Evolutionary Parasitology

Evolutionary Parasitology

The Integrated Study of Infections, Immunology, Ecology, and Genetics

Second Edition

Paul Schmid-Hempel

Emeritus Professor, Institute of Integrative Biology (IBZ) and Genetic Diversity Centre, ETH Zürich, Switzerland

OXFORD
UNIVERSITY PRESS

OXFORD
UNIVERSITY PRESS

Great Clarendon Street, Oxford, OX2 6DP,
United Kingdom

Oxford University Press is a department of the University of Oxford.
It furthers the University's objective of excellence in research, scholarship,
and education by publishing worldwide. Oxford is a registered trade mark of
Oxford University Press in the UK and in certain other countries

First Edition published in 2011
Second Edition published in 2021

Impression: 1

Published in the United States of America by Oxford University Press
198 Madison Avenue, New York, NY 10016, United States of America

British Library Cataloguing in Publication Data
Data available

Library of Congress Control Number: 2021934827

ISBN 978–0–19–883214–0 (hbk.)
ISBN 978–0–19–883215–7 (pbk.)

DOI: 10.1093/oso/9780198832140.001.0001

Printed and bound by
CPI Group (UK) Ltd, Croydon, CR0 4YY

Preface

Parasites and infectious diseases are everywhere around us and have affected the ecology and evolution of organisms since the early days of life on this planet. In fact, this second edition of *Evolutionary Parasitology* was finished during the Corona year, 2020. The pandemic brought grief and misery to many people, not to speak of the enormous economic costs. At the same time, this pandemic is an impressive illustration of the pervasive influence of parasitism that affects virtually all aspects of the hosts' lives. The field of evolutionary parasitology, therefore, cuts across many disciplines for a more comprehensive approach to studying hosts and parasites, to appreciate the mechanisms that guide their interactions and to identify the selective forces that shape their biology.

As before, I am using the generic term 'parasite' to cover various other names, such as 'pathogen' or 'parasitoid', which are more common in fields like medicine or agriculture. However, parasitism is the core ecological relationship towards which all scientific endeavours in the larger field gravitate. This relationship is based on molecular and physiological processes, on probabilities of contacts, on binding between surfaces and specific molecules, but also results in more or less success of either party. Hence, the relationship is also under selection and has evolved and co-evolved over the aeons and still continues to do so. In some cases, we see fast evolutionary changes, as with the rise of antibiotic resistance in bacteria, whereas the conserved nature of some elements in immune defence systems points to their deep ancestry across organisms. Indeed, immune systems are among the most complex natural systems that have evolved and, doubtlessly, parasitism was a major driver along this way. But parasites are not just the passive partners, as their typical organismal simplicity would suggest. Instead, parasites have evolved mind-boggling mechanisms and strategies to evade, overwhelm, and manipulate their hosts in their own favour—this is even true for viruses that undermine their hosts' defence systems in amazing ways. Therefore, to unravel these fantastic processes and to clarify the evolutionary reasons for the enormous diversity of host defences and parasite strategies is an endlessly captivating venture.

This is a completely rewritten update of *Evolutionary Parasitology*. It contains a number of tables that cannot be a comprehensive review of the respective topics. Such an attempt would be close to impossible, given the enormous range of activities in this huge area. Rather, and as in the previous edition, the tables should illustrate typical studies, while giving an impression of the variety of study subjects and approaches. As before, I must also apologize to the plant world that my examples are primarily zoonotic in origin. Similarly, social parasites such as inquiline ants or brood parasites in birds are not considered in much detail. Nevertheless, the principles guiding those host–parasite interactions are also the topics of this book.

Looking back, it is astonishing to see how much has happened in the broader field within the decade since the original book appeared. Three elements contributed in important ways. Firstly, the advance in molecular technologies is breathtaking. What once used weeks, is now done in a day, and at a fraction of the cost. Sequencing technologies, for example, have sparked a new age for virology, allowing an ongoing epidemic to be traced almost in real time. Discoveries based on mechanisms in immune defence systems, such as RNAi or CRISPR–Cas, allow the genotypes of organisms to be changed in unprecedented ways. And with mRNA technology a next toolbox is already on the horizon that not only makes for a new generation of vaccines but can help to further dissect the mechanisms underlying host–parasite interactions. A second methodical element that has contributed to the advance in the field is the progress in mathematical algorithms and computing power, often lumped together as bioinformatics, that makes it possible to use large amounts of information and to analyse these with improved statistical techniques. Reconstructing the molecular epidemiology of viral diseases is just one of the applications of these powerful methods. Finally, the field has progressed in its concepts, which is the ultimate aim of any scientific exploration. For instance, the early phases of infection have come into focus, as did concepts to predict the outcome of an infection based on measures of host status at certain stages of the process. Clearly, evolutionary parasitology has matured, but it will not end soon—too diverse and intriguing are its subjects, too riveting the study of these, and too important the practical implications for matters of agriculture, conservation biology, medicine, and public health.

The daily work of a scientist often is a very lonesome activity, but the process of doing science is not. Therefore, this book also rests on the work of many others. I have been blessed to meet so many outstanding colleagues and to have the chance to discuss questions at the forefront of their respective fields, all of which has influenced this book perhaps more than is visible. To pick just a few, I am grateful for the extended contacts with Janis Antonovics, Mike Boots, Sylvia Cremer, Dieter Ebert, Steve Frank, Andrea Graham, Andrew Read, Jens Rolff, David Schneider, and many others. David Schneider's concept of the disease space has been a particularly illuminating addition and is used in this book as a guide through the different sections—in the hope that it will always show the relationship between the underlying mechanisms and the ecologic and evolutionary outcome of a parasitic infection. My own scientific home in the Institute of Integrative Biology (IBZ) has been an enormously fruitful setting over the years; the interactions with the groups of Sebastian Bonhoeffer and Roland Regoes especially helped me to reach out into the theoretical domains. Moreover, good fortune has brought many outstanding students and postdocs to my own research group. Working together on topics of host–parasite interactions has been enriching, and a real pleasure. From the more recent past, I just mention Boris Baer, Seth Barribeau, Mark Brown, Jukka Jokela, Hauke Koch, Joachim Kurtz, Yannick Moret, Kathrin Näpflin, Oliver Otti, Livia Roth, Ben Sadd, Rahel Salathé, Yuko Ulrich, Maze Wegner, Lena Wilfert, without any disregard for all the others that have contributed in many other ways. The administrative and technical help of Rita Jenny, Roland Loosli, Christine Reber from IBZ, and Aria Minder from the Genetic Diversity Centre kept many a burden off my table. Of course, my wife Regula has not only shared the ups and downs during writing, but has also helped in many and important ways, both scientifically and with technical support. Finally, a number of colleagues have volunteered to read through the earlier drafts. I am thus very grateful for the valuable input given by Seth Barribeau, Mark Brown, Austin Calhoun, Roger Kouyos, Elyse McCormick, Andrew Read, Roland Regoes, Bryan Sierra Rivera, Jens Rolff, Ben Sadd, and Logan Sauers. A special thanks goes to Louis du Pasquier who had already helped with the first edition, and whose critical advice was essential for the discussion of immune defences. The remaining errors are, of course, mine. Last but not least, I thank Ian Sherman and Charles Bath from Oxford University Press for their generous support and unobtrusive coverage of the entire process. May the efforts aid the field of evolutionary parasitology and advance our scientific understanding of nature.

Paul Schmid-Hempel
November 2020
ETH Zürich, Institute of
Integrative Biology (IBZ), and
Genetic Diversity Centre at ETH, Switzerland

Contents

xiv CONTENTS

List of common acronyms

Acronym	Name	Description
AGO	Argonaut	Binds to short RNA (siRNA) in the antiviral defence of invertebrates.
AID	Activation-induced cytidine deaminase	Enzyme involved in gene conversion, somatic hypermutation, class switching.
AIDS	Acquired immune deficiency syndrome	Disease caused by HIV.
AMP	Antimicrobial peptide	Effector protein with antimicrobial activity.
APC	Antigen-presenting cell	Cells that can bind to and present parasite peptides to passing, for example, CD4+ T-cells.
CD4+, CD8+	T-cell with CD4, CD8 protein	Helper cells.
CDV	Canine distemper virus	Paramyxoviridae (ssRNA−).
CHIKV	Chikungunya virus	Togaviridae (ssRNA+).
CoV	Coronavirus	Coronaviridae (ssRNA+).
CRISPR	Clustered regular interspaced palindromic repeats	Genetic loci of bacteria and archaea that store viral sequences from previous encounters to provide memory and defence.
CTL	Cytotoxic T-cell (lymphocyte)	An activated CD8+ T-cell able to destroy an infected host cell.
DAMP	Damage-associated molecular pattern	Biomolecules, e.g. DNA fragments, that indicate the presence of a parasite.
DENV	Dengue virus	Flaviviridae (ssRNA+).
Dicer	Dicer	Enzyme involved in the (antiviral) RNAi system.
Dscam	Down syndrome adhesion molecule	Recognition protein, primarily of arthropods.
ED	Effective dose	Dose of a pathogen or substance that causes an effect.
EMP	Erythrocyte membrane protein	Surface protein on red blood cells, recognized by the immune system. Encoded by the parasite (*Plasmodium*).
ESS	Evolutionarily stable strategy	Strategy that cannot be beaten by rare, alternative strategies.
FREP	Fibrinogen-related protein	Recognition protein of molluscs.
GFG	Gene-for-gene	A genetic host–parasite interaction model.
GWAS	Genome-wide association study	Association of a phenotype over all loci in the genome.

Acronym	Name	Description
HBV	Hepatitis B virus	Orthohepadnaviridae (dsDNA).
HIV	Human immunodeficiency virus	A Lentivirus, retrovirus (+ssRNA).
HLA	Human leukocyte antigen	Binding site, known as the MHC in other jawed vertebrates.
HRV	Human rhinovirus	Picornaviridae (ssRNA+).
IED	Individual effective dose	Dose of a pathogen or substance that causes an effect in a given individual.
IFN	Interferon	Cytokines, signalling protein for viral infections, e.g. IFN-γ.
Ig	Immunoglobulin	Recognition protein; in particular, the antibodies of jawed vertebrates.
IL	Interleukin	Cytokine that signals between leucocytes.
Imd	Immune-deficiency pathway	A canonical immune-signalling pathway of insects and some other arthropods.
IRAK	IL-1R associated kinases	Signalling protein in innate immune pathway.
IUCN	International union for the conservation of nature	Membership union of governments and other parties to protect nature.
IVA	Influenza A virus	Orthomyxoviridae (ssRNA-).
JAK/Stat	Janus kinase-signal transducer activator of transcription	A canonical immune-signalling pathway of insects and some other arthropods.
LD	Lethal dose	Dose of a pathogen or substance that causes death.
LD	(Genetics) Linkage disequilibrium	Association of alleles in a genotype that deviates from random.
LPS	Lipopolysaccharide	A component of the cell wall of Gram-negative bacteria.
LRR	Leucin-rich repeat	A common outward domain of transmembrane receptors in the immune system.
MAC	Membrane attack complex	Formed with complement activation to destroy microbial cell walls.
MDV	Marek's disease virus	Herpesviridae (dsDNA).
MERS	Middle East respiratory syndrome	Disease caused by the MERS virus.
MHC	Major histocompatibility complex	A large genomic region in jawed vertebrates that codes for molecules binding to parasite-derived peptides.
MV	Measles virus	Paramyxoviridae (ssRNA-).
MyD88	Myeloid differentiation primary-response gene 88	A transducing protein associated with TLR receptors.
NF-κB	Nuclear factor-κB	A transcription factor in the immune signalling cascade of animals.
NK	Natural killer cell	A phagocytic cell of the innate immune system.
NLR	NOD-like receptor	Intracellular sensors that detect intracellular parasites (PAMPs) or the associated damage (DAMPs).
NOD	Nucleotide-binding and oligomerization domain	An intracellular receptor involved in regulation of defence.

Acronym	Name	Description
OTU	Operational taxonomic unit	A taxonomic group defined by minimal sequence divergence.
PAMP	Pathogen-associated molecular pattern	A molecular pattern (epitope) that is recognized by an innate receptor (PRR).
PGRP	Peptidoglycan receptor protein	An innate immune receptor.
PO; proPO	Phenoloxidase	Key enzyme in the defence cascade of arthropods; precursor to PO.
PRR	Pattern-recognition receptor	Binds to general molecular patterns (epitopes), e.g. on bacterial cell walls.
QTL	Quantitative trait locus	A genetic locus statistically associated with a phenotypic trait.
RAG	Recombinase-activating gene	Involved in somatic recombination of genetic elements for lymphocytic receptors.
Relish	Relish	Transcription factor, e.g. activated by the Toll pathway in insects.
RISC	RNA-induced silencing complex	Protein complex binding and cleaving RNA strands. Part of the antiviral defence of invertebrates.
RNAi	Interference RNA	A system that silences genes by degrading the transcribed RNA. Antiviral defence in invertebrates.
ROS	Reactive oxygen species	Non-saturated oxygen molecules with high reactivity; toxic for microbes.
SARS	Severe acute respiratory syndrome	Disease caused by SARS-CoV-1 virus.
SIR	Susceptible–infected–recovered	Refers to standard model of epidemiology.
SNP	Single-nucleotide polymorphism	Variation at a single nucleotide position in a population.
SR	Scavenger receptor	Receptors that trigger removal of modified molecules (lipids) from the cell, but also have immune functions.
TCR	T-cell receptor	A receptor on the surface of a T-cell, e.g. the CD4 protein.
TEP	Thioester-containing protein	A class of phagocytic opsonization factors in insects.
TGIP	Transgenerational immune priming	The phenomenon that offspring of challenged (parasite-exposed) parents are better protected.
Th1, Th2	T-helper cells type 1, type 2	Helper cells that produce various cytokines, involved in defence against bacteria and viruses (Th1) and helminths (Th2).
TLR	Toll-like receptor	A family of key receptors in the innate immune system.
TNF	Tumor necrosis factor	Membrane-bound cytokine of the immune defence, e.g. inflammation, but also with many other functions.
VLR	Variable lymphocyte receptor	Receptors at the surface, e.g. on B-cells.
VSG	Variable surface glycoproteins	Polymorphic surface molecules (epitopes) recognized by the immune system, e.g. in trypanosomes.

Glossary

Adaptive immunity Immune defence that adapts to ongoing infections by becoming more specific and stronger.

Aetiological agent The agent (parasite) causing a particular disease. For example, HIV causes AIDS.

Affinity Strength of binding, usually between receptor and ligand.

Affinity maturation The process by which B-cells that bind more strongly to a given parasite (antigen) become more common, based on somatic hypermutation.

Allele An alternative variant of a gene at a given locus.

Allograft A foreign tissue that is transplanted onto (or comes in contact with) a host individual.

Alternative splicing A process during gene expression that results in different mRNAs and proteins derived from a single gene.

Anergic An immune cell (lymphocyte) that is unresponsive to an antigen.

Antagonistic pleiotropy Pleiotropic genes affect several phenotypic characters. Antagonistic pleiotropy is often used for a gene that has a positive effect early in life but a negative effect late in life.

Antibiotic resistance Acquired resistance of microbes to antibiotic agents. Also known as antimicrobial resistance, or drug resistance.

Antibody A secreted immunoglobulin (Ig) that binds to a parasite epitope.

Antigen A parasite molecule (or other foreign substance) that stimulates an immune response.

Antigen-presenting cells (APC) A heterogeneous group of immune cells that process and present parasite molecules (antigens) at their surface for other immune cells.

Antigenic drift A change in the antigenic properties of a parasite that results from mutation accumulation in a population, e.g. in an infecting viral population.

Antigenic shift A change in the antigenic properties of a parasite that results from the expression of different stored variants of the individual parasite, or by recombination among different co-infecting strains of viruses, for example.

Antigenic variation Scheduled or random variation of recognized molecules on the surface of parasites (epitopes) to evade the host immune system.

Antimicrobial peptide (AMP) A short protein (peptide) that is able to destroy a (microbial) parasite. Also effective against protozoans. AMPs differ in the exact mechanism of how they damage the parasite. AMPs are effectors of the innate immune system.

Apoptosis Programmed cell death.

Attenuation The process of a parasite losing virulence over generations.

Bacteriocin Molecules produced by bacteria to suppress competing bacteria.

Basic reproductive number, R_0 This is the number of newly infected hosts resulting from one already infected host in a population of all susceptible hosts.

Bateman's principle The observation that males vary more in their reproductive success than females.

Biofilm A dense aggregation of bacteria embedded in a matrix of biopolymers.

Bridge host Used in the study of zoonoses to characterize a host that mediates between background reservoir and the target species.

Candidate gene A gene that is suspected to play a role in a given function. For example, the peptidoglycan recognition genes are very likely to act as recognition molecules for certain kinds of pathogens.

Case mortality rate Mortality rate per diagnosed case, i.e. the probability of host death once infected.

Central tolerance (Immunological) The establishment in lymphocytes of tolerance towards own tissues during maturation of the B- and T-cell populations in the primary lymph organs.

Chemokine A chemical attractant, a molecule, in the immune defence system.

Class switching A process during which an immunoglobulin (antibody) changes its class, e.g. converts from an IgD to an IgE type.

Clearance The process by which the parasite is removed (cleared) from the host; the host becoming uninfected again.

Clonal expansion The process during which B- and T-cells of the vertebrate adaptive immune system multiply in numbers (and mature) to fight a specific infection.

Co-infection Often used to denote the infection of a host by more than one different parasite species or by otherwise very different types. Also more commonly used as a generic term meaning multiple infection by different parasite species or variants.

Coalescence The convergence of two phylogenetic lines at some time in the past.

Constitutive defence A defence that is present and active even before an infection. It can therefore act immediately, should an infection occur.

Copy number variation Variation in the number of copies of a gene within a genome.

Critical community size (Epidemiology) Critical population size to endemically maintain an infectious disease.

Cytokine A signalling protein for immune cells. Helps to orchestrate the immune response.

Cytokine storm An unregulated, massive release of cytokines.

Cytoskeleton The internal skeleton of a cell that allows it to keep and change shape or to move. A highly dynamic structure consisting of protein filaments and microtubules.

Cytosol The fluid components of the plasma inside a cell.

Defensins A class of small cysteine-rich cationic antimicrobial peptides (15–20 residue). They are found in all animals and some higher plants.

Dendritic cell (DC) A type of haemocyte that patrols the body and is able to present antigens (in a MHC–peptide complex) to passing helper T-cells.

Deuterostomes Animals that develop through a 'mouth second' scheme, i.e. the first opening of the embryo becomes the anus, and the mouth develops from a sperate opening. These includes a few advanced invertebrate groups, such as the echinoderms, and the chordata, including the vertebrates.

Digenic A parasite having two hosts in its life cycle. Sometimes this term also covers three and more hosts. Synonym: dixenic.

Dioecious Male and female parasites use different host species (e.g. in some Strepsiptera).

Dixenous Having two hosts, or a host and a vector in the life cycle.

Domain (protein) A domain in a protein is a region with a conserved amino acid sequence and thus of tertiary structure, which defines its function.

Dose The number of parasite cells, cysts, etc. needed to cause a response to infection.

Drift (genetic) With drift, alleles and genotypes are lost by chance, the effect being stronger in small populations.

Drug resistance The same as antibiotic resistance.

Dysbiosis Loss or change of the normal structure (and/or microbes) of the microbiota.

Effective population size Population size that, in terms of population genetics, functions like a standard outbred, diploid population.

Effector Any molecule or process at the end of the immune response cascade that actually affects the parasite. Examples are antimicrobial peptides, encapsulation, cytotoxic lymphocytes.

Emerging disease An infectious disease, not present before, that appears in a host population. Typically, by transfer from a reservoir.

Encapsulation An important effector mechanism in invertebrate immune systems. A parasite thereby gets surrounded by melanizing haemocytes; eventually the parasite is completely enclosed in a sealed capsule and becomes killed.

Endemic A persistent infection in a population in the absence of novel infections coming from the outside.

Endemic threshold Minimum host population size to endemically maintain an infection.

Endocytosis Ingestion of macromolecules by specialized cells such as macrophages.

Endotoxins Compounds associated with the pathogen itself, e.g. located on the bacterial cell, and which cause damage to the host while helping the parasite to infect or spread.

Epidemic An infection in a population that shows a dynamic course starting from a few cases, e.g. a new infection that is spreading.

Epidemiology The study of host–parasite dynamics with population biology and population genetics. In medicine, 'epidemiology' is a field that identifies statistical associations between the occurrence of disease and putative causal factor.

Epistasis An effect on the phenotype (e.g. the fitness of an organism) that is due to the particular combination of genes (alleles) at two (digenic epistasis) or more loci. More strictly defined as the deviation in fitness from an additive effects model due to combination of genes.

Epitope A molecular pattern on the surface of a parasite that is recognized by a receptor or ligand of the host.

ESS (Evolutionarily stable strategy) A strategy that, if adopted by all individuals in the population, cannot be invaded by a rare mutant.

Exon Any part of a gene that is finally transcribed into mRNA.

Exotoxin Proteins released by a pathogen such as a bacterium and which can take effect far from the site of infection.

Extant Still existing today, e.g. a currently living species.

Fecundity In ecology, the average per capita number of offspring in a population.

Final (definitive) host For a parasite with several hosts in its life cycle, the final host is judged to be the most 'important'. Often this is the host where the parasite sexually reproduces, but this is not always so.

Force of infection The rate at which an exposed uninfected individual becomes infected by transmission from infected hosts. In a mass action model (such as in the standard SIR model), this is proportional to the product of the number of infected individuals and transmission rate (i.e. infection probability per encounter).

Gene conversion A process that happens during a (homologous) recombination event where the 'donor' gene remains the same but the 'receptor' gene acquires the recombined sequence. This leads to an altered gene; i.e. the gene has converted into a new one.

Genome The entire genetic sequence of an organism.

Genomics The study of genomes.

Gram-negative bacteria A heuristic category for bacteria that appear red or pink after the Gram stain process and subsequent safranin treatment. Gram-negative bacteria have two membranes—a thin peptidoglycan layer and an outer layer of lipopolysaccharides—separated by the periplasmic space.

Gram-positive bacteria A heuristic category for bacteria that appear blue or violet during the Gram staining process. Gram-positive bacteria have a thick cell wall but only one membrane layer.

Haematopoiesis Cell development (of immune cells).

Haemolymph The circulating body fluid in insects (or arthropods more generally); the 'blood' of insects.

Helper T-cell Same as CD4+ T-cell. Functions to provide a signal necessary to stimulate the antibody or cytotoxic lymphocyte response (CTL).

Herd immunity A population is protected from an infectious parasite by herd immunity when a critical fraction of the population can no longer become infected, e.g. by vaccination. The effect results from the epidemiological dynamics of host–parasite interactions so as to lower the basic reproductive rate of the parasite to a value less than one.

Heterologous immunity Immunity directed against a different (heterologous) parasite than what originally caused it.

Heteroxenic In parasite life cycles: using different host species.

Horizontal (gene) transfer (Lateral gene transfer) The transfer of genes from one phylogenetic line or species to another. Different processes can be involved.

Horizontal transmission The transmission of parasites between hosts of the same population and generation. i.e. not to own offspring.

Host range The list of host species used by a parasite. Sometimes called 'host spectrum'.

Host reservoir Where an infection usually resides endemically and away from the host under scrutiny.

Hypermutation A process of somatic mutation in vertebrate adaptive immunity where mutation rates are increased during the maturation of lymphocytes.

Hypersensitive response In plants: a non-specific, early and fast immune response. It is characterized by a rapid induction of apoptosis in cells around the infection site. An oxidative burst occurs.

Immune priming Used to denote the phenomenon of an immune memory in invertebrates.

Immunocompetence The capacity to mount an immune response to a challenge. Sometimes also defined more loosely as the ability to withstand infection and disease. Originally considered to be a summary measure for all possible immune responses.

Immunodominance A response dominated by a few epitopes, triggering affinity maturation.

Immunogen A stimulus, such as a foreign object or substance, able to trigger an immune response.

Immunoglobulin (Ig) Globulins in serum with antibody activity. There are five major classes: IgG, IgM, IgA, IgE, IgD.

Immunological tolerance A process during the maturation of lymphocytes whereby self-reactive cells are eliminated or modified.

Immunopathology Pathological effects caused by the immune system itself.

Incubation period Time from infection to first signs of the disease.

Index case The first identified case (i.e. infected host) in an epidemic.

Induced defence A defence that is activated upon infection. The defence therefore needs to be built up before it can take an effect.

Infective dose Dose needed to start an infection.

Inflammasome A large, cytosolic protein complex formed during inflammation.

Inflammation An early, innate immune defence where immune cells such as macrophages or monocytes are recruited (by cytokines) to the site of infection.

Innate immunity A collection of diverse defence systems in all animals or plants, e.g. phagocytosis, complement, or TLR pathways. Essentially based on germline-encoded molecules with no specific somatic modifications.

Inoculum The population of parasites used for (experimentally) infecting a parasite. From the Latin word 'inoculare' ('to graft a scion').

Integument The outer shell of an animal's body. Examples are the mammalian skin or the insect cuticles.

Intensity (infection) The number of parasite cells or parasite individuals within a given host individual (parasite load, parasite burden).

Interferon A cytokine active in the context of antiviral defence.

Interleukin A cytokine that signals between leukocytes.

Intermediate host A host where the parasite passes through obligatory developmental steps.

Isolate A sample of parasites that has been obtained from an infected individual (primary sample) or from a restricted number of infected individuals. An isolate contains a population of parasites that is affected by the host from where it has been isolated.

Isotype In immunology, denotes different classes of immunoglobulins, such as IgM or IgA.

Jumping genes Genes that can transfer to another location within the same genome.

Lamellocytes An important class of plasmatocytes (haemocytes) in arthropods, invertebrates. Contribute to the encapsulation of a parasite.

Latent period The time interval between infection and the appearance of symptoms (incubation period in the medical sense), or the interval between infection and when the parasite is transmissible (latency in the epidemiological sense) by having developed into the corresponding transmission stage.

Leukocytes The 'white' blood cells of vertebrates, i.e. a family of cells of the immune system circulating in the body (lymphocytes, eosinophils, basophils, polymorphonuclear cells).

Life history The temporal schedule of birth, development, reproduction, and death in organisms.

Ligand A usually smaller molecule that is able to bind to another larger biomolecule and form a complex.

Linkage (disequilibrium) (LD) In population genetics, the alleles at two (or several) loci are said to be in linkage disequilibrium (LD) if the observed frequency at which certain genotypes (combinations of alleles at different loci) deviates from the expected frequency at which the same genotypes would occur if the alleles were combined at random; i.e. there is a statistical association between alleles at different loci. With positive LD, the genotypes are more common, with negative LD they are rarer than expected by chance.

Locus In genetics, the (idealized) location within the genome where a gene is found. In reality, a gene consists of several exons that can be distributed widely along the genomic sequence.

Lymphoid cells One of two families of blood cells in vertebrates. They function as effectors of immune defences (lymphocytes, natural killer cells).

Lysis The process of destruction of a bacterium by damage of the external cell membrane and where the cell contents spills out. This might be caused by viral phages that thereby kill the host cell to release the newly produced phage daughter virions.

M-cells (microfold cells) Part of the specialized gut epithelium that overlies Peyer's patches.

Macrophage A vertebrate blood cell that is specialized for phagocytosis; a 'professional' phagocyte. One of two (myeloid) monocytes. Macrophages act as presenters of MHC–peptide complexes to signal infection to passing helper T-cells.

Mass action principle A model of transmission that assumes infected and susceptible are perfectly mixed and meet at random. In this case, the number of encounters is given by the product of the number of infected and susceptible individuals. Infection then occurs in a fraction of these encountered, quantified by the transmission rate.

Metazoa Multicellular organisms; sometimes more specifically referred to as those with differentiated tissues.

MHC Multi-histocompatibility complex. A set of tightly linked loci that are important in immune defence by coding for recognition proteins.

Microbiota, Microbiome The ensemble of microbes that live together in a more or less structured community in or on a host, e.g. in the gut or on skin.

Molecular epidemiology Epidemiology based on genes and genomic sequences. Typically used to reconstruct the spread of an infectious disease, but also for risk assessment.

Monoxenic (monogenic) Having only one host in the life cycle.

Multidrug resistance Resistance that is directed against several drugs (antibiotics); for example, in multiply resistant bacteria.

Myeloid cells One of two families of blood cells in vertebrates. Some function as effectors of immune defences (e.g. granulocytes, mast cells, monocytes).

Natural antibody Antibodies secreted by non-activated (naïve) B-cells. They are mostly of IgM type.

Necrosis A process during which cells die and an entire tissue gets destroyed.

Neutrophil A kind of granulocyte that is important in the inflammatory response of the vertebrate immune system.

Obligate killer A parasite that needs to kill its host for the completion of its life cycle. Examples are many parasitoids, fungi, and microsporidia.

Opsonin, opsonization A molecule deposited on the surface of a foreign particle by the host and which acts to facilitate binding for the subsequent process of phagocytosis or destruction. Examples are antibodies or components of the complement system. The parasite is thus marked for destruction (opsonized).

Ortholog A homologous gene (or part thereof) in different species, as a result of a speciation event.

OTU (Operational Taxonomic Unit) Referring to groups of sequences from individual probes that are sufficiently different, usually with more than 3 per cent sequence divergence, to be classified as a different 'taxon'. OTUs might represent species.

Oxidative burst (respiratory burst) Also called a respiratory burst. This is the rapid release of reactive oxygen molecules (reactive oxygen species, ROS) in cells and serves as a defence against microbes, which can be killed by these reactive oxygen species.

Pandemic An epidemic that has spread to a large part of the globe.

Paralog A homologous gene (or part thereof) that arose by gene duplication.

Parasitaemia The condition where parasites are found in the bloodstream. Often with a measure such as the density of parasite cells per unit volume of blood.

Parasite load, parasite burden A loosely defined generic term. Usually it refers to the number, or the diversity, of parasites that an individual host or a host population carries. For macroparasites, the number of individual parasites in a host.

Parasitoid Lives parasitically when young (a larva) but free-living as an adult, e.g. parasitic wasps.

Paratenic host A host where no parasite development occurs, but infectious stages can accumulate. Typically an incidental infection that serves as an intermediate host. It can aid the spread of the parasite and might evolve into a regular host.

Pathogen Used interchangeably with 'parasite'.

Pathogen-associated molecular pattern (PAMP) A classical term that refers to any structural feature of microbes that is recognized by the immune system.

Pathogenicity island Blocks of genes in the genome of pathogens (like bacteria) that code for virulence factors. Pathogenicity islands are often formerly mobile genetic elements that have been transferred into the pathogen's genome.

Pattern-recognition receptor (PRR) A generic term, usually meaning a host receptor able to recognize a generalized motif of a parasite surface (such as a PAMP).

per os Via the mouth; this is one route along which parasites can infect hosts.

Peripheral tolerance Screening of B- and T-cells to remove lymphocytes that are too strongly self-reactive; occurs peripherally, in lymphoid tissues.

Peyer's patch Focal accumulation of lymphoid tissue in the wall of the vertebrate intestine. An entry site for many bacterial pathogens.

Phage A virus that parasitizes bacteria (bacteriophages).

Phagocyte, Phagocytosis A cell that is specialized for phagocytosis (i.e. engulfing a foreign object such as a bacterium).

Phagosome A within-cell vesicle, specialized to receive, contain, and destroy internalized particles during phagocytosis.

Phase variation (Bacteria) Frequent, reversible changes in a (bacterial) phenotype, often by gene expression.

Phoresy, Phoresis The phenomenon of being carried by a transport host. From Greek 'phorein, pherein' meaning to bear.

Phylodynamics A combination of immunology, epidemiology, and evolutionary biology to study the effects of host immune responses on parasite population structure and epidemiological patterns, typically using the phylogeny of infections in host populations.

Plasmatocyte A cell in the haemolymph (plasma) of insects (or arthropods more generally); a 'blood cell'. Plasmatocytes are differentiated into several types, e.g. lamellocytes and crystal cells.

Plasmid A piece of DNA that is separated from chromosomal DNA. Plasmids are typically circular strands of DNA and are particularly common in bacteria. Plasmids can be transferred between bacteria.

Polymorphism The simultaneous presence of several variants (morphs, genotypes, alleles) in a population.

PPO cascade The prophenoloxidase (PPO) cascade is a central defence cascade of the invertebrate immune system. It leads to the production of melanin and cytotoxic compounds. The cascade is triggered by a wide range of elicitors.

Prevalence The prevalence of an infection is the fraction (percentage) of infected host individuals in a population.

Primary response Response by the immune system upon a first encounter with a parasite (see also: Secondary response).

Proteasome A large intracellular protein complex that degrades proteins derived from an intra-cellular parasite (e.g. a virus) into peptides.

Proteomics The study of expressed proteins in an organism.

Proteostome Animals that develop through a 'mouth first' scheme, i.e. the first opening of the embryo becomes the mouth. These include most of the invertebrates.

Pseudogene A genetic sequence that is very similar to a gene but not functional. Typically, a sequence of a former gene that has deteriorated.

Quasispecies Descriptive term for the set of very similar (viral) sequences 'surrounding' a master sequence, typically from within an infection.

Quorum sensing The ability of microbes (bacteria) to respond to their population density by gene regulation.

Red Queen dynamics A process caused by time-lagged antagonistic co-evolution between hosts and parasites, such that host and parasite genotype frequencies fluctuate over time. This dynamic is hypothesized to provide an advantage for sexual reproduction and recombination.

Refractory Describes a host that is difficult or impossible to infect.

Reservoir Host species infected by a parasite that potentially can jump over to another host population (e.g. to humans).

Resistance Generalized term implying a host defence capacity to reduce and clear an infection.

Secondary response Response by the immune system upon a second or further encounter with the same parasite. This response is faster and/or more efficient if a memory is present. Memory is known from vertebrates, but analogous patterns are also observed in insects or crustaceans.

Selective sweep A rapid increase of a genetic variant, e.g. an allele, in a population with, typically, eventual dominance and fixation.

Septic shock A severe medical condition resulting from infections characterized by tissue perfusion (i.e. influx of liquids) and massive release of cytokines, triggering an over-response of the immune system.

Serial interval In epidemiology, the average time interval from infection of the first to infection of the subsequent hosts.

Serial passage Successive infections of the same host types.

Serovar, Serotype A group of a parasite that can be distinguished by antigenic properties; these properties set them apart from other such serotypes. Serotyping is classically done by antigen–antibody reactions, but can also be based on various other factors, such as virulence type, genotyping, etc.

Sexual selection Selection with respect to traits that affect mating and reproduction (fecundity selection).

Sexual transmission Transmission during mating and sexual activities.

Signal transduction The conversion of a primary stimulus (e.g. an antigen binding to a receptor) into a signal that is passed to the next element of a cascade.

SIR model A standard model for the epidemiology of infectious diseases tracking the number of susceptible (S), infected (I), and recovered (R) hosts in a population. The number of parasites is thereby not modelled explicitly.

SNP Single-nucleotide polymorphism. Denotes variation in a single position of the genomic sequence among individuals in a population.

Social immunity Collective (immune) defences in cooperative social groups, e.g. allogrooming.

Strain A generic term denoting a parasite variant that can be separated from others by their properties, e.g. infectivity or virulence, and/or by genetic markers. In the extreme, a strain is a clone. In most practical cases, however, a strain refers to a parasite variant that remains as such a separate entity for the purpose and time scale under consideration.

Superinfection Often used to denote the infection of a host by more than one strain of the same parasite, leading to strain diversity within the host.

Superspreader A host individual that transmits an infection further disproportionally often.

Tissue tropism The propensity of infecting parasites to migrate to a preferred tissue.

Tolerance (Evolutionary parasitology) The capacity of a host to reduce and minimize fitness loss when infected.

Tolerance (Immunology) The prevention of a reaction to self, by absence of self-reactive lymphocytes.

Toxin (Mostly) proteins released by pathogens that typically cause damage to the host, but are also essential for the success of the infection.

Transcription The conversion of genetic information (DNA) into RNA.

Transgenerational immune priming (TGIP) The phenomenon that offspring of challenged (parasite-exposed) parents are better protected when they encounter the same or similar parasites.

Translation The conversion of messenger information (mRNA) into proteins.

Transmission The passage of a parasite from one host to the next.

Transmission mode The (physical) method of transmission, e.g. by air.

Transmission route The actual (physical) path taken from one host to the next, e.g. the faecal–oral route.

Transposon A generic term, meaning genetic material that can move to different positions within the genome.

Variable region (V) A variable region of an immunoglobulin (membrane-bound or antibody) that defines the binding specificity.

Vector A temporary animal vehicle that carries the parasite to a next host, e.g. mosquitoes.

Vertical transmission The transmission of parasites to own offspring.

Virion The replication products of the virus that leave the host cell.

Virulence A generic term denoting the capacity of a parasite to inflict damage and a reduction in host fitness (in interaction with the host's response).

Virulence factor An element of the pathogen (e.g. a protein coded by a specific genetic element) that is essential for the successful completion of the pathogen's life cycle in the host. If successful, virulence emerges. If factor is absent, the parasite is attenuated or non-pathogenic.

Virulence gene The gene(s) that code(s) for a particular virulence factor. Often located on pathogenicity islands.

Zoonosis An epidemic in human populations that has its origin in animals.

CHAPTER 1

Parasites and their significance

1.1 The Panama Canal

On New Year's Day of 1880, the young Fernanda Lesseps stood on board a steam launch in the mouth of the Rio Grande, some 15 km east of the recently founded town of Colón, on the Caribbean coast of the Isthmus of Panama, then a province of neighbouring Colombia. She put a shovel full of sand into a box that had been emptied of its champagne bottles, to symbolically mark the start of the construction work for the Panama Canal. Fernanda was performing on behalf of her father, Count Ferdinand Lesseps (1805–1894). He was in his seventies and a public hero. Lesseps was the architect of the Suez Canal, which had opened on 17 November 1869. The Viceroy of Egypt eventually convinced Giuseppe Verdi (1813–1901) to write a piece for the opening of the new opera house erected to celebrate the Canal. Verdi's *Aida* premiered in Cairo on 24 December 1871, with pomp and glamour. Ferdinand Lesseps, therefore, had every reason to be confident that he would also succeed in constructing the long-desired maritime shortcut from the Atlantic to the Pacific Ocean. To finance the work, he had just founded a new company, the *Compagnie Universelle du Canal Interocéanique*. In January 1881, around 200 engineers from France and other European countries, together with 800 labourers, had arrived in Colón to start the work. Count Lesseps could not foresee that it would take 34 years from this day, and a second attempt by American companies and engineers, to finish the project. Eventually, the canal opened on 15 August 1914, with the passage of the vessel SS *Ancon*. An estimated 80 000 people had worked on the canal, and more than 30 000 lost their lives in the effort. The engineering was an extraordinary challenge, but the parasites of the hot and humid lowlands of Panama proved to be the most challenging problem to overcome.

The canal work began in 1882 along the route of the Panama Railroad. This railroad connected the Atlantic with the Pacific coast and had opened in 1855. Lesseps started by erecting moorings, roads, and barracks for the labour force. However, the lowland tropics were different from the Arabian deserts of the Suez Canal. Social insects proved to be the first problem and the gateway to disaster. Termites quickly destroyed the wooden constructions for the workers' housing. Heavy trafficking by ants inside the barracks was not only a nuisance but a problem for hygiene. Lesseps therefore decided to put housings and storage facilities on wooden stilts. Stilts were placed in large, water-filled drums to prevent termites and ants from gaining access to the buildings and destroying the wooden structures. This countermeasure was a success—termites and ants were no longer a problem. However, the tropics were far from defeated. The water drums soon attracted hordes of mosquitoes that used the pools as their breeding grounds. While this created the additional nuisance of insect bites, the real threat emerged with the arrival of yellow fever, for which mosquitoes are a vector. By the end of 1881, some 2000 men were at work. In 1882, 400 deaths from yellow fever occurred, and in 1883 a total of 1300 men had died from the disease. Probably as much as one-third of the labour force were infected at any one time. By December 1888, rampaging yellow fever, together with the ever-increasing cost of the construction, led to the financial collapse of Lesseps' company. It dissolved in February 1889. An invisible parasite had stopped the ambitious work (Wills 1996).

Evolutionary Parasitology: The Integrated Study of Infections, Immunology, Ecology, and Genetics. Second Edition.
Paul Schmid-Hempel, Oxford University Press. © Paul Schmid-Hempel 2021.
DOI: 10.1093/oso/9780198832140.003.0001

Yellow fever is caused by a single-strand, positive-sense RNA virus belonging to the Flaviviridae (group B arboviruses). This family of viruses includes dengue, hepatitis C, and Zika virus (Chippaux and Chippaux 2018). The virus is haemorrhagic—that is, by damaging blood vessels it can lead to uncontrolled bleeding ('haemorrhage'). The first symptoms appear three to six days after the infection, with swellings and cell death. In the majority of cases, the infection is short and intense, and patients fully recover and acquire a long-lasting immunity against the virus. However, in a minority of cases (around 15 per cent of patients), the infection develops into a severe problem. Sudden high fever, a yellow tint in the eyes, jaundice, and bleeding that leads to 'black vomit' are the typical symptoms. In the process, the liver cells are destroyed, which leads to acute liver failure (and so to jaundice). Severe infections, if untreated, are associated with high case-mortality rates of 30–60 per cent. The blood remains infective and can be transmitted further by mosquitoes during a period from the first to the third day of fever (Cook and Zumla 2008).

Yellow fever originated in West Africa, where the virus has a reservoir in wild animals, especially in monkeys. With the increasing trade connections between Africa and the Caribbean in the sixteenth and seventeenth centuries, and the slave trade in particular, yellow fever spread to the New World (Powell et al. 2018). It was first recorded in 1648 in the Yucatan peninsula and the Spanish settlement of Havana, Cuba. Twenty years later, in 1686, yellow fever had reached Brazil. Following the trading routes, yellow fever subsequently jumped back from the Americas to the European continent, where it caused an outbreak in Cádiz, Spain, in 1730. Later, such outbreaks happened in Marseilles, France, and England (1878). After its first introduction in Central America, yellow fever had established an animal reservoir there, too, mainly in howler monkeys. Epidemic outbreaks in howler monkeys had repeatedly occurred, starting in Panama and spreading along the east coast of Central America to Guatemala. In 1914, Sir Andrew Balfour (1873–1931), then the founder of the Wellcome Museum of Medical Science (and, in 1923, the first director of the London School of Hygiene and Tropical Medicine), noted

that a yellow fever epidemic in Trinidad had led to a 'silent forest'. All howler monkeys had died from the infection (Balfour 1914; Cook and Zumla 2008). This so-called 'sylvatic' or 'forest cycle' of yellow fever is still a reservoir of infections for the human population (Klitting et al. 2018; Moreira-Soto et al. 2018).

Yellow fever was one of the most feared diseases in the eighteenth and nineteenth centuries—a threat that, for the last century, has been distant for people in modern Western civilizations. Furthermore, yellow fever's historical consequences were remarkable. It was not only a prime factor in the depopulation of tropical America at these times, but also affected America's history. In around 1800, the French controlled large territories in the Caribbean, Central America, Mexico, Louisiana, and Canada. In 1791 a rebellion by slaves under the Black leader General Toussaint Louverture (1743–1803), himself a descendant of African slaves, started in the French colony of Santo Domingo (now Haiti). By 1801, Louverture had declared a sovereign Black state. Napoleon was forced to send his brother-in-law, General Charles Leclerc, to subdue the rebellion. However, over 27 000 troops, including Leclerc himself, died from yellow fever within months of arriving in Santo Domingo. At the same time, yellow fever had little effect on the Black African rebels whose ancestors came from West Africa, the region where yellow fever had been around for a very long time and where the population had become less susceptible to the infection. One consequence of the epidemic was that the French withdrew from the Americas and sold Louisiana to the United States (Oldstone 1998). Others suffered, too. In Philadelphia in 1793, the American capital at the time, the disease claimed over 10 per cent of the population (around 40 000 people). From 1793–1796, the British Army in the Caribbean lost some 80 000 men, over half of them to yellow fever. Even in the peaceful period between 1817 and 1836, the annual death rate of British soldiers in the West Indies was six to ten times higher than at home, primarily due to diseases such as yellow fever. West Africa itself became nicknamed the 'White Man's Grave', mostly because of the widespread presence of yellow fever in this area; the associated mortality was 30 times as high as in the homeland (Crosby 1986). Up to the

Figure 1.1 A fumigation car for the control of yellow fever in Panama City, 1905. Such control measures were used in preparation of the construction of the Panama Canal by American companies. Control of mosquito populations was first introduced by the US Army medical scientist Walter Reed and his team in Havana, Cuba, after 1900. Photo: Panama Canal Museum (Canalmuseum.com).

early years of the twentieth century, massive yellow fever epidemics repeatedly swept through the Caribbean and up the North American coasts, regularly terrifying people.

In 1881–1882 Carlos Juan Finlay, a Cuban physician based in Havana, suggested that yellow fever was a mosquito-borne disease. A next step occurred with the Spanish-American War of 1898, where the United States backed the rebels fighting for the independence of Cuba from Spain. As the United States sent in troops, yellow fever decimated many times more men than the combat itself, especially also during the occupation period that followed the initial fights. The US Army medical scientist Walter Reed (1851–1902) and his team ('The Reed Commission') staged bold experiments in research stations just outside Havana where volunteers were exposed to mosquitoes. The experiments finally proved, in 1900, that the yellow fever mosquito, *Aedes aegypti*, indeed vectors yellow fever. The insight allowed for successful campaigns to destroy the mosquitoes' breeding grounds. Only with such control measures (Figure 1.1) did it became possible to complete the construction of the Panama Canal during the years of 1906–1914. Finally, by 1928, the South African virologist Max Theiler (1899–1972) and his Harvard mentor Andrew Sellards showed that the agent of yellow fever is a virus. In 1937, Theiler, then working at the Rockefeller Institute, developed a safe and successful vaccine that is still

in use today. He was awarded the Nobel Prize in 1951 for the discovery. Nevertheless, yellow fever remains a health problem in tropical America and Africa (Barrett 2018; Douam and Ploss 2018) (Figure 1.2).

1.2 Some lessons provided by yellow fever

This dramatic piece of history illustrates several issues that this book covers. First of all, we might ask: what is a parasite? Numerous definitions exist. Here, we use a pragmatic view—parasites are organisms (including viruses) that live in or on another organism, from which they obtain resources (e.g. nutrition, but also shelter): ecologically, the two parties are antagonistic to one another, with parasites gaining fitness at the expense of the hosts, and vice versa. Note that, throughout the book, the terms 'parasite' and 'pathogen' are used interchangeably, with 'pathogen' typically referring to viruses, bacteria, or protozoa that cause disease (i.e. are pathogenic); yet, in their ecological relationships, pathogens are parasites.

1.2.1 Parasites have different life cycles and transmission modes

Yellow fever is a parasite that needs a vector—a more or less passive transport vehicle—to get from one host to the next. Not all parasites transmit in this way. Many can jump directly; for example, via air in close contact (e.g. influenza virus, SARS), by transfer of body fluids (e.g. HIV or Ebola virus), or by water over more considerable distances (cholera, typhoid bacteria). Some parasites have evolved to utilize an intermediate host where necessary developmental steps occur. For example, the digenean trematode *Schistosoma mansoni* (causing bilharzia) uses the freshwater snail *Biomphalaria glabrata* as its intermediate host, from where it transfers to the (final) human host. In the final host, *Schistosoma* reproduces sexually. The parasite's eggs penetrate the host's veins, intestines, or bladder, where they cause harm. A few parasites have incorporated more than two hosts in their life cycle, such as the lancet liver fluke (*Dicrocoelium dendriticum*) that passes through hosts of three very different phyla:

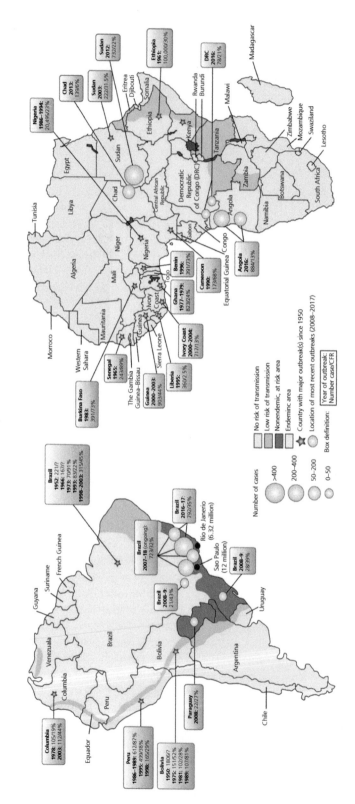

Figure 1.2 Yellow fever today. The figure shows the geographical distribution of the endemic and high-risk areas in Africa and South America. The boxes indicate major outbreaks since 1950 and until 2018. See legend for further explanations. CFR: case fatality rate. Reprinted from Douam and Ploss (2018), with permission from Elsevier.

Number of cases

>400
200–400
50–200
0–50

No risk of transmission
Low risk of transmission
Nonendemic, at risk area
Endemic area
Country with major outbreak(s) since 1950
Location of most recent outbreaks (2008–2017)

Box definition:
Year of outbreak:
Number case/CFR

**Chad
2013:**
139/65%

**Sudan
2012:**
732/2.2%

**Sudan
2003:**
222/31.5%

**Ethiopia
1961:**
100,000/30%

**DRC
2016:**
78/21%

**Nigeria
1986–1994:**
20,495/23%

**Benin
1996:**
391/73%

**Cameroon
1990:**
173/68%

**Angola
2016:**
884/13%

**Burkina Faso
1983:**
391/73%

**Senegal
1965:**
243/69%

**Ghana
1977–1979:**
823/24%

**Ivory Coast
2000–2004:**
717/13%

**Guinea
2000–2003:**
903/44%

**Liberia
1995:**
360/2.5%

Rwanda
Burundi
Kenya
Malawi
Tanzania
Zambia
Angola
Namibia
Botswana
Zimbabwe
Mozambique
Swaziland
Lesotho
South Africa
Madagascar
Congo
Gabon
Equatorial Guinea
Democratic Republic of Congo (DRC)
Central African Republic
Sudan
Ethiopia
Eritrea
Djibouti
Somalia
Chad
Niger
Nigeria
Togo
Sierra Leone
Guinea
Guinea-Bissau
The Gambia
Mali
Mauritania
Western Sahara
Morocco
Algeria
Libya
Egypt
Tunisia

**Columbia
1978:** 105/19%
2003: 112/44%

**Peru
1986–1989:** 612/87%
1995: 499/38%
1998: 165/29%

**Bolivia
1950:** 180/67
1975: 151/52%
1981: 102/28%
1989: 107/81%

**Brazil
1952:** 221/?
1966: 167/?
1973: 70/91%
1993: 83/22%
1998–2003: 315/45%

**Brazil
2016–17:**
792/55%

**Brazil
2007–18 (ongoing):**
723/32%

**Brazil
2008–9:**
28/39%

**Brazil
2008–9:**
21/43%

**Paraguay
2008:** 22/27%

Rio de Janeiro
(6.32 million)

Sao Paulo
(12 million)

Columbia
Ecuador
Venezuela
Guyana
Suriname
French Guinea
Peru
Brazil
Bolivia
Paraguay
Argentina
Uruguay
Chile

snails (Mollusca), insects (Arthropoda), and then to a vertebrate (Chordata). Finally, a large number of insect species have evolved to become parasitoids. The larval stages of parasitoids live inside or on the surface of a host, from which they extract resources. The adult insect is free-living, searches for mates, and females eventually lay eggs or larvae to invade a next host. These differences in life cycles and transmission modes have various consequences for the ecology and evolution of host–parasite interactions, as well as for their control.

1.2.2 Not all host individuals, and not all parasite strains, are the same

Only some people infected by yellow fever progress to the second more dangerous stage of the disease. Similarly, West Africans were generally more resistant to yellow fever than French or British soldiers. In other words, there is within- or among-population variation in the susceptibility of hosts to a given parasite. On the other hand, not all yellow fever viral strains are the same, either. Today, epidemiologists distinguish between urban yellow fever that is transmitted by the mosquito *Aedes aegypti*, and which is prevalent in tropical urban areas; and sylvatic or jungle yellow fever, which is the same parasite but a variant that primarily causes a disease of monkeys in the tropical forests of South America and Africa, and where humans only occasionally become hosts. An infected female mosquito can also transmit the yellow fever virus to its offspring, from where it can again infect another monkey or human. Differences not only exist between urban and jungle forms of yellow fever. There are also more or less virulent strains. For example, the standard yellow fever vaccine (YF-VAX) is based on strain 17D, which was initially isolated from a patient named Asibi. The properties of this strain allowed Max Theiler to maintain it in cell culture, where it attenuated to become a safe, live vaccine.

1.2.3 Physiological and molecular mechanisms underlie the infection

Consider the 'problems' a yellow fever virus has to solve to be successful. First, it must reach a new host through the bite of an infected mosquito. Once inside the bloodstream of the human host, it must enter a suitable target cell and multiply. Finally, it has to again be present in the bloodstream to be taken up by another mosquito and become transmitted to the next host. All of these steps require processes that unfold at the physiological, biochemical, and molecular levels. For example, the virus gains entry by receptor-mediated endocytosis, i.e. it manages to get ingested by the host cell. Later, the synthesis of new viral RNA occurs in the host cell cytoplasm, while the synthesis of viral proteins (that form the capsule) happens in the host endoplasmic reticulum. In the subsequent assembly of new viruses (the virions), protein C binds RNA to the viral nucleocapsid and thus ensures proper packaging of the genetic information (RNA) into the (protein) capsule. The protein NS1 affects the release from the host cells, and so forth. The host's immune system, in turn, responds to infection by activating signalling cascades and expressing the genes responsible for antiviral defence; this includes the recruitment of lymphocytes that can recognize virus-infected cells and destroy them. Furthermore, parasites like yellow fever, requiring a vector for transmission, have to deal not only with the human (vertebrate) immune system but also with that of the mosquito (an insect). These systems are exceedingly complex and will be illustrated in Chapter 4.

Together, these physiological and molecular mechanisms produce macroscopic phenomena that we know as parasite 'virulence', or host 'resistance'. They are based on genes that become differentially expressed at various stages of parasite infection, replication, and transmission. We therefore distinguish between the mechanisms that lead to a particular outcome of the infection, the underlying genetic basis for these mechanisms, and the function of parasite virulence or host resistance; that is, their value for survival and reproduction (the fitness) of host and parasite. Indeed, virulence and resistance are macroscopic traits that show phenotypic and genotypic variation within populations and are thus able to evolve. We must therefore expect that these traits have been shaped by natural selection to increase the fitness of their carriers. It is necessary to investigate the underlying physiological and molecular mechanisms. At the same time, the knowledge of mechanisms cannot answer questions about adaptive value and fitness—and, vice versa, we cannot infer mechanisms from the knowledge of the adaptive function of a trait alone.

1.2.4 Parasites and hosts are populations

The medical sciences, for obvious reasons, focus on hosts as individuals. Here, hosts and parasites are interacting populations. On the ecological scale, a population dynamics process, for example, a yellow fever epidemic, unfolds from this interaction. Its details depend, for example, on susceptibility and clearing of infections by the hosts, or perhaps on medical intervention. Nevertheless, an epidemic is also an ecological process in which population densities, frequency dependence, or other factors produce the changes in the level of infection over time, and both populations might also change as they evolve over ecological time scales. An epidemic is as much dependent on molecular mechanisms as on the laws of ecology, population dynamics, and the evolutionary process.

1.2.5 Parasites can be controlled when we understand them

The control of mosquito breeding grounds first achieved the control of yellow fever. Later, a vaccine also became available. Vaccination works because of a protective immune memory, but not all vaccines and not all parasites allow for such safe and long-lasting protection as is the case for yellow fever. Vaccination not only has consequences for the individual host that gains protection, but also alters the ecology and selection regime for the parasite population as a whole. For example, vaccination decreases the number of available hosts for the parasite. If enough hosts are so removed from the population, 'herd immunity' can be reached, such that the parasite is unable to find a sufficient number of new hosts. The parasite population declines and will eventually be eliminated. On the other hand, the vaccine-associated selection pressure on the parasite may lead to adaptations that are desired, or could lead to unwanted effects in the long term (Gandon et al. 2001). Again, we are reminded that the study of host–parasite interactions is not possible without an integrated approach that spans all levels, from molecules to ecology and evolution.

1.3 Parasites are not a threat of the past

Conservative estimates and the regular reports by the World Health Organization suggest that, currently, hundreds of millions of people are infected by parasites worldwide. Very many thousands die because of infections every year. Moreover, parasites are not only present in less developed countries, but should also be a source of disquiet for industrialized countries with high living standards and modern medical services. Hence, the threat by parasites and the associated diseases is real, persistent, and potentially devastating (see Figure 1.2). It probably needed the recent SARS-CoV-2/Covid-19 pandemic to remind the world at large of these facts. By comparison with yellow fever (up to 15 per cent overall), the fatality rate of Covid-19 (probably around 0.5–1 per cent of cases overall) is relatively low. The plague (or 'Black Death' during 1347–1353) caused by the bacterium *Yersinia pestis* killed one-third to one-half of the population in Europe (DeWitte 2014). It was an essential factor in the historical turning point that ended the Middle Ages.

With the SARS-CoV-2 pandemic, we also see that neglecting and underestimating any parasite with pandemic potential, such as a highly transmissible virus, is never a good strategy and will cost vastly more in terms of money and human lives than staying prepared. It is an issue where a holy grail of evolutionary biology is touched—is it possible to predict the future evolution? The practical need is obvious: because it takes years even to develop, and then months to produce an available vaccine in large quantities, it would be an enormous advantage to be able to predict, for example, which viral strain is likely to cause the next seasonal influenza or the next major pandemic (Du et al. 2017). However, even for a well-studied pathogen like the influenza virus, this is far from trivial. There is no simple relationship between genetic sequence (readily screened at large scales in a population) and the phenotype that determines the antigenic properties of the virus (Qiu et al. 2017). Nevertheless, predicting useful targets for a vaccine against the influenza virus may be within reach (see Figure 11.12). For a range of related questions, predicting evolution will probably remain the unattainable grail forever, since chance events can push the processes in very different directions.

A large proportion of the progress in human welfare and personal happiness is mostly due to improving public health by sanitation and hygiene, alongside the discovery of new medication to escape the grip of parasites. Not only do living organisms have physiology and follow the laws of genetics or molecular biology, but they also interact with their environment and thus are subject to evolution by natural selection in a given ecological context. The traditional boundaries between fields are not helpful for this necessarily integrating approach and must be put aside. The terms 'host' and 'parasite' are probably the universal ones throughout this book. They capture the notion that the ultimate job is to understand how and why they interact in the way we see it, regardless from which field our wisdom comes from, and regardless of whether we take 'parasite' to mean a virus, nematode, or parasitic insect.

Furthermore, we will benefit from the methods used in ecological and behavioural field studies, laboratory experiments, molecular and genomic techniques, mathematical modelling, and computing, and require a good sense for what might be going on between hosts and parasites. Studying parasites and their ways has often been equated with the work of a detective (De Kruif 1926). Indeed, much of the fascination of this study subject comes from the vast and still mostly unexplored terrain, where (to cite George E. Hutchinson, 1903–1991) hosts and parasites act out their evolutionary play in the ecological theatre.

CHAPTER 2

The study of evolutionary parasitology

In this book, we will consider phenomena such as immunological mechanisms, molecular processes, genomic patterns, or the ecological dynamics of interacting host and parasite populations, to mention some topics. However, the overarching theme for all of these questions is evolution by natural selection. It is, therefore, necessary to first look at the basic principles of this process.

2.1 The evolutionary process

Evolution is a dual process that unfolds in genotypic and phenotypic space (Figure 2.1). A phenotype develops from the 'program' laid down in the genotype, in interaction with effects from the environment. Thus, different genotypes produce different phenotypes, but the same genotype can also produce different phenotypes, depending on the environment (a phenomenon called 'plasticity') and within a specific range of possibilities (the 'reaction norm'). In the case of host–parasite interactions, selection may be through hosts being resistant to parasites or, vice versa, by the detrimental effects of infection on the hosts. Phenotypes that survive these challenges can eventually reproduce and leave progeny that carry the corresponding, 'successful' genes by inheritance into the next generation. Notably, selection only 'sees' the phenotypes, regardless of how they come about. For example, a parasite can infect a host because it has a 'non-resistant' genotype; alternatively, the host just happens to be in bad condition. In both cases, the selective event is the same (i.e. the host loses reproduction), but the evolutionary consequences are quite different. Since there is no genetic basis for the latter, no evolutionary change occurs.

The evolutionary process is governed by the short-term success of phenotypes and will produce traits best fitted to the organism's current environment. Hence, selection is also blind to the long-term consequences. Currently, successful variants will accumulate in the population, regardless of whether the population can no longer deal with a future change in the environment. Darwin's postulates for evolution by natural selection summarize these principles. They stipulate four basic observations that can independently be verified:

1. Individuals in a population vary in their phenotype. For example, individual hosts vary in their susceptibility to infection (Variation).
2. Parents produce more offspring than eventually can survive and reproduce themselves, leading to competition for resources (Competition). For example, the human parasitic fluke, *Schistosoma mansoni*, lays around 300 eggs every day over many years of infection. This parasite is obligatorily sexual, and, hence, in a stable population, only two out of hundreds of thousands of eggs will, on average, give rise to an adult pair of worms.
3. Some offspring happen to be better suited for current conditions. These are more successful in surviving and reproducing than less suitable types (Selection). For example, a few *Schistosoma* offspring may be more likely to escape the immune responses of the intermediate host, a freshwater snail. These individual flukes will be

Evolutionary Parasitology: The Integrated Study of Infections, Immunology, Ecology, and Genetics. Second Edition.
Paul Schmid-Hempel, Oxford University Press. © Paul Schmid-Hempel 2021.
DOI: 10.1093/oso/9780198832140.003.0002

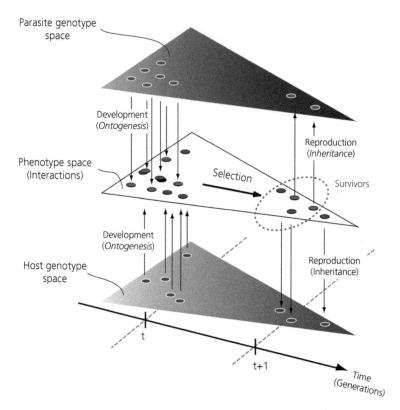

Figure 2.1 The process of host–parasite co-evolution. The dots represent the collection of different genotypes in the population of hosts and parasites in their genotype spaces (space characterized by triangular plane at top, parasites; at bottom, hosts). The genotypes produce a corresponding phenotype (in interaction with the environment) which is mapped into the phenotype space (dots in middle plane), and where host and parasite phenotype meet and interact (the processes and molecules involved in this interaction can be characterized as the 'interactome'). Some host phenotypes are resistant against a given parasite phenotype, and some parasite phenotypes can infect a given host phenotype (a process of natural selection). Successful phenotypes survive and eventually reproduce. The dotted oval shows this region of survivors. Here, it is assumed that non-infected hosts survive (blue dots) and reproduce; likewise, parasites that have infected a host (red dots) can reproduce, too. The genes of reproducing hosts and parasites are passed on to offspring (by inheritance; arrows to the right) and are represented in the respective genotype spaces of hosts (bottom) and parasites (top) of the next generation ($t + 1$). Modified from original rendering by Lewontin (1974).

more likely to survive to reach the final host, a human.

4. Part of the variation among all phenotypes is heritable (Inheritance). For example, some *Schistosoma* eggs might inherit genes that code for surface proteins not easily recognized by the snail's immune system. As a result, these genes code for a favourable phenotype that has an edge in the competition.

If these four conditions hold, evolution will become inevitable. As selection removes the less-suited forms, the better-suited ones become more common—the population phenotypically adapts to its environment (Adaptation). Because phenotypes are (partially) heritable, in the modern wording, evolution is a change of gene frequencies in the population, which results in the change of phenotypes over generations (Figure 2.1). Correspondingly, a prerequisite for an effect of selection is the presence of genetic variation in the population, from which selection can choose. The rate of evolutionary change is directly proportional to the fraction of genetic variation that is heritable (i.e. 'additive heritability'; see Chapter 10). Note that if selection were to continue unchanged, the genetic variation would finally become exhausted, and evolution would grind to a halt even though selection persisted. Therefore, continued evolution in a population requires either the appearance of new genetic variants (by mutation or immigration) or a change in selection regimes (see the discussion on the Red Queen scenario in Chapter 14).

Because selection can only choose among existing variants, the universally best adaptation does not necessarily result, simply because the existing variants may not contain the best solution. Instead, the evolutionary process is based on small 'improvements' on the existing variants by adding or changing something available. For example, the immune defence can become more efficient by merely reusing an existing molecule for a new purpose, e.g. a change of functions in immunoglobulins. Alternatively, a new regulatory element can modify the existing immunological cascade. Overall, the immune defence (the phenotype) does adapt in this way over the generations. Nevertheless, the underlying 'construction' inherits the past solutions. It resembles more the work of a tinkerer that adds whatever is within reach than the carefully considered construction an engineer would realize from scratch.

Because phenotypic and genotypic variation is needed, evolution is not something for the individual but something that happens in populations. Indeed, evolutionary considerations rest on 'population thinking'—which means numbers and probabilities. Evolutionary processes are, in fact, stochastic ('randomized') in many respects, which is an inevitable consequence of the numerous, varying factors that act on organisms and their environment. Therefore, events are not precisely predictable for any given case, but nevertheless follow an underlying probability distribution that can be determined. Hence, even against a 'noisy', stochastic background, phenotype A may still have a slightly higher probability of surviving and reproducing than phenotype B. Because this selective difference comes into play on many occasions in the population (i.e. with many individuals of type A or B, respectively), on average, more descendants of type A than B will eventually be produced and carry the responsible genes into the next generation. As a result, 'genes' of type A will accumulate at the expense of type B. The strength, speed, and direction of evolutionary change, therefore, depend on various

parameters and involve several mechanisms at the same time. None of them contradicts the basic Darwinian scenario. Rather, these additional processes add to the range of phenomena that govern the course of evolution (Box 2.1). The above describes the process of 'microevolution', which is at the heart

---

Box 2.1 The basic evolutionary forces

The modern theory of evolution considers several different processes. The combination of these evolutionary processes determines the size, direction, and speed of evolutionary change.

- *Mutation*: This is a summary term that covers several different processes such as point mutations, translocations, deletions, insertions, or gene duplications. Mutations arise *de novo* and change the existing genetic information in a gamete. A mutation is the ultimate source of novel genetic information on which all other evolutionary forces can act. Gene duplication, in particular, is an essential source of major evolutionary novelties.

- *Selection*: This is the differential survival and reproduction of individuals, genes, or genotypes within a population, depending on how well the associated phenotype fits the current environment. Selection is the most powerful evolutionary force.

- *Genetic drift*: A chance process, especially noticeable in small populations and caused by the sampling from a given distribution of frequencies. Genetic drift occurs when only a limited number of gametes or offspring can survive and reproduce that together cannot carry all possible genes and their combinations. Drift can lead to chance loss of genes and genotypes.

- *Gene flow*: When individuals migrate to another population and reproduce there, new genes immigrate into this population. Continuous gene flow can homogenize gene frequencies among populations.

- *Inbreeding*: Inbreeding occurs when offspring are produced by individuals that carry more similar genes than expected by chance in the population. Inbreeding promotes homozygosity and happens when relatives mate with each other.

- *Recombination*: Recombination rearranges existing genes into new combinations but does not change existing genes. Combinations of genes become relevant when selection on genes depends on the presence of others somewhere else in the genome (epistasis).

of evolutionary change. Extending these processes in space and time leads to cumulative changes that sum up to 'macroevolution', notably the formation of new species. Despite this complexity, fields such as population or quantitative genetics have developed powerful toolboxes to analyse evolutionary change, and fields such as behavioural or evolutionary ecology have defined concepts and procedures to study the adaptive values of traits.

2.2 Questions in evolutionary biology

Evolutionary biology wonders about the diversity of organisms and their characteristics. Therefore, questions are about the adaptive value of traits and how these have evolved from their origins. A trait is a characteristic of an organism that we can delineate, define, describe, and measure. In the current context, these traits may be host resistance, tolerance to infection, or parasite virulence. Readers less used to the evolutionary discourse should note that when characterizing an adaptive strategy, one often speaks 'as if' animals or a parasite would be able to think and decide. This cognitive ability is, of course, not meant by this. The wording simply is a very powerful shorthand. Instead of saying something like 'during evolution by natural selection, parasite genotypic variants with a phenotype causing intermediate virulence were transmitted more often and therefore left more progeny. As a consequence, the genetic information for this trait accumulated in the population. Extant parasites, therefore, show intermediate virulence, which, compared to other levels of virulence, is associated with the highest probability of survival and reproduction in the current host population', we might say, for short, 'parasites choose an intermediate level of virulence to maximise fitness'. This shorthand does not imply that a virus, for example, can 'think' and 'decide'; instead, the process of evolution by natural selection has produced a result that follows rational terms.

To study a trait, the Dutch ethologist Niko Tinbergen (1907–1988) (Tinbergen 1951) suggested that four different questions are relevant:

1. *Mechanism*: How does a trait work? This is a question about the physiology and molecular biology behind an observed trait or phenomenon.

For example, which molecules are synthesized by hosts for defence against infection? Irritatingly, this is called a study of 'function' rather than of 'mechanism' in immunology or molecular biology.
2. *Function*: What does a trait serve for? This asks for the value of a trait in terms of survival and reproduction of its carrier (the adaptive value). We may thus ask whether the production of, for example, antimicrobial peptides is efficient for defence against a particular infection, or whether other defences would be better.
3. *Ontogeny*: What is the development of a trait from the egg/zygote to the mature individual? For example, from which stage onwards can a host even produce antimicrobial peptides and respond to an infection?
4. *Phylogeny*: When and how did this trait appear during the historical course of evolution of organisms? Have ancestors produced antimicrobial peptides, and when in history did it happen?

These points illustrate some essential elements for the study of adaptive traits. In this book, we will also use the concept of the disease space, developed by David Schneider (Schneider 2011; Torres et al. 2016), to particularly highlight the intrinsic link between the within-host mechanisms and the phenomena seen from the 'outside', such as the variability in defence characteristics among individual hosts or the evolution of parasite virulence. The disease space characterizes host status with parasite load and contains the trajectory that an infecting parasite population takes through this space (Box 2.2). The disease space aids in connecting the different topics discussed in the book's chapters.

2.3 Selection and units that evolve

We have treated Darwin's four postulates as reflecting the advantages for individuals rather than for groups or species. This distinction touches the question of levels of selection and what units can evolve. The 'individual' here is a shortcut for the genetic information (laid down in the DNA or RNA sequence of the genome) that codes, for example, for a virulent parasite (the phenotype). If beneficial,

Box 2.2 The disease space

The disease space is a concept initially developed by David Schneider and co-workers (Torres et al. 2016). The space plots the individual host status against the parasite load. The disease space can be rendered in many ways, depending on the measures used. For example, host status could be the current body mass, the titre of red blood cells, or the level of fat reserves. In contrast, parasite load can mean infection intensity (the number of parasite cells in the host), parasite body size, or the current growth rate of a viral population. We will here use a generalized representation showing host condition vs parasite infection intensity. The disease space itself does not explain why an individual host status occurs, or why parasites have multiplied to a given load. However, it serves as a tool to illustrate the dynamics and consequences of infections that unfold in individual hosts.

In this general form, host status in disease space has four domains—Healthy, Sick, Recovery, and Death (Figure 1). These are broad classes of host condition and refer to the medical interest in host health. A typical course of infection follows a path through disease space (the 'infection trajectory', see arrow in Figure 1), which starts when a healthy host acquires an infection at point (1). The parasite then establishes and

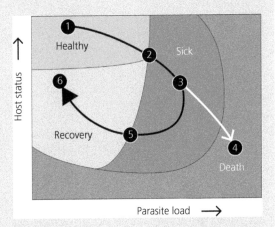

Box 2.2 Figure 1 The disease space. Adapted from Torres et al. (2016) under CC BY.

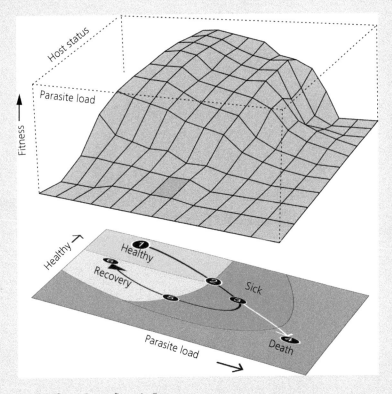

Box 2.2 Figure 2 Host fitness in disease space.

continued

Box 2.2 *Continued*

grows or multiplies within the host. As the parasite load increases, health typically decreases, and the host eventually becomes sick (2). In this area, a decisive point (3) is reached, where the infection trajectory either leads to a fatal condition (4) and the host dies, or where the infection can be controlled and cleared such that the host condition allows recovery (5). If cleared, the host will eventually return to a healthy status (6). Numerous variations of this basic scenario are possible, depending on strategies of host resistance and tolerance, or of parasite transmission to a next host.

Within each domain, the host condition is not steady but gradually changes as we move along the surface. The situation becomes clearer when host fitness values are added to each point of the space, as in Figure 2, forming the host's fitness surface associated with the disease space. Fitness values are assumed highest when the host is healthy and furthest from a fatal condition. As the host status moves into the sick domain, fitness decreases and becomes zero when the host is dead. A few points are worth mentioning. The

fitness values characterize the statistically expected future fitness of a host that has reached a certain point in disease space. This is similar to the residual reproductive value in life history theory, and which is associated with a given age. The value takes into account the population background, e.g. the age structure or future mortality risks. Strictly speaking, host death is also not necessarily equal to zero fitness; for example, when fitness also results indirectly through relatives (kin selection).

Similarly, Figure 2 shows host fitness as a static surface. We might also imagine this surface to change dynamically as the host–parasite interaction unfolds, as environmental conditions change, and so forth. Furthermore, we could also plot a fitness surface of the parasite in this disease space, which would—by definition—not be congruent to the host's surface. Probably only a perfect symbiont would match the host's surface, suggesting how differences between the fitness surfaces of host and parasite translate into selection pressures at different points in disease space.

this genetic variant becomes more frequent in the population, and more individual parasites carry this genetic variant. The individuals themselves cease to exist at the end of their life, and only the genetic information they pass on to their offspring persists over the generations.

So, what entities can evolve in the first place? We require that such entities have long-term persistence and are heritable, but not immutable. Individuals and groups do not qualify because, in each generation, they develop anew. Genotypes (the ensemble of all genes in a genome) do not automatically qualify, because they are typically destroyed by recombination in each generation, too. The 'genes' (i.e. the genetic information) are the most fitting units that can evolve—genes are passed on to the next generation and can persist indefinitely. At the same time, they can occasionally change by mutation such that new variants come into existence. Note that the 'gene' in this sense is an abstract concept that ignores how a coding gene is physically structured (e.g. in one stretch or several parts).

Now consider a situation where parasites of type A and B co-infect the same individual host. Three parties are now involved, and selection acts for all

three of them at the same time. The host defends itself, whereas the parasites counteract the immune responses, and each parasite competes with the other. Here, selection acts at different levels, for example on the single individuals, on the pair of parasites, on the host, but also on the entire set of hosts and parasites. Whereas it is often straightforward to identify selective forces at such various levels, it is typically more challenging to understand what the evolutionary response to these forces will be.

In many cases, for instance, the effects of selection extend beyond the individual. When parasites A and B are related to one another (e.g. being descendants of the same parents), they share a certain fraction of their genes due to common ancestry. A trait of A that selectively benefits B will also favour the same genes of type A residing in B, at least with a certain probability (a contribution given by the degree of relatedness of B as viewed from A). Therefore, looking at parasite A in isolation does not fully account for the total evolutionary effect of selection. In this case, the calculations will have to follow William D. Hamilton's (1936–2000) concept of 'inclusive fitness' (Hamilton 1964). In biology, relatedness by common descent is the prime process that generates similarity of genes. However, it is the statistical associations between the genotypes of the interacting parties, no matter how they come about, in combination with how selection affects the different levels, which predicts the evolutionary change that takes place. Evolution in such correlated landscapes can follow its own course. Technically, this can be analysed with George R. Price's (1922–1975) covariance equation, which applies to all forms of hierarchically organized units and all forms of selection (Frank 1995, 1997).

2.4 Life history

The idea of a 'life history' is a powerful concept when studying adaptations (Stearns 1977, 1992). For a free-living organism, the life history starts with its birth and ends with its death. By analogy, the life history of a parasitic infection starts with infection and ends with the infection disappearing from the host. The life history concepts, therefore, add a time axis and a 'lifetime achievement' to the study of parasitism (Figure 2.2). On this time axis, it becomes apparent that fitness in the evolutionary ('Darwinian') sense results from both survival and

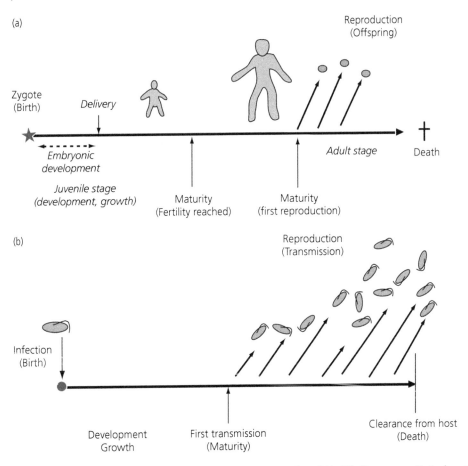

Figure 2.2 The life history framework. (a) Life history in a free-living, sexual organism. The individual life history starts with the formation of the zygote. In the case of mammals, embryonic development is ended by the delivery (physical birth) of the young. The juvenile stage lasts until physiological, sexual maturity is reached. Sometime thereafter, reproduction takes place, typically spaced out over different episodes. The life history ends with the death of the individual. (b) Life history of a parasitic infection. Here, the life history starts at the moment of infection ('birth' of the infection). In a first phase, the parasite develops as it establishes itself, invades host tissues, grows in size (e.g. helminths), or multiplies in number (e.g. bacteria). The time of first transmission marks the 'maturity' of the infection, which is followed by bouts of transmission ('reproduction'). The life history in this example ends with the clearance of the infection from the host ('death').

reproduction over a lifetime. Fitness is zero if an organism does not survive to reproduce, and also zero if it survives but does not reproduce. In particular, fitness results from adding up the number of offspring an individual is producing over its lifetime (ignoring further factors such as quality of offspring). In practice, this is often cumbersome or not feasible. Therefore, proxies are being used instead, such as survival to reproduction, competitive ability, and several other measures. The choice of proxies depends on the questions asked but remains embedded in the life history perspective. The principle is illustrated with the formal treatment of the expected mean fitness of a population, W. This adds the contributions of different age classes (x) to reproduction (b_x), weighted by the probability to reach this age (l_x), such that:

$$W = R_0 = \sum_{x=A}^{k} l_x b_x \qquad (2.1)$$

where A is the age at maturity, and k symbolizes the end of an average lifetime. Fitness W yields the growth of a population; more precisely, this is a ratio and also known as the 'reproductive number', R_0; that is, a factor by which the population multiplies per generation. In epidemiology, R_0 describes the fitness, that is, the ability to spread, of a parasite that enters a host population (see Chapter 11). In the long run, most populations are stable, that is, $R_0 = 1$. Hence, with Euler's equation (named after the Swiss mathematician Leonhard Euler, 1707–1783), we have:

$$1 = \sum_{x=A}^{k} e^{-rx} l_x b_x \qquad (2.2)$$

where r is the intrinsic (instantaneous) rate of increase of the population (or of a particular genotype). These equations are useful to gain an understanding of how changes at age class, x, affects lifetime fitness, and how the different parameters are connected to each other.

The life history framework also illustrates the basic principle of a 'trade-off'. For example, an individual might put more resources into reproduction at an early age, yet at a cost for survival later in life—thus changing the values of b_x and l_x along the

age classes in eq. (2.1). Such trade-offs between different components of fitness (b_x, l_x) are at the heart of the life history theory. Limitations might set a trade-off in the physiological capacity of the organism; for example, when a strong defence requires time and resources that are then no longer available for reproduction (see Chapter 6). A trade-off can also be based on a gene that affects both aspects. For example, the expression of a cytokine gene can stimulate one type of mammalian immune cell (e.g. Th1 cells) but at the same time suppresses another one (Th2). This case is 'pleiotropic antagonism'—where the same cause (the expression of a gene) increases one trait but decreases another. Antagonistic pleiotropy can also connect the expression of traits at different stages in the life history; for example, the expression of the same gene might increase reproduction early in life but have detrimental effects later. Pleiotropic antagonisms play an essential role in thinking about how different tasks should be timed in the life history in order to maximize the eventual lifetime fitness. Formally, this can be analysed with eqs (2.1) and (2.2).

2.5 Studying adaptation

2.5.1 Optimality

The concepts of 'optimality' and 'evolutionarily stable strategies' (ESS) are among the most powerful tools for studying the adaptive value of traits. In each case, the question is, what strategy might provide the maximum possible fitness for an organism, i.e. a host or a parasite? A 'strategy' is a set of decision rules of the form: 'when in situation A, take action B'. For example, migratory locusts follow the strategy: 'when in a dense population, up-regulate the immune system' (termed 'density-dependent immunoprophylaxis'; Wilson et al. 2002), which serves to lower the risk of succumbing to infections where the risk of contagion is high.

A given trait or strategy is said to be 'optimal' when the associated phenotype (e.g. a parasite strain or a host genotype) achieves the highest possible fitness in a given environment. An optimum is only defined within the boundary conditions of the problem (the 'constraints') and with the relevant trade-offs. By definition, constraints cannot be changed by the individual during its lifetime because they are deeply rooted in the evolutionary history of the organism and genetic variation for this trait is now lacking. For example, some microsporidian parasites have evolved to kill their insect hosts to become transmitted. Spores cannot leave the host other than when the host corpse decays, e.g. *Nosema whitei*, which infects the larvae of flour beetles (*Tribolium* spp.). Within these constraints, the trade-offs become essential. Trade-offs assume that a trait is 'plastic': it can be changed within the lifetime of an individual—at a cost to another trait. In this sense, a microsporidium can 'choose' to balance the production of new spores against their harmful effects on the survival of the host. If parasite multiplication is too fast, the host is killed too early, and not enough spores are produced overall. If multiplication is too slow, the total number of spores remains too small when the host dies. Only the 'right' trade-off between these parameters will maximize the number of spores that the parasite produces. We will return to these considerations in the discussion of parasite virulence (see Chapter 13).

2.5.2 Evolutionarily stable strategies (ESS)

When the environment is ignorant of a strategy adopted by an organism, finding some kind of optimum strategy is feasible. However, if the environment responds to a chosen strategy and changes its strategy accordingly, this is no longer possible. This is particularly relevant for host–parasite interactions. For example, a defence strategy by the host is countered by an immune evasion strategy of the parasite (see Chapter 8). More generally, whenever the costs and benefits of a given strategy chosen by an actor (such as a host) depend on the strategy chosen by the recipient (i.e. the parasite), the situation is a 'game', and the toolbox of game theory is needed. Here, the most crucial concept is the ESS. An ESS, when adopted by the majority of actors in the population, cannot be bettered by any other alternative strategy (within the set of known constraints and

trade-offs); hence, an ESS is stable against the invasion of rare actors that adopt an alternative (mutant) strategy. An ESS must not be a simple strategy, such as a fixed point in time where the host should be killed. Instead, an ESS can be a complicated 'recipe' in itself, such as a probability distribution for the day the host should be killed, which could additionally depend on the environmental conditions. We will make frequent use of the ESS-strategy terminology without going into any of the mathematical details that allow their calculations.

2.5.3 Comparative studies

Evolutionary parasitology is studied in various ways. Methods include observation, experimentation, and theory (as with optimality and ESS analysis). However, comparative studies are an additional powerful tool. These studies explore the fact that nature has already made 'experiments' for us. Different lineages that have originated from the same ancestor have evolved in different ways and therefore can reveal why traits are different. At least some of these differences reflect adaptations for different environments, including the effects of co-evolution. A plausible hypothesis suggests that older lineages of parasites have acquired more host species. Comparing parasite lineages of different ages will provide a test for this hypothesis. In this case, the evidence is weak, and there is no universal trend towards a broader host range in older taxa (see section 7.2).

Comparative studies have their problems, of course. In particular, lineages with the same ancestor are not independent events as required for statistical analyses. How to eliminate these statistical dependencies is part of the methodical toolbox. Comparative studies, furthermore, rely on the reconstruction of phylogenetic relationships. Often, these phylogenetic relationships are uncertain, which can prompt alternative interpretations of the same data. The large toolbox developed for comparative studies and for understanding the tree of life more generally can now be used to study how an epidemic unfolds (see section 11.7).

Important points

- Evolution by natural selection unfolds in populations and corresponds to a change in gene frequencies. The process maps to the genotype space, which has a correspondence in the phenotype space. Selection affects phenotypes, but evolution only results when there are genetic variation and inheritance.
- Tinbergen's four questions are a useful framework for studying host–parasite interactions. Notably, the question for the underlying mechanism and for the function (the adaptive value) of a trait complement each other.

- The primary unit of evolution is the gene that 'acts' alone or in concert with others, and whose success is defined by the fitness of its carriers. The concept of life history considers events at different points in the lifetime of an organism. Their balance affects the overall fitness of an organism.
- Several methods are available for the study of adaptive traits. These include optimality analysis, the identification of ESS, and comparative study.

CHAPTER 3

The diversity and natural history of parasites

3.1 The ubiquity of parasites

The majority of all living organisms are parasites (Windsor 1988). This sounds like an exaggeration. However, any single-host species can be infected by numerous parasite species, and any given individual host can carry many different infections at the same time. For example, in a study in several lakes in Finland, it was found that at least 42 parasite species used the perch (*Perca fluviatilis*; Figure 3.1) as their host, and at least 38 species used the roach (*Rutilus rutilus*) (Valtonen et al. 1997). Very similar

numbers were found in other studies, for example in freshwater fish of Poland (Morozinska-Gogol 2006). These studies concentrated on macroparasites (Box 3.1). They did not include the wealth of microparasites, such as bacteria, viruses, or fungi, which are also known to parasitize fish.

Fish are typically large individuals and therefore may harbour many parasites. Nevertheless, small-bodied hosts rival their larger counterparts in the number of parasite species. Individual honeybees, for example, are smaller in size than perch or roach.

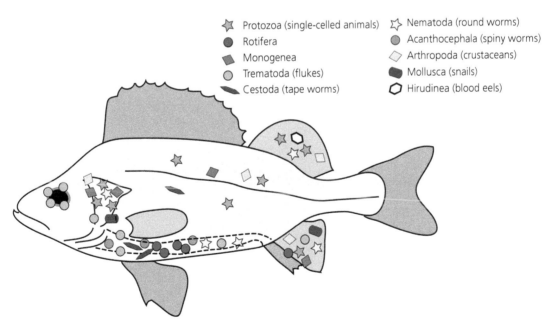

Figure 3.1 The diversity of parasites. The sketch shows the typical sites of infection for a number of different macroparasites of the European perch (*Perca fluviatilis*). On average, a single fish harbours 99.4 individual macroparasites from eight different taxa. The data are taken from Valtonen et al. (1997).

Evolutionary Parasitology: The Integrated Study of Infections, Immunology, Ecology, and Genetics. Second Edition.
Paul Schmid-Hempel, Oxford University Press. © Paul Schmid-Hempel 2021.
DOI: 10.1093/oso/9780198832140.003.0003

Box 3.1 Types of parasites

Parasites are assigned to different functional categories, independent of their taxonomic status.

Macroparasites: These are usually small parasites (up to a few hundred μm in size), typically referring to the viruses, bacteria, fungi, and protozoa. Microparasites have short generation times and can multiply to large numbers within their host. Their central feature, however, is that in epidemiological analyses (SIR-model) hosts can be considered infected or not infected. The parasites are not taken into account individually; growth of each individual parasite is typically neglected, and only replication is considered.

- *Macroparasites*: Usually large parasites, typically referring to helminths (the worm-like parasites: nematodes, cestodes, trematodes), parasitic insects, and some other groups (Acanthocephala, Hirudinea). Macroparasites often have generation times similar to their hosts and are found in low numbers in their hosts, sometimes with just a single individual. In epidemiological analyses, each parasite is considered individually; the analysis of body growth is as essential as reproduction.
- *Parasitoids*: Parasites that have a free-living stage while the juvenile lives in or on the host animal. This category typically consists of parasitic insects (such as ichneumonid wasps). However, it might also include species such as the mussel *Glochidium*, whose larva lives on the gills of fish, whereas the adult is free-living. Often, but not always, parasitoids kill their hosts before progressing to the next stage in their life cycle.
- *Endoparasites*: Parasites that live inside the host. Most microparasites fall into this category. Also, most para-

sitoid larvae are found inside a host. Vice versa, some microparasites live on the surface of the host, for example fungi that infect the skin.

- *Ectoparasites*: Parasites that live on the host or are attached to it. Typical examples are mites, ticks, lice, or parasitic groups such as suckerfish.
- *Social parasites*: Parasites that exploit the social life and structure of a host group. These are typically macroparasites such as insects, or other higher eukaryotes. Sometimes it is not easy to distinguish their actual functional status. Social parasites can be commensal (species that mostly live on debris or food remains without any noticeable harm, e.g. some beetles that live in the nests of ants). They can also be cleptoparasites (species that rob food from their hosts, e.g. some spiders that live in the web of their host spider) or brood parasites (e.g. cuckoo species that seduce the host to raise the parasite's eggs). Brood parasites, in particular, are often species related to their hosts. For example, closely related bumblebee species lay their eggs in the nests of other bumblebees (all within the genus *Bombus*), and some geese dump their eggs into the nests of conspecifics, which gives rise to intraspecific parasitism. Social parasites are not the focus of this book.

Several other terms exist, but strict definitions are not always helpful. As with any categorization of living systems, a specific organism may sometimes fall in between the definitions. Therefore, the above categories are a good guideline but cannot replace careful consideration in any given case.

However, honeybees are also social animals and thus offer additional opportunities for parasites to exploit the bee colony that might contain several tens of thousands of potential host individuals. More than 70 parasite species are described for this host species. In this case, the list also includes the microparasites such as viruses, bacteria, and fungi (Schmid-Hempel 1998; Chantawannakul et al. 2016). Most of the knowledge on parasites comes from hosts that have a commercial value for humans, such as freshwater fish or honeybees, or from pest species that cause much damage to human cultures or installations, such as the fall armyworm (*Spodoptera frugiperda*). The voracious larvae of this

moth are significant pests for many crops, but it turns out that this moth has numerous natural parasites. For example, 148 parasitoid species parasitize the larva of *Spodoptera*, 20 species attack the pupa, and a further eight species parasitize its eggs (Molina-Ochoa et al. 2003). There are undoubtedly many more that are not yet known. Some of these natural enemies might become useful adversaries to protect cultures against *S. frudiperda*. Hence, the situation illustrated in Figure 3.1 is probably rather conservative. It is also evident that almost all parts of a host body are exploited, and few spared. In all, for every host species, there are dozens or even hundreds of parasite species, and not every parasite, in turn, uses a similar number of host species. Indeed, there are more parasitic species than host species living on our planet.

3.2 A systematic overview of parasites

The diversity of parasites is genuinely enormous, with parasites varying across a wide size range (Figure 3.2) and coming from many different lineages (Goater et al. 2014; Morand et al. 2015). A short overview of the most important or biologically most exciting groups follows next. The diversity of parasitic plants, such as vines and root parasites, is not discussed in this book. Note that the taxonomy and systematics of many of the following groups are subject to ongoing changes, especially in the age of genomic approaches. However, classification is not the focus here. Instead, the groupings should primarily help to grasp the diversity and

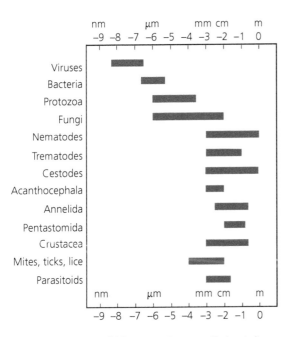

Figure 3.2 Body sizes of different parasite groups. The bars indicate a characteristic size range on the log scale (10^{-x}). Note that in any group there are a few very large or very small species that go beyond the typical range shown here. For example, paramyxoviruses can be up to 14 µm long, which is the size of a protozoan; in whales, nematodes can reach 7 m in length, cestodes up to 30 m. Fungi, too, are extremely variable in size; the length of their hyphal network in trees may reach dozens of metres, or even kilometres, in length.

characteristics of organisms that have evolved a parasitic lifestyle.

3.2.1 Viruses

Viruses are extremely reduced life forms that have no metabolism of their own. They must exploit the host cells to carry out crucial tasks for their survival and replication. Some scientists maintain that viruses are not a life form at all. Not surprisingly, therefore, all viruses are either parasites or live as symbionts within their hosts. The viral life cycle includes the attachment to a host cell and the gaining of entrance into the cell, followed by the uncoating of the virus and insertion of its genetic information into the host's genetic program. Its expression leads to the subsequent production and assembly of new viruses that eventually leave the host cell to infect a next one. A completed virus particle outside the host cell is a 'virion'. It consists of a nucleic acid molecule that carries the genetic information (DNA or RNA), with its supporting proteins and a protein shell (the capsid). The capsid allows for the morphological distinction of viruses. Also, some viruses (e.g. HIV) sequester some of their host's cell membrane as an additional outer lipid hull (the viral envelope), which presumably serves as additional protection against the host's defence responses. Finally, some viruses possess rather sophisticated morphologies, such as a complex outer wall. Bacteriophages, viruses that parasitize bacteria, possess complex tail-like structures. They act like 'legs' to land on and attach to the surface of the host cell.

The genetic material of viruses is either DNA or RNA. Sometimes, both types are present, such as in cytomegalovirus, where there is a core of DNA and several fragments of RNA. Furthermore, the nucleic acid molecules can be either single- or double-stranded (labelled ss and ds, respectively), and are organized either linearly or in loops. In RNA viruses, the genetic information is coded either in positive-sense (identical to the sequence of the corresponding mRNA) or in negative-sense (i.e. complementary to mRNA). The latter, therefore, needs to be translated into a positive-sense sequence first). In the so-called ambisense RNA viruses, stretches of positive- and negative-sense ssRNA (or ssDNA) are located on the same strand. These

Table 3.1 Viruses according to the Baltimore classification system, with examples.

	Group[1]	Families (examples)	Virus (example)	Remarks
I	dsDNA	Baculoviridae	Granulovirus	Mostly in insects.
		Myoviridae	Phage T4	Parasitic on enterobacteria.
		Papillomaviridae	Human papillomavirus (HPV)	On body surfaces, mouth, genital tract.
		Poxviridae	Orthopoxvirus (Variola major)	Agent of smallpox.
II	ssDNA	Geminiviridae	Maize streak virus	Insect-transmitted maize parasite.
		Parvoviridae	Parvovirus B19	Childhood infection, involved in arthritis.
III	dsRNA	Brinaviridae	Drosophila X virus	Mostly transmitted vertically.
		Cystoviridiae	Pseudomonas phage F6	A lytic phage that infects plant-pathogenic bacteria.
		Reoviridae	Blue tongue virus	Disease of livestock, vectored by insects.
		Totiviridae	Saccharomyces cerevisiae virus	A virus of fungi (yeast).
IV	(+) ssRNA	Coronaviridae	SARS	Respiratory diseases of mammals.
		Falviviridae	Hepatitis C virus	Liver infection leading to inflammation and cirrhosis.
		Picornaviridae	Poliovirus	A major human disease leading to paralysis.
		Togaviridae	Rubella virus	A classical childhood disease.
V	(−) ssRNA	Filoviridae	Ebola virus	A highly virulent haemorrhagic virus for humans.
		Paramyxoviridae	Measles virus	Long association with humans, causes epidemics.
		Rhabdoviridae	Rabies virus	Causing encephalitis, highly virulent.
VI	ssRNA-RT	Retroviridae	HIV	Agent of AIDS.
VII	dsDNA-RT	Hepadnaviridae	Hepatitis B virus	'Serum hepatitis'; liver inflammation and cirrhosis.

[1] ds: double stranded; ss: single stranded; RT: reverse transcription virus; (+): positive-sense RNA; (−): negative-sense RNA.

stretches are each transcribed on their own. As the few examples in the list of Table 3.1 impressively demonstrate, the diversity of viruses alone is bewildering, and every single virus has its own specificities and a life cycle that varies according to the group.

A special note is due to retroviruses. These are related to reverse-transcribing elements in the genome ('retrotransposons'). Their reverse transcriptase (RT, a polymerase) is a key feature of retro-elements and allows the (reverse) transcription of RNA into DNA (Finnegan 2015); they can also integrate themselves into a genome. Retrotransposons behave in several ways; i.e. as mobile elements that 'jump' around to different locations in the genome only to become degraded over time, or as plasmids, but also, essentially, as viruses. In this latter case, these elements pack their DNA or RNA into a virion and cycle through integration into the host's genome for their replication (Koonin et al. 2015).

Retroviruses are incredibly abundant in the genome of eukaryotes (Finnegan 2015), notably also in vertebrates (Xu et al. 2018) that harbour thousands of ancient retroviral sequences (the 'endogenous retroviruses') (Johnson 2019). Their effect on hosts is still mostly unknown, but substantial (e.g. causing cancer) and manifold for the cases studied so far. Notably, retroviruses can add novel genes and regulatory elements to the host's repertoire. For example, endogenous retroviruses may have shaped aspects of the innate immune system, such as the regulation of interferons (Chuong et al. 2016).

3.2.2 Prokaryotes

Prokaryotes are the most basic living organisms that have their own metabolism. However, they lack a cell nucleus and are typically unicellular. Prokaryotes are grouped into two major systematic domains—the Archaea and the (Eu-)Bacteria. The latter domain contains some of the most essential and abundant pathogens of animals, plants, and fungi.

3.2.2.1 Archaea

The Archaea (earlier somewhat misleadingly called the Archaebacteria) are among the oldest extant and most spectacular organisms on our planet. Their unique place in the tree of life was recognized only in the late 1970s (Woese and Fox 1977). Archaea superficially resemble bacteria, but biochemically and genetically they are very different. Most interestingly, the Archaea are probably more closely related to the higher organisms (the Eukaryotes) than the bacteria are. Archaea typically live in extreme habitats where no other organisms can survive. Examples are highly saline waters (e.g. *Halobacterium* dwells in salt lakes) and hot springs (*Thermoproteus* lives in the acidic hot water around Icelandic geysers), or the deep sea with its chemically extreme conditions (e.g. *Methanocaldococcus* is associated with deep-sea vents in the Pacific). New research shows, furthermore, that Archaea are also abundant in the plankton of the open seas. It is remarkable, however, that almost no parasites have so far been found among the Archaea (e.g. *Nanoarchaeum equitans*; Waters et al. 2003; Moissl-Eichinger and Huber 2011).

3.2.2.2 Bacteria

This is one of the most important parasitic groups. Until recently, the number of described bacteria (and Archaea) was relatively small, with a count of several thousand (Staley 2006). However, there is a general difficulty in identifying a bacterial species in the first place (Rosselló-Móra and Amann 2015; Munson and Carroll 2017)—a problem that also exists with viruses and other essentially non-sexual prokaryotes. Today, identification typically uses 6S rRNA gene sequences. Probably some 250 000 species of Bacteria and Archaea are yet to be described (Rosselló-Móra and Amann 2015). For most purposes, however, it is essential to distinguish and characterize different bacterial lines, regardless of their taxonomic status (a species or not), especially if these lines correspond to different functional types (Cohan 2006) and show different host specificities or degrees of pathogenicity.

A simple diagnostic tool to classify bacteria is the Gram staining process (named after the Danish bacteriologist Hans Christian Gram, 1853–1938). 'Gram-positive' bacteria retain the violet dye of the stain and appear blue or violet under the microscope. In contrast, 'Gram-negative' bacteria do not retain the dye and appear pink or red. Gram-positive bacteria have only one (inner) cell membrane that is covered with a thick layer of peptidoglycans. Peptidoglycans are important molecular signatures detected by immune systems to recognize a potential infection. Gram-negative bacteria possess an additional outer membrane. It overlays the peptidoglycan layer and contains lipopolysaccharides (LPSs), which are also detected by immune systems. Between the inner membrane and the outer layer, there is a periplasmic space. Therefore, Gram-negative bacteria have evolved several sophisticated secretion systems that allow transporting molecules from the inside of the bacterium to the external environment across this space. These secretion systems (e.g. the type III system) are essential in host–parasite interactions. Among the Bacteria, the Mollicutes lack a proper cell wall.

Bacteria are generally small (typically, some μm in length; Figure 3.2) and have no nucleus, which defines them as prokaryotes. They come in different shapes—spherical (Cocci), spiral (Spirochaetes), or rod-shaped (Bacilli). Bacteria have very different lifestyles and might dwell as individual cells or in dense aggregates, such as in biofilms (an example of which is dental plaque). Some bacteria are also able to form spores that can survive in the environment for a very long time. A drastic example is the spores of *Bacillus anthracis* (the cause of human anthrax). During the Second World War, British military scientists in 1942 conducted a test of biological warfare with sheep and a highly virulent strain of *B. anthracis* on Gruinard Island, off the Western coast of Scotland. The sheep died within a day, but the infective spores had entered the soil, and the subsequent decontamination work was unsuccessful. Therefore, Gruinard Island was put under quarantine and remained a forbidden island for many decades. Only in 1990, after an intensive decontamination program, and 48 years after the trial, was Gruinard Island reopened.

Bacteria reproduce asexually by fission. However, they can exchange genetic material among each other, even between different species. For example, the exchange of plasmids is the central process that contributes to the rapid evolution of antibiotic resistance, and a mechanism for the evolution of new, virulent strains and bacterial species (Gal-Mor and Finlay 2006). Bacteria have indeed evolved to infect virtually every other group of organisms except the viruses. Bacteria of one genus, *Bdellovibrio*, even attack other bacteria. They swim very fast, collide, and penetrate the cell wall of their victims to consume the cytoplasm. *Bdellovibrio* then grow inside their host, form new cells, and leave after the host cell wall is dissolved. Hence, *Bdellovibrio* behaves like a virus or a parasitoid insect. In general, parasitic bacteria mostly infect through the mouth and digestive tract. Less commonly, they infect via the integument (skin) or in the trachea and lung. Bacteria infecting the respiratory tract, however, include some of the most dangerous pathogens of humans (e.g. *Bacillus anthracis*, *Clostridium tuberculosis*).

Systematically, the (mostly) Gram-positive cluster of the Firmicutes is very diverse and among the largest groups of parasitic bacteria. The Actinobacteria are also Gram-positive bacteria with high G+C content. They mostly live as decomposers of organic material, e.g. *Actinomyces*, but with some being important pathogens (e.g. *Mycobacterium*, *Corynebacterium*). The restricted taxon Firmicutes (the low G+C group) includes the classes Bacilli, Clostridia, and Mollicutes, containing many feared pathogen genera such as *Bacillus*, *Staphylococcus*, *Streptococcus*, *Enterococcus*, *Listeria*, or *Clostridium*. Many species can form durable capsules resistant to desiccation and so can survive extreme conditions. The outer wall of these capsules consists of polysaccharides and is highly antigenic for the host.

The class of Mollicutes is an unusual group of bacteria, evolutionarily close to the Firmicutes. Mollicutes have a minimal genome (580–2200 kb) and no cell walls (and thus do not respond to Gram staining), and also lack some other components, such as fibrils, but are nevertheless able to actively 'swim' (Browning and Citti 2014). The Mollicutes include the (order) mycoplasms that can move by gliding on surfaces. Mollicutes presumably split from the Gram-positive Firmicutes around 65 million years ago. They progressively became more specialized to the parasitic lifestyle and lost some of their

genome in the process. Altogether, there are probably 200 extant species (Trachtenberg 2005). In humans, they are mostly extracellular and live on epithelial tissue in the respiratory and urogenital tract. The constituent group of Spiroplasms infects plants and their insect vectors such as bees.

The Spirochaetae are elongated and spiral-shaped bacteria with a characteristic corkscrew-like appearance. Most are free-living and can tolerate quite extreme conditions, similar to the Archaea; some have become parasitic. Due to their shape and a particular system of locomotion propelled by endoflagella, Spirochaetae adapted to move through highly viscous media, such as mucus, and to penetrate tissues that other bacteria cannot. *Treponema pallidum*, the causative agent of syphilis, is probably one of the best-known representatives of this group. Chlamydiae are very small (sometimes smaller than viruses) obligatory intracellular parasites in eukaryotic hosts. Examples are *Chlamydia trachomatis*, which causes eye trachoma, and *C. pneumoniae*, responsible for lung infections.

The large group of Gram-negative Proteobacteria include a wide variety of well-known pathogenic bacteria and further divide into the subgroups of Alpha- to Epsilon-Proteobacteria, showing different characteristics. The Alpha group consists mostly of phototrophic species but also contains dangerous pathogens such as *Rickettsia*. This group probably gave rise to eukaryotic mitochondria. The Beta group are important bacteria in soil, e.g. for nitrogen fixation, but also contain the pathogenic genera *Neisseria* and *Burkholderia*. Gamma-Proteobacteria contain some of the most significant human parasites, such as the Enterobacteria, the genera *Vibrio* (causing cholera), *Salmonella*, *Yersinia* (bubonic plague), *Escherichia coli*, and the Pseudomonadaceae. Finally, the Epsilon group is relatively small, and most species seem to dwell in the digestive tract of animals. Among those, *Helicobacter pylori* is an important and widespread human pathogen inhabiting the stomach.

3.2.3 The basal eukaryotes

Diplomonads are basal, flagellated eukaryotes lacking mitochondria and having two nuclei. The group is small (50–60 species) but includes, for example,

Giardia lamblia, which can be contracted via contaminated water and causes gastrointestinal infections in humans. Similarly, Trichomonads are mostly parasitic, primitive eukaryotes. They include *Trichomonas vaginalis*, which inhabits the human vagina, often found in otherwise weakened patients. Trichomonads are typically transmitted by direct contact.

3.2.4 Protozoa

Protozoa are a highly diverse group of small, single-celled organisms that otherwise possess all the basic traits of the higher metazoa (the multicellular organisms) and, in particular, are heterotrophic like animals or fungi (Cheng 1986). The vast majority of protozoa are free-living, but among those that have adopted a parasitic lifestyle, we find some important pathogens of humans and their livestock. The parasitic forms are not only very diverse in their morphology, life cycle, and habits, but also use a wide range of hosts (Table 3.2).

Protozoa are eukaryotes, i.e. they have nuclei containing the (nuclear) genetic material. However, the genetics of protozoa can be quite complicated. For example, some groups have one nucleus; others have two nuclei that resemble each other (e.g. the Sarcomastigophora, Apicomplexa, Myxospora). Further groups (e.g. the Ciliophora) have one micro- and one macronucleus. The protozoa reproduce by fission. It remains unclear, though, how many protozoa reproduce strictly asexually and so form clonal lines, or whether they occasionally reproduce sexually ('intermittent' or 'epidemic' sexuality). Also, whether sexual reproduction is common, and whether the reproduction involves the standard laws of meiotic recombination, remains unclear (Tibayrenc and Ayala 2002). Note that sexual reproduction means biparental reproduction, i.e. two parents contribute genes to any given offspring. Protozoa feed by several methods, such as by 'phagotrophy' (i.e. engulfing solid food such as amoebae). Alternatively, protozoa can also directly absorb nutrients through the body wall (saprozoic feeding), as in amoebae or flagellates (e.g. *Crithidia*). Many protozoa are capable of forming a cyst that can survive unfavourable conditions. Cysts are also transmission stages and can, furthermore, specialize concerning the attachment to the host's surface;

this is the case for the ciliate *Ichtyophthirius* that infects the epithelium of fish.

How to group and classify the different protozoa is a matter of ongoing debate, although, with the increasing use of genomic methods, protozoan taxonomy and systematics have become more accessible than before. In the most straightforward scheme, the protozoa are divided into four primary groups—the flagellates (Mastigophora), amoebae (Sarcodina), sporozoans (Sporozoa, Apicomplexa), and the ciliates (Ciliophora). Alternative classifications at all levels of the taxonomy exist. For example, the Sarcodina and Mastigophora are also sometimes grouped in the phylum Sarcomastigophora.

Each group has some important and remarkable parasitic species. For example, several genera of trypanosomes (*Herpetomonas*, *Phytomonas*, *Leptomonas*— all grouped as Mastigophora) parasitize important crops (e.g. tomato). The trypanosomatids are primarily in invertebrates, and most are monoxenic, i.e. are directly transmitted with one host in the life cycle, such as *Crithidia* in bees. Some trypanosomes are dixenous, and use insects as vectors (tsetse flies, triatomine bugs); they are important pathogens of humans and animals. Examples are tropical diseases such as sleeping sickness (*Trypanosoma brucei*), Chagas disease (*T. cruzi*), or leishmaniasis (*Leishmania*). The taxonomy of the Trypanosomatidae is debated. Currently, 14 genera of monoxenous (e.g. *Herpetomonas*, *Leptomonas*, *Crithidia*) and five genera of dixenous (e.g. *Phytomonas*, *Leishmania*, *Trypanosoma*) species are recognized (Kaufer et al. 2017).

The most prominent group among the Sarcodina are the amoeba (belonging to the Rhizopoda). *Entamoeba histolytica* is a well-known human pathogen. Its feeding stage ('trophozoite', the actual amoeba) is found in the colon and rectum of humans and other primates, but dogs or cats might also become infected. It feeds by engulfing fragments of the host epithelial cells, but also blood cells in the bloodstream, or bacteria that inhabit the same host. Sometimes, *E. histolytica* leaves the intestines and invades other tissues, notably the liver, where the trophozoite feeds on cells and causes severe damage (leading to a liver abscess). The infective stage of *E. histolytica* passes out as a cyst in the faeces of hosts. Cysts in water survive for several weeks, and many days under adverse conditions, such as dryness,

Table 3.2 Protozoan parasites.

Group	General characteristics	Parasitic habit	Examples	Remarks
Mastigophora	With flagellum to move through fluids. Trypanosomes with streamlined body shape to move in bloodstream.	In animals and plants. Mostly in digestive tract, lymphatic system; some in specific tissues.	- Kinetoplastidae: important parasites of humans, livestock, e.g. *Trypanosoma*, *Leishmania*, or in agriculture (*Phytomonas*). - Dinoflagellates, e.g. on gills of fish.	Kinetoplast contains separate genetic material, in macro- and very many microcircles. Trypanosomes can change their antigenic surface (antigenic variation).
Sarcodina	Surface with very flexible membrane. Can change shape and form pseudopodia. Move 'amoeba-like', glide on surfaces. Typically reproduce by fission.	Often in alimentary tract, associated lining of intestines. Feeding form is the 'trophozoite', that can be transmitted. Typically form durable cysts at transmission stages that can survive for weeks under adverse conditions.	- Actinopoda: only a few parasitic species. Radiolaria typically not parasitic. - Rhizopoda: Amoeba as major parasitic group. Human pathogenic *Entamoeba histolytica*.	Some free-living, sexual species have meiosis and flagellated gametes. Foraminifera are free-living Rhizopoda and can become very large (marine Xenophyophores in syncytium, up to 20 cm in diameter).
Sporozoa	Several groups of important pathogens with severe effects, notably the Apicomplexa.	- Gregarine: nutrition through the body wall. Majority is sexual (e.g. the Eugregarinidae). - Coccidia: epithelia lining of intestinal tract, vertebrates and invertebrates. Alteration of sexual/asexual cycle. - Plasmodiidae: use vectors.	- Gregarines parasitize invertebrates, mostly arthropods. - Coccidia: *Eimeria tenella* in poultry. Other *Eimeria* spp. in rabbits, grouse, pigs, sheep and goats, cats, dogs, cattle, and trout. - Haemosporida: Plasmodiidae with *Plasmodium* (malaria).	Some Coccidia (e.g. *Toxoplasma*) cycle through two hosts (cats, invertebrates). *T. gondii* spreads in leucocytes to target organs such as the nervous tissue or the eye.
Ciliophora	Heterokaryotic, with micro- and macronucleus. Complex infra-cilial structure associated with external cilia that can cover entire body (e.g. *Paramaecium*). Divide by lateral fission.	Most free-living; important symbionts in herbivores (digest cellulose). Parasitic forms in many host organisms, e.g. in host intestines, or on external surfaces.	*Balantidium coli* in pigs, humans. Causes dysentery, diarrhoea. A similar direct life cycle is known from *Ichthyophthirius multifiliis* in the external surface of fish ('white spot disease'); wide variety of freshwater hosts (carp, trout, catfish).	*B. coli* transmitted as cysts. Parasite attaches to sediments and becomes encapsulated in a gelatinous covering, develops further to form 'theronts' that attach to passing fish.

high temperatures (up to 50 °C), or weak concentrations of chlorine (Cheng 1986). Because *E. histolytica* does not cause pathogenic effects in all host individuals, asymptomatic hosts can carry the infection and spread cysts in the population without being noticed.

Among the Sporozoa, the Apicomplexa contain the Gregarines, Coccidia, Haemosporia, and Piroplasmea (e.g. *Babesia*, a parasite of cattle). Species in all of these groups can cause severe pathologies. Some have a complex life cycle. In the sexually reproducing Eugregarinida, for example, infection usually occurs by ingestion of spores that typically contain eight sporozoites which then independently penetrate the host epithelial cells. Sporozoites develop into trophozoites (the feeding stage), and eventually into gamonts; these associate as pairs and encyst together as a gametocyst. Within each gametocyst, each gamont undergoes cell division to form 'male' or 'female' gametes, respectively. The gametes then fuse to form a zygote, such that multiple zygotes develop within a gametocyst. Subsequently, the gametocyst further develops into a secondary cyst, the oocyst (sporocyst). These sporocysts are shed in the host's faeces to be transmitted to another host (Cheng 1986). Of feared repute, *Plasmodium* (Haemosporia) is the cause of human malaria., but also infects lizards, frogs and toads, birds, rodents, and other non-human primates. *Plasmodium* is a vectored parasite that depends on certain species of mosquitoes.

The Ciliophora contain probably 8000 species, of which the non-parasitic genus *Paramaecium* is the best-known. Ciliophora generally divide by lateral fission, but some also engage in sexual reproduction with conjugation, during which two individuals exchange their genetic material. The macronucleus is polyploid and contains genes responsible for the cell's metabolism. It is involved in phases of asexual reproduction. The micronucleus is active in sexual reproduction. In some cases, the micronucleus produces a new macronucleus after conjugation.

3.2.5 Fungi

Fungi are a species-rich group (with perhaps more than 1.5 million species) grouped into four different phyla: the Chytridiomycota, Zygomycota, Ascomycota, and Basidiomycota, with an inferred evolutionary history and phylogeny that follows this ordering. Microsporidia, formerly thought to belong to the Myxozoa, are highly reduced parasitic fungi that probably diverged very early in the evolution of the group (Keeling and Fast 2002; James et al. 2006). This incidentally underpins the fact that fungi are closer to the animals than they are to the plants.

Parasites occur in all of the major fungal groups, and it is therefore difficult to generalize. It will suffice to give a few examples. Among the Chytridiomycota, *Batrachochytrium dendrobatidis* infects amphibians (e.g. frogs), where it can cause a deadly disease (chytridiomycosis), and is partly responsible for the worldwide decline in amphibian diversity (Stuart et al. 2004). Among the Zygomycota, the Entomophthorales contain widespread pathogens, mostly of insects, but some can infect mammals (e.g. *Basidiobolus*). Infection by these fungi sometimes causes spectacular changes in the behaviour of the infected hosts (Schmid-Hempel 1998; Hughes et al. 2011; de Bekker et al. 2014). Infected wood ants (*Formica*), for example, tend to climb on grasses in the evening and stay there. The fungal hyphens then grow out of the body and fix the ant onto the substrate. The growing fungus then kills the ant. The fungal spores disperse from the cadaver that remains fixed in this vantage point (Marikovsky 1962; Loos-Frank and Zimmermann 1976). Other Entomophthorales produce spectacular spore stalks in many different shapes that protrude from the insect host and facilitate the dispersal of the parasite's spore (Hughes et al. 2011).

The Ascomycota are the largest and most diverse group of fungi comprising the yeasts and many of the fungal partners in lichen. Most Ascomycota are parasitic on plants, e.g. the Phyllachorales. The Erysiphales typically overwinter inside plant buds as a mycelium (a fungal thread). As the buds open in spring, the sexual ascospores disperse by wind and rain and land on the surface of a new host. During the growing season of the fungus, asexual spores (the conidia) disperse the infection. After spore germination, the fungal hyphae enter the plant tissue (the epidermal cells), utilizing haustoria, specialized tips of the hyphae that can penetrate the host cell surface by enzymes that degrade the cell wall. An infection is typically visible by the whitish or grey powdery appearance of the leaf surface ('powdery mildew'). This infection affects many crops such as grapes, wheat, barley, vegetables (onions, cucurbits), and valuable horticultural plants (e.g. apples, roses). A full infection can spread throughout the entire plant. Damage results because leaves and buds die off. Further Ascomycete fungi belonging to the Laboulbeniales are parasitic on insects and other arthropods. Within the Pneumocystidomycetes, *Pneumocystis* causes lung infections, typically in immunocompromised human patients.

The Basidiomycota represent the other large fungal group in this kingdom (the 'higher fungi'). Most are free-living, but some, like the rusts (Pucciniomycota) or smuts, parasitize plants or, sometimes, other fungi (Tetragoniomyces). Other species such as *Cryptococcus* are pathogens of animals. Fungal taxonomy is complicated by the fact that the same organism changes its morphology dramatically during different life stages. The same fungus might, for example, have a yeast-like (anamorphic) form during one stage but grow with hyphae in another (e.g. the equivalence of *Cryptococcus* being *Filobasidiella*). This reflects the enormous diversity of lifestyles and complex life cycles.

Finally, the Microsporidia are highly specialized parasitic fungi, with around 1500 described species. They primarily infect insects but also occur in fish, crustaceans, and occasionally in other animals, including mammals. *Nosema apis*, for example, is a parasite of the honeybee and can cause severe problems for honeybee breeding. Symptoms include dysentery and sluggish behaviour, e.g. a weak sting reflex. The colony thus becomes considerably weakened. Furthermore, infected queens may become superseded by a new one. Some Microsporidia are hyperparasites of trematodes; that is, they are parasites on another parasite. For example, *Nosema legeri* and *N. spelotremae* are parasitic on trematodes that infect marine bivalves (Cheng 1986).

Microsporidia lack mitochondria and structures for active locomotion and form spores with a wall that contains chitin. The most remarkable feature of microsporidia is the polar tube that is coiled up inside the spore (Figure 3.3). Upon contact with a host cell, the tube is explosively discharged and penetrates the cell wall. Subsequently, the sporoplasm

of the parasite enters the host cell via this tube. Once inside the host cell, it develops by growing and dividing before eventually producing new spores for further transmission. While spores are relatively easy to spot even under the light microscope, the stage of infection during which only the sporoplast is present is difficult to detect. Because there can be some time between infection and the formation of new spores, many infections go unnoticed, unless they start to produce detrimental effects on the host. Microsporidia can have complex life cycles, with sexual and asexual phases, and some species are also known to have distinct, polymorphic spores with presumably different functions.

3.2.6 Nematodes (roundworms)

Nematodes are a diverse group—perhaps as many as one million species (Blaxter and Koutsovoulos 2015)—with a remarkably simple body plan yet one that is remarkably successful in virtually all habitats. *Caenorhabditis elegans* has become one of the standard model organisms of biology and was the first metazoan whose genome was sequenced in 1998. Within the nematodes in the broadest sense, the Nematomorpha ('hairworms', *c.*300 species) are obligate parasites of terrestrial and marine arthropods. Their larvae are parasitic in the body cavity of the host, whereas the adult stage is free-living. This qualifies them, ecologically speaking, as parasitoids. Furthermore, Nematomorpha is a sister group to the class Nematoda (the 'actual' nematodes) (Dunn et al. 2008), which in turn contains three subclasses with free-living and parasitic species, with simple and complex life cycles in all habitats.

Nematodes are often important parasites in domestic animals, such as in poultry or sheep, where they infect the intestines, lung, muscles, or the eye. However, nematodes are also of medical relevance. For example, the guinea worm (or medina worm, *Dracunculus medinensis*) is a nematode that has been known since antiquity, causing dracunculiasis in humans. It is contracted by contaminated water and infects the deep connective and subcutaneous tissues and also the skin. One to two years after infection, a fully developed dracunculiasis has formed painful blisters on the skin at places where the female worm will emerge (mostly

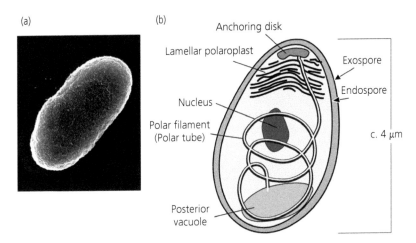

Figure 3.3 Morphology of microsporidia. Microsporidia are highly specialized parasitic fungi. The transmission stages are spores. (a) *Nosema bombi*; the spore is a few μm long (electron scanning image by Experimental Ecology, ETH; Boris Baer) (b) Internal morphology of a spore. Infection occurs by an explosive discharge of the polar tube that penetrates the host cell. From Cali and Owen (1988), adapted by permission from Springer Nature.

on limbs). The blisters burst, and one tip of the nematode emerges. Infected people typically attempt to alleviate the pain and burning sensation by putting the affected limb into water. When in the water, the parasitic female releases large numbers of tiny larvae that can infect an intermediate host; this is a crustacean where male and female worms mate. The ancient (and still used) treatment is to slowly extract the parasite through the opening of the blister utilizing a small stick onto which the worm is wound. Some historians believe that this method is the origin of the medical symbol with a snake and stick (the Rod of Asclepius) that we use today.

3.2.7 Flatworms

'Helminths' and 'flatworms' are not valid taxonomic or systematic units. However, the terms are widely understood to describe the set of 'worm-like' parasites, such as the nematodes, trematodes, and cestodes. The dispute over the taxonomic validity and classification of various helminths is not new. Genomic methods are now helping to clarify the evolutionary relationships (Pérez-Ponce de León and Hernández-Mena 2019). Among the problems encountered, an uneven representation of mitochondrial genomes hampers inferences of

phylogenies (Zhang et al. 2017). Hence, there is an ongoing debate about the systematics of 'flat-worms'. Generally, the groups discussed in this section belong to the phylum Platyhelminthes—the 'flatworms'. This phylum contains the Turbellaria (planarians), the Trematoda—the flukes, including the classes Digenea (with life cycles having at least two hosts) and Monogenea (life cycles having one host)—and the Cestoda (tapeworms). As a group, the Platyhelminthes are bilaterally symmetrical but non-segmented animals that have no coelom (body cavity). Some are free-living, but most are parasitic. Flatworms reproduce sexually, or asexually. Most are hermaphrodites, and many have evolved mechanisms to avoid self-fertilization. Some groups, such as the Digenea, can have remarkably complicated life cycles with several specialized life history stages. Digenea is the largest group within the subphylum Neodermata, with well over 10 000 species (Cribb et al. 2003). Some prevalent flatworm infections are extremely irritating or damaging (e.g. *Schistosoma* and the liver flukes).

The Cestodes (tapeworms) are parasites of vertebrates and of some freshwater oligochaetes. They have no digestive tract or mouth but absorb their nutrients through their body wall. Cestodes, however, possess sophisticated structures (the 'scolex' or 'holdfast' at the anterior end, formed of either hooks or suckers) to fix themselves in the host's intestinal tract. Right after the scolex, there is a single specialized segment of the body, followed by a repetitive multisegmented body part. Each of these repetitive segments is a 'proglottid'. Each proglottid typically has its own reproductive organ—with immature segments near the scolex and mature ones near the posterior end of the animal. Ripe, gravid proglottids have uteri filled with eggs but can release sperm as well. The fertilized egg (zygote) develops into the larvae (the 'oncosphere'), which is the transmission and infection stage. The oncosphere is ingested, for example, by an invertebrate intermediate host. The larva then develops into a 'metacestode'. When a predator eats the invertebrate prey, this stage reaches the final vertebrate host to become again a sexually mature cestode.

Diphyllobothrium latum is the largest tapeworm that can infect humans. It generally causes few symptoms except for potentially dangerous anaemia. *D. latum* can grow to a size of 20–25 m and a diameter of almost 2 cm (Marquardt et al. 2000). The mature parasite may have up to 4000 proglottids. Its life cycle requires several hosts and is associated with freshwater habitats. The 'oncospheres' use a variety of cladocerans (e.g. *Cyclops*) as the first intermediate host. These develop into a 'procercoid' within two weeks. The infected cladoceran must then become prey to a planktivorous fish. Once ingested by the fish, the procercoid migrates into a diversity of tissues (mainly muscle, ovaries, liver) and develops into the 'plerocercoid' that is already some 6 to 10 mm long. As this fish becomes eaten by a still larger predatory fish, the plerocercoid simply continues its existence in the new host and accumulates in numbers. A large pike, for example, may harbour around 1000 such parasitic forms (Marquardt et al. 2000). At this point, the large fish can be consumed by the final host, such as a human or perhaps a dog (but a bear, for example, would not be suitable). Subsequently, successful infections in humans are due to insufficiently cooked food. The parasite grows about 5 cm a day in the human intestine and starts to shed eggs after 30 days. Such infections—if untreated—can persist for as long as 30 years. Infections by *D. latum* are known from industrialized countries, too. For example, a survey in Sweden

showed that, in the study area, 100% of fish were infected. Furthermore, 20 people in the vicinity of these lakes and rivers were diagnosed with this tapeworm (Von Bonsdorff and Bylund 1982; Marquardt et al. 2000).

3.2.8 Acanthocephala

The Acanthocephala (spiny-headed worms) are a small (c.1300 species) group of entirely parasitic species. They are often discovered in the final hosts such as fish, reptiles, birds, and, sometimes, in mammals. The intermediate hosts of Acanthocephala are invertebrates, such as isopods or insects that are eaten by the final host, for example a duck. A prominent feature of this group is the retrievable proboscis armoured with hooks (from which the phylum receives its name) that serve to fix the adult animal in the intestines of its final host. Acanthocephala are sexual, with larger females and smaller males.

A typical life cycle starts with the female producing eggs that, when fertilized, leave the female as ova with a developing larva (the 'acanthor') inside. Once ingested by an appropriate intermediate host, the acanthor pierces the host's gut wall and migrates into the haemocoel, where it develops into the next stage, the 'acanthella'. The acanthella eventually becomes encysted in its intermediate host, thus forming the 'cystacanth'. When its carrier is eaten by a predator, this form is now ready to infect the final host. Cystacanths can stay alive and remain infective for the final host even when they reside in a transport or paratenic host. In the final host, the cystacanth drops its hull, attaches to the gut wall, and develops into the adult form.

Acanthocephala have no digestive tract. Nutrients are absorbed through the body wall. Because Acanthocephala possess many intriguing and unique characteristics, their placement in the tree of life has always been unclear. For some time, they were in a position within the Aschelminthes—a large summary group of very diverse animals. Molecular evidence now suggests that Acanthocephala are a sister group to the rotifers (Garey 2002; Wallace 2002; García-Varela and Pérez-Ponce de Leon 2015). Within the Acanthocephala, there are several orders with different characteristics.

3.2.9 Annelida

By far the majority of Annelida (the segmented worms) are free-living, such as its best-known representative, the earthworm (*Lumbricus* sp.). However, some annelid species are parasites, with leeches as the most prominent example. Annelids are very advanced organisms with a closed circulatory system, a complete digestive tract, and respiration via the integument or gills.

The (class) Polychaeta are marine worms, with some species being endo- or ectoparasitic on fish. For example, *Ichthyotomus sanguinarius* attaches to the fins of eels and uses an apparatus composed of protruding stylets, a sucking pharynx, a large gut, and proteins that prevent coagulation to feed on its host's blood (Marquardt et al. 2000). Among the approximately 500 species in the (class) Hirudinea (the leeches), about one-quarter consume blood from vertebrate hosts in different habitats. After a blood meal, a leech drops off the host and develops its eggs elsewhere before it attaches to the new host. Leeches have annual life cycles and mate in the early season. They can also serve as vectors for other parasites such as trypanosomes or gregarines (protozoa). The medical leech (*Hirudo medicinalis*) was used to cure several diseases since antiquity. The first applications go back to ancient India, and Nicander of Colophon (*c.*130 BC) gave detailed instructions on how to use them in ancient Greece. Even today, leeches are used for some specific purposes. *H. medicinalis* is a source of hirudin, a potent anticoagulant, and is, for example, used in treatments involving tissue grafts and reattachment surgery. Its saliva contains a range of other compounds that increase the blood flow and act as local anaesthetics.

3.2.10 Crustacea

The large group of crustaceans contains a considerable number of parasitic species (Table 3.3), although the distinction between mutualism and parasitism is sometimes tricky to check (Poulin 2004; Rohde 2005a). The Pentastomatida have traditionally been classified as a separate phylum. According to newer morphological and molecular data, they are more likely to be crustaceans, perhaps close to the fish-ectoparasitic Branchiura. The example of *Porocephalus crotali* illustrates their life cycle. The developing embryos leave the host via nasal or oral secretions to release the primary larva, which has small leg-like appendices. The larva is then ingested by an appropriate intermediate host, such as a mouse. When a final host, e.g. a rattlesnake, preys on the intermediate host, the nymphs penetrate the intestines and migrate to the lungs, where they eventually lodge. The primary larva then moults several times to develop into a nymph that becomes encapsulated in the host's tissue. Later, the mature eggs pass out into the nasal secretions, from where the developing primary larva leaves.

Copepods are small crustaceans of marine and freshwater habitats; some 13 000 species are known. Probably half of them are parasitic, typically as ectoparasites (Table 3.3). Those parasitizing fish are the best known, with the 'fish lice' (family Caligidae, *c.*500 species; Dojiri and Ho 2018) not only being prevalent in wild fish but causing considerable economic damage in fish farms (Overton et al. 2019). Natural protection is by the immune system (Fast 2014), but also surface mucus that contains antimicrobial effectors (Reverter et al. 2018), or by eventual removal by cleaner fish (Overton et al. 2020).

The order Isopoda is another large group with *c.*10 000 species. From a total of 95 recognized families of isopods, seven families are parasitic (Bopyridae, Cryptoniscidae, Cymothoidae, Dajidae, Entoniscidae, Gnathiidae, Tridentellidae; Smit et al. 2014), and around 6 200 species are found in marine environments (Poore and Bruce 2012). Some are ectoparasites of other crustaceans, both on the host larvae and adults (Williams and Boyko 2012; Yu et al. 2018). The parasitic habit seems very old, e.g. the family Cymothoidae is reaching back at least to the Jurassic. In the typical life cycle, the female carries the brood. The early, active larval stage is then released and capable of swimming and orientation. It seeks out new hosts where they attach (Hata et al. 2017).

Several other groups of crustacea are less numerous but have also evolved a parasitic lifestyle (Table 3.3). For example, the Branchiura are obligatory

Table 3.3 Crustacean parasites.

Group	General characteristics	Parasitic habit	Examples	Remarks
Pentastomida	A very small group (c.100 species) of endoparasites. No system for circulation, excretion, or respiration; outer chitinous cuticle. Sexual, with males and females; eggs are fertilized internally.	Live in the respiratory tract of vertebrates. Also called 'tongue worms', since they can often be found lodged in the nasopharyngeal area of the host (e.g. *Linguatula serrata* in the nose of dogs).	*Linguatula serrata* is the cause of halzoun or marrara syndrome in humans: heavy swelling of affected tissues in the throat or nasal ducts, which can lead to suffocation. There is also a risk of secondary bacterial infections.	Poor cooking, or traditional dishes (Middle East), pose risks to humans.
Copepods	A large group of small crustaceans. Half of them are parasitic, associated with many marine and freshwater host organisms such as sponges, molluscs, echinoderms, or fish. Their morphology and body shapes are varied, but adapted to attach and cling to the fish surfaces. Some copepods have a two-host life cycle.	Parasitic copepods damage hosts by their attachment mechanisms (claws and suckers), and by feeding. Parasites of fish feed by rasping material from the fish's surface or by sucking blood and other host fluids. In the latter case, they clearly deserve their nickname as 'sea lice'.	Around 30 families of copepods contain species that lodge in fish gill chambers, the nostrils, on fins, and around eyes.	Cleaner fish can remove ectoparasites such as sea lice. Can be a problem in commercial fish farming.
Isopods	A large group of crustaceans, with several parasitic families, and habits similar to copepods. Often, thickened and calcified cuticle as protection. Attach themselves with their hooked legs. Some forms also commensal. Marine and freshwater habitats.	Primarily ectoparasitic on fish and other crustaceans. Dwell in gill chamber, mouth, fins, on the body surface, sometimes also in abdominal muscles. Damage through the attachment and consumption of host tissue and fluids.	Suborder Epicaridea (c.800 species) are ectoparasites of crustaceans. Cymothoidae were among the first described isopods.	Often very specific sites of attachment on host. Isopods parasitic on other crustaceans normally attach to the cephalo-thorax of their host.
Branchiura	About 200 species; also known as 'fish lice'. Free-swimming larval forms seek out hosts. Can move or keep position on host skin with sucker disks.	Attach to host; adults use a tube-like mouth structure to pierce host integument and suck blood.	*Argulus* spp. is the most widespread genus, mainly in freshwater.	Can impose severe threats and economic losses to farmed fish. A few forms are parasitic on amphibians.
Other	Several other groups parasitize other crustaceans, and a variety of other host groups. - Rhizocephala (Cirripedia) are highly specialized parasites (c.300 species). - Tantulocarida, c.30 species on other crustaceans, in freshwater and estuaries. - Ascothoracida, c.100 species on echinoderms.	A free-swimming larva (nauplia, cyprid) settles on a host. After moulting, it penetrates host integument and develops inside the host, where it extracts resources.	*Sacculina* spp., infecting crabs.	Cirripedia are generally sessile feeders with a highly modified morphology, fixed 'upside down' with gland-produced cement. Parasitic form further modified and extremely reduced.

fish ectoparasites mainly in fresh water or estuaries (Neethling and Avenant-Oldewage 2016; Suárez-Morales 2020). As with other reduced and specialized parasitic forms, it is often difficult to assign relationships based on morphology alone (Jenner 2010). Larval stages disperse and seek out a host, aided by sensory and behavioural adaptations (Mikheev et al. 2015). The Cirripedia (barnacles, c.1000 species; Pitombo 2020) are widely known as dwellers of rocky shores and are often depicted in nature films as attached in large numbers to whales. However, the parasitic Cirripedia (Rhizocephala) are highly evolved parasites (Høeg et al. 2020) that use a wide range of hosts, such as polychaetes, crustaceans, dogfish, sea anemones, Echinoidea, or crabs (Høeg et al. 2015). Most of these are barely recognizable as crustaceans anymore. Among those, *Sacculina carcini* is among the most bizarre parasites more generally (Figure 3.4). It infects several crab species. The female larva (the cyprid) settles on the host surface and uses its stylet to penetrate the cuticle. Subsequently, a specialized stage (the vermigon) is injected into the host haemocoel. After a period of growth, the parasite appears externally as a small virgin female that attracts male parasites. The female grows further to become sexually mature and to reproduce. Eventually, the parasitic female has filled its host with a network of body appendages that drain the host of its resources, whereas the reproductive parts are outside at a location where the healthy host females usually carry their brood. Male hosts also become feminized, and in the process can change their body shape to resemble females. Many of these parasitic crustaceans have rather complex and sometimes bizarre life cycles as well as strongly modified morphologies.

3.2.11 Mites (Acari), ticks, lice (Mallophaga, Anoplura)

Mites, ticks, and lice together are an immensely species-rich group of perhaps one million species. The mites (Acari) infect plants (spider mites, gall mites) as well as animals (on the skin, in hairs, and in feathers). Taxonomically, the mites are currently grouped into the Acariformes, comprising the mites (Trombidiformes, Astigmata), parasitic on very different kinds of organisms, the Parasitiformes, including the Mesostigmata, parasitic on birds and insects, the ticks (Ixoda), and the Opilioacariformes (mites resembling harvestmen). Correspondingly, their lifestyles and adaptations vary tremendously. In some sense, however, the mites are for terrestrial hosts what the parasitic crustaceans are for aquatic ones. Typical for all of them is that these species are ectoparasitic or lodge in body openings (e.g. the tracheal mites of honeybees), where they consume host tissues and body fluids. In many cases, however, it is not clear whether they are genuinely parasitic, commensalistic, or only phoretic, e.g. use an insect or a bird as a transport vehicle from one site to the next. It is known, however, that this group of parasites also serve as vectors for microbial diseases. For example, ticks are vectors for Lyme disease, caused by the bacterium *Borrelia burgdorferi*, and they also carry and transmit the meningitis virus. Similarly, the mite *Varroa destructor* vectors and activates acute bee paralysis virus as well as

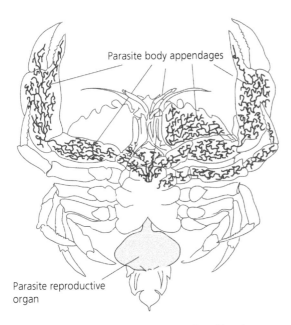

Parasite body appendages

Parasite reproductive organ

Figure 3.4 A bizarre parasite. This drawing, adapted from the nineteenth-century zoologist Ernst Haeckl (*Die Kunstformen der Natur* (The forms of art in nature), 1904), shows a crab parasitized by the barnacle *Sacculina carcini* (Cirripedia). The parasite is the ramified network visible in the illustration (parasite body appendages, red lines). These are body extensions of the parasite that function to extract resources from the host. The reproductive organ of the parasite is located where healthy host females normally carry their brood.

several other viruses in honeybees (Schmid-Hempel 1998; Levin et al. 2016).

3.2.12 Parasitic insects (parasitoids)

There are a large number of insects that parasitize other arthropods, notably other insects. Typically, these are parasitoids; that is, the adult stage is free-living, whereas the larval stage is parasitic. Several hundreds of thousands of hymenopteran species from diverse families, such as the Braconidae, Ichneumonidae, Chalcididae, Pteromalidae, have evolved this parasitic lifestyle. Also, several highly diverse groups of flies (Diptera) have evolved to become parasitoids (e.g. Phoridae, Conopidae, and Sarcophagidae). Also, the enigmatic group of the Strepsiptera comprise ten extant families that are parasitoids of a large number of insect families. They show many extreme adaptations in their biology and lifestyle (Kathirithamby and Michael 2014; Kathirithamby et al. 2015). For example, in some species, males and females use the same host species ('monoecious'); in others, the two sexes use different host species ('dioecious'). Several detailed accounts exist that can provide an introduction to the biology and special significance of this large group (Godfray 1994; Jervis 2005; Wajnberg et al. 2008; Wajnberg and Colazza 2013). Finally, a special tribute is due to the countless insect species, among them many flies and mosquitoes, which act as vectors for many important diseases. Even though they are not themselves parasitic in the strictest sense, insect vectors are an essential group of organisms for parasitology.

3.3 The evolution of parasitism

To be a parasite offers several advantages over a free-living lifestyle. The benefits include ready access to nutrition from the host, free transport to other places, and shelter from adverse environmental conditions or from predators. The evolutionary path from a free-living form to parasitism typically exploits some of these advantages. At the same time, the new parasite must evolve ways to deal with new problems, that is, find a host and deal with its defences, and to ensure transmission.

Many species are only facultatively parasitic, which suggests that evolution towards parasitism is never in one big leap. Instead, it unfolds in steps and stages, each step with some advantages over the previous step, resulting in the highly specialized parasitic forms we see today.

The routes from a free-living form to parasitism vary among the different groups. Some of the more common ones are as follows:

1. Facultative parasitism evolves into obligate parasitism. The possibility of being parasitic at least some of the time can provide advantages at all steps of the evolutionary process and is a vital preadaptation to evolve towards higher specialization and a permanently parasitic lifestyle (Luong and Mathot 2019). In the process, free-living and parasitic forms can co-exist for some while, sometimes at different stages of the life cycle. For example, some obligatory ectoparasitic crustaceans (cymochoid isopods) must have evolved from facultative parasites of fish, as suggested by their phylogeny (Brusca 1981).
2. Phoresy turns into parasitism. Small organisms that routinely attach to larger ones for dispersal into new areas ('phoresy') evolve to exploit their former transport host and so become 'true' parasites. This route is likely for mites and some nematodes (Walter and Proctor 1999).
3. Escape from variable environments. Hosts provide a steady environment that becomes exploited. Steady hosts offer an alternative to dispersal in space or time (such as diapause or durable stages) when the environment changes. Parasitism may have evolved along this route in some crustaceans (copepods) (Hairston and Bohonak 1998). In some groups, the pre-existence of durable stages would have facilitated this route of evolution (Crook 2014).
4. A parasite or associate in a prey species might evolve to survive in the predator and become parasitic. This route is involved in the evolution of complex life cycles (see section 3.4). However, it might also be helpful for the evolution of parasitism in the first place. For instance, the ciliate *Lambornella clarki*, commonly a free-living form in a pond, responds to the presence of a predator (such as mosquito larvae, *Aedes sierrensis*) by

developing into a parasitic cell able to attach and penetrate the mosquito's cuticle; so it is a parasite in this host (Washburn et al. 1988).

Parasites are often morphologically simpler than their free-living relatives—a pattern that, for example, is found in the Myxozoa (Canning and Okamura 2004). However, this pattern is far from universal. Parasitic nematodes are superficially not much different from free-living ones, and yet there are gains of characters, such as buccal teeth in hookworms, or changes in segmentation. Hence, character losses are readily balanced by evolutionary gains of characters in parasitic forms. In some cases, the size of the nuclear genome of the parasitic forms is reduced when compared to their free-living cousins, e.g. in the amoeba, fungi, mites, and nematodes. The difference may not be massive, though, as in flatworms (Platyhelminthes) (Poulin and Randhawa 2015). Significant reduction, however, is known for bacteria, especially the loss of non-functional DNA sequences, compared to non-parasitic species (Mira et al. 2001; Moran 2002) or in the Microsporidia that have a very compacted genome as compared to other fungi (Nakjang et al. 2013). Again, genomic reduction is not universal either. For example, the kinetoplastid, *Bodo saltans*, is the closest free-living relative of the fully parasitic trypanosomatids. The comparison suggests that this group shows no reduction in the size or functionality of the genomes. Instead, an expansion of many gene families has occurred (Jackson 2015; Jackson et al. 2016). A pattern of gene births and gene family expansions is also prevalent in nematodes and flatworms (International Helminth Genomes Consortium 2019).

Furthermore, a reduction can affect the organelles, notably the mitochondria, which are either maintained or lost in several taxa (Poulin and Randhawa 2015), with few exceptions. In all, changes in the genome are due to losses of genes that code for functions not necessary for parasites as compared to their free-living relatives, e.g. homeoboxes in tapeworms (Tsai et al. 2013). However, gains and expansions of gene families that cover essential functions are also commonplace, such as heat shock genes in tapeworms or surface proteins in trypanosomes (Jackson 2015).

Hence, the transition to parasitism involves reduction but also adds complexity. Therefore, parasitism is not just a degenerated lifestyle but a highly specialized way of life that necessitates evolutionary changes of many kinds. Just as there are many different kinds of parasites, so should we expect that there is considerable variation in the route to parasitism and the associated morphological and genomic changes.

3.3.1 Evolution of viruses

To retrace the actual evolutionary history can be treacherous, as the parasite's current morphology and genetic endowment may not yield many hints as to where it originated. For example, viruses are remarkably adapted to a parasitic lifestyle and are of ancient ancestry; most of the traces have thus been lost in time. Currently, two major views prevail as to their origin. With the 'cell-first' hypothesis, viruses are thought to have evolved from cells that lost their components and genes, and so became obligatory parasites of other cells; this is conceivable for large DNA viruses. Viruses could also have evolved from mobile genetic elements that can move within the genome to different locations (i.e. transposable elements, 'jumping genes'). Eventually, such elements managed to escape their genome and became freely moving viruses. In fact, non-viral, mobile genetic elements are involved in many aspects of virus evolution, especially for viruses of eukaryotes (Koonin et al. 2015).

The 'virus-first' hypothesis proposes an origin from precellular life, with self-replicating units based on RNA (Pressman et al. 2015). These units may have started to infect the emerging primitive cells, to acquire the cell's coating proteins, and to evolve into what are essentially RNA viruses, which then later diversified (Koonin et al. 2015; Zhang et al. 2018b). In this scenario, the emergence of viruses predated the existence of the last universal cellular ancestor (known as LUCA), from which the three major extant groups (the archaea, bacteria, and eukarya) have descended. Retroviruses depend on genomic DNA (into which they are retro-transcribed), and therefore originated later in life, likely in a marine environment of the early Palaeozoic (Aiewsakun and Katzourakis 2017; Xu et al. 2018).

Apart from viruses, many other extant parasitic groups consist entirely of parasitic species. Examples are the Trypanosomatida, Trematoda, or Cestoda. This makes it difficult to understand how parasitism evolved in these groups in the first place. But parasitism is an old phenomenon, as shown by the fossil record; for example, there are traces of infection in trilobite hosts 570 million years ago (Conway Morris 1981). Such traces are extending to the early Palaeozoic in brachiopods (Bassett et al. 2004). Parasitism has also repeatedly and independently emerged in very many lineages (e.g. in lice; Johnson et al. 2004), especially also concerning the transition from free-living forms to parasites (Blaxter et al. 1998; De Meeûs and Renaud 2002; Whitfield 2003).

3.3.2 Evolution of parasitism in nematodes

Nematodes are an excellent example to illustrate the path to parasitism. Extant forms have a wide variety of lifestyles; some are free-living, others phoretic or facultatively or obligatory parasitic. The parasitic Nematomorpha and Mermithidae (Nematoda), for example, have independently evolved a parasitoid habit, with a free-living adult and parasitic larval stages. Phylogenetic reconstruction of nematode evolution shows that parasitism has evolved at least 18 times independently. Nematodes that parasitize plants have three origins, those that infect invertebrates have ten, and vertebrate parasites have five (Blaxter and Koutsovoulos 2015). Not all of these evolutionary routes could have been the same.

One likely route for some nematodes starts with free-living forms that are dung dwellers and utilize organic waste (Figure 3.5). Initially specialized to feed on the dung of mammals, they came into contact with insects that also visit dung and lay their eggs in this resource, such as dung beetles or dung flies. At a later stage, these insects can serve as vehicles by which the nematode becomes transported to another dung patch—a phoretic relationship. The nematodes thus evolved an efficient strategy for their dispersal. Among extant nematode species, transport associations (phoresy) are well known, often with arthropods and molluscs. Associations are sometimes quite specific; for

example, nematodes lodge at specific locations on the transport host, e.g. under the elytra of beetles (Blaxter 2003), and synchronize their life cycle with the host. Such associations generally entail negligible fitness costs to the carrier insect (Richter 1993). The evolution of true parasitism can progress when the insect–nematode transport relationship takes a new turn. For instance, the nematodes might incidentally start to feast on the carcass of their transport host in the case of its uncorrelated death. This lifestyle gives access to a new resource that is not competed for by other dung-dwelling organisms. From this stage, and if this new resource is profitable enough, a further step can occur when the nematode actively promotes the death of its vehicle to generate the food source. It thus becomes an active parasite. Incidentally, in many unrelated

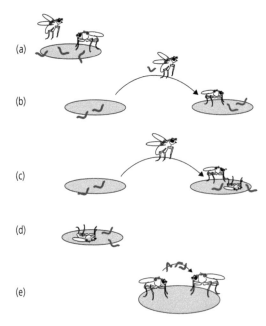

Figure 3.5 Evolution of parasitism in nematodes. In this hypothetical scenario, parasitism evolves via several stages. (a) Free-living nematodes (red symbols) feed on mammalian dung. They come into contact with dung-dwelling insects such as dung flies. (b) The nematodes associate with the insect as transport vehicles to new dung patches (phoresy). (c) The nematodes benefit from incidental host death and start to feed on the dead insects in the new place. (d) The nematodes increase the likelihood of host death by various means. It is possible that around this evolutionary stage, bacterial symbionts are recruited for the production of toxins. (e) The nematodes have become obligatory parasites that might transfer from insect to insect host on dung patches. Scenario after Blaxter (2003).

groups of nematodes, such a host death is caused by symbiotic bacteria of the nematode. These bacteria produce the toxins that eventually kill the insect host (Blaxter 2003).

Whether any particular species has followed this sequence of events is typically challenging to elucidate. However, pre-existing associations of free-living forms with other animals certainly facilitated the evolution towards parasitism. Other features of nematodes may have been additionally helpful. For example, nematodes develop through different larval stages that involve moulting. Thus, at each larval stage, new surface structures can be expressed that can be adapted for the next stage; indeed, moulting coincides with steps in the life cycles of nematodes more generally (Viney 2017). In addition, the third larval stage often becomes developmentally arrested, such that it turns into a 'dauer-stage' (German for a permanent stage). Arrestment can be considered a preadaptation for the evolution of an infective stage (Crook 2014). Whereas these properties are favourable to the evolution of parasitism, the differences in the respective genomes do not necessarily match the differences in such characteristics among nematodes. For example, a conspicuous acquisition of more proteases is known for *Strongyloides*, but these have other plausible primary functions in addition to initiating a dauer-stage (Viney 2017, 2018).

Few extant nematodes exploit their hosts by living on the host's surface or by consuming it after its death. Most parasitic nematodes are gut-dwellers, or they live in tissues (e.g. the guinea worm). Gut-dwellers exploit the food resources and the associated bacterial flora of the gut. Those that invade tissues must have undergone additional, essential shifts in their exploitation strategies. Not least, they must first reach their specific target tissue. Many of these species also eventually migrate to the gut for their reproduction and the release of eggs. Finally, many parasites of vertebrates have evolved to utilize vectors or intermediate hosts for transmission to the next host. The most common vectors are arthropods, mainly insects, which suggests that the association with arthropods rather than with vertebrates is the ancestral stage. This scenario may be real for some groups but may not be the general pattern (Blaxter 2003).

3.4 The diversity and evolution of parasite life cycles

Just like any other organisms, parasites must survive and reproduce to gain (Darwinian) fitness. As we will see, to precisely analyse how parasites achieve fitness is an important key to understand some seemingly unrelated problems, such as parasite virulence, host switching, strategies of immune defences, and the like. Moreover, in turn, the life cycle is key to analysing how parasites survive and reproduce.

3.4.1 Steps in a parasite's life cycle

Looking at the typical life cycle of a parasite, we can see the following steps (Figure 3.6).

1. *Finding a host*: Before any parasite can grow, multiply or reproduce, it has to find a suitable host. Unfortunately, hosts can be far apart for a small parasite such as a virus or a protozoan—organisms typically of a micron scale that live in a large animal. To get to a new host is, therefore, a significant problem. Not surprisingly, parasites have evolved

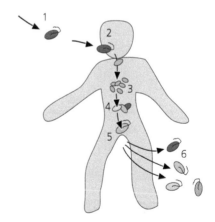

Figure 3.6 Basic steps in the life cycle of a parasite. These consist of: (1) Finding a host. (2) Infection and establishment in the host. (3) Growth or multiplication of the parasite inside the host. (4) Reproduction, e.g. by exchange of genetic material between co-infecting strains. (5) Development of transmission stages. (6) Transmission to the next host. The different colours symbolize the changes in the stages and genotypes of an infection. The scheme could apply to an infectious disease such as cholera, caused by the bacterium *Vibrio*, where infection is by ingestion, change by pathogenicity islands, and transmission via faeces.

many different ways to increase the chances of successfully finding a host. For example, the production of an enormous number of offspring increases the chance that at least one propagule successfully reaches a new host, but utilizing a vector is also an efficient strategy.

Parasites can utilize (a) *passive dispersion* to reach a new host. With this simplest mode of movement, the propagules become transported by air or water, or via surfaces. Examples are human rhinovirus (HRV), or the new SARS-CoV-2 virus. In both cases, transport starts with the shedding from the host by coughing, sneezing, or breathing. The virus particles (the virions) passively disperse in the airflow, in droplets or aerosols, until another host is encountered. When hosts are near to each other, when air flows in the right direction, when the virus particles do not degrade quickly outside the host, or when the new host is more receptive, e.g. in cold weather, the chances of finding a host are therefore higher. In winter, HRV is more infective, since human mucosal layers dry out, with many small ruptures and thus entry points. Similarly, the bacteria that cause cholera (*Vibrio cholerae*, *V. eltor*) are released via faeces into the water and so are transported passively by currents and human water use. The extremely durable spores of a *Bacillus anthracis*, the agent of anthrax, are found on surfaces or in the soil, or reside in animal carcasses until a new host comes into contact with the contaminated material. Ecologically, *B. anthracis* behaves like a 'sit-and-wait predator', except that its chances of finding a host are minute compared to a predator with highly evolved sensory capacities. The ability to survive a long time in the environment is therefore essential. Incidentally, *B. anthracis* is also an example of where dispersal by air can be highly successful under the right conditions: labourers in woolsheds or in factories can suffer from an anthrax infection known as 'wool-sorter's disease'. In this case, exposure is to spore-containing dust when animal hair and hides are handled.

In contrast to passive dispersion, (b) *active host finding* is more efficient. Parasites can use a vector, for example, to become transported to suitable hosts with more precision. Active finding typically also involves the manipulation of host and vector behaviours in the parasite's favour. Active host finding is also an option for actively moving, mobile parasite stages. For example, the females of parasitic wasps are free-living insects and can locate a new host on their own. Females of *Rhyssa*, whose larvae parasitize wood-boring beetles, can readily locate their victims across centimetres of solid wood. Furthermore, many parasites that do not use a vector have nevertheless evolved ways to bias the behaviour and sometimes the morphology of their current hosts to increase their chances of finding a new host.

2. *Infecting and establishing in the host*: Once a host is encountered, the parasite must overcome the outer barriers (e.g. the skin or the cuticle) to enter the host's interior (in the case of endoparasites). It can also become attached to the host's surface, such as on the gills of fish (ectoparasites). The infective form is thereby often different from the form in transit between hosts or that the parasite assumes inside a host. The most direct way to infect is through body openings such as airways but, in particular, also *per os*; that is, when hosts ingest spores, cells, or any other infective form of the parasite. Infection in this way bypasses the external body wall but must still overcome internal protective layers (which, histologically, are partly 'outer' layers, too). A vast number of parasites enter their hosts in this way. Infections by ingestion can end up in the airways, but more generally in the digestive tract. In the gut of a mammalian host, the so-called M-cells of Peyer's patches are used to subsequently invade the host's internal organs (especially by bacteria). The mucosa lining the gut wall in Peyer's patches is particularly thin. These patches act as sentry points for the gut's immune system; parasites, therefore, gain direct access to the underlying tissues or to the lymph draining system that allows for efficient dispersal to other parts of the host body. In fish, some parasites arrive via the water flow around the gills and settle there. Alternatively, a parasite may also enter through other body openings such as the anus, the cloaca, or the trachea in arthropods. Another route for infection is directly through the integument. This route is illustrated by fungi, by nematodes, but also by the specialized cercariae of trematodes. Penetration of the integument is not always easy, however. An insect cuticle, for example, is a solid protective outer hull which is

difficult to cross. Some fungi have therefore evolved specialized structures (the appressoria) that allow the growing fungus first to gain a steady hold on the surface. With this hold, it penetrates the cuticle by mechanical (pressure by the growing tip) and chemical means (chitinases that digest the cuticle).

However, for most parasites, reaching the host's interior, or being anchored safely to its surface, is not yet sufficient for further development. Endoparasites will have to reach the right kind of organ or tissue. Hence, they need to migrate through the host's body. Ectoparasites, too, need to get at their resources. For example, ticks need to sink their stylet into the host's blood vessels at suitable points, and crustacean fish parasites need to be able to grasp host scales or other tissue at appropriate places. After infection, some ectoparasites firmly establish by embedding themselves into the outer layer of the host's skin or even induce specialized structures that hold them (e.g. the larvae of gall wasps). These additional steps often require different mechanisms from the initial infection of the host.

3. *Growth, multiplication*: Once the parasite has reached its target location or organ, it can grow in size (e.g. in cestodes or parasitoid larvae) or multiply. The latter characterizes infecting populations of parasites such as viruses, bacteria, or protozoa, which multiply in numbers within the host. Thus, uptake of nutrition and other essentials becomes of paramount importance. This phase is associated with the extraction of host resources, while preventing the host from clearing the parasite. We will discuss the manifold consequences of these processes in later chapters.

4. *Reproduction*: In the general sense used here, reproduction is associated with, but conceptually separated from, multiplication, since it involves the transmission of genetic information to offspring (or among co-infecting lineages) without necessarily increasing numbers. With sexual reproduction, the genetic material transferred to one offspring comes from two parents (biparental reproduction) and therefore yields new genotypes. Similarly, horizontal gene transfer via plasmids is changing the genetic endowment of bacterial cells. Reproduction can thus modify genotypes. Note that this is also the case with the accumulation of new mutations in a replicating infecting population. Mutations can occur whenever the genetic material is duplicated or copied.

5. *Formation of transmission stages*: Multiplication and reproduction by a parasite may also involve the formation of special transmission stages that are either mobile in the external environment (e.g. the actively swimming cercariae of trematodes) or able to survive difficult environmental conditions for extended periods (e.g. spores in many bacteria or fungi).

6. *Transmission*: This is the transition from one host to the next, and, therefore, it connects up with step 1; that is, finding a host. Transmission is the point of the life cycle that is most amenable to a quantitative analysis of parasite fitness, because it represents a bottleneck that is observable from the outside. Transmission can also include long intervals of no particular activity, e.g. in durable spores.

3.4.2 Ways of transmission

Transmission is a crucial step in a parasite's life cycle and discussed in section 9.1. In brief, parasites become transmitted in various modes (the physical means, e.g. droplets) and along different routes (e.g. the faecal–oral route). Transmission can happen (Figure 3.7) with (a) *direct transmission*, where the parasite transfers to a new host without any other intervening stage or organism. For example, HIV gets transmitted to a recipient by direct contact of blood from a donor, and the cholera bacterium *Vibrio* spreads through water. The intestinal trypanosome, *Crithidia bombi*, transmits to a new host when bumblebee workers pick up infective cells from a flower. These were previously deposited by another infected individual (Durrer and Schmid-Hempel 1994). Using a medium, such as air or water, or using a highly mobile host (such as a bee) aids in the dispersal of the parasite to new places and hosts. The direct transmission also happens through time, with durable stages (spores, cysts).

Transmission with (b) *paratenic hosts* (a transport host) happens with an optional substitute host. There, the parasite does not develop, grow, or multiply, no morphological changes occur, and no special effects on the host occur. Paratenic hosts

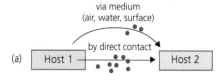

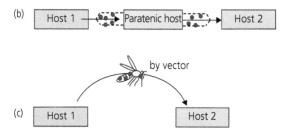

Figure 3.7 Ways of transmission. (a) Direct transmission upon contact between hosts, or as mediated by water, surfaces, etc. Examples are *Vibrio* (cholera), where contact with contaminated water is needed, or trypanosomatids that are picked up by bees from contaminated flowers. (b) Transmission via an opportunistic (paratenic) host. For example, tapeworms can infect when humans consume raw fish that are normally not hosts. The paratenic host might prey on host 1, and in turn be consumed by host 2, while the parasite survives. (c) Transmission via vectors, e.g. *Plasmodium* (malaria), uses bloodsucking insects such as mosquitoes.

typically acquire the parasite by ingestion of the original host. An example would be the broad fish tapeworm (*Diphyllobothrium latum*) that can also infect humans (Marquardt et al. 2000). Its regular hosts are plankton-feeding fish, where it develops. When a predatory fish consumes an infected fish, the infective tapeworm stages remain passive in the predator. If they survive in this novel environment, they are ready to infect a next, more suitable host.

A most elaborate way is (c) *vector transmission*. Vectors are other organisms, such as mosquitoes, ticks, or leeches that allow efficient transport to the next host. A vector can cover large distances (at the scale relevant for a single parasite) and is much more specific in getting to a next suitable host than transport by a paratenic host. Parasites living in the bloodstream, such as *Plasmodium*, are obvious candidates for vectoring by bloodsucking insects, ticks (e.g. *Borrelia* bacteria), or leeches (as in trypanosomes of fish). In contrast to the transport via a paratenic host, vector transmission also involves many more physiological interactions, sometimes rather complex

ones, between the parasite and its vector. Hence, the vector is more than just a transport vehicle. Inside the vector, the parasite can develop further. For example, the amastigote form of *Leishmania*, ingested by a sandfly, develops into procyclic promastigotes that multiply by binary fission. They attach themselves to the midgut epithelium of the fly. Finally, promastigotes develop further into metacyclic promastigotes, ready for transmission to a new host. In some cases, parasites also undergo reproductive episodes in their vectors. When several strains of *Trypanosoma brucei* infect the same tsetse fly (*Glossina* spp.), the parasites can additionally exchange genetic material, and so manage to generate new, infective genotypes. Finally, horizontal transmission (among hosts of different lineages) is distinct from vertical transmission (i.e. within the same lineage, such as to offspring; see Figure 9.1).

3.4.3 Complex life cycles

Some parasites have seemingly absurd life cycles, with several different host species, which furthermore need to be infected in the right sequence. In the intermediate host(s), the parasite undergoes obligatory steps in its development. The distinction between the role of an intermediate host and a vector is sometimes vague. Typically, parasites stay longer in intermediate hosts than in vectors and undergo distinct developmental stages. In the final (definitive) host the parasite reproduces sexually, but this is not a universal feature either. A final host is often the one considered to be the most important one for a variety of reasons, notably, when of medical importance for humans.

The best examples of complex life cycles are in the cestodes and trematodes. However, other parasitic groups, such as the Microsporidia or Coccidia, have also evolved complex life cycles with several hosts; for example, the coccidian *Aggregata eberthi* has a cycle with a squid and a crustacean as hosts (Cheng 1986). The utilization of intermediate hosts is associated with striking examples of host manipulation to increase the chances of transmission to the next host in the sequence. For example, the cestode, *Anomotaenia brevis*, infects the larvae of forest-dwelling ants of the genus *Leptothorax* and persists into the host's adult stage. In the adult worker ant, it induces a sluggish behaviour together with persistent foraging activity outside the nest. Also, a conspicuously yellow colouration develops (these ants usually are brownish). All of this makes the ant more visible and more comfortable to catch for predators (e.g. blackbirds, woodpeckers) that are the next (final) host for the cestode.

Many aspects of the phylogeny of helminths are still under debate (Pérez-Ponce de León and Hernández-Mena 2019). Such details are important to understand along which route the exceedingly complex life cycles of the Digenea and Cestoda might have evolved. For example, the Monogenea show a basic pattern and are typically ectoparasitic, with a simple one-host life cycle. They infect poikilotherm vertebrates, such as teleost fish or sharks, typically lodging on gills, scales, or fins. Monogenea also parasitize amphibia and reptiles, and a few aquatic mammals, crustaceans, or cephalopods. Against this rule, a few species are true endoparasites, such as *Dictyocotyle* in the body cavity of a ray, and *Polystoma integerrium* in the bladder of amphibians (Cheng 1986). Ectoparasitic Monogenea possess a highly sophisticated structure (the 'opisthaptor'), by which they attach themselves to the surface of the host. The one-host life cycle of the Monogenea involves the 'oncomiracidium' stage (a larva that develops from eggs). The oncomiracidia are the free-swimming stages that seek out a new host. They will attach and develop into the adult form. Mating partners, once found, often form a tight association. For example, in *Diplozoon*, which parasitize fish, the larva becomes attached to the gills, and two such larvae of opposite sexes become attached to one another and remain paired throughout their life. Even their reproductive systems eventually become connected to form a single physiological unit. In endoparasitic forms (e.g. *P. integerrium*) the host ingests the parasite *per os*. The parasite larva then migrates through the host's alimentary canal to eventually become lodged in the bladder of the frog or toad where it reaches maturity (Cheng 1986).

By contrast, the Digenea are endoparasites of vertebrates and have life cycles with two or more hosts (Figure 3.8). In the vertebrate (final) host, they lodge in the alimentary and urinary tracts, in the blood, the oesophagus, and most other major organs. In contrast to the Monogenea, they possess two prominent suckers on their body surface, of which one typically serves to hold the animal in place and the other, to extract nutrition from the host (in the monostomes only one sucker is present, at least in the early stages). Some of the Digenea have the most complex life cycles among all parasitic animals. Examples are *Schistosoma mansoni* (with two hosts), *Paragonimus westermani* (with three), and *Halipegus occidualis* with four (!) hosts (Esch et al. 2002) (Figure 3.8). All forms are sexual, but some can self-fertilize. Flukes lodging in the blood are of great medical and veterinary importance (Table 3.4).

On an interesting historical note, the Italian poet and physician Francesco Redi (1626–1698), after whom the redia stage in the digenetic life cycle is named, described several 'worms' from many different animals (among them, the first Acanthocephala). However, it is unclear what precisely all of these forms were (Andrews 1999). However, he published the first illustration of the liver fluke (*Fasciola hepatica*) after extracting a specimen from the liver of a castrated ram. During the second half of the seventeenth century, Francesco Redi acted as a physician to the Grand Dukes Ferdinand II and Cosimo III of Florence. In 1668, he experimented in the modern sense of the word to disprove that maggots arise out of rotting meat spontaneously; he showed that unless flies are given access to lay their eggs, maggots will not appear. This was a first significant proof against the then widely accepted theory of spontaneous generation of life forms and an essential step towards the Darwinian concept of evolution by common descent.

3.4.4 The evolution of complex parasite life cycles

The association of trematodes and helminths with their hosts is very ancient and probably dates back to the late Cambrian (Parker et al. 2003). But how can such complex life cycles evolve by natural selection? Moreover, in such life cycles, why is parasite reproduction delayed until a specific (final) host, but suppressed in intermediate hosts? There are, in fact, several very plausible scenarios.

One possible pathway to complex life cycles—involving different hosts (i.e. a 'heteroxenic' life

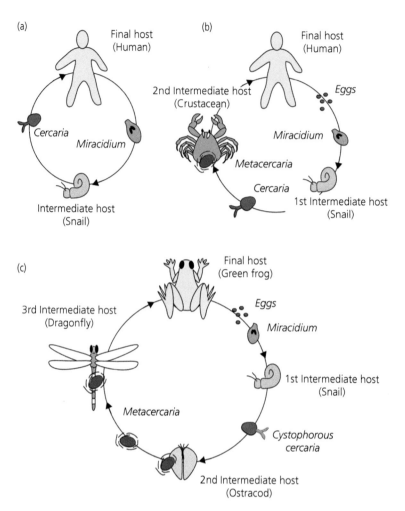

Figure 3.8 Complex life cycles in Digenea. (a) *Schistosoma mansoni* infects humans as the definitive (final) hosts. Aquatic snails are the first intermediate hosts. (b) *Paragonimus westermani* infects humans and cats as definitive (final) hosts. Aquatic snails are the first intermediate, and crabs the second intermediate host where the metacercaria encysts. (c) *Halipegus occidualis* uses the green frog (*Rana clamitans*) as its final host, where it always sits under the tongue. The first intermediate host is a pulmonate snail (*Helisoma anceps*), where cystophorous cercariae develop. Cercariae infect the second intermediate host, an ostracod. In nature, the cercariae do not yet fully develop, but must infect a dragonfly to encyst as metacercariae that can then be ingested by the predatory final host, the frog. Drawn after descriptions in Esch et al. (2002).

cycle)—is by 'upward incorporation' (Figure 3.9a). It assumes that an infected host containing a parasite becomes consumed by a predator (Parker et al. 2003, 2015a). Usually, the parasite would then perish, along with its host. In some cases, however, the parasite may manage to survive and so find itself in a new host with new resources. Such chance events are indeed observed; for example, in nematodes normally parasitizing fish (Moravec 1994) If the parasite manages not only to survive but also to utilize these new resources, the evolution towards a

new life cycle gains momentum. If the parasite has the option to delay its maturity, it can continue growing in this new host, rather than initiating reproduction. Moreover, because predators (the new host) are typically larger than their prey (the former host)—the parasite might attain a larger body size than what would have been possible in the former, original host. Larger body size, in turn, generally correlates with lower mortality and higher fecundity (Parker et al. 2009). At this stage, reproduction would still be possible in the original

Table 3.4 Some digenetic trematodes ('flukes') of medical and veterinary importance.[1]

Trematode	Principal hosts	Location	Disease
Schistosoma mansoni *S. japonicum, S. haematobium*	Humans, livestock.	Blood.	Schistosomiasis (debilitation, damage to liver and internal organs).
Fasciola hepatica, F. magna, F. buski	Livestock, humans (*F. hepatica*).	Liver, bile ducts, duodenum.	Fascioliasis (liver rot).
Dicrocoelium dendriticum	Sheep, cattle, goats, horses, pigs, dogs, rabbits, humans.	Liver, bile ducts.	Dicrocoeliasis (diarrhoea).
Opisthorchis felineus	Cats, sometimes humans.	Biliary and pancreatic ducts.	Opisthorchiasis (diarrhoea, jaundice).
Parametorchis complexus	Cats.	Bile ducts.	Liver cirrhosis may result.
Heterophyes heterophyes	Humans, cats, dogs, foxes.	Small intestine.	Heterophyiasis (mild, but can cause severe haemorrhagic diarrhoea).
Metagonimus yokogawai	Humans, dogs.	Small intestine.	Metagonomiasis (diarrhoea, abdominal pains; possible migration of eggs into heart and brain).
Sphaeridiotrema globulus	Ducks, swans.	Small intestine.	Ulcerative enteritis.
Echinostoma ilocanum	Humans, rats, dogs.	Small intestine.	Colic, diarrhoea.
Paragonimus westermani	Humans, cats.	Encapsulated in lungs.	Paragonimiasis (chronic inflammation of the lungs).
Nanophyetes salmincola	Dogs, foxes, coyotes, cats, bobcats, racoons, humans.	Intestines.	'Salmon poisoning'.
Collyriclum faba	Chickens, turkeys.	Encysted in skin.	Emaciation and anaemia, fatal.
Watsonius watsoni	Humans.	Intestines.	Severe diarrhoea.
Typhlocoelum cymbium	Ducks, geese.	Tracheae, bronchi, air sacs.	Suffocation with heavy infections.

[1] After Cheng 1986. *General parasitology.* Academic Press.

host. By exploiting the new host and continuing to grow, the parasite can increase its reproductive success compared to variants that only use the original host. This requires that the associated drawbacks (e.g. loss of infectivity for the original host, longer generation times) are not too large. In further evolutionary steps, the parasite would become better adapted to the new host, might become larger, produce more offspring for the transfer back to the original host, and so forth. A larger body size and delaying maturity would still be expressed in the original host, however. Therefore, when the parasite has grown to such a large body size that the current host can no longer sustain it, it would become advantageous to become transmitted to the new host and to entirely rely on the new, larger host for maturation and reproduction. Now, the original host has become the intermediate host and the new

host the final host. Upward incorporation is facilitated by the benefits of increased fecundity and lower expected mortality in the new host. Moreover, the ability to reproduce by selfing (or asexually) also facilitates the evolution down this path. These conditions may explain why complex life cycles are primarily found in organisms that grow as individuals (such as helminths) and therefore benefit from larger body size. By contrast, parasites that only multiply (e.g. viruses, bacteria) and where body size does not relate to fecundity cannot benefit.

With 'downward incorporation', a new host, located lower in the trophic chain than the original host, is added to the parasite's life cycle (Figure 3.9b). Finally, with 'lateral incorporation' (Figure 3.9c), a new intermediate host or a new final host is utilized. Here, the addition of hosts is in 'parallel' rather than 'in series', and does not change the length of the life

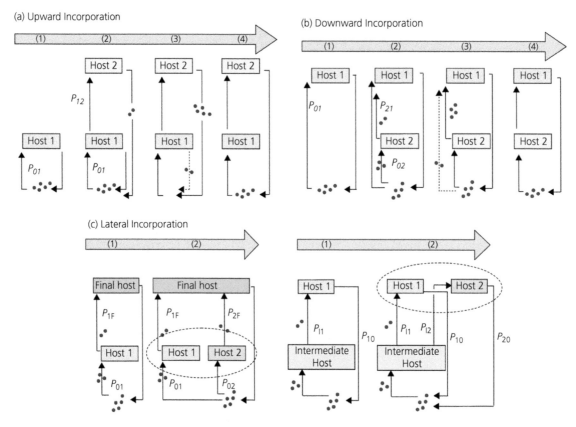

Figure 3.9 The evolutionary steps towards complex (heteroxenic) life cycles. (a) Upward incorporation. Originally (1), parasite propagules (red dots) enter host 1 with probability P_{01}. At some stage (2), Host 1 is consumed by its predator (host 2), but the parasite survives with probability P_{12}. This contributes a few propagules that again infect host 1. When predation is common (3), the parasite might evolve to grow better in host 2. Hence, the relative contribution to success via host 1 declines. Finally (4), the contribution by Host 1 disappears, the direct transmission between host 1 therefore becomes lost, and a two-host cycle has established. (b) Downward incorporation. Originally (1), parasite propagules enter the original host with probability P_{01}. At some stage (2), the parasite propagules accidentally enter Host 2 with some probability P_{02} and survive. Host 2 is also consumed by the same predator (host 1) and, therefore, some parasites are transferred to host 1 with probability P_{21}. At this stage (3), both paths can be used. But if predation is common, the relative contribution via host 2 to overall parasite fitness increases, and the direct contribution via host 1 decreases. Finally (4), the contribution of direct transmission to host 1 is lost and a two-host cycle has established. (c) Lateral incorporation. Two scenarios are possible. *Left*: Originally (1), the life cycle includes an intermediate (host 1) and a final host, with transmission probabilities P_{01} and P_{1F}, respectively. At some stage (2), the parasite increases its range of intermediate hosts to include host 2, with transmission probabilities P_{02} and P_{2F}, respectively. *Right*: Originally (1) the life cycle includes an intermediate and a final host (host 1) that is reached with probability P_{11}. At some stage (2), the parasite increases its range of final hosts to include host 2, with transmission probabilities P_{12}, while reaching the intermediate host with probabilities P_{10} and P_{20}, respectively. In both cases, ecological over lap between old and new hosts makes this more likely. With lateral incorporation, the length of the life cycle remains the same, but the host range increases. Evolution of complex life cycles depends on the relative weights of the P-values and the associated costs and benefits. Evolutionary steps (numbers 1 to 4) marked in the arrow of time. Panels (a) and (b) adapted from Parker et al. (2003) adapted by permission from Springer Nature, (c) adapted from Parker et al. (2015a) by permission of John Wiley and Sons.

cycle but just expands the host range. Lateral incorporation is more likely to happen when the existing and the added hosts overlap in their ecological niches and can easily be reached. These three hypotheses refer to ecological relationships and

growth benefits in hosts. It is also conceivable that the need to find a mating partner drives evolution towards a more complex life cycle. Similar to the idea of upward incorporation, predators that consume infected prey could act as 'concentrators' of parasites within their bodies and so allow finding a mate much more efficiently (Brown et al. 2001). Furthermore, the presence of several co-infecting lineages of the parasite in 'concentrators' allows for genetic exchange. Hence, a possible genetic linkage between managing to survive in such a new host and the benefits of outcrossing may also select for upward incorporation (Milinski 2006a).

3.4.5 Example: trypanosomes

In contrast to the nematodes, all living members of the Trypanosomatida are parasitic, which makes it difficult to compare with a free-living condition. In all, parasitism in this group has evolved four times independently (Lukeš et al. 2018). The trypanosomatids descended from within the Bodonid group, with *Bodo saltans* being the closest free-living species (Figure 3.10). Today, free-living Bodonids (the Neobodonida, Parabodonida, Eubodonida) live in a wide variety of aquatic habitats. From current evidence, parasitism had emerged roughly coincidently with the common trypanosomatid ancestor. This early form had evolved from a phagotrophic feeder, i.e. a protozoan that engulfed particles or other organisms as a whole, similar to feeding by amoeba (Jackson et al. 2016). The change to parasitism led to a loss of many of these capacities but replaced them with new functions and reorientations of previous faculties. It is likely that the ancestral trypanosomatids first exploited insects or other invertebrates as hosts and were transmitted by contamination, predation, or faecal discharge. Only later did they become parasitic on vertebrates by the blood-feeding habits of their original hosts. For example, *Blastocrithidia culicis* only infects insects such as mosquitoes but is found in the salivary glands. This would make a transmission to a vertebrate by a mosquito bite very likely (Nascimento et al. 2010).

Some forms may have managed to survive the consumption of their current host by a vertebrate predator ('upward incorporation'). Furthermore, some may have managed to lodge in various host body cavities, e.g. in the reproductive organs of snails, fish, or leeches. Some of the modern-day *Cryptobia* are flagellated trypanosomes and ectoparasites on gills, in the gut, but also in the bloodstream of fish;

they also form durable cysts. Living in the bloodstream must have provided the early forms with advantages in terms of a steady supply of nutrition, more or less homeostatic conditions, and shelter from enemies (Stevens et al. 2001). In turn, these early forms had to evolve ways to deal with the host's defence responses. Durable stages able to survive outside the host and achieve transmission to a new host certainly added to the success.

At the level of the genome, genes encoding for processes to degrade large molecules have become reduced, in line with the move away from phagocytosis. There is also a reduction in the level of redundancy of functions ('streamlining'; Jackson et al. 2016), perhaps as a result of living in a more stable habitat inside a host. On the other hand, adaptive radiation of membrane functions happened, notably in the diversity of major surface proteases, which are essential for the infection process or to evade the host's immune defences. Only later in the evolution of trypanosomatids did the variable surface glycoproteins (VSG) emerge, which are the basis of the now-characteristic surface antigenic variation seen in this group (Jackson et al. 2016).

An 'insect-first' scenario is conceivable for the monoxenous trypanosomes. This lifestyle is the supposed ancestral state. Current phylogenies suggest that the derived dixenous life cycle (two hosts) has likely evolved independently in each of the genera *Trypanosoma*, *Leishmania*, and *Phytomonas* (Maslov et al. 2013). Along the way, the monoxenous trypanosomes evolved the capacity to survive in a warm-blooded vertebrate by upward incorporation. At the same time, the former host was gradually relegated to act as the vector. Perhaps the ancestral species in the genus *Trypanosoma* were first parasites of vertebrates that subsequently exploited, for instance, bloodsucking invertebrates by downward incorporation (i.e. a new host). These were insects in terrestrial and leeches in aquatic environments that later served as vectors (Stevens et al. 2001).

The phylogeny of extant Trypanosomatida shows a mixture of free-living and parasitic, as well as a variety of monoxenous and dixenous species (Figure 3.10). The route to parasitism was therefore very varied, and in each case, the evolution of the genome (gains and losses of genes) had its different

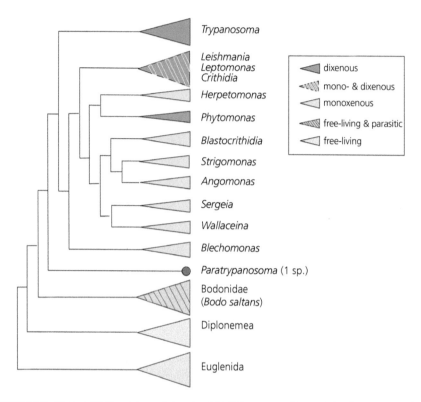

Figure 3.10 Parasitism in trypanosomatids. The phylogeny is based on SSU rRNA sequences. Extent of triangles approximately proportional to the number of sequences (species) that were considered. Free-living and parasitic lifestyles varies among genera, according to legend. *Bodo saltans* is the closest free-living extant species to the rest of the group. Redrawn from Lukeš et al. (2014), with permission from Elsevier.

pattern (Lukeš et al. 2014; Jackson 2015). At a large scale, the phylogeny of kinetoplastids suggests that parasitism must have evolved several hundred million years ago in this group. They acquired novel host groups as they evolved over the ages. Extant monogenetic insect parasites could only have emerged after the appearance of the first insects. The Orthoptera (emerging *c.*300 million years ago) are the most basic insect group harbouring trypanosomes today. Similarly, only after their potential vertebrate hosts would have colonized the land, could a shift to arthropod vectors have occurred, and only after the appearance of bugs (Hemiptera, *c.*250 million years, and Diptera, 200 million) would these specific bloodsucking vectors have been available. What is typically much more difficult to understand, and where research is often lacking, is which selection pressures have

driven a particular evolutionary step and led to a corresponding change in the genome.

3.4.6 Example: helminths

Different groups of helminths (the worm-like parasites) have also followed different evolutionary paths to complex life cycles. As already mentioned above, trematodes could have used arthropods or aquatic molluscs as their hosts first, followed by upwardly incorporating vertebrate hosts later (Cribb et al. 2003). The alternative of 'vertebrate-first' suggests that intermediate hosts were added by downward incorporation. Molecular evidence suggests that Monogenea are basal to the Neodermata (also including the Trematoda and Cestoda); because Monogenea are typically ectoparasites of vertebrates, the scenario of downward incorporation is also plausible.

Downward incorporation may also have been the path to adding intermediate hosts in many Digenea (Figure 3.11). There are species where alternative paths of transmission are known (direct vs indirect), as expected under this scenario. For example, the cercaria of the liver fluke, *Fasciola hepatica*, can be transmitted either directly via the water to the final, vertebrate host or, indirectly, by first encysting on leaves of a water plant that may then become eaten by a vertebrate herbivore (Mas-Coma and Bargues 1997; Choisy et al. 2003). The likely common ancestor had a miracidium that penetrated a mollusc (a snail). From the snail, the fork-tailed cercariae emerged that were eaten by the final host, for example a fish or a bird (Cribb et al. 2003). From this stage, second intermediate hosts were added independently to generate a three-host life cycle (as in the Diplostomida). Forked tails of cercariae allow finding and pursuing a host with much more efficiency than simple tails and are considered to be an ancient (plesiomorphic) trait in Digenea (Marquardt et al. 2000; Cribb et al. 2003). That fork-tailed cercariae infect the final host suggests that this was likely the original host and that the mollusc was added downwards at a later stage as an intermediate host.

The blood flukes (Schistosomatoidea), by contrast, have probably shortened their more ancestral three-host into a two-host cycle (Poulin and Cribb 2002; Cribb et al. 2003). Initially, the parasite went from the final host to a first intermediate host, then to a second intermediate host, and back again to the final host. Precocious reproduction in the second intermediate host could truncate the life cycle by dropping the need to infect the former, final host for maturation. The ancestral second intermediate host eventually becomes the new final host (Figure 3.11). Some species of *Alloglossidium* (Macroderoididae), where all parasites produce eggs in their second intermediate as well as in the final host (crustaceans or leeches), are examples of such truncation (Smythe and Font 2001). Similarly, some individuals of *Coitocaecum parvum* (Opecoelidae) produce eggs as metacercariae in the amphipod host, while some other individuals only produce eggs in the final host, a fish (Poulin and Cribb 2002). Truncation of the life cycle can thus occur by neoteny; that is, by

the acceleration of reproduction until reproduction occurs in a formerly juvenile stage.

An alternative way of truncating a host from the life cycle is when cercariae do not leave the first intermediate host but directly develop into metacercariae, which then transfer to the final host (Figure 3.11). Cercariae are therefore no longer free-swimming, and the second intermediate host becomes dropped from the life cycle, as in several digenean parasites (Poulin and Cribb 2002). *Gymnophallus choledochus* truncates the life cycle seasonally. In summer, the parasite has its normal, three-host cycle, while in winter, a two-host cycle occurs (Poulin and Cribb 2002). As above, the shortening of the life cycle occurs when the second intermediate host becomes the final host. In this case, the metacercariae are retained in the second intermediate host, and then develop into the adult forms that mate in the same host (now a final host). An example is *Haplometra cylindracea* (Macroderoididae), where metacercariae infect the buccal cavity of their frog host and become encysted in the epithelium. After a few days, the cysts burst, and the juvenile worms migrate to the frog's lung to become adults and mate. Sometimes the metacercaria stage is bypassed (Grabda-Kazubska 1976, cited in Poulin and Cribb 2002). Hence, a shortening of the life cycle can occur by upward (as in the second and third scenario) or downward (as in the first scenario) transfer of maturation.

Downward incorporation is a likely evolutionary scenario for cestodes, too (Hahn et al. 2014). For example, the larval stages of *Bothriocephalus barbatus* and *B. gregarius* use a planktonic copepod as their intermediate host. The copepod is eaten by the final host, a flatfish, where the parasite sexually matures. However, in addition to this two-host cycle, *B. gregarius* typically uses a second intermediate host, a goby, which then becomes the prey of a flatfish. Gobies are more common prey items for the flatfish than the copepods. Hence, the addition of this new (initially paratenic) second intermediate host—which yields a three-host life cycle—increases transmission efficiency as compared the two-host life cycle (Robert et al. 1988; Morand et al. 1995).

When parasites expand their life cycle by upward incorporation, they also ascend in the trophic

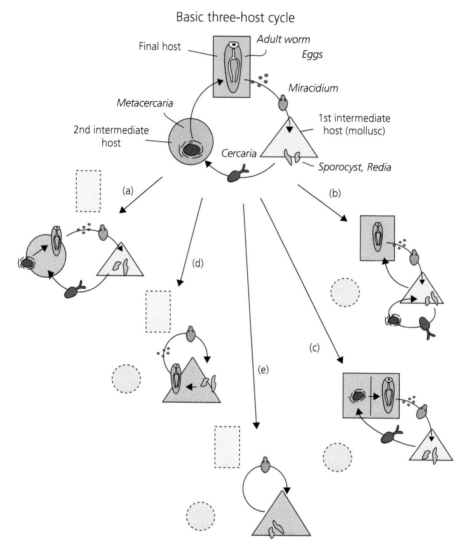

Figure 3.11 Variation in digenean life cycles. Top panel. In the basic three-host life cycle, the adult worm lives in the final host (rectangle) and releases eggs (red dots), which develop into miracidia. These infect the first intermediate host (a mollusc) (triangle). After asexual multiplication inside sporocysts (sometimes rediae), the parasite develops into mobile cercariae that seek out and infect the second intermediate host (circle). There, the parasite encysts as a metacercaria. The final host then ingests the second intermediate host, which leads to development into the adult worm. Modifications of this basic pattern are found in the following cases: (a) The adult worm develops in the second intermediate host ('progenesis') and the original final host is skipped (dashed rectangle). (b) The metacercaria already develops and encysts in the first intermediate host. The second intermediate host is skipped (dashed circle). (c) The metacercaria has already developed into the adult worm in the first intermediate host. This host simultaneously becomes the final host. Original second intermediate host is skipped (dashed circle). (d) The adult worm develops already in the first intermediate host, which now acts as the only (final) host. Second intermediate (dashed circle) and previous final host (dashed rectangle) are both skipped. (e) Sporocysts develop directly into miracidia that can infect this single host. No other stages and adult worm appears. Second intermediate (dashed circle) and previous final host (dashed rectangle) are both skipped. Colours symbolize functional intermediate (green) and final or only hosts (blue).

cascade. Utilizing a high-level predator (usually a large-bodied species) offers a protected, stable environment that allows for growth and reproduction; this is, therefore, a highly advantageous host to reach. At the same time, high-level predators have little niche overlap with, for example, free-living helminths. Intermediate hosts connect these trophic levels across the food web and allow the transmission to higher levels of the food chain. To what extent this gap between trophic levels exists, called the 'trophic transmission vacuum' (Benesh et al. 2014), is an empirical question. From comparative data, the observed transmission chains—from the lowest to the highest trophic level of potential hosts in an ecosystem—are longer than the shortest possible route (i.e. directly from lowest to highest level). They are also longer than expected with random chains, but still somewhat shorter than the maximum-length route, suggesting that trophic gaps exist to some degree (Benesh et al. 2014; Parker et al. 2015a). The 'trophic transmission vacuum' is an important selective factor that maintains the use of intermediate hosts, rather than just being a selective force for its evolution in the first place.

Important points

- Parasites are more numerous than non-parasitic species and have evolved in virtually all groups of organisms, such as viruses, prokaryotes (bacteria), protozoa, fungi, nematodes, flatworms, acantocephalans, annelids, crustaceans, and arthropods (crustacea, mites, ticks, insects).
- A parasite life cycle goes through the steps of infection, establishment, growth, and transmission. Transmission can be vertical or horizontal, direct or indirect (by vectors), and involve one or several hosts in the cycle.
- Parasitism can evolve from a non-parasitic lifestyle by various routes, such as phoresy and later consumption of the former transport host, or through an initially commensalistic relationship that becomes exploited by the parasite.
- The evolution of complex life cycles can be by upward incorporation, i.e. adding a predator of the original host. With downward incorporation, a new host is accidentally acquired that is the prey of the original, single host. With lateral incorporation a new host is added in parallel, thus expanding host range. Bridging trophic levels with the help of intermediate hosts to reach a 'safe' host at the top level affects food webs.
- Examples of life cycle evolution are trypanosomes that have added a vector, and Digenea that evolved complex life cycles with two or more hosts. The latter illustrates how extant life cycles evolved further via different routes by skipping some hosts.

CHAPTER 4

The natural history of defences

4.1 The defence sequence

Host defences are processes that minimize fitness loss due to parasitism. They start with avoidance of infection in the first place and continue with mechanisms that reduce the negative effects when an infection has become unavoidable. Hence, the defence is not just immunology but includes a wide range of possibilities such as changes in behaviour, life history, and even morphology (Parker et al. 2011). Defence comes at a cost, visible in other components of the host's fitness: for example, reproduction or

feeding activity. For that reason alone, the 'chosen' level of defence is always a compromise between different needs, and rarely maximal. Host defences are under selection by parasites (and the other host needs) and change over the generations in an evolving population. By contrast, the defences discussed here are under the control of an individual during its lifetime. Such responses are 'plastic' in the sense that the host can respond to changes in the environment. Host responses are ordered along a defence cascade—with the processes occurring from pre-infection to post-infection (Figure 4.1).

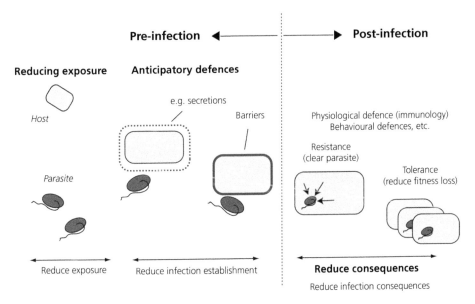

Figure 4.1 The defence sequence. Pre-infection defences reduce exposure to infection (behavioural avoidance, repelling parasites, change of life history, group life, etc.), use of anticipatory defences (upregulation of the immune system, vaccination, change of the chemical milieu or of physical conditions), and the existence of barriers (skin, cuticle, epidermis, endothelia). Post-infection defences include physiological mechanisms (immune defence, changes in chemical or physical conditions, behavioural defences). The strategy can be to resist (essentially, to kill and clear the parasite) or to tolerate the infection by reducing fitness losses (behavioural changes, medication, changes in life history, social support, etc.).

Evolutionary Parasitology: The Integrated Study of Infections, Immunology, Ecology, and Genetics. Second Edition.
Paul Schmid-Hempel, Oxford University Press. © Paul Schmid-Hempel 2021.
DOI: 10.1093/oso/9780198832140.003.0004

4.1.1 Pre-infection defences

Pre-infection defences minimize exposure to or reduce the risk of successful entry and establishment of a parasite by anticipatory defences or barriers. There are several ways to defend oneself before an infection occurs (Table 4.1). Indeed, there are an astonishing variety of phenomena, although most of the supporting evidence is by correlation rather than by experiment.

4.1.1.1 Avoidance behaviour

Such defence includes avoiding localities that are associated with parasites or migrating away from such places. Changing habitat might also reduce the risk of parasitism. Habitat choice is therefore likely to be a compromise between benefits (e.g. food availability) and different risks, such as between predation and parasitism (Poulin and FitzGerald 1989; Decaestecker 2002). Similarly, avoiding infected conspecifics helps to prevent infection, for example by social avoidance of group members (Poirotte et al. 2017; Kavaliers and Choleris 2018). Diurnal and seasonal activity patterns can also be important (Westwood et al. 2019). For example, malaria is a periodic fever which results from the synchronous cycles of reproduction of all parasite cells (*Plasmodium*) within a host; a characteristic scheme of days between fevers occurs. Also, *Plasmodium*-infected blood cells all burst around midnight and release the gametocytes that can be taken up by blood-feeding mosquitoes (Mideo et al. 2013b). This timing is partly matched by the biting activity of the mosquito vectors that also peaks during the night times. Also, mosquito populations have shifted their biting activity towards day times in areas where protective bed nets for the local people were introduced (Moiroux et al. 2012).

Avoidance can also reduce the risk of foodborne infections. Such risks range from bees becoming infected by consuming nectar (Durrer and Schmid-Hempel 1994), to risks for human health from eating fish (Gauthier 2015) or meat (Fredriksson-Ahomaa 2018; Rukambile et al. 2019). Avoidance behaviour may differ between the sexes, with females often being more conservative than males, and also differs among individuals, as observed in primates (Poirotte and Kappeler 2019). The feeling of 'disgust'

is likely an evolved psychological mechanism that guides a consumer towards safer foods (Curtis et al. 2011).

4.1.1.2 The selfish herd and group-living

When an individual joins a group, it will incur a risk of contracting a new infection. However, the risk of becoming infected can also decrease as the group increases in numbers. This 'dilution' of risk has been termed the 'selfish herd' effect; i.e. individuals live in groups for the selfish benefit of lowering their individual risk (Hamilton 1971; Mooring 1992). Reduced risks of parasitism in social groups are indeed known for a wide range of animals, e.g. monkeys (Snaith et al. 2008), rodents (Bordes et al. 2007), and birds (Krebs et al. 2014). However, the effect is not always clear-cut (Wilson 2003). For instance, for highly mobile parasites (i.e. those actively seeking a host) group size is not a good correlate for infection risk (Côté and Poulin 1995). On the other hand, from the group members' point of view, imported infections should be avoided. Thus, keeping out infected individuals becomes essential. This strategy may point to an evolutionary root of xenophobic behaviours observed in animal societies as well as in humans (Schaller 2011; Curtis 2014; Kavaliers and Choleris 2018).

4.1.1.3 Anticipatory defences

The collection and application of appropriate chemical substances for protection is a prophylactic behaviour, different from self-medication, which acts as a cure. Many groups of animals, from primates (Carrai 2003; Fowler 2007; Huffman 2016) to insects (de Roode et al. 2013), show this behaviour. The best-known examples in humans are preventive antimalarial drugs or vaccination. In several insect species, individuals that congregate in dense aggregates upregulate their immune system and activate the phenoloxidase (PO) cascade—one of the general canonical defence lines in invertebrates. This anticipation of an increased infection risk has been termed 'density-dependent prophylaxis' (Wilson 2003). Even transgenerational protection by parental prophylaxis, i.e. passing the protective effects to offspring, occurs (Wilson and Graham 2015).

Table 4.1 Pre-infection defences.

Defence element	Description, examples	Source
Avoid unsafe places	Great tits (*Parus major*) avoid nests experimentally infested with fleas. Ladybirds avoid contact with soil and leaves harbouring fungal spores (*Beauveria*). Water fleas balance risk of fish predation (upper water column) and risk of parasitism by Microsporidia (near bottom) by choosing an intermediate position in the water column. In reindeer, migrating herds have fewer flies than herds that stay in the calving grounds.	8, 11, 15, 27, 29, 30
Migration to escape parasites	Reindeer and other ungulates' migratory and foraging behaviour reduces parasite load. Long-distance migration of monarch butterflies leaves parasites behind.	4, 16, 27
Shift activity times	Ants shift activity times to avoid parasitoids such as phorid flies.	13, 14, 25, 31
Avoid unsafe food	Disgust in humans as a psychological mechanism to avoid risky food. Feral Soay sheep avoid grass when experimentally associated with risk of nematode infections.	10, 20
Avoid infected group members Join small and safe groups	Infection levels by various parasites are lower in smaller groups in a range of social mammals and birds. Solitary living animals have lower loads of many but not all parasites.	1, 5, 12
Repel parasites, camouflage	Elephants use tools (twigs) to repel biting flies. Zebra stripes might be less attractive for biting flies (tsetse) than uniformly coloured dark skin.	17, 36
Hygiene	Some insects defecate away from their daily routines. In social insects, waste is dumped at locations that are kept separate from the rest of the nest.	18, 42
Choose a healthy mate	Mice can detect the infection status of a mating partner from urine and subsequently avoid mating.	21, 33, 34
Anticipatory upregulation of immune system	Locusts in dense aggregates upregulate antibacterial activity and the PO cascade even when not infected ('density-dependent prophylaxis').	3, 43, 44, 45
Protect surfaces	Over 250 bird species practice 'anting', i.e. they rub crushed insects etc. into their plumage. This transfers antimicrobial compounds for protection. Thin layers of AMPs on body surface prevent establishment of fungal spores on ants. AMPs protect the surface of the human eye.	9, 10, 24, 32
Prophylactic self-medication	Sifakas (*Propithecus verreauxi*) increase intake of tannin against helminths. African great apes suspected to use leaves as medicines. Many lepidopteran larvae take up plant compounds that are toxic to their parasitoids. Wood ants collect resins that provide protection against bacteria and fungi and increase survival of larvae and adults.	6, 7, 19, 22, 28
Hygienic behaviour	Mutual grooming in social groups, application of antimicrobial secretions or compounds.	26
Polyandry	Multiply mated females of bumblebees, honeybees, etc. have colonies with lower parasite loads and higher fitness, perhaps due to more concerted production of AMPs.	2, 23, 37, 40
Adjust recombination rate	In *D. melanogaster*, infected mothers plastically increase recombination rate, or have a surplus of gametes with recombination events, to diversify and thus protect offspring.	38, 39
Transgenerational immune priming	If infected, protect direct offspring by a signal to increase their immune activity; reported from beetles, lepidopterans, bees, crustaceans.	35, 41

Sources: [1] Altizer. 2003. *Annu Rev Ecol Syst*, 34: 517. [2] Baer. 1999. *Nature*, 397: 151. [3] Barnes. 2000. *Proc R Soc Lond B*, 267: 177. [4] Bartel. 2011. *Ecology*, 92: 342. [5] Bordes. 2007. *Biol Lett*, 3: 692. [6] Carrai. 2003. *Primates*, 44: 61. [7] Chapuisat. 2007. *Proc R Soc Lond B*, 274: 2013. [8] Christe. 1994. *Anim Behav*, 52: 1087. [9] Clayton. 1993. *Auk*, 110: 95. [10] Curtis. 2014. *Trends Immunol*, 35: 457. [11] Decaestecker. 2002. *Proc Natl Acad Sci USA*, 99: 5481. [12] Fauchald. 2007. *Oikos*, 116: 491. [13] Feener. 1988. *Behav Ecol Sociobiol*, 22: 421. [14] Folgarait. 1999. *Ecol Entomol*, 24: 163. [15] Folstad. 1991. *Can J Zool*, 69: 2423. [16] Gunn. 2003. *Wildl Soc Bull*, 31: 117. [17] Hart. 1994. *Anim Behav*, 48: 35. [18] Hart. 2001. *Behav Ecol Sociobiol*, 49: 387. [19] Huffman. 2003. *Proc Nutr Soc*, 62: 371. [20] Hutchings. 2001. *Ecology*, 82: 1138. [21] Kavaliers. 1995. *Proc R Soc Lond B*, 261: 31. [22] Krief. 2004. *Antimicrob Agents Chemother*, 48: 3196. [23] Lee. 2013. *Naturwissenschaften*, 100: 229. [24] McDermott. 2009. *Ophthalmic Res*, 41: 60. [25] Mehdiabadi. 2002. *Proc R Soc Lond B*, 269: 1695. [26] Meunier. 2015. *Philos Trans R Soc Lond B*, 370: 20140102. [27] Nilssen. 1995. *Can J Zool*, 73: 1024. [28] Nishida. 2002. *Annu Rev Entomol*, 47: 57. [29] Oppliger. 1994. *Behav Ecol*, 5: 130. [30] Ormond. 2011. *FEMS Microbiol Ecol*, 77: 229. [31] Orr. 1992. *Behav Ecol Sociobiol*, 30: 395. [32] Ortius-Lechner. 2000. *J Chem Ecol*, 26: 1167. [33] Penn. 1998. *Trends Ecol Evol*, 13: 391. [34] Penn. 1998. *Ethology*, 104: 685. [35] Roth. 2010. *J Anim Ecol*, 79: 403. [36] Ruxton. 1982. *Mamm Rev*, 32: 237. [37] Simone-Finstrom. 2016. *Biol Lett*, 12: 20151007. [38] Singh. 2015. *Science*, 349: 747. [39] Stevison. 2017. *Philos Trans R Soc Lond B*, 372: 20160459. [40] Tarpy. 2003. *Proc R Soc Lond B*, 270: 99. [41] Tidbury. 2011. *Proc R Soc Lond B*, 278: 871. [42] Weiss. 2006. *Annu Rev Entomol*, 51: 635. [43] Wilson. 2003. *J Anim Ecol*, 72: 133. [44] Wilson. 2001. *Ecol Lett*, 4: 637. [45] Wilson. 2002. *Proc Natl Acad Sci USA*, 99: 5471.

4.1.1.4 'Genetic' defences

In animals (or plants), where the success of a female depends on the genetic diversity among its off-spring, additional options exist. For example, social insect females (the queens) can mate multiply with several unrelated males and so reduce the parasite load in their colonies (Baer and Schmid-Hempel 1999; Tarpy 2003; Lee et al. 2013b). How this effect comes about is often unclear. Alternatively, and in anticipation of parasitism, mothers might plastically increase the recombination rate. Recombination increases genotypic diversity in offspring. Indeed, recombination rate differs between the sexes, varies with age, and plastically changes in response to temperature or starvation in various organisms, e.g. *Drosophila* (Jackson et al. 2015; Stevison et al. 2017) or yeast (Abdullah and Borts 2001). Modular organisms, such as corals or plants, have further options. Plants, for example, start to systemically increase recombination rate in their tissues when infected (Abdullah and Borts 2001) Although it is often unclear whether recombination changes in response to parasites, changes in response to stress—a typical correlate of any infection—are well documented.

4.1.2 Post-infection defences

When exposure or actual infection by a parasite has become inevitable, hosts use a variety of different means to reduce the expected fitness loss or to clear the infection altogether. Among these, the most obvious response is by the immune system. Its response profile can be depicted in the disease space as well (Box 4.1). Nevertheless, there are also large numbers of behavioural defences or physiological responses (Table 4.2).

4.1.2.1 Behavioural changes

Infected animals can change their time budget and direct more activities towards defence behaviour. Infected hosts might also change their life history and start to reproduce earlier than when uninfected ('fecundity compensation'; see Chapter 15). Such shifts are known for snails infected by trematodes.

Box 4.1 Disease space: defences

The disease space illustrates how the infection trajectory moves through different host states, and thus reflects the 'internal workings' of the host (immune) defences (Box 2.1). Immune defences can kill and clear a parasite or nudge the infection trajectory into a corner of the disease space where the parasite can be tolerated, i.e. where damage to the host is manageable. Tolerance can be more efficient than incurring the cost of removing the parasite altogether. For the parasite, being nudged may be better than having to resist clearance; both parties may thus achieve higher fitness with tolerance.

The effects of different defence efforts can be illustrated as changing the shape of the various regions in disease space (Figure 1). Alternatively, one can depict defences by a change in the infection trajectory that keeps it away from a dangerous zone. When clearing the infection, the trajectory would regain the healthy zone as quickly as possible.

The shape and structure of the disease space itself is a result of the immune mechanisms that become activated at a given point along the infection trajectory (Figure 2). Hence, a sequence of immune responses unfolds—initially by the constitutive defences, then by the early and late innate, and eventually, by the adaptive system (numbered points in Figure 2). Only the trajectory through disease space is noticed from the 'outside', but not the mechanisms that nudge it along a particular path. The trajectory also depends on the parasite, and the strategies of the two parties generate the actual trajectory.

We expect that early innate mechanisms (such as complement) are active in a region of disease space where the host may still appear healthy, or is on the verge of becoming sick (point 1 in Figure 2). Later innate mechanisms (e.g. production of AMPs) follow afterwards (point 2), whereas adaptive cascades become active after a delay (point 3) and can clear the infection. Because the infection trajectory may take a different course for a different parasite, different regions in disease space are thereby touched. As a consequence, different mechanisms can also become activated; the same would be true if the same parasite infected another host. Therefore, different hosts are likely using different mechanisms to combat the same infection—adding to the staggering variation in host responses.

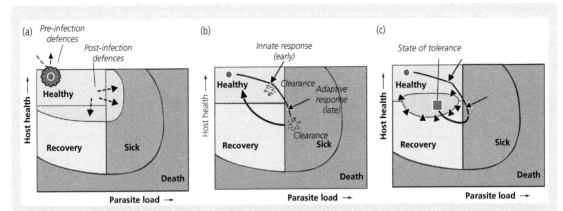

Box 4.1 Figure 1 Effects of defences in disease space. (a) Pre-infection defences avoids infection (blue shield and broken arrow). Some post-infection defences compare to expanding the healthy zone (green zone, broken arrows) to prevent an infection from moving into the sick or deadly region. (b) As an infection takes its course, immune defences become activated. The innate response is an early response that can change the course of an infection (the kink in the trajectory) or eliminate an infection at an early stage without necessarily leaving the healthy zone, even though a reduction in fitness may occur (see fitness landscape in Box 2.1). The adaptive response comes later, and may clear the infection when the host is already sick. The host then returns to the healthy state. A further possibility is that hosts tolerate an infection. In this case, innate and adaptive defences are recruited, but the host returns to a 'tolerance status', presumably close to healthy (a 'grey zone'). This state is maintained by continuous responses (black triangles) to keep the infection from leaving the grey zone and moving the host back to a sick or deadly status. The red dot indicates host status at time of infection.

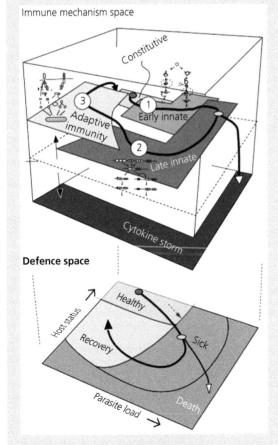

Box 4.1 Figure 2 Immune mechanisms and disease space. *Upper panel:* The infection trajectory (thick black line) leads through different domains of the immune mechanism space. It thereby moves 'upwards' from one defence level to the next, according to mechanism (from constitutive, to early and late innate, and to adaptive defence; small insets symbolize the different immunological cascades at points 1 to 3). The trajectory finally drops back to a healthy host status (green zone). In some cases, the trajectory may break out (at yellow dot) and fall down to a lethal condition of a cytokine storm. The projection of the trajectory through the different immune mechanisms is the infection trajectory through the different zones of disease space, as shown in the lower panel. Immune mechanisms only very loosely correspond to the different zones of the disease space. Red point is start of infection.

Table 4.2 Post-infection defences.

Defence element	Description	Source
Change places and activity	Ant colonies infected by fungi dislocate more often. Great tits reduce sleeping time in favour of increased nest sanitation when infected by ticks.	5, 24
Reduce food intake (anorexia)	Infection-induced anorexia helps to combat infections by increasing resistance or tolerance. Found in many animals.	2, 3, 11, 27, 29
Self-medication (cure); also for medication of others (kin)	Many animals use a wide variety of means to cure themselves, such as certain leaves, antibiotic substances, secondary plant metabolites, toxins, etc. Also across generations to protect offspring.	1, 7, 8, 14, 18, 22, 25
Behavioural fever (warm or cold)	Many ectothermic animals raise temperature by seeking out appropriate places (e.g. sunny locations). Raised body temperature damages and clears parasites ('resistance') or alleviates symptoms ('disease tolerance'). Preference for lower temperatures impedes development of endoparasites in bumblebees, *Drosophila*, and snails.	7, 9, 16, 17, 23, 28, 31, 32
Grooming behaviour, hygiene, social immunity	Self- and allogrooming is widespread among animals. Effective against threats such as fungal spores, deposited eggs, larvae, and ectoparasites more generally. Nest cleaning in birds. Cleaning behaviour by specialized fish, shrimps, or birds. Several additional mechanisms in social animals.	4, 6, 7, 12, 21, 30, 34, 37
Change life history, tolerance mechanisms	Fecundity compensation: Acceleration of reproduction before onset of parasite-induced castration in snails, sticklebacks, *Drosophila*, mosquitoes, etc.	13, 20, 26, 35
'Adaptive suicide'	Controversial hypothesis. Infected individuals sacrifice their life to remove the parasite from a group or their neighbours. Beneficiaries must be close kin to compensate for the cost. Bacteria have self-abortive Abi systems to remove phage infections.	15.
Fever by internal heat	Endothermic animals can produce heat internally and raise body temperature. 'Normal' fever as part of the immune response.	10, 33, 36
Reduce food intake (anorexia)	Less caloric intake can increase resistance. In crickets, based on trade-off in use of lipoproteins for immune defence vs digestion.	2
Activate immune system	Immune system responds in many ways when parasite is recognized.	see Chapter 4
Genetically diversify cells	Infected plants increase recombination rate in leaf cells even away from infection site.	19

Sources: [1] Abbott. 2014. *Ecol Entomol*, 39: 273. [2] Adamo. 2010. *Anim Behav*, 79: 3. [3] Ayres. 2009. *PLoS Biol*, 7: e1000150. [4] Barber. 2000. *Rev Fish Biol Fish*, 10: 131. [5] Christe. 1994. *Anim Behav*, 52: 1087. [6] Cremer. 2007. *Curr Biol*, 17: R693. [7] de Roode. 2012. *Insects*, 3: 789. [8] de Roode. 2013. *Science*, 340: 150. [9] Elliot. 2002. *Proc R Soc Lond B*, 269: 1599. [10] Evans. 2015. *Nat Rev Immunol*, 15: 335. [11] Exton. 1997. *Appetite*, 29: 369. [12] Hart. 1991. *Neurosci Biobehav Rev*, 12: 123. [13] Heins. 2012. *Biol J Linn Soc*, 106: 807. [14] Huffman. 2003. *Proc Nutr Soc*, 62: 371. [15] Humphreys. 2019. *Biol Lett*, 15: 20180823. [16] Hunt. 2016. *J Anim Ecol*, 85: 178. [17] Lefcort. 1991. *Parasitology*, 103: 357. [18] Lefevre. 2010. *Ecol Lett*, 13: 1485. [19] Lucht. 2002. *Nat Genet*, 30: 311. [20] Minchella. 1985. *Parasitology*, 90: 205. [21] Mooring. 2004. *Biol J Linn Soc*, 81: 17. [22] Morrogh-Bernard. 2017. *Sci Rep*, 7: 16653. [23] Müller. 1993. *Nature*, 363: 65. [24] Oi. 1993. *Fla Entomol*, 76: 63. [25] Parker. 2011. *Trends Ecol Evol*, 26: 242. [26] Polak. 1998. *Proc R Soc Lond B*, 265: 2197. [27] Povey. 2014. *J Anim Ecol*, 83: 245. [28] Rakus. 2017. *Dev Comp Immunol*, 66: 84. [29] Rao. 2017. *Cell*, 168: 503. [30] Sazima. 2010. *Biota Neotrop*, 10: 195. [31] Schieber. 2016. *Pathog Dis*, 74: [32] Stahlschmidt. 2013. *Naturwissenschaften*, 100: 691. [33] Stearns. 2008. *Evolution in health and disease*, 2nd ed. Oxford University Press. [34] Vaughan. 2017. *Fish Fish*, 18: 698. [35] Vézilier. 2015. *Biol Lett*, 11: 20140840. [36] Wojda. 2017. *J Therm Biol*, 68: 96. [37] Zhukovskaya. 2013. *Insects*, 4: 609.

Snails will also shift their reproduction if only exposed to a chemical signal that is associated with trematode infective stages in the water (Minchella 1985). Numerous studies have shown that hygienic behaviour, such as grooming, is effective against infecting ectoparasites or helps to remove spores from the body surface. In social groups, additional options exist (see section 4.1.3 on 'social immunity'). A unique, interspecific form of allogrooming is cleaning behaviour, known from specialized cleaner fish, or birds, e.g. oxpeckers. Hence, hygienic behaviour comes in many forms and has evolved in virtually every organism (Curtis 2007).

4.1.2.2 Physiological responses

A change in body temperature has many consequences for the host metabolism but also for the infecting parasites. Endotherms actively regulate their body temperature by producing heat and thus can use temperature as a defence mechanism. Such physiological fever seems to be a phylogenetically ancient trait (Evans et al. 2015; Harden et al. 2015). By contrast, ectotherms cannot metabolically regulate body temperature, but a change of behaviour achieves a similar result. By moving into warmer or sunnier places, the body temperature increases and, in most cases, prolongs survival, reduces infection levels, or increases performance in other respects. Such 'behavioural fever' is widespread and has been reported in virtually all groups of ectothermic vertebrates, such as amphibians, reptiles, and fish (Stahlschmidt and Adamo 2013; Schieber and Ayres 2016), but also in snails (Wang et al. 2018) and many insects (Campbell et al. 2010; Catalán et al. 2012; Rakus et al. 2017). Some flying insects have evolved ways to heat their bodies with their large flight muscles and without actually moving the wings. In bees, for example, this behaviour achieves temperatures above 35 °C. To some degree, therefore, this allows them to regulate body temperature independently of ambient temperature. This mechanism is used by honeybees to combat infections through cooperative warming of the colony as a whole ('social fever'; Starks et al. 2000). Finally, sleep may also be one element of the defence against parasites. Sleep does affect various aspects of immune functions, and infection does change sleep patterns, with beneficial effects on host status and health. However, it remains open to what extent sleep is an adaptive response to contain, tolerate, or even clear infections (Krueger and Opp 2016).

4.1.3 Social immunity

Socially living organisms have additional options to defend themselves against parasites and to combat infections. 'Social immunity' is the result of the collective, behavioural, and physiological defences that add to the individual defences of colony members; it also includes the consequences of colony organization on the spread of a parasite. These mechanisms are prominent in the social insects but are also present in social groups more generally (Cremer et al. 2007; Cotter and Kilner 2010; Cremer et al. 2018). Mechanisms include nest hygiene or sanitary care. At some point, the most efficient defence from the group's point of view is to remove, rather than to care for, individuals. Removing

infecteds is an option for highly integrated animal social groups; the strategy is termed the 'care–kill dichotomy' (Cremer and Sixt 2009). However, despite the similarities between workers of social insect colonies and cells in an individual body, there are many more conflicts of interest among group members than among cells of a body (West et al. 2002). Social immunity is thus vulnerable to cheating. Not every individual might contribute equally to the social defence, be this the individual contribution towards heating the nest, as with social fever in honeybees, or the costly upregulation of an individual's immune system to prevent further spread of a pathogen in the group (herd immunity).

Many different elements of individual, behavioural, and physiological defences contribute and combine to provide social immunity in different groups of animals (Table 4.3). For example, the individual immune response to an experimental challenge is stronger in cooperatively breeding species of birds as compared to their solitary counterparts (Spottiswoode 2008). However, this relationship is not universal (Wilson 2003). Ant workers avoid contaminated areas or food (Tranter et al. 2015), prevent infected individuals from entering the nest (Drum and Rothenbuhler 1983), sanitize the nest with antimicrobial substances or potent chemicals (Brütsch et al. 2017), remove the infection or sources of contamination from the nest (Diez et al. 2014), and so forth (Cremer et al. 2018). Hygienic behaviour, such as allogrooming, leading to lower parasite loads, is known from an extensive range of taxa, e.g. social bees (Waddington and Rothenbuhler 1976; Drum and Rothenbuhler 1985), ants and termites (Hughes et al. 2002), but also primates (e.g. howler monkeys, Sanchez-Villagra et al. 1998). Grooming frequency, likely to reduce parasite loads, often increases with group size, e.g. in primates (Dunbar 1991) or ants (Schmid-Hempel 1998). In social insect colonies, chemical signals most likely help to organize these defences. For example, when honeybees are immune-challenged (i.e. mimicking an infection by injecting an antigen), the profile of their cuticular hydrocarbons changes (Richard et al. 2012). These changes relate to altered worker behaviour towards infected nestmates (Richard et al. 2008). Fungal-infected pupae of the ant *Lasius neglectus* emit a 'sickness cue' that leads to their

Table 4.3 Social immune defences in social insects and other social animals.[1] Some defences can serve in different contexts.

Defence element	Description	Type[2]	Mode[3]	Host type[4]
Avoidance behaviours	Avoid cannibalizing infected corpses.	Post	Behav	Ants, Termites
	Avoid contaminated food, corpses, etc.	Pre	Behav	Social insects, Group-living (crickets, cockroaches), Solitary (ladybirds, bugs, moths)
	Guard foraging trails.	Pre, Post	Morpho, Behav	Social insects
	Avoid contaminated habitats.	Pre	Behav, Spatial	Social insects, Group-living (burying beetles, mole crickets), Solitary (beetles, ladybirds, mosquitoes, bugs)
	Guard nest entrance.	Pre, Post	Behav	Social insects
	Non-overlapping foraging ranges.	Pre	Behav, Spatial	Social insects
Organization	Division of labour: only some individuals forage and are exposed.	Pre, Post	Behav	Social insects
Hygiene	Self-medication: collect antimicrobial substances to coat nest.	Pre	Behav	Social insects
	Produce chemical secretions, e.g. in metapleural gland, sternal glands, venoms.	Pre	Behav, Physiol	Social insects, Group-living (spruce beetles, pine beetles, beewolf with bacterial symbiont)
	Faecal material.	Pre	Physiol	Social insects, Group-living (cockroaches, burying beetles)
Waste management	Waste management: removal of corpses; keep waste deposits and 'graveyards' separated (also to reduce spread within group).	Pre	Behav, Spatial	Social insects, Group-living (cockroaches, burying beetles, web spiders, footspinners (Embiidina), caterpillars), Solitary (grasshopper)
	Encapsulation of parasites, infectious propagules, or infected individuals ('walling').	Post	Behav, Spatial	Social insects
Remove/avoid parasite	Remove parasite: mechanical removal by self- and allogrooming.	Post	Behav	Social insects, Group-living (earwigs, ambrosia beetles)
	Remove or cannibalize infected individuals.			Social insects
	'Sacrifice': infected individuals leave nest (controversial).			Social insects
	Pathogen alarm (vibrational displays in termites).	Post	Behav	Social insects
	Abandon infected nest areas. Nest relocation.	Post	Behav, Spatial	Social insects
Protective applications	Chemical substances (antimicrobials) placed on brood, or fed to brood.	Post Pre	Behav	Social insects, Group-living (earwigs, beewolf, wasp, cockroach, monarch butterfly), Solitary (house fly)
Organizational immunity	Behavioural structuring (by age and caste), spatial nest compartmentalization. Protect queen. Reorganize colony after infection.	Pre	Behav	Social insects.
Increase genetic diversity	Multiple mating and/or multiple queens.	Pre	Behav, Genetic	Social insects.
	High recombination rate.	Pre	Genetic	Social insects.

Defence element	Description	Type[2]	Mode[3]	Host type[4]
Facultative immunity	Immunity transfer by social interaction.	Post	Behav, Physiol	Social insects.
	Transgenerational transfer of immunity, such that colony members or direct offspring are protected.	Pre, Post	Physiol	Social insects, Group-living (burying beetles), Solitary (beetles)

[1] Sources: Cremer. 2007. *Curr Biol*, 17: R693; Meunier. 2015. *Philos Trans R Soc Lond B*, 370: 20140102.

[2] Type is either a defence that generally occurs pre-infection (Pre) or post-exposure (Post).

[3] Mode is either a defence based on behavioural change (Behav), physiological change (Physiol), spatial organization (Spatial), difference in morphology of acting individuals (Morph), or genetic structure of colonies (Genetic).

[4] Host type broadly grouped as social insects, group-living, or solitary animals.

removal and destruction by the attending nurse workers (Pull et al. 2018)—a process that is equivalent to the presentation of signals by infected cells in a body, which also leads to their destruction.

Instead of resisting and clearing an infection at a cost, its consequences might be tolerated (Medzhitov et al. 2012). In social insects, for example, resilience against worker loss (Müller and Schmid-Hempel 1992; Straub et al. 2015) or adaptive shifts in task allocation (Mersch et al. 2013) compensate for deficiencies caused by incapacitated workers (Natsopoulou et al. 2016). Finally, the organization of the social group or colony does affect how easily pathogens can spread. For example, new pathogens are routinely picked up from outside the group or nest. Hierarchical social groups should then prevent a disease from reaching the most valuable individuals, such as the queen in social insects (Schmid-Hempel and Schmid-Hempel 1993). 'Organizational immunity' (Schmid-Hempel and Schmid-Hempel 1993; Schmid-Hempel 1998; Stroeymeyt et al. 2014) is, therefore, an implicit effect in social groups and defined by the network of interactions among individuals, points of contact among group members, or simply by host density that is usually higher in advanced societies. Sophisticated schemes of a division of labour, such as found in most insect societies, can impede transmission and eventually block the spread of a disease (Schmid-Hempel 1998; Hock and Fefferman 2012; Sah et al. 2018). Social networks are, furthermore, plastic, and can therefore be adapted to contain an infection. For instance, when experimentally infected workers are added to a colony of the ant *Lasius niger*, individual behaviours and social contacts are rearranged to increase separation and modularity among worker groups (Stroeymeyt et al. 2018). Organizational immunity is quite general and known for social insects as well as social mammals (Altizer et al. 2003).

4.2 Basic elements of the immune defence

The immune system has several different functions, including the surveillance against aberrant cells (e.g. those causing cancer), to control 'damage' by wounding or cell death, or to remove debris, including debris from cells that undergo apoptosis during the development of organs and structures. Such damage generates 'danger-associated molecular patterns' (DAMPs), for example, double-stranded DNA freed from damaged cells, and which can be sensed by receptors of the immune system. DAMPs also reveal the presence of a parasite and its damaging actions. In the current context, the associated defence against parasites is the primary function of the immune system.

Parasites are either extracellular, such as most bacteria, or intracellular, such as viruses. Figure 4.2 is a cartoon of how the immune system responds to these two kinds of parasites. Although this simple scenario captures the basic working of the immune system, the actual processes in terms of receptors, pathways, and effectors are immensely complex. A specific immunology textbook provides the details, while keeping in mind that the standard textbooks

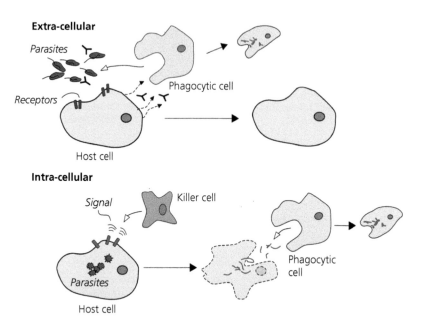

Figure 4.2 Two basic kinds of parasites and the corresponding responses. *Top panel*: Extracellular parasites are recognized by host cells through surface receptors (dark blue rods). A B-cell, for example, secretes antibodies (Y-shaped symbols) that can bind to the parasite. Eventually, phagocytic cells are recruited by signals, and the parasites engulfed and eliminated. *Lower panel*: Intracellular parasites are recognized by internal process and their presence inside the host cell is signalled at the surface (purple rods). This recruits killer cells that destroy the infected host cell and its dangerous contents. Eventually, phagocytic cells are recruited to the site to gobble up the debris. In both cases, the humoral components of the immune response are involved, and tightly integrated with the cellular defences. These simple response schemes are implemented in extremely complex pathways and very diverse arrays of molecules. Sketch courtesy of Sebastian D. Fugmann.

typically cover the immune system of the (jawed) vertebrates, but less frequently those of insects, other invertebrates, or plants. Immunology is also a vast and active field of research, where knowledge expands and concepts keep changing by the year, and sometimes even by the month. Furthermore, the immune system components are classified according to several criteria. In particular, the cellular and humoral arms of the defence, or the innate and adaptive systems. These distinctions refer to differences in the underlying mechanisms. In fact, within the entire immune system, all arms of the defence are tightly integrated in order to function correctly. The molecules and pathways of the immune system also belong to different compartments, such as recognition, the signalling cascade, or the effectors.

4.2.1 Humoral defences

Humoral defences involve non-cellular components, such as serum proteins and enzymes. These nevertheless depend on cells that associate with or secrete these molecules.

4.2.1.1 Immunoglobulins

Serum globulins with antibody activity, collectively called immunoglobulins (Ig's), are essential humoral responses in gnathostomes (jawed vertebrates). Immunoglobulins group into five major classes: IgG, IgM, IgA, IgE, and IgD. Immunoglobulins bind with different specificities to foreign objects (the antigens). Even with identical binding specificities, the different classes of immunoglobulins fulfil different functional roles in the defence. Immunoglobulins are discussed in more detail in section 4.4.1.

4.2.1.2 Complement

The complement system consists of a set of soluble proteins of the blood serum that 'complement' the cellular components and complement the functions of the 'ordinary' antibodies. Complement is one of the phylogenetically oldest parts of the immune system, and probably originated in the common ancestor of all Metazoa (Nonaka 2014). Essential complement components exist in deuterostomes (such as sea urchins and hemichordates; Nonaka 2001; Zarkadis et al. 2001). Some elements are also known from protostomes, e.g. from early arthropods, corals (C3-like genes), or sea anemones, suggesting that this defence cascade is at least one billion years old (Flajnik and du Pasquier 2013). These early components seemed to have been lost and were presumably replaced by other thioester-containing proteins (e.g. TEPs of insects, *Drosophila*, *Anopheles*; Blandin and Levashina 2004) that function as opsonins. In humans and mammals more generally, complement consists of around 30 to 60 different proteins, which includes the soluble and cell-membrane-associated molecules. Complement is one of the very first lines of defence against invading pathogens and is found extra- as well as intracellularly. Its activation leads to the destruction of the invader by direct lysis of the target cell. Complement is furthermore 'marking' foreign material for destruction ('opsonization') via phagocytic cells, such as by polymorphonuclear cells (PMNs) and neutrophils. Complement also activates specialized immune cells and adaptive immunity (Reis et al. 2019) and stimulates inflammation by anaphylatoxins.

Complement becomes activated by one of three different pathways, the so-called 'classical', 'lectin', and 'alternative' pathways, as Figure 4.3 shows in more detail. In these three pathways, the presence of an invader is detected by different means. In all of them, the key complement protein, C3, is involved, a glycoprotein that not only allows keeping the system in a 'standby' state but also is a point of convergence of the activating cascades (Ricklin et al. 2016). In the key process, C3 becomes cleaved into C3a (an inflammation mediator) and C3b (an opsonin), the two most critical upstream components of any activation. Well-defined recognition molecules trigger the classical and lectin pathways (Ig's, MBLs, etc.). The alternative pathway has no such definition but depends on the background production and subsequent recruitment of C3b proteins to general motifs of the parasite surface; this pathway thus continuously monitors for the presence of parasites.

Complement activation results in the formation of a membrane attack complex (MAC) that punches the parasite's cell wall and destroys it by lysis, but also can induce the inflammation process. The attack complex can also lead to the destruction of its own, aberrant cells. Complement furthermore connects innate immune defences (e.g. cell lysis by the attack complex) with adaptive immunity, e.g. the production of antibodies, stimulating B- and T-cells, and responses by antigen-presenting cells (APCs). The CR2 receptor recognizes C3b on B-cells. This step supports eventual immune memory formation. Because of its ever-ready state and high reactivity to cell surfaces, complement also needs to be tightly regulated to avoid self-damage. Interestingly enough, some pathogens (such as the bacterium *Porphyromonas gingivalis*) actively interfere with complement to avoid their clearance. In the process, C5 is enzymatically cleaved, which eventually stimulates the degradation of MyD88, an essential element in the innate Toll and Toll-like receptor (TLR) pathways which is required for clearance (Maekawa et al. 2014). Beyond defence, complement also has several other functions that are basic for the organism (Kolev et al. 2014; Hajishengallis et al. 2017). Complement is now understood, more generally, as a global immune regulator and mediator of tissue homeostasis (Hajishengallis et al. 2017; Reis et al. 2019). Complement can even regulate the microbial flora (the microbiome, section 4.8; Chehoud et al. 2013).

4.2.1.3 Other humoral components

A range of other compounds circulate outside cells and thus are part of the humoral arm of the response. Among those, the antimicrobial peptides (AMPs; see section 4.4.2) exist in virtually all organisms. Various other humoral compounds are needed for the inflammatory, clotting, coagulation, or melanization responses. Clotting, for example, aids in limiting tissue damage due to parasite activities or self-damage by the immune system. Clotting requires humoral components like hemolectin or Fondue in *Drosophila*, and a variety of factors (called IX, XI, etc.) in mammals (Sheehan et al. 2018). In insects, the fat body—a loose collection of cells rich in lipids and glycogens, lining the rim of the haemocoel—is essential for the production and secretion of AMPs and other factors in the humoral defence. Similarly, midgut and salivary glands release humoral factors that kill and control microbial infections (Hillyer 2016). Furthermore, proteolysis is a process of degradation of proteins by specialized humoral enzymes, the proteases.

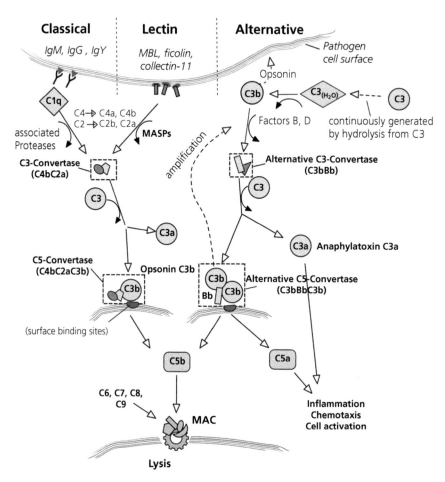

Figure 4.3 Complement in mammals. Complement is activated by one of three pathways that respond to different signals. *Classical pathway.* Complement protein C1q binds to complement-fixing antibodies (IgM, IgG, IgY) that have recognized and attached to epitopes on the pathogen's cell surface (red triangles). C1q-associated proteases cleave proteins C4, C2, which then leads to formation of the protein (enzyme) complex C4bC2a, also known as classical pathway C3-convertase (dashed rectangle). C3-convertase activates the key protein C3, eventually generating C3a, and opsins C3b that bind to pathogen surface sites (red ovals). Opsonins mark the pathogen surface and are part of the complex of C4bC2aC3b, which forms the C5-convertase that in turn cleaves C5 into C5a and C5b. C5b is necessary to form—together with complement proteins C6, C7, C8, C9—the Membrane Attack Complex (MAC). MAC forms pores in cell walls and thus leads to the destruction of the parasite by lysis. *Lectin pathway.* Mannose-binding lectin (MBL), ficolin, and collection-11 (blue rods) bind to carbohydrates (red ovals) on the pathogen surface. The mannose-binding protein-associated serine proteases (MASPs) then activate/cleave C4 and C2 to form the lectin pathway C3-convertase, which is the same as the classical one (C4bC2a). These subsequent steps converge with the classical pathway. *Alternative pathway.* Even when not infected, the key protein C3 is continuously hydrolysed into $C3(H_2O)$, which generates fragment C3b with the help of factors B, D and at a low background level. C3b continuously monitors for the presence of invaders. In fact, C3b is highly reactive and binds to generally responsive chemical groups on the pathogen surface; it can thereby act as an opsonin. In addition, factors B and D are recruited to process C3b into the alternative pathway convertase C3Bb (dashed rectangle). This convertase cleaves C3 into the anaphylatoxin C3a and, again, C3b. With another C3b, the alternative C5-convertase (C3bBbC3b, with opsonin; dashed rectangle) is formed that cleaves C5 into C5a and C5b. C5a is involved in, for example, the inflammation process. C5b converges with the same product from the classical/lectin pathways and becomes part of MAC. Adapted from Freely et al. (2016) by permission of John Wiley and Sons.

Proteolytic cascades are triggered by the binding of soluble host recognition proteins to a corresponding parasite motif. The best studied such cascade actually is complement.

4.2.2 Cellular defences

The cellular defence is based on immune cells. The defences based on B- and T-cells in the jawed vertebrates have received most of the attention and will be discussed in more detail below (section 4.6.3). Other important cellular components include monocytes and macrophages. These are recruited to the site of infection in the process of inflammation (see section 4.3.1). The release of cytokines is a significant function of immune defence cells and serves to orchestrate the cellular response. Cells infected by viruses produce 'interferons' (IFN; cytokines) to alert nearby cells to this threat.

Invertebrates have no closed blood circulation system. Their cellular response is based on a variety of specialized cells that are freely circulating in the body (e.g. plasmatocytes, lamellocytes, and crystal cells in the haemocoel of insects). Such cells are known from all Metazoa, even though ontogeny and functions may differ. However, the diversity of immune cell types in invertebrates is generally much lower than in vertebrates. In arthropods, crystal cells store a precursor molecule (PPO) for a key enzyme (phenoloxidase, PO) of a major defence cascade. This cascade results in the release of toxic molecules and eventual melanization. Crystal cells readily rupture and release their contents into the haemolymph when activated. Other circulating cells perform functions such as phagocytosis, encapsulation, nodule formation, and clotting, or the production of a range of effector molecules. Activation of the cellular defence is by the recognition of epitopes on the surface of parasites, or via opsonization. Alas, there is a confusing nomenclature. The circulating cells are called 'haemocytes' in the arthropods, but 'coelomocytes' or 'amoebocytes' in annelids, molluscs, and echinoderms (Flajnik and du Pasquier 2013). Even more generalized nomenclatures exist (Hartenstein 2006).

4.2.2.1 Haematopoiesis (cell development)

All specialized immune cells originate from the haemopoietic stem cells. These are totipotent and self-renewing. During haematopoiesis stem cells differentiate into various types, such as macrophages or T-cells, and often increase in their specific binding properties along the different routes (Figure 4.4). The molecular processes of haematopoiesis seem relatively conserved across animal lineages (Hartenstein 2006). However, the location where the development happens—the haematopoietic tissues (or the 'haematopoietic stem cell niche')—differs among taxa. For example, crustaceans use lobes in the head, connective tissues of thorax and abdomen, and tissues at the base of each limb (Söderhäll 2016); insects use haematopoietic pockets in each larval segment and a specialized, anterior lymph gland (Gold and Brückner 2015; Hillyer 2016). Gut-associated tissues are an active site in all vertebrates, whereas a fully developed thymus and spleen appear with the jawed vertebrates. In addition, bone marrow has evolved to be an essential stem cell niche for various cell types, including immune cells. The most prominent haemopoietic tissues are the lymphoid organs. In mammals, B-cells and various other cell types mature in bone marrow (Crane et al. 2017), whereas in birds, B-cell maturation is in the bursa of Fabricius. T-cells also start their development in bone marrow but eventually migrate to the thymus glands for final maturation (see section 4.6.3). Secondary lymphatic organs are located in the periphery and harbour the mature, functional lymphocytes. Different kinds of secondary lymphatic organs exist at various sites of the body, e.g. lymph nodes, spleen, tonsils, or in the mucosa lining the digestive (Peyer's patches), respiratory, or the genital-urinary tracts. The architecture of the lymphatic system varies among organisms; most species do not have lymph nodes (Boehm and Bleul 2007).

Interestingly, stem cell niches and myeloid-lineage immune cells of colonial tunicates (a primitive chordate) share characteristic similarities with those of mammals, suggesting a common evolutionary origin (Rosental et al. 2018). The stem cell niches typically also change as the individual develops from embryo to juvenile or larval stage, and finally into an adult. Vertebrate haematopoiesis follows two pathways with the lymphoid or myeloid progenitors (Figure 4.4a). It occurs in distinct waves as the individual develops. In humans and mice, the early wave involves the placenta and other organs, including tissue associated with blood vessels. The later waves occur in tissues such as liver, thymus, spleen, and bone marrow (Wang and Wagers 2011; Crane et al. 2017). The various immune cells then assume different functions—macrophages become phagocytes, NK-cells destroy virus-infected cells, and B- and T-cells are part of the adaptive immune

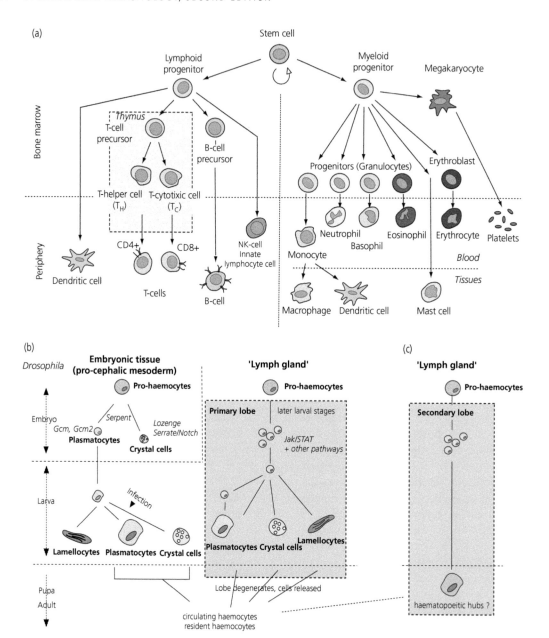

Figure 4.4 Immune cell development (haematopoiesis). (a) Haematopoiesis in vertebrates. Totipotent stem cells develop into two families of blood cells (myeloid and lymphoid cells). In the thymus, T-cells differentiate into T-helper cells (with receptor CD4+) and cytotoxic T-cells (CD8+). Adapted from Coico and Sunshine (2015) by permission of John Wiley and Sons. (b) Haematopoiesis in *Drosophila*. In insects, haematopoiesis also follows two (myeloid) lineages that share characteristics with the vertebrates. Furthermore, insects typically develop through different stages (larvae, pupae in holometabolic groups) that have their own adaptations. Basically, blood cells develop from embryonic mesoderm as pro-haemocytes that differentiate into plasmatocytes (the large majority of blood cells; functionally similar to vertebrate macrophages) and crystal cells (similar to vertebrate granular cells; in insects they are the reservoir of PPO). Plasmatocytes reside in tissues and also proliferate and self-renew. Upon infection, plasmatocytes can also differentiate into lamellocytes. It seems controversial whether the secondary lobe of the lymph gland transforms into a haematopoietic organ in the adult fly. Adapted from Gold and Brückner (2015), with permission from Elsevier. (c) Development in secondary lobe after description in Hillyer (2016). Some important transcription factors and pathways are indicated in italics.

response. Some of these cells are freely circulating in the bloodstream, and others are resident in tissues, from where they are quickly recruited to a site of infection or injury.

In insects, haemocytes ('blood cells') are the prime immune cells and have mainly been studied in *Drosophila*, mosquitoes, and the silkworm. Given the enormous number of extant insect species, our knowledge is therefore minimal. In the model species, *Drosophila*, immune cells develop from self-renewing stem cells along two routes and differentiate into plasmatocytes, crystal cells, and, upon infection, into lamellocytes (Figure 4.4b) (Gold and Brückner 2015; Vlisidou and Wood 2015; Hillyer 2016). Plasmatocytes represent the lion's share of all cells. They develop in the embryo as they spread through the body. In the larval stage, haemocytes are produced in the 'lymph gland', a specialized haematopoietic tissue around the anterior, dorsal blood vessel. This gland degenerates during the pupal stage, and the cells are released. The crystal cells originate in the larval stage from plasmatocytes within aggregates of blood cells (Leitão and Sucena 2015). Among all insects, and also among the vast group of other invertebrates, there are many variations to this theme. Nevertheless, the primary pathways and cell types share similarities and are partly conserved (Hartenstein 2006; Grigorian and Hartenstein 2013). Typically, invertebrates have several subpopulations of cells with different functions, but their nomenclature is not standardized.

4.2.2.2 Phagocytosis

This is a universal and evolutionarily conserved cellular response. Phagocytosis removes cellular debris, foreign particles, and parasites and starts within seconds of infection. It ends with the internalization of a particle by the phagocytic cell, followed by its destruction (Stuart and Ezekowitz 2008; Lim et al. 2017). Historically, phagocytosis is also one of the first active immune defence mechanisms discovered. In around 1900, Elie Metchnikoff pricked the larvae of a starfish with a thorn and noted how cells migrated to the site of the wounding (Gordon 2016a).

Phagocytosis follows a general scheme found in virtually all animals (Figure 4.5a) (Gordon 2016b;

Hillyer 2016; Lim et al. 2017). The process starts with the recognition of a (non-self) pathogen, or of an 'altered self', e.g. the molecular signature of a necrotic cell. For this, receptors bind to opsonized particles (e.g. the C3b opsonin of complement) or directly to parasite surface components with pattern-recognition receptors (PRRs) such as scavenger receptors (SRs), Nimrod proteins, or peptidoglycan-recognition proteins (PGRPs). In insects, soluble factors such as TEPs can mark a pathogen for later destruction, whereas in mammals antibodies (IgGs, secreted by plasma cells), sensed by Fc receptors, also mark a particle (Figure 4.5a). The particle is internalized by a phagocytic cell, which by rearrangements of the cytoskeleton can move and change shape. The phagocytic cell forms pseudopods that facilitate uptake of the particle into a phagosome. As the parasite is internalized, large tracts of new membrane are synthesized by the phagocytic cell and delivered to the surface to compensate for the engulfed part. For example, mammalian phagocytes can replace the equivalent of more than their own surface within half an hour (Greenberg and Grinstein 2002). Subsequently, the internalized parasite becomes transported to a specialized vesicle, the phagosome. Phagosomes subsequently undergo a process of maturation, during which they may split (fission) and fuse with endosomes and lysosomes into a phagolysosome. The resulting chemical environment leads to the degradation and destruction of the engulfed particle or microorganism. The type of phagosome that will develop depends on several factors, such as the involved receptors, the kind of particle swallowed, or the nature of the membrane. The phagosome is, therefore, a rather complex organelle (Gordon 2016b; Nazario-Toole and Wu 2017). For example, 600 different proteins are involved in *Drosophila melanogaster*, of which 70 per cent are orthologues of the mammalian phagosome (Stuart and Ezekowitz 2008). Eventually, the parasite is also killed by oxidative mechanisms that produce reactive molecules in a process called 'respiratory burst', or 'oxidative burst'. In this process, cells release reactive oxygen species (ROS: superoxides, hydrogen peroxides) and nitric oxide, which are toxic to microorganisms. The oxidative burst is especially prominent in the hypersensitive responses of plants but also occurs

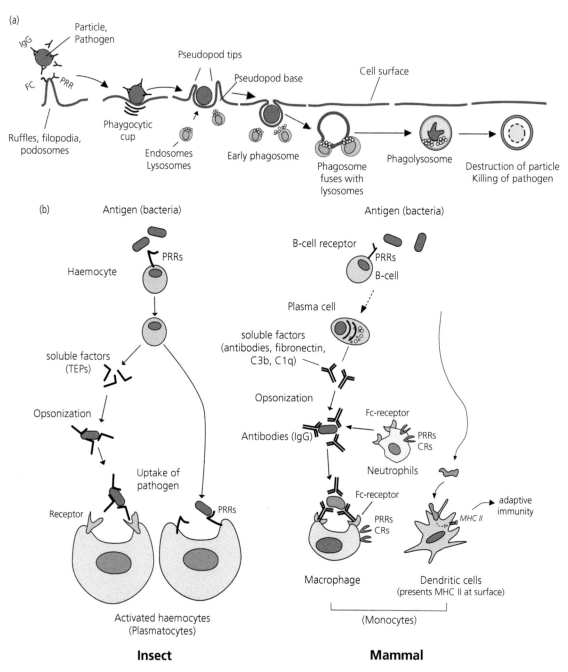

Figure 4.5 The mechanism of phagocytosis. (a) In the general scheme, a foreign particle (such as a parasite, red) is delivered from the surface to the interior of the cell, and into a phagosome. To find foreign material, the phagocyte explores its surroundings by filopodia, etc. Material can be marked ('opsonized') with antibodies (IgGs, recognized by Fc receptors), or is recognized by general PRRs (pattern-recognition receptors, e.g. recognizing bacterial cell wall components). Recognition triggers an ordered series of events where, first, the cytoskeleton acts to form pseudopodia that help engulf the particle. By fusion with endosomes and lysosomes (green), the new vacuole matures into a phagolysosome; the acidic and hydrolytic milieu destroys the parasite. (b) Simplified schemes for 'professional phagocytes'. *Left*: In insects (*Drosophila melanogaster*), the pathogen (red) is recognized by cell-surface PRRs of haematocytes or by soluble factors (e.g. TEPs, thioester-containing proteins). With opsonization, the pathogen (red) is recognized by activated haemocytes (e.g. plasmatocytes), followed by phagocytosis (Hillyer 2016; Nazario-Toole and Wu 2017). *Right*: In mammals, the recognition is also directly through cell-surface PRRs of B-cells (green). Soluble factors opsonize a pathogen, e.g. with immunoglobulins (from activated B-cells) and complement factors (C3b, C1q). The opsonized particle is recognized and subsequently phagocytosed by various cells that arrive at the site of infection at different times. The phagocytic cells (neutrophils, macrophages) express a variety of receptors, such as Fc, PRRs, or complement receptors (CRs). Dendritic cells are an example of 'non-professional' phagocytes that have narrower targets and are less efficient (Gordon 2016a; Lim et al. 2017). Sketches after descriptions in, and adapted from, Stuart and Ezekowitz (2008), with permission from Springer Nature.

in insects and mammals. These mechanisms are very ancient and conserved across phyla.

Tissue-resident 'professional phagocytes', such as macrophages, are the first responders, recruited to the site of infection by chemical signals. In mammals, neutrophils are the first cells recruited from the bloodstream to the site of infection, guided by chemo-attractants released by the pathogen itself or by the affected host cells. Later arrivals are the monocytes that differentiate into macrophages or dendritic cells. 'Non-professional' cells are other cell types (e.g. dendritic cells, epithelial cells) that can phagocytose but have a narrower range of targets: they are less efficient, are slower, or have a more specialized function. Phagocytosis by dendritic cells provides a link to the adaptive arm of immunity. For the defence against intracellular parasites (e.g. protozoans, bacteria, or viruses), programmed cell death ('apoptosis') can remove infections by the self-destruction of the cell. Apoptosis follows various modes and uses different and complex pathways that can be regarded as safeguards against parasite interference (Jorgensen et al. 2017). In all, phagocytosis is not only crucial for defence against pathogens but also for body maintenance (homeostasis) (Arandjelovic and Ravichandran 2015). Macromolecules (rather than entire parasitic cells) can be engulfed by specialized cells, such as macrophages, in a process called endocytosis. Endocytosis is unspecific by simple membrane fold ('pinocytosis') or based on the binding of the macromolecules by specific receptors, followed by the internalization of the foreign material. In either case, the vesicle containing the macromolecules fuses with endosomes (containing acidic components) and with lysosomes (containing degrading enzymes, as in phagocytosis). This leads to the destruction of the ingested material.

4.2.2.3 Melanization, encapsulation

The prophenoloxidase (PPO) cascade is found in invertebrates but is a major defence response in arthropods (Eleftherianos and Revenis 2011; Hillyer 2016) and is absent from vertebrates (Flajnik and du Pasquier 2008). The recognition of a parasite by a PRR (e.g. lipopolysaccharide (LPS)-binding proteins, PGRPs, or Gram-negative binding proteins (GNBPs)), activates a serine protease cascade, which eventually cleaves the pro-form, PPO, into the active enzyme

phenoloxidase (PO). Through a series of steps, PO converts tyrosine into melanin. Melanin is involved in the hardening and darkening of insect cuticles during development or wound healing (Bilandžija et al. 2017). It is also a part of the defence against parasites through melanization (Hillyer 2016). With the cascade, cytotoxic intermediates are also produced, such as phenols, quinones, and reactive oxygen species (Nappi and Christensen 2005; Strand 2008), which can directly kill pathogens. In insects, the major reservoir of the precursor molecule, PPO, is in the circulating crystal cells. These represent around 5 per cent of all plasmatocytes; their degranulation releases the PPO. The PPO cascade generates a range of aggressive compounds. Therefore, regulatory proteins (serpins) control and localize the melanization process to avoid self-damage (Shakeel et al. 2019). Small parasites, such as bacteria, become melanized in a capsule and are phagocytosed. The circulating plasmatocytes recognize larger invaders, such as eggs or larvae of parasitoids, and attach to their surface. Subsequently, a large number of lamellocytes, plasmatocytes, and granulocytes are attracted. The lamellocytes form a multilayered capsule around the invader that eventually melanizes into a tight capsule, sealing off the parasite from the rest of the host body in a process known as 'encapsulation' (Lemaitre and Hoffmann 2006; Cerenius et al. 2008; Hillyer 2016). The PPO cascade is again central to this process. The cascade crossreacts with clotting and nodulation, and eventually leads to melanization that is visible from the outside by the darkening of the encapsulating cells.

4.2.2.4 Clotting, nodule formation

In vertebrates, local blood clotting is a process concurrent with the activation of complement and inflammation. The resulting blockage of the blood flow draining from the site of infection impedes the spread of infecting microorganisms to the rest of the body (Markiewski et al. 2007). However, with the potential to form thrombi, to spread systemically, and to disable blood flow, blood clotting also carries a risk of inflicting secondary damage to the host. It is, therefore, also under the control of inhibitors in the form of anticoagulation systems (Markiewski et al. 2007). The open circulatory system of arthropods poses additional challenges, as invading microorganisms can spread in the haemocoel and throughout the entire body without efficient barriers. Local clotting is a crucial response by the arthropod immune system. The formation of a matrix of cells during clotting additionally facilitates wound sealing and healing (Theopold et al. 2004). Nodule formation in insects is a further cellular defence response during which invading microorganisms such as bacteria become trapped inside an aggregation of haemocytes.

4.3 Basic defences by the immune system

4.3.1 Inflammation

Inflammation is an early, diverse, and complex response by the innate arm of the immune system of vertebrates—combining humoral and cellular components. Inflammation is directed against all kinds of harmful stimuli, including injury, cell debris, foreign particles, or parasites. Inflammation contributes to eliminating the challenge and to the repair of the affected tissue. With inflammation, fluids (plasma) and blood cells (leukocytes, especially neutrophils, granulocytes) rapidly move to the site of infection, differentiate, and interact with local resident macrophages in the affected tissue. For this to happen, a cascade of biochemical events ensures that blood flow to a site of infection increases, and that blood vessels become permeable such that cells can infiltrate the tissue surrounding the site of infection or injury. Typically, inflammation is visible from the outside as swelling and an increased reddening of the skin. Inflammation is triggered by PRRs, such as TLRs, located on tissue-resident macrophages or mast cells that act as sentinels, which recognize epitopes on a pathogen's surface. Recognition triggers a signalling cascade that eventually results in the production of type-I IFNs and proinflammatory cytokines, which attracts leukocytes. These immune cells, together with the proteins contained in blood plasma, e.g. from the complement system, act to contain and eliminate the infection before it can spread further. Soon after inflammation begins, the sequence to terminate the process is initiated, and it ends when granulocytes are removed and the macrophages have been drained by the lymphatic system (Serhan and Savill 2005).

A key step in inflammation is the formation and activation of the 'inflammasome'. These are large protein complexes in the cell's cytosol; their specific assembly and composition are governed by the kind of receptor that is stimulated. Several different inflammasomes with their activating pathways have so far been identified. For example, the inflammasome of type-'NLRP1' assembles when infected by *Bacillus anthracis*, which releases a lethal toxin; type-'NLRP3' results from detecting bacterial LPS by a TLR4 receptor. Inflammasomes differ subtly in their protein composition, but typically include a NOD-like receptor (NLR) with an adaptor protein. They converge in the eventual production of the enzyme caspase-1 (or other caspases, depending on inflammasome; Broz and Dixit 2016), which mediates the release of proinflammatory cytokines. Activation can also lead to inflammatory cell death ('pyroptosis') (Bauernfeind and Hornung 2013; de Zoete et al. 2014; Guo et al. 2015; Broz and Dixit 2016). If excessively expressed, inflammation can damage the organism's own tissue. In fact, chronic inflammation is associated with various 'modern' diseases of humans, e.g. inflammatory bowel diseases, arthritis, diabetes, or cancer (Okin and Medzhitov 2012). It is, therefore, a tightly controlled process (O'Connell et al. 2012). In the gut, the resident microbiota can modulate, or dampen inflammation, even at sites distant from the intestines (Blander et al. 2017).

4.3.2 Innate immunity

'Innate immunity' is a collection of various and quite diverse defence systems, such as complement, phagocytosis, or TLR pathways, which were added and modified in the course of evolution. One could crudely consider 'innate' immunity to be germline-encoded, i.e. using genes as they are present in the genome. By contrast, 'adaptive' immunity is based on somatic adaptations, such as somatic recombination or alternative splicing that leads to deviations in proteins, as compared to their 'original' encoding in the genome. Adaptive immunity, therefore, allows for tuning and individualizing the defences more precisely towards an actual challenge (see section 4.6.3). However, adaptive immunity is a delayed response. Innate immune defences, by contrast, are either constitutively expressed (e.g. complement), or become active with a short delay. Innate defences are therefore crucial as a first and generalized defence and are prominent in physical barriers such as skin or mucosa. In vertebrates, the innate immune system also contributes to the activation and orchestration of the adaptive response (Iwasaki and Medzhitov 2015).

Innate immunity contains many different systems that are evolutionarily old. Innate immunity also has a limited number of receptors that recognize parasite surface molecules (epitopes). These typically represent molecular motifs conserved across a range of microbial taxa (Kimbrell and Beutler 2001), for example deep and crucial structural elements of a parasite's cell wall (Beutler 2004), and seem difficult to change. Host receptors that can recognize those general features of a microbial parasite are the PRRs, and the features that they recognize are parasite-associated molecular patterns (PAMPs). In fact, PAMPs are not different from any other epitope that an immune system might recognize, and PRR is a term that is synonymous with a recognition molecule more generally. As the study of innate immunity progresses, also more and more elements are discovered that are functionally similar to those of the adaptive system, for example, specific responses and the formation of a 'memory' (discussed under many different names) (Gourbal et al. 2018; Netea et al. 2019) (see section 4.7.2).

4.3.3 Adaptive (acquired) immunity

'Adaptive' has a somewhat different meaning in immunology as compared to evolutionary biology. In the latter, it means a trait that positively contributes to fitness in the environment where the organism lives. In immunology, 'adaptive' means that the immune system responds increasingly specifically to an infection within the lifetime of an individual. For this purpose, the adaptive immune system acquires information about the ongoing infection and adapts the response to the particular parasite. As a consequence, an adaptive immune response is not only more specific but also inevitably delayed. Response times typically require days rather than minutes to hours as with innate defences. The adaptive system, in the narrower sense, has evolved in the vertebrates. It appears, therefore, later in the

history of life. Nevertheless, some somatic adaptations already exist in a few invertebrate phyla. Regardless, innate and adaptive systems interact closely; an adaptive response would not be possible without a previous innate response. Note that there are two versions of the adaptive immune system—that of the jawless vertebrates (Agnatha, i.e. hagfish and lampreys) and that of the jawed vertebrates (Gnathostomata: the higher vertebrates). These two versions show analogous as well as homologous features, with differences in receptors but conserved cell lineages, and partly shared mechanisms of cell-dependent antibodies. A hallmark of adaptive immunity in the higher (jawed) vertebrates is the presence and clonal expansion, upon infection, of highly variable lymphocytes (B- and T-cells) with antigen-specific functions, directed by the major histocompatibility complex (MHC), as well as the formation of derived memory cells (see section 4.7). Variable lymphocyte receptors (VLRs) have a similar role in jawless vertebrates, for example in the lamprey. Adaptive immunity is discussed in more detail in section 4.6.3.

4.3.4 Regulation of the immune response

The immune system is a biochemical-molecular network that is tightly regulated to generate an appropriate response in the right place and at the right time. The system recognizes the presence of an infection. The recognition event is subsequently converted into a signal ('signal transduction') that stimulates the downstream signalling cascade. The cascade itself is activated or inhibited by a variety of concurrent processes that act in parallel or are part of overlapping molecular cascades. Eventually, the cascade yields a variety of further signals. These are either transcription factors that initiate

1. the production of effectors (e.g. AMPs) and secreted signals to other cells (e.g. cytokines)
2. activation of specialized immune cells (e.g. cytotoxic T-cells),

or they are signals that trigger apoptosis. Although the signalling cascades are somewhat different in different organisms, the basic scheme of 'recognition-signalling effectors' is universal. By contrast, the signalling molecules themselves rapidly diversify and show considerable genetic divergence among lineages (Waterhouse et al. 2007).

A key to regulation is the production of signalling molecules that, for example, can form or disassemble protein complexes needed for the further passing of signals. Among those, cytokines are essential and produced by almost all cells of the immune system to communicate with other cells. Cytokines that signal between leukocytes have been named interleukins (ILs); cytokines that interfere with viral replication are IFNs, especially type-I IFNs (type-II IFN generally activates macrophages). Cytokines are secreted (e.g. IFN-γ, IL-2), or remain membrane-bound (tumor necrosis factors, TNF-α, TNF-β). But their effect on cells depends on the target cells, which possess cytokine receptors that are specific to some cytokines but not others (Cho and Kelsall 2014; Ivashkiv and Donlin 2014; Boraschi et al. 2017). Some cytokines act antagonistically with one another, such as the ones inducing either T_H1- or T_H2-cell development. Hence, the combination and concentration of different cytokines recognized by a cell ultimately decide its activity in the immune response. Cytokine molecules do not persist for very long and are soon degraded to ensure swift regulation.

A paradigmatic case is the signalling cascade of the mammalian TLR pathways that are important for innate and adaptive immunity. Several different types of TLRs are expressed on different cells, and in different compartments inside these cells. These are sensitive to different kinds of stimuli, e.g. lipopolysaccharides (LPS), dsRNA (from viruses). As Figure 4.6 shows, regulation of the TLR pathway occurs at many points. For instance, the constitutive CHIP protein affects a relatively early step in the cascade, whereas the receptor-interacting protein RIP140 regulates the transcription factor NF-κB further downstream. Often, the assembly and disassociation of critical molecule complexes (e.g. IRAK, TAK1) are themselves the means of regulation, such that the signal is passed on, or stopped. In the example, CHIP and RIP140 stimulate the cascade. A wide variety of regulators, e.g. for complement, include soluble or membrane-bound molecules and their respective receptors (Zipfel and Skerka 2009). The primary regulatory mechanisms can be characterized as follows.

4.3.4.1 Regulation by protein–protein interactions

Proteins can act at key steps to attenuate a response. An example is IRAK-M (IRAK-3), an inducible negative regulator of the NF-κB pathway. It directly interacts with other proteins, and its main effect seems to be the binding to protein TRAF-6, which inhibits the latter's 'normal' interaction with IRAK-1 (Rothschild et al. 2018).

4.3.4.2 Regulation by miRNAs

Regulatory micro-RNAs (miRNAs) are the most dramatic discovery during the last one or two decades. miRNAs target a particular RNA sequence, which is a transcript of a particular gene that needs to be regulated. Functional miRNAs emerge from their primary genomic templates through a series of events and become active in appropriate molecular complexes. They pair with their coding mRNA sequence and degrade it, which results in the silencing of the respective gene (Rothschild et al. 2018). Hence, miRNAs suppress the targeted element. But miRNAs always affect their immediate target. Therefore, the overall response can be either down- or upregulated. For example (Figure 4.6), miRNA19 dampens the zinc finger protein A20 that, in turn, suppresses TRAF-6, an important element of the TLR cascade. Hence, the net result is an upregulation of the response. The repertoire of immune-regulatory miRNAs now numbers in the hundreds or even thousands, and they affect every corner of the immune system. This includes the basic, canonical innate immune cascades (He et al. 2014; Forster et al. 2015), the regulation via cytokines and inflammation (Contreras and Rao 2012; Forster et al. 2015), apoptosis, phagocytosis (Zhou et al. 2018), adaptive immunity such as T-cell function (Keck et al. 2017), antiviral responses (Ojha et al. 2016), immunological tolerance (Chen et al. 2013), and immune cell maturation (Mehta and Baltimore 2016). Regulatory miRNAs are not restricted to mammals but also found, for example, in insects (Hussain and Asgari 2014), or plants (Weiberg and Jin 2015).

4.3.4.3 Regulation by post-translational modification

Several processes can modify proteins and thus change function (Rothschild et al. 2018). These

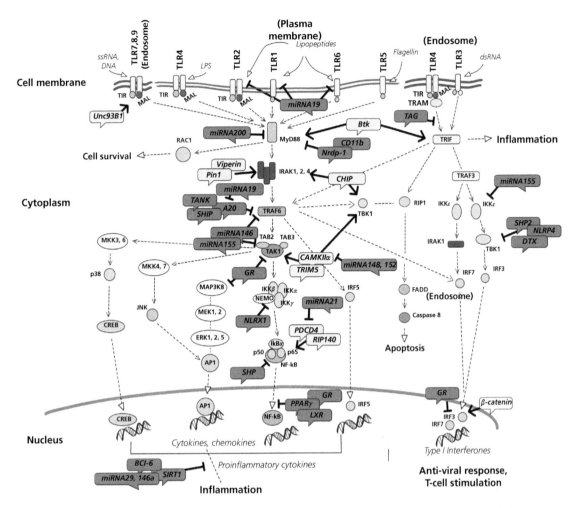

Figure 4.6 The complexity of immune response regulation. The sketch shows the TLR pathways, e.g. in a mammal, starting with TLR receptors (TLR1 to TLR9, blue rods) that recognize different challenges (*ssRNA*, *LPS*, etc.), and are located on different membranes. Recognition triggers signalling cascades (components with different colours). Eventually, transcription factors (green) are produced that initiate defences such as inflammation, apoptosis, antiviral responses, or T-cell stimulation. Broken lines illustrate the signalling cascades. Call-out boxes show additional, explicit factors that either stimulate (thick arrows, yellow boxes) or dampen (blocking symbols, red boxes) particular steps in the cascade. Note that regulation also occurs by assembly or disassociation of molecular complexes such as the IRAKs, which is not shown here. For clarity, by no means are all regulating factors, or all adaptors, cascade elements, and their interactions, shown here. In particular, a considerable number of regulatory miRNAs are not shown. The full names for the acronyms are not relevant here, but see text and Glossary for further information. Pathway schemes adapted and redrawn after Gay et al. (2014b), with permission from Springer Nature; regulators following descriptions in Qian and Cao (2013).

include proteolysis, methylation, acetylation, glycosylation, phosphorylation, or ubiquitination. For example, phosphorylation causes a conformational change of the protein (i.e. a change in its three-dimensional shape), and thus to a different function. Typically, it is an activating signal, for example IRAK and IκB kinase (IKK) complexes, activating the NF-κB cascade (Rothschild et al. 2018). Ubiqui-

tination involves ubiquitin, a small peptide in eukaryotic cells that can attach to proteins and thus regulates their functions. In the NF-κB pathway, ubiquitin can bind to NEMO ('NF-κB essential modulator'); ubiquitin also marks proteins for later degradation, as is the case for IκBα (Yang et al. 2015) (Figure 4.6). In these cases, the cascade is downregulated. Ubiquitination also regulates T-cell functions

(Versteeg et al. 2014; Barbi et al. 2015; Ivanova and Carpino 2016). Other post-translational modificatory processes have a similarly wide spectrum of targets and effects.

4.3.4.4 Negative regulation

An unregulated response may become too strong and damage own tissue. For example, when inflammation is not sufficiently regulated or becomes sabotaged by a parasite, the process can eventually progress to a massive, systemic circulation of cytokines ('cytokine storm') that results in fatal systemic sepsis (Tisoncik et al. 2012). Hence, downregulation or negative regulation of an immune response prevents self-damage ('immunopathology'). Besides damage to self-tissue, an overshooting response also wastes time, energy, and nutrients, which could otherwise be used for other vital functions that the host needs to maintain.

There are different evolutionary reasons for negative regulation (Figure 4.7). In particular, negative regulation may be a primary consequence of a pathogen detection system that is selected for a rapid response and thus has high responsiveness towards a non-self signal. Alas, high responsiveness inevitably generates many 'false alarms', which are 'corrected' by shutting down the response as soon as possible. Life history indeed predicts that early mechanisms to stop a response in the case of a detection error should be under stronger selection than those that are selected for their consequences to avoid immunopathology later in life (Frank and Schmid-Hempel 2008) (see Chapter 12). As of today, little is known about them, but negative regulators are extraordinarily diverse and numerous in any cascade of the immune system. In particular, this is the case for the very first responders, that is, complement and innate immunity.

4.4 Immune defence protein families

Immune systems would not work without suitable molecules. By and large, immune systems are based

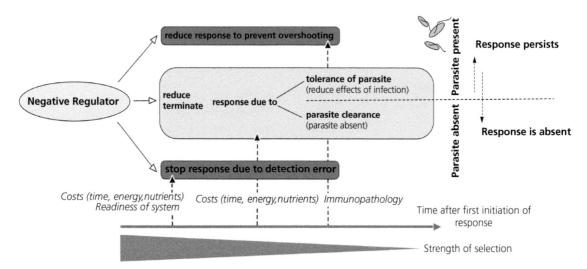

Figure 4.7 Negative regulation. Downregulation of an immune response can exist for different adaptive reasons. The most obvious is the dampening of a response to prevent immunopathology, i.e. damage to own tissue because of an overshooting response (such as with excessive inflammation). A second reason is that the response is no longer needed because the parasite's effects can be tolerated, or because it has been cleared. Finally, the response may be stopped because it was triggered by a detection error. In the first two cases, the parasite remains present and the response may still be active, albeit at a lower level. In the latter two cases, the parasite is absent and the response should therefore be absent, too. The selective factors shaping these scenarios are the immunopathological effects on host fitness, the costs wasted on an inappropriate response, or possibly the loss in the state of readiness for a next, true infection. Note that these costs typically appear at different times after the initiation of the response. In particular, immunopathology is a relatively late effect after a response has been initiated, whereas costs associated with false detection appear immediately.

Table 4.4 Protein families in immune defences.

Family	Structure	Functions	Remarks, source
Immunoglobulin superfamily (IgGs)	One to several domains, each with β-strands, forming two sheets in 3D. Variable domain folding defines binding specificity.	Five important classes (human, mouse): IgG (in blood), IgA (mucosal surface), IgM (on immature B-cells), IgD (B-cell surface), IgE (mast-cell receptors).	IgM binds to complement, is early responding. IgE involved in inflammation. Classification by makeup of domains: V (variable), I (intermediate), C (constant).
Leucin-rich repeats (LRRs)	Two to several dozen motifs (20–30 bp). Typical horseshoe-shaped form; inner, concave part binds to carbohydrates or proteins. Often flanked by cystein-rich domains.	Extra- or intracellular, soluble, or transmembrane. Protein–protein interaction, cell adhesion, signal transduction, DNA repair, recombination. In immune defences: antigen recognition, receptor (TLR), cytokine, control of motility of lymphocytes (vertebrates), haemocytes (insects).	Found in plants (NBS-LRR) to Metazoa. In immune defence: antigen recognition, receptor (TLR), cytokine. Motility of lymphocytes (vertebrates), haemocytes (insects). In humans, c.370 different LRRs. 1, 2, 20
Toll, Toll-like receptors (TLRs)	Three domains: intracellular TIR, transmembrane, extracellular LRR domain. Toll binds to ligand Spaetzle (a self-peptide), resulting from upstream cleavage of precursor in recognition (by PGRPs).	Extra- and intracellular. LRR domain provides binding to ligand. TIR and transducing protein MyD88 activates signalling cascade. Toll involved in insect development and activates Toll defence pathway.	In plants, protozoa, early Metazoa (sponges, cnidarians), to mammals. 16
Lectins	C-type lectins with similar hydrocarbon-recognition domains. Can also recognize peptides.	Soluble or membrane-bound. Bind to carbohydrates (sugars, glycoproteins) on surface of microbes. C-type lectins activate complement (lectin pathway), and T-cells; also involved in dendritic cells, immunological synapse, etc.	Found in many phyla, from early metazoans. Mouse NK-cells with C-type lectin as (peptide) receptor on membrane. Controls MHC expression. 6, 11, 15, 18
Cytokines	Includes chemokines, interferons, interleukins, lymphokines, tumor necrosis factors (TNFs). Chemokines with unique 'serpentine' shape in their extracellular part. TNFs have variable cytoplasmic tails but conserved extracellular parts binding ligands. Some TNFs membrane-bound. Tails include death, decoy, or activating receptors.	Extra- or intracellular. Interact with receptors on wide range of immune cells (neutrophils, mast cells, macrophages, B- and T-cells, etc.). Chemokines bind to other chemokines or parasites, e.g. *Plasmodium* to Duffy, HIV to CR4 receptors of immune cells. Essential for almost any immune response.	A large and diverse group produced by a wide range of immune cells. Duffy blood group antigen is a chemokine receptor. 9, 13, 27
PGRPs, GNBPs	Peptidoglycan-recognition proteins. Gram-negative binding proteins (GNBPs), which recognize β-1,3 glucans. Various types.	On blood cells, or soluble. Bind to microbial (bacterial) ligands. Activate Toll, Imd pathways, proteolytic cascades, leading to AMPs, melanization, phagocytosis.	In invertebrates, somatic diversity by alternative splicing. PGRP homologues in vertebrates, with antibacterial activity. 17, 23
NOD-like intracellular sensors (NLRs)	LRRs at the C-terminal of the molecules are the binding sites. Four subfamilies with different functions.	Defence inside cells; damage and stress responses. Variation in LRR domains provide binding specificity. Involved in immune signalling pathways, inflammation (inflammosome), apoptosis. Also for development.	Well represented in plants. NOD-like receptors diversified in bony fish, sea urchins. 7, 10, 22, 26
Scavenger receptors (SRs)	Transmembrane proteins. Extracellular domain isoforms as soluble forms. Subdivided into ten classes (A to J).	Recognize modified lipids to reduce damage, but also bind to proteins and parasites (function as PRRs), e.g. receptors in phagocytosis of bacteria.	Class-C scavenger receptor I (dSr-CI) in *Drosophila* one of the first recognized SRs. 8, 25

(Continued)

Table 4.4 Continued.

Family	Structure	Functions	Remarks, source
Dscam	(Ig superfamily). Membrane-bound, transmembrane proteins with extracellular receptor region of Ig domains and intracellular, diverse cytoplasmic tail.	Organizer in neuronal development (axon guiding). Role in immune defences (e.g. in phagocytosis) currently not fully understood. Could be a modifier for recognition.	In vertebrates (e.g. humans), invertebrates. Arthropod, *Drosophila* Dscam with large diversity of binding specificities, generated by alternative splicing. 3–5, 19, 24
Fibrinogen-related proteins (FREPs)	Large protein family with a fibrinogen-related domain, connected to one to two immunoglobulin domains.	Invertebrate immune defence, e.g. the snail *Biomphalaria glabrata* against trematodes (*Schistosoma*). Capacity for somatic diversification; many isoforms in snails and mussels.	First found and described in molluscs. Conserved in animals. 12, 14, 21
Antimicrobial peptides (AMPs)	Short peptides with different secondary structures and polarity. See Table 4.5.	Important effectors on surfaces, respiratory and urinary tracts. Delivered by lymphocytes. Generally poor specificity, but highly effective against microbes.	Different AMPs can have synergistic effects.

Sources: [1] Akira. 2004. *Semin Immunol*, 16: 1. [2] Akira. 2006. *Cell*, 124: 783. [3] Armitage. 2017. *Front Immunol*, 8: [4] Armitage. 2015. *Dev Comp Immunol*, 48: 315. [5] Brites. 2015. In: Hsu, eds. *Pathogen–host interactions*. Springer. [6] Brown. 2018. *Nature*, 18: 374. [7] Buckley. 2015. *Dev Comp Immunol*, 49: 179. [8] Canton. 2013. *Nat Rev Immunol*, 13: 621. [9] Chu. 2013. *Cancer Lett*, 328: 222. [10] Dangl. 2004. *Nature*, 411: 826. [11] Drummond. 2013. *PLoS Path*, 9: e1003417. [12] Gordy. 2015. *Fish Shellfish Immunol*, 46: 39. [13] Griffith. 2014. *Annu Rev Immunol*, 32: 659. [14] Hanington. 2012. *PLoS Negl Trop Dis*, 6: e1591. [15] Hardison. 2012. *Nat Immunol*, 13: 817. [16] Lindsay. 2014. *Dev Comp Immunol*, 42: 16. [17] Lu. 2020. *Dev Comp Immunol*, 102: 103468. [18] Mayer. 2017. *Histochem Cell Biol*, 147: 223. [19] Ng. 2015. *Dev Comp Immunol*, 48: 306. [20] Ng. 2011. *Proc Natl Acad Sci USA*, 108 Suppl 1: 4631. [21] Pila. 2017. *Trends Parasitol*, 33: 845. [22] Platnich. 2019. *Arch Biochem Biophys*, 670: 4. [23] Wang. 2019. *Curr Opin Insect Sci*, 33: 105. [24] Watson. 2005. *Science*, 309: 1874. [25] Zani. 2015. *Cells*, 4: [26] Zhang. 2010. *Immunogenetics*, 62: 263. [27] Zlotnik. 2012. *Immunity*, 36: 705.

on a limited set of protein (super)families, but the members of these families show a high degree of diversification and specialization across organisms. Moreover, during evolutionary history, members of the same family have been put in the service of many different immune defence functions (Flajnik and du Pasquier 2013; du Pasquier 2018). This pattern suggests that the general structure of these protein families makes them valuable and flexible enough for key functions such as recognition and binding, or as potent effectors. Table 4.4 shows an overview of the most important such families.

4.4.1 The major families

Proteins of the *immunoglobulin superfamily* (IgSF) are among the most important ones. They represent a vast number of different molecules with a broad range of functions, from cell adhesion in the nervous system, receptors for antigens, and antigen presenters, to co-stimulators, 'classic' antibodies, or T-cell receptors. The surface of vertebrate lymphocytes, for example, can show more than 30 different

types of IgSF receptors (Barclay 2003). Immunoglobulin domains can also associate with other proteins, such as *Leucine-rich repeats* (LRRs) or fibronectins, to form mixed proteins (chimaeras) that assume an additional variety of functions. The genetic coding of immunoglobulins is quite complex, because different gene segments can encode various parts of their β-strands. For example, in T-cell receptors, the (variable domain) V-gene exon encodes for strands in both sheets, while the (joining) J-segment encodes only one strand (the G-strand) in one of the two sheets of the V-domain. Within each of the regions, each of the various domains, V, C (constant), and I (intermediate), show diversification across the organisms. The V-domain alone, or in combination with other V-domains, acts to recognize antigenic epitopes and is, therefore, the most prominent part. The binding specificity of this domain results from different underlying amino acid sequences that lead to different three-dimensional structures of the folds.

Another large and important group are the LRRs, found in plants and all metazoan animals (Ng et al.

2011; Ng and Xavier 2011). There are several classes (subfamilies) in this protein superfamily. These classes differ in length of their sequence, and in the molecular motifs that make up the leucine-rich repeat domains within these proteins. LRR proteins can also associate with other domains intra- and extracellularly. Many protein families are part of (innate) PRRs. PRRs include different groups, e.g. the TLRs, C-type lectins (CTLs), NLRs, RIG-1-like receptors (RLRs), or absent-in-melanoma-like receptors (ALRs). The first two are membrane-associated, the remainder are intracellular receptors. *Toll and Toll-like receptors* (TLRs) are found from plants and protozoans to mammals, and are structurally very similar. They activate fairly conserved cascades, but the outer, extracellular part varies and works differently among the various taxa. The recognition part uses LRRs. Parasites are recognized directly, or by a signal peptide that acts as a ligand. For example, Toll in insects binds to a signalling peptide (Spaetzle) and not directly to the parasite; it is therefore not a receptor in the strict sense (Lindsay and Wasserman 2014). While the number of different Toll receptors in invertebrates is limited, the group of TLRs in vertebrates is varied and specific. Remarkably, however, echinoderms (sea urchins) have hundreds of different TLRs, compared to 18 in bony fish, ten in birds, and 13 in mammals (Flajnik and du Pasquier 2008; Buchmann 2014). *Lectins* can bind to carbohydrates (Mayer et al. 2017), which are often important surface elements of microbial parasites, e.g. the variable glycoprotein surface molecules of African trypanosomes. C-type lectins (CTLs) are a particularly important group that is also more generally involved in homeostasis (Brown et al. 2018). In the context of the MHC, lectin receptors recognize peptides. Lectins activate complement, or T-cell responses (Drummond and Brown 2013; Mayer et al. 2017; Brown et al. 2018). The diversity and polymorphism of lectins within any given population is typically large (Flajnik and du Pasquier 2013; du Pasquier 2018).

Peptidoglycan-recognition proteins (PGRPs) come in various types—and their somatic diversity is increased by alternative splicing (leading to isoforms). PGRPs activate canonical insect defence pathways, such as Toll, or Imd. *Gram-negative-binding proteins* (GNBPs) can bind to Gram-negative bacteria and fungi, but also to Gram-positive bacteria (e.g. in *Drosophila*). GNBPs recognize β-1,3 glucans and play a vital role in activating appropriate defence pathways (Flajnik and du Pasquier 2013). *NOD (nucleotide-binding oligomerization domain)-like receptors* (NLRs) are intracellular sensors. They detect parasite signatures, such as PAMPs and DAMPs, amid a multitude of intracellular processes and self-molecules. NLRs have diversified enormously in groups such as bony fish and sea urchins (Zhang et al. 2010; Buckley and Rast 2015). *Scavenger receptors* (SRs) are membrane-bound receptors that bind to lipoproteins, and various other ligands, such as endogenous proteins, or to parasite epitopes (Canton et al. 2013; Zani et al. 2015). *Nimrods* are phagocytic receptors in *Drosophila* that contain Nimrod (NIM repeats), and are also active in coagulation, adhesion, or recognition. An example from *Drosophila* is the receptor Eater (Melcarne et al. 2019). *Down syndrome cell adhesion molecules* (Dscam) typically are membrane-bound, transmembrane proteins with an extracellular receptor region of Ig domains. Dscam lends identity to neurons to guide neuronal development (e.g. human Dscam). It is involved in immunity, but its precise role is still unclear and controversial. Dscam may function as a phagocytic receptor, or perhaps specifies the identity of a haemocyte (Brites and Du Pasquier 2015; Armitage et al. 2017). As proteins, Dscam can have many isoforms. In the early arthropods, such as the chelicerates (e.g. spiders), Dscam isoform diversity reflects the dozens of Dscam genes in the genome. In other basic groups (e.g. centipedes) diversity results from gene variants plus alternative splicing in one genomic cluster of exons. D (e.g. crustaceans, insects) generate a large number of isoforms (more than 10 000) by somatic, alternative splicing from a single gene (*Dscam1*). Dscam is expressed on the surface of haemocytes. *Fibrinogen-related proteins* (FREPs) also contain immunoglobulin domains and are a large family of conserved proteins found in all animals. They, too, can become somatically diversified to yield many isoforms (Gordy et al. 2015). There are numerous other families of defence molecules that show different structures from the ones mentioned above or that are composed of domains coming from different families.

Signalling molecules are essential for virtually all defence responses. *Cytokines* are a diverse and extensive group of signalling molecules and include the chemokines (Zlotnik and Yoshie 2012), interferons, interleukins, and lymphokines. A broad range of immune cells produces cytokines (Griffith et al. 2014). The *tumor necrosis factor* (TNF) family contains particularly important cytokines (Chu 2013), characterized by their variable cytoplasmic tails. TNFs are involved in the process of cell apoptosis, as proinflammatory cytokines (TNF-α), and in many other functions. One of the most prominent TNFs is the CD40 co-receptor of B- and antigen-presenting cells (Clark 2014).

4.4.2 Effectors: antimicrobial peptides

AMPs are a very diverse group of small proteins that defend against the microbial parasites. They occur in all multicellular organisms, from prokaryotes to humans (Hancock et al. 2016; Zhang and Gallo 2016; Ageitos et al. 2017; Keehnen et al. 2017).

Typically, they protect surfaces, such as mucosa lining the gut, respiratory, or urinary tracts. AMPs derive from larger precursor molecules. In some cases, the germline encodes multiple copies of the same AMP, perhaps to increase the speed at which a large number of peptides can be transcribed and synthesized.

The number and diversity of AMPs are quite impressive (Table 4.5). Several thousands of AMP sequences from a wide range of organisms are deposited in various data repositories, such as APD (Antimicrobial Peptide Database; Wang et al. 2016) and CAMPR (Collection of Antimicrobial Peptides; Waghu et al. 2015). Specific databases also exist for different taxa, e.g. shrimps (PenBase; Gueguen et al. 2006), plants (PhytAMP; Hammami et al. 2008). Other databases list functions (e.g. DBAASP; Pirtskhalava et al. 2016), or focus on patented peptides and clinical aspects (e.g. DRAMP; Kang et al. 2019). AMPs often are classified according to their secondary structure (as in Table 4.5). They can also be ordered according to polarity (cationic, anionic),

Table 4.5 Antimicrobial peptides (AMPs). AMPs are often classified according to their secondary structure.[1]

Class (secondary structure)	Characteristics	Examples in invertebrates	Examples in vertebrates
α-helical	Most abundant AMPs. Typically form helical structure when in contact with membranes. Cationic and hydrophobic, anionic. Typical action is by inducing membrane defects in pathogen. Active against Gram-negative and -positive bacteria.	Cecropin, andropin, moricin, ceratotoxin (insects), melittin.	Cathelicidin LL37 (humans), human lactoferrin, PMAP-36 (pig), seminal plasmin, BMAP, SMAP, PMAP (cattle, sheep, pigs), CAP18 (rabbits), magainin, dermaseptin, esculentin, brevinin-1, buforin II (amphibia), lysenin, dermicidin (humans).
β-sheet (one or several bonds)	Often a cyclic molecule with sheets, stabilized by disulphide bonds.	Tachyplesin, polyphemusin-1 (horseshoe crab), defensin, drosomycin (insects).	Defensins: α-, β-defensins. HNP-1 (human α-defensin), HBD-1 (human β-defensin), protegrins PG-1 (pig), bactenecins (sheep, goats),
Extended	No particular secondary structure, but can fold into amphipathic structures in contact with membranes. Defined by a high content of peptides, such as glycine, histidine, arginine, tryptophan.	Abaecin, apidaecin (honeybee), hymenoptaecin (honeybee), coleoptericin, holotricin (beetles). Drosocin (*Drosophila*, bugs), pyrrhocorin (bugs).	indolicin (cattle), tritripticin, PR-39 (pig).
Other	Loops		Maximin H5 (amphibia). Dermidicin (human).

[1] Simplified scheme based on: Brogden. 2005. *Nat Rev Microbiol*, 3: 238; Kumar. 2018. *Biomolecules*, 8: biom8010004; Mahlapuu. 2016. *Front Cell Inf Microbiol*, 6: Article 194; Wang. 2019. *Medicinal Research Reviews*, 39: 831; Zasloff. 2002. *Nature*, 415: 389; Zhang. 2016. *Curr Biol*, 26: R14.

peptide enrichment (e.g. with enriched proline), by activity spectrum (e.g. against Gram-negative bacteria), or by their mechanism of action (e.g. poreforming). The α-helical AMPs are the most abundant ones, with defensins among the best studied (Mahlapuu et al. 2016). For many (e.g. defensins, cathelicidins)—but not all groups—the amino acid sequences in some genomic regions are fairly conserved, indicating common constraints on their production, delivery, or mode of action (Zasloff 2002; Mahlapuu et al. 2016; Wang et al. 2019a).

AMPs have broad activities and can quickly kill (within seconds) protozoans, fungi, bacteria, or viruses, but may also function as immunomodulatory proteins (Zasloff 2002; Brogden 2005; Kumar et al. 2018). AMPs often disrupt the integrity of microbial membranes, e.g. by forming pores (Brogden 2005; Bahar and Ren 2013; Lee et al. 2016). Other AMPs target intracellular elements and inhibit the synthesis of proteins, DNA, and RNA, disrupt protein folding, and impede cell division, cell wall synthesis, or lipid metabolism (Le et al. 2017). Some peptides (e.g. buforin II, apidaecin) permeate the parasite's cell membranes and accumulate in its cytoplasm, where they inhibit nucleic acid, protein synthesis, or enzyme activities (Brogden 2005). In other cases, a primary AMP forms pores in the membrane (e.g. hymenoptaecin in Gram-negative bacteria), followed by the entry of a second AMP through these pores into the cell where it takes effect (e.g. abaecin interacts with chaperones; Rahnamaeian et al. 2015). Individual hosts often produce a 'cocktail' of different peptides when infected (Zasloff 2002), which can potentiate the effects of each AMP. Variation in the composition of the AMP 'cocktail' might be of great significance to combat a diversity of parasites (Yan and Hancock 2001; Ganz 2003; Rosenfeld et al. 2006; Riddell et al. 2009; Rahnamaeian et al. 2015; Marxer et al. 2016). Such variable, synergistic action might similarly prevent microbes from quickly adapting and evading the effect of AMPs. Bacteria, in fact, have mechanisms to resist AMPs. These include a reduction in AMP-binding capacities or the modification of membranes; often, anionic/cationic changes are involved. Resistance can be transferrable among bacteria by plasmids (Andersson et al. 2016).

4.5 The generation of diversity in recognition

Host–parasite co-evolution is an antagonistic interaction. Such systems tend to become diversified over evolutionary time—hosts evolve new defences and parasites adapt and evolve new attack strategies, and vice versa. As parasites diversify, hosts must cope with a diversified set of the respective molecular motifs that need to be recognized. Any single infecting parasite can have thousands of such epitopes, and almost the entire surface of a parasite presents many overlapping molecular motifs that might be recognized as 'antigens'. How defence systems generate capacities to recognize and attack diverse parasites is therefore a central theme in immunology. At the same time, we observe that across the vast kingdoms of chordates, invertebrates, and plants, very different solutions as to how to diversify the repertoire of recognition have evolved. Such diversity can result from having a diversity of germline genes that code for the relevant proteins, such as LRRs. This kind of diversity will change as the population evolves under the effects of selection by parasites. Alternatively, and in addition, diversity is generated somatically, during the lifetime of an individual, and from a given repertoire of genetic elements that the individual has at its disposal.

A major problem of any defence system is to distinguish 'self' from 'non-self' against a background of highly diverse parasite motifs. Three major strategies can be characterized (Medzhitov and Biron 2003):

1. *Recognition of microbial non-self*: This is a universal strategy, certainly among the animals. It relies on the detection of conserved motifs that reveal the presence of a pathogen (e.g. the PAMPs, epitopes on microbial surfaces).
2. *Recognition of missing self*: This strategy is followed by, for example, NK-cells in higher vertebrates, which respond to the absence of a signal that identifies a cell as 'self'. Cells lacking this signature are treated as 'non-self' and destroyed. This strategy is especially effective against intracellular parasites that otherwise might hide from the immune defences.

3. *Aberrant activities*: This involves mechanisms that detect the aberrant consequences of pathogen activity in the host body. Examples are parasite virulence factors that change normal processes, for example in plants (the 'guard hypothesis', Holt III et al. 2003). Similarly, the activity of the symbiotic microbiota, for example, within the gut of an animal, can be monitored. When a pathogen enters this 'ecosystem', the regular functioning changes and the immune system detects this, as in *Anopheles gambiae* (Meister et al. 2009).

4.5.1 Polymorphism in the germline

With genetic polymorphism in the narrow sense, several variants of the same gene are present in different individuals of a population. Alternatively, polymorphism can refer to different genes at different loci within the same individual genome ('polylocism'). Such genes are part of the germline and persist over generations, slowly changing under the effects of evolutionary processes. This is also true for immune protein families, where mutation and selection generate a birth–death process in different gene families. Additionally, gene conversion and gene duplication, followed by the subsequent divergence of the copies, can lead to clustered genes within the same neighbourhood. But translocation to other sites within the genome is also known. As a result, different lineages end up with a richer or poorer repertoire of immune genes that are arranged in the genome in various ways.

Examples of genetic polymorphisms are many. In the genome of *Caenorhabditis elegans*, around 180–280 different C-type lectin-like protein domains (receptors, Table 4.5) have been found. However, the functions are not known for most of them (du Pasquier 2006; Pees et al. 2016). Similarly, in insects or crustaceans, recognition and opsonization molecules such as PGRPs, GNBPs, TEPs, and the recognition segments of their TLRs (the LRRs) are genetically polymorphic. This is the case, too, with variable chitin-binding proteins (VCBPs), lectins, and LRRs/TLRs of the early chordates (the 'lancelet', *Amphioxus*). In jawless (with their VLRs) and the higher (jawed) vertebrates, many essential elements are encoded by polymorphic genes and become further diversified by somatic processes. This includes the immunoglobulins, TLRs, complement factors, and the MHC. The underlying genes can be species-specific, but some are orthologues in congeneric species, or orthologous across larger groups (Figure 4.8). Nevertheless, one of the immunological conundrums is that the number of possible host genes is still much lower than the enormous diversity of parasites and the epitopes they possess. Several diversifying, somatic processes overcome this limitation.

4.5.2 Somatic generation of diversity

4.5.2.1 Alternative splicing

With alternative splicing, a single gene codes for multiple proteins. In this regulated process, the various exons of a gene are either included or excluded from the final, processed mRNA by various mechanisms. The process yields multiple mRNA templates (the alternatively spliced mRNAs) for protein synthesis from the same gene. Alternative splicing is a fundamental process of gene expression in eukaryotes. Somatic diversification by alternative splicing (Figure 4.9a) gives rise to randomly individualized and changing repertoires; this includes FREPs in many molluscs, such as in the snail *Biomphalaria glabrata* (Zhang et al. 2004; Adema 2015; Gordy et al. 2015). FREPs bind to carbohydrate motifs on cellular surfaces of pathogens (Hanington et al. 2012; Pila et al. 2017). Further examples include scavenger receptor cysteine-rich (SRCR) proteins in sea urchins where thousands of isoforms are expressed based on a few genes (Pancer 2000; Hibino et al. 2006). Arthropod Dscam is a protein whose receptor part is composed of different IgSFs. With the exception of the Dscam of humans, the receptor part is hypervariable due to alternative splicing; thousands of isoforms are generated in any one individual (Brites et al. 2008; Armitage et al. 2012; Chiang et al. 2013). Its role in immune defence is still unclear, however. Many other receptors, from plants to the higher animals (e.g. LRR domains in mammals), are diversified by alternative splicing. This also includes the diversification of effectors (e.g. human defensins, du Pasquier 2006; *SpTrf* in molluscs).

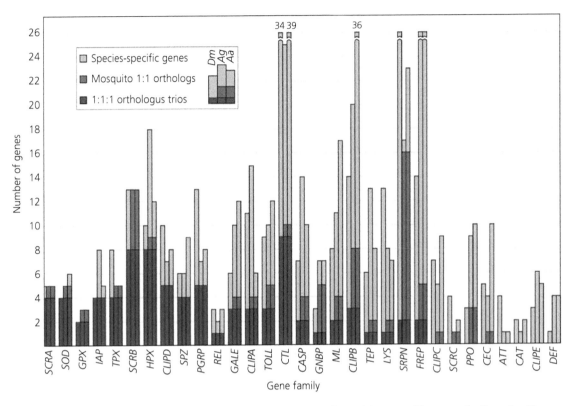

Figure 4.8 Diversity of the immune gene repertoire. The graph shows the number of genes belonging to different gene families, as found in two species of mosquitoes (Ag: *Anopheles gambiae*, Aa: *Aedes aegypti*) and in *Drosophila melanogaster* (Dm). Dark red: Orthologous genes in all three species. Orange: Orthologous genes for mosquitoes. Light red: Genes only found in one species. Some gene names are mentioned in text and Glossary; see source for full descriptions. Reproduced from Waterhouse et al. (2007), with permission from the American Association for the Advancement of Science.

A related mechanism is RNA editing. In this process, a specific nucleotide of the mRNA is changed catalytically. In contrast to alternative splicing, therefore, the transcripts are not cut and reassembled in different ways. Instead, the information of the mRNA is altered in some places. RNA editing seems to be very old, as it exists in single-celled organisms; it may occur in some higher organisms, too (du Pasquier 2006, 2018).

4.5.2.2 Somatic DNA modification

'Rearrangement' refers to the RAG-mediated recombination found in the jawed vertebrates (see section 4.5.3). In jawless vertebrates (hagfish and lampreys, e.g. *Petromyzon marines*), alternative processes ensure that germline-encoded DNA segments somatically come together in new combinations. Two types of

T-like lymphocyte populations express VLRAs and VLRCs—with unclear specificities—and one B-like lymphocyte lineage expresses VLRB antibodies that bind to epitopes of a protein or carbohydrate nature (Boehm et al. 2018). VLRs are membrane-bound proteins consisting of N-terminal (LRRNT) and C-terminal (LRRCT) caps, a constant stalk region that anchors the molecule to the membrane, plus a hypervariable region that provides binding specificities. The variability in the binding region is generated in a very unusual way (Figure 4.9b). Single lymphocytes have a monotypic expression but—as a population—vary in their binding specificities, which are analogous and comparable in size to the lymphocyte repertoire of the higher (jawed) vertebrate. Precise estimates of how many different isoforms exist are still lacking, though (Boehm et al.

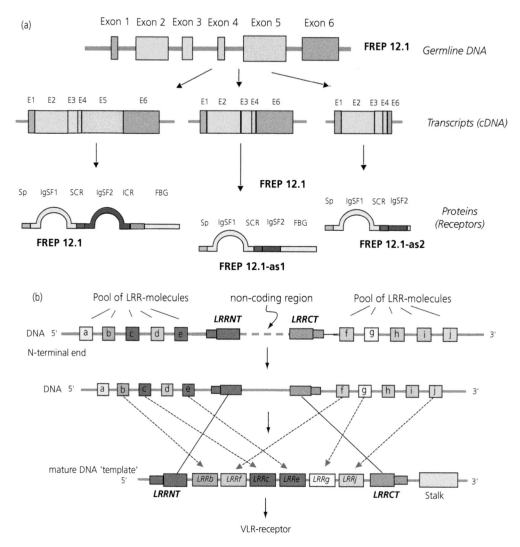

Figure 4.9 Mechanisms of somatic diversification for receptors. (a) Alternative splicing of FREPs in the snail *Biomphalaria glabrata*. The genome (germline DNA) encodes for six exons (E1 to E6), transcribed into mRNA in alternative ways. In this example, FREP12.1 generates three different transcripts (here shown as cDNAs). These are translated into three polypeptides (proteins), which fold into different three-dimensional structures and, hence, have different binding properties. Protein domains are signal peptide (Sp), immunoglobulin superfamily (IgSF1, IgSF2), small connecting region (SCR), and fibrinogen β/γ sheet (FBG). Adapted from Zhang and Loker (2003), with permission from Elsevier. (b) Somatic rearrangement in VLRs of jawless vertebrates (lamprey). The germline DNA contains an N-terminal (*LRRNT*) and C-terminal (*LRRCT*) sequence (leucin-rich repeats), separated by a non-coding region. During cell maturation, this non-coding region is replaced by a variable number of LRR sequences that are recruited from a large pool of LRR cassettes. The cassettes are located in the germ line to either side of the core module (LRRNT, LRRCT); (see elements a–j in the sketch). The process is complex and has a random component; it resembles the early stages of a gene conversion with enzymes CDA1, CDA2 (distantly related to AID; Trancoso et al. 2020). Homologies in the LRR modules allow the synthesis of the final DNA template. The template yields a functional VLR. Adapted from Schatz (2007), with permission from Springer Nature. (c) Somatic hypermutation and gene conversion in the jawed vertebrates. The germline encodes a number of *V*, *D*, and *J* elements. These are re-arranged on the genomic sequence by RAG-dependent recombination. Subsequently, somatic hypermutation (asterisk) changes residues in the sequence. Also, gene conversion can copy and convert a pseudogene (*ΨV*) into the sequence. In both cases, the enzyme AID initiates the process by changing residues to uracil (U), which leads to error-prone repair (mutation) or conversion of a pseudogene (gene conversion). Adapted from Schatz (2007), with permission from Springer Nature.

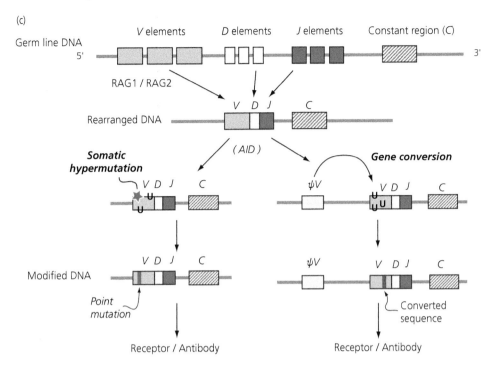

Figure 4.9 Continued.

2018). Comparing the T- and B-cell mechanisms in the jawed with VLRs and their generation in jawless vertebrates suggests that the dichotomy of B- and T-cells traces back to a common ancestor, but that the process by which these receptors become somatically diversified uses different mechanisms (Trancoso et al. 2020).

4.5.2.3 Somatic (hyper-)mutation, gene conversion

Somatic mutation to diversify recognition molecules is an evolutionary old process and plays an important role in the lymphocytes of vertebrates. 'Gene conversion' is a large mutation event caused by asymmetric recombination that leads to the transfer of a gene sequence from one genomic set to another; the receiving sequence is thereby modified. In the vertebrate immune system, gene conversion is mediated by the key enzyme Activation-Induced Cytidine Deaminase (AID) (Figure 4.9c). Gene conversion leads to the diversification of vertebrate immunoglobulins, e.g. in B-cells of rabbits (Winstead et al. 1999), or birds (V-region of light chain, McCormack et al. 1991; Flajnik 2002).

Mutation and gene conversion are also involved in the diversification of FREPs in molluscs. AID seems an ancient all-purpose enzyme that is involved in somatic hypermutation, gene conversion, class switching, and RNA editing in many lineages (Papavasiliou and Schatz 2002; Honjo et al. 2004; Odegard and Schatz 2006; Schatz 2007).

4.5.3 Variability and B- and T-cells

In the jawed vertebrates, populations of leucocytes, the B- and T-cells, undergo clonal expansion upon infection to orchestrate the 'adaptive response'. This process involves a cascade of events—which can vary among lineages (Flajnik and du Pasquier 2008; du Pasquier 2018)—that lead to an enormous diversification of the associated receptor molecules.

4.5.3.1 B-cells

A membrane-bound form of immunoglobulins (Ig) is present on the surface of B-cells. When it recognizes an antigen (i.e. an epitope on a foreign cell), the B-cell becomes 'activated'. Immunoglobulins will then be secreted, known as 'antibodies'. The structure of immunoglobulins has several features which make them uniquely suited to act as recognition molecules. On the one hand, the molecule folds into a three-dimensional structure that allows binding to an antigen, with a specificity defined by the so-called 'variable region'. On the other, this region is somatically diversified by several processes (Figure 4.9c). Although any single B-cell has its own specificity, the population of all B-cells in a given host individual represents an enormous range of different specific recognition capacities.

The basic structure of a B-cell immunoglobulin consists of four polypeptide chains (Coico and Sunshine 2015; Punt et al. 2018). The four chains of an antibody are made up of two identical 'light chains' and two identical 'heavy chains', forming a symmetrical globular molecule, which is usually sketched as a Y-shape (Figure 4.10). In mammals, the heavy chains come in five different classes (the isotypes: the IgA, IgM, IgG, IgE, and IgD forms)—an individual host has all five types. Among taxa, these classes vary, e.g. classes IgA, IgM, IgY for birds, IgM, IgD, IgY IgX for amphibia, IgM, IgD,

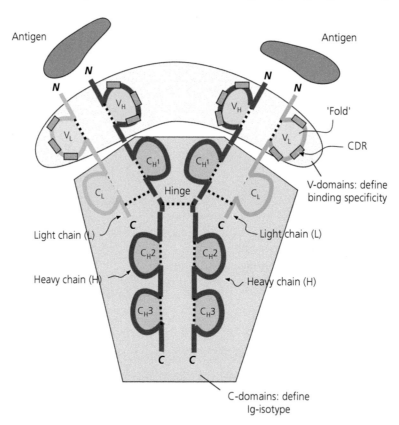

Figure 4.10 Structure of immunoglobulin (Ig) receptors. Shown is an antibody secreted by B-cells (plasma cells). Two light chains (L; light blue) and two heavy chains (H; dark blue) form the backbone of the Ig. The chains contain within-chain disulfide bridges (dotted black lines) that lead to the formation of three-dimensional Ig-fold domains ('folds' or 'loops'). The N-terminal loop of each chain is variable (V_H, V_L); additional variability comes from three hypervariable regions (complementary-determining regions, CDRs; orange rectangles). These are the prime binding elements for the antigen (symbolized by the red shapes). There are two loops in the light, and four to five loops (or more in different species) in the heavy chains. Together, the V-domains of the L- and H-chain form the antigen-binding site (towards the top) and determine the binding specificity of the antibody (blueish areas within loops). Towards the C-terminal ends, there are less variable 'constant domains' (C; greenish areas within loops), numerically dominated by the heavy chains; these determine the isotype (class) of the Ig. Disulfide bridges also keep the chains together. The hinge region (purple) allows the molecule to change the spread of the distal ends, so as to accommodate antigen-binding sites with slightly variable geometries. The amino acid sequence determining the binding region of the molecule is varied by somatic V(D)J-gene rearrangement (hypermutation, gene conversion, etc.). Adapted from Coico and Sunshine (2015), with permission from John Wiley and Sons.

IgZ for bony fish (Flajnik and du Pasquier 2008). Ig classes are further divided into subclasses, for example into IgA_1, IgA_2, and the IgG subclasses. The different types have the same binding specificity but fulfil different biological functions in the immune system. For example, IgM is involved in activating the complement, whereas IgG can agglutinate and precipitate antigens. Consequently, the Ig-isotype forms vary in their abundance in different tissues and at different times of the immune response.

In the germline, the light chain is encoded by a variable (V) and a small, joining (J) genetic segment (both derived from a common presequence that is split and rearranged). The variable part of the heavy chain is encoded by gene segments for the V- and J-regions, plus one extra segment for the D-region. Together, the VDJ domains determine the variable region of the molecule. To give a rough indication, for the heavy chain, humans have around 50 genes, six genes for the J-segment, and some 20 genes for the D-region of the heavy chain. For the light chain, there are 30 to 40 genes for the V- and one to five genes for the J-segment (Coico and Sunshine 2015). There are two types of light chains (κ and λ) in mammals, but up to four types in other groups such as shark or amphibians, whereas birds have only λ-chains (Criscitiello and Flajnik 2007). Both of the κ and λ-chains are present in an individual host; they are encoded by genes located on different chromosomes and by slightly different numbers of genes.

As the B-cells mature and the chains are synthesized, the gene segments encoding the two (sometimes, three) domains of the V-genes are somatically rearranged to generate a novel genetic sequence for the molecule. This process is referred to as 'VDJ recombination'. This modified and novel genetic sequence defines a specificity for the antibody. The process is under the control of the recombination-activating genes (*RAG1, RAG2*) that are typical for the jawed vertebrates. Interestingly, the RAG process exists in echinoderms and *Amphioxus*, but so far seems not to be involved in immunity. Given the number of genes, there are approximately 50 x 20 x 6 = 6 000 different VDJ combinations in the heavy chains, and 40 to 200 VJ combinations in the light chain. As the two chains can by themselves associate in different ways, the total number of combinations is in the order of 10^6 different specificities—all

derived from around 150 genes. Nevertheless, this enormous number is still below the estimated 2.5×10^7 antigenic specificities known to circulate in the blood and lymph systems of a human body.

Additional processes generate diversity: (1) *Junctional diversity* results from the fact that the VJ-preform of the genetic sequence is split to generate the later V and J forms. This split is not always precisely in the same position. When the segments are subsequently rejoined, these split variations lead to alterations of the amino acid sequence in the respective fragments because the reading frame have changed (Hsu et al. 2006). (2) With *N-region diversity*, a small number of amino acids are inserted in the joining regions of V and D, and of D and J in the heavy chain. (3) With *P-diversity*, palindromic (P-) nucleotides are added, or deleted, at the junction of gene segments (Wuilmart et al. 1977; Di et al. 2009). (4) *Somatic hypermutation* affects the genes coding for the V-regions of light and heavy chains, and thus binding specificity. These mutational events are stimulated by binding to antigens and happen over the lifetime of the B-cell; these are around 10^5 times more frequent than the background mutation rate in the germline. This is also the key process for 'affinity maturation', during which the B-cells increase their binding capacity towards a prevalent antigen. Mutations are more frequent in older cells, and these are therefore more important for the secondary response; that is, when the host re-encounters the same antigen. (5) *Class switching* occurs after gene rearrangement (determining specificity), and when the heavy chain (isotype) undergoes alternative splicing in the constant region. This yields different classes of the immunoglobulin forms, for example, IgM and IgD, which can be expressed by the same B-cell. With class switching, specificities for a recognized parasite transfer to other classes of immunoglobulins with different biological functions. By contrast, B-cells that have not yet been stimulated (the naïve cells) produce immunoglobulins of broad specificity, called 'natural' antibodies (typically IgM). Their presence characterizes a kind of background that circulates independently of any infection. Other groups of jawed vertebrates may use other processes to diversify their VDJ specificities. In birds and rabbits, for example, AID-dependent gene conversion leads to somatic diversification of B-cell specificities. In the process, a stretch of DNA is copied into a receptor strain; only this gene is converted into a new sequence. As we begin to understand more of the immune system of vertebrates, it becomes apparent that there is a remarkable diversity in how variation in the immunoglobulin recognition molecules is generated.

4.5.3.2 T-cells

T-cells are the other major category of lymphocytes. They cooperate with B-cells and fight parasites that have managed to infiltrate host cells, e.g. viruses, bacteria, but also many protozoa. T-cells mature in the thymus and are always membrane-bound. T-cells have T-cell receptors (TCRs) with antigen-specific IgSF receptors. Each clonal line of T-cells has a different specificity, analogous to the situation in the

B-cells. Together, the population of T-cells in a single host represents a repertoire of a vast number of different specificities. Diversity is generated by the same VDJ gene recombination rearrangement as described above (such rearrangement is not known from the jawless vertebrates, Flajnik and du Pasquier 2008). However, TCRs never change by hypermutation or class switching; an exception is sharks, where mutations occur (in their γ, δ chains) (Flajnik and du Pasquier 2008).

Structurally, TCRs show similarities with the immunoglobulins of B-cells, although they consist of only two polypeptide chains (α and β), again linked by disulfide bridges (Figure 4.11). TCRs preferentially recognize small protein fragments (peptides) that are between approximately 8 and 17 amino acids long. These fragments result from antigenic (parasite) proteins, broken down to such

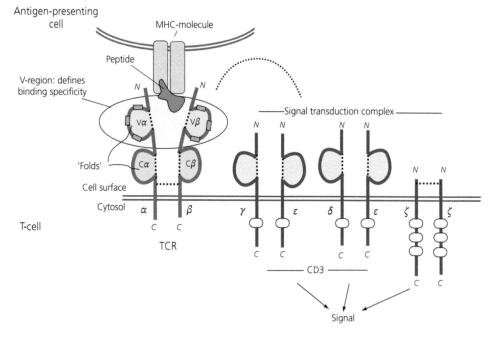

Figure 4.11 Structure of the T-cell receptor complex. The T-cell receptor (TCR) consists of α- and β-chains, roughly similar to antibodies. TCRs bind with their variable regions (Vα, Vβ; blueish area within loops) to a peptide (red shape) that is presented by the MHC molecule (yellow) on the surface of, for example, an antigen-presenting cell (APC). The variable region also has three hypervariable regions (orange rectangles). The C-terminal region anchors the molecule in the T-cell surface. The TCR complex furthermore contains the signal transduction complex, CD3, assembled from γ, ε, and δ-chains, plus the two ζ-chains connected by disulfide bridges (dotted lines), with long tails into the T-cell interior. The Ig folds of CD3 are invariable (grey areas within loops). Only the TCR binds to the peptide. The binding to CD3 activates the signal transduction complex that generates the signal. Each such transducing polypeptide chain has several adaptor molecules at their intracellular end (white ovals). The T-cell surface, furthermore, contains co-receptors (such as CD4, CD8; not shown here) that bind to the MHC molecule. Adapted from Coico and Sunshine (2015), with permission from John Wiley and Sons.

peptides, which bind specifically to one out of a large repertoire of MHC molecules. The MHC–peptide complex then moves to the host cell surface, where the peptide is 'presented' to passing T-cells; the T-cell binds if it has the appropriate specificity (Figure 4.11). Multiple copies of MHC molecules are present on the cell surface at any one time, and each presents a peptide that was 'sampled' from the cell interior. Recognition by a T-cell indicates that an infection or something foreign is around and that the immune system needs to respond. In contrast to other Ig molecules that can directly bind to parasite surfaces, TCRs recognize the MHC–peptide complex. MHC class I molecules present peptides of intracellular parasites and bind to C-type lectins (with CD8 co-receptors). MHC class II molecules present peptides of extracellular parasites and bind to CD4+ T-cells (with CD4 co-receptors); these subsequently release cytokines. MHC class II molecules are constitutively expressed by APCs but can be induced on other cell types, too. On the surface of T-cells, however, the TCRs are always expressed within a receptor complex with CD3, forming the signal transduction complex, which generates the actual signal for the activation of an immune defence cascade (Figure 4.11). Because the organization of the genes coding for these regions is similar to the immunoglobulins of the B-cells, the two recognition molecules likely evolved from the same ancestral form.

4.6 The diversity of immune defences

This section is to remind the reader of the vast diversity of immune defence systems across the various groups of organisms. The range of different systems is truly enormous, and the diversity mind-boggling and fascinating at the same time. Although only a small fraction of this diversity is investigated so far, some common themes emerge. For example, new molecules emerge from mutation or horizontal transfers, while existing molecules or cascades are reused for new functions. The evolution of immune systems was anything but orderly and directed; it was more the work of a tinkerer than of a brilliant engineer (Jacob 1977). Out of this patchwork of existing and novel molecules, of old and new mechanisms, solutions of very different kinds evolved.

Some defence mechanisms withstood the test of time and are very old, forming large-scale homologies across taxa. Sometimes molecules and functions appeared and were lost again in a lineage. Hence, modern complex mechanisms did not appear suddenly. Bits and pieces were in place before, perhaps serving a different purpose. These were eventually coerced for new functions in the immune defence. Nevertheless, there are particularities in the defence systems that are characteristic of certain groups of organisms rather than others. One of the most remarkable, only recently discovered 'immune' defences is the CRISPR–Cas system of prokaryotes (Bacteria, Archaea), targeted against viral infections (bacterial phages) or harmful mobile genetic elements, as illustrated in Box 4.2. An analogous system, albeit working with entirely different mechanisms, are the 'virophages', recently discovered in protists. Both systems also generate a 'memory' of past infections (Mougari et al. 2019).

4.6.1 Defence in plants

Plants face the same challenge as any other organism. They have evolved an effective immune system that can protect them against a variety of pathogens, such as viruses, bacteria, fungi, nematodes, or helps to control a plethora of herbivore insects, such as wasps, lepidopteran larvae, and leaf beetles. The characteristic plant defence is the hypersensitive response, including a rapidly induced cell death at and around the site of infection (Table 4.6). Plants have many genes that encode for the appropriate receptors, such as NLRs (nucleotide-binding leucine-rich repeat receptor, NB-LRR), the largest family of plant resistance genes. Extracellular receptors (e.g. FLS2 sensing a component of bacterial flagellin) mediate the recognition event. After transduction of the signal via the transmembrane domains, this activates a MAPK (mitogen-activated protein kinase)-dependent cascade that leads to the release of WRKY-transcription factors, which induce the production of effector proteins (Zipfel and Oldroyd 2017). Transcription factors such as WRKY are unique to plants and, vice versa, the transcription factors found in insects or mammals are not present in plants. Hence, it appears that the plant immune system, despite

Box 4.2 Adaptive immunity in prokaryotes: the CRISPR–Cas system

The CRISPR–Cas system defends archaea and bacteria against their viral parasites ('phages') (Horvath and Barrangou 2010; Doudna and Charpentier 2014; Jiang and Doudna 2017). CRISPR stands for 'Clustered Regularly Interspaced Short Palindromic Repeats' and describes the structure of this defence locus in the genomic sequence. CRISPR consists of a series of short direct repeats that are inter-

spaced with short and variable sequences, the so-called 'spacers' (Figure 1). These spacers are DNA sequences acquired from viruses or plasmids during earlier infection events. Therefore, the library of spacers is a memory of past infections. Furthermore, the spacer sequences encode corresponding 'non-coding interfering CRISPR-RNAs' (crRNAs) for later defence (by RNA interference), as they can match a

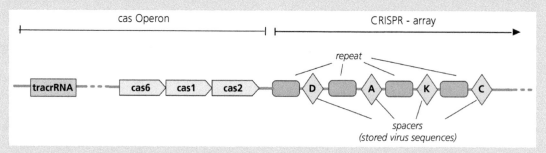

Box 4.2 Figure 1 The structure of the CRISPR–Cas locus. On the genomic sequence, the CRISPR array contains a series of sequence repeats (blue cartouches) interlaced with short, but variable spacer sequences (D, A, K, C, ...; greenish diamonds) that have been acquired from previous infections. A functional system also contains several Cas proteins (...Cas6, Cas1, Cas2) plus tracrRNA needed for the maturation of the transcribed target RNAs when activated for defence. These operon cassettes vary among species (Koonin et al. 2017).

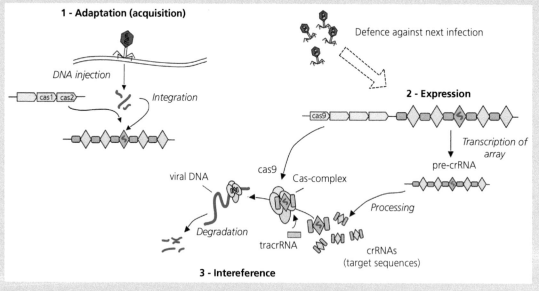

Box 4.2 Figure 2 The functioning of the CRISPR–Cas system. In the first phase (1) 'Adaptation': A new viral (or plasmid) sequence is acquired (red) and inserted as a spacer (green diamond) into the bacterial CRISPR array; for this step Cas1 and Cas2 are required. (2) 'Expression': With a next infection, the CRISPR array is activated. The stored array (incl. green diamond with red sequence) is transcribed into a precursor (pre-crRNA). This is processed to form mature crRNAs. These act as guides to the corresponding sequences in a new infection. For this purpose, the crRNAs are loaded into an effector complex (Cas complex, whereby tracrRNA is needed). The complex can target a new invader. (3) 'Interference': The Cas complex binds to the sequence of the new invader to cleave and degrade it, when it contains the targeted crRNA motif. This sketch represents a type II system, as seen in *Streptococcus thermophilus*.

new infection with an already known one for an efficient response. The fully functional CRISPR–Cas system contains an operon with Cas genes ('CRISPR-associated genes') that encode the Cas proteins.

The system works in three phases (Figure 2) (Hille et al. 2018). With the short generation time of bacteria, it is unlikely that the same individual bacterial cell becomes reinfected by the same virus a second time. However, the archive is passed on during cell division and protects offspring. This defence system, therefore, has a pronounced 'Lamarckian' component.

CRISPR–Cas systems vary among taxa and according to function. Currently, two classes are recognized, Class 1 and Class 2, which are divided further into six types and many subtypes (Koonin et al. 2017; Shmakov et al. 2017; Mohanraju et al. 2018). Across this diversity, the acquisition step is relatively uniform, with Cas1 and Cas2 proteins to form the spacers. For interference, Class 1 systems (found in bacteria and archaea, with types I, III, IV) use multi-subunit effector complexes, containing several Cas proteins (notably Cas5, Cas6, Cas7). Class 2 systems (found almost only in bacteria: types II, V, VI) use only one protein ('single-effector' complex). The best-known system is Class 2 type II, with the Cas9 protein, derived from an ancestral system that presumably was similar to the extant type III (Koonin et al. 2017; Mohanraju et al. 2018). Different types of CRISPR–Cas systems can interfere with different targets. For example, Cas9 type II addresses dsDNA from viruses, whereas Cas13a type VI interferes with ssRNA. Viruses, in turn, have responded to the CRISPR–Cas systems of their hosts by evolving anti-CRISPR mechanisms. So far, more than 20 quite different viral protein families, with activity against type-I and type-II CRISPR–Cas systems, have been found. They prevent, for example, the binding of the effector complex to DNA by direct interference with various Cas proteins, prevent the docking of additional components needed for cleavage, or even mimic dsDNA, to deflect the effector complex from the real target (Pawluk et al. 2018).

Recall that dsDNA is the standard genomic configuration in eukaryotes. Because the CRISPR–Cas system can induce double-stranded DNA breaks at a precise position defined by the crRNA target sequence, it is developed into a genomic engineering tool, notably the single-effector class systems involving Cas9 (Strich and Chertow 2019). After a break has occurred, one of two DNA-repair pathways is activated by the cell: the non-homologous and end-joining (NHEJ) or the homologous-directed repair (HDR) process. With the HDR process, a piece of DNA with homology to the region around the break will be integrated into the targeted genomic DNA; this region is known from the target sequence. However, this piece can also carry a piece of a new sequence. Hence, it becomes possible to insert the desired sequence at a desired position of the genome. A given sequence has thus been 'edited' and rewritten with a new code. In technical applications, the whole process is not error-free but has already become a powerful genomic tool (Doudna and Charpentier 2014; Waddington et al. 2016).

Table 4.6 Immune defence systems in various groups.[1]

Group	Some characteristics	Mechanisms, molecules	Sources, further information
Prokaryotes (Bacteria, Archaea)	CRISPR–Cas system (Box 4.3) as defence against viruses (phages) and mobile genetic elements.	Target sequence RNA (crRNA) is transcribed from genomic archive. Also serves as a memory.	23, 26, 30, 32, 46
Plants	Hypersensitive response, leading to apoptosis of infected cells and tissues.	Pathogen effector molecule ('avirulence factor', *avr*) recognized by plant receptor. *avr* are essential factors for parasites. System only partly homologous to animals.	1, 8, 36, 54
Unicellular eukaryotes	Organisms such as marine protists, acanthamoeba, etc., have 'virophages' integrated into their genomes. When virophages replicate or insert themselves into infecting viral genomes, they suppress replication of pathogenic (giant) viruses. A 'memory' of encountered viral sequences.	Different systems. For example, mavirus infection is integrated in host genome; mediates resistance against later Cafeteria roenbergensis virus (CroV) infection. MIMIVIRE works with co-infection of, e.g., mimivirus A and Zamilon virophage, which is later integrated in viral sequence and renders it dysfunctional. Viruses have evolved countermeasures.	29, 33

(Continued)

Table 4.6 Continued.

Group	Some characteristics	Mechanisms, molecules	Sources, further information
Placozoa (basal forms of free-living metazoans)	'Proto-MHC' involved in intracellular defences; the genomic region associated with stress-response genes.	Rich repertoire of TLR, NOD-like, SR, fibrinogen-related genes. Pathways deviate partly from other invertebrates. System is also basal to the Bilateralia. Patterns consistent with early genome duplication event in metazoan evolution.	28, 47
Porifera (Sponges)	Epithelium as defence barrier. Many extracellular (TLRs, IL-R1-like receptors), and intracellular receptors (NLRs, RIG-like). Signalling cascades (e.g. rudimentary NF-κB). No recognizable specialized immune cells. Histocompatibility responses. Phagocytosis (amoebocytes).	Rich immune gene repertoire. Highly variable allorecognition factor genomic region; additional variation by RNA editing. FREPs involved in coagulation. Effectors include reactive ROS, NOS.	7, 10, 35, 37, 41, 53
Cnidaria (incl. corals, anemones, *Hydra*)	Epithelium as defence barrier. Extracellular TLRs, NOD-like (NLRs) receptors. SRs, IL-R1-like, RIG-like (extra- and intracellular receptors). Specialized immune cells. Histocompatibility responses. TLR pathway shows conserved elements with higher groups.	Polymorphic allorecognition allelic system (FuHC locus—fusion and histocompatibility). PO/melanization pathway. Rich repertoire of AMPs, also reactive cytotoxic molecule species.	2, 5, 6, 34, 38, 44
Nematodes	No cellular defences, but autophagy present. Recognize different kinds of microbes, but mechanisms unclear. Conserved p38/MAPK pathway for antimicrobial defence. Other pathways include DAF-2/insulin-like receptor (ILR) pathway, and the TGFβ-related pathway.	Antiviral defence involves Dicer and argonaut proteins, vaguely similar to other invertebrates and plants (see Box 4.3). MyD88 and NF-κb transcription factors missing. Effectors include C-type lectin-containing proteins, which may act as specific defences, proteins with a CUB domain (C1s/Clr complement components) or AMPs. Model organism is *Caenorhabditis elegans*.	11, 14, 15, 31, 40, 42 49
Molluscs	FREPs as variable receptors. Group is taxonomically highly diverse (snails, mussels, cephalopods, chitons, etc.). Apoptosis, autophagy.	Enormous radiation and functional diversification of immune protein families. Interferon-like response to viral infections. Knowledge mainly comes from snails and oysters.	17, 18, 20, 45, 50
Crustacea	Major canonical pathways (Toll, Imd, JAK/STAT) and canonical receptors (Toll, lectins, GNBPs, β-glucan-binding NOD-like, SRs, Dscam, TEPs).	PO/melanization, AMPs as major defences. Antiviral pathways based on small RNAs (miRNA, siRNA, piRNA). Hypervariable Dscam with unclear function.	9, 21, 24, 27
Insects	Major canonical pathways and receptors present. Mammalian TLR pathway resembles insect Toll pathway.	Molecules and cascades show many homologies to vertebrates. Yet, enormous diversification and richness in protein families among insects,	See text for further details.
Echinodermata (sea urchins)	A very rich and complex pattern-recognition receptor repertoire (TLRs, SpTLR, SRs, NOD-like).	*SpTrf* as hypervariable effector molecules.	See text for further details.
Cephalochordata	Rich repertoire of receptors. Over 70 TLR genes, hundreds of C-type lectins and LRRs. Proto-RAG gene with similar functions as in the VDJ rearrangement of higher vertebrates; an ancestral version of the vertebrate RAG transposon. MyD88-independent pathways probably originated here.	Canonical pathways with NF-κb, or JNK. Greatly expanded, germline-encoded repertoire of adaptor proteins (MyD88, TIRs). Putative alternative splicing in receptor domains (*Branchiostoma*). Essential knowledge refers to the lancelet, *Amphioxus*.	13, 25, 51, 52

Group	Some characteristics	Mechanisms, molecules	Sources, further information
Urochordata (tunicates)	The major conserved mechanisms (complement, PO cascade, apoptosis, inflammation) and receptors (PGRPs, C-type lectins, TLRs) and VCBPs (studied in *Ciona*). No evidence for MHC, RAG genes, AID enzymes. Some elements later co-opted for MHC activity.	In *Botryllus schlosseri*, inflammation upon contact with neighbours triggered by the polymorphic *FuHC* locus: Alternative splicing generates a diverse repertoire of putative FuHC receptors. Closest living relatives of the vertebrates.	3, 12, 13, 16, 39, 43
Agnatha (jawless vertebrates)	VLRs are highly diverse and important receptors. Have similar roles to variable lymphocytes of jawed vertebrates.	Lymphocytes (VLRs) diversified by processes similar to gene conversion, mediated by AID-like enzymes (CDA1, CDA2). *RAG1, RAG2* genes absent.	4, 19, 22, 48
Gnathostomata (jawed vertebrates)	Adaptive system based on B- and T-cell lymphocytes.	Large families of Igs.	See text for further details.

[1] See glossary for immunological acronyms.

Sources: [1] Alamery. 2018. *Crop Pasture Sci*, 69: 72. [2] Augustin. 2010. In: Söderhäll, ed. *Invertebrate immunity*. Springer. [3] Azumi. 2004. *Immunogenetics*, 55: 570. [4] Boehm. 2018. *Annu Rev Immunol*, 36: 19. [5] Bosch. 2014. *Trends Immunol*, 35: 495. [6] Brown. 2015. *Sci Rep*, 5: 17425. [7] Buchmann. 2018. In: Cooper, ed. *Advances in comparative immunology*. Springer. [8] Christopoulou. 2015. *G3-Genes Genom Genet*, 5: 2655. [9] Clark. 2016. *Integr Comp Biol*, 56: 1113. [10] Degnan. 2015. *Dev Comp Immunol*, 48: 269. [11] Dierking. 2016. *Philos Trans R Soc Lond B*, 371: 20150299. [12] Dishaw. 2016. *Nat Comm*, 7: 10617. [13] du Pasquier. 2018. In: McQueen, ed. *Comprehensive toxicology*, Vol 11. [14] Engelmann. 2010. In: Söderhäll, ed. *Invertebrate immunity*. Landes Bioscience, Springer. [15] Ermolaevaa. 2014. *Semin Immunol*, 26: 303. [16] Franchi. 2017. *Front Immunol*, 8: Article 674. [17] Green. 2015. *J Gen Virol*, 96: 2471. [18] Green. 2018. *Viruses*, 10: 133. [19] Gunn. 2018. *J Mol Biol*, 430: 1350. [20] Guo. 2016. *Philos Trans R Soc Lond B*, 371: 20150206. [21] He. 2015. *Mol Immunol*, 68: 399. [22] Hirano. 2013. *Nature*, 501: 435. [23] Horvath. 2010. *Science*, 327: 167. [24] Huang. 2020. *Dev Comp Immunol*, 104: 103569. [25] Huang. 2016. *Cell*, 166: 102. [26] Jiang. 2017. *Ann Rev Biophys*, 46: 505. [27] Jin. 2013. *Fish Shellfish Immunol*, 35: 900. [28] Kamm. 2019. *BMC Genomics*, 20: 5. [29] Koonin. 2017. *Curr Opin Virol*, 25: 7. [30] Koonin. 2017. *Curr Opin Microbiol*, 37: 67. [31] Kuo. 2018. *Autophagy*, 14: 233. [32] Makarova. 2020. *Nat Rev Microbiol*, 18: 67. [33] Mougari. 2019. *Viruses*, 11: [34] Nicotra. 2019. *Curr Biol*, 29: R463. [35] Nicotra. 2011. *J Immunol*, 186: 170.4. [36] Niehl. 2019. *Mol Plant Pathol*, 20: 12013. [37] Oren. 2013. *Immunobiology*, 218: 484. [38] Palmer. 2018, and [39] Parrinello. 2018. In: Cooper, editor. *Advances in comparative immunology*. Springer. [40] Pees. 2016. *J Innate Immun*, 8: 129. [41] Pita. 2018. *Sci Rep*, 8: 16081. [42] Pradel. 2004. *Annu Rev Genet*, 38: 347. [43] Satake. 2019. *Cell Tissue Res*, 377: 293. [44] Schröder. 2016. *mBio*, 7: e01184. [45] Schultz. 2017. *Dev Comp Immunol*, 75: 3. [46] Shmakov. 2017. *Nat Rev Microbiol*, 15: 169. [47] Suurväli. 2014. *J Immunol*, 193: 2891. [48] Trancoso. 2020. *Curr Opin Immunol*, 65: 32. [49] Viney. 2005. *Int J Parasitol*, 35: 1473. [50] Wang. 2018. *Dev Comp Immunol*, 80: 99. [51] Xu, editor. *Amphioxus immmunity*. Elsevier; 2018. [52] Yuan. 2016. In: Xu, editor. *Amphioxus immmunity*. Elsevier. [53] Yuen. 2014. *Mol Biol Evol*, 31: 106. [54] Zipfel. 2017. *Nature*, 543: 328.

some similarities and the presence of similar protein families, is not entirely homologous.

Defence against intracellular parasites, notably viruses, follow similar principles as in many other organisms. For example, antiviral RNA silencing is one such defence, and is similar to that found in insects (Niehl and Heinlein 2019). The second major pathway in plant antiviral defence is R-gene dependent. A major class of these genes encodes NLRs. These receptors presumably do not recognize viral epitopes or viral products directly. Instead, they detect when specific host proteins, which act like sentinels, change under the action of the infecting virus (the 'guard hypothesis') (Jones and Dangl 2006); such surveillance can be in different forms (e.g. as guards, baits, decoys). The signalling cascades activate a variety of different compounds, such as salicylic acid, jasmine acid, ethylene, nitric oxides, and reactive oxygen species

(Ausubel 2005; de Ronde et al. 2014). These pathways also have points of mutual interactions and crosstalk to one another (Moon and Park 2016).

4.6.2 Defence in invertebrates

As a group, invertebrates make up the vast majority of all of the extant animals. Given the enormous number of species that have conquered almost any habitat on this planet, their immune defences are, perhaps not surprisingly, enormously diverse; Table 4.6 provides a rough overview.

4.6.2.1 Insects

Insects are the most species-rich group and a prime study subject for innate defences. In particular, studies in *Drosophila* (Imler 2014; Parsons and Foley 2016; Mussabekova et al. 2017) and the mosquito *Anopheles* (Bartholomay and Michel 2018)

Box 4.3 Antiviral defence of invertebrates

RNA interference (RNAi) is an ancient antiviral mechanism of invertebrates. Activities of RNA viruses, for example, typically lead to the production of dsRNA (double-stranded RNA) in the host cell, which is degraded or blocked by RNAi from being translated into a protein.

Three classes of RNAi mechanisms can be distinguished (Figure 1) (Mussabekova et al. 2017; Leggewie and Schnettler 2018):

1. *Micro-RNA pathway (miRNA)*: In *Drosophila*, hundreds of miRNAs are known that affect gene expression. After recognition of viral mRNA, pri-miRNA is transcribed in the form of stem-loops that are processed in the cytoplasm (insects) or the nucleus (plants); stem-loops are generally important during development in animals and plants. After binding to Dicer-1 (an enzyme that cleaves RNA), small pieces of duplex RNA (21, 22 nt long) are generated. These can be recognized and become loaded to

argonaut-1 (AGO1). Together, this forms the miRNA-programmed RNA-induced silencing complex (miRISC) that can bind to the corresponding, virus-induced target mRNA sequence. This blocks translation and leads to the slicing and degradation of viral mRNA. The pathway seems to be present in most animals.

2. *Small interfering RNA pathway (siRNA)*: In this pathway, viral dsRNA is directly recognized and bound by dicer-2 (Dcr2). This cuts the dsRNA into smaller (21 nt) fragments of duplex RNA, the siRNAs. The fragments are recognized by argonaut-2 (AGO2) and, together with co-factor R2D2, bind to form a pre-RISC complex. The precomplex matures into the RISC complex by ejecting one strand of siRNA (the 'passenger strand'); the other 'guide strand' remains as a template and binds to the viral mRNA that corresponds to its template siRNA. This process leads to viral mRNA degradation. The siRNA pathway in nematodes is very similar. In the latter, Dicer-related helicase 1

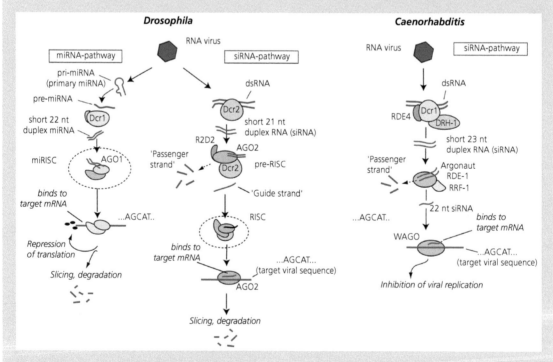

Box 4.3 Figure 1 Antiviral defence by RNAi interference in insects (*D. melanogaster*) and nematodes (*C. elegans*). In *Drosophila*, primary RNA (red), derived from the virus, is processed to pre-miRNA (left-hand side) in the miRNA pathway. It binds to Dicer (Dcr1) to form the silencing complex (RISC) that targets the virus-induced mRNA and degrades it. The siRNA pathways of *Drosophila* and *Caenorhabditis* are very similar. Further descriptions in text. Schemes after descriptions in Xu and Cherry (2014), Paro et al. (2015), Gammon 2017; Mussabekova et al. (2017) and Schuster et al. (2019b).

(DRH-1) may act as a sensor for viral RNA. Short duplex siRNA then loads onto an argonaut protein (RED1). Subsequently, one strand of siRNA is expelled; the other strand associates with worm-specific argonaut (WAGO) to bind to the target (virus-induced) mRNA and degrades it.

3. *Piwi-associated RNA pathway (piRNA).* This path is independent of Dicer and involves additional argonaut proteins. piRNAs are small (24–27 nt long) non-coding RNAs. They associate with piwi and proteins aubergine and argonaut-3 (AGO3) to form active complexes. The piRNA pathway is well known for acting against transposable elements in the genome, including viral elements. It may additionally be involved in defence against viruses, but this seems not to be the case in *Drosophila*. The pathway exists in flies and vertebrates.

Different groups of organisms differ in the particular proteins that take part in these pathways. For example, the micro-RNA pathway of insects contains two forms of Dicer. In vertebrates, only one Dicer seems present. Similarly, organisms differ in the kinds of co-factors needed for Dicer to function.

Argonaut proteins are central and exist in different forms in plants, insects, and nematodes. Generally, argonauts are enzymes that specifically target RNA templates and slice them into pieces. They are guiding a template within the RISC towards complementary viral mRNA sequences. The Imd–JNK, JAK–STAT, and Toll pathways all seem additionally involved in antiviral defences (Mussabekova et al. 2017; Chow and Kagan 2018; Mondotte and Saleh 2018). Hence, the antiviral defences involve several pathways, but the RNAi-silencing mechanism is a centrepiece in plants and invertebrates.

Viruses, in turn, have evolved various suppressors of the RNAi-silencing mechanisms (VSR, viral suppressor of RNA silencing). For example, FHV (flock house virus) and DCV (Drosophila C virus) of *Drosophila* have suppressors for the dsRNA/dicer step (e.g. proteins 1A and 340R from DCV). Suppression generates strong selection on host proteins to escape this interference. Indeed, antiviral proteins are generally fast-evolving, and in *Drosophila*, Dcr2, AGO2, and co-factor R2D2 are among the most rapidly evolving genes (Obbard et al. 2006; Kemp and Imler 2009; van Mierlo et al. 2014).

have been helpful to understand insect immune systems more generally (Hillyer 2016; Marques and Imler 2016). The insect immune system shows several distinct pathways, notably the canonical Toll, Imd, and JAK/STAT pathways. Insects also have an astonishingly diverse range of different receptors that serve different purposes. Examples include receptors for phagocytosis (e.g. integrins, perhaps Dscam) or cytokine regulation and AMP expression (Toll, PGRPs, domeless) (Ferrandon et al. 2007; Wang et al. 2019b). The Toll pathway responds to signatures from fungi and bacteria, but the Toll receptor itself binds to its ligand Spaetzle instead of a parasite epitope itself. Strictly speaking, it is thus not a receptor. The insect Toll pathway resembles the mammalian signalling cascades, such as those associated with the TLR, or the interleukin-1 receptor (IL-1R) pathway. The Imd pathway responds primarily to Gram-negative bacteria. It has a transmembrane PGRP and depends on a signalling complex with FADD (Fas-associated protein with death domain), DREDD (Death related Ced-3/Nedd2-like caspase), and the Imd protein. The signal results in the production of the transcription factor Relish that activates genes for AMPs such as Diptericin. As with almost any of the pathways, there is crosstalk; that is, JAK/STAT can be activated by Imd. Insect

defence against viruses is illustrated in Box 4.3 as an example of antiviral defences in invertebrates more generally. Signalling pathways of insects and mammals share similarities and are homologous in many of their elements. Nevertheless, it should be kept in mind that *Drosophila* and *Anopheles* are not paradigmatic for all insects. For example, honeybees (*Apis mellifera*) have only around one-third of the immune genes compared to other insects (Evans et al. 2006), which is an ancestral trait in bees (Barribeau et al. 2015), and aphids seem to lack the Imd pathway altogether (Gerardo et al. 2010).

4.6.2.2 Echinoderms

This is a group of highly evolved animals. Judging from the genomes of several echinoderms, these animals have a surprisingly complex immune system (Smith et al. 2018a; Oren et al. 2019). Their repertoire of PRRs is extraordinarily complex, for example in the sea urchin, *Strongylocentrotus purpuratus*. Its Toll-like receptors (SpTLRs) have expanded to more than 250 genes, grouped in 11 subfamilies; they are differentially expressed in larvae or adults. Furthermore, the sea urchin's NOD-like receptors (SpNLRs) have more than 200 genes, and the scavenger receptors contain multiple cysteine-rich domains (SpSRCR) with more than 1000 domains in more than 200 genes (Buckley and Rast 2015; Smith et al. 2018a). The SpSRCR proteins can form more than 1000 domains by alternative splicing (du Pasquier 2006; Hibino et al. 2006). These numbers contrast with the typical 5–20 genes in the corresponding protein families of, for example, insects. The SpTLRs consist of an extracellular LRR domain and a transmembrane region with a cytoplasmic Toll/IL-1R (TIR)-signalling domain that elicits the cascade inside the cell. Furthermore, a large set of over 200 cytoplasmic recognition proteins (NACHT domain and leucine-rich NLR proteins, resembling plant LRRs) are encoded in the germline. The extraordinary richness of genes that code for different immune receptors and their close relatedness among each other seems to be the result of an evolutionarily recent expansion in the number of genes (Flajnik and du Pasquier 2008). Moreover, the genome of *S. purpuratus* contains genes (SpRAG1-like, SpRAG2-like) with compelling similarities to the RAG genes of vertebrates (which function in the recombination rearrangement of receptors); these are thus suspected of assuming similar roles in the adult sea urchin (Smith et al. 2018a). Immune responses in echinoderms are based on immune cells, the coelomocytes of various types, such as phagocytes, spherule cells, haemocytes, and crystal cells. To date, their functions are not well known.

The echinoderm (*S. pupuratus*) *SpTransformer* (*SpTrf*, formerly *Sp185/333*) gene family has no sequence similarities elsewhere. They show two exons, the larger with various types of repeats, that also occur in copy numbers. Each individual sea urchin has a unique *SpTrf* gene repertoire; hence, genes of an identical sequence are not shared among individuals in a population, which therefore contains a vast diversity of such genes. The expressed *SpTrf* proteins result from the 'mosaic' arrangement of genetic elements, but also RNA editing or alternative splicing (Oren et al. 2019). The genes are rapidly upregulated upon infection; their exact function is still unclear, but they may disrupt parasite membranes (Lun et al. 2017; Oren et al. 2019).

4.6.3 The jawed (higher) vertebrates

The higher (jawed) vertebrates, too, use the innate immune system as a first responder. Three major families of PRRs exist that together recognize a wide range of pathogens: the intracellular NLRs and RLRs, and the extracellular TLRs (Cao 2016). Among those, the TLRs are a well-studied example. Signal transduction (as in other cases) occurs by a conformational change of the transmembrane TLR molecules; the change recruits adaptor proteins, such as MAL or MyD88, and leads to signalling along a cascade; Figure 4.6 shows some details. The different TLRs respond to different stimuli and have slightly different signalling architectures (De Nardo 2015; Cao 2016). For example, TLR3 senses double-stranded RNA (dsRNA) that results from the replication of positive-stranded RNA viruses (e.g. West Nile virus). Other cascades result from other intracellular viral receptors, activating key kinases, for example IKK or TANK-binding kinase (TBK), and eventually the same transcription factors. Innate immune signalling cascades in mammals thus show complexity, redundancy, and various degrees of convergence of key elements. Furthermore, they share similarities with those of invertebrates; for example, the Toll and TLR pathways are partly conserved between *Drosophila* and humans (Lindsay and Wasserman 2014).

Together with the evolution of the MHC, a hallmark of the adaptive immune system of jawed vertebrates is the B- and T-cell lymphocyte populations that undergo clonal expansion when stimulated by the recognition of an antigen and a T-helper cell. During haematopoiesis of mammalian B-cells (see Figure 4.4), the B-lymphocytes develop from stem cells in the bone marrow and end up as immune cells circulating in the blood and the lymphatic system. As the cells mature, a huge diversity of different binding specificities in the Ig receptor domains are generated by the various processes shown in Figure 4.9c. Some of those might accidentally recognize and react against self-tissue. Hence, before they are entering the defence repertoire, B-cells first undergo a process of negative selection against self-reactive variants by being exposed to self-molecules. If self-reactive, the cells induce their apoptosis and die, or they can become tolerant to self by further receptor modifications through gene rearrangements ('receptor editing'). Furthermore, some cell lines no longer express self-reactive antibodies on their surface (they become 'anergic').

Together, these are processes that yield 'central tolerance'—a key element of 'immunological tolerance'. Self-reactive B-cells that nevertheless reach the periphery are silenced by apoptosis or become anergic, leading to 'peripheral tolerance'. Note that 'immunological tolerance' is not to be confused with 'tolerance' in the sense of a defence strategy (see section 5.5).

Mature B-cells secrete their Ig receptors as antibodies that can recognize and bind to parasite epitopes. In 'naïve' (not activated) B-cells, these are 'natural' antibodies of type IgM (Figure 4.12a). As the B-cells circulate in the host's body, their membrane-bound Ig receptors might recognize and bind to epitopes of an infecting parasite, which leads to the secretion of more IgM antibodies (Figure 4.12b). However, to start the full response, the B-cell also needs to be recognized by an already activated T-helper cell. The entire process involves antigen presentations on the surface of APCs and the B-cells, as sketched in Figure 4.12c. Eventually, the binding to the parasite epitopes and to the T-helper cell activates the B-cell and starts cell division and proliferation; a clonal expansion to large numbers of this particular B-cell subpopulation with this receptor specificity occurs. B-cells that happen to bind more strongly (because their receptors match the epitopes better) divide more rapidly. By somatic changes (hypermutation) the cell population further fine-tunes its binding specificities ('affinity maturation'). At the same time, the secretion of antibodies of other types (IgA, IgG) with the same specificity increases. In the process, affinity, i.e. the binding strength towards this recognized single epitope, can be approximately 30 000 times higher than in the beginning. This change results from the substitution of amino acids that changes the shape of the binding pocket. The particular parasite epitopes that have triggered this process cause 'immunodominance'; that is, a response dominated by these few epitopes. Macroscopically, the immune system adapts to an infecting parasite by recruiting more and more cells that specifically target this infection. An important consequence of this expansion is that some of these activated B-cells develop into memory cells that maintain the trace of the former infection and, upon exposure to the same parasite, can produce larger numbers of matching antibodies

much faster than on the first encounter. B-cell activation happens in specialized organs, that is, in the lymph nodes of mammals, where the B-cells with their load migrate to and find a matching T-helper cell. Lymph nodes have a high density of T-cells, and so the frequency of contacts is much higher than if the B-cell had to find its T-helper cell somewhere else in the body.

TCRs can bind to various targets, but primarily to small fragments (the peptides). T-helper cells, in particular, have the co-receptor CD4 on their surface (the CD4+ T-cells; Figure 4.11b), and engage with the APCs that circulate in the blood and lymphatic system, as well as with the B-cells. The CD4+ T-helper cells can furthermore differentiate into various subsets, notably the T_H1- and T_H2-cells (Coico and Sunshine 2015). This difference is essential, as one subset inhibits the development of the other set. Differentiation is affected by cytokines that are released mainly from the dendritic cells, mast cells, and NK-cells. T_H1-cells require the cytokine IL-12 to develop, which typically is generated in response to bacterial and viral infections. T_H1-cells subsequently release the key cytokine, IFN-γ, which inhibits T_H2-cells, but also activates NK-cells and macrophages. T_H2-cells, by contrast, are stimulated by the cytokine IL-4. This cytokine is produced, for example, in response to parasitic worm infections. T_H2-cells subsequently produce IL-4 and IL-5, which stimulates B-cell growth and promotes IgE and IgG synthesis but suppresses T_H1-cells. Because of the mutual suppression of these T-cell populations, their differential regulation in response to different infections has been implicated in various autoimmune diseases, notably the development of asthma (Coico and Sunshine 2015).

In contrast to B-cells and antibodies, which patrol the exterior environment in the blood and lymphatic system, the T-cells also fight parasites inside host cells. In the process, the proteins of an intracellular, invading parasite are processed and cut into small peptides by the host cell's proteases. These proteases are part of the proteasome, large intracellular protein complexes that degrade damaged own or foreign proteins (such as antigens) and cut them into fragments (the peptides). In further steps, the transporter associated with antigen processing

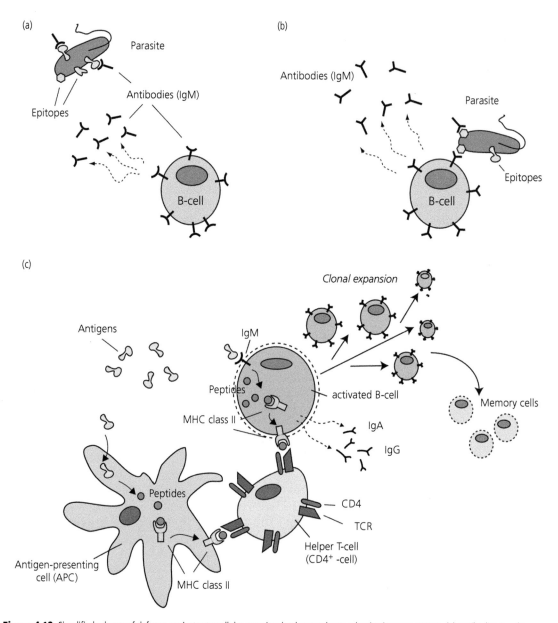

Figure 4.12 Simplified scheme of defence against extracellular parasites by the vertebrate adaptive immune system. (a) Antibodies are the secreted Ig receptors of B-cells and circulate to bind on epitopes (antigens; reddish shapes) of the parasite. 'Naïve' B-cells secrete 'natural' antibodies of type IgM. (b) The same Ig receptors are also membrane-bound on B-cells (green) and recognize the same epitopes. If a B-cell binds to epitopes, it becomes activated to secrete more antibodies (IgMs). (c) B-cells (green) become activated for clonal expansion when additionally binding to a T-helper cell (blue). This happens when, firstly, parasite-derived molecules (antigens; reddish shapes) are recognized and internalized by APCs. Inside the APC, the antigens are degraded into pieces (peptides) that can be recognized and bind to MHC class II proteins (yellow). The MHC–peptide complex subsequently is presented on the surface of the APC to passing T-helper cells. If the T-helper cell (here a CD4+-cell) has the matching specificity for the peptide—as determined by its TCR-receptor complex (green/grey; see Figure 4.11)—the T-helper cell binds to the APC. Meanwhile, the B-cell similarly internalizes and processes the antigen to peptides, which eventually are presented with the MHC class II complex on the B-cell surface. The same matching T-helper cell that is activated by the binding to the APC can therefore also bind to the B-cell. This binding yields a signal to the B-cell, such that it becomes activated to proliferate for clonal expansion and affinity maturation. In the process, antibodies of other types (IgA, IgG) are secreted instead of IgM's. Besides, binding to the inducing antigens becomes stronger. In the further course of the response, some B-cells might develop into memory cells that can be reactivated more quickly and specifically on a secondary challenge by the same parasite. The binding of APCs and B-cells with T-cells leads to the formation of an 'immunological synapse' that contains many molecules and signals.

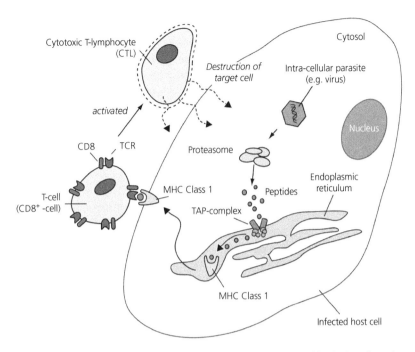

Figure 4.13 Simplified scheme of defence against intracellular parasites (such as a virus; red shape) by the (jawed) vertebrate adaptive immune system. The proteasome (a large intracellular protein complex) degrades the proteins derived from a parasite. The resulting peptides (red) are transported into the endoplasmic reticulum by TAP (transporter associated with antigen processing), where they are recognized and bind to MHC class I proteins (yellow). The MHC–peptide complex is then presented on the cell surface, where it is recognized by a passing T-cell with a matching TCR (see Figure 4.11) and co-receptor (CD8 in this case). If binding occurs, the T-cell becomes activated and transforms into a cytotoxic T-cell (CTL) that starts to actively destroy the infected host cell.

(TAP) complex moves these peptides into the endoplasmic reticulum, where they are loaded onto MHC class I molecules to be presented on the cell surface to passing T-cells (CD8+ T-cells; Figure 4.13). This binding activates the T-cell and transforms it to become a 'cytotoxic T-cell' (CTL). The cell will start to destroy the host cell that has signalled as being infected. CD8+ T-cells thus recognize infected cells by the MHC–peptide complex that signals non-self.

During haematopoiesis, T-cell precursors move from the bone marrow to the thymus via the bloodstream, where they further develop and differentiate (Figure 4.4a). In this complex multistep process (Coico and Sunshine 2015), the machinery to somatically diversify the receptor specificities is activated first and generates a specificity for a cell. Subsequently, these 'pre-T-cells' (thymocytes) start to express their receptor complex, followed by the expression of the CD4+ and CD8+ co-receptors

('double-positive cells'). At this point, a process known as 'thymic selection' sets in. In a first episode, double-positive T-cells that can bind to MHC class I and class II molecules presenting self-peptides become 'educated' to recognize these self-MHC molecules, i.e. survive and proliferate ('positive selection'). Here, the recognition of self-MHC is of the essence, regardless of what peptides would later be presented by these MHCs. Thymocytes not able to bind to self-MHC (an estimated 90 per cent or more of all thymocytes) will die at this stage, leading to the 'self-MHC restriction' of the T-cell response. Those T-cells that have survived the positive selection step may have too high an affinity to self and will thus present a considerable risk for self-damage. Therefore, these cells induce apoptosis and become eliminated during 'negative selection', aided by the transcription factor AIRE (autoimmune regulator, Mathis and Benoist 2007). As a consequence, only

T-cells with some low to intermediate affinity to self will survive this two-step process. In a final step ('lineage choice'), the doubly positive thymocytes commit to developing into a CD4+- or a CD8+-cell and leave the thymus as mature T-cells to begin their work. In humans, there are around 10^{11} T-cells that are patrolling the body, but this represents only a minor fraction (estimates are less than 1 per cent, Müller and Bonhoeffer 2003) of all thymocytes that have developed in the thymus. Like in the case of the B-cells, thymic selection has led to central tolerance, which is dominated by the elimination of unwanted cells early in development. Self-reactive T-cells that nevertheless make it to the periphery will subsequently become inactivated in the process of 'peripheral tolerance'. Similar to B-cells, the host retains a memory of an earlier encounter, and when a second exposure to the same epitopes occurs, the matching T-cells can multiply more rapidly. In the interaction with MHC molecules, TCRs utilize short peptide motifs that are likely to be around eight to ten amino acids long in MHC class I, and 13–17 amino acids long in MHC class II (Janeway et al. 2001). These lengths are surprisingly short. Nevertheless, even with these lengths and given that there are 20 different amino acids, an enormous number of combinations are possible: with eight amino acids we have $20^8 = 25.6 \times 10^9$ different peptide sequences, and with 17 amino acids we get $20^{17} = 1.3 \times 10^{22}$ different sequences, a combinatorial space that covers almost all realistically possible peptide variants, also for those encountered during parasitic infections (Burroughs et al. 2004).

4.7 Memory in immune systems

'Memory' means that after the first encounter with a parasite, the 'primary challenge', a second or subsequent encounter with the same parasite, or a variant of the same parasite, induces a faster, more robust, or more efficient response (Figure 4.14). Its discovery has a long history (Box 4.4), and it has become clear that memory is not only a property of the adaptive arm of the immune system but also a property of the innate immune system, even in higher vertebrates.

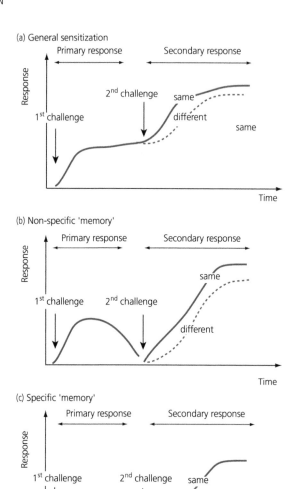

Figure 4.14 Immune responses and immune 'memory'. A first challenge by an antigen elicits the primary immune response (y-axis: strength of the response). A subsequent second challenge with either the same (solid line) or a different (broken line) antigen elicits a secondary response that is generally stronger, faster, lasts longer, etc. Several cases can be distinguished. (a) General sensitization (priming) of the immune response: the primary response does not subside; a secondary response builds on the remaining activity and is enhanced. (b) Non-specific memory. The primary response subsides (the immune system is quiescent) and the secondary is stronger, but not specific, as it responds similarly to the same or a different challenge. (c) Specific immune memory. The primary response subsides, the secondary is stronger and primarily directed against the same but not towards a different antigen. This is the recall response that is typically associated with a bona fide immune memory.

Box 4.4 Priming and memory

The epidemic that struck Athens in 430 BC—now considered smallpox, the plague—probably was the first documented case where humans took notice of immune 'memory'. An eyewitness, the Athenian general and historian Thukydides (454–*c*.396 BC) left the now-classical account, where he noted that survivors of the infection did not get the disease a second time (Littman 2009). Later, turning such observations into vaccination started in sixteenth-century China, and culminated a first time in 1796 with Edward Jenner (1749–1823). He inoculated a boy with extracts from a dairymaid that had cowpox symptoms, and another time with material from smallpox; the latter led to complete protection (Riedel 2005). It was still a long while before the

modern vaccines that we know today were developed (Plotkin 2011). However, the success story of vaccination is a powerful proof that immune systems have a 'memory', such that a second infection of the same type (the same parasite, or the same strain of a parasite) elicits a response that is faster, stronger, or more efficient. In the case of vaccination, this results in better protection of the individual. Although this general feature is undisputed, there is much discussion about what precisely a 'memory' is, and what mechanisms may be responsible for it.

Until recently, it was conventional wisdom that only the jawed vertebrates, with their B- and T-cell lymphocytes, and—given the results of transplantation experiments—

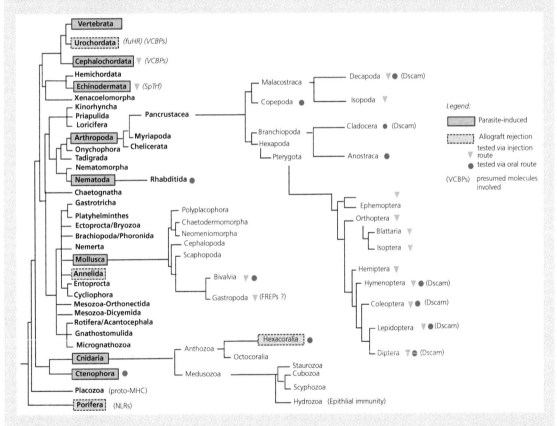

Box 4.4 Figure 1 Phylogenetic overview of immunological memory phenomena. These are based on innate systems; vertebrates also have an 'adoptive' memory. Boxes indicate parasite-induced memory, candidate molecules in parentheses, etc. (see legend). This crude overview is continuously changing as new evidence accumulates. The patchy distribution of memory suggests that the phenomenon is probably almost universal across the tree of life. Adapted from Milutinovic and Kurtz (2016), with permission from Elsevier.

continued

Box 4.4 *Continued*

perhaps also the jawless vertebrates, could have an immuno-logical memory. However, studies in insects dating back many decades (Paillot 1920; Metalnikov 1921; Chorine 1929; Gingrich 1964), along with most work done during the 1980s and 1990s (Karp and Rheins 1980; Faulhaber and Karp 1992), already suggested that insects possess some kind of immune 'memory', too, and are protected the second time. In plants, memory phenomena were already described in the 1930s (Chester 1933). Not all of these studies were perfectly controlled and designed. However, collectively they suggested that an immune memory does not depend on B- and T-cells, and that some kind of 'innate immune memory' must exist—a veritable paradigm shift in the field (Netea et al. 2015). Hence, the current view is more extensive, as there is now good evidence for memory in vertebrate NK-cells, macrophages, and monocytes (Sun et al. 2014; Hamon and Quintin 2016; Pradeu and Du Pasquier 2018). The cells of the innate immune system may undergo epigenetic changes

upon a primary infection that leads to 'trained' immune cells for a secondary challenge (re-exposure) ('trained immunity'; Netea et al. 2011, 2015, 2016). Now, 'memory' phenomena are known from many groups of invertebrates, for example earthworms, tunicates, and corals (Pradeu and Du Pasquier 2018) (Figure 1).

Of special note are the memory systems in protozoa, based on the CRISPR–Cas mechanisms (see Box 4.2). These systems can 'remember' a repertoire of encountered viral sequences (Sternberg et al. 2016; Nussenzweig et al. 2019). A second special note is deserved for parents providing immune protection to their offspring (Moret and Schmid-Hempel 2001)—now known as 'transgenerational immun-ity' (Sadd et al. 2005) (see Chapter 7). The wealth of new terms should, however, not distract from the fact that some kind of 'immune memory' is widespread among organisms (Figure 1), and conveys more benefits than costs in a world full of parasites.

4.7.1 Memory in the adaptive system

The 'adaptive immune memory' is the prime reason why vaccination is possible—a major achievement by the medical sciences that has saved thousands of lives and reduced human suffering in unprecedented ways. It is based on B- and T-cells. A small fraction of these cells turn into memory cells, often considered to be the 'stem memory cells'. In T-cells, they further develop and multiply into 'effector memory cells' (alternatively, a direct development into effector memory cells is conceivable). The entire system is more complex, though, as at least three different groups of memory T-cells exist (Mueller et al. 2013). The maintenance of the memory cells is still controversial, but cytokines seem to induce occasional cell divisions in a process known as 'homeostatic proliferation'. Similarly, B-cells can turn into memory cells. A first group involves B-cells that are recruited early during a response and mainly have the IgM type of antibodies. A second group are those B-cells that circulate later and have already undergone affinity maturation. They express other classes of Igs, such as IgA or IgG, with better binding specificities than the IgMs of B-cells in the first group. Both groups are associated with

the formation of the germinal centres—sites within a lymph node or spleen where B-cells proliferate (a B-cell follicle); these establish within days of infection and remain active for some weeks. Other descendants of the B-cell germinal centre lead to plasmablasts, perhaps only after a second challenge, and establish in the bone marrow (or in the mucosa, and elsewhere), where they can produce specific antibodies for a very long time ('long-lived plasma cells', LLPCs). The recall of the memory is in itself a complex process, with various cues for stimulation of the respective cell populations (Hoyer and Radbruch 2017; Chang et al. 2018; Khodadadi et al. 2019).

4.7.2 Memory in innate systems

An immunological memory exists in invertebrates, plants (Reimer-Michalski and Conrath 2016), and even in protists and protozoa (see CRISPR–Cas in Box 4.2) that lack B- and T-cells. Here we use the lump term 'immune priming' (Little and Kraaijeveld 2004) to separate these phenomena from the lymphocyte-based vertebrate adaptive memory, in an attempt to avoid conflicting meanings as well as to account for the fact that the underlying mechanisms are very different. Lumping is an overt oversimplification, because the innate immune system is not homogeneous. Instead, it is a collection of very different mechanisms, each one probably having the capacity to evolve a memory component (Pradeu and Du Pasquier 2018). In fact, the mammalian innate immune system also shows memory, which involves macrophages, NK-cells, and some other cell types (Hamon and Quintin 2016). As in any rapidly developing field, there is considerable confusion with concepts and terminology because the memory/priming phenomena vary with respect to duration, specificity, and strength, and are observed in a wide range of taxa that may or may not use quite different mechanisms to achieve the same goal. Nevertheless, the central insight of the last decade is the recognition that the 'memory' phenomenon is widespread, gradual, and diverse, and must have evolved several times due to the apparent benefits it provides for organisms that repeatedly may encounter the same or similar parasitic challenges (Box 4.4).

Many examples come from insects (Lanz-Mendoza and Garduño 2018). Worker bumblebees, for example, develop protection against a second, lethal dose of the same bacterial infection that lasts for the worker's lifetime (Sadd and Schmid-Hempel 2006). Similar findings were made for *Drosophila* (Pham et al. 2007), the flour beetle, *Tribolium castaneum* (Roth et al. 2009), and for the silkworm, *Bombyx mori* (Wu et al. 2015). In the latter two, 'memory' might even be specific to the level of different bacterial species and perhaps strains. However, not all pathogens elicit a memory effect, for reasons which are still enigmatic. Immune priming is also known for viruses in, for example, oysters (Green and Speck 2018), shrimps (Syed Musthaq and Kwang 2014), and mosquitoes (Ligoxygakis 2017). Nevertheless, priming does not work for every case. For example, populations, different stages and the two sexes of flour beetles (*Tribolium*) vary in the level of immune priming (Khan et al. 2019), and experimentally evolve specific priming only when treated with high doses of bacteria (Khan et al. 2017b). In *Tenebrio*, (non-specific) priming seems primarily elicited by challenges with Gram-positive bacteria (Dhinaut et al. 2018).

The biochemical and molecular mechanisms that generate an immune memory are generally poorly known, especially for the invertebrates. In the simplest case, a first challenge leads to a primary response, which leads to a generally increased sensitivity of the immune system to subsequent challenges of any kind (Figure 4.14). Such a 'memory' could simply be associated with a primary response that does not disappear completely. In contrast, a 'memory' in the narrower sense is not dependent on keeping the immune system active between successive antigenic challenges. Instead, the information is laid down somewhere, for example, in specialized cells of the immune system. A secondary response is faster and stronger than the primary response, even though no immune activity is occurring in the meantime. Moreover, the secondary response might extend to antigens different from the primary challenge, in which case the memory is cross-reactive and not very specific. Note that these definitions are in functional categories and independent of the underlying mechanisms.

There are some hints as to what mechanisms may be important, though (Gourbal et al. 2018). In plants, the primary infection causes epigenetic changes in immune defence genes. Examples are DNA methylation or histone modification, and higher expression levels of receptors, signalling enzymes, and differences in transcription factors have been described, which allow for a better response the second time (Reimer-Michalski and Conrath 2016). A priming activator and its receptor have been described (Durrant and Dong 2004; Conrath et al. 2015; Ramirez-Prado et al. 2018). In snails, gene expression and gene knock-out studies suggest that a response to a second challenge shifts from cellular to humoral defences (Pinaud et al. 2016). In mosquitoes exposed to *Plasmodium*, a haemocyte differentiation factor (lipoxin–lipocalin complex) is crucial for the formation of immune priming. The penetration of the gut (by the parasite's ookinetes) during the primary infection induces the priming (Ramirez et al. 2015). Epigenetic processes that affect gene expression upon a second exposure are perhaps some of the main mechanisms that generate immune priming (Wang et al. 2009; Zhao et al. 2013; Pinaud et al. 2016; Greenwood et al. 2017). This is also the case for the innate memory in mammals, where parasite-derived compounds such as β-glucans seem to initiate the reprogramming of monocytes for memory formation (Hamon and Quintin 2016). Also, changes in the abundance or the ongoing persistence of regulating RNAs (miRNA) after a primary infection could be important mechanisms (Gourbal et al. 2018). Finally, the microbiota contributes to the formation of an innate immune memory, although the underlying mechanisms are still unclear (Futo et al. 2016).

4.8 Microbiota

A large number of diverse microorganisms, notably bacteria and protozoans, form persistent assemblies on the surface of their hosts, e.g. on vertebrate skin, insect cuticle, or inside the gut, various body openings, and inside cells, e.g. within the specialized bacteriocytes (Qin et al. 2010; Spor et al. 2011; Weinstock 2012; Douglas 2015). Collectively, these microbial assemblies are called the 'microbiome' or 'microbiota'. Of those, the gut microbiota has received most of the attention.

The gut microbiota is known to provide essential functions for its host, such as for the uptake of nutrition, gut homeostasis, and development, and thus has quite generally positive (or negative) effects for health (Knights et al. 2014; Levy et al. 2017; Sommer et al. 2017; Jackson et al. 2018). There is crosstalk between the gut microbiota and distant organs, allowing the microbiota to relate to the immune and hormonal systems, as well as to brain functions in the so-called 'gut–brain axis'. This involves a range of bacteria-derived or bacteria-stimulated molecules (Schroeder and Bäckhed 2016). Furthermore, the microbiota also affects autoimmune diseases (Bach 2018) and has been implicated as a basis of the so-called 'hygiene hypothesis', although the underlying mechanisms are not yet fully understood (Blander et al. 2017; Bach 2018). This hypothesis refers to observations that people in Western, industrialized countries more frequently acquire allergies, as compared to people in the developing world. The hygiene hypothesis proposes that this results from an almost complete absence of common parasites, notably helminths, in modern societies, with their more hygienic conditions. With a lack of parasitic infections, immune responses are

more frequently to allergenic substances instead of 'real' challenges. A major contributing factor, presumably, is the differences in the structure of the microbiota in humans of developed vs developing countries (Villeneuve et al. 2018).

The effect of microbiota on the resistance to colonization by gut pathogens has been known since at least the early 1950s. For example, the weakening of the microbiota by antibiotics facilitated *Salmonella* infections in mice (Bohnhoff et al. 1954). The presence or absence of certain bacteria (*Prevotella*) in the human gut microbiota is, in fact, predictive for the infection success of *Entamoeba histolytica* or the associated disease symptoms (diarrhoea) (Burgess et al. 2017). The microbiota interacts with parasites and pathogens in various ways—either directly or indirectly, e.g. by the mobilization of the immune system. On the other hand, it remains a challenge to understand how the host defence system can distinguish between beneficial and harmful microbes. This could also be an adaptation by the microbes themselves. For example, in mosquitoes (*Aedes aegypti*), the gut microbiome protects itself against host defences with C-type lectins to escape the effects of AMPs (Pang et al. 2016). Because of the many effects of microbiota on the host, combined with technological advances in measuring the microbiota, several initiatives have established data banks that hold genomic and other information about these associated microbes.

4.8.1 Assembly, structure, and location of the microbiota

The microbiota represents a few per cent of the host's body mass. Taxonomically, a microbiota contains bacteria, archaea, fungi, ciliates, and flagellates, depending on the host group and host species. Viruses are also present, but very little is known about them as of today (Abeles and Pride 2014). Bacteria are the most numerous. In humans, an estimated 3×10^{13} bacteria colonize the body, which is twice as many as the number of human cells themselves (Qin et al. 2010). The composition of the microbiota is characterized by the identity and the number of species, and the members of a microbiota are typically identified by their genetic sequences. These usually refer to hypervariable parts of the 16S rRNA gene, generally resulting in millions of gene sequence reads for any given probe. These sequences are grouped based on a critical amount of sequence divergence (usually 3 per cent). The resulting sequence clusters are called 'Operational Taxonomic Units' (OTUs); it is not generally clear whether these are species, nor would most OTUs have ever been described as a species before. Typically, the bacterial component of the microbiota of a vertebrate host would contain hundreds or perhaps thousands of OTUs (Spor et al. 2011; Hird et al. 2015). Insects, by contrast, and with the conspicuous exception of the termites (Brune and Dietrich 2015), have only a few dozen OTUs. In humans, the microbiota resides at various sites that vary in the relative frequencies of the dominant bacterial phyla. The gastrointestinal tract, for example, is rich in Bacteroidetes, whereas skin and nostrils are rich in Actinobacteria (Spor et al. 2011). Various factors affect this distribution and composition, and the microbiota, therefore, varies not only among sites of the body, but also among individuals, among populations, and between species (Nishida and Ochman 2019). Across species, the location of the microbiota in the gut can also be quite different (Figure 4.15).

How a microbiota is assembled, i.e. what determines its composition and how and why microbial pathogens can end up in the host gut in the first place, are important elements, as they set the stage for host–parasite interactions. Microbes—pathogenic or not—are acquired by a host through transmission from a source, and opportunities of getting in contact with sources of microbes vary, for example, by diet (Muegge et al. 2011; David et al. 2014), social contacts (Keiser et al. 2018), and the like. Hence, the transmission dynamics affects microbiota composition, and partly explains why certain microbes are either permanently associated with a host or remain transient (Douglas 2015). Typically, a mixture of both is present: permanently associated microbes can become symbionts. The commonly present microbes are known as the 'core microbiome' of a particular host species or group. As such, the core microbiome varies with host taxonomy, e.g. as seen in neotropical birds (Hird et al. 2015) or bees (Kwong et al. 2014).

However, transmission and contact with a host do not necessarily lead to the incorporation of a particular microbial taxon into the resident microbiota. For example, microbiota composition varies with host genotypic background (Luca et al. 2018). In some cases, a single-locus SNP (a variant amino acid in the gene sequence) is associated with a rich or poor microbiota, e.g. the human *MEFV* locus that encodes for pyrin (Spor et al. 2011). In fact, genetic variation in humans and mice correlates with the composition of the microbiome (Org and al. 2015), and some microbial taxa show significant heritability in studies of twins (Goodrich et al. 2014). Differences

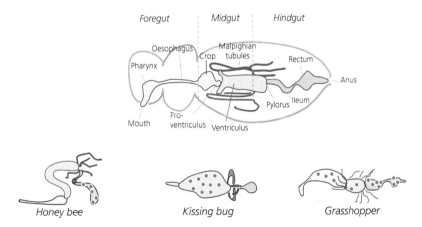

Figure 4.15 Location of the gut microbiota in insects. The top panel shows a simplified scheme of an insect gut; the sketch would be as found in a bee (the crop being the honey stomach). Below are three examples of where the microbiota is located in different insect groups (red dots). In the honeybee, this is in the hindgut and rectum; in the kissing bug, in the midgut/ventriculus; in the grasshopper, the microbiota is found in all parts of the gut at varying densities. Adapted from Engel and Moran (2013), with permission from Oxford University Press.

in general physiology can override the effects of the diet even when the ecological niches overlap, e.g. in primates (Amato et al. 2018) and humans living in the same city (Deschasaux et al. 2018). Nevertheless, diet is often an important determinant and can be stronger than genetic factors, e.g. in small rodents (Carmody et al. 2015). Hence, a variety of factors affect the composition of the microbiome and often are difficult to tease apart. Besides, interactions among different members of the microbiota will also shape its structure and location. These processes can be analysed within the framework of classical community ecology, keeping in mind that interactions within the microbiota can be rather complex, especially among bacteria. Bacteria produce common goods (e.g. siderophores for iron sequestering) or have quorum sensing (see Box 12.5). Finally, the microbiota can be seriously affected by the presence of pathogens (Reynolds et al. 2015; Burgess et al. 2017; Cortes et al. 2018).

4.8.2 Mechanisms of defence by the microbiota

In some cases, the defensive effect is due to a single microbial species. For instance, the bacterium

Hamiltonella defensa protects its aphid host from a parasitic wasp through the production of a toxin (Degnan and Moran 2008; Vorburger and Perlman 2018), and *Wolbachia* protects mosquitoes from viruses, e.g. dengue or yellow fever (Teixeira et al. 2008). By contrast, the defensive effect of an entire community of microbes—the microbiota—is more complex, and the exact mechanisms by which the protection occurs are still poorly known. Several correlates exist that link the composition and structure of the microbiota with susceptibility to infection, e.g. as shown for the effect of a more diverse microbiota against trypanosomes in bees (Mockler et al. 2018) or some diseases in humans (Luca et al. 2018). Nevertheless, this does not always prove causation, and how the microbiota affects these processes remains of prime interest.

The defensive effects of the microbiota can be direct or indirect (Table 4.7). Direct, active interference with other resident microbes or invaders involves, for example, the release of bacteriocins that differ considerably in the spectrum of other bacteria against which they work. Not only the gut microbiota but also the nasal and other microbiota provide colonization resistance by the production

Table 4.7 How microbiota affects host defences.[1]

	Organism	Mechanism, effect of microbiota	Source
(a) Direct effects			
Competition for nutrients	Mouse, *C. rodentium*, Humans, *E. coli*	Resident microbiota outcompetes pathogen with respect to carbohydrates (sugars) required for growth. Commensal *E. coli* outcompete pathogenic strains for sugar utilization. 'Passive competition'.	5, 15, 19, 26
Biotoxins	Honeybee	Bacteria in ileum (gut) produce biotoxins to kill competitors. 'Interference competition'.	2
Bacteriocins	Mouse	Lactobacillales species produce various substances, such as nisin (against Gram-positive bacteria), Abp118, lactocillin. *Clostridium thuringiensis* produces thurin, primarily affecting *C. difficile*.	14, 21, 26
Antimicrobial proteins		*Bacteroides* spp. secrete antimicrobial proteins that inhibit growth of other, similar species.	23
Reactive oxygen species (ROS)	Mosquito, *Plasmodium*	*Enterobacter* produces ROS and/or inhibits parasite defence system. Also *Wolbachia* infections induce ROS and immune effectors.	9, 11, 16
Bile salts	Mouse, *Clostridium*	Metabolic activity of microbiota enriches bile salts, which inhibit bacterial growth.	1, 5, 26
Direct cell contact	Humans	Type-VI secretion systems allow interference upon cell contacts. Resident communities of Bacteroidales share same types of system and resist colonization by others.	28
Peritrophic matrix	Mosquito	Microbiota supports building of matrix that hinders dissemination of bacteria into body cavity.	22

	Organism	Mechanism, effect of microbiota	Source
(b) Indirect effects			
Increase basal response level	Mosquito, *Plasmodium* (Malaria)	Stimulation of AMP production.	13
Increase basal response level	*T. gondii*	Priming of response to *T. gondii* by TLRs.	3
Treg cells, Anti-inflammation, Th2 response	Humans	Microbiota-derived histamines suppress IL-18 and AMPs that predispose to gut inflammation. Capsular polysaccharides of bacteria stimulate Treg (regulatory CD4+T-cells) to produce cytokines. Induction of Treg cells support Th2 response against nematode infections.	12, 18, 20
Cellular activity	*H. polygyrus*	Induction of Treg responses.	see 17, p.4
Inflammatory environment	Protozoa	Microbiota-derived metabolites stimulate inflammosomes to expel pathogens.	8
Antibodies	Humans, mouse	Epithelium-associated microbes stimulate production of IgA.	6
Antiviral signalling	*Drosophila*	Microbiota-derived peptidoglycans from bacteria stimulates signalling cascade and leads to resistance.	24
Th1 response	Mouse, *T. gondii*	Microbiota provides signals that trigger IL-12, IFN-g and thus the Th1 response.	3
Gut homeostasis	Humans, mouse	Microbiota contributes to regulate the gut environment, for example, as epithelial cells sense the presence of microbes, which in turn prevents inflammation or provides colonization resistance. Changes of acidity (pH value).	27
Permeability of gut mucosa	Vertebrate gut	Microbiota-driven change in the synthesis or permeability of mucosa; makes attachment to or penetration of this barrier more difficult for a parasite.	17,
Host gene expression	Mouse	Microbiome affects host gene expression, also beyond the gut, including effects on infection resistance. The exact pathways are largely unknown.	27
Tolerance effects	Mouse	Stimulation by *E. coli* reduces muscle wasting and fat loss upon *Salmonella* infection.	25

[1] See also reviews in refs 4, 7, 10, 26, 27.

Sources: [1] Allain. 2018. *Front Microbiol*, 8: 2707. [2] Anderson. 2017. *Curr Opin Insect Sci*, 22: 125. [3] Benson. 2009. *Cell Host Microbe*, 8: 187. [4] Blander. 2017. *Nat Immunol*, 18: 851. [5] Buffie. 2015. *Nature*, 517: 205. [6] Bunker. 2015. *Immunity*, 43: 541. [7] Burgess. 2017. *Infect Immun*, 85: e100101. [8] Chudnovskiy. 2016. *Cell*, 167: 444. [9] Cirimotich. 2011. *Science*, 332: 855. [10] Cortes. 2018. *Trends Parasitol*, 34: 640. [11] Dennison. 2016. *Malaria Journal*, 15: 425. [12] Donaldson. 2016. *Nat Rev Microbiol*, 14: 20. [13] Dong. 2009. *PLoS Path*, 5: e1000423. [14] Donia. 2014. *Cell*, 158: 1402. [15] Kamada. 2012. *Science*, 336: 1325. [16] Kambris. 2009. *Science*, 326: 134. [17] Leung. 2018. *Front Microbiol*, 9: 843. [18] Levy. 2015. *Cell*, 163: 1428. [19] Maltby. 2013. *PLoS ONE*, 8: e53957. [20] Ohnmacht. 2015. *Science*, 349: 989. [21] Rea. 2010. *Proc Natl Acad Sci USA*, 107: 9352. [22] Rodgers. 2017. *PLoS Path*, 13: e1006391. [23] Roelofs. 2016. *MBio*, 7: 1. [24] Sansone. 2015. *Cell Host Microbe*, 18: 571. [25] Schieber. 2015. *Science*, 350: 558. [26] Sorbara. 2019. *Mucosal Immunol*, 9: 1. [27] Thaiss. 2016. *Nature*, 535: 65. [28] Verster. 2017. *Cell Host Microbe*, 22: 411.

of bacteriocins (Zipperer et al. 2016). The indirect effects of the microbiota on host defence are subtler but also more diverse than the direct effect (Table 4.7). During an infection, the host's immune genes are expressed in an orchestrated manner, and this can be modulated by the microbiota (Näpflin and Schmid-Hempel 2016). Such effects on host gene expression are presumably widespread (Sorbara and Pamer 2019) and, in parts, connected to the sensing of microbes by host innate immune receptors (Sommer et al. 2015). Differential expression of transcription factors, the binding to chromatin, and methylation of histones are likely processes in this context, presumably mediated by metabolites derived from microbiota (Thaiss et al. 2016).

Hosts themselves regulate their microbiome (Thaiss et al. 2016). For instance, human or mouse epithelial cells that line the intestine express an extensive repertoire of immune receptors recognizing microbes (Pott and Hornef 2012). Their expression and signalling pathways ensure the integrity of the epithelial barrier, which prevents effects such as spontaneous inflammation and a possible breach, followed by dissemination of pathogens through the gut wall. Regulation can be complex, as illustrated by the presence of the protein NLRP6 in epithelial cells. NLRP6 regulates the interaction of host and microbiota with the production of IL-18, the expression of AMPs, and the secretion of mucus. With NLRP6-deficient mice, the composition of the microbiota changes and makes the host more susceptible to enteric infections (Elinav et al. 2011). The protein is involved in antiviral defence, too (Wang et al. 2015b). As mentioned, the host genotype is associated with the composition of its microbiota. Similarly, the expression of host genes depends on regulatory loci that respond to the presence of the microbiome. This corresponds to a kind of microbiome–host genotype interaction (Luca et al. 2018). Understanding the protective function of microbiota remains a dynamic field and a great challenge (Dheilly et al. 2019).

4.9 Evolution of the immune system

4.9.1 Recognition of non-self

As soon as the first cellular organisms had evolved, the problem of distinguishing self from non-self must have emerged. Mechanisms that can defend a single-celled organism by internal defences (e.g. with RNAi's and CRISPR-like mechanisms) and requiring intracellular receptors would probably have evolved first. Simple multicellular (metazoan) organisms would have relied on more localized cellular defences. Mobile cells that can phagocytose foreign material were presumably among the earliest such defence systems. At this point, the recognition of non-self had to be delegated to receptors located at the cell surface. Receptors in diverse taxa do indeed share deep molecular homologies, whereby the lectins, LRRs, Igs, EGFs, and complement proteins are the most widespread and presumably among the oldest receptor families (du Pasquier 2018).

Because the challenges to early organisms were already pressing and rather diverse, the immune defences were likely to already be fairly sophisticated early in life. For example, a fundamental process of allorecognition is intolerance to foreign tissue, e.g. the rejection of grafts, which is already present in very early Metazoa (e.g. sponges). Their histocompatibility ('tissue compatibility') response is not based on the same MHC system as in vertebrates, however (Flajnik and du Pasquier 2008). Nevertheless, sponges and Cnidaria already possess a complex histocompatibility system with polymorphic genes (Flajnik and du Pasquier 2008). The Bryozoa show allorecognition, too, with partial tolerance for kin-related tissue, as in *Celleporella hyaline* (Hughes et al. 2004b). Allograft rejection was also the original observation that led to the discovery of the vertebrate MHC-linked mechanism, notably from allografts given to pilots suffering from severe burns during the Second World War.

During the evolution of organismic diversity, the mechanisms of allorecognition and the ability to distinguish self from non-self must have diversified, too. On the one hand, the relatively constant PAMPs on the parasites are matched by generalized PRRs by the host. But as diversification of epitopes on parasites progressed—selected by hosts that could recognize them—the host's receptors had to be diversified, too. In the process, gene duplication events led to the evolution of multigene families (e.g. the TLR superfamily, Ig superfamilies) that could keep up with the diversifying parasites. The same principles appear to apply to the evolution of effector molecules, such as reactive oxygen species or families of AMPs. In many cases, the immune system must also have evolved to control and regulate the microbiota. This is suggested, for example, in the case of Cnidaria (Bosch 2014). The members of the microbiota are non-self, but not all of them should be rejected. The problem of distinguishing between friend and foe among the associated microorganisms was most likely a vital selection pressure already early in the history of life, and affecting various parts of the immune system (Lee and Mazmanian 2010; Maynard et al. 2012).

4.9.2 The evolution of signal transduction and effectors

How the evolution of the signal cascades exactly happened is not well understood. However, successful pathways were preserved in many different organisms across large kingdoms, and also among different functions. For example, various variants of MyD88 are common adaptor proteins in vertebrates or insects, and the NF-κB pathways are likewise shared—in immune defences or developmental pathways. In fact, the signalling pathways as such are relatively conserved, even though the same pathways have been recruited for different functions in the various organisms. On the other hand, the participating molecules have changed and diversified (du Pasquier 2018). Furthermore, immune defences increasingly were delegated to specialized cell populations in the various tissues, such as haemocytes, monocytes, macrophages, or B- and T-lymphocytes, which share deep homologies in their development.

4.9.3 The evolution of adaptive immunity

The innate immune system is a loose collection of independent units (e.g. complement, phagocytosis, TLR pathways, and so forth), showing enormous diversity among organisms, and yet able to communicate with one another for an efficient response. By contrast, the advanced adaptive immune system is a tightly integrated unit, centred on B- and T-lymphocytes and directed by the MHC (du Pasquier 2018). Perhaps, for this reason, the adaptive arm of the immune system of the jawed vertebrates has been the archetypical immune system until recently. However, today it is more appropriately seen as 'just another', although very remarkable, addition to the defence repertoire of organisms. So, when did this remarkable component appear in evolutionary history? Unfortunately, neither cells, molecules, nor immunological mechanisms fossilize as readily as do skulls or bones. Instead, the evolutionary events have left traces in the genome and the organization of immune systems of extant organisms. The evidence for the evolutionary steps in the immune system, therefore, comes from comparative immunology that has made enormous progress with the advance of genome sequencing projects (Flajnik and du Pasquier 2008, 2013) (Figure 4.16).

A common theme among the vertebrates is the evolution of somatically diversified lymphocytes as the basis for receptor diversity, for instance from lampreys and hagfish (the jawless fish, Agnatha), to the gnathostome cartilaginous fish (the Chondrichthyes), and to the higher vertebrates. In the jawless

vertebrates, receptor diversity is by VLRs. How the VLR lymphocyte specificities are selected during cell development remains unclear, though. Furthermore, the key molecules of the advanced adaptive immune response, such as MHC class I and MHC class II molecules, as well as T- and B-cell receptors (TCRs) only appear with the jawed vertebrates (du Pasquier 2018; Kaufmann 2018). Hence, depending on vertebrate class, diversification of the receptor repertoire is based on different mechanisms (Schatz 2007; Yuan et al. 2014). In particular, the integration of *RAG1/RAG2* genes (recombination-activating genes) by horizontal transmission from RAG transposons into the vertebrate line, was a crucial event in the evolution towards the gnathostome vertebrate adaptive immune system. RAG sequences from the genomes of molluscs, sea urchins, and cephalochordates (*Amphioxus, Branchiostoma*) suggest that these genes are ancient. The origin of the modern adaptive immune system, therefore, follows from a long previous history (Flajnik and du Pasquier 2008; du Pasquier 2018; Zhang et al. 2019), rather than in an 'immunological big bang'. Note that the transposon activity of RAG genes can threaten genome integrity and is therefore also a dangerous acquisition. The RAG transposon was 'domesticated' to become a useful RAG recombinase in the jawed vertebrates but kept its transposon activity in the protochordates and other invertebrates (Zhang et al. 2019).

The first adaptive immune system based on RAG activities appears in the early Chondrichthyes (the cartilaginous fish, e.g. sharks), some 450 million years ago (Figure 4.16). Virtually all elements were in place at that time. They had evolved gradually before and were then newly recruited for the immune function (Flajnik and du Pasquier 2008). A similar picture arises for the other key process of adaptive immunity. For example, the thymus became the important lymphoid organ only in the jawed vertebrates, whereas an anatomically homologous region in the pharynx of lampreys assumes similar functions for the VLRs. With the amphibians, the canonical process of Ig class switching evolved, allowing for a further finetuning of the response in different tissues and various infections. Gene conversion in the heavy and light chains starts with birds, and then with mammals. Specialized

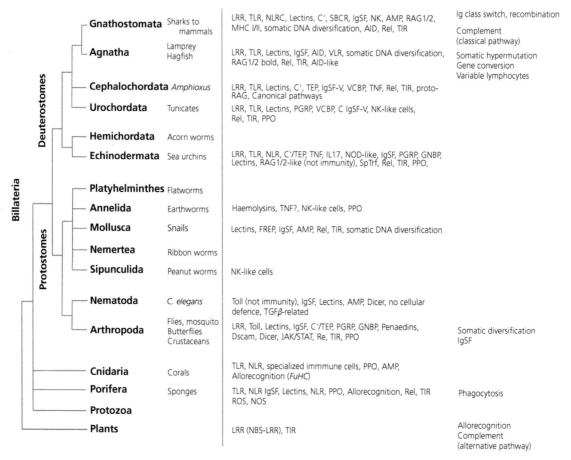

Figure 4.16 The evolution of immune systems in animals. The phylogeny of large groups of organisms shows the different elements of immune defences (molecules, genes, cells). Terms can indicate the existence of some kind of somatic change and diversification. Others are elements found in all groups. On the right, the approximate emergence of some elements is indicated. For complement system (with the three pathways—alternative, lectin, classical) information from Nonaka (2001); other elements after Flajnik and du Pasquier (2008). Note that not all groups necessarily have kept the same system in their evolutionary history, even though the system might have come into existence earlier. For immunological acronyms, see Glossary. Redrawn from Flajnik and Du Pasquier (2004), with permission from Elsevier.

germinal centres for B-cell maturation finally evolved in mammals, too (Flajnik and du Pasquier 2004; du Pasquier 2018). Among all gnathostome vertebrates, the organization of the MHC locus is remarkably different (Kaufmann 2018), which makes it difficult to trace the origin and evolutionary history of the MHC. It appears that a backbone of the MHC is an ancient, linked region with similarities in insects (*Drosophila*), cephalochordates (*Amphioxus*), and even placoderms (*Trichoplax*, Suurväli et al. 2014). Two rounds of genome dupli-

cations occurred basal to the line leading to the jawless vertebrates, leaving MHC paralogue regions elsewhere in the genome. These events might thus have freed up copies of linked regions that could be coerced for the adaptive arm of the immune defences, but this hypothesis remains controversial (Flajnik and du Pasquier 2008; Kaufmann 2018). It is, however, fair to say that many gaps in our understanding of this history persist and leave many questions unanswered (du Pasquier 2018; Kaufmann 2018).

Important points

- Hosts avoid infections by changes in behaviour, diet, habitat choice, and by having body walls. Post-infection defences include changes in behaviour or life history. However, immune responses are a major defence in all organisms.
- Immune systems have a humoral and a cellular arm, and an innate and adaptive arm. Whereas the innate arm is a collection of various defence systems, the adaptive response of a jawed vertebrate is a tightly integrated unit. In all cases, signalling cascades ensure that a recognized infection leads to a suitable response. Cascades, but not the participating molecules, are relatively conserved.
- Immune systems are based on a limited number of protein families, such as IgSF, LRRs, lectins, and AMPs. Molecules within these families are divergent among organisms and sometimes have diversified very fast.
- A high diversity of recognition molecules must cope with very diverse parasites. Diversity is encoded in the germline as genetic polymorphism. However, somatic diversification of receptor specificities, via gene conversion, rearrangement, somatic hypermutation, alternative splicing, RNA editing, and copy choice, dramatically increase the repertoire.
- The microbiota adds to defence in various ways, either through direct or indirect effects of defensive microbes. This protects the gut and various surfaces from infection.
- Memory for a past infection that allows for a more efficient response the second time is highly evolved in the adaptive arm, but also exists in innate immune systems. Memory is probably ubiquitous and an old achievement.
- The primary immune defence capacities evolved very early in the history of life. Evolution of all systems has been gradual and often coerced existing molecules and processes for novel functions. The most advanced adaptive immune system, based on B- and T-cells and the MHC, evolved with the gnathostome vertebrates.

CHAPTER 5

Ecological immunology

5.1 Variation in parasitism

It is a common observation that not all individuals of a given population carry the same loads of parasites: some individuals are heavily infected while others remain uninfected, which, in statistical terms, means that parasites are typically 'overdispersed', or 'clumped' (Anderson and May 1985). For example, in a study of the incidence of infection by malaria and other febrile diseases among schoolchildren from the coastal district of Kilifi, Kenya, considerable variation was observed (Mackinnon et al. 2005). Using analysis of variance (ANOVA), the observed differences could be statistically partitioned according to the contribution of different factors to provide insight into the possible causes for this variation (Figure 5.1). The significance of factors differed somewhat between cases with mild or severe symptoms, but 'household' and 'genetics' always explained a large part of the variation. A large part of the variation remained unexplained (factor 'Unknown'), which reflects the fact that no study can ever consider or explain all possible effects. The study suggests, perhaps not surprisingly, that whether or not an individual is infected depends on many factors, some with more and some with less weight. To partition the infection status of a set of individuals into different known or assumed factors is a general principle and useful to guide further research into causation (Figure 5.2). Furthermore, variation in infection and health status among individuals is reflected in the differences among the trajectories that the same infection takes through the disease space of various individuals. Such differences can result from differences in resource allocation to the defence (Box 5.1).

5.1.1 Variation caused by external factors

External factors include exposure, environmental conditions, behaviour, or other general host characteristics (Table 5.1). For example, potential hosts can reduce the risk of exposure to parasites by avoiding specific locations, habitats, or times of day (section 4.1.1). Besides, parasite loads vary according to sex, social status, food supply, nutritional status, or environmental conditions (e.g. chemical milieu, pH, humidity). Among those, temperature is probably the most important single factor. This is especially true for ectotherm animals, where body temperature typically follows ambient temperature (Thomas and Blanford 2003; Catalán et al. 2012; O'Connor and Bernhardt 2018). Temperature can affect not just infection success, but also the effect that the parasite has on its host, the parasite's 'virulence' (see Chapter 13). For example, the bacterial parasite *Pasteuria ramosa* is very virulent for its host, *Daphnia magna*, at an ambient temperature of 20–25 °C, where infection leads to sterilization. In a colder environment (10–15 °C), the infection is benign, and almost all infected females can produce offspring (Mitchell et al. 2005).

Note that 'temperature' is often a summary term that contains the effects of many temperature-dependent processes in a given environment. As a consequence, temperature not only scales up or down but also has non-linear effects on the outcome (Thomas and Blanford 2003). Social environment, social status (e.g. rank in a dominance hierarchy), mating strategy, group size, crowding, or sex (male, female) are additional parameters that co-vary with parasite load. Finally, parasite load varies with genetic factors, for example known genetic variants

Evolutionary Parasitology: The Integrated Study of Infections, Immunology, Ecology, and Genetics. Second Edition.
Paul Schmid-Hempel, Oxford University Press. © Paul Schmid-Hempel 2021.
DOI: 10.1093/oso/9780198832140.003.0005

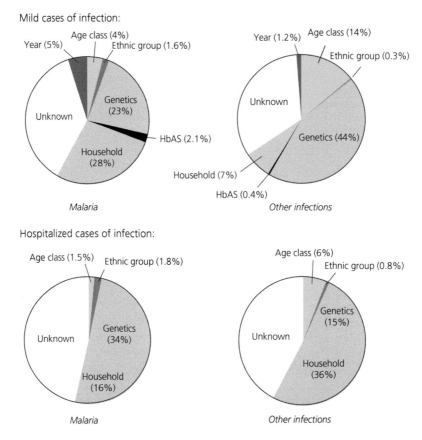

Figure 5.1 Variation in the incidence of malaria and other febrile diseases (e.g. measles, meningitis, gastroenteric, and acute respiratory infections) among schoolchildren in Kenya. Children were classified by age class, household (typically, three to six adjacent houses involving different fathers and mothers), year of study, ethnic group, and genetic parameters. *Top row:* observed cases with mild symptoms from a total of 640 children (aged ten or under) from 77 different households monitored for approximately three years. *Bottom row:* Hospitalized cases (i.e. severe symptoms) from a cohort of 2914 children observed over four years. The graphs show the proportion of variance explained by the different factors. For example, the household from where the child comes explains 28 per cent of the differences (having malaria or not) among all cases of malaria in the study of the top row. Factor HbAS refers to allelic variants of the haemoglobin S-gene. Redrawn from Mackinnon et al. (2005), under CC BY.

or genetic markers like single-nucleotide polymorphisms (SNPs). This was the case of Kenyan children (Figure 5.1) but is also observed in many other host populations, such as Soay sheep and their helminth parasites (Brown et al. 2013).

5.1.2 Variation in immune responses

Environmental conditions affect immune defences in many ways. Experimental infections under controlled conditions are therefore needed to control for these confounding effects. Experiments show that immune responses vary consistently with

environmental parameters such as temperature, condition (e.g. stress, nutritional status), social position (e.g. high vs low ranking), sex, age, or geographic region, to mention only some influences (Table 5.2). Some of these parameters affect the parasites more than the host—for example, by having low infectivity under high temperature—or they reflect a combination of effects, e.g. temperature combined with differences in nutrient supply.

The analysis of variation in immune defences with environmental conditions and, in particular, as a consequence of the underlying individual investment 'decisions', is a domain of 'ecological immunology'

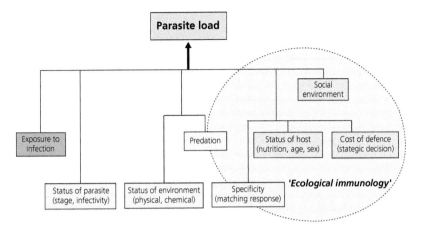

Figure 5.2 Parasite load varies due to a number of factors. The sketch shows how various effects can contribute towards the observed parasite load. Effects might be hierarchically partitioned, such as when predation and environment act together. Any of these factors can in principle interact with others. The analysis of factors in the grey encircled area crudely delimits the domain of 'Ecological Immunology'. This field investigates, for example, what causes lead to differences in infection levels between the sexes, and how this might depend on the sex-specific differences in the immune defence strategies.

Box 5.1 Disease space and costs of defence

A central argument in evolutionary parasitology is that the evolution and deployment of immune defences (as well as any other defence) is associated with costs. These costs are paid because the respective resources are not available for another need of the organism, e.g. growth or reproduction. The best defence strategy should thus balance these costs against the benefits of defence while compromising on other needs. The costs are visualized by a change in the various domains of the disease space. For example, a host organism can invest in defence mechanisms that allow recovery from infection. The investment can be made visible by an enlargement of the 'recovery' domain in disease space. If so, it becomes more likely that a given infection trajectory reaches this area and brings the host back to a healthy status. On the other hand, recovery becomes more unlikely when the 'recovery' domain gets smaller because the corresponding

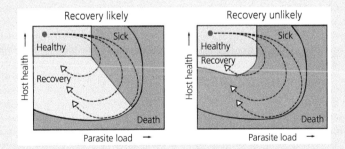

Box 5.1 Figure 1   Different investment into recovery mechanisms. *Left panel*: Hosts that invest much into defence mechanisms for recovery will enlarge the 'recovery' domain in disease space. It thus becomes more likely that an infection trajectory (broken lines) reaches the recovery zone. *Right panel*: Low investment into recovery mechanisms makes it less likely that an infection trajectory leads to recovery. Red dot is start of infection.

continued

Box 5.1 *Continued*

mechanisms have not received many of the resources (Figure 1). Note that it could also be assumed that the allocation of more resources into fast recovery shows up as a higher speed in traversing the domains, leading towards 'recovery'. Regardless, costs and investment into defence will affect the structure of the disease space.

The importance of reaching a given region of the disease space is illustrated by mice experimentally infected with the bacterium *Listeria monocytogenes* (Lough et al. 2015). The

mice came from different stocks—that is, were genetically different—but each individual mouse showed a trajectory through disease space that followed the same general pattern. A significant difference, however, was that some mice reached a zone that resulted in surviving the infection. In contrast, others crossed a boundary that led them to the 'death' domain in which they eventually perished (Figure 2). Investment into defences that would keep a host away from this boundary will therefore lead to resistance or tolerance to infection.

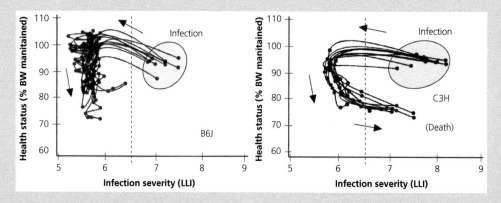

Box 5.1 Figure 2 Infection trajectories and eventual outcome. Each trajectory is for a single mouse and starts with infection (top-right of panels); dots mark days post-infection. After infection, the mice move first to the left in disease space (i.e. towards lower infection severity) before dropping down towards lower health along the health status axis, and finally back towards higher infection severity again. However, the outcome of the infection is different, depending on whether the mouse remains left of a boundary (at roughly 6.5 LLI; broken line), as in the left panel (all of the B6J-strain mice), or whether the boundary is crossed (right panel) upon which the individual will eventually die (all of the C3H-strain mice). Incidentally, these data also show the importance of the genetic background of mice (B6J, C3H) in their chances of surviving this infection. Health status is given as percentage body mass maintained (y-axis). Infection severity, LLI, is measured by the current light intensity of the (engineered) bioluminescent bacteria; this is a measure of infection intensity. Adapted from Lough et al. (2015), with permission of The Royal Society.

(Folstad and Karter 1992; Sheldon and Verhulst 1996) (Figure 5.2). For example, controlled experiments with nestlings of collared flycatchers (*Ficedula albicollis*), cross-fostered among biological and surrogate parents, showed that 5.7 per cent of the variation in the immune response was due to the environment, that is, due to differences in the nests, and 18.2 per cent due to genetic differences among the biological parents. From the remainder, the largest part (69.9 per cent) was unexplained variance, i.e. due to factors not quantified in the study (Cichon et al. 2006). The pattern is similar to the results of

Figure 5.1. A general difficulty of such studies remains in that there are many different immune responses. Without at least some knowledge about the underlying mechanisms, the relevant response may not be found (Forsman et al. 2008).

Among all effects, nutrition and food intake are important for immune responses. Poor nutrition generally increases susceptibility to disease in humans (Kelley and Bendich 1996), or in farm animals such as poultry (Klasing 1998) or sheep (Kahn et al. 2003). Changes in nutrition can also be directly associated with the responsiveness of the

Table 5.1 Infection status, parasite load, and inferred causal factors.[1]

Host[2]	Finding	Source
Daphnia magna (E)	Infection levels (*Pasteuria ramosa*) and within-host parasite multiplication reduced with a high C/P-nutritional ratio (carbon: phosphorous).	5
Agricultural plants (O, E)	Different nutrients such as N, Zn, P, etc. have varying effects on disease resistance and tolerance. High levels of N lead to more severe symptoms of infection by obligate parasites but weaker symptoms with facultative parasites.	4
Tengmalm's owl (O, E) (*Aegolius funereus*)	In poor years, all owls infected by blood parasite (*Trypanosoma avium*). In intermediate years, females in poor condition and parents with large broods had higher parasite loads. Experimental supplementation of diet reduces parasite load.	6
Pea aphid (E) (*Acyrthosiphon pisum*)	Susceptibility of aphid clones to fungal infections (*Erynia neoaphidis*) varies with temperature, but in a non-linear fashion.	3
Amazonian primates (O)	Species that sleep in sheltered tree holes have lower rates of malaria infection than species sleeping in the open, even when controlled for group size and phylogeny.	8
Olive baboon (*Papio cynocephalus anubis*) (O)	Troops with contact to humans have higher loads of schistosomes.	9
Bumblebee (*Bombus terrestris*) (E) Honeybee (*Apis mellifera*) (E)	Multiply mated females have colonies with lower parasite loads. Worker patrilines differ in susceptibility.	1, 2, 10, 11
Drosophila melanogaster (E)	Infection success with fungus (*Beauveria bassiana*) depends on genotype and varies with site of origin (Africa vs temperate region).	12
Water python (*Liasis fuscus*) (O)	Load of blood parasites (*Hepatozoon* spp.) varies with polymorphism of MHC alleles (typed with restriction *fragment* length polymorphism—RFLP—fragments), and with presence of a specific C-fragment. Lowest parasite load is with an intermediate number of fragments.	7
Sticklebacks (*Gasterosteus acculateus*) (O)	Diversity of infecting parasites and MHC types and diversity varies across lakes. Intermediate MHC allelic diversity correlates with minimum parasite load.	13

[1] Factors considered: Nutrition, diet, temperature, exposure, mating strategy, genetics. [2] O: observational, E: experimental study.

Sources: [1] Baer. 1999. *Nature*, 397: 151. [2] Baer. 2003. *Ecol Lett*, 6: 106. [3] Blanford. 2003. *Ecol Lett*, 6: 2. [4] Dordas. 2008. *Agron Sustainable Dev*, 28: 33. [5] Frost. 2008. *Ecology*, 89: 313. [6] Ilmonen. 1999. *Oikos*, 86: 79. [7] Madsen. 2006. *J Evol Biol*, 19: 1973. [8] Müller. 2001. *Folia Primatol*, 72: 153. [9] Müller-Graf. 1997. *Parasitology*, 115: 621. [10] Palmer. 2003. *Naturwissenschaften*, 90: 265. [11] Seeley. 2007. *Proc R Soc Lond B*, 274: 67. [12] Tinsley. 2006. *Parasitology*, 132: 767. [13] Wegner. 2003. *J Evol Biol*, 16: 224.

immune system, such as with food-supplied anti-oxidants (Beck et al. 2004; Sorci and Faivre 2009). Likewise, hosts can respond to infection by increasing their intake of energy (Moret and Schmid-Hempel 2000) or important nutrients (Ponton et al. 2011a) to compensate for the extra cost of the infection. Nutrition and immunity are intertwined in complex ways, and many nutrients are known to affect disease susceptibility and defences (Noland and Noland 2020). The host's immune system depends on the availability of energy, proteins, and nutrients, which are shared with the parasites, and therefore have direct effects on the infection. Furthermore, the gut microbiota is important for host defences (section 4.8) and is affected by nutrition,

too (Ponton et al. 2013). The framework of 'nutritional immunology' therefore offers new perspectives for understanding variation in defence levels and predicting the combination of nutritional components that would provide the best defence level. There is also considerable potential for medical applications (Ponton et al. 2013; Raubenheimer and Simpson 2016; Noland et al. 2020). As the examples in Table 5.2 illustrate, many other factors affect immune defences. For example, males generally have higher parasite loads and weaker immune defence (Roved et al. 2017; Kelly et al. 2018) (Chapter 6). There are also social environment matters, mediated by social rank, dominance, or density of aggregations (section 4.1). The sheer presence of

Table 5.2 Variation in immune response and inferred causal factors.[1]

Host[2]	Immune response[3]	Observation, causal factors	Source
Autumnal moth (*Epirrita autumnata*) (E)	Encapsulation, PO activity.	Encapsulation response lower when larva reared on high-quality plant leaves. No change in PO activity.	12
Kittiwake (*Rissa tridactyla*) (E)	Immunoglobulins, phytohemagglutinin (PHA) test.	Immunoglobulins decrease seasonally due to shortage of food supply. Seasonal decrease occurs in PHA test but not affected by food supply.	10
Sand martin (*Riparia riparia*) (E)	PHA test.	Food plentiful: body mass correlates positively with immune response. Food scarce: body mass correlates negatively with immune response.	5
Snail (*Lymnaea stagnalis*) (E)	Phagocytosis.	Predator attacks mimicked by touching the snail. Retraction into shell leads to haemolymph being expelled. When then exposed to water containing parasites, survival and fecundity decreased. Repeated predator avoidance behaviour reduces ability to defend against parasites.	19
Damselfly (*Lestes viridis*) (E)	Haemocyte count, PO activity.	Photoperiod experimentally manipulated to mimic short or long season, in combination with high or low food supply. Short season reduced PO activity. Low food reduced both measures of immune defence.	20
Damselfly (*Lestes viridis*) (E)	PO activity.	Experimental manipulation of hatching date (early vs late), photoperiod (normal vs delayed), and predation risk (fish vs no fish). PO activity reduced with short developmental time (as given by hatching date and photoperiod) and under predation risk.	24
Deer mice (*Peromyscus*) (O, E)	Clearance of experimental bacterial infection. Antibody response to novel antigen.	Across species, the two responses are negatively correlated. Complex relationship to life history parameters such as age at maturity, litter size.	15
Ground squirrel (*Spermophilus tridecimlineatus*) (O)	Parameters of intestinal immunity (lymphocytes, leukocytes).	Percentage of B-cells higher, T-cells lower in hibernating squirrels compared to summer. Higher levels of cytokines (TNF-α, IL-10, mucosa response).	13
Neotropical bird species (O)	Natural antibody titre, complement.	Across 70 species, higher antibody titre in birds with longer development times (incubation period). Complement activity correlates with clutch size (short incubation period).	14, 18
Barn swallow (*Hirunda rustica*) and European passerines (O)	Humoral immune response (antibody titre). Cellular response (PHA test).	Investment in nest building (volume of nest material) correlates with immune response within *H. rustica* and across bird species.	22
Sparrows (Emberizidae) (O)	Acute-phase response upon injection of LPS.	Seasonal variation in acute-phase response of males, especially with respect to mating and breeding season.	17
Drosophila melanogaster (O, E)	Resistance (survival) to experimental infection with different bacteria.	Resistance varies with origin of population. In particular, individuals from areas with species-rich bacterial communities showed higher resistance.	7
Sticklebacks (*Gasterosteus aculeatus*) (E)	Oxidative burst, granulocytes, lymphocyte proliferation.	Two ecotypes (river, lake) differ. Immune response of river type higher when exposed to lake instead of river conditions, but also harbour higher parasite loads in lakes than the lake types in same habitat.	21
White-crowned sparrow (*Zonotricha leucophrys*) (E)	Antibody titre after challenged with diphtheria–tetanus vaccine (humoral). PHA test for cell-mediated immune response.	Even in same experimental photoperiods, males from lower latitudes show stronger humoral response. No difference in cellular response.	16
Bees (O, E)	Antimicrobial activity of body surface against standard bacteria.	Increased antimicrobial response in species with large colony sizes and close within-nest relatedness.	25
Ant (*Cataglyphis velox*) (O)	PO-enzyme activity.	Foraging workers have higher activity of a key enzyme than nest workers. Perhaps as prophylaxis to exposure to infection.	4

Host[2]	Immune response[3]	Observation, causal factors	Source
Mealworm (*Tenebrio molitor*) (O)	Melanization, PO cascade.	Living in dense groups (crowding) leads to more active PO cascade and cuticle melanization.	2
Zebra finch (*Taenipygia guttata*) (E)	PHA test. T-cell-mediated immune response.	Low-ranking males have lower immune responses.	11
Cooperatively breeding birds (O)	PHA test.	Among 66 species, the 18 cooperative breeders have higher cell-mediated immune responses.	23
Small rodents (O, E)	Lymphocyte titres, IgG.	Social stress reduces immune functions; studies have varied interaction patterns, housing as individuals or in groups.	3
Yellow throat (*Geothlypis trichas*) (O)	PHA test. T-cell-mediated immune response.	Nestlings from extra-pair fathers have higher response than their half-sibs from within-pair father. Effect only visible in the colder year.	9
Cichlid fish (*Pundamilia nyererei*)	Antibody response to sheep red blood cells.	When rival males are present, nuptial colouration more intense; more colourful males have lower antibody response as compared to males with no rivals.	8
Leafcutter ants (*Acromyrmex echinatior*) (O, E)	Antibacterial response.	Males have lower immune response, also after experimental infection by fungus.	1
Collared flycatcher (*Fidecula albicollis*) (E)	PHA test.	Besides environmental factors, there is an effect of origin, i.e. from which biological parent a nestling descends, on the strength of the immune response.	6

[1] Factors considered: Diet, life history, predation, habitat, social factors, mating strategy, sex. [2] O: observational, E: experimental study. [3] Swelling after injection of PHA (phytohaemagglutinin). PO: Phenoloxidase cascade.

Sources: [1] Baer. 2005. *Insectes Soc*, 52: 298. [2] Barnes. 2000. *Proc R Soc Lond B*, 267: 177. [3] Bartolomucci. 2007. *Front Neuroendocrinol*, 28: 28. [4] Bocher. 2007. *J Evol Biol*, 20: 2228. [5] Brzek. 2007. *J Exp Biol*, 210: 2361. [6] Cichon. 2006. *J Evol Biol*, 19: 1701. [7] Corby-Harris. 2008. *J Anim Ecol*, 77: 768. [8] Dijkstra. 2007. *Behav Ecol Sociobiol*, 61: 599. [9] Garvin. 2006. *Mol Ecol*, 15: 3833. [10] Gasparini. 2006. *Funct Ecol*, 20: 457. [11] Gleeson. 2006. *Aust J Zool*, 54: 375. [12] Klemola. 2007. *Entomol Exp Appl*, 123: 167. [13] Kurtz. 2007. *Dev Comp Immunol*, 31: 415. [14] Lee. 2008. *J Anim Ecol*, 77: 356. [15] Martin. 2007. *Ecology*, 88: 2516. [16] Owen-Ashley. 2008. *Brain Behav Immun*, 22: 614. [17] Owen-Ashley. 2007. *J Ornithol*, 148: S583. [18] Palacios. 2006. *Oecologia*, 146: 505. [19] Rigby. 2000. *Proc R Soc Lond B*, 267: 171. [20] Rolff. 2004. *Am Nat*, 16: 559. [21] Scharsack. 2007. *Proc R Soc Lond B*, 274: 1523. [22] Soler. 2007. *Behav Ecol*, 18: 781. [23] Spottiswoode. 2008. *Behav Ecol Sociobiol*, 62: 963. [24] Stoks. 2006. *Ecology*, 87: 809. [25] Stow. 2007. *Biol Lett*, 3: 422.

predators can affect the investment in immune defences and modulate responses in various ways, too. For example, when female damselflies (*Coneagrion puella*) were exposed to their (caged) predators (the dragonfly *Aeshna cyanea*), haemocyte density and PO (phenoloxidase) activity increased (two measures for the baseline activity of the immune system), but this was dependent on the presence or absence of a parasite (a water mite) (Joop and Rolff 2004).

5.2 Ecological immunology: The costs of defence

5.2.1 General principles

Immune responses entail a cost, visible as a reduction in another component of the host's fitness, e.g.

reproduction. This trade-off concept is an essential argument for ecological immunology: because of limited resources, the different needs have to be weighed against each other. Therefore, the defence is a compromise, too, and not necessarily expressed at its maximum. Figure 5.3 shows a useful classification of costs, where costs are classified according to the cause (evolution, maintenance, deployment) and their implementation (genetic, physiological). The scheme captures the idea that immune defences require a physiological 'machinery' to function. 'Machinery' is a colloquial term, summarizing the molecular, biochemical, or physiological processes that together form the immune defence system—without implying that this machinery is simple, or understood in all cases. Just as any other trait, the defence machinery evolves. It thereby attains a certain degree of perfection at the detriment of another fitness-relevant function of the organism. Furthermore, once the machinery is in place, it has to be maintained even when not used. Both requirements can be considered a cost of possessing and keeping the immune defence machinery in a ready state. Hence, there is a cost of having evolved the immune defence summarized as 'evolution and maintenance' in Figure 5.3, even though these two costs may be analysed separately from one another. Once an infection has occurred, the machinery is activated to generate an immune response. Activation comes at a cost to the organism; for example, a reduction in performance of another function due to competing demands on resources. These kinds of costs can be considered costs of deployment.

Furthermore, the costs are implemented in essentially two different ways—genetic or physiological. Note, however, that genetic costs are ultimately also based on physiological processes. The difference

refers to whether the genotype fixes the costs, or whether they can be altered at the individual, physiological level, e.g. by using more or fewer resources for defence (the reality being somewhere in between). Costs entrenched in the genetic makeup of a population represent a framework of constraints for a given individual and its genotype. Any changes affecting this cost can only occur through the process of selection and evolution. Such costs are visible as negative genetic covariances between different traits. For example, some genotypes might be able to mount a robust immune response when challenged, yet they have low fecundity when no infection occurs; other genotypes might show the reverse pattern instead (McKean et al. 2008). Across the genotypes in a population, immune response and fecundity are thus negatively correlated, indicating an evolutionary cost of evolving a high capacity to respond to an immune challenge. Nevertheless, genotypic variation

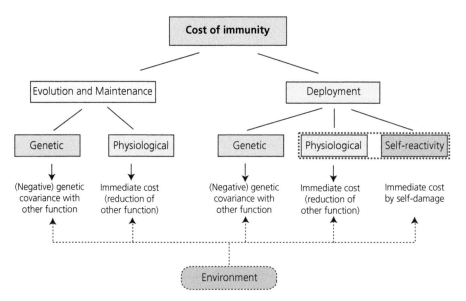

Figure 5.3 Costs of immune defences. Costs can be classified according to cause (evolution, maintenance, deployment) or according to implementation (genetic, physiological). For example, the immune defence 'machinery' must evolve in the first place. This can cause genetic costs, since an increased response capacity can only evolve at the price of a reduced performance in another fitness component (e.g. a negative genetic covariance with fecundity). The machinery can also cause physiological costs of maintenance even when not used. The system is activated upon infection (deployment); hence additional costs emerge. These again can be genetic costs, i.e. using the defence reduces another fitness component (e.g. predator avoidance). Activation can also induce physiological costs, e.g. energy consumption that is not available for reproduction. Furthermore, the costs are affected by the environment. Self-reactivity (leading to immunopathology) is, strictly speaking, a physiological cost. However, it is distinct enough to warrant its own category. Physiological costs are plastic (the individual can 'decide'), whereas genetic costs are constrained by the genotype.

can be maintained. For example, during times or at locations where parasitism is rare, the weak responders (with high fecundity) will outcompete the strong responders with low fecundity. When parasitism is common, however, strong responders are favoured, even with lower fecundity, as parasites remove the weak responders. The variation in parasitism across time and space thus sets the conditions for the maintenance of polymorphism.

Other kinds of defence costs are physiological in nature; these costs are plastic, because they arise

with an individual's 'decisions'—the bases of which, of course, have a genotypic component, too. In other words, the individual can 'choose' to allocate resources into defence or any other concurrent need, e.g. to produce eggs. Such decisions arise not only when the immune system responds after infection but also for the maintenance of defences in a state of readiness. Physiological costs of maintenance become visible, for instance, when the demand for another fitness component increases. For example, an increase in sexual activity during the

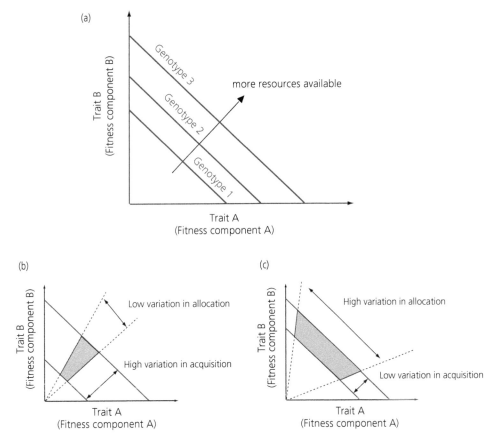

Figure 5.4 Trade-offs between different components of fitness. (a) Due to limited resources, allocation to a fitness-relevant trait A is negatively correlated with allocation to a fitness-relevant trait B, as indicated by the lines having negative slopes. However, genotypes (genotypes 1…3) might differ in their intrinsic capacity to acquire resources (arrow), and thus show trade-offs (lines) at different resource levels. (b) In this population, the genotypes vary widely in their resource acquisition capacity but show little variation in how they allocate their resources. As a result, the population-wide observation is a positive correlation between the allocations to two competing fitness components (as indicated by the reddish area). Nevertheless, this positive association is based on the same negative trade-off as in (a). (c) The situation is reversed in this population. The population-wide observation is a negative correlation between fitness components (reddish area). This is known as the 'car–house paradox', i.e. some people have large houses *and* large cars simply because they are rich. Adapted from Van Noordwijk and De Jong (1986), with permission from University of Chicago Press.

mating season would reduce the immune system's responsiveness. Whereas maintenance costs exist even in the absence of infection, additional physiological costs result from the actual deployment of immune defences. The magnitudes of these costs are generally not independent of each other. For example, the immune defence machinery might evolve towards higher capacity, thus requiring more resources for maintenance but causing lower physiological costs when deployed.

The term 'cost of defence' can be used very generally. However, to make the concept useful, benefits and costs have to be measured in the same currency, and for all steps in a defence cascade (Figure 4.1). The ultimate currency is Darwinian fitness, a combination of survival and production. However, many practical problems make it difficult or impossible to measure these elements directly. Therefore, in any specific case, the measures used are proxies for Darwinian fitness, given by practicability and the question of interest.

In any given population, genotypes might also differ in their capacity to acquire the resources, as they vary in their overall adaptation to the current environment. A well-adapted genotype will thus have more resources at its disposal than a less well-adapted one. Such variation can produce counterintuitive results. For example, even with a consistently negative relationship between resources available for immune defence (component A) or reproduction (component B), the population-wide relationship can be different (Figure 5.4). Ideally, therefore, the environment and the genotypes should be controlled wherever possible in any such study, e.g. for *Drosophila* (Schwenke et al. 2015), but this is often difficult or impossible in practice.

5.2.2 Defence costs related to life history and behaviour

Because defences can occur at different steps (Figure 4.1), defence costs are not always related to the immune system. For example, snails (*Biomphalaria glabrata*) infected by *Schistosoma* start to reproduce earlier than when non-infected, to avoid the consequences of the later parasite-induced castration (Sorensen and Minchella 2001). Although this tactic prevents a complete loss of reproduction, it

nevertheless comes at the cost of a reduced number of eggs being produced in comparison to a regular cycle (Minchella 1985). Such a strategy of 'fecundity compensation' (see section 15.1.2) also exists in nematodes infected by bacteria (Pike et al. 2019), and may perhaps occur in humans, too (Blackwell et al. 2015). Body size is another life history trait that can be protective. For example, small larvae of *D. melanogaster* are better protected against parasitoids and thus survive better to adulthood. Nevertheless, smaller larvae also have reduced fecundity as adults (Kraaijeveld and Godfray 2003).

Changing behaviour can also reduce the risk of parasitism at a cost. For example, lines of the honeybee that are bred for hygienic behaviour to contain foulbrood show reduced larval growth (Sutter et al. 1968). Similarly, lines of *Drosophila nigrospiracula* evolved towards lower fecundity (with no difference in longevity) when selected for behavioural responses to reduce parasitism by the mite *Macrocheles subdadius* (Luong and Polak 2007a). The effect was more substantial in warmer environments, and a genetic covariance between defence and fecundity was found (Luong and Polak 2007b; Luong and Polak 2007a). Not in all cases can a cost of behavioural defences easily be detected. For example, field crickets (*Acheta domesticus*) from populations with many parasitoids (and predators) not only had a better defence by staying inside the burrow for longer after an enemy attack but also higher levels of encapsulation, suggesting the absence of a trade-off between predation and defence against parasites (Kortet et al. 2007).

5.2.3 Cost of evolving an immune defence

5.2.3.1 Genetic costs associated with the evolution of immune defences

This issue refers to a 'hard-wired' trade-off between different fitness components; that is, a negative genetic covariance between different traits. For instance, the same genetic basis (i.e. particular alleles in a genotypic line) associates with an increase in the immune response and a decrease in fecundity. These costs become visible in the absence of parasitism. Then, the benefits of better defence are not realized, while the costs are still there. An example is a test with 40 hemiclones of *D. melanogaster*, constructed from a natural population. In hemiclones, one-half of the genome, inherited by one of the parents, is identical for the entire population, whereas the other half varies. Among hemiclones, a negative correlation between the fecundity of females in the absence of parasites and resistance to experimental infection when no reproduction takes place was observed (Figure 5.5). Therefore, a genotype that can resist an infection has a cost in the form of lower fecundity, which appears when no parasites are around. Note that the consequence of activating the immune system for fecundity, a measurement that would indicate the costs of deployment (see Figure 5.3), is not under scrutiny here. In this case, the trade-off is only visible in a poor environment. Genetic costs may, therefore, not always become visible as long as the environment remains sufficiently favourable. Similar costs of evolution and maintenance have been demonstrated in other studies of *Drosophila*, and in many other systems (Table 5.3).

Individuals that grow faster or end up with a larger body size generally have an advantage by, for example, being less vulnerable to predation, or having higher fecundity. In the poultry industry, chicken and turkey breeding lines are selected for higher growth rates, and sometimes for increased resistance against common diseases. In a meta-analysis of 14 studies, birds selected for larger body mass showed a significant decrease in immunocompetence (measured in various ways). On the other hand, where birds were selected for an increase in immunocompetence, the associated effects on body mass seem to be quite variable, and often but not always negative (van der Most et al. 2010).

Furthermore, a genetic cost of defence, resistance (or tolerance; section 5.5) and components of immune responses should have heritability, as is indeed observed for many cases (Table 5.4). In the example of *D. melanogaster* (Figure 5.5), the heritability of resistance to infection by *Providencia rettgeri* is 13 per cent in the food-limited environment and 8 per cent in the good environment (McKean et al. 2008). Note that—as in the example—the degree of heritability depends on the environment in which it is measured (see Box 10.3). Furthermore, defences are dynamic and unfold over various stages throughout an infection (Hall et al. 2017). In an early phase, preventing an infection from establishing should be a prime function of the defence, whereas in a later stage, the premium can be on keeping the infection under control. Therefore, genetic correlates associated with defence are also not necessarily the same for every stage (Hall et al. 2017).

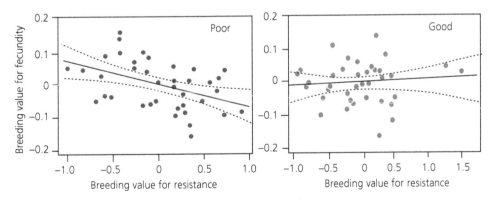

Figure 5.5 The evolutionary genetic cost of immune defence. Across different genotypes (each dot represents a hemiclone) of *D. melanogaster*, an evolutionary cost emerges as the negative correlation between fecundity in the absence of parasitism (breeding value for fecundity) and resistance to infection by the bacterium *Providencia rettgeri* (breeding value for resistance). The negative slope of the line indicates negative genetic covariance. The extent of covariance depends on the environment (*Left*: Poor; *Right*: Good environment). Adapted from McKean et al. (2008), under CC BY 2.0.

Table 5.3 The cost of evolution and maintenance of immune defences.

Cost	Organism	Immune performance[1]	Source
Selection for resistance against bacteria (*Pseudomonas aeruginosa*) (E)	*Drosophila melanogaster*	Survival post-infection as measure of resistance. Reduced longevity and egg/larval viability with higher resistance, but faster rate of development, perhaps due to pleiotropic effects.	12
Selection for delayed reproduction (E)	Mosquito (*Aedes aegypti*)	Higher encapsulation response.	6
Slower larval growth (B)	Honeybee (*Apis mellifera*)	Selected for increased resistance to bacterial disease (foulbrood).	10
Slower development, lower egg viability, but increased pupal mass (E)	Indian meal moth (*Plodia interpunctella*)	Selected for increased resistance to granulosis virus.	3
Selected for increased body mass (B)	Turkey (*Meleagris gallopavo*)	Reduced immune function.	2, 9
Lower survival of larvae (E)	Fruit fly (*Drosophila melanogaster*)	Increased encapsulation response towards virulent larval parasitoid (*Leptopilina boulardi*).	4
Fewer eggs (E)	Mosquito (*Aedes aegypti*)	Increased resistance to nematode infections.	5
Longevity, number of eggs	Flour beetle (*Tenebrio molitor*)	No visible cost of higher capacity to encapsulate a parasite.	1
Reduced competitiveness for food (E)	Fruit fly (*Drosophila melanogaster*)	Increased encapsulation response towards common larval parasitoids (*Asobara tabida*).	7, 8
Reduced competitiveness; growth rate (E)	Bacteria (*Streptococcus thermophila*)	Strains with resistance against phage 2972 have lower growth rates compared to CRISPR–Cas-deficient variants. A cost of maintenance.	11

[1] O: observational; E: experimental study; B: breeding programme in animal husbandry.

Sources: [1] Armitage. 2003. *J Evol Biol*, 16: 1038. [2] Bayyari. 1997. *Poultr Sci*, 76: 289. [3] Boots. 1993. *Funct Ecol*, 7: 528. [4] Fellowes. 1998. *Proc R Soc Lond B*, 265: 1553. [5] Ferdig. 1993. *Am J Trop Med Hyg*, 49: 756. [6] Koella. 2002. *Evolution*, 56: 1074. [7] Kraaijeveld. 1997. *Nature*, 389: 278. [8] Luong. 2007. *Heredity*, 99: 632. [9] Nestor. 1996. *Poultr Sci*, 75: 1180. [10] Sutter. 1968. *J Invertebr Pathol*, 12: 25. [11] Vale. 2015. *Proc R Soc Lond B*, 282: 20151270. [12] Yixin. 2009. *PLoS Path*, 5: 1.

5.2.3.2 Physiological costs associated with the evolution (maintenance) of immune defences

The costs of having a piece of immune defence machinery are present even if not activated. Such maintenance costs should be minimized, for example, with a general downregulation when not needed, but upregulated when environmental conditions become more difficult. During winter, many organisms experience energy shortages and have to reduce certain functions, and some have a winter diapause like bears or marmots. Perhaps to maintain defences even under low resources, a seasonal upregulation of the immune system is found in small rodents or fish during winter and when days are short (Nelson et al. 1998; Bowden et al. 2007). Similarly, mucosal IgA levels are increased in torpid and hibernating ground squirrels as compared to the summer population (Kurtz and Carey 2007).

During such 'true' hibernation, the immune system is not active, but parameters known to correlate with the capacity to respond (Ig titres, haemocyte density) show high levels, typically regulated by hormones (Martin et al. 2008). In these cases, the physiological costs are unknown, but the programmed changes make it likely that costs drive these changes. As these examples make clear, a high level of defence is not always the better solution, because a higher investment into defence also has costs. For example, immune defences might also be downregulated to reduce the risk of self-damage when the risk of infection is low. Hence, to maximize fitness, an intermediate investment into defence is likely to be superior ('optimal defence').

Balancing the activities of the innate vs the acquired immune system may be another way to minimize the overall costs. For example, the costs of evolving an adaptive immune system are probably

Table 5.4 Heritability of immune response and general resistance traits.

Organism	Response measures	Heritability (mean ± SE)	Remarks	Source
Slash pine (*Pinus elliotti*)	Resistance to infection by fungal rust (*Cronartium quercuum*).	0.21 (maximum value under 72 per cent infection prevalence).	Large data set analysed. Dominance variance low compared to additive genetic variance.	4
Egyptian cotton leafworm (*Spodoptera littoralis*)	PO activity in haemolymph (optical density units).	PO activity: 0.65 ± 0.11. Antibacterial activity: 0.63 ± 0.11. Haemocyte density: 0.36 ± 0.08.	Haemocyte density negatively co-varying with antibacterial and positively, with PO activity.	3
House cricket (*Acheta domesticus*)	Haemocyte density.	0.20 ± 0.12		8
Scorpionfly (*Panorpa vulgaris*)	Phagocytic activity.	0.83 ± 0.28		6
Drosophila nigrospiracula	Resistance to ectoparasite (*Macrocheles subbadius*)	0.152 ± 0.014	No differences between the sexes.	7
Velvetbean caterpillar (*Anticarsia gemmatalis*)	Resistance to nucleopolyhedro-virus (NPV).	means: 0.218, 0.469, 0.657	Selection regimes for increased resistance (survival) with three cycles of parasite pressure and relaxation. Heritability means calculated at each cycle.	5
Chicken (broilers)	Antibody titre against vaccines, wing web assay, clearance rate.	range: 0.06…0.53	Estimated from sire variance components. Variance among immune defence components generally negative.	1
Humans	Resistance to infectious diseases (tuberculosis, leprosy, poliomyelitis, hepatitis B).	Concordance among monozygotic twins: 32–65 per cent.		2

Sources: [1] Cheng. 1991. *Poultr Sci*, 70: 2023. [2] Cooke. 2001. *Nat Rev Genet*, 2: 967. [3] Cotter. 2004. *J Evol Biol*, 17: 421. [4] Dieters. 1996. *Silvae Genet*, 45: 235. [5] Fuxa. 1998. *J Invertebr Pathol*, 71: 159. [6] Kurtz. 1999. *Proc R Soc Lond B*, 266: 2515. [7] Polak. 2003. *J Evol Biol*, 16: 74. [8] Ryder. 2001. *J Evol Biol*, 14: 646.

higher than those for the innate system, but using the adaptive system has lower costs than with the innate system (Råberg et al. 2002; McDade et al. 2016); an evolutionary trade-off between these two systems can indeed be observed (Wegner et al. 2007). Such considerations are also part of the 'pace-of-life hypothesis'. It proposes that 'fast-living species' (with low survival rates but high fecundity, generally having high basal metabolic rates) should invest more in non-specific and innate defences. 'Slow-living species' (with high survival rates and low fecundity, low metabolic rates), by contrast, should invest more in specific and acquired immune defences (Lee 2006). Theoretical predictions about the life span and investment in immune defences are involved, but life span sometimes does seem to affect the best investment in either of the two systems (Miller et al. 2007; Sandmeier and Tracy 2014).

5.2.4 Cost of using immune defences

5.2.4.1 Genetic costs associated with the deployment of immune defences

In the example of *D. melanogaster* (Figure 5.5), infected females also lay fewer eggs, i.e. have reduced fecundity after using the defence. This reflects the costs of deployment. However, there were no differences between genetic lines in the magnitude of this reduction and, hence, no measurable genetic cost of deployment. However, challenged females show lower body mass when measured sometime after the treatment, and this difference varies according to genetic line (McKean et al. 2008). This represents a genetic cost of the immune response as measured by a loss in body mass. Similarly, larval *D. melanogaster* that better survive infection by parasitoids (*Asobara tabida*) have less resistance to desiccation and starvation when tested as adults. This capacity varied among different genetic lines (Hoang 2001), again suggesting a genetic cost of defence deployment.

5.2.4.2 Physiological costs associated with the deployment of immune defences

Physiological costs reveal themselves either when another demanding activity temporarily reduces the capacity of the immune defence system, or when activating the immune system leads to a loss in some other fitness component. This type of cost is the easiest to measure and has, therefore, attracted most of the attention (Schwenke et al. 2015). Both approaches demonstrate physiological fitness costs of using the immune defences (Tables 5.5 and 5.6). An example is workers of the bumblebee, *Bombus terrestris*. They typically do not reproduce by themselves but work for the colony, for example by foraging for energy and nutrients. On the one hand, when foraging was made more costly, the immune defence was compromised (Figure 5.6a). On the other, when the immune defence was experimentally activated, the workers died earlier (Figure 5.6b). Many other examples show a trade-off between using immune defences and reproductive activities

(Schwenke et al. 2015). Physiological costs of immune defences are therefore always present. However, these will sometimes not become visible, such as when animals compensate by feeding more (Moret and Schmid-Hempel 2000; Schmid-Hempel 2003).

An infection might also lead to a drastic loss of metabolic reserves in a process called 'wasting'. In *D. melanogaster* infected by *Mycobacterium marinum*, insulin activity regulates the degree of wasting (Dionne et al. 2006). These processes might lead to effects such as disease-induced morbidity but could also function in mobilizing more resources for defence. Physiological costs are also of relevance in 'exercise immunology', i.e. when investigating the effect of excessive physical activity on immune responses, such as is the case in athletes (Gleeson et al. 2004; van Dijk and Matson 2016). Typically, intensive training decreases measures of innate and adaptive responses but is not necessarily associated with higher disease levels (Walsh and Oliver 2016). Physical exercise might improve health status via altering of the microbiota (Monda et al. 2017). Moreover, how episodes of exercise are patterned seems a significant determinant for the eventual outcome (Simpson et al. 2015a).

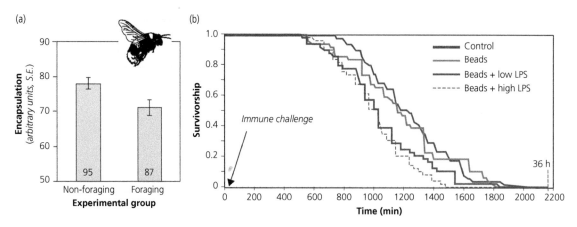

Figure 5.6 Physiological costs of immune system deployment in the bumblebee, *Bombus terrestris*. (a) A high workload reduces the strength of the immune response. Here, workers that could forage showed lower levels than their counterparts that did not work as hard ('non-foraging'). Adapted from König and Schmid-Hempel (1995), with permission from The Royal Society. (b) Activating the immune response reduces survival. In this case, the immune system was challenged either with high or low doses of LPS (initiating the antibacterial response), with sephadex beads (triggering the encapsulation/phagocytosis response), or with a combination thereof (see legend). As compared to sham-injected controls (heavy grey line), any immune challenge significantly reduced survival. Survival after an initial challenge was monitored for 36 h. The animals were food restricted (withholding sugar water). Adapted from Moret and Schmid-Hempel (2000), with permission from the American Association for the Advancement of Science.

Table 5.5 Physiological cost of using immune defences: demand on another fitness component increased.

Induced cost[1]	Organism	Observed effect on immune defences	Source
Clipping wing feathers (E)	Tree swallows (*Tachycineta bicolor*) in the wild	Higher workload: lower humoral response (and later egg laying) with clipped wings.	6
Mated vs unmated beetles (E)	Mealworms (*Tenebrio molitor*)	PO-enzyme activity reduced with mating activity.	13
Males exposed to female scent (E)	Mouse (*Mus musculus*)	Exposed males lose more body mass during an infection by *Salmonella enterica*, even when equally able to clear infection and with same diet as control males. Suggests energetic costs of controlling infection.	14
Enlarged clutch size (E)	Flycatcher (*Ficedula spp.*) in the wild	Lower cell-mediated immune response (PHA test) in females with larger broods. Reduced antibody titre against antiviral vaccine. Higher loads of blood parasites.	2, 7, 10, 11
Decreased brood size (E, O)	Great tit (*Parus major*)	Lower loads of blood parasites (malaria) with smaller clutches. Higher nestling body mass correlates with more blood parasites.	12
Mechanical disturbance during 15 min (E)	Oyster (*Crassostrea gigas*)	A decrease in several immune response parameters during stress periods; increases again 30–40 min afterwards.	8
Chronic stress (O)	Arctic ground squirrel (*Spermophilus parryii plesius*)	Lower immune response of males during mating and breeding season.	1
Stress due to captivity (O)	Zebra finch (*Taenopygia guttata*)	Wild birds have lower cellular immune response (PHA test), but higher leucocyte titres than freshly caught birds kept in confinement for the first ten days.	3
Intense physical activity by chasing birds around.	Zebra finch (*Taenopygia guttata*)	Delayed response to PHA test when birds are chased shortly before injection, but no difference after 2 h.	4
Protein-rich vs protein-poor diet (E)	House sparrows in captivity (*Passer domesticus*)	Increased cellular but decreased humoral response with protein-rich diet.	5
Food restriction vs *ad libitum* food (E)	Deer mice (*Peromyscus maniculatus*)	With low food intake: testing in a second challenge shows lower antibody titre (IgG), thus reduced immune memory.	9

[1] O: observational, E: experimental study.

Sources: [1] Boonstra. 2001. *Ecology*, 82: 1930. [2] Cichon. 2000. *Oecologia*, 125: 453. [3] Ewenson. 2001. *Naturwissenschaften*, 88: 391. [4] Ewenson. 2003. *Anim Behav*, 66: 797. [5] Gonzalez. 1999. *J Anim Ecol*, 68: 1225. [6] Hasselquist. 2001. *Behav Ecol*, 12: 93. [7] Ilmonen. 2002. *Oecologia*, 130: 199. [8] Lacoste. 2002. *Dev Comp Immunol*, 26: 1. [9] Martin. 2007. *Am J Physiol Regul Integr Comp Physiol*, 292: R316. [10] Moreno. 1999. *Proc R Soc Lond B*, 266: 1105. [11] Nordling. 1998. *Proc R Soc Lond B*, 265: 1291. [12] Ots. 1996. *Proc R Soc Lond B*, 263: 1443. [13] Rolff. 2002. *Proc Natl Acad Sci USA*, 99: 9916. [14] Zala. 2008. *Behav Ecol Sociobiol*, 62: 895.

5.2.4.3 Costs due to immunopathology

An important cost of an immune response to infection is the risk of self-damage (Graham et al. 2005; Sorci and Faivre 2009). For example, the insect PO cascade protects against invading parasites such as eggs and larvae of parasitoids. However, the activation of the cascade leads not only to the production of melanin used in the encapsulation process but also to the release of cytotoxic intermediates such as quinones or reactive oxygen species, which can seriously damage own tissue (Nappi and Ottaviani 2000; Sugumaran et al. 2000; Wang et al. 2001). Self-reactivity is also observed away from the actual site

of a challenge, e.g. when beetles respond to an immunogenic challenge (Sadd and Siva-Jothy 2006).

Self-reactivity is well known in the form of auto-immunity in vertebrates. As a defence response unfolds, self-reactive oxygen and nitrogen species are produced, similar to the situation in insects. These reactive molecules are typically kept in check by antioxidants, but failure to do so results in oxidative stress and potential damage to own tissue (Halliwell and Gutteridge 1999; von Schantz et al. 1999; Sorci and Faivre 2009). Consequently, adding antioxidants (especially carotenoids) to the diet of greenfinches (*Carduelis chloris*) alleviates the oxidative stress of these birds (Horak et al. 2007), and has

Table 5.6 Physiological cost of using immune defences: immune response stimulated, another fitness component surveyed.

Immune challenge[1]	Organism	Consequences for other fitness component	Source
Injection of LPS or Sephadex beads (E)	Bumblebee (*Bombus terrestris*)	Survival rate reduced but only under food-limited conditions.	5
Infection by parasitoid wasp (*Asobara tabida*) (E)	*Drosophila melanogaster*	Larvae that have survived attack show reduced capacity to withstand desiccation and starvation as adults. Genetic (isofemale) lines vary in this capacity.	3
Injection of diphtheria–tetanus vaccine (E)	Blue tit (*Parus caerulus*)	Increased basal metabolic rate. Lower antibody titre in birds also exposed to cold temperatures.	9
Injection of antigen (sheep red blood cells) (E)	Great tit (*Parus major*)	Increased basal metabolic rate and leukocyte stress index. Loss of body mass.	6
Injection of tetanus vaccine (E)	Pied flycatcher in the wild (*Ficedula hypoleuca*)	Reduced foraging effort and thus fewer offspring.	4
Experimental infection with microfilariae (E)	Mosquito (*Armigeres subalbatus*)	Reduction of egg development.	2
Inducing encapsulation, or wounding (E)	Various insects	Increase in metabolic rate when encapsulating.	1
Stimulating immune defences by elicitors, or pathogens	Various insects	Reduction of fecundity (fewer eggs, etc.) in most studies.	8
Injection of diphtheria–tetanus vaccine (E)	Blue tit (*Parus caerulus*)	Females reduce feeding of nestlings.	7
Injection of antigen (sheep red blood cells) after first brood (E)	Starling (*Sturnus vulgaris*)	No effect on parameters of reproductive success for a second brood.	10

[1] E: experimental study.

Sources: [1] Ardia. 2012. *Funct Ecol*, 26: 732. [2] Ferdig. 1993. *Am J Trop Med Hyg*, 49: 756. [3] Hoang. 2001. *Evolution*, 55: 2353. [4] Ilmonen. 2000. *Proc R Soc Lond B*, 267: 663. [5] Moret. 2000. *Science*, 290: 1166. [6] Ots. 2001. *Proc R Soc Lond B*, 268: 1175. [7] Råberg. 2000. *Ecol Lett*, 3: 382. [8] Schwenke. 2015. *Annu Rev Entomol*, 61: 239. [9] Svensson. 1998. *Funct Ecol*, 12: 912. [10] Williams. 1999. *Proc R Soc Lond B*, 266: 753.

similar effects in crustaceans (Babin et al. 2015). Also, lymphocyte-based adaptive immunity, absent in invertebrates, is a major source of autoimmune reactions in the vertebrates (Sarvetnick and Ohashi 2003). As discussed in section 4.6, to avoid such complications, lymphocyte lines with affinities to self that are too high are eliminated before they can exert damage. Due to many limitations in the process, elimination is not perfect, thus leaving some potentially dangerous lymphocytes circulating in the bloodstream (Müller and Bonhoeffer 2003). The generality of autoimmune reactions has led to the hypothesis that the immune system, rather than being inactive as the default status, is, in fact, active but needs to be continuously and actively suppressed when not needed, to avoid self-reactivity.

5.3 The nature of defence costs

Although costs of immune defences exist, it is often challenging to identify them (Lochmiller and Deerenberg 2000; Martin et al. 2008). In many cases, when immune responses are activated, no change in energy consumption, behaviour, growth, fecundity, or many other observable traits occurs (Williams et al. 1999; Lozano and Ydenberg 2002; Verhulst et al. 2005; Colditz 2008; Penley et al. 2018), and deployment costs only appear under very adverse conditions (Figure 5.6b).

5.3.1 What is the limiting resource?

The nature of the limiting resources that cause the defence costs are often unclear. On the one hand,

Table 5.7 Estimates of the energetic cost of immune defences.

Organism	Immune response or challenge[1]	Estimated cost[2]	Source
Cabbage butterfly (*Pieris brassicae*)	Nylon filament implanted in diapausing pupae.	Standard metabolic rate increases by 8 per cent on day 3 after challenge.	4
Insects (*Tribolium, Acheta, Cotinis, Periplaneta*)	Antigenic challenge, wounding.	Metabolic rate increases up to 28 per cent. Degree of encapsulation and metabolic rate positively correlated among individuals.	1
Collared Dove (*Streptopelia decaocto*)	Humoral response upon SRBC challenge.	8.5 per cent increase in basal metabolic rate (BMR) (on day 7) and loss of body mass. By comparison: cost of thermoregulation = 5.3 per cent per 1 °C.	3
Great tit (*Parus major*)	Overwintering birds in the wild: humoral response upon SRBC challenge.	8.6 per cent increase in BMR (on day 7). Loss of body mass.	11
Blue tit (*Parus caeruleus*)	Diphtheria–tetanus vaccine.	8–13 per cent increase in BMR (on day 7), generally low costs.	13
Green finch (*Carduelis chloris*)	Humoral response upon SRBC challenge.	No effect on BMR, but birds fed *ad libitum* in aviary.	6
House sparrow (*Passer domesticus*)	Cell-mediated.	29–32 per cent increase over resting metabolism.	8
Zebra finch (*Taenopygia guttata*); Guinea pig (*Cavia porcellus*); White-footed mouse (*Peromyscus leucopus*)	Humoral response.	No effect; metabolic rate lower.	2, 9, 12, 14
Shore birds	Experimental challenge by antigens (tetanus, diphtheria)	13 per cent increase in BMR in Red Knots during secondary antibody response; but 15 per cent reduction in Ruffs.	10
Laboratory rat	IL infusion. Inflammation.	18 per cent, 28 per cent increase over resting metabolism	7
Laboratory mouse	KLH challenge (keyhole limpet haemocyanin injection).	27–30 per cent increase over resting metabolism.	7
Gerbils (*Gerbilus*)	Infection by fleas.	Infection leads to 3 per cent (*G. andersoni*) to 16 per cent (*G. dasyurus*) increased average daily metabolic rate.	5

[1] SRB: Sheep red blood cell injection. [2] BMR: Basic metabolic rate.

Sources: [1] Ardia. 2012. *Funct Ecol*, 26: 732. [2] Derting. 2003. *Physiol Biochem Zool*, 76: 744. [3] Eraud. 2005. *Funct Ecol*, 19: 110. [4] Freitak. 2003. *Proc R Soc Lond B*, 270: S220. [5] Hawlena. 2006. *Funct Ecol*, 20: 1028. [6] Horak. 2003. *Can J Zool*, 81: 371. [7] Lochmiller. 2000. *Oikos*, 88: 87. [8] Martin. 2003. *Proc R Soc Lond B*, 270: 153. [9] Martin. 2008. *Philos Trans R Soc Lond B*, 363: 321. [10] Mendes. 2006. *J Ornithol*, 147: 274. [11] Ots. 2001. *Proc R Soc Lond B*, 268: 1175. [12] Pilorz. 2005. *Physiol Behav*, 85: 202. [13] Svensson. 1998. *Funct Ecol*, 12: 912. [14] Verhulst. 2005. *J Avian Biol*, 36: 22.

genetic costs result from negative genetic covariances among traits. For example, a genotype that ensures effective immune defence may not tolerate desiccation—as in the example of *Drosophila* selected for resistance against parasitoids (Hoang 2001). The negative genetic covariance by itself does not, however, reveal the nature of the limiting resource or the limiting process.

5.3.1.1 Energy

The prime limiting resources for physiological costs are either nutrients or energy. In the example of the

bumblebee, *B. terrestris* (Figure 5.6b), the only difference between workers that survived the immune challenge and those that died prematurely was access to sugar water. Given the physiology of bees (where sugars are the primary energy source), this points to energy as the limiting resource (Moret and Schmid-Hempel 2000).

Many studies have tried to quantitatively estimate the energetic cost associated with immune responses by measuring metabolic rates, especially in vertebrates (Table 5.7). These estimates vary widely, from observing no effect to an increase of

30–50 per cent over the resting metabolism, and perhaps up to 200 per cent or more with severe illness in human patients. Even mild challenges, such as those caused by vaccination, can increase the basal metabolic rate by 15–30 per cent in humans. Also, for every 1°C of fever, an increase of 5–13 per cent is observed (Baracos et al. 1987). Studies in birds have typically found an approximate 8 per cent increase in metabolic rate for humoral responses (e.g. antibodies against an injected antigen), similar to moderate levels of thermoregulation. In the collared dove, the energetic cost of mounting a humoral response against sheep red blood cells (SRBC), summed up until day 15 post-injection, was estimated to be 69.15 kJ, approximately the same as the energy content of an average egg (70 kJ) (Eraud et al. 2005). Studies on costs in the wild are scarce but suggest considerable costs. For example, shags (*Phalacrocorax aristotelis*) with high loads of endoparasites have a 10 per cent higher energy consumption and spend 44 per cent less time in flight (Hicks et al. 2018). Not all studies, however, have found deploying an immune response to have significant energy demands (Table 5.7). Approximately half of these energy costs are probably due to protein synthesis as the immune response unfolds (Lochmiller and Deerenberg 2000). To fuel the energy metabolism, 'glucogenesis' is upregulated during infections. Glucogenesis is the production of glucose from metabolizing proteins, carbohydrates, and lipids. Energy is also needed to mature the immune system and convert it into a state of readiness. The demand is not well known and might be small. The mass of lymphocytes in birds, for example, accounts for only about 1 per cent of the total body mass, yet cell populations renew at a high rate. Experiments with lymphocyte-deficient mice showed a decrease in the basal metabolic rate by approximately 10 per cent (Råberg et al. 2002).

5.3.1.2 Food and nutrients

These can be limiting resources in their own right. For example, vertebrates must synthesize proteins such as immunoglobulins. The body's stores partially cover this demand. In the process, muscles become proteolysed to recruit proteins needed to deploy and sustain an immune response. Although protein synthesis increases upon infection, protein use (catabolism) also increases, e.g. in septic rats by 40 per cent; in humans, N-excretion is 160 per cent of normal. Carbohydrates and lipids, too, are used to produce glucose (via glucogenesis), and they eventually must be replaced as well. These demands explain why the activation of the immune response is often associated with a loss of body mass. Severe infections in humans are associated with a loss of 15–30 per cent of body mass and a 20 per cent reduction of skeletal muscle mass. Similar values are found for pigs (body mass loss of 21 per cent when vaccinated), or chickens (18 per cent loss when exposed to injections of antigen) (Lochmiller and Deerenberg 2000). Illness-induced anorexia leads to body mass loss, too, but at the same time may shift metabolic pathways to free resources (lipids) for immune defence (Adamo et al. 2010).

The overall balance of, or the presence of, essential nutrients is a critical issue. For example, humans that develop fever require vitamin A in higher quantities to compensate for the loss by urinary excretion (Lochmiller and Deerenberg 2000).

Larvae of the moth *Spodoptera littoralis*, given a high-quality protein (supplementary casein) diet, survived better and had more heavily melanized cuticles than their counterparts on a low-quality diet (with zein instead of casein as a supplement) (Lee et al. 2008b). In fact, an optimized nutrient balance now emerges as one of the key elements for health and the defence against parasites (Cotter et al. 2011; Leulier et al. 2017). Animals themselves seem able to adjust their diet. For example, when larvae of *Spodoptera* could choose their diet, those that increased their protein intake were more likely to survive a viral infection than those that did not (Povey et al. 2014); a similar pattern was observed for their resistance against opportunistic bacterial pathogens (Povey et al. 2009). An active change to a protein-rich diet upon infection is quite commonly observed, as seen also in flour beetles (Catalán et al. 2011).

Pollinating insects can acquire critical compounds from flowers. For example, honeybee larvae can increase resistance to fungal disease by being fed with pollen from particular plants (Foley et al. 2012), and bumblebee workers prevent infection by dangerous trypanosome through intake of callunene from heather (Koch et al. 2019). Such active changes in diet are an example of 'self-medication' (see section 4.1), whereby hosts protect or cure themselves by changing their nutritional intake (Ponton et al. 2011b, 2013). Similar effects of nutrition, energy demand, and regulation of intake are also of importance to the vertebrate immune system and, thus, to human health (Wolowczuk et al. 2008). These observations also fit more general considerations on an optimal diet to meet various demands (Simpson et al. 2015b; Machovsky-Capuska et al. 2016), also applicable to human health (Raubenheimer and Simpson 2016).

5.3.2 Regulation of allocation

Different fitness-relevant tasks must be met at the same time, and they compete for limited resources. For instance, two tasks might use the same physiological cascade (Figure 5.7). For example, the same PO cascade determines size and darkness of the melanized wing spot in damselfly males (a trait important for attracting females), and the strength of the immune response (encapsulation, toxic molecules). The fractions of resources allocated to one or the other task are visible from the ratio of melanin

in wings vs the level of PO activity in haemolymph (or parasite load), respectively. Among males of the damselfly, *Calopteryx splendens*, a negative correlation between these quantities exists, suggesting that a (genotypically) fixed allocation pattern regulates how the resources are distributed from within the same biochemical cascade (Siva-Jothy 2000). Alternatively, each task could depend on its own physiological cascade that is competing for the same resource. In this case, the animal can 'decide', by up- or downregulation of the cascades, what fraction of the resources is used by either one of them (Figure 5.7). The allocation of resources shapes the trajectory an infection takes through disease space (Box 5.1) and, therefore, the observed variation in outcome among individuals.

What mechanisms might be involved in resource allocation is often not known. In crickets (*Gryllus texensis*), for example, lipid-transporting molecules (apolipophorin III) in haemolymph are used by both the immune system (e.g. affecting susceptibility to bacterial infections) and the flight muscles (for energy). Competition for lipid transporters thus affects this balance between conflicting needs (Adamo et al. 2008). Several hormones are known

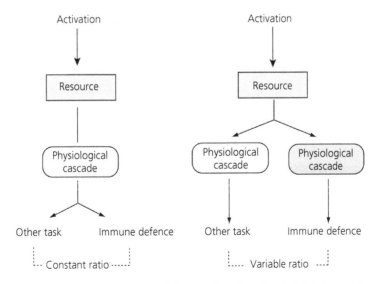

Figure 5.7 Two ways of allocating limited resources to immune defence vs other tasks. *Left*: A physiological cascade is used by both tasks and supplies each with resources in a constant ratio. The physiological cascade can only be changed by selection and evolution, but not within an individual's lifetime (allocation is by a 'constraint'). *Right*: Each task has its own physiological cascade. The individual can 'decide' on how to allocate the resources to one of the two cascades ('trade-off'). As a result, the two tasks are either performing in a constant or variable ratio. In reality, a given allocation mechanism might be a combination of both processes.

Table 5.8 Hormones that regulate immune function but also other needs.

Hormone	Function	Remarks	Source
Melatonin	Various roles. Can increase in immune function, e.g. affect proliferation of cells. Also affects reproduction. Can regulate seasonal activities.	Produced at night. A prime signal for daylength. Melatonin is present in most organisms. Important, e.g. for migratory birds.	5, 6, 10
Glucocorticoids (steroid hormones)	Regulates metabolic, cardiovascular, reproductive, and immunological functions. Can stimulate or suppress immune functions. Affects inflammation (the DTH-response; delayed-type hypersensitivity). Also affects reproduction.	Cortisol, corticosterone as examples. In vertebrates.	3, 4, 16
Gonadal steroids	Key regulators of reproduction and connects with immune function; neuro-endocrine axis. Typically immunosuppressive, but exact role often unclear. Prolactin a potential antagonist in stress responses.	Testosterone (androgens) as example. Suggested as mediator of the 'immunocompetence handicap'-hypothesis. In vertebrates.	8, 12, 14
Leptin (peptide hormone)	Stimulates humoral and cellular immune functions; generally downregulating and involved in seasonal changes. Also regulates energy stores, and reproductive activities. Secreted by adipose tissue.	In mammals.	7, 11, 12
Tyrosine derivates	Important regulators in stress responses; generally downregulate immune function.	Octopamine, norepinephrine as examples. In many organisms, e.g. chordates, molluscs, arthropods.	1, 2
Insulin (polypeptide hormone)	Key role in carbohydrate and fat metabolism. Regulates allocation of energy to different needs.	Insulin/insulin-like growth factor signalling important in insects.	9, 13, 17, 18
Juvenile hormone (JH; sesquiterpenoid hormone); 20-hydroxy-ecdysone (20E; steroid hormone)	JH downregulates immune function. Together with 20E for development, metamorphosis, egg maturation.	JH a key hormone in insects, regulating many physiological functions.	15, 17, 18

Sources: [1] Adamo. 2008. *Invertebrate Surviv J*, 5: 12. [2] Adamo. 2017. *Horm Behav*, 88: 25. [3] Bowers. 2015. *Am Nat*, 185: 769. [4] Cain. 2017. *Nat Rev Immunol*, 17: 233. [5] Calvo. 2013. *J Pineal Res*, 55: 103. [6] Carrillo-Vico. 2013. *Int J Mol Sci*, 14: 8638. [7] Demas. 2005. *Proc R Soc Lond B*, 272: 1845. [8] Folstad. 1992. *Am Nat*, 139: 603. [9] Garcia. 2010. *Proc R Soc Lond B*, 277: 2211. [10] Goldmann. 1993. In: Reiter, ed. *Melatonin: biosynthesis, physiological effects, and clinical applications*. CRC Press. [11] Lord. 1998. *Nature*, 394: 897. [12] Martin. 2008. *Philos Trans R Soc Lond B*, 363: 321. [13] Odegaard. 2013. *Science*, 339: 172. [14] Roberts. 2004. *Anim Behav*, 68: 227. [15] Rolff. 2002. *Proc Natl Acad Sci USA*, 99: 9916. [16] Sapolsky. 2000. *Endocr Rev*, 21: 55. [17] Schwenke. 2017. *Curr Biol*, 27: 596. [18] Schwenke. 2015. *Annu Rev Entomol*, 61: 239.

to modulate the activity of the immune system, and this can come at the expense of other tasks (Table 5.8). This modulation can be fast acting, such as observed for stress-induced immunosuppression, or act more in the long term; for example, with seasonal up- or downregulation of immune activity (Martin et al. 2008).

5.4 Measurement and fitness effects of immune defence

There is little doubt that immune defence is advantageous for the individual when infected by parasites. However, to which qualitative or quantitative degree the capacity for immune defence relates to eventual lifetime fitness is not easy to study. Often, the capacity of an individual to mount an immune response to a parasitic challenge is termed 'immunocompetence'. Nevertheless, keep in mind that immune responses are complex and that the different arms of the immune system are neither independent nor always positively correlating with one another. Because of the complexity, measuring host defence levels or an immune response in a standardized way is not straightforward, and various measures can be defined and used (Box 5.2). In practice, immunocompetence is often measured by responses involving the humoral and cellular arm in various

Box 5.2 Measures of host defence

Defence against parasites can have many components, and the meanings adopted in different studies can vary widely. Examples for commonly used terms are:

- *Resistance*: A generic term indicating the host's capacity to resist a parasite by various means. Resistance is typically measured as a gain in units of host fitness when exposed to parasites or, more narrowly, when infected.
- *Immunocompetence*: The capacity to mount an immune response to a challenge. Sometimes also defined more generally as the ability to withstand infection and disease. Originally considered to be a summary measure for all possible immune responses.
- *Susceptibility*: The failure to resist an infection. The term sometimes is used to denote any capacity to withstand a parasite or to avoid infection in the first place. These definitions would include elements of behaviour or habitat choice.
- *Recognition*: The host's capacity to recognize a parasite. This is typically given as a probability that the host's recognition system can identify an arbitrarily presented parasite (epitope) with a small error margin. Recognition by itself is necessary to trigger a response.
- *Recovery*: The ability of the host to survive and recover from the infection. This occurs as the host's (immune) defence system clears the parasite. Often, it is assumed that during the recovery process, the host becomes immune against reinfection (at least for a specific time period) (e.g. in SIR models; see Box 11.2). Recovery would also be observed when the host keeps the parasite below a certain damage level.
- *Clearance*: The host's ability to eliminate the parasite from its body. On average, the clearance rate is the inverse of the duration of an infection.
- Reducing virulence: The host's ability to suppress the parasite or its effects in order to reduce the infection-induced damage.
- *Robustness*: The host's capacity to mount an effective response in the face of parasite subversion and manipulation of the immune system. A robust response is buffered against the failure of completing single steps in a given defence cascade.

manners, such as antibody titres, cellular responses in birds, the activity of the prophenoloxidase (PPO) cascade, or encapsulation rate in insects. Advances in technology now expand the toolbox and allow the expression of immune- or defence-related genes to be measured at large scales. This typically reveals the up- or downregulation of hundreds of genes but is followed by the question of which ones may be the most relevant ones. Quite generally, the relationship between measures of immunocompetence and fitness is a difficult topic due to many confounding factors, and the matter is sensitive to the measures used (Viney et al. 2005).

Although a seemingly obvious relationship, empirical studies therefore need to show that there is a benefit from being able to mount a strong immune response. Data to test these expectations suffer from many confounding variables, for example the age structure of populations, wrong identification of the relevant immune responses, or the effects of immunosuppression by the parasite. Such confounding variables might explain that sometimes even a positive association of immune response and parasite load exists. Nevertheless, many studies show the expected negative correlation. For example, higher pre-infection levels of immunoglobulins (IgE) reduce the success (fecundity, number of eggs released) of hookworms infecting humans (Quinnell et al. 2004). Finally, a higher immune response is also expected to provide higher fitness. For example, in damselflies, PO activity (indicative of the strength of the melanization–encapsulation response) correlates positively with the number of copulations (i.e. fitness) relative to the population mean (Rolff and Siva-Jothy 2004). Also, the survival of overwintering blue tits correlated with higher antibody titre against experimentally injected diphtheria–tetanus vaccines (Råberg and Stjernman 2003), again suggesting a fitness advantage for strong responders. In a comparative study across 25 bird species, clutch size correlated positively with adult immune responsiveness (Martin et al. 2001), suggesting that a response capacity, at least during certain life history stages (and summarized as 'immunocompetence'), has positive fitness effects and is likely under selection by parasites. One inherent problem with such observational studies is that they cannot distinguish between the effects of high immunocompetence per se and the effects of generally good quality of the individual; for example, a strong-responder bird also being an efficient forager.

5.5 Tolerance as defence element

Hosts cannot only 'resist' an infection, e.g., by clearing the parasite, but may instead limit the damage of infection without necessarily preventing parasite replication or clearance. This defence strategy is known as 'tolerance' (Råberg et al. 2007; Ayres and Schneider 2012; Råberg 2014) and has initially been developed in plant pathology (Caldwell et al. 1958; Schafer 1971; Roy and Kirchner 2000). Tolerance is also an example of a 'reaction norm', which defines the variation in the phenotype (i.e. a decrease in health) in various environments (with increasing parasite load) that is associated with a given genotype (Stearns 1992). Alternatively, 'tolerance' can be considered a dose–response curve, which is a more familiar concept in parasitology (Ayres and Schneider 2012).

5.5.1 Defining and measuring tolerance

Tolerance describes how parasite-induced fitness loss varies with the parasite load. 'Parasite load' can be measured as the number and density of parasites within the host (infection intensity, I), and therefore needs to be combined with the effect on host fitness (fitness, W). Hence, a measure of tolerance describes how these quantities co-vary, as formally given by the term Cov (W, I). In the simplest case, the relationship is a line with a slope and a starting point for the uninfected host with their fitness value (sometimes called 'vigour'). Instead of a linear relationship, empirical data also suggest a sigmoidal curve that can be characterized by at least three parameters (Figure 5.8). The exact shape of the curve does not matter for the concept of tolerance. However, it makes a difference for its measurement and for inferring possible causes of the decline; for example, is it caused by a microbial load or by damage from the immune response (Louie et al. 2016)? 'Resistance', by the way, would be proportional to the inverse of the average parasite load ($\sim 1/\bar{I}$), since more resistant hosts have fewer parasites.

Measuring parasite load in the tolerance concept is not as straightforward as it seems, because one could use infection dose, infection intensity at a

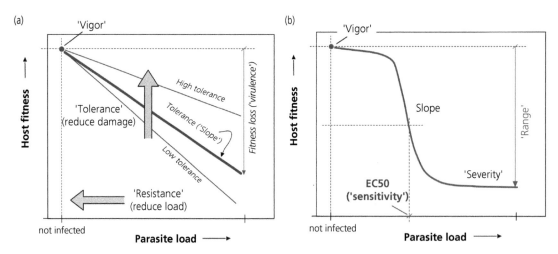

Figure 5.8 Resistance and tolerance. (a) Host fitness, e.g. measured with some indicator of host health, typically declines with an increase in the parasite load, e.g. with infection intensity (thick red line). Here, the relationship is a straight line, characterized by 'vigour' (host fitness when not infected), and 'slope' (the decrease in host fitness with increasing parasite load). The loss of vigour when heavily infected can be considered the 'virulence' effect on this host. To avoid fitness loss, the host can either reduce parasite load with eventual clearance, which is defined as 'resistance' to infection, or reduce damage and thus limit fitness loss, which is defined as 'tolerance'. Strong resistance reduces the parasite load effectively. A high tolerance keeps the negative effects of an increasing parasite load small (shallow slope of curve), low tolerance does not limit effects well (steep slope). (b) Sometimes, the tolerance curve (thick red line) is sigmoidal in shape. In this case, tolerance can be characterized by 'vigour', 'slope' (the decrease in fitness around the 'sensitivity' point, where half of the vigour is lost), 'severity' (host fitness at high parasite loads), and 'range' (the loss of vigour when substantially infected). Adapted from Louie et al. (2016), under CCBY 4.0.

certain time after infection, the maximum level achieved, or an integral of intensities over all time points (Ayres and Schneider 2012; Louie et al. 2016). These measures will be useful in different contexts. Various established methods, such as ELISA, can be made parasite-specific and offer useful solutions to study wildlife immunology (Garnier and Graham 2014). Likewise, host fitness (*W*) can also be measured in various ways—for example, by a measure of host health such as body weight or anaemia—and these measures can also be taken at different points in time. In *D. melanogaster*, for instance, host fitness was defined by the survival rate of flies (or time-to-death) for a given infective dose relative to uninfected flies (Ayres and Schneider 2008). A measure of host status was anaemia, and parasite load is parasitaemia of malaria parasites experimentally infected in mice. Different mouse strains vary in their response to the parasite burden, with some strains showing less anaemia when parasitaemia increases, i.e. showing higher tolerance according to the definition (Råberg et al. 2007).

Any such measure eventually depends on the particular host–parasite system, because infection intensity or virulence is a result of the host–parasite interaction rather than of the actions of the host or parasite alone. Furthermore, the host defence elements—resistance and tolerance—can vary independently from one another but, biologically, are typically correlated. Also, the fitness of an uninfected host is not independent of the tolerance level, since evolving a higher degree of tolerance might come at the price of lower fitness when uninfected. This can be seen in a negative genetic correlation between host fitness in the absence of infection and a measure of tolerance (Simms and Triplett 1994; Stowe et al. 2000). Moreover, increased tolerance might come at a cost to resistance so that more tolerant host genotypes might be less resistant (Stowe et al. 2000). In theory, at least, such a negative correlation between tolerance and resistance is expected to evolve under certain conditions (Fornoni et al. 2004; Restif and Koella 2004). Strains of mice that show higher resistance to experimental infections by *P. chabaudi* do indeed show lower tolerance (Råberg et al. 2007). Together, the relationship of tolerance with resistance, and the effects of different

genotypes on these characteristics, can produce a range of different outcomes (Soares et al. 2017) (Box 5.1; Figure 5.9).

5.5.2 Mechanisms of tolerance

Whereas the mechanisms of resistance largely are explained by immunological responses, the mechanisms that result in tolerance are still poorly understood. At the immunological level, tolerance mechanisms are, in fact, very diverse (Ayres and Schneider 2012). In *Drosophila*, the capacity to phagocytose mediates tolerance against bacterial infections (Shinzawa et al. 2009), whereas cytokines and inflammatory processes are important in birds (Adelman et al. 2013) or mammals (Medzhitov et al. 2012; Lippens et al. 2016). Moreover, the immune response can be weak or suppressed when the parasitic challenge is minor, which is not only an efficient use of the defence system but can also result in tolerance. Lack of an appropriate receptor and compartmentalization by separating microbes from the immune system (Hooper 2009) are possible ways to avoid unnecessary activation. When the TLR4 receptor of mice that recognizes LPS is experimentally blocked (as in mutant strains), higher doses of LPS can be tolerated (Qureshi et al. 1999). Furthermore, mice can reduce the concentration of LPS outside the cell by enzymatic degradation with alkaline phosphatase. Both mechanisms can change tolerance or resistance, depending on the circumstances (Ayres and Schneider 2012). Not least, infected and non-infected individuals follow different trajectories through disease space (Box 5.1), observations that can be used to elucidate tolerance mechanisms (Lough et al. 2015).

A further option is to control the damage inflicted by an infection, which involves conserved stress and damage responses (Soares et al. 2017), but also the use of specific compounds. For example, antimicrobial peptides (AMPs) are effective against microbes but may spare the host cells (Welling et al. 2001). Similarly, to control levels, not only effectors are activated but also their antagonists, as with, for example, reactive oxygen species and their antioxidant scavengers (Lambeth 2007). Moreover, repair mechanisms can restore tissue and thus limit damage (Soares et al. 2017). Also, diet can yield tolerance

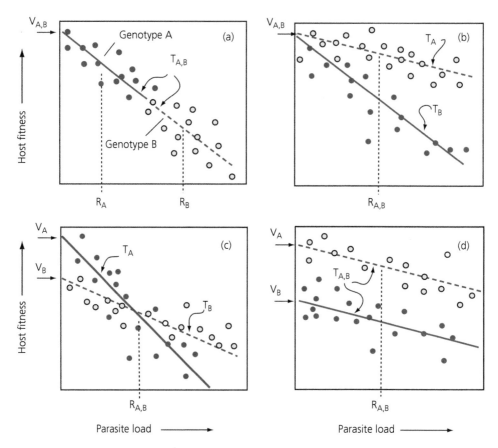

Figure 5.9 Tolerance as a defence element. In this hypothetical examples, two host genotypes (Genotype A: solid line, red symbols. Genotype B: broken line, blue symbols) are assumed. The dots represent measurements for individual hosts of a given genotype, observed in situations where they have different parasite loads. Lines characterize how host fitness depends on parasite load; this relationship could also be curvilinear. The position and slope of a line depend on the combination of various defence elements; notably, general host status or 'vigour' (V; the fitness of uninfected hosts), host resistance (R), and host tolerance (T; slopes) for a genotype (A, B). Note that the situation shown here also would depend on the parasite, since different parasitic infections might change how V, R, and T are shaped for the same genotype. The four panels show how the overall relationships depend on how the defence elements vary. (a) The genotypes differ by resistance ($R_A > R_B$) but have the same vigour ($V_{A,B}$) and tolerance ($T_{A,B}$). (b) The genotypes differ by tolerance ($T_A > T_B$) but have the same vigour ($V_{A,B}$) and resistance ($R_{A,B}$). (c) The genotypes differ by vigour ($V_A > V_B$) and tolerance ($T_B > T_A$) but have the same resistance ($R_{A,B}$). This case, incidentally, illustrates a cost in general vigour as a result of increased tolerance. (d) The genotypes differ by vigour ($V_A > V_B$) but have the same resistance ($R_{A,B}$) and tolerance ($T_{A,B}$). The consequences for the evolution of parasite virulence or host defences are different in the four cases. Adapted from Råberg et al. (2007), with permission from the American Association for the Advancement of Science.

rather than using the immune mechanisms for this purpose. For example, burying beetles (*Nicrophorus*) infected by pathogenic bacteria survived much longer on a fat-rich and protein-poor diet without any noticeable change in immune activity or bacterial load, suggesting that nutrients and energy supply lead to tolerance effects (Miller and Cotter 2018). Not least, the microbiota is also involved in generating tolerance along various routes (Soares et al.

2017). In mice, the commensal *E. coli* mediates the activation of the NLRC4 inflammasome (see section 4.3.1), which initiates a cascade that eventually suppresses the expression of proteins that otherwise would have led to muscle atrophy and loss of body mass, which are typical damages of an infection (Schieber et al. 2015). Increasing host tolerance is also a way for the commensal microbes to increase their survival and further transmission (Soares et al.

2017). Several additional mechanisms for tolerance are conceivable, including the manipulation of parasites by the host or an appropriate redistribution of energy use upon infection (Ayres and Schneider 2012).

5.5.3 Selection and evolution of tolerance

Tolerance, as a trait, just like resistance, shows genetic variation, as the steepness of the slope that defines tolerance varies among genotypes (Figure 5.8). For example, different genetic strains of mice or *D. melanogaster* vary in tolerance to the same infection (Råberg et al. 2007; Ayres and Schneider 2008). Genomic scans reveal several loci that are associated with tolerance, many of them involved in the regulation of gene expression,

especially the downregulation of immune responses (Howick and Lazzaro 2017). In humans, heterozygosity of an allele in the MHC locus (HLA-B) provides higher tolerance against HIV infections (Regoes et al. 2014).

However, tolerance is not always measured in the same way, and the same overall outcome in terms of host fitness and parasite load could follow from different defence elements (Figure 5.9). In *D. melanogaster*, a single mutant gene changes the relative contribution of tolerance or resistance to infections by various pathogens. This gene encodes for a protease that is involved in the melanization cascade (Ayres and Schneider 2008). In the unmanaged Soay sheep population on the remote Scottish island of St Kilda, tolerance (body condition vs parasite load) varied among individuals, mostly imposed by

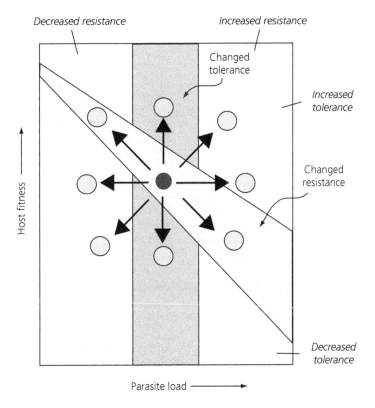

Figure 5.10 Evolution of resistance and tolerance. From an original type (dark blue circle), evolution can move the host phenotype towards one of nine basic patterns (arrows and light blue circles). The orange region corresponds to changes in resistance, with increased resistance reducing the parasite burden (x-axis) and improving host fitness (y-axis). The yellowish region corresponds to changes in tolerance. The remaining regions (grey) are based on changing both properties at the same time. Changes in any one direction can result from one or few mutations. The combined effects of resistance and tolerance are under selection in a given environment. Adapted from Ayres and Schneider (2008), under CCBY.

environmental factors. Nevertheless, longitudinal data over many years eventually showed that individuals with higher tolerance also had higher lifetime reproductive success, a sign of positive directional selection on tolerance (Hayward et al. 2014b). At the same time, the standing heritable genetic variation was small, in line with theoretical expectations (Roy and Kirchner 2000; Miller et al. 2005).

Tolerance of infection could give the parasites an advantage. It might lead not only to an increase in parasite virulence under a wide range of conditions—because the host imposes no defence and thus there is no cost for the parasite—but also to more parasites being transmitted. So how can host tolerance evolve, despite the apparent adverse effects for the host population? Theoretical studies show that this is possible under a range of reasonable conditions (Roy and Kirchner 2000). Studies in *D. melanogaster* furthermore suggest that only a few mutations of key genes might be required to shift the combination of resistance vs tolerance in various directions. Each such change would have different effects on the outcome in terms of host fitness and parasite load, which would prompt the parasite population to adapt (Figure 5.10).

However, the evolution of host tolerance also depends on the exact mechanisms of tolerance and can result in different outcomes (Miller et al. 2006). For example, when the tolerance mechanism is only able to reduce the effects of infection up to a particular parasite load, host and parasite might evolve towards apparent commensalism. No apparent effects of the infection appear at this point; nevertheless, the mechanisms of tolerance itself remain costly to the host. Seemingly paradoxical outcomes are possible in theory, too. When tolerance reduces the effect of the infection for the host, the total number of hosts dying from parasites increases in the population at large because more hosts become infected. Parasite fitness also increases when tolerance reduces host mortality rate but does not change when tolerance reduces the loss of host fecundity (Best et al. 2008). The models also suggest that tolerance, unlike resistance, may not even evolve towards stable polymorphism or a cyclic change of tolerance levels, but to multiple states that can suddenly shift. At the same time, parasites may evolve towards high virulence in non-tolerant

hosts (Best et al. 2014). Tolerance, therefore, can have complex evolutionary consequences at the population level, and the consequences vary with the exact mechanisms.

5.6 Strategies of immune defence

5.6.1 General considerations

An optimal immune defence strategy would be one that provides the best defence at the minimum cost (Viney et al. 2005). However, immune defences could have been selected for different reasons. For example, they may impede the establishment of an infection, eliminate an infection, ensure fast recovery, reduce the pathogenic effects, avoid self-damage, tolerate infections, or be robust against parasite manipulation, and can serve many other purposes. Moreover, evolution happens in an ecological setting, where the dynamics of hosts and parasites depend on the defence strategies in effect (Cappuccio et al. 2015). Differences also emerge when parasites are assumed to co-evolve with their hosts as compared to a more static scenario. Host defences should also become adapted to how frequent parasites are in the host's environment and what effect they may have. In such large-scale considerations, the particular kind of defence that evolves can be different. For example, innate defences might evolve when parasites appear sufficiently regularly (Mayer et al. 2016).

'Mathematical immunology' theoretically investigates the dynamics of immune cells, effectors, and so forth, to understand how immune responses unfold and what the best host response would be (Castro et al. 2016; Eftimie et al. 2016; Perelson 2018) (see also Chapter 12). Models based on evolutionary considerations, in particular, analyse various responses (e.g. constitutive or induced), consider their effect, and assign a cost to them. Such costs are reduced fecundity or a lower intrinsic growth rate of the host population. A large number of such models were formulated, and predictions for many detailed scenarios have been generated (Table 5.9). Parasites, in turn, are assumed to gain more transmission by replicating faster inside the host, which is assumed to cause more harm to the host (increased 'virulence'; see Chapter 13). Furthermore, the success

Table 5.9 Models of optimal investment in (immune) defence.[1]

Scenario, main question	Finding, predictions	Remarks	Source
Population dynamic (SIR-) model with resistant hosts growing more slowly in numbers (cost of defence in absence of parasite).	Outcome depends on how defence increases with investment, i.e. the costs of defence. With accelerating costs, a single ESS exists (higher resistance with lower parasite virulence). With decelerating costs, some hosts are maximally resistant and others are undefended. More resistance expected when hosts do not compete much ecologically.	Analysis of invasion into wild type population. Directly transmitted parasite. No acquired immunity, no recovery. Infected hosts do not reproduce and die earlier. No co-evolution by parasites.	1
Population dynamic (SIR-) model where hosts defend by being able to recover and parasites co-evolve in their virulence (related to increased transmission). Host defence reduces host reproductive rate (fecundity), and parasite has lower transmission rate.	When parasites cannot co-evolve, optimal recovery rate is maximal for intermediate force of infection. With high force, recovery does not pay and host resources are invested into reproduction. When parasites co-evolve, optimal recovery rate increases for low rates of host reproduction (fecundity) but bifurcates (defence polymorphic with no defence) for intermediate reproduction rates and goes to zero for high rates.	SIR model, where effect of defence affects population parameters and so generates feedback via force of infection.	5
SIR model to find ESS investment in immune defence as a function of average host life span.	Relationship of investment into defence and life span is complex and also depends on kind of defence. (a) Investment to better recovery: increase with life span up to maximum level. (b) Investment to reduce virulence effects: No defence for short life spans; then bistable solution—either none or increase as life span increases. (c) Investment to reduce susceptibility: peak investment at intermediate life spans, then decrease.	No co-evolution by the parasites assumed.	6
Optimal allocation of resources to avoid starvation and disease (by investment in immune defence) in relation to levels of reserves.	Individuals with low reserves should invest less in defence than those with high reserves; investment increases monotonically with level of reserves. When resource availability varies, individuals should invest more into defence at periods when resources are abundant.	State-dependent model of resource allocation.	3
What is best investment in recovery from infection? Two distinct costs of defence assumed: resources and immunopathology. Resource costs can be constitutive or when using the immune system. Measured as loss of birth rate, and also as mortality for immunopathology.	Benefits of recovery vanish when increasing immunopathology costs, leading to chronic infections. Also, low recovery when parasites are virulent.	An epidemic (SIR) model. Cost functions defined. Immunopathology an important cost that drives recovery/chronic infections.	2
Trade-off of investment into immune defences, resistance, tolerance, and development of larval stages in insects, especially energy allocation budgets.	Developmental interference (i.e. allocating resources to development instead of other needs) favours tolerance and higher levels of constitutive immunity. But predictions depend on age at infection.	Within-host dynamic model with compartments for resources, development, immune function, parasite numbers.	4

[1] Defence typically means resistance against parasites in various forms. None of the models explicitly consider multiple infections.

Sources: [1] Boots. 1999. *Am Nat*, 153: 359. [2] Cressler. 2015. *Proc R Soc Lond B*, 282: [3] Houston. 2007. *Proc R Soc Lond B*, 274: 2835. [4] Tate. 2015. *Am Nat*, 186: 495. [5] Van Baalen. 1998. *Proc R Soc Lond B*, 265: 317. [6] Van Boven. 2004. *Am Nat*, 163: 277.

of a given host or parasite strategy depends on the level of defences in all other hosts of a population. For example, if the average host in the population reduced its defence, the parasite would become more prevalent. If so, a higher investment into an individual's defence is worth more and should be selected for. The dependence of the success of one's own strategy on the strategy of all other individuals in the interacting population defines a game situation, as already mentioned in section 2.5.2. Game theory seeks to identify the 'evolutionarily stable strategy' (ESS) for host defence (or parasite virulence). Whatever the exact assumptions and scenarios, models can yield testable assumptions and predictions, such as those described in Table 5.9. In many other cases, models are a tool to explore possibilities and to structure the thinking on a given issue (Best and Hoyle 2013a; Boots et al. 2013).

5.6.2 Defence and host life span

Intuitively, short-lived organisms should not have much use for an elaborate adaptive immune system and immunological memory. Alas, intuition can be misleading, and this is more a myth than a justified claim (Boots et al. 2013). For example, early vertebrates were presumably short-lived creatures, and despite this, essential elements of an adaptive immune system did evolve (Rolff 2007). The argument also ignores epidemiological processes. Short life span also means a high birth rate; therefore, many new susceptible host individuals enter the host population at any one time and are available for infection. As Chapter 11 discusses, parasites can thus persist more easily and can become more prevalent. Hence, the likelihood of any individual

becoming infected and having to defend itself are higher, too. The same happens when individuals are not refractory to a second infection. As a consequence, there will be more hosts available, the parasite will spread to higher levels, and hosts will have to invest more into defence. The relationship between average host life span and the evolution of costly defences or memory is therefore not as straightforward as it might seem.

Theoretical models suggest that host life span or the timing of events in life history does relate to the choice of best immune defences, although in complex ways (Table 5.10). In particular, and contrary to the classical wisdom, short life span does not necessarily select against memory or costly defences. For example, the theory suggests that if such immune protection wanes fast, investment into defence by recovery from infection (e.g. faster clearance) should indeed increase with host life span, but not without bounds. Moreover, intermediate rather than long host life spans should select for a maximum investment (Figure 5.11). Among other things, such predictions are sensitive to epidemiological processes; for example, when immune protection is weak, parasite numbers increase and hosts are more frequently infected, which makes defences more valuable.

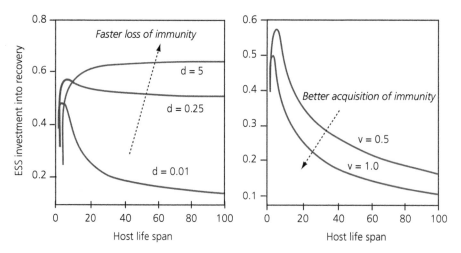

Figure 5.11 Optimal immune defence and memory. Shown is the evolutionarily stable (ESS) investment in defence for recovery from infection. In both cases, an immune memory can form. *Left:* When immune protection ('memory') is lost fast (model parameter values *d*), the ESS investment in recovery generally increases and saturates with increasing life span of the host. If memory is persistent (low value of *d*), long-lived hosts should not invest much into recovery. *Right:* Long-lived hosts should not invest much in recovery even when the likelihood that an immune memory can be formed is low (low values of model parameter *v*). Maximum investment into recovery is to be expected for hosts with intermediate life span. ESS investment and host life spans are in arbitrary units. Adapted from Miller et al. (2007), with permission from John Wiley and Sons.

Table 5.10 Examples of models for optimal investment in various kinds of defence.[1]

Scenario, main question	Finding, prediction	Remarks	Source
Find optimal B-/T-cell repertoire that minimizes probability that parasite escapes recognition vs risk of self-reactivity in B-cell repertoire.	Repertoire size (diversity of B/T-cell specificities) primarily determined by avoiding risk of self-reactivity and not by diversity of parasites.	Probabilistic model where response depends on having at least one lymphocyte that matches a given epitope.	1
Modified Lotka–Volterra dynamics (tracking different host and parasite genotypes), where hosts invest in defence at a cost to their intrinsic rate of increase; parasites pay a cost in growth (cost of virulence). Hosts have general resistance (for all parasite types) or specific resistance (according to genotype).	(a) General defence. With accelerating costs high variation in defence and lower mean defence in cases. With decelerating costs of defence, low variation and intermediate defence levels. (b) Specific defence. Large variation if parasite must pay high virulence cost to infect broad host range. A negative correlation of general vs specific defence always evolves.	Specific resistance assumed as gene-for-gene or matching alleles interaction. Emphasis on observable measures such as variation in host defence within population.	2
Allocation of resources to development or immune defences.	Depending on the developmental stage where the parasite attacks, optimal allocation varies. But if development is often prioritized, constitutive rather than induced defences should be favoured.	Developmental interference i.e. when allocation to development takes priority over other needs, is as a key regulatory process.	6
Constitutive and delayed, induced defence reduces parasite growth rate. Costs are reduction in host growth and reproduction, and self-damage. Modelled as number of defence proteins that are either stored (constitutive) or newly produced (induced).	The optimal combination of defence minimizes total cost to host with (a) constitutive defence only, favoured by fast-growing, virulent parasites, long delay to induced, cost-effective constitutive defence; (b) induced defence only, favoured by low effectiveness of constitutive defence; (c) a combination of both; if fast-growing, virulent parasites, cost-effective induced response. Model with proteins: number of stored proteins decreases with higher variation in parasite dose, i.e. uncertainty about parasite threat.		4, 5
Within-host model of immune dynamics. Constitutive defence is fixed but varies among individuals; induced (but not specific) defence produced in proportion to parasite numbers and with limited duration. Defence increases parasite mortality rate by sum of two defence types. Costs to host come from parasite load and from constitutive or induced defence, but no a priori assumptions about cost of defence.	Parasite types only differ by growth rate but not by specificity of response. With a constant type, only constitutive defences just large enough to clear infection are needed. With variable parasite types, constitutive defence increases with increasing mean (within-host) growth rate of parasite. With increasing variation in parasite growth rates, in addition to constitutive defence.	No differential costs required for explaining deployment of both defence types. In invertebrates, constitutive defence represent by PO cascade. A numerical threshold assumed, below which parasites go extinct.	3
In bacteria, constitutive defences with permanently expressed receptors. Induced response via the CRISPR–Cas system.	High resource availability selects for high constitutive, low for high induced defences.	Costs of CRISPR–Cas likely due to increased mortality or reduced replication.	7

[1] Defence typically means resistance against parasites in various forms. None of the models explicitly considers multiple infections.

Sources: [1] De Boer. 1993. *Proc R Soc Lond B*, 252: 171. [2] Frank. 2000. *J Theor Biol*, 202: 283. [3] Hamilton. 2008. *Proc R Soc Lond B*, 275: 937. [4] Shudo. 2001. *J Theor Biol*, 209: 233. [5] Shudo. 2002. *J Theor Biol*, 219: 309. [6] Tate. 2015. *Am Nat*, 186: 495. [7] Westra. 2015. *Curr Biol*, 25: 1043.

The best defence strategy is also affected by the degree of specificity stored in the memory, and how large or responsive the memory is. These parameters could depend on the number of memory cells that are stored away, and that can be reactivated when needed. The costs of producing and maintaining the immune cells, or the risk of self-reactivity, in turn, yield the costs for a memory. Under some conditions, rather than clearing a secondary infection by memory-induced responses, the infection is better restricted to a level of parasite tolerance (Shudo and Iwasa 2004). Comparative studies or test assays can help to address these predictions. Some studies confirm the hypotheses (Tieleman et al. 2005; Martin et al. 2007; Lee et al. 2008a; Previtali et al. 2012; Sandmeier and Tracy 2014), mostly where more specific measures or challenges are used (Boots et al. 2013). Tests are inconsistent or unclear in other taxa or contexts (Tella et al. 2002; Versteegh et al. 2012).

Host life span ('longevity') is by itself a trait that evolves and adapts to the organism's environment and the pattern of age-dependent extrinsic mortality (Stearns 1992). Hence, a relationship between life history and immune function is to be expected. The decline in immune functions with age, i.e. ageing or senescence, is, in fact, a common observation in almost any organism (Bueno et al. 2017). It affects different tissues—e.g. the lymph nodes and spleen (Turner and Mabbott 2017), lung (Boe et al. 2017), or skin (Chambers and Vukmanovic-Stejic 2020)—and can be caused by various mechanisms, e.g. immunopathology (Khan et al. 2017a), or differences in diet and nutrition (Simpson et al. 2017). Immunosenescence itself is a trait that can evolve. For example, Drosophila selected for longer life spans also change the age-dependent regulation of immune function. Longer life spans, in particular, alleviate immunosenescence (Fabian et al. 2018). This correlated change shows that immune defences are under selection by the overall life history pattern, especially the onset of reproduction and subsequent life span.

5.6.3 Specific vs general defence

A general immune response works against many different types of parasites, whereas a specific response is directed against a few. Examples of non-specific defences are complement or the PPO cascade of invertebrates. Specific defence occurs with the diversified B- and T-cell repertoire of the jawed vertebrates. General and specific defences complement each other to defend the host against parasites, and both strategies have their benefits and costs. General resistance is assumed to be more costly maintain (see above, section 5.2.3). Specific defence, on the other hand, is more efficient, as it

targets particular parasite types at lower costs. Theoretical analyses suggest that the two kinds of defence might evolve to correlate negatively with one another. The negative correlation between general and specific defence varies depending on the underlying genetics; a strong correlation is predicted for gene-for-gene interactions and a weak one for matching allele interactions (Frank 2000) (see section 10.5.2). What balance of general vs specific defences evolves depends on many environmental and genetic factors.

5.6.4 Constitutive vs induced defence

A defence is constitutive when it remains in a state of readiness even before the infection occurs, such as with the alternative pathway of complement (Figure 4.3). The infection activates an induced response and thus needs to be built up before it can have an effect. The time delay until the induced defence takes effect is an important parameter, and hosts are assumed to 'decide' to either respond quickly (constitutively) or accept some delay after the infection (induced). This choice becomes prominent when different time delays differ in costs, or in the range of parasites that are attacked. A purely constitutive response, for example, has a time delay of practically zero and the costs are primarily determined by maintaining the state of readiness; e.g. the evolutionary costs resulting from having the system in place. In turn, an induced defence causes additional costs when it is activated. An optimal strategy of defence might minimize the total costs—that is, the costs of maintenance plus the costs of responding—provided the parasite is thereby eliminated or controlled (tolerated). The best choice should also depend on the kind of parasite. For example, bacteria or viruses can multiply rapidly when not controlled and quickly reach a density that causes severe effects on the host.

Theoretical models, perhaps not surprisingly, typically propose a constitutive response should be used when the parasite infects with a large dose, can rapidly multiply within the host, or potentially causes severe damage for other reasons. By contrast, a combination of constitutive and induced responses is optimal: for example, when the latency to the induced response is not too long, and the

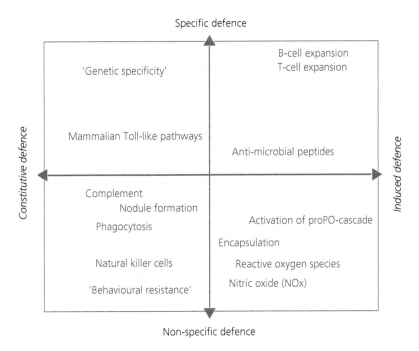

Figure 5.12 The defence chart. This hypothetical chart shows different immune defence mechanisms. These are placed according to how general or specific they are (y-axis), and according to whether they are always present (constitutive) or induced upon infection (x-axis). The actual position of the various mechanisms is not precisely known and is shown here tentatively. Adapted from Schmid-Hempel and Ebert (2003), with permission from Elsevier.

parasite does not multiply too fast (Shudo and Iwasa 2001). Investment in both kinds of defence should increase with the damage done by the parasite ('virulence'). However, the predictions for the best mix of constitutive and induced defences depends on many assumptions and many factors (Table 5.10). There is not, therefore, a simple answer. Ideally, the models should, at least, yield predictions that are empirically testable.

These considerations take the different immune responses as evolved defence strategies. Immunologists classify responses by the mechanisms, e.g. whether or not B- and T-lymphocytes are involved. However, the molecular mechanisms of an immune response are not directly 'seen' by selection. Instead, it is the consequences of a given defence for the host's fitness that matter for evolution. From this point of view, the various immunological, molecular mechanisms are not the prime issue. Rather, defences should be arranged in a 'defence chart', for example with the functional elements 'delay to response' and 'specificity of the response' as its two

principal axes (Figure 5.12). How such a chart exactly looks is not known, but it would provide a framework within which the evolution of different defence mechanisms can be meaningfully analysed.

5.6.5 Robust defence

Probably all parasites have evolved strategies to subvert, suppress, manipulate, and evade host defences (see Chapter 8). These strategies have typically evolved with specialized molecules that target specific elements of the host defence system. A 'robust' immune defence would reduce the success of these parasite interventions and nevertheless ensure an effective defence. A robust defence requires a secured direct detection of the parasite, e.g. by pattern-recognition receptors that recognize PAMPs. The recognition subsequently should trigger a defence cascade that is robust against parasite interventions. A robust defence can also be indirect. For example, perturbations of host homeostasis and damage to host cells as a consequence of the

Box 5.3 Structurally robust immune defences

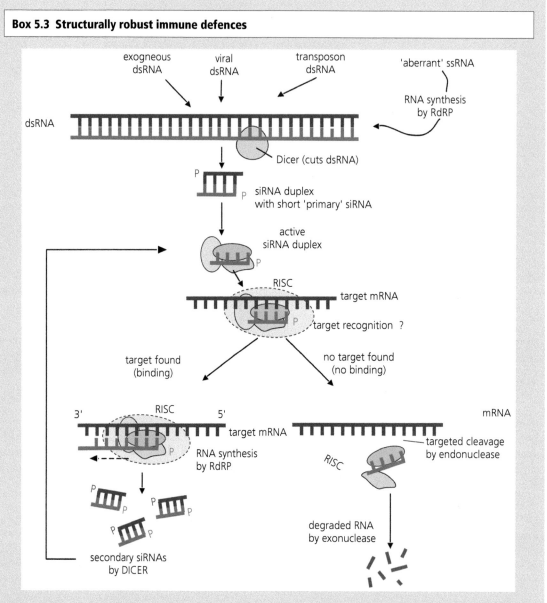

Box 5.3 Figure 1 Robust defence against viral RNA. After recognition of a double-stranded RNA (dsRNA), the enzyme Dicer cuts the original dsRNA into shorter pieces ('primary siRNA'). These are stabilized within a RISC complex. This complex can bind to a target sequence (mRNA). *Bottom left*: After the RISC/siRNA complex has bound to a target sequence, the process continues with RNA synthesis by a host-encoded RNA polymerase (RdRP). The resulting product becomes cleaved again to yield short secondary siRNAs. The whole cycle is self-amplifying, as it produces ever shorter copies of the original siRNA that are templates for further RISC complexes. Eventually, the pieces become too short and the process ends when the RISC/siRNA complex no longer finds a sequence to bind to. In this case, the process stops, and the complex plus RNA is degraded (bottom right). Various sources for dsRNA are destroyed by this system. The pathways correspond to the RNAi defence system (see Box 4.3). This scheme might be found in the nematode *Caenorhabditis elegans*. Following Bergstrom and Antia (2006); drawing adapted from Plasterk (2002), with permission from the American Association for the Advancement of Science.

Some elements of the RNAi-based immune defence against viral infections elucidate how robust immune systems can work (Bergstrom and Antia 2006). Viral infections are often associated with the presence of high doses of double-stranded RNA (dsRNA) in the host cell's cytosol. This is sensed by the immune system to initiate a cascade of events that eventually leads to the degradation of the viral genetic information. On the one hand, the combination of these cues (dose, location) is hard to avoid for the parasite. On the other, the unique features of the degradation process make it quite robust against manipulation by the virus, while keeping the risk of self-damage low.

In the RNAi-defence cascade, viral dsRNA is bound in a siRNA-duplex complex. Counterintuitively, viral RNA sequences are later copied and amplified rather than eliminated at this step (Figure 1). However, one crucial element is that the RNA polymerase (RdRP, Figure 1) generally works in a unidirectional mode. That is, RdRP only copies the genetic sequence from the 5' end to the 3' end. Because the copying process cannot start at the very end of the template sequence, each new, synthesized piece is necessarily shorter than the template. By the principle of unidirectionality, each siRNA is converted into a shorter siRNA than before. After several rounds of such amplification, the remaining siRNA in the cell corresponds to the short sequences at the downstream end of the original template viral dsRNA. At some point, these pieces can no longer bind to the mRNA, nothing is copied anymore, and all viral RNA has become degraded.

This mechanism of unidirectional amplification is one of several elements of the defence that are hard to circumvent by a viral intervention. Indeed, viruses do sabotage and interfere with the RNAi system (Li et al. 2002; Li et al. 2004b; Kemp and Imler 2009). However, this specific amplification and degradation process has its own, process-dependent stopping point that is difficult to subvert by a virus. An additional safety element exploits the fact that viruses must enter through the cytoplasm of the eukaryotic host cell before they can interact with the nucleus, their final target. However, when passing through the cytoplasm, the viral RNA inevitably triggers RNAi-defence mechanisms. The RNAi defence, furthermore, needs a certain threshold to be activated, and only viral dsRNA is normally present in large enough numbers to surpass this threshold. The complicated mechanism does not seem to be energetically very efficient but is a good safeguard against viral manipulation (Bergstrom and Antia 2006).

infection could be elicited by the detection of damage-associated molecular patterns (DAMPs). Thus, several checkpoints for sensing infections exist (Blander and Sander 2012). A somewhat controversial scenario for sensing infections by damage or by a general cue for a threat is the 'danger model' (Matzinger 2002). It seems appropriate for so-called 'sterile inflammation', where inflammation occurs in response to internal signals rather than by an actual infection (Stuart et al. 2013). However, the damage is not always a trigger in the context of infections. Therefore, danger theory is now met with some reservations (Pradeu and Cooper 2012; Józefowski 2016). Nevertheless, immune systems could become robust against parasite manipulation by being able to respond to some kind of perturbation that cannot be avoided or concealed by the parasite. For example, parasites typically use effector molecules to recruit resources, or to manipulate the host's immune system; the host could detect these. Such 'effector-triggered immunity' exists in plants, where intracellular recognition mechanisms, based on polymorphic NLR receptors, can detect parasite effectors that interfere with host defence; this activates the immune response (Cui et al. 2015). In humans, a receptor (leukocyte immunoglobulin-like receptor A2, LILRA2) has been found that can sense antibodies as they become degraded by microbes. In response, it activates monocytes and neutrophils to maintain an immune response despite manipulation by the parasite (Hirayasu et al. 2016). 'Effector-triggered' immune responses are probably widespread in animals (Stuart et al. 2013). By contrast to sensing an actual threat, immune systems could also be structurally robust, i.e. having mechanisms that make it difficult or impossible for a parasite to manipulate the defence. An example is some elements of the RNAi-based immune defence against viral infections (Box 5.3).

From these studies, a few principles emerge that allow for a robust defence (Bergstrom and Antia 2006):

1. Redundancy: a structural organization ensuring that when one component fails the whole system nevertheless can work. Such redundancy might be at the heart of the exceedingly complex regulatory

networks that are so typical for immune systems (Soyer and Bonhoeffer 2006; Salathé and Soyer 2008).

2. Distributed processing of information, e.g. along different receptors routes and cascades, can guard against total failure and is a general hallmark of robust systems.

3. 'Commitment' and 'robust action': a defence response, once started (and once having passed the stage where negative regulation could set in), should not be halted until it has run its due course. For example, infected cells are destroyed (a full commitment) rather than repaired. A repair might make it easier to subvert the system.

4. Using multiple inputs to cross-validate the state of the system before any action is taken. Models of signalling network indeed suggest that robust defences possess many independent pathways with few connections between them (Schrom et al. 2017). For this reason, the diverse toolbox of loosely connected constitutive and innate response systems should be more robust against parasite interference than the tightly integrated system of the B- and T-cell based adaptive response.

Important points

- The rate of parasitism and the level of immune response typically vary among individuals and populations. Variation is explained by ecological, genetic, and immunological factors.
- Immune defences have costs of evolution and maintenance and costs of using. These can be genetic, i.e. a negative genetic covariance between defence and other fitness components. Costs can also be physiological, e.g. consuming resources that are lacking for another task. Limiting resources typically are energy, food, and nutrients. Various hormones orchestrate the regulation of allocation.
- In addition to resistance (e.g. clearing the infection), hosts can also tolerate parasites and mitigate against the induced damage (tolerance). Both components can evolve and interact with one another.
- The capacity to immune defend relates to host fitness. Defence elements, such as specificity, delay to response, recovery from infection, or immunological memory, should evolve for the most efficient response mix. Theoretical models can predict optimal defences and allocation strategies. In general, short host life spans do not select for low investment into defence.
- Immune defences should be selected for robustness against parasitic manipulation, too. This implies structural properties such as redundancy, committed action, multiple inputs, or structural robustness.

CHAPTER 6

Parasites, immunity, and sexual selection

6.1 Differences between the sexes

6.1.1 Differences in susceptibility to parasites

In general, males age more rapidly and die earlier (Clutton-Brock and Isvaran 2007). At the same time, males are more susceptible to parasitism than females (Klein 2000; Zuk and Stoehr 2010). This holds true for almost all organisms studied so far, including humans (Figure 6.1). Not only are males more likely

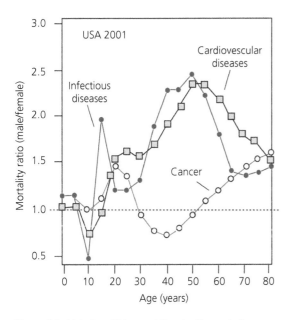

Figure 6.1 Males have higher mortality rates. The graph shows mortality rates of men relative to women (ratio > 1 means males die more frequently) for the population of the United States in 2001. Many more men than women die from infectious disease in the age classes of 30–65 years. Adapted from Kruger and Nesse (2006), with permission from Springer Nature.

to contract an infection, but they are also more likely to suffer greater infection intensity and more severe symptoms than do females. Examples to the contrary, i.e. where females are more susceptible or suffer from worse consequences of parasitism are less common. For instance, female mice are more susceptible than males to *Toxoplasma gondii*, *Schistosoma mansoni*, or *Taenia crassiceps* (Klein 2004). Similarly, comparing 33 bird studies, the common malaria-like blood parasite (*Haemoproteus*) was more prevalent in females than in males, yet only in females of polygamous species (McCurdy et al. 1998). Sex differences also extend to plants (Kaltz and Shykoff 2001; Williams et al. 2011). Some of the most pronounced differences are found in insects (Kelly et al. 2018) (Table 6.1), which suggests that specific physiologies do not constrain the pattern.

Such a male bias may be due to differences in hormones and the genetic background, or due to behavioural differences that lead to higher exposure to parasites. Males of many species partake in more risky behaviours, have larger body sizes—especially in mammals—and might have a different diet. These factors are not only associated with increased risk of accidents but also with more frequent exposure to infection by parasites (Zuk and McKean 1996). However, risky behaviour and exposure do not explain all cases (Table 6.1).

6.1.2 Differences in immune function

A sex bias in immune functions results when males or females express the same or different genes of the immune system but in different ways (Winterhalter and Fedorka 2009). Sex differences in immune

Evolutionary Parasitology: The Integrated Study of Infections, Immunology, Ecology, and Genetics. Second Edition.
Paul Schmid-Hempel, Oxford University Press. © Paul Schmid-Hempel 2021.
DOI: 10.1093/oso/9780198832140.003.0006

Table 6.1 Sex differences in parasitism.

Organism	Observation	Remarks	Source
Plant: White campion (*Silene latifolia*)	Infection by fungus (*Microbotryum violaceum*) more successful in female flowers.	Effect due to longer-lived female flowers and thus increased opportunity for fungal growth.	5
Arthropods	No significant bias across different arthropod–parasite (*n* = 61) systems; male bias in infection intensity in a few cases only (*n* = 31 systems).	Various parasites. No correction for phylogeny.	13
Insects: House cricket (*Acheta domesticus*); Flour beetle (*Tribolium* spp.); Tsetse fly (*Glossina moritans*); Japanese damselfly (*Mnais costalis*)	Higher susceptibility to parasites in males. Damselflies: Higher parasite loads in orange-winged males than in females. But clear-winged males (no sexual signal) have lower loads than either group.	Experimental infections, respectively, with bacteria (*Serratia marcescens*), rat tapeworm (*Hymenolepis diminuta*), *Trypanosoma rhodesiense*. Natural infections with Eugregarines in damselflies.	2, 3, 14, 15
Fruit fly (*Drosophila melanogaster*)	Survival upon infection by viruses, bacteria, or Microsporidia not universally biased to one sex.		1
Birds, mammals	In two-thirds of all helminth infections (nematodes, cestodes) males generally harbour more and slightly larger parasites (*n* = 48 host–parasite systems).	Corrections done for sample sizes.	11
Mammals: laboratory mouse, rat; free-living rodents	Small but consistent male bias in infection by arthropod but not helminth parasites (*n* = 149 host species). Males more susceptible to various parasites (protozoa, fungi, viruses, bacteria). IFNγ levels: higher in females when infected by *Leishmania*, but lower with *Toxoplasma*.	Both observational and experimental studies. Effect stronger for experimental studies, where associated with circulating steroid hormone concentration. Lower levels of IFNγ explains higher susceptibility to toxoplasmosis. In a few cases, females more susceptible.	6, 7, 12
Mammals (many species)	Male bias with higher parasite loads in most mammalian orders (infections by viruses, bacteria, fungi, protozoa, or helminths). Male bias found in 49 out of 58 studied parasitic infections.	Male bias larger for polygamous species. In many studies, pattern confirmed by experimental manipulation of sex hormones. Various studies under different conditions. In a few cases, females more susceptible. Males suffer higher mortality rates and recover more slowly from infection.	7–10
Humans	Ten major parasitic diseases studied. Most showed a male bias, except for severe dengue, mild leprosy, and typhoid fever. Moreover, sex-specific differences in behaviour did not generally explain the patterns, indicating a role for hormonal differences.	A study in Brazil during 2006–2009, taking into account age classes. Based on notification reports of the health authorities.	4

Sources: [1] Belmonte. 2020. *Front Immunol*, 10: Article 3075. [2] Burtt. 1946. *Ann Trop Med Parasitol*, 40: 74. [3] Gray. 1998. *J Invertebr Pathol*, 71: 288. [4] Guerra-Silveira. 2013. *PLoS ONE*, 8: e62390. [5] Kaltz. 2001. *J Ecol*, 89: 99. [6] Klein. 2000. *Neurosci Biobehav Rev*, 24: 627. [7] Klein. 2004. *Parasite Immunol*, 26: 247. [8] Kruger. 2006. *Human Nature*, 17: 74. [9] Moore. 2002. *Science*, 297: 2015. [10] Nunn. 2009. *Philos Trans R Soc Lond B*, 364: 61. [11] Poulin. 1996. *Int J Parasitol*, 26: 1311. [12] Schalk. 1997. *Oikos*, 78: 67. [13] Sheridan. 2000. *Oikos*, 88: 327. [14] Siva-Jothy. 2001. *Physiol Entomol*, 26: 1. [15] Yan. 1995. *J Parasitol*, 81: 37.

functions have been viewed with some controversy (Klein and Flanagan 2016; Roved et al. 2017), and average differences may be small, if present beyond confounding factors (Foo et al. 2017; Kelly et al. 2018). Nevertheless, differences do exist but are also very heterogeneous. Innate immune responses often seem stronger in females, whereas males sometimes have a high adaptive response. However, these patterns also change throughout the lifetime of an organism (Klein and Flanagan 2016), are not universal, and show many exceptions (Table 6.2).

In some mammalian species (including humans or rodents), females have a more robust inflammatory response, more phagocytosis, higher production of cytotoxic nitric oxides, more efficient antigen-presenting cells, and stronger adaptive

Table 6.2 Sex differences in immune defences.

Organism	Finding	Measures	Source
Field crickets (*Teleogryllus* spp.)	Males show stronger encapsulation response than females.	Encapsulation.	12
Crickets (*Gryllus texensis*)	Males have lower PO activity and resistance to bacterial infections. Difference occurs only during reproductive period.	PO enzyme activity, parasite load.	1
Wellington tree weta (*Hemideina crassidens*)	Field-caught males have stronger immune response than females. Females presumably using resources for egg production instead of defence.	Encapsulation response.	5
Cabbage white butterfly (*Pieris rapae*)	Males generally have stronger response than females. Perhaps due to necessity to produce dark wing spots with melanin.	Encapsulation response.	10
Monarch butterfly (*Danaus plexippus*)	Late instar larvae tested. Females have higher density of circulating haemocytes when not infected. When experimentally infected with a protozoan (*Ophryocystis elektroscirrha*), males have higher haemocyte concentrations.	Haemocyte concentration. But measure not necessarily related to lower adult parasite loads.	9
Fruit fly (*Drosophila melanogaster*)	After mating, expression differs between sexes. Males express more Gram-negative defences. Overall investment similar. Expression differs among paralogues of major immune gene families.	Immune gene expression.	11
Fruit fly (*Drosophila melanogaster*)	Basal activity of melanogenesis higher in males. Higher expression of Toll-pathway components in females.	Sex differences in immunity are a complex phenomenon. Mechanisms remain mostly unknown.	4
Scorpion fly (*Panorpa vulgaris*)	Stronger immune defences in females.	Haemocyte number, phagocytosis, PO activity, lysozyme-like activity.	7, 8
In many taxa (sea urchins to humans)	In females, particularly the innate immune function is higher, whereas adaptive functions (T-cells) often higher in males.	Various.	6
Humans	Stronger immune reactivity in females associated with higher incidence of autoimmune diseases (diabetes, multiple sclerosis, rheumatoid arthritis, thyroid diseases, etc.). Women have stronger immune responses (antibody titre, cell-mediated defences) than men after infection or vaccination.	Associated with differences in sex hormones. Similar findings in mice, rats, dogs. Autoimmune responses in females show remission during pregnancy.	2, 3, 6

Sources: [1] Adamo. 2001. *Anim Behav*, 62: 417. [2] Ahmed. 1985. *Am J Pathol*, 121: 531. [3] Beagley. 2003. *FEMS Immunol Med Microbiol*, 38: 13. [4] Belmonte. 2020. *Front Immunol*, 10: Article 3075. [5] Kelly. 2009. *Physiol Entomol*, 34: 174. [6] Klein. 2016. *Nat Rev Immunol*, 16: 626. [7] Kurtz. 2001. *J Invertebr Pathol*, 78: 53. [8] Kurtz. 2000. *Dev Comp Immunol*, 24: 1. [9] Lindsey. 2009. *Evol Ecol*, 23: 607. [10] Stoehr. 2007. *Ecol Entomol*, 32: 188. [11] Winterhalter. 2009. *Proc R Soc Lond B*, 276: 1109. [12] Zuk. 2004. *Can J Zool*, 82: 627.

response than males (Klein 2004) (Table 6.2). Conversely, males might sometimes mount higher levels of proinflammatory cytokines and show higher levels of NK-cell activity than females. At the same time, the incidence of autoimmune diseases is also generally higher in females than in males. In humans, for example, women are more likely to develop inflammatory rheumatic diseases than men. Similar female biases towards autoimmune-related diseases exist in mice, rats, and dogs (Klein 2000; Beagley and Gockel 2003; Klein 2004). This difference may reflect the cost of a more active immune system in females. For most cases, the molecular underpinnings of sex differences in immune function are not known. In *Drosophila*, for example, differences in the Toll pathway seem responsible for the higher susceptibility of females to bacterial infections, with sex differences in constitutive and infection-induced expression (Duneau et al. 2017b).

6.1.3 The role of sex hormones

Sex hormones can affect the functioning of the immune system. This topic has mainly been studied in vertebrates. Their effects are manifold, and hormones can up- or downregulate the components of the immune system (Table 6.3). Androgens are among the most important male sex hormone regulators of immune activity, and testosterone is the biologically most relevant androgen. Testosterone reduces the activity of NK-cells and the synthesis of proinflammatory cytokines such as TNFα, while increasing the production of anti-inflammatory cytokines such as IL-10. In general, testosterone is considered immunosuppressive. However, studies have produced mixed results. Experimental tests suggest a general, albeit only moderate suppressive effect, while observational data tend to yield small or even no effects after controlling for many confounding factors (Foo et al. 2017; Kelly et al. 2018). In females, oestrogens affect the innate and adaptive (acquired) immune response in various ways and are generally immune stimulating. However, the effects depend on the immune measure used (Kelly et al. 2018). In general, cell-mediated immunity is enhanced by oestrogens, and small effects on humoral defences are also found. This effect is also reflected in lower parasite loads when oestrogens are increased (Foo et al. 2017).

A different sex hormone class, the progestins (especially progesterone), are essential for reproductive

activities and immune defences, where they act to both stimulate and suppress immunity. In females, progesterone enhances humoral (Th2-type) but reduces cell-mediated (Th1-type) responses, whereas the Th2-type response is suppressed in males (Roved et al. 2017). Finally, glucocorticoids are generally important for the development of sexual dimorphism. At the same time, glucocorticoids have immunosuppressive effects; for example, on proinflammatory cytokines (Table 6.3). Glucocorticoids are found in higher concentrations in females than in males.

It is plausible that different titres of sex hormones are a crucial mechanism that leads to differences in sex-specific immunity and susceptibility to parasites. After all, sex hormones have immunosuppressive and immunostimulating effects. Hormone levels are furthermore affected by stress, and males typically are subject to higher stress levels than females during the mating season, e.g. due to increased competitive activities. Again, the exact role of these hormones in shaping sex differences in susceptibility to infection is not fully understood. Apart from these sex hormones, titres of many other hormones such as prolactin or growth hormone differ

Table 6.3 Vertebrate sex hormones and their effects on components of the immune system.[1]

Hormone	Upregulated components[2,3]	Downregulated components[2,3]
Androgens:		
Dihydrotestosterone (DHT) IL-4		IL-4, IL-5, IL-6, IL-1β, CD4+ T-cells, antibody production, Fc-receptor expression.
Testosterone	TGF·β, IL-4, IL-5, IL-10, IL-13, CD4+ T-cells, CD8+ T-cells.	Eosinophil degranulation, macrophages, nitrite, TNFα, IL-4, IL-5, antibody production, Fc-receptor expression.
Oestrogens:		
Oestradiol	Eosinophil degranulation, macrophage phagocytosis, NK, MAPK, TGF·β, IL-1, IL-2, IL-4, IL-5, IL-6, IL-10, IL-13, IFNγ, TFNα, T-cell apoptosis, B-cell activity, antibody production, Fc-receptor expression, wound healing.	Nitrite, NK, NF-κB, TNFβ, IL-1, IL-2, IL-6, IFNγ, CD8+ T-cells, lymphocyte activation markers.
Progestins:		
Progesteron	Eosinophil degranulation, TFNα, IL-2, IL-4, IL-5.	Nitrite, nitric oxide, NK activity, NF-κB, TNFα, IgA, IgG.
Glucocorticoids:		
Cortisol, Corticosterone	TGFβ,· IL-4, IL-10, DTH, antibody production, lymphocyte and monocyte apoptosis.	NOS activity, NK, NF-κB, IL-1, IL-2, IL-6, IL-12, TNFα, STAT-4, TNFβ, IFNγ, CD8+ T-cells, circulating lymphocytes and monocytes, lymphocyte proliferation.

[1] Adapted from: Klein. 2004. *Parasite Immunol*, 26: 247, with permission of John Wiley and Sons.

[2] Production or activity, may also indicate mRNA levels.

[3] For immunological acronyms, see glossary.

between the sexes (Klein 2004). More generally, the effects on immune responses are a result of sex-specific different allocations of resources. Moreover, the levels of host hormones can also be actively altered by the parasite, with the corresponding effects on defences (see Chapter 8). Finally, sex differences in hormone titres need to be connected to the adaptive value of sex differences more generally.

6.2 Parasitism and sexual selection

'Sexual selection' refers to selection concerning reproductive activities in the broadest sense (also called 'fecundity selection'). Reproduction is episodic, typically occurring during a limited time in a year. For the rest of the time, selection affects the organism beyond reproduction. It is known as 'survival selection' or 'natural selection'—the latter term being somewhat misleading, as all selection in the wild is 'natural'. Sexual selection affects males and females differently because they have different reproductive strategy fitness optima, which can also lead to sexual conflict over reproductive decisions (Trivers 1972; Parker 1979). In short, females almost always invest more in each individual offspring (eggs and dependent young) than males (sperm). As a result, eggs are a limiting resource, whereas males can potentially fertilize many more eggs than a single female can produce. Therefore, males must compete for access to female eggs, and females can, in principle, choose among potential mates. Furthermore, individual males differ in their abilities to secure mating opportunities. These differences lead to an outcome where the reproductive success among males varies much more than among females. This observation is known as 'Bateman's principle'. The evolutionary study of sex differences in parasitism and immunity can therefore be based on sexual conflict theory, which has been one of the most active areas of research over the last decades (Box 6.1) (Arnqvist and Rowe 2005; Zuk and Stoehr 2010).

According to sexual conflict theory, sex differences have emerged independently of the causal mechanism, such as the presence of testosterone in mammals. In fact, insects lack testosterone, but insect females express higher levels of immune

functions, too (Nunn et al. 2009). Moreover, sex differences in immune function are quite heterogeneous, which fits a background of sexual selection that is somewhat variable, given the diversity of life histories and underlying mechanisms among organisms (Foo et al. 2017; Kelly et al. 2018).

6.2.1 Female mate choice

Because females and their eggs are in short supply, females, in general, can choose among males ('female mate choice') rather than vice versa. A wide range of hypotheses has fathomed the many advantages of female choice (Figure 6.2). In particular, female mate choice potentially links sexual selection with parasitism, as it can affect infection and immunity for the female and her offspring. For example, females could avoid sexually transmitted diseases, secure better help to raise young, or receive favourable genes for their offspring when mating with healthy males. Furthermore, female choice can take place at different stages of the male–female interaction (Arnqvist and Rowe 2005; Andersson and Simmons 2006), all of which are likely to be influenced by the immune response or parasite infections.

1. *Premating*: This includes female choice of males whose secondary characters promise that offspring inherit paternal genes, providing resistance against the prevailing parasites (Hamilton and Zuk 1982).
2. *Prezygote*: Once mating has happened, selection can act on the received sperm (or pollen in plants) for access to the female's egg. Male seminal fluids contain several components that affect the female's immune system directly and therefore undermine this kind of female selection (Arnqvist and Rowe 2005; Lawniczak et al. 2006).
3. *Postzygote*: After fertilization, zygotes and embryos are nourished by the female. The female can thus also abort an embryo and select against the associated male contribution. Postzygotic conflicts between the female (who cares for developing young) and the male (who donated only their sperm) affect the eventual outcome. This sexual conflict may become visible as increased female susceptibility to disease.

Box 6.1 Sexual selection

Sexual selection is a potent force of evolution. It is responsible for most trait differences between the two sexes; for example, excessive male ornamentation such as colourful feathers and elaborate behavioural display in birds. Several hypotheses have been suggested to explain the evolutionary forces driving sexual selection. A particular focus is female choice and elaborate male ornaments:

- *Direct benefits*: Females may benefit from mating with males carrying elaborate ornaments, since they can reflect the male's ability to provide resources for them or their offspring, e.g. to defend a good territory. Alternatively, such choice might reduce direct costs otherwise imposed by males such as through harassment or manipulatory seminal fluids (Arnqvist and Rowe 2005).
- *Fisher's Runaway Process*: The process starts when females show a (heritable) preference for a (heritable) male trait that is correlated, even weakly, with fitness (e.g. being more resistant to parasites). Natural selection (fitness of male trait) and sexual selection (female choice) operate together to produce somewhat fitter offspring, since they are more likely to carry the father's advantageous trait. As a result, both female preference and male advantageous trait would spread in a population and become genetically associated. As the process unfolds, males carrying the trait would start to gain an additional advantage because of the increasing proportion of females with a preference for the trait (a sexually selected advantage). At this point, the process becomes self-enforcing—male trait and female preference 'run away' to become ever more exaggerated. This evolutionary progress becomes stabilized when excessive exaggeration is checked by natural selection; that is, when large ornaments become a hindrance to survival. The statistical association between the male trait and female preference is a hallmark of Fisher's hypothesis (Fisher 1930; Pomiankowski and Iwasa 1998). This scenario has sometimes also been labelled the 'sexy son hypothesis'.
- *Indicator models*: Indicator hypotheses suggest that a male's traits signal his genetic quality and that females choose a mate by these signals (Mead and Arnold 2004). A significant problem for this class of hypotheses is to understand how genetic variation for such traits and the associated genetic quality can be maintained (the 'lek paradox'). In 'indicator models', male trait and female preference can co-evolve even in the absence of a genetic linkage, but male trait and female preference become associated in the process. In contrast to Fisher's runaway

process, the indicator models require that the trait signals fitness beyond sexual selection, i.e. has a real advantage under natural selection. Two cases are especially interesting:

- o *Hamilton's good genes model*: Females benefit from choosing the fittest males for their offspring, and males would be able to signal their fitness by the expression of secondary sexual signals (the 'Hamilton–Zuk hypothesis'). Hamilton's good genes model suggests, in particular, that the indicator signals resistance against the prevailing parasites (Fisher 1915; Hamilton and Zuk 1982).
- o *Zahavi's handicap model*: Male traits that are preferred by the females might at the same time have a disadvantage to the carrier (Zahavi 1975), e.g. long tail feathers that disrupt an efficient bird flight. Only the fittest males could fully express such disadvantageous traits and carry the cost that comes with it. The indicator is, therefore, an 'honest' signal for male quality. Female preference for such male traits would spread because of the assured good genetic background that these males provide for her offspring. The handicap model is a variant of the more general indicator model with an emphasis on honest signalling due to the high cost of expressing a trait that is disadvantageous except for female choice.
- *Genetic compatibility*: Non-additive genetic benefits could result from choosing a mate with alleles that complement those of the choosing sex (usually the female). Such benefits are likely for the MHC locus, where complementary alleles add to those of the female in generating the most parasite-resistant offspring (Milinski 2006b). By adjusting the number and perhaps kind of alleles that offspring receive, sexual selection could lead to 'immunogenetic optimality'. Compatibility may extend beyond resistance to parasites and provide benefits more generally (Zeh and Zeh 1996; Mays and Hill 2004).
- *Sensory bias*: This hypothesis focuses on the kind of signal that might become the target of sexual selection. Neuronal systems process information in particular ways and have been targeted by selection outside of sexual selection (Ryan 1998). For example, in a given habitat, sensitivity for the red light in the visual system might be advantageous. Such a pre-existing bias could affect the process of sexual selection and lead to male traits that emphasize red colours, as females are intrinsically more responsive to red than to any other colour (Endler and Basolo 1998; Fuller et al. 2005).

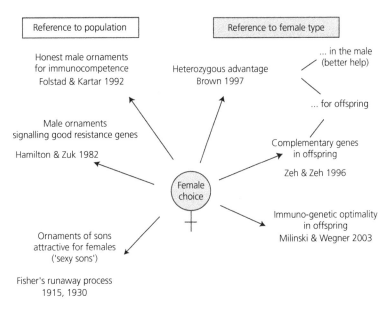

Figure 6.2 Hypothesized advantages of female mate choice. The advantages can refer to the population at large, e.g. more elaborate male ornaments signalling an above-average male quality as compared to the rest of the population (left side). Alternatively, advantages only refer to the female's own type (or status; right side). For example, a particular male may have a number of MHC alleles different from the female's. Parasites select for offspring with optimal MHC diversity (immunogenetic optimality). Sources for each idea are indicated.

Most studies have focused on premating choice, i.e. female mate choice. According to 'indicator models', females should base their choice on indicators that reveal the male's quality concerning relevant fitness components, such as resistance to parasites. For example, female satin bowerbirds (*Ptilonorhynchus violaceus*) choose males able to hold a decorated bower at display sites. These males show increased resistance against ectoparasites (the louse, *Myrsidea ptilonorhynchi*) during a critical life stage as juveniles (Borgia et al. 2004). With 'non-indicator models', female choice of males has consequences for, for example, the genetic makeup of offspring. In both cases, female choice determines the fitness of her offspring for all traits, including immunity to infection.

In insects, for example, mating and reproduction have a range of immunological consequences that illustrate the presence of postmating conflicts (Schwenke et al. 2015). For example, courtship and mating in *D. melanogaster* is associated with the up- or downregulation of a large number of immune genes (Lawniczak and Begun 2004). Mating activities often reduce immune responses, albeit not always. Mated females in beetles show reduced PO activity (Rolff and Siva-Jothy 2002), and courting and mating *Drosophila* males are less able to clear bacterial infections (McKean and Nunney 2001). Females of *D. melanogaster* show postmating immune suppression. In this case, males transfer a sex peptide (Acp70a) during mating, which activates juvenile hormone synthesis in the female. This releases additional energy for the developing offspring but reduces resistance to bacterial infections (Schwenke and Lazzaro 2017). By contrast, mated female crickets show no decrease in immune capacity and are more resistant to experimental bacterial infections as compared to virgin females (Shoemaker et al. 2006).

In some species, males wound the female while mating. In bed bugs (*Cimex* spp.), males penetrate the female abdomen at a specific site, the spermalege, and inject sperm into the body cavity, rather than using the genital tract (Reinhardt and Siva-Jothy 2007; Siva-Jothy 2009). In the process, microbes will also be transferred into the female's body (Bellinvia et al. 2020). In turn, the spermalege contains a high titre of haemocytes and shows humoral immunity. These defences will limit the spread of a microbial infection (Siva-Jothy 2009). Inside the female body, sperm eventually migrate through haemolymph to the receptory organs and ovaries. During this journey, phagocytosis of sperm might mediate postmating cryptic female choice.

6.2.2 Males indicate the quality of resisting parasites

The most apparent differences between the sexes are in the secondary sexual characters, such as the colourful plumage of male birds. A range of ideas has formulated to explain how such differences may relate to parasitism (Box 6.1).

6.2.2.1 The Hamilton–Zuk hypothesis

This hypothesis has been critically important to develop the field of sexual selection and has, in particular, set the stage for investigating how parasitism might affect patterns of sexual dimorphism (Hamilton and Zuk 1982). It suggests that secondary sexual characters ('ornaments'), such as colourful plumage and red skin, signal male health status, and that females will choose the healthiest males according to these signals (an indicator model). Females so garner 'good genes' for their offspring that ensure resistance against the prevailing parasites. Such female choice will only be favoured when heritable variation for male quality and signalling is maintained in the population. However, the evolution of any polymorphism subject to directional selection (such as when females choose the best males) will generally reduce additive genetic variance and thus heritability. At this point, the female choice would no longer have any impact (this is known as the 'lek paradox'). Antagonistic host–parasite co-evolution can rescue this paradox. With this process, the frequency of genes in both parties may cycle continuously and thereby maintain additive genetic variation for the respective traits (see section 14.2).

The Hamilton–Zuk hypothesis predicts that females choose males according to secondary sexual characters, that attractiveness of these characters are determined by parasitic infection, and that both the characters and parasite resistance are heritable, thus providing advantages for the female's offspring. Testing these predictions is difficult because of many methodological limitations, such as measuring the ornament's 'strength' or the influence of environmental conditions and nutritional status on ornaments, as well as choice of host species and parasites (Balenger and Zuk 2014). In a number of cases, females indeed choose males with fewer parasites, e.g. in sticklebacks (*Gasterosteus acculateus*) (Milinski and Bakker 1990), pheasants (Hillgarth 1990), or bowerbirds (Borgia and Collins 1990). On the other hand, female guppies (*Poecilia reticulata*) choose males with larger orange spots, but the size of this ornament is unrelated to parasite load (Martin and Johnsen 2007). The situation can also be more complex; for example, mating with uninfected males was more likely in western bluebirds, but this may reflect female choice or the propensity of infected males to pursue fewer extra-pair copulations (Jacobs et al. 2015). Exactly what kind of ornament matters is sometimes difficult to assess. For example, the bright plumage of males considered to be relevant in jungle fowls (Zuk et al. 1990) was later shown not to be important for female choice (Ligon and Zwartjes 1995). By and large, females often show choice for larger ornaments, males with larger or more conspicuous ornaments are often less affected by parasites, and this choice benefits the female and her offspring (Arnqvist and Rowe 2005). When a male trait that is favoured by females has heritable variation, it can also become subject to the Fisherian runaway process (see Box 6.1). Hence, the two processes—traits selected by (somewhat arbitrary) female preference and as an indicator of good quality (as with Hamilton–Zuk)—are not mutually exclusive (Pomiankowski 1987; Pomiankowski and Iwasa 1998).

In invertebrates, too, secondary sexual characters often, but not always, correlate with measures of immune defence, and females often choose accordingly (Lawniczak et al. 2006) (Table 6.4). These secondary sexual characters include male pheromones, horn length in beetles, wing spots, or 'attractive' characteristics of male songs, such as specific motifs or song duration. Furthermore, there is some (limited) evidence that the sexual selection and the respective signal and measures of immunocompetence or parasite resistance are

Table 6.4 Female choice in invertebrates.

Organism	Finding	Relation to parasitism and immunity	Source
Wolf spider (*Hygrolycosa rubrofasciata*)	Males increase mating success with higher drumming rate. More mobile males have higher lytic activity and are also more likely encountered.	High drumming rate correlates with higher encapsulation response. At the same time, lytic activity in haemolymph reduced, suggesting a cost of sexual advertisement.	1, 2
Wolf spider (*Schizocosa ocreata*)	Symmetry of male foreleg ornaments associated with bacterial infection intensity.	Ornaments indicate male resistance, but females mate indiscriminately.	4
Australian cricket (*Teleogryllus commodus*)	Females show slight preference for males with certain song motifs.	Preferred song motifs relate to higher encapsulation response.	14
Field cricket (*Gryllus bimaculatus*)	Females prefer male songs with high click rate and duration.	Preferred song components correlate with high encapsulation response but low lytic activity of haemolymph. Encapsulation response heritable.	8, 12, 13
House cricket (*Acheta domesticus*)	Females prefer males with more syllables per chirp.	Preferred song motifs correlate with higher haemocyte titres, a heritable component of immune defence.	10, 11
Banded agrion damselfly (*Calopteryx splendens*)	In the field, males with darker, more homogeneous wing spots have lower parasite loads (eugregarines).	After immune challenge, only males with light, heterogeneous wing spots increased PO activity. Encapsulation response uncorrelated, or correlates positively with encapsulation response in other study.	7, 15
Beetle (*Tenebrio molitor*)	Females prefer pheromones of immunocompetent males.	Preferred males have higher encapsulation response and PO activity in haemolymph. PO activity is condition dependent.	6, 9
Burying beetle (*Nicrophorus vespilloides*)	Sex pheromone varies among males and attracts females to males and their carcasses.	Emission of sex pheromone unrelated to male immunocompetence, and not indicator for attractiveness for females.	3
Horned beetle (*Euoniticellus intermedius*)	Females show higher PO activity; no difference in encapsulation rate	PO activity correlates positively with male horn length, and encapsulation rate with elytra size.	5

Sources: [1] Ahtiainen. 2002. *Behav Ecol*, 15: 602. [2] Ahtiainen. 2005. *J Evol Biol*, 18: 985. [3] Chemnitz. 2017. *Sci Nat Heidelberg*, 104: 53. [4] Gilbert. 2016. *Anim Behav*, 117: 97. [5] Pomfret. 2006. *Behav Ecol*, 17: 466. [6] Rantala. 2002. *Proc R Soc Lond B*, 269: 1681. [7] Rantala. 2000. *Proc R Soc Lond B*, 267: 2453. [8] Rantala. 2003. *Biol J Linn Soc*, 79: 503. [9] Rantala. 2003. *Funct Ecol*, 17: 534. [10] Ryder. 2000. *Proc R Soc Lond B*, 263: 1171. [11] Ryder. 2001. *J Evol Biol*, 14: 646. [12] Simmons. 1995. *Behav Ecol*, 6: 376. [13] Simmons. 1992. *Anim Behav*, 44: 1145. [14] Simmons. 2005. *Anim Behav*, 69: 1235. [15] Siva-Jothy. 2000. *Proc R Soc Lond B*, 267: 2523.

genetically correlated. As an example, resistance to parasitoids relates to mating success in *D. melanogaster* (Rolff and Kraaijeveld 2003). On the other hand, when males of *D. melanogaster* were selected for resistance to bacterial infections, their success in fathering offspring declined, suggesting a negative genetic correlation between these quantities (Kawecki 2020).

When looking at variation across different species, the Hamilton–Zuk hypothesis predicts that male secondary characters should be more exaggerated (more showy males) in species with many parasites, as in the example of subfamilies of birds (Figure 6.3). Indeed, a large number of studies in insects, fish, amphibians, reptiles, birds, and

mammals have investigated this prediction. Such interspecific comparisons yield some support for the hypothesis (Garamszegi and Møller 2012). Examples include ornaments in passerine birds from North America (109, 114, and 131 species studied), Europe (113 species, Read 1987; Read and Weary 1990), and New Guinea (64 species; Pruett-Jones et al. 1991), song repertoires in European birds (six species, Garamszegi 2005), or sexual dichromatism in Irish freshwater fish (24 species, Ward 1988). By contrast, no relationship of plumage characteristics and parasitism was found, for instance, among wood warblers in Canada (ten species, Weatherhead et al. 1990). On the critical side, some provisos therefore remain. For example, it is

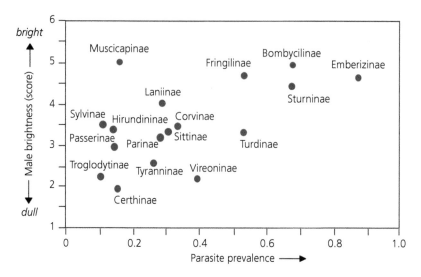

Figure 6.3 Male brightness increases with parasite prevalence. Each dot refers to a different bird (sub-)family. Male brightness was scored on a six-point scale (y-axis). Parasite prevalence refers to different kinds of parasites, as reported in the literature. The positive correlation is independent of the effects of body size, geographical latitude of breeding grounds, or mating system. However, male brightness and the complexity of the male's song (sexually selected traits) in these data are not independent of phylogenetic relationship. Re-analyses of this pattern have thus produced ambiguous results, albeit with some support in favour of the Hamilton–Zuk hypothesis (e.g. Garamszegi and Møller 2012). Adapted from Read (1988), with permission from Elsevier.

difficult to separate a pattern from the effects of phylogenetic relatedness that would link similarities in parasite spectrum and prevalence on the one hand, with the expression of male characters on the other. When phylogenetic similarities are factored out, song characteristics (another sexually selected trait) do not correlate with the prevalence of malaria-like parasites (Read and Weary 1990). Thus, many such relationships are likely confounded by phylogenetic dependencies. Furthermore, whereas spectrometry can standardize the measurement of plumage colour (Montgomerie 2006), the studies were not consistent in their choice of methods, parasites, time of year, or stage in host life history. With all of these difficulties taken into account, the relationship of parasitism to the expression of secondary sexual characters typically becomes moderate at best and more likely weak. Regardless, male secondary sexual characters remain important for mating strategies.

6.2.2.2 The immunocompetence handicap hypothesis

The resources that an individual can invest limit the expression of elaborate male ornamentation. In

turn, resources additionally are demanded by other tasks, such as immune defences. Accordingly, the 'immunocompetence handicap' hypothesis suggests that only good-quality males have enough resources to maintain an effective immune defence and simultaneously express elaborate ornaments. Females choosing males with elaborate ornaments would therefore also acquire good-quality genes for their offspring (Folstad and Karter 1992). Together with Hamilton–Zuk's concept, this hypothesis has been highly influential in fostering the field of ecological immunology. It is a variant of Zahavi's handicap hypothesis (Box 6.1), since it assumes that elaborate ornaments are costly; expressing them, therefore, is an honest signal. The primary mediators of this process are thought to be hormones, in particular testosterone in vertebrate males, and limited resources such as carotenoids.

To test the hypothesis, studies in vertebrates have measured or experimentally altered the level of testosterone in males (Table 6.5). A naturally high level of testosterone or an experimentally increased level indeed often—but not invariably—correlates with the size of male secondary sexual traits. For example, testosterone-implanted males of the

Table 6.5 Immunocompetence and sexual selection in vertebrates.

Organism[1]	Sexual character	Immune defence measure[2]	Finding	Source
European starling (*Sturnus vulgaris*) (O)	Male song	PHA test, antibodies against SRBC.	Testosterone correlates with song and negatively with immune responses.	5, 6
White-tailed deer (*Odocoileus virginainus*) (O)	Male antler size	Helminth infections.	Males with large antlers have more parasites. A gene from the MHC complex correlates with testosterone level.	4
Various vertebrate taxa (amphibia, birds, reptiles, mammals) (O)	Colouration, morphology, vocalization	Various stress measures as mediators for sexual signals.	Meta-analysis shows no relationships.	8
Humans (O)	Male facial attractiveness	Antibody response to hepatitis B vaccine.	High testosterone correlates with attractiveness and stronger immune response. Presumably moderated by cortisol (a stress hormone).	11
Humans (O)	Male androgen levels	Various measures of innate and adaptive immunity.	No evidence for immunosuppressive effect of androgens. No evidence for immunocompetence handicap hypothesis.	10
Red grouse (*Lagopus lagopus scoticus*) (E)	Testosterone level, sexual behaviour	Parasite infection.	Implant that increases testosterone level but does not change associated behaviour leads to increased parasite loads in wild birds.	9
Blackbird (*Turdus merula*) (E)	Bill colour in males	Males injected with either PBS (phosphate-buffered saline solution; controls) or SRBC. Carotenoids as important immune defence factors. Infection by intestinal parasites.	Birds immunized with SRBC show duller beaks (loss of yellow colour) because carotenoids are used for defence. Males supplemented with carotenoids maintain bill colour even when infected.	1, 7
Siberian hamsters (*Phodopus sungorus*) (E)	Sex hormones (testosterone, oestradiol)	Antibody titre, lymphocyte proliferation, PHA test. Tests *in vivo* and *in vitro* (cell extracts).	Animals intact, gonads removed, and replaced by adding hormones. Hormones enhance immune functions in both sexes.	2
Mouse (*Mus musculus*) (E)	Scent attractiveness of males	Experimental infection with nematodes.	Factorial experiment with testosterone injection (yes/no) and infection (yes/no). Testosterone 'compensates' for infection-induced loss of attractiveness. But corticosterones not involved.	12
European tree frog (*Hyla arborea*) (E)	Visual and acoustical	PHA test.	Signals testosterone dependent, but PHA injection does not suppress signals in testosterone-supplemented males.	3

[1] Observational (O), experimental (E) study.

[2] Phytohaemagglutinin (PHA) injected and tissue swelling measured. Sheep red blood cells (SRBC) induce the formation of antibodies (humoral response).

Sources: [1] Baeta. 2008. *Proc R Soc Lond B*, 275: 427. [2] Bilbo. 2001. *Am J Physiol Regul Integr Comp Physiol*, 280: R207. [3] Desprat. 2015. *Behav Ecol*, 26: 1138. [4] Ditchkoff. 2001. *Evolution*, 55: 616. [5] Duffy. 2002. *Proc R Soc Lond B*, 269: 847. [6] Duffy. 2000. *Behav Ecol*, 11: 654. [7] Faivre. 2003. *Science*, 300: 103. [8] Moore. 2015. *Behav Ecol*, 27: 363. [9] Mougeot. 2005. *Am Nat*, 166: 158. [10] Nowak. 2018. *Sci Rep*, 8: 7392. [11] Rantala. 2012. *Nat Comm*, 3: 694. [12] Zhang. 2014. *Can J Zool*, 92: 817.

dark-eyed junco (*Junco hyemalis*) outcompete control-implanted males in courtship displays and are more attractive to females (Enstrom et al. 1997). Also, the expression of secondary sexual morphological characteristics often is more pronounced, with high levels of testosterone. For example, testosterone positively relates to badge size in house sparrows (*Passer domesticus*) (Gonzalez et al. 1999), and to size and colour of the frontal head shield in male moorhens (*Gallinula chloropus*) (Eens et al. 2000). However, in several studies, no (or a negative) correlation between the size of the secondary sexual character and immune responses has been found (Table 6.5). Overall, the empirical evidence for the immunocompetence handicap is therefore mixed, ranging from studies yielding no support, mainly in mammals (Roberts et al. 2004; Nunn et al. 2009; Nowak et al. 2018), to moderate support, e.g. in birds (Roberts et al. 2004). One possible reason for these contradictory findings is the large degree of heterogeneity among studies. For example, experimental treatments appear to have effects in reptiles and birds, but not in mammals. Where males were not castrated, the experimental effects of testosterone implants were significant. Nevertheless, where males were castrated and then implanted, no significant effects emerged (Roberts et al. 2004). Observational studies can furthermore be confounded, as high-quality males have larger parasite loads and nevertheless remain fitter than low-quality males with fewer parasites (Getty 2002). Furthermore, corticosteroids like testosterone can redistribute activities among different arms of the immune system rather than just up- or downregulate immune responses (Kurtz et al. 2000; Poiani et al. 2000).

Alternative hypotheses have assumed that a significant function of the immunosuppressive effect of sex hormones is, indeed, to suppress immune responses. When mating activities generate additional stress, immunosuppression may simply reduce the risk of immunopathology (Råberg et al. 1998; Westneat and Birkhead 1998). Other considerations suggest that, in vertebrates, a stress response leads to the release of glucocorticoids, which redistributes resources to short-term stress-associated priorities and away from long-term needs. Through various mechanisms, this can affect the expression of sexual signals (Bortolotti et al. 2009; Moore et al. 2016). This may happen, for example, through a shared physiological pathway embedded in the mitochondrial functions as a mechanistic basis for the hypothesized relationships (Koch et al. 2017). Correcting for these effects may thus bring the immunocompetence handicap hypothesis back in line with the ambiguous observations (Leary and Knapp 2014). Nevertheless, the evidence remains inconclusive, and meta-analyses could not find a significant relationship between the physiological correlates of stress and sexual signalling (Moore et al. 2016). Newer technologies, such as the study of transcriptomes in response to stress,

have not changed the picture (Wenzel et al. 2013). Ironically, therefore, the 'immunocompetence handicap' hypothesis was very fruitful, despite being probably wrong.

6.2.3 Male genotypes and benefits for resistance

Theoretical considerations suggest that females could benefit from mating with males that are genetically dissimilar to themselves (Colegrave et al. 2002). Dissimilarity can provide two exclusive genetic advantages—heterozygosity and complementary genes (Mays and Hill 2004; Neff and Pitcher 2005). For example, dissimilar males might not become infected by the same diseases, or have complementary properties that extend the range of options for the pair, e.g. better parental care. Dissimilarity might also benefit the female by ensuring heterozygous offspring, which are generally fitter than homozygous offspring. However, male 'quality', as indicated by ornaments, and the advantages through genetic dissimilarity do not necessarily match up. Genetic dissimilarity can be hard to gauge, and any signals of genotype would not necessarily be the same as those indicating male quality more generally (Arnqvist and Rowe 2005). The genotype of a male could be revealed, for example, by odour cues, e.g. peptide ligands for major histocompatibility complex (MHC) in sticklebacks (Milinski et al. 2005; Gahr et al. 2018) or humans (Milinski et al. 2013). At the same time, to sense male dissimilarity, at least, the females have to make a choice based on a reference to the self, as is known for sticklebacks (Aeschlimann et al. 2003).

6.2.3.1 Heterozygosity advantage

There is considerable evidence for a general fitness advantage of heterozygotes, including lower susceptibility to parasites. Examples of heterozygous advantage can be found, for example, in birds (Zelano and Edwards 2002; Voegeli et al. 2012), Soay sheep (Coltman and al. 1999), and sickle cell anaemia—the classic example in humans (Aidoo et al. 2002). In song sparrows (*Melospiza melodia*) the size of the male song repertoire (a sexually selected trait) was positively correlated with its ability to mount an immune response to an experimentally induced challenge, and with male heterozygosity (Reid et al. 2005). Heterozygosity at the MHC locus reduced rather than increased the fitness of mice experimentally infected with *Salmonella enterica* in large population enclosures (Ilmonen et al. 2007) but provided higher resistance against multiple bacterial infections. For further discussion of the effects of MHC heterozygosity on parasitism, see section 10.4.

Females might sometimes be able to sense cues for heterozygosity in males. In the blue tit (*Parus caerulus*), heterozygous males have crown feathers (a secondary sexual signal) with a brighter chroma (an ultraviolet component of colouration). These heterozygous males sired offspring with higher fledging success and ultimately produced more recruits for the local population (Foerster et al. 2003). Several other traits are also positively associated with heterozygosity at functional loci, which can potentially be used by females (Ferrer et al. 2015). In house mice (*Mus musculus*), females prefer heterozygous outbred over inbred males; this preference was even more pronounced when the choosing female was inbred herself. Infecting the males with bacteria did not change the preference, but inbred males were also not more susceptible to infections. The preference was mediated by odour signals (Ilmonen et al. 2009). Females can also increase heterozygosity of their offspring through extra-pair copulations, which counteracts inbreeding imposed by social mate choice and by male mate guarding. In the blue tit, young sired by an extra-pair male are more heterozygous than their within-social pair half-sibs (Foerster et al. 2003). However, across many studies and for general fitness effects (i.e. not just parasitism), the genetic benefits of extra-pair mating have been put in doubt (Akçay and Roughgarden 2007).

6.2.3.2 Dissimilar genes

Mouse females, for example, prefer males that mark more frequently (which correlates with male dominance), but also those genotypically more dissimilar. They also base their choice on the variability among available males (Roberts and Gosling 2003). Females can also choose genotypically dissimilar males, based on postmating choice ('cryptic female choice') (Mays and Hill 2004). A well-known example is a selection against their own pollen in flowering plants (Hiscock and Tabah 2003). Postcopulatory mate choice is also observed in animals, e.g. by sperm selection (Tregenza and Wedell 2002; Bernasconi et al. 2004; Gahr et al. 2018; Kekäläinen and Evans 2018).

Genes at the MHC locus are an essential element of the defence against parasites in the (jawed) vertebrates. As discussed in Chapter 4, MHC class I genes are involved in defence against intracellular, and MHC class II, against extracellular parasites. In both cases, the MHC molecules present self- or parasite-derived peptides to other cells. The primary locus consists of three tightly linked loci and is highly conserved. During evolution, however, this genomic region has been duplicated several times, such that there are three, six, nine, or more loci. Hence, in a single individual, three loci would represent a total of six alleles; with six loci a total of 12 alleles (such as in fish, Milinski 2014), and with nine loci (e.g. in humans; Horton et al. 2004) a total of 18 alleles are present, which together define the MHC genotype. In a human population, however, each locus can have hundreds or more alleles (Marsh et al. 2000; Janeway et al. 2005). MHC allelic frequencies can vary among populations and change across generations, as in Soay sheep (Charbonnel and Pemberton 2005). Those MHC loci that harbour high allelic diversity have received most of the attention, such as MHC class IIB. Why such high genetic diversity is maintained in the face of ongoing selection is a general question in population genetics (Eizaguirre et al. 2012b). In the current context, female choice selecting for dissimilar genes is a process of disassortative mating that can preserve genetic diversity (Hedrick 1992).

Many studies support the idea that females prefer males with an MHC genotype that can complement or add MHC alleles to their own, although not in all cases (Table 6.6) (Kamiya et al. 2014b). Hence, female mate choice seems to lead to higher diversity and more dissimilarity of genes at the MHC locus, e.g. in vertebrate animals, even though the effects may statistically often be small (Kamiya et al. 2014b). Selection for intermediate diversity, as discussed below, could be one reason for weak relationships found with statistical procedures.

Table 6.6 Examples of female choice and MHC type in vertebrates.

Organism	Finding	Cue	Remarks	Source
Atlantic salmon (*Salmo salar*)	Dissimilar MHC males; is not inbreeding avoidance.	Not tested	Dissimilarity especially at peptide-binding region (peptide ligands). MHC-dissimilar fish are less infected.	2, 12
Brown trout (*Salmo trutta*)	Males with intermediate MHC dissimilarity sired more offspring. No such effect in females.	Not tested	Parentage analysis in free-living fish.	7
Stickleback (*Gasterosteus aculeatus*)	Dissimilar MHC males. Males more attractive when offering suboptimal numbers of peptides, less attractive with superoptimal numbers.	Male odours mimicked by offering peptides	Especially dissimilar peptide ligands. Females also choose the optimal number of MHC alleles that corresponds to their local habitat.	8, 15
Stickleback (*Gasterosteus aculeatus*)	Postmating selection on sperm carrying different genotypes. MHC-complementary male gametes are more successful.	Not tested	Most successful sperm yield the offspring MHCgenotype that correlates with highest parasite resistance.	13
Tawny dragon lizard (*Ctenophorus decresii*)	No evidence of MHC-associated mating.	Male throat colour	Lower parasite loads with certain MHC types. Could be maintained by rare allele advantage/fluctuating selection.	9
House sparrow (*Passer domesticus*)	Breeding failure when combined number of MHC alleles in offspring was too low or too high.	Not tested	Non-random pair formation with respect to MHC type. Not an effect of inbreeding avoidance.	1
Yellowthroat (*Geothlypis trichas*)	Males with more MHC alleles more resistant to malaria. Extra-pair mating could give females advantages.	'Mask' size (colouration)	'Mask' size larger in males with more MHC alleles.	4
Great snipe (*Galingao media*)	Mating not according to MHC type. No female preference for males with many or rare MHC alleles.	Not tested	Some allelic lineages more common in successful males.	5
House mouse, sheep, humans	Review of studies: evidence of MHC-dependent mating preferences in mice and humans.	Odour	Not all studies find evidence for mating preferences.	18
Tuco-tuco (*Ctenomys talarum*)	Successful males more heterozygous and carry distinct, but among each other similar alleles at MHC, compared to random males.	Odour	Laboratory and field data.	3
Soay sheep (*Ovis aries*)	No evidence for MHC-related mate choice.		MHC heterozygosity correlates with fewer parasites.	16, 17
Grey mouse lemur (*Microcebus murinus*)	Males more heterozygous and dissimilar MHC types are more likely to sire offspring. Disassortative mate choice for MHC class II genes (DRB, but not DQB).	Not tested	DRB is under strong diversifying selection.	10, 19
Alpine marmot (*Marmota marmota*)	Social pairs with higher MHC dissimilarity than random. Extra-pair pairings higher when social pair has low dissimilarity.	Not tested	MCH and neutral alleles considered.	6
Comparative study with 112 mammal species	MHC nucleotide diversity increases with parasite richness in bats and ungulates; decreases in carnivores.	Not tested	MHC nucleotide diversity increases with testes size in all taxa. Various confounding factors taken into account.	21
Comparative study with fish, amphibia, birds, reptiles and mammals (116 effect sizes)	Support for female choice for MHC diversity, and for dissimilarity across multiple loci.	Not tested	Little difference among vertebrate taxa, but also a problem of statistical power.	11
Human (*Homo sapiens*)	Females prefer males with dissimilar MHC type. Pleasantness of odours increases with dissimilarity of MHC alleles.	Male odour	No simple effect of heterozygosity.	14, 20

Sources: [1] Bonneaud. 2006. *Proc R Soc Lond B*, 273: 1111. [2] Consuegra. 2008. *Proc R Soc Lond B*, 275: 1397. [3] Cutrera. 2012. *Anim Behav*, 83: 847. [4] Dunn. 2012. 67: 679. [5] Ekblom. 2004. *Mol Ecol*, 13: 3821. [6] Ferrandiz-Rovira. 2016. *Ecol Evol*, 6: 4243. [7] Forsberg. 2007. *J Evol Biol*, 20: 1859. [8] Gahr. 2018. *Proc R Soc Lond B*, 14: 20180730. [9] Hacking. 2018. *Ecol Evol*, 8: [10] Huchard. 2013. *Mol Ecol*, 22: 4071. [11] Kamiya. 2014. *Mol Ecol*, 23: 5151. [12] Landry. 2001. *Proc R Soc Lond B*, 268: 1279. [13] Lenz. 2018. *Evolution*, 72: 2478. [14] Milinski. 2006. *Annu Rev Ecol, Evol Syst*, 37: 159. [15] Milinski. 2005. *Proc Natl Acad Sci USA*, 102: 4414. [16] Paterson. 1997. *Proc R Soc Lond B*, 264: 1813. [17] Paterson. 1998. *Proc Natl Acad Sci USA*, 95: 3714. [18] Penn. 1999. *Am Nat*, 153: 145. [19] Schwensow. 2008. *Proc R Soc Lond B*, 275: 555. [20] Wedekind. 1997. *Proc R Soc Lond B*, 264: 1471. [21] Winternitz. 2013. *Proc R Soc Lond B*, 280: 20131605.

Scent composition reveals the MHC type of a male, and therefore signals immunological dissimilarity from the female (Penn 2002; Milinski et al. 2005). Detailed tests in sticklebacks, in particular, have demonstrated that the active components of the odour are peptide ligands that reveal the specificity of the receptors in recognizing antigens; these are encoded by MHC class I locus genes (Milinski et al. 2005). The importance of scent associated with the peptide-binding region of the MHC locus has also been demonstrated in Atlantic salmon (Landry et al. 2001), mice (Leinders-Zufall et al. 2004), and humans (Milinski et al. 2013). The parasites themselves can alter the scent mark of infected males. For example, microbial communities such as the microbiota in mice have known sources of odours. Infection by nematodes, for instance, changes the odour profile and leads female mice to choose uninfected males (Ehman and Scott 2001). Furthermore, the microbial community composition is affected by genetic background and the MHC type of the mouse, and hence the host immunological type also sets the scent mark (Lanyon et al. 2007; Ehman and Scott 2001).

Female choice for MHC types could serve different possible adaptive functions (Howard and Lively 2004). These include increasing offspring heterozygosity (increasing the diversity of genes), procuring 'good', dissimilar, or rare alleles for offspring that might be advantageous against rapidly co-evolving parasites (Milinski 2006b). Female choice for MHC is not completely equivalent to Hamilton and Zuk's version of the 'good genes' hypothesis. The reason is that 'good' may very well differ according to whoever is choosing. In the stickleback example, the peptide ligands by themselves (and thus the genes responsible for it) are neither attractive nor discouraging for female choice. Rather, their choice value depends on the MHC composition of the mating pair (that is, male MHC plus female MHC) and therefore is based on female reference to self. Hence, the major effect is the dissimilarity aspect rather than the garnering of 'good genes' in a pure sense (Neff and Pitcher 2005). Observations and tests show, furthermore, that

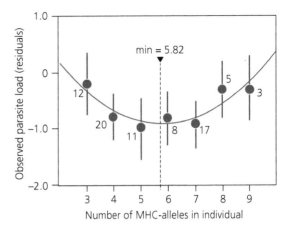

Figure 6.4 Experimental test of the dissimilarity hypothesis. Experimentally raised three-spined sticklebacks (*Gasterosteus aculeatus*) were infected with helminth parasites (*Diplostomum spathaceum*, *Camallanus lacustris*, and *Anguillicola crassus*) and the resulting parasite load determined. The established parasite load varied with the number of expressed MHC class IIB molecules (alleles at the diploid locus) of the individual (x-axis). Parasite load is given as residuals from a General Linear Model including dose as covariate, and sibship of fish as random factor. The data are statistically best described by a quadratic function (red curve) with a minimum parasite load at 5.82 alleles. Values are means ± S.E. Sticklebacks have six loci in this MHC class and thus carry a maximum of 12 alleles (Milinski 2006b). Adapted from Wegner et al. (2003a), with permission from the American Association for the Advancement of Science.

females choose males that are neither too dissimilar nor too similar to their own type (Aeschlimann et al. 2003). The choice thus goes for males that, together with the female's own MHC type, will ensure that offspring carry an intermediate diversity of alleles at their MHC loci. Such intermediate diversity is associated with the lowest parasite load and the highest expression of elements of innate immunity. This status has been called 'immunogenetic-optimality' (Wegner et al. 2003a; Wegner et al. 2003b; Kurtz et al. 2004; Milinski 2006b) (Figure 6.4). An upper limit to diversity is presumably set by the increasing risk of self-reactivity if an ever more diverse MHC locus can recognize too many peptides. Immunogenetic optimality is also reported from other fish (trout; Forsberg et al. 2007), birds (Bonneaud et al. 2004), or voles (Kloch et al. 2010).

Important points

- The sexes differ in parasite load and immune defences, with males being generally more susceptible and having lower immune responses than females. Immune functions are, among other things, affected by sex hormones.
- Sex differences result from different life histories of males and females and the resulting sexual conflicts. Females are generally the more contested sex and males compete for mating opportunities.
- Females choose their mates based on indicators that signal the male's health or quality. Females might choose genetically heterozygous or dissimilar males to provide offspring with good genes or with more compatible genes, which has benefits in terms of parasite resistance.
- Males might indicate their quality with elaborate ornamentation (secondary male sexual characters). Females prefer males with more elaborate ornaments, and such choice can benefit their offspring.
- In (jawed) vertebrates, female mating preferences use odours linked to the MHC locus. An intermediate allele number at the MHC is associated with lowest parasite loads and more active immune functions.

CHAPTER 7

Specificity

It is a remarkable fact that most host species seem resistant to most parasites—even to those that are not encountered commonly—a phenomenon called 'non-host resistance'. Hosts become more resistant or tolerant by co-evolution with their parasites (where parasites change, too). Resistance can also evolve by one-sided evolution (where parasites do not change evolutionarily), or as a side product. The latter may occur because the parasite specializes on other hosts (Antonovics et al. 2013). All of these components vary among host species, populations, or genetic variants, and they add up to 'host specificity', which is the variation in resistance (or tolerance) towards different parasites. If a parasite gains little or no fitness by counteradaptations (a one-sided scenario), host resistance presumably evolves in different directions to limit the damage from an infection.

Host defences should become more specific when the parasite co-evolves. Specialized hosts can defend themselves against a small proportion of parasites and are susceptible to most. A corresponding situation exists for parasites—they, too, become specialized to various degrees. 'Parasite specificity' thus relates to the differences in performance in different hosts (Box 7.1). Similar to host resistance, parasite specificity can have very different evolutionary reasons. Host and parasite specificities affect the ecological and evolutionary dynamics of host–parasite systems. For example, resource partitioning by specialization is an essential determinant of the structure of ecological communities. However, specificities are also crucial for disease control and eradication strategies in medicine or agriculture.

Box 7.1 Specificity in defence space

Several 'filters' can characterize how specifically hosts interact with their parasites; primarily, 'ecological' and 'physiological' filters (see Figure 7.4 in the main text; Combes 2001). Various mechanisms shape these filters. In particular, an ecological filter can be set by the host's chances to encounter a parasite. In contrast, the physiological filter can result from a series of successive defence mechanisms that may or may not limit the establishment or growth of an infection inside the host (Schmid-Hempel and Ebert 2003). The parasite, in turn, should breach these filters and succeed to infect, establish, and transmit further to a next host. If successful, this host is within the parasite's host range.

When the parasite can use this host, the infecting parasite population moves along the infection trajectory in disease space, where different outcomes are possible (Figure 1). The outcomes are a mechanistic way to look at host ranges and specificity.

In Figure 1, for example, parasites A and B would infect, but the host can immediately resist the infection and quickly recover to a healthy state. An observer would probably conclude that these parasites are unable to infect the host and that this host is not part of the parasite's host range. Parasite

continued

Evolutionary Parasitology: The Integrated Study of Infections, Immunology, Ecology, and Genetics. Second Edition.
Paul Schmid-Hempel, Oxford University Press. © Paul Schmid-Hempel 2021.
DOI: 10.1093/oso/9780198832140.003.0007

Box 7.1 *Continued*

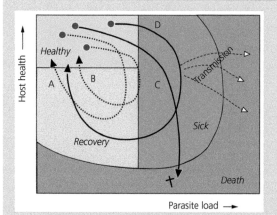

Box 7.1 Figure 1 Specificity in disease space. The trajectories generate the host range and the specificity of host defences (see text).

C can infect, make the host sick, or even kill it, but does not transmit further. In this case, this host would be considered susceptible to this parasite but not a part of the normal host range, because no transmission happens; the parasite has no fitness, and the dead-end host may be an accidental one. Parasite D uses this host as a part of its range since it develops along a 'normal' infection trajectory and gains fitness by transmission. Hosts can keep out parasites A and B (here, by immediate clearance) but not C and D. If so, these hosts would be considered as having specific defences against A and B. In 'paratenic' hosts, by definition, the parasite can reside and also transmit further, but no development occurs; i.e. the parasite load (the x-axis of defence space) does not increase. This case is not explicitly shown here.

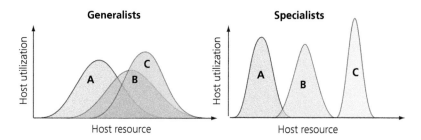

Figure 7.1 Generalist and specialist parasites. The graph shows how different parasites (A, B, C) utilize the range of available host resources. Resources for a parasite might be different host species, different host genotypes, or different target organs. Host utilization is the probability that a given parasite uses a particular resource; the area under each curve symbolizes the 'niche' of the parasite.

7.1 Parasite specificity and host range

Parasites are regarded as 'specialized' when they can successfully infect only a narrow range of host species, whereas 'generalist' parasites can infect many hosts (Figure 7.1). Hence, the 'host range' of a parasite is roughly inverse to its 'specificity'. Host specificity and host range are key defining characteristics of any parasite, but, despite its relevance, parasite specificity is not easy to define and measure. Several measures (Box 7.2) and approaches are in use.

7.1.1 Measuring parasite specificity and host range

7.1.1.1 Observation of infections

This is the classical and still most practical way to determine host specificity. It merely compiles all observations into a list of host species per parasite species observed in the wild (a host range; Figure 7.2). The count also yields an index of 'host specificity' (HS; the number of host species parasitized) (Caira et al. 2003). Such lists typically show that most parasites use one or a few hosts, with only

Box 7.2 Host specificity indices

Various indices exist to characterize host specificity with a single number. These indices combine the different components that describe specificity, and, in particular, the relative abundance of each parasite species and parasite species number.

Rohde's indices: Rohde (1980) suggested several different indices:

(a) *Density index (S_i)*: This index considers how evenly parasites are distributed among host species:

$$S_i = \left[\sum \frac{x_{ij}}{n_j h_{ij}}\right] / \left[\sum \frac{x_{ij}}{n_j}\right] \quad (1)$$

where x_{ij} is the number of individuals of parasite species i in host j and n_j the number of individuals of host species j that are examined. h_{ij} is the rank of host species j based on diversity of its infections (as given by x_{ij}/n_j), with the highest diversity being ranked top (rank 1). This index does not take into account the total number of host species in the sample.

(b) *Frequency index (S_i)*: This index also uses eq. (1) above. But x_{ij} is now the number of individuals of host species j infected with parasite species i, and h_{ij} is the rank of host species j, based on the frequency of infections with the highest frequency ranked top (rank 1).

(c) *Probability index*: This index checks the probability of finding a host species infected by parasite species i out of a set of n_j examined host species. It is assumed that enough hosts are screened to detect the infections. If a total of n_i host species are infected by parasite species i, then this probability is $P_{ij} = n_i/n_j$. From this, a host specificity index for parasite species i can be defined as

$$S_i = \frac{1}{p_{ij}n_j} = \frac{1}{n_i} \quad (2)$$

or, alternatively:

$$S_i = 1 - P_{ij} \quad (3)$$

In the first case (eq. (2)), specificity is the inverse of P_{ij} corrected for the number of host species examined. In the second case (eq. (3)), specificity is the probability of not finding an infected host species. This index ignores the frequency distribution of infections and is strongly affected by the number of species examined.

Rao's entropy index (Q) (Rao 1982)*; Host breadth index (HB)* (Fallon et al. 2005): This index weighs the phylogenetic distinctiveness of host species used by parasites; it is calculated from the host perspective as

$$Q, HB = \sum_{i=1}^{s} \sum_{j=1}^{s} (d_{ij} p_i p_j) \quad (4)$$

where d_{ij} is the taxonomic or phylogenetic distance between host species i and j. The values p_i, p_j are the prevalences of the parasite in host species i and j, respectively, in the set of s host species. This index is sensitive to host-species richness, as the expected average distance increases with more host species.

Poulin–Mouillot index S_{TD} (Poulin and Mouillot 2003): S_{TD} is among the most used indices. This index takes account of the taxonomic distance, d_{ij}, between host species i and j in a set of s host species used by a given parasite. The difference, d_{ij}, could be estimated by the number of nodes in the phylogenetic tree that must be passed to reach species j from species i, and vice versa.

$$S_{TD} = 2 \frac{\sum\sum_{i>j} d_{ij}}{s(s-1)} \quad (5)$$

A high value of S_{TD} indicates that host species are, on average, distant from one another and that the parasite is not specific. S_{TD} is not computable for parasites with only one host species. The same index is also described as SPD_i (Poulin et al. 2011).

Poulin–Mouillot improved index S_{TD*} (Poulin and Mouillot 2005): This index combines the prevalence of a parasite in different hosts with the latter's taxonomic similarity:

$$S_{TD*} = \frac{\sum\sum_{i<j}(d_{ij} p_i p_j)}{\sum\sum_{i<j}(p_i p_j)} \quad (6)$$

where d_{ij} is the taxonomic distance between host species i and j. The values p_i, p_j are the prevalences of the parasite in host species i and j. The summation runs over all host species ($1 \ldots i, j \ldots s$). The index has been further modified by using the variance in the taxonomic distances, $Var(S_{TD})$ between host species (Hellgren et al. 2009):

$$S^* = S_{TD} + \frac{s-1}{1 + Var(S_{TD})} \quad (7)$$

Host skew: A given parasite species occurs at different relative frequencies in the different hosts of a set of host species. If a parasite uses these hosts at random, in proportion to their availability, no skew exists in the distribution of relative frequencies. When host species are used non-randomly, the distribution of species in the parasite's host

continued

Box 7.2 *Continued*

spectrum becomes skewed. The skew in the host spectrum can measure specificity. With respect to the phylogenetic relatedness of hosts in a parasite's spectrum, the Colless index, I_C (Blum and Francois 2005), is zero if the phylogeny of hosts is fully matching the parasite's host spectrum; the value of I_C approaches one when the parasite's host spectrum deviates from the phylogenetic relationships of hosts (Krasnov et al. 2008).

There are several other indices—often recruited from classical community ecology—that have been used in various studies (Pojmańska and Niewiadomska 2012). Examples include Lloyd's index (Novotny and Basset 1998), Hurlbert's

index (Hurlbert 1971), and the (spatial) 'index of beta-specificity' (β_{SPF}; referring to the replacement of species when moving from one site to another) (Krasnov et al. 2011). Several indices have also been proposed for 'phylogenetic specificity', i.e. the tendency to utilize only closely related hosts. Examples include the 'net relatedness index' (Cooper et al. 2012a), and an index of phylogenetic specificity, PS_{ji}, that measures the total length of branches in the phylogenetic tree that links species i with j, and which can be corrected for the number of host species. The effects of host phylogeny and differential host use at different localities can be captured with an index of phylogenetic β-specificity (PBS_i; Poulin et al. 2011).

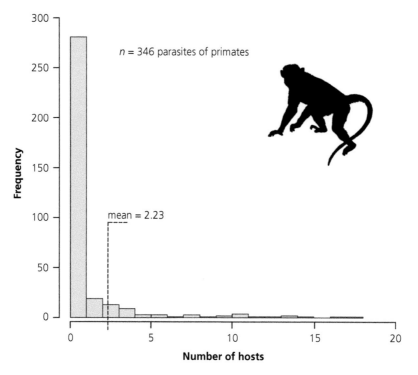

Figure 7.2 Host ranges in parasites of primates. (a) Shown are the frequencies of host numbers (x-axis) per parasite for 38 primate species. Parasites are bacteria ($n = 33$), fungi ($n = 3$), helminths ($n = 180$), protozoa ($n = 78$), and viruses ($n = 52$). The average number is 2.23 hosts per parasite species. In this sample, the phylogenetic similarity among pairs of parasites correlates with similarity among pairs of hosts. Silhouette from phylopic.org. Based on data in Cooper et al. (2012a).

a minority present in many host species. Lists also show that the degree of host specificity varies among parasite groups.

The method has some problems. For example, an observed infection may be a transient, chance

infection, e.g. in a paratenic host. Also, the likelihood of detecting a host depends on sampling effort, because more intensive studies will reveal more hosts and vice versa (Byers et al. 2019). Study intensity is highest for larger animals (e.g. vertebrates), for more widespread and visible taxa (e.g. birds), for those of commercial interest (such as cattle), and for parasites that are of interest to humans. Moreover, the parasite type plays a role, e.g. a vectored pathogen, a virus, or any other such category. Measures that take account of missing coverage with too small sample sizes (completeness of data), or of uneven sampling (sample heterogeneity), can reduce these errors or at least yield an estimate of its effects. Examples are methods for regressing the number of hosts on the number of studies, or rarefaction methods as used in ecology. Also, the 'abundance-based coverage estimator' (ACE) is such a (non-parametric) estimator of species richness in a sample and can estimate host range (Chao and Lee 1992). Using ACE in combination with the Global Mammal Parasite Database showed that accuracy of host range estimation varies with parasite group and that a large fraction (20 to 40 per cent) of the host range is commonly unknown (Dallas et al. 2017). Based on the rate of discoveries over the last decades, some current estimates suggest that perhaps half of all parasites may have been found (Costello 2016). This seems a somewhat optimistic view, given the many cryptic species and microparasites such as protozoa, viruses, or bacteria that have barely been covered in depth so far.

7.1.1.2 Screening with genetic tools

This method is essentially the same as used for the classical list, except that the presence and identity of a parasite are established by molecular genetic tools, e.g. by testing for the presence of a partial sequence of the *Cyt b* gene. Often, these analyses reveal numerous cryptic species of parasites that could not have been distinguished by the classical methods (Poulin and Morand 2004; Miura et al. 2006). As a result, host specificity tends to increase (Seghal et al. 2005).

7.1.1.3 Experimental infections

This approach removes the ecological filter, e.g. when there are no natural encounters, by controlled infections of different host–parasite combinations. Experimental infection can test novel combinations and study the potential for host jumping and the risk of emerging diseases (Figure 7.3). For example, it asks: how restricted is a parasite by the phylogeny and physiology of its hosts (Ruiz-Gonzalez and

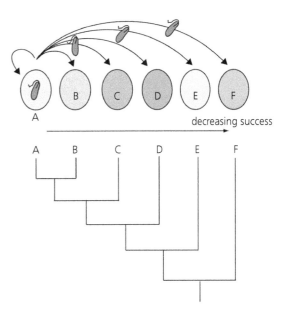

Figure 7.3 Cross-infecting parasites. A parasite found in host A is experimentally infected to other hosts (B, …, F) and, as a control, into the same host (A). In this example, infection success decreases from host B to F. The decrease might reflect the phylogenetic distances between host A and the test hosts, as indicated by the hypothetical phylogeny shown underneath. The distance to which the parasite remains infective reveals its compatibility with respect to host taxa. Alternatively, the tree might represent geographic distance, habitat, or immunological similarity. Adapted from Poulin and Keeney (2008), with permission from Elsevier.

Brown 2006; Solter 2006; Kuris et al. 2007), or is it mainly constrained by ecological factors (Detwiler and Janovy 2008)? Due to the practical difficulties, the experimental approach is not feasible for the vast majority of host–parasite combinations.

Host specificity of parasites varies considerably (Table 7.1), and host ranges are typically right-skewed; that is, have long tails towards high host numbers (Figure 7.2). In general, when the phylogenetic dependencies among hosts are accounted for, host choice seems much more opportunistic; i.e. host taxonomic groups are chosen according to their ecological availability (Poulin 2007b). On the other hand, the discovery of cryptic species suggests that most parasites are more host-specific than what immediately meets the eye (Poulin and Keeney 2008). The genetic analysis also helps to identify the life history stage at which parasites are most specific in their host choice (Randhawa et al.

Table 7.1 Host specificity of parasites.

Parasite	Host	Specificity[1]	Remarks, measures	Source
Various macroparasites (17 spp.)	Fish	Adult helminths show high degree of specificity. (O)	Parasite community dominated by parasites of non-native fish. Species richness, Shannon–Wiener index.	1
Ticks (2 spp.)	Rodents (8 spp.), birds (4 spp.)	Not very specific as measured by index. More uneven among hosts when measured by prevalence and intensity. (O)	S_{TD*} index, prevalence, intensity.	2
Ticks (61 spp.)	Mammals (95 spp.)	Essentially generalists, especially juveniles, whereas adults are more specialized. Using intensity yields more specialization. (O)	S_{TD*} index, intensity.	7
Gregarines (6 spp.)	Beetles (5 spp.) (Tenebrionidae)	All parasites infected larvae and adults of several host species. Range not determined by host phylogeny. (E)	Host range in the field must be largely determined by ecological factors. Mean abundance.	4
Water mites (Unionicolidae)	Fresh water mussels	Highly host-specific with one or two hosts per parasite species. (O)	Affected by host phylogeny. Mites on gills more specific than mites on mantle. Number of host species, and S_{TD*} index.	5
Helminths	Marine animals	Wide range of different host specificities. (O)	Various measures (review of studies).	8
Helminths (29 spp.)	Fish (in six communities)	Host specificity not much affected by host phylogeny. (O)	Parasites are opportunistic in what they infect. Intensity, prevalence relative to main host.	12
Helminths (73 spp.)	Antarctic fish	Not very specific. (O)	Broad range for intermediate hosts, but narrower range for final hosts. Species lists.	13
Various (437 spp.): viruses, helminths, protozoa, bacteria, fungi	Primates (128 spp.)	Most parasites are generalists, as they infect not only closely related hosts. (O)	Net relatedness index (NRI).	3
Eugregarines	Cockroaches	Almost completely host-specific. (E)	Tested by experimental infection in gut homogenates for combinations of hosts and parasites. Establishment in host.	14
Haemosporidia: *Haemoproteus, Plasmodium, Leucocytozoon*	Birds of prey, owls	Broad range of host species but specificity varying substantially among parasite species. (O)	Prevalence established by molecular methods.	9

			$S_{TD}*$ index. HB index.	16
Haemosporidia: *Haemoproteus, Plasmodium*	Passerine birds (nine spp.)	Parasite lineages restricted to set of host species. Generalist parasites most widespread. (O)		
Haemosporidia: *Haemoproteus, Plasmodium*	Birds (14 spp.)	Not specific, share many hosts. (O)	Degree of specificity varies among parasite strains. Measure: degree to which a single parasite infects a single host line.	15
Global avian haemosporidian database (2451 spp.)	Birds	Among larger parasite clades, differences according to host phylogeny occurs. But sister lineages can nevertheless infect unrelated hosts.	Presence/absence and DNA species fingerprint with molecular tools (Cyt b). Phylogenetic signal estimated by various measures: Pagel's λ, Blomberg's K, Moran's I.	6
Phytophagous insects (900 spp.)	Plants (51 spp.) in New Guinea	Most phytophagous insects feed on several, but closely related host plants. However, many parasites shared even among distant host taxa. (O)	Parasites are not very host-specific overall and thus not strongly separated by niches. No difference between temperate and tropical systems. Percentage of herbivores feeding on a single host species.	10, 11

[1] O: observations, E: experimental infections.

Sources: [1] Choudhury. 2004. *J Parasitol*, 90: 1042. [2] Colombo. 2018. *Ticks Tick-borne Dis*, 9: 781. [3] Cooper. 2012. *Ecol Lett*, 15: 1370. [4] Detwiler. 2008. *J Parasitol*, 94: 7. [5] Edwards. 2006. *J Parasitol*, 94: 7. [6] Ellis. 2018. *Int J Parasitol*, 48: 897. [7] Espinaze. 2016. *Parasitology*, 143: 366. [8] Hoberg. 2002. *Parasitology*, 124: S3. [9] Krone. 2008. *J Parasitol*, 94: 709. [10] Novotny. 2005. *Proc R Soc Lond B*, 272: 1083. [11] Novotny. 2002. *Nature*, 416: 841. [12] Poulin. 2005. *Parasitology*, 130: 109. [13] Rocka. 2006. *Acta Parasitologica*, 51: 26. [14] Smith. 2008. *Comp Parasitol*, 75: 288. [15] Szymanski: 2005. *J Parasitol*, 91: 768. [16] Ventim. 2012. *Parasitology*, 139: 310.

2007). Finally, the degree of host specificity is also affected by the measure or index used to describe it. Sometimes, formalized indices suggest higher degrees of generalist patterns than simple observations (Espinaze et al. 2016).

7.1.2 Characteristics of a host

A parasite that uses a particular host has fitness benefits and costs, especially with respect to specialization vs generalization. Organisms rarely can evolve to specialize on one resource without losing their efficiency on another—this is known as the 'niche breadth' or 'trade-off' hypothesis. This hypothesis has been a central concept to explain how diverse parasites can be (Price 1980), and proposes that host specialization should be favoured. Advantages for specialists indeed exist. For example, specialization means a better establishment in the host (e.g. for parasitoids, Rossinelli and Bacher 2014), higher parasitaemia (malaria in primates, Garamszegi 2006), or achieving a higher prevalence (blood parasites in birds, Medeiros et al. 2014). In fact, generalist parasites typically have lower prevalences and infection intensities as compared to specialists, which fits the hypothesis. Similarly, parasites that are serially passaged through a novel host generally lose their ability to efficiently infect or multiply within the original host (Agrawal 1997; Ebert 1998). This pattern suggests a negative genetic correlation of performances in different hosts—a trade-off that is the basis for the evolution of more specialized parasites (Duffy et al. 2007; Henry et al. 2008).

Nevertheless, a negative correlation of parasite success on different hosts is not the general rule. For example, compared to specialists, generalist helminths (Poulin 1999) achieve higher prevalence, or are at least similarly successful in colonizing new regions (Drovetski et al. 2014). The same holds for some generalist malaria parasites of birds (Hellgren et al. 2009). Even for the most generalist parasite species, there is considerable variation in their success and their effects on the different hosts (Bielby et al. 2015). However, these may not always correlate negatively in a simple way. From observations, therefore, although most parasites use only a few hosts (Figure 7.2), many others infect several hosts, and many are rather opportunistic. The fungus

Batrachochytrium dendrobatidis, for example, infects over 500 amphibian species (Bielby et al. 2015). A broader host range is also observed for many parasites infecting humans, some of which have animal reservoirs and opportunistically jump across species barriers (Cleaveland et al. 2001b; Woolhouse et al. 2001). Furthermore, specialization is not irreversible, and parasites are therefore capable of evolving a narrow or broad host range regardless of their ancestral state (Johnson et al. 2009a).

7.1.3 Evolution of parasite specificity and host range

Host range is not a simple trait and is, therefore, determined by a combination of factors. Nevertheless, the basic concepts are apparent. Host range evolves through natural selection, where ecological factors (the 'ecological filter') such as niche and shared habitats affect the likelihood of an encounter between host and parasite species. Physiological factors (the 'physiological filter') determine whether a parasite that has encountered a host can infect, establish, and transmit further from a host (Figure 7.4). This filter depends on the molecular-biochemical repertoire of host and parasite to resist infection or to gain entry into the host and the target tissues, respectively. Phylogenetically close hosts are physiologically similar, too. They should thus have similar physiological filters and attract the same parasites. Nevertheless, many parasites use several hosts to complete their life cycle, which may even belong to different phyla. Examples are parasites that use insects, snails, and crustaceans, but eventually also infect vertebrates. Therefore, jumping across vast phylogenetic distances does not seem a major universal constraint, even though these life cycles have evolved in several steps.

Adaptation to a new host is generally associated with genetic changes. Sometimes, the acquisition of a genetic element (e.g. a new virulence factor) by the parasite allows expansion of the host range. Furthermore, local adaptation likely leads to specialization, as parasites adapt faster when they can infect similar genotypes (Lively 1999). Conversely, sizeable genetic diversity within a host population can limit the adaptation of parasites to a given host, e.g. in bacteriophages and their bacterial hosts (Morley et al.

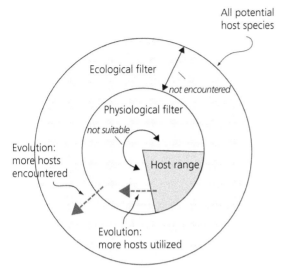

All potential
host species

Ecological filter

not encountered

Physiological filter

not suitable

Host range

Evolution:
more hosts
encountered

Evolution:
more hosts utilized

Figure 7.4 Ecological and physiological filters determine parasite host range. The outer circle represents all potential host species. Of all hosts, some are never encountered (grey area), of those that are encountered ('ecological filter'; inner circle), some are not suitable as hosts ('physiological filter'; yellow area). Hosts that are encountered and suitable form the host range (green area). An evolutionary change in the host range (red arrows) can result from a change in the ecological filter, leading to more hosts being encountered, or a change in the physiological filter, leading to more hosts being utilized. Note that the physiological filter, in particular, is a result of the host–parasite interaction and, therefore, affected by the characteristics of both parties. In the plant literature this scheme is often shown as a 'disease triangle' where host, parasite, and environment are the three dimensions of the problem (Barrett et al. 2009). Adapted from Combes (2001), with permission from University of Chicago Press.

2017). Furthermore, constraints to adaptation can be set by how host resistance and parasite infectivity are genetically encoded, e.g. by allele matching or other processes (Chapters 10 and 14).

An important ecological constraint for the evolution of parasite specificity is the opportunity to encounter different hosts; this factor relates to the problem of transmission. Transmission depends on the specificity of a vector, if one exists, the spatial proximity of hosts, or the ability of the parasite to disperse in space and time by durable stages. For example, some arthropod vectors are generalist feeders, whereas others specialize on a few host species; parasites that largely are restricted to humans use vectors that primarily feed on humans (Woolhouse et al. 2001). Spatial proximity is suggested by ectoparasites (fleas) that parasitize colonially nesting birds

and show narrower host ranges than fleas infecting birds nesting in their own territories. Parasite mobility (estimated from the morphology of macroparasites, e.g. body size, leg length) and geographical range likewise decrease with more aggregation in the hosts (Tripet et al. 2002). However, although there is an association of transmission strategy and host range within each parasite group, there is no consistent direction in this association (Pedersen et al. 2005). The ecological filter also reflects historical accidents, e.g. when a parasite or a host species is accidentally transferred to a new geographical region. The subtle role played by ecological factors is illustrated by cross-infection experiments that remove the effects of transmission barriers.

7.2 Factors affecting the host range

A large number of explicit hypotheses have explicitly postulated factors that could have determined the evolution of the host range.

7.2.1 Biogeographical factors

7.2.1.1 Parasite geographic distribution

Parasites that have a broad geographic distribution will typically also encounter a more extensive range of potential hosts as compared to parasites with a restricted distribution. Nevertheless, even in a broad range, only a few of the suitable hosts may be found. Hence, widely distributed parasites might initially evolve higher specificity, which could then result in local differentiation. On the other hand, to exploit all resources, widely distributed parasites could evolve towards a broader host range while being less well adapted to each of their various local hosts. Ultimately, the direction that specificity evolves in widely distributed parasites may depend on gene flow among localities in host and parasite. There is evidence for several of these expectations. For example, a broader host range correlates with less local adaptation in several experimental cross-infection tests (Lajeunesse and Forbes 2001). Also, more widely distributed parasites had broader host ranges for fleas infesting small mammals (Krasnov et al. 2008). In this case, fleas were more opportunistic in their host choice if widely distributed

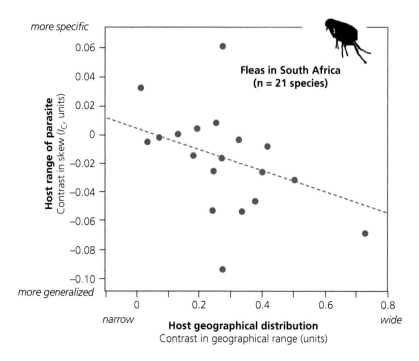

Figure 7.5 Parasite host range vs host geographical distribution. Shown are the data for 21 flea species infesting small mammals in South Africa. The values are corrected for phylogenetic dependencies among flea species (i.e. phylogenetic contrasts instead of raw values). Widely distributed parasite species show low skew (index I_C; y-axis) in their host range (x-axis: phylogenetic contrasts); i.e. they infect host taxa according to their availability. Narrowly distributed parasite species show high skew; i.e. the host range deviates from taxonomic availability and is thus more specific. Silhouette from phylopic.org. Adapted from Krasnov et al. (2008), with permission from Springer Nature.

(Figure 7.5; this pattern was not seen for a second region in North America, though). Similarly, among a sample of 866 parasite species of mammalian hosts, more widespread parasites had more host species. Presumably, there are more opportunities to expand the range if more host species can be infected (Byers et al. 2019). However, the reverse is also observed—some trematodes have broader host ranges on islands compared to their counterparts on expansive continents where the parasites range far and wide (Combes 2001).

Finally, the geographical latitude is a correlate to many other parameters, such as average temperatures, precipitation, day length, or biodiversity in a community. Not surprisingly, therefore, the number of host and parasite species typically varies with latitude; for example, viruses in rodents decline as one moves away from the equator (Bordes et al. 2011). For the host range, there are mixed results. For example, digenean parasites of fish are more

specialized in warmer tropical waters (low latitude), but in monogenean parasites no relationship with latitude is observed (Rohde 2002).

7.2.1.2 Spatial heterogeneity

A heterogeneous environment, e.g. locally differentiated host populations, might make it more difficult for parasites to adapt to any particular population. Therefore, a broad host range and generalism should be more common. Experimentally, phages do indeed show less specialization in experimental heterogeneous, spatially structured populations of bacterial hosts (Hesse et al. 2015). Different locations or populations could also form a geographic mosaic of co-evolution. As each location follows its host–parasite co-evolutionary trajectory, different degrees of specificity may exist across the geographic mosaic at large. Local adaptation is one possible outcome (Gandon and Nuismer 2009). Effects of geographical mosaics on specialization are found, for example, in rust (*Melampsora*) infecting flax (*Linum*) (Laine et al. 2014), as well as in other systems.

7.2.2 Phylogeny and available time

7.2.2.1 Constraints by host phylogeny

Phylogenetic constraints shape characters due to common ancestry. A parasite may not be able to infect a host, entirely because it belongs to a group that has never evolved the necessary mechanisms. For example, many cestodes and Monogenea specialize in specific groups of fish and birds, although host switching across different phylogenetic lineages does occur (Combes 2001; Mendlová and Šimková 2014; Braga et al. 2015). Weak, yet significant correlations of host ranges between pairs of helminth species indicate that parasite sister species often show similarly narrow or broad host ranges (Mouillot et al. 2006). Several fish parasites tend to have more host species when hosts belong to a species-rich taxonomic group, suggesting that there are many hosts available that are similar enough to be infected (Poulin 1992; Sasal et al. 1998). Phylogenetic differences between hosts sometimes are apparent. For ectoparasitic lice, feathers of birds are a different microhabitat from hairs of a mammal. For most parasites, homeothermic hosts are physiologically different from poikilothermic hosts. Moreover, immune defences differ among taxonomic lines. Such differences may drive broad patterns of parasite specificity, determined by host phylogeny, but further factors must act in combination to determine the host range of a parasite. Many parasites of primates, for example, infect closely related hosts less than expected and instead are generalists (Cooper et al. 2012a).

In all, however, the effect of phylogenetic proximity among hosts, also called 'phylogenetic inertia', is fairly common. It has been found—to different degrees—in many different groups, such as in several marine host–parasite systems (Hoberg and Klassen 2002), freshwater fish (Braga et al. 2015; Vanhove et al. 2016), malaria in birds (Clark and Clegg 2017), or for various parasites of bats and rodents (Presley et al. 2015) (see also Table 7.1); sometimes, host and parasite phylogeny are even largely congruent (see Chapter 14).

7.2.2.2 Phylogenetic age of groups

The evolution of host–parasite associations may need time. A simple hypothesis states that in the course of the evolutionary history of a parasite group, the host range does expand (towards generalism), because parasites continue to evolve to use more and more host species. For example, in *Cichlidogyrus*, a species-rich group of monogenean parasites of cichlid fish, specialization is the ancestral state, and some species have evolved to become generalists (Vanhove et al. 2016). Using the number of branching events in a phylogeny (as a ranked proxy for the phylogenetic age), the lineage age among 297 species of fleas, parasitic on small mammals, was indeed positively related to the number of host species used, and as expected from the expansion-over-time hypothesis (Poulin et al. 2006). In a study of 20 congeneric monogenean ectoparasites of fish, however, the advanced taxa were not found to differ in host specificity as compared to the more basal taxa, and contrary to expectation (Desdevises et al. 2002). Phylogenetic age seems, therefore, a conceivable but weak factor for the explanation of host range. Instead, host range appears to evolve rapidly and under selection independently of the age of the lineage.

7.2.2.3 Constraints by parasite group

Different groups of parasites differ in their evolved repertoire of using hosts. For example, viruses can be less specific than helminths in primates (Pedersen et al. 2005). However, parasite taxonomic status necessarily is confounded with many other physiological and ecological factors. Therefore, taxonomic status does not explain much by itself when it comes to parasite specificity.

7.2.3 Epidemiological processes

7.2.3.1 Transmission opportunities

The chances for transmission should affect the range of hosts that are infected. This 'transmission hypothesis' seems relevant in many, but not all, host–parasite associations. Nevertheless, more frequent transmission within the same host species could also lead to more specialization instead. In avian malaria, for instance, hosts of specialist parasites were more abundant (i.e. offering more transmission opportunities) than those of generalists (Svensson-Coelho et al. 2016). For many ectoparasites—such as fleas, mites, or ticks—good transmission opportunities only appear when their hosts come into close physical contact with one another, or when a suitable host encounters the parasite's dormancy stage. If these opportunities occur infrequently, the limited transmission restricts the parasite's dispersal, which can lead to well-structured parasite populations, as found in the mites of South African mice (Mathee and Krasnov 2009). Failure of transmission and the inability to reach new hosts also restricts the host range; it is said that parasites 'miss the boat'.

When there are frequent transmission opportunities, a higher parasite diversity in any one host species is found (Morand 2015), e.g. in cichlid fish and their helminth parasites (Hayward et al. 2017). This is not necessarily associated with a broader host range of the parasite, though. Similarly, even when encounters between hosts are frequent and transmission possible, the genetic structure of parasite populations can show that host specialization due to other causes nevertheless exists, as in fleas of mice and voles (Withenshaw et al. 2016). Often, host characteristics (e.g. body size, longevity) and phylogeny can override the effects of transmission opportunities; this is the case for several parasites infecting desert rodents (Dallas and Presley 2014). A particular determinant of transmission opportunities is social life and host group size. Empirically, group size does sometimes, but not universally, relate positively to parasite prevalence and intensity (Schmid-Hempel 2017), and very often seems to negatively—or to not—relate to parasite diversity or their specialization (Ezenwa et al. 2006b; Patterson and Ruckstuhl 2013). Section 15.1.3 discusses this topic further.

7.2.3.2 Differences in host predictability

When the availability of hosts is predictable, hosts represent a large and persistently exploitable resource; specificity should thus evolve (Combes 2001). This hypothesis has some support. For example, fleas studied in North America and South Africa are consistently more specific if they infest bird or mammalian hosts with larger body masses, assumed to represent a more stable resource; for

example, large organisms are longer-lived (Krasnov et al. 2006). In some cases, haemosporidian parasites of birds are more specialized on more predictable (i.e. more abundant, longer-lived) host species (Svensson-Coelho et al. 2016). Similarly, the proportion of specialized parasites of fish, and fleas of mammals, increases with host abundance, while parasites and hosts are randomly associated otherwise (Vazquez et al. 2005). Monogeneans infecting cyprinids are more specialized on more predictable host species (Šimková et al. 2006) but not in cichlids (Mendlová and Šimková 2014). At the same time, lower host-species diversity facilitates the spread of more specialized parasites (Mitchell et al. 2003). As the examples show, in many studies the proxies for host predictability are host body size, longevity, or abundance (Sasal et al. 1999; Šimková et al. 2006; Mendlová and Šimková 2014).

7.2.3.3 Transmission mode

Whether parasites become transmitted by direct contact or by vectors should conceivably affect the host range. For example, free-living parasitic stages are often mobile and actively seeking hosts; they can thus actively avoid unsuitable hosts and thus become more specific (Noble et al. 1989). The evidence for this hypothesis is mixed, however. Among many different parasites of 119 primate hosts in the wild, transmission mode (direct, vectors, intermediate hosts) was significantly associated with the degree of parasite specificity (direct transmission less specific), although this varied among parasite groups (viruses, protozoa, helminths) (Pedersen et al. 2005). However, with phylogenetic distances among hosts added, no effect of transmission mode was found in another study with 128 primate species (Cooper et al. 2012a). Contrary to expectations, ectoparasites such as parasitic flies that can disperse widely on their own often show high host specificity (Dick and Patterson 2007). Hence, transmission mode may affect the evolution of the host range, but not always in the expected direction.

7.2.4 Constraints set by life history

7.2.4.1 Host body size and longevity

Apart from the aspect of predictability, large and long-lived hosts can be considered equivalent to large islands in biogeographical theory, offering more resources and diverse niches. On the one hand, within-host parasite abundance and/or parasite richness does often positively correlate with host size, e.g. in primates (Nunn et al. 2003), terrestrial mammals (Hechinger 2013), ungulates (Ezenwa et al. 2006b), rodents (Bordes et al. 2011; Kiffner et al. 2014), bats (Bordes et al. 2008), and fish (Garrido-Olvera et al. 2012; Hayward et al. 2017); in many cases after correcting for host phylogeny. But no relationship of parasite richness or specificity with host body size is also commonly observed, e.g. for fleas on rodents (Krasnov et al. 2004; Kiffner et al. 2014; Morand 2015), monogeneans in fish (Sasal et al. 1999; Šimková et al. 2006), or avian malaria (Svensson-Coelho et al. 2016). Hence, whereas host body size sometimes matters, a general effect of body size is not the rule. Depending on assumptions, theory even suggests that large-bodied hosts should have more rather than fewer generalist parasites if directly transmitted but a varied outcome for trophically acquired parasites, with mixed evidence for these expectations (Walker et al. 2017).

7.2.4.2 Complexity of the life cycle

The utilization of several hosts in a life cycle can either limit or expand the range of hosts. An inherent problem for the parasite is the need to cope with, for instance, immune defences of insect vectors during one stage, and the adaptive immune system of vertebrates after this. Free-living stages of parasites encounter considerable challenges, too, e.g. predation, desiccation, UV light, and so forth. Given the trade-offs for such different needs, this should make it more difficult for parasites to acquire new hosts and thus favour specialization. However, directly transmitted parasites of primates have fewer, rather than more, host species than parasites with an intermediate host (Pedersen et al. 2005). Also, similar species richness independent of transmission mode is found among nematodes (Morand 1996).

7.2.4.3 Selection regimes within the parasite's life cycle

Parasites that have intermediate and final hosts will interact differently with their hosts at every such stage. For parasites that do not interact closely, e.g. when the parasite uses an encysted form during some stages, the host range should be broader, as

selection is likely weaker on cysts than it would be on the active form. Conversely, a narrow host range could result when a stage actively develops in a host. Host range might therefore also be determined by the parasite's capacities to evade the host's immune system (van Baarlen et al. 2007). Unfortunately, the strength of selection on different parasite life history stages is very poorly known, and the merits of this hypothesis are therefore hard to judge.

7.2.5 Virulence and defence

7.2.5.1 Virulence of the parasite

Parasites can cause damage to host tissues (i.e. cause virulence; see Chapter 9). Typically, parasites also use different tissues or sites within their host. Where the colonized tissues are prone to damage by an infection, the effect may limit host range, provided this is associated with lower transmission success (see Chapter 13). Hence, under certain conditions at least, the host range for more virulent parasites (i.e. inducing more damage to the host) should be narrower. This hypothesis also assumes that higher parasite fitness due to higher virulence in one host correlates negatively with fitness in others. In other words, 'mismatched' parasites are not efficient in exploiting a host and would be selected against (Lievens et al. 2018). Serial passage experiments indeed show that the virulence in the original host decreases as the parasite adapts to a new host (Ebert 1998; Barrett et al. 2009). Among different species of malaria infecting primates, a negative correlation between host range and peak parasitaemia (a proxy for virulence) exists; that is, virulent parasites are more host-specific than more benign ones (Garamszegi 2006). More generalist fish parasites also attained lower infection intensities and prevalence (Poulin 1998). The opposite is observed for helminths infecting birds (Poulin 1999) or different strains of the rust fungus, *Melampsora lini*, which produce fewer spores (are less fecund) when they infect a wide range of hosts as compared to the more specialized lineages (Thrall and Burdon 2003).

Alternatively, generalist parasites have more opportunities to infect and therefore can become more virulent, since they do not pay a cost for their virulence; e.g. they can 'escape' to different hosts (Kirchner and Roy 2002). *Bacillus thuringiensis*, for example, is generally virulent but can also infect an extensive range of different hosts. Similarly, parasites with durable stages or those causing opportunistic infections escape the direct costs of their virulence. However, the relationship of virulence and host range is not straightforward, even if the underlying trade-off of virulence and host specificity might be present (Gandon 2004). One reason for this discrepancy is that the evolution of virulence itself is a non-trivial process (see Chapter 13).

The manipulation of hosts by their parasites (see Chapter 8) can set additional constraints on the host range. Manipulation is only possible if the different hosts have similar biochemistries or molecular pieces of the defence machinery that govern the targeted processes. For example, parasites that lodge close to the central nervous system (e.g. trematodes, some fungi) can manipulate host behaviour more precisely. At the same time, this strategy is more demanding and should restrict the range of hosts that can be used in this manner. Indeed, compared to such parasites, those only debilitating their hosts or changing their behaviour unspecifically have broader host ranges (van Baarlen et al. 2007; Fredensborg 2014). This 'manipulation hypothesis' certainly warrants more attention.

7.2.5.2 Immune defences and defensive symbionts

Immune defences are part of the physiological filter (Figure 7.4) and can affect host ranges by their specificity. Hosts that are well defended can force parasites to become host specialists. The defence includes defensive symbionts (e.g. the microbiota; section 4.8) that affect host range and specificity of parasites (Ford et al. 2017; Vorburger and Perlman 2018). For example, aphids harbour defensive symbiotic bacteria (*Hamiltonella*, *Regiella*). These provide specific resistance to different parasitoid species (McLean and Godfray 2015; Parker et al. 2017) or parasite taxonomic lineages (Cayetano and Vorburger 2013). There is indeed increasing evidence that host endosymbionts restrict parasitoid host range. This happens through a variety of processes, e.g. by reducing larval parasite survival

or by reducing the fitness of emerging adults (Monticelli et al. 2019). Vice versa, the parasitoids express putative venom toxins and 'symbiotic' virus particles when infecting aphids. The expression differs among parasitoid species and may underlie the aphid-lineage-specific success of these wasps on their hosts (Dennis et al. 2017).

Together, all of the above hypotheses have some arguments in their favour, but none explains all the observations. One hypothesis usually fits well in a given system, but the same hypothesis fails in another. Many of the factors that determine host suitability are also entangled with one another and hard to separate. It is therefore plausible that the diversity of explanatory factors simply reflects the diversity of evolutionary routes that have led to host–parasite associations more generally.

7.3 Specific host defences

Specific host defence can occur at any level of discrimination between species of parasites or between variants ('strains') of the same parasite. As for host defences more generally, specificity is not necessarily due to the response of the immune system itself. Several other processes also contribute to this pattern, such as behavioural or physical defences.

7.3.1 Specificity beyond the immune system

7.3.1.1 Behavioural defences

Behavioural defences can take several forms: avoidance behaviour to prevent infection, to reduce parasite growth, to facilitate clearance, or to increase tolerance to infection (de Roode and Lefèvre 2012; Curtis 2014; see also section 4.1). Hence, a rich repertoire of behavioural changes before or after an infection is documented, and a long list of behavioural changes is known to avoid specific groups of parasites (Moore 2002; de Roode and Lefèvre 2012). A spectacular example is the 'minor' workers in leafcutter ants (*Acromyrmex* spp.). Minors 'hitchhike' on the leaves carried back to the nest by the larger workers (the 'majors'). The minors behaviourally ward off various parasitic flies (phorids) that attack the transport workers on their trails

(Wetterer 1995). Another example is the long-distance migration by reindeer (*Rangifer tarandus*), which lowers the number of warble fly (*Hypoderma tarandi*) attacks (Folstad et al. 1991). Similarly, great tits behaviourally avoid boxes that are infested by ticks (Christe et al. 1994). Female water striders (*Aquarius paludum*) oviposit in deep water to avoid infections by egg-parasitic wasps (Amano et al. 2008). By and large, avoidance behaviours generally act to exclude entire groups of parasites rather than to exclude certain species or types, e.g. they are generally not very specific at the adequate level of within-population differences.

7.3.1.2 Other non-immunological defences

This category includes a variety of physical and chemical barriers, and typically comprises non-specific defences that act against a wide range of parasites. For example, a suitable integument will provide resistance against fungal spores. Hairs (trichomes) on the surface of plants illustrate another mechanism to prevent infections by parasites more generally (Levin 1973). Skin and the respiratory tract of mammals are surfaces with glands that secrete antimicrobial peptides (AMPs) with a broad spectrum of activity (Agerberth and Gudmundsson 2006). Broadly acting lysozymes are found in eye fluids (Hankiewicz and Swierczek 1974); saliva and breast milk are also known to contain a range of antimicrobial compounds (Moreau et al. 2001). Metapleural glands of ants continuously secrete AMPs, forming a thin film over the entire body surface that is highly effective against microbial infections (Beattie et al. 1986). Hence, defences with barriers and antimicrobial compounds on surfaces are widespread, but these are acting in an unspecific manner.

7.3.2 Specificity of immune systems

The specificity of the adaptive immune system of jawed vertebrates is based on the differential binding of antibodies and of B- and T-cell receptors to different epitopes of the parasite (see section 4.5). Similarly, there is specificity in the response of the innate system, also in connection with the adaptive arm. For instance, complement is activated by the binding of antibodies to an invading parasite in the classical pathway (see Figure 4.3). This eventually activates proteases that kill the invader; besides this, several other processes unfold, such as opsonization of the pathogen.

Invertebrates, too, specifically interact with their parasites. For example, when different clones of the crustacean *Daphnia magna* are exposed to different clones of the bacterium *Pasteuria ramosa*, the likelihood of infection varies according to the combination (Carius et al. 2001) (see Box 10.4). Indeed, many examples show specificity in the innate immune defences of invertebrates (Table 7.2). Moreover, specificity in the innate immune response can also evolve, as shown in experiments with *Tribolium* infected by bacteria (Khan et al. 2017b; Ferro et al. 2019).

7.4 Memory, transgenerational protection

The ability of the immune system to 'learn' from a previous challenge and to increase the efficiency of the second immune response to a subsequent challenge characterizes memory, as discussed in section 4.7. However, genuine memory and immune priming implicitly depend on the capacity for specific defences in hosts.

7.4.1 Evolution of memory and immune priming

Having a memory (in its broadest sense) has clear advantages but also costs in terms of resources, energy, or the potential of self-damage. Such costs should become visible when primed individuals (i.e. those having been exposed to the parasite before) do not become exposed or infected a second time. For example, female mosquitoes primed against malaria (*Plasmodium berghei*) experience a loss of fecundity (their egg-hatching rate declines) compared to non-primed females, while they cannot benefit from protection when not infected again (Contreras-Garduño et al. 2014). Such costs may help to explain why innate immune priming seemingly has not evolved against all pathogens or is not always very specific. Memory and its components—the likelihood of establishment and the duration of memory—evolve against this background of costs, benefits, and the threat imposed by parasites.

Table 7.2 Empirical studies suggesting specificity and/or immune memory in invertebrates.[1]

Organism	Test	Result	Type[2]	Ref.[3]
Sponge: *Callyspongia diffusa*	Tissue transplant	Allografts rejected.	I, S, M	
Amphimedon queenslandica	Genomic analyses	Specificity inferred by diversity of NLR receptors.	(S)	
Ctenophora: *Mnemiopsis leidyi* (sea walnut)	Exposure to bacterial epitopes (Gram-positive, Gram-negative)	Gene expression caries according to primary exposure.	I, S, M	
Cnidaria: *Eunicella stricta* (gorgonia)	Colonial contact	Incompatibility (rejection) of xenogenic and allogenic tissue.	S	
Montipora verrucosa (stone coral)	Tissue transplant	Allografts rejected. Reaction is rapid with specific but slow with unspecific second sets.	I, S, M	
Exaiptasia pallida (sea anemone)	Challenge with bacteria (*Vibrio corallilyticus*)	Protection (for <6 weeks) upon secondary exposure. *V. corallilyticus* is a serious pathogen of corals.	I, S, M	
Nematoda: *Caenorhabditis elegans*	Challenge by bacteria (*Serratia marcescens*)	Specific upregulation of genes (lectins, lysozymes, etc.).	I	
Annelida: *Lumbricus terrestris*, *Eisenia foetida* (earthworms)	Tissue transplant	Xenografts rejected (specificity at level of species recognition).	I, (S), (M)	
Mollusca: *Biomphalaria glabrata* (snail)	Challenge by trematodes (*Schistosoma mansoni*)	Selection possible for strain-specific susceptibility. In other tests, reinfection inhibited.	S I	
Crassostrea gigas (oyster)	Challenge with bacteria (*Vibrio splendidus*)	Haemocyte titre higher and earlier upon secondary challenge. Highly specific.	I, S, M	
Crustacea: *Daphnia magna* (water flea)	Challenge by bacteria (*Pasteuria ramosa*)	Host-clone vs parasite-strain specific infection probability. In other tests, maternal transfer of strain-specific susceptibility.	S (I), S, (M)	
Macrocyclops albidus (copepod)	Challenge by tapeworm (*Schistocephalus solidus*)	Reinfection varies according to sibship of tapeworm, with more protection to same sibship.	I, S, M	
Penaeus monodon (shrimp)	Challenge by white spot syndrome virus (WSSV), viral proteins	Oral vaccination with virus envelope protein leads to lower infection rates.	I	a
Porcellio scaber (woodlouse)	Challenge by heat-killed bacteria	Phagocytic activity of primed hosts higher in homologous combinations, especially for priming with *B. thuringiensis*.	I, S, M	b
Insects: *Galleria monella* (wax moth)	Challenge by bacteria	Hosts can protected by previous injection with sublethal dose.	I	
Periplaneta americana (cockroach)	Challenge by venoms (honeybee, snake)	Lethality of second dose depends on primary challenge. No cross-design.	I	
Zootermopsis angusticollis (dampwood termite)	Challenge by bacteria (*Pseudomonas aeruginosa*), fungus (*Metarhizium anisopliae*).	Mortality reduced upon second challenge with same parasite as compared to controls. A first demonstration of generalized immune memory or increased unspecific sensitivity. Social transfer of resistance against fungus (by altered behaviour).	I, (S), M	c

Species	Challenge	Notes	Response
Drosophila melanogaster	Challenge by bacteria (*A. cloaca*)	Protection when primed but also against other bacteria.	I
Drosophila melanogaster	Challenge by bacteria (*Streptococcus pneumoniae*)	Sublethal challenge protects coarsely specifically against later second lethal challenge; persists for lifetime of fly. Phagocytosis and Toll pathway involved. Not all bacteria can be primed against.	I, S, (M), N [d]
Drosophila melanogaster	Exposure to Drosophila C virus (DCV)	No protection achieved.	N
Anopheles gambiae (mosquito)	Challenge by bacteria (*E. coli, M. luteus*)	Bacterial challenge reduces susceptibility to malaria, *Plasmodium*.	I
Bombus terrestris (bumblebee)	Challenge by bacteria	Survival longer and clearance of infection more likely with homologous as compared to heterologous challenges.	I, S, M [e]
Tenebrio molitor (flour beetle)	Challenge by lipopolysaccharide (LPS)	Long-lasting protection against the fungus *Metarhizium anisopliae*.	I
Tribolium castaneum	Challenge by bacteria	Survival longer with homologous as compared to heterologous challenges, specific to different strains of same bacterium. Not all bacteria can be primed against.	I, S, M, N [f]
Formica selysi (wood ant)	Challenge by fungus (*Metarhizium anisopliae*)	Survival does not differ in challenged and unchallenged ants.	N
Echinodermata: *Lytechinus pictus* (painted sea urchin)	Tissue transplant	Allografts rejected. Foreign grafts rejected.	I, S
Tunicates: *Botryllus* spp. (star tunicate)	Colonial contact, allografts	Fusion or rejection of neighbours based on a single genetic locus.	S [g]
Cephalochordata: *Branchiostoma belcheri* (lancelet)	Challenge by inactivated bacteria	Within-individual priming concluded from differential gene expression.	I, S, M

[1] Modified after Kurtz. 2005. *Trends Immunol*, 26: 186; Milutinović. 2016. *Semin Immunol*, 28: 328; Contreras-Garduño. 2016. *Ecol Entomol*, 41: 351.

[2] Response can be induced (I), is specific (S), shows evidence for memory (M), or no protection is observed (N).

[3] Additional references are: [a] Syed Musthaq. 2014. *Dev Comp Immunol*, 46: 279; [b] Roth. 2009. *Dev Comp Immunol*, 33: 1151; [c] Rosengaus. 1999. *Naturwissenschaften*, 86: 588; [d] Pham. 2007. *PLoS Path*, 3: e26: 1; [e] Sadd. 2006. *Curr Biol*, 16: 1206; [f] Roth. 2009. *Proc R Soc Lond B*, 276: 145; [g] Dishaw. 2016. *Nat Comm*, 7: 10617.

Theoretical analyses show that the conditions favouring a more specific memory are manifold. For example, the value of a specific memory saturates as its 'repertoire' increases, i.e. as more and more specific responses are possible. Furthermore, the lower the cost of having a given memory repertoire, the larger the size of the repertoire that will eventually evolve. Nevertheless, the 'optimal' memory size and specificity will depend more on the probability of reinfections and its protective value than on the composition of the parasite community itself (Graw et al. 2010). Similarly, when faced with highly virulent parasites, specific memory is favoured over unspecific defences (Best et al. 2013). Section 5.6.2 discusses the possible influence of life span on memory evolution. The various conditions under which memory and, in particular, a highly specific memory evolves are far from understood.

7.4.2 Transgenerational immune priming (TGIP)

In many different groups of animals, the immunogenic experience of parents is transferred to offspring and can protect them against infection (Roth et al. 2018)—a phenomenon already demonstrated in around 1900 (Grindstaff et al. 2003). Such 'transgenerational immune priming' (TGIP) is known for a large number of vertebrates (Table 7.3) and an ever-increasing number of invertebrates (Table 7.4) (Roth et al. 2018; Tetreau et al. 2019). TGIP is as a case of extended parental care, even though the protective effect may sometimes extend beyond the next generation. In mammals, antibodies produced after an immune challenge by the mother are transferred through the placenta to the foetus. After birth, mothers can transfer antibodies via breast milk (Hanson 1999). This maternal help is valuable, since newly born mammals, like many other vertebrates, have limited abilities to produce their own antibodies. Therefore, passive protection via the mother is an alternative to bridge the time until the young can develop their defences. The persistence of maternal antibodies lasts anything from a few days in fish to months in humans, and transfer of maternal antibodies probably serves as a signal to upregulate the immune system more generally. The circulation of maternal antibodies

subsides in concert with the increasing production of antibodies by the offspring (Grindstaff et al. 2003; Lemke et al. 2003). In some cases, e.g. in mice, rats, and insects, protection extends for a very long period after birth, and can last up to a lifetime.

Transgenerational effects occur in invertebrate taxa as far apart as crustaceans or molluscs, all of which have no antibodies to transfer (Table 7.4). As a simple mechanism, a general, induced upregulation of immune defences may protect bumblebee workers from queens (mothers) exposed to bacterial antigens (Moret and Schmid-Hempel 2001). In this case, protection associates with an (unknown) factor that is present in the queen's eggs and that is acquired from the biological mother (Sadd et al. 2005; Sadd and Schmid-Hempel 2007) (Figure 7.6a). However, TGIP often is more specific. In bees and beetles, for example, workers become better protected against homologous (by the same bacteria) than against heterologous (by different bacteria) secondary infections (Figure 7.6b). Specific protection through TGIP is now reported for several invertebrate taxa (Table 7.4). Furthermore, in flour beetles, the protective effect can also be transferred via the genetic father, irrespective of the mother (Roth et al. 2010; Eggert et al. 2014) (Figure 7.6b). This fact is remarkable, as fathers only contribute sperm to offspring—they do, however, also transfer seminal fluids during copulation.

The mechanisms of transgenerational protection in invertebrates are largely unknown, but several mechanisms have been suggested (Roth et al. 2018; Tetreau et al. 2019). A simple mechanism may be the transfer of a 'signal', e.g. bacterial fragments passed on by the mother, which would then stimulate the offspring's immune defences. Egg yolk vitellogenin, a ubiquitous insect protein, has been tentatively implemented as a carrier for such fragments in honeybees (Salmela et al. 2015). Alternatively, the effector molecules themselves might be transferred, similar to antibody transfer in mammals. This mechanism might occur in insects where AMPs are present in eggs from challenged but not from naïve mothers. Examples are tenecin-1 in *Tenebrio molitor* (Dubuffet et al. 2015), or gloverin in *Manduca sexta* (Trauer-Kizilelma and Hilker 2015). Eggs of several molluscs (mussels and snails) contain receptors (lectins) and antibacterial factors that are effective

Table 7.3 Examples of transgenerational transfer of protection in vertebrates.

Organism	Test	Result	Remarks	Source
Pleuronectes platessa (plaice)	Females immunized with rabbit red blood cells.	Haemagglutinin activity found in egg yolk.	Transfer of immunity in eggs to offspring.	2
Ophioblennius, Salario (blennies)	Brooding behaviour observed in wild.	Male anal glands produce mucus with antimicrobial compounds. This is rubbed on the nest to protect offspring.	A sexually dimorphic trait. An inducible response?	3
Syngnathus typhle (pipefish)	Injection of heat-killed *Vibrio* or *Tenacibacter* in parents (males, females). Offspring challenged with homo- or heterologous bacteria. Gene expression, haemocyte counts in offspring.	Older (four months) offspring primed against *Vibrio* but not *Tenacibacter*.	Gene expression observed. These pipefish have sex-role-reversed brood care (males have a placenta-like structure).	1
Astatotilapia burtoni (cichlid, mouthbreeder)	Injection of heat-killed *Vibrio anguillarum* in parents. Offspring challenged with same or different (*Tenacibacter*) heat-killed bacteria.	No explicit signs for TGIP, but challenged mothers can invest less into mouth breeding and produce lower quality offspring.	Expression of immune genes in parents and offspring observed. Lower expression in offspring from challenged mothers.	7
Hyalinobatrachium colymbiphyllum (glass frog)	Observation in the wild.	Antimicrobial skin peptides and defensive microbiota transferred to offspring. Protects against fungus.	An inducible response?	14
Gopherus agassizii (desert tortoise)	Comparing mothers sero-positive for bacteria (*Mycoplasma*) and sero-negative.	Offspring of sero-positive mothers without infections; presumably protected by higher antibody titres in egg and blood of young. Transfer of IgM and specific IgG antibodies.	Effect also seen in one-year-old young.	13
Parus major (great tit)	Ectoparasite (flea) loads of mothers manipulated during egg production.	Offspring of ectoparasite-exposed mothers have lower infestation rates of fleas.	Also increased growth rate and survival in the field.	5
Ficedula hypoleuca (pied flycatcher)	Females (mothers) challenged with LPS.	Brood challenged with LPS vs controls. Offspring of challenged mothers have higher LPS-specific antibodies and higher antibody titres (IgG), especially when older (14 days post-hatching).		4
Calonectris borealis (Cory's shearwater)	Vaccination of female against Newcastle disease virus (NDV).	Mother transfers antibodies to offspring up to five years after vaccination. Persists for several weeks in offspring.	Second vaccination boosts levels in offspring. A long-lived bird (over 30 years).	12
Chicken (broilers)	Hens are immunized with various antigens from *E. coli*.	Chicks tested with pathogenic *E. coli*. Offspring from immunized hens survived better than controls at age 7 days or 14 days. No protection when infected by heterologous strain.	Older young (up to 45 days) showed higher resistance.	6
Mouse	Mothers immunized with antibody against viruses (RSV) after giving birth.	Offspring produce antibodies against same virus at higher rate.	Protection is mediated via breast feeding.	11
	Mothers infected with nematodes (*Heligmosomoides polygyrus*).	Offspring of parasitized mothers better able to clear nematode infection.	Protected offspring also have higher growth rates.	8

(*Continued*)

Table 7.3 Continued.

Organism	Test	Result	Remarks	Source
	Female mice immunized with various antigenic compounds.	Mothers transfer antibodies. Protection not only for neonates but extends much longer.	Maternal antibodies act as epigenetic activators for defence of offspring over lifetime.	9
Rat	Mothers inoculated with hantavirus.	Foetuses of pregnant mothers show antibodies. Newborns from immunized mothers infected by virus survive better than those from control mothers.	Transfer of antibodies during pregnancy.	15
Human	Maternal antibodies against measles after vaccination or natural infection of mother.	Antibodies transferred through placenta. Interplay of maternal antibodies and vaccination of offspring important for protection.	A review of literature. Recommendations for public health programmes.	10

Sources: [1] Beemelmanns. 2016. *Ecol Evol,* 6: 6735. [2] Bly. 1986. *Comp Biochem Physiol A,* 84A: 309. [3] Giacomello. 2006. *Biol Lett,* 2: 330. [4] Grindstaff. 2006. *Proc R Soc Lond B,* 273: 2551. [5] Heeb. 1998. *Proc R Soc Lond B,* 265: 51. [6] Heller. 1990. *Avian Pathol,* 19: 345. [7] Keller. 2017. *BMC Evol Biol,* 17: 264. [8] Kristan. 2002. *J Exp Biol,* 205: 3967. [9] Lemke. 2003. *Vaccine,* 21: 3428. [10] Leuridan. 2007. *Vaccine,* 25: 6296. [11] Okamoto. 1989. *J Immunol,* 142: 2507. [12] Ramos. 2014. *Am Nat,* 184: 764. [13] Schumacher. 1999. *Am J Vet Res,* 60: 826. [14] Walke. 2011. *Biotropica,* 43: 396. [15] Zhang. 1988. *Arch Virol,* 103: 253.

Table 7.4 Examples of transgenerational transfer of protection in invertebrates.

Organism	Test	Result	Remarks	Source
Nematodes: *Caenorhabditis elegans*	Initial exposure to flock house virus (FHV) via induced expression in transgenic worms carrying viral DNA.	Small virus-interfering RNAs (viRNAs) are transmitted up to fourth generation when initial experience is made. Tested by heat shock-induced expression.	viRNAs expressed when virus is expressed and dampens viral replication.	11
Molluscs: *Crassostrea gigas* (oyster)	Protective chemical poly (I:C) injected into parental oysters. Test with ostreid herpesvirus (OsHV-1).	Offspring (larvae) from parents primed with poly (I:C) show higher survival when exposed to virus.	Oysters treated with poly(I:C) show protection against virus. Protection only via father.	5, 7
Chlamys farreri (scallop)	Primed with heat-killed *Vibrio anguillarum*. Gene expression, survival of eggs and larvae.	Expression and level of immune protein higher in eggs; higher larval survival.		16
Branchiopoda: *Artemia* (brine shrimp)	Progenitors (clonal reproduction), and the offspring larvae exposed to *Vibrio campbelli*.	Larvae from primed progenitors survive better.		10
Crustacea: *Daphnia magna* (water flea)	Mothers exposed to different strains of the bacterium, *Pasteuria ramosa*.	Offspring receiving homologous challenges (same strain) have higher fecundity than those with heterologous challenges (different strain).		2, 8
Penaeus monodon (shrimp)	Females ready to spawn injected with β-glucans from yeast.	Offspring from challenged mothers survived better when exposed to virus as compared to controls.	WSSV, a pest in aquacultures. General protection by glucans.	6
Insects: *Bombus terrestris* (bumblebee)	Mother queens challenged by heat-killed bacteria (*Arthrobacter globiformis*) prior to colony founding.	Workers from colony tested by challenge with LPS. Eggs, workers from challenged mothers show higher antibacterial activity than controls. Transcriptome differs in challenged offspring from primed others.	No difference in phenoloxidase (PO) activity levels. Cross-fostering experiment shows that effect is due to biological mother, not colony environment.	1, 13, 14
Tenebrio molitor (flour beetle)	Injection of LPS into larvae (parents); tested with response to LPS injections in larval offspring.	Antimicrobial activity higher in offspring of challenged parents.	No effect on PO activity.	9
Tenebrio molitor (flour beetle)	Injection with heat-killed bacteria (*Arthrobacter*, *Bacillus*) in mothers. Offspring survival, etc., with/without exposure to bacteria.	Eggs of primed mothers show higher antibacterial activity. Larvae of challenged mother survive better, irrespective of whether also exposed or not.	Egg size smaller but hatching rate higher for challenged mothers. Not all bacteria can be primed well.	3
Tribolium castaneum (flour beetle)	Mothers or fathers challenged by heat-killed bacteria. Offspring tested for survival under same and different bacterial challenges.	Both parental and maternal protective effects observed.	Paternal reduces offspring fecundity.	12
Trichoplusia ni (cabbage semilooper)	Parents as larvae kept under diets with or without bacteria.	Three immune parameters (PO-enzyme activity, protein expression in haemolymph, transcript abundance) elevated in offspring from bacteria-diet parents.	No paternal effect.	4
Manduca sexta (tobacco hornworm)	Bacterial extract (*Micrococcus luteus*) injected into pupae (parents). Then, eggs tested and exposed to parasitoid (*Trichogramma*).	PO and antimicrobial activity higher in non-parasitized eggs of challenged parents. Also, infection by parasitoid less successful.	No effect on PO activity in non-parasitized eggs.	15

Sources: [1] Barribeau. 2016. *PLoS ONE*, 11: e0159635. [2] Ben-Ami. 2010. *Am Nat*, 175: 106. [3] Dhinaut. 2018. *J Anim Ecol*, 87: 448. [4] Freitak. 2009. *Proc R Soc Lond B*, doi:10.1098. [5] Green. 2016. *Mol Immunol*, 78: 113. [6] Huang. 1999. *Dev Comp Immunol*, 23: 545. [7] Lafont. 2019. *Dev Comp Immunol*, 91: 17. [8] Little. 2003. *Curr Biol*, 13: 489. [9] Moret. 2006. *Proc R Soc Lond B*, 273: 1399. [10] Norouzitallab. 2015. *Fish Shellfish Immunol*, 42: 426. [11] Rechavi. 2011. *Cell*, 147: 1248. [12] Roth. 2010. *J Anim Ecol*, 79: 403. [13] Sadd. 2005. *Biol Lett*, 1: 386. [14] Sadd. 2007. *Curr Biol*, 17: R1046. [15] Trauer-Kizilelma. 2015. *Dev Comp Immunol*, 51: 126. [16] Yue. 2013. *Dev Comp Immunol*, 41: 569.

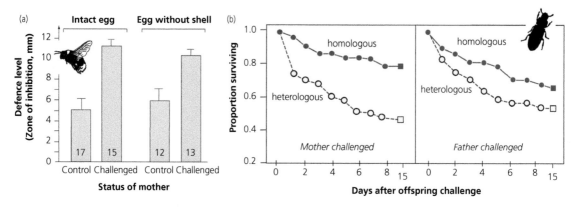

Figure 7.6 Transgenerational transfer of immune protection in insects. (a) Immune protection of offspring in a social insect, *Bombus terrestris*, is mediated by a factor that mothers transmit inside their eggs. The graph shows the strength of the antibacterial response (zone-of-inhibition assay; y-axis) of intact eggs (left) and internal egg extracts (right). In each case, antibacterial activity is restricted to eggs coming from challenged mothers. Adapted from Sadd and Schmid-Hempel (2007), with permission from Elsevier. (b) Mothers or fathers of the beetle *Tribolium castaneum* were challenged (primed) with bacteria (*E. coli*, *B. thuringiensis*). Later, their worker offspring were challenged either with the same (homologous) or different bacteria (heterologous), and survival was observed. For both parents, offspring lived longer with the (primed) homologous challenge as compared to any other combination of challenges. Adapted from (Roth et al. 2010), with permission from John Wiley and Sons.

against infections (Wang et al. 2015a). Also, the direct transfer of vitellogenin, which acts as a modulator of immune defences, might be a way to protect offspring. However, an inherent problem of transferring the effectors is that they may not be sufficient in quantity, nor long-lasting enough, to effectively protect young. Epigenetic modifications are more likely candidates for transgenerational immune protection. Such modifications include the acetylation of histones or the methylation of immune genes. Epigenetic modifications explain priming that lasts into the adulthood of offspring in the absence of an immune challenge (Barribeau et al. 2016). Such epigenetic markers are not necessarily removed in next-generation offspring. This could explain why, in some cases, priming lasts over several generations (Norouzitallab et al. 2015). Currently, the knowledge about such epigenetic mechanisms in invertebrates is still limited.

Protection by TGIP is advantageous for organisms such as social insects, where most of the (female) offspring stay at home and, therefore, are exposed to the same parasites as their mother. In non-social organisms, advantages exist when the dispersal of young is limited, e.g. for flour beetles that typically live in dense sympatric populations. However, immune protection of offspring not only

provides benefits but also has costs. Workers of bumblebees, for example, show higher levels of protective antibacterial activity when the same bacteria had challenged their mother. At the same time, however, they become more susceptible to infection by a prevalent trypanosome (Sadd and Schmid-Hempel 2009). In neonatal vertebrates, very high levels of antibodies are disadvantageous, as they seem to block the development of the neonatal immune system, thus lowering later resistance against infections (Grindstaff et al. 2003; Staszewski and Siitari 2010). As with innate immune priming itself, TGIP also does not work for all parasites. In flour beetles, TGIP against Gram-positive bacteria was more efficient than against Gram-negative bacteria (Dhinaut et al. 2018). Similarly, mosquito offspring altogether showed no effect after mothers were challenged (Voordouw et al. 2008). In other cases, the transgenerational effect may even have affected offspring fitness negatively (Vantaux et al. 2014; Littlefair et al. 2017). Theoretical considerations suggest that TGIP is more likely to evolve in species that have limited dispersal from the natal site, are longer-lived, or may encounter more predictable parasite communities (Metcalf and Jones 2015; Pigeault et al. 2018). The selective advantages of, and thus the evolution of, transgenerational protection will be as diverse as the organisms and their lifestyles themselves (Roth et al. 2018).

7.5 Adaptive diversity and cross-reactivity

The adaptive immune response of jawed vertebrates uses a diversified repertoire of lymphocytes with different specificities to recognize parasite epitopes and parasite-derived peptides. These specificities are not strict, however, but a particular lymphocyte specificity cross-reacts with several different epitopes ('heterologous immunity'). This tails with the general question of how diverse and how specific the defence repertoire should be.

Consider the binding regions for parasite-derived peptides in the MHC class I complex. This region is around nine amino acids (9-mers) long. Is this small number sufficient to reliably distinguish between self and non-self? Calculations show that this is, in fact, the case. Humans probably have around 30 000 different own proteins. This number translates into around 10^7 possible different 9-mers that belong to own proteins. This set needs to be distinguished from the set of possible foreign 9-mers coming from all parasites. To calculate the latter number, all 9-mers that are possible in a typical set of known bacteria and viruses can be enumerated. Surprisingly, all possible sequences that come from bacteria and viruses (and are based on a 9-mer recognition motif) are still sufficiently different from the set of possible 9-mers characterizing human proteins. These two sets overlap in only 0.2 per cent of cases. Shortening the length of the recognition motif to 7-mers would lead to 3 per cent overlap, and 11-mers would overlap in around 0.1 per cent of cases (Burroughs et al. 2004). Hence, although longer sequences than 9-mers would increase the reliability of discrimination, the typical 9-mers are already amazingly well suited to distinguishing pathogens against non-self, with more than 99.5 per cent probability. This capacity incidentally reflects the sizeable evolutionary distance that separates

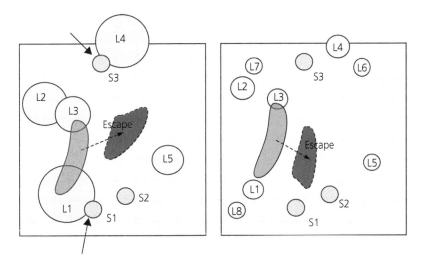

Figure 7.7 Antigenic space. The two grey squares symbolize the encountered antigenic space, i.e. the set of all possible antigens that the immune system encounters, plotted in two dimensions. Different lymphocytes ($L_1 \ldots L_8$; green and blue circles) have different specificities and different degrees of cross-reactivities. This is shown here by the position of lymphocytes in antigenic space, where the size of the circle indicates the degree of cross-reactivity; left is a case with extensive cross-reactivity (green circles), right is a case with narrow cross-reactivity (blue circles). The yellow circles ($S_1 \ldots S_3$) are antigens that represent self. With extensive cross-reactivity (left), fewer lymphocytes are needed to cover the antigenic space but more lymphocytes react to self (arrows) and cause autoimmune failures. The red shapes represent the antigens occupied by a parasite population. With extensive cross-reactivity (left), the parasite population is recognized by two lymphocyte specificities (L_1, L_3), whereas a single specificity (L_3) matches under narrow cross-reactivity (right). The parasite population evolutionarily responds by escape into a region of antigenic space not surveyed by the host population. The direction of escape will differ depending on how lymphocytes survey the antigenic space. Adapted from Fairlie-Clarke et al. (2009), with permission from John Wiley and Sons.

the proteomes of bacteria and viruses from the 'self-proteome' of vertebrates.

Cross-reactivity might seem like a design failure of the immune system. However, calculations show that a certain degree of cross-reactivity is essential to discover a novel antigen, simply because the number of possible encountered antigens (perhaps in the order of 10^{13} different varieties) is much larger than the repertoire of immune cells (Van den Berg et al. 2001; Borghans and De Boer 2002). Cross-reactivity is even needed to cover the antigenic space adequately, i.e. the set of all possible antigenic motifs. With limited cross-reactivity, many foreign antigens are missed, leading to 'holes' in the defence repertoire that can be exploited by parasites to escape recognition. With extensive cross-reactivity, there is a risk of self-damage by 'false alarms'. The optimal level of cross-reactivity should therefore be small for potentially self-

reactive cells and more extensive for those recognizing foreign antigens (Van den Berg and Rand 2007; Fairlie-Clarke et al. 2009). Exactly how extensive cross-reactivity should be is not easy to predict, except that it should vary depending on circumstances (Figure 7.7). In some cases, cross-reactivity depends on the similarity of amino acids in the epitopes of the parasites, e.g. for HIV viruses targeted by T-cells (Frankild et al. 2008). Hence, single amino acid changes can lead to immune escape by the parasite, and allow movement into a part of the antigenic space that is not surveyed by the hosts. Such moves are essential for the emergence of repeated epidemics in cases such as the influenza virus (Ferguson et al. 2003; Koelle et al. 2006; Wikramaratna and Gupta 2009). As yet, we know little about the precise ways cross-reactivity selects for parasite diversification, host range, or parasite speciation.

Important points

- The host range of parasites typically is right-skewed, with most parasites having few host species. Parasite specificity and host range are measured by various methods, and several indices are used for characterization.
- Host range is determined by an ecological and a physiological filter. A number of hypotheses seek to explain what selective pressures shape these two filters. Although all of the hypotheses agree with some observations, none can satisfactorily explain all aspects.
- Parasite specificity is crucial for the ecological and evolutionary dynamics, and specific defences by the host are a

corollary to parasite specialization. Specificity results from behavioural, chemical, or immunological defences.
- Parents can protect offspring by transgenerational immune priming (TGIP). Such transfer can happen via mothers and fathers, but at a cost. TGIP is found in many organisms.
- As an example, specificity and cross-reactivity of lymphocytes in the adaptive immune system of jawed vertebrates must cover the entire antigenic space. An optimal cross-reactivity balances between the detection of parasites and the risk of self-reactivity.

CHAPTER 8

Parasite immune evasion and manipulation of host phenotype

8.1 Parasites manipulate their hosts

A parasitic infection often spreads without any obvious reaction by the host's immune system, even when the infection eventually leads to severe damage. The case of *Bacillus anthracis* (Box 8.1) is just one of very many examples that illustrates the situation. Immunologists and microbiologists have unravelled a fantastic and sometimes bizarre variety of mechanisms by which parasites escape or modulate the host's immune response in their favour. In this way, pathogens such as *B. anthracis* can replicate to high numbers to become a fatal threat without being checked by the defences.

Parasites do not merely hold out against the host's immune defences—they actively evade, sabotage, or manipulate the immune system and alter the host phenotype (e.g. by changing host behaviour) for their benefit. Here, these diverse phenomena are summarized as 'immune evasion', i.e. when host defences are blocked, and 'manipulation', when the host phenotype is changed. The latter also refers to the 'adaptive host manipulation hypothesis' (Holmes and Bethel 1972). At the same time, 'manipulation' is a generic term covering parasite interference with host defences more generally. Evasion and manipulation are widespread and evolved in all major groups of parasites, although not every parasite has such effects (Lafferty and Shaw 2013).

Some parasites even manage to utilize and feed on the host's immune response directly. For example, *Leishmania* use host cytokines as growth factors and tapeworms utilize antibodies as nutrients (Damian 1997). The principle itself was discovered more than 100 years ago. Paul Ehrlich (1854–1915), one of the fathers of immunology, reported on the 'disappearance of receptors' during infections by African trypanosomes in his Nobel Lecture of 1908. This phenomenon is now known as 'antigenic variation' (Bloom 1979; Damian 1997) and is one of the mechanisms by which trypanosomes escape recognition by the host's immune system.

As a note of caution, most current studies are done in vitro, or with the standard model systems (e.g. the mouse), and may thus not be representative for other systems. It is also often not clear whether an observed change is beneficial for the parasite in the first place (Box 8.2). What is clear, however, is that the manipulations should increase the fitness components of a parasite—avoid clearance by the host's defences (i.e. increase parasite survival), and increase the chances of transmission to the next host (i.e. increase parasite reproduction). On the one hand, manipulation mechanisms can therefore be classified by the mode of action (passive or active) and by the molecular targets that are addressed by the parasite. On the other, they can be classified by the targeted function (clearance, transmission).

Evolutionary Parasitology: The Integrated Study of Infections, Immunology, Ecology, and Genetics. Second Edition.
Paul Schmid-Hempel, Oxford University Press. © Paul Schmid-Hempel 2021.
DOI: 10.1093/oso/9780198832140.003.0008

Box 8.1 Immune evasion by *Bacillus anthracis*

Bacillus anthracis is a Gram-positive, spore-forming bacterium that infects mammals and is a potential bioweapon. Spores reside in the soil and can survive for decades in the environment. They enter a host body through skin lesions, lungs, or the gastrointestinal tract (Goel 2015). Eventually, high parasitaemia (a high number of bacterial cells) in lymph nodes and the bloodstream results, which—combined with other effects—is a lethal condition for the host. Based on studies of chromosomal genes, *B. anthracis* belongs to the *B. cereus* group, containing at least six 'species', which should be considered distinct strains of *B. cereus* rather than separate species (Helgason et al. 2000). Among these strains, *B. anthracis* can produce anthrax toxin, which is encoded by (extrachromosomal) plasmid genes (pXO1) (Mock and Fouet 2001). Anthrax toxin is a major virulence factor that mediates immune evasion. The other key virulence factor, the poly-γ-D-glutamic acid capsule, is encoded by another plasmid, pXO2; it protects the bacterium against phagocytosis (Moayeri et al. 2015). Note that 'toxins' are not poisons, as the name might imply, but are parasite-produced molecules that, typically, are finely tuned to disarm the host's immune response repertoire. The high number of new spores produced and their dissemination from a dead host into the environment ensure the further transmission of this bacterium.

Anthrax toxin consists of three components—lethal factor (LF), edema factor (EF), and protective agent (PA). These already are expressed at the spore stage and by newly germinated spores (Moayeri et al. 2015). During infection, PA first binds to host cell receptors, primarily those of the immune system. By complex mechanisms, pores in lipid bilayers (such as in a cell membrane) are formed, without provoking an immune response. The process aids the transport LF and EF to their targets inside the host cell, where they end up in intraluminal vesicles protected from host proteases (Abrami et al. 2005). The combinations of PA and LF (also called the 'lethal toxin', LT), and of PA and EF ('edema toxin', ET), are released into the host cell cytoplasm. They target multiple host functions, notably those of universal importance for immune defence in almost all cell types. Even at low doses early in the infection, LT evades several host immune responses by suppressing MEK (mitogen-activated protein kinase kinase) signalling, which disables the mitogen-activated protein kinase (MAPK) pathways. These pathways would otherwise trigger several crucial responses, including the release of proinflammatory cytokines and a variety of cellular stress responses. LT can also lyse macrophages, induce apoptosis of endothelial cells, and interfere with antigen presentation by dendritic cells in mice and humans. Similarly, ET primarily elevates intracellular cAMP levels; this causes a variety of additional effects, e.g. the upregulation of receptors for PA (Moayeri et al. 2015). Different bacterial strains appear to vary in their capacities and perhaps even in the precise mechanics of immune suppression (Abrami et al. 2005; Moayeri et al. 2015). For example, one strain of *B. anthracis* (strain 9131) has only one extracellular secreted protein, an inhibitor of metalloproteases (InhA1), which is probably involved in the degradation of antibacterial peptides (Gohar et al. 2005).

As the bacterial population grows, the increasingly higher dose of the toxin causes severe pathogenic effects and eventual host death. The pathogenic effects follow from damage induced by LT and ET to vital systems, e.g. the liver and the cardiovascular system, as well as with the developing sepsis in the bloodstream. Among other things, LT induces vascular leakage, systemic hypoxia, and a shock-like collapse. ET, on the other hand, affects the liver cells (hepatocytes), inducing liver oedema and fluid influx into the intestinal lumen, and seemingly damaging other tissues as well. The entire cascade eventually leads to host death; LT and ET are sufficient to produce such symptoms of anthrax infection. Mutants lacking these elements are attenuated and do not cause such severe damage (Moayeri and Leppla 2004).

8.2 The diversity of immune evasion mechanisms

8.2.1 Passive evasion

Evasion is 'passive' when no parasite-borne molecules are secreted. Passive evasion happens in several ways (Table 8.1). For example, a major strategy is to avoid recognition by the host's immune system in the first place. For this purpose, parasites can produce decoy molecules to divert host receptors or change and hide their molecular surface motifs to escape recognition. A further widespread strategy is to continuously change identities, notably antigenic variation as in African trypanosomes (see section 12.3.2); this was Ehrlich's original observation. Furthermore, two different, co-infecting strains of viruses often recombine to form novel and different subtypes, a phenomenon called 'antigenic shift'. A change of identity can also result from the mutation of epitopes ('mutant escape'). In such cases,

---

Box 8.2 Is manipulation adaptive, and for whom?

The infection by a parasite leads to several changes in the expression of the host immune defences and the visible phenotype of the host, e.g. its behaviour. However, are the observed changes in the host indeed under the control of the parasite? Several possibilities exist. (1) A manipulation has evolved for the benefit of neither the host nor the parasite. It could merely be an unavoidable side effect of the infection, although it accidentally also yields benefits for host and parasite. For example, infection typically causes sickness behaviours, such as fatigue or reduced mobility (Hart 1988), which incidentally frees up resources in favour of the parasite. (2) A change is beneficial to the parasite, but it has evolved as a coincidental side effect of other needs. In this case, selection might eventually lead to the incorporation of this side effect into the parasite's repertoire; it would then become an active strategy. The initial trait is an 'exaptation', i.e. a fortuitous benefit from other adaptations (Gould and Vrba 1982). For example, eye flukes (trematodes infecting the eye) benefit from the fact that the eye's interior is an immune-privileged site, i.e. a tissue with low activity of the immune system. The host's visual acuity is thereby impaired, and it becomes more likely to be a predated by a next host. This initial side effect can be co-opted for the new purpose of increasing the chances of transmission. The distinction between exaptation and adaptation is based on a judgement of what the 'true meaning' of a trait is, i.e. what selection pressure maintains it at present and how has it historically evolved? (3) Finally, the change has evolved due to benefits for the parasite; it is a proper adaptive parasite strategy.

Several criteria have been suggested to identify genuine parasite adaptations (Poulin 1995; Lefèvre and Thomas 2008). For example, an adaptation is more plausible if the same or a similar mechanism has evolved independently in different lineages; e.g. infections by acanthocephalan parasites cause a change in the intermediate host's behaviour (these are gammarids) such that infected individuals become more susceptible to predation by the final host (waterfowl and shorebirds). Infections by many species of Acanthocephala, but also infections by trematodes, cause similar changes that make predation by a next host more likely. These are sometimes quite specific in targeting particular hosts (Fredensborg 2014; Bakker et al. 2017).

Also, measuring the presumed benefits makes the existence of a parasite adaptation more likely. Compared to uninfected controls, gammarids infected by Acanthocephala are indeed eaten more often by birds. The effects are non-specific concerning the predator species, but the transmission is indeed more likely (Jacquin et al. 2014). Furthermore, a higher level of manipulation should lead to more success for the parasite. For instance, malaria-infected mosquitoes show a changed schedule of blood meals as compared to controls (i.e. more frequent but smaller meals). This change leads to increased survival, more host contacts, and thus higher transmission potential for the parasite. Notably, the same changes occur with an immune stimulation alone and are therefore not bound to *Plasmodium* per se (Cator et al. 2015). Such studies are challenging to carry out, especially in the natural setting. The lack of compelling evidence for a benefit to the parasite for any given case has thus prompted some critical reviews of the subject (Poulin 1995, 2000; Hurd 1998; Thomas et al. 2005; Poulin and Maure 2015). Nevertheless, the observed changes are often those that increase parasite transmission or survival within the host. Such plausibility cannot be proof, but it can lead to a reasonable working hypothesis to investigate a case in more detail.

---

the host immune system eventually may fail to track the changing parasites during an infection. Finally, parasites can make themselves unavailable for any recognition or response by the host's immune defences. With quiescence, for example, bacteria have little or no metabolic activity, no cell division and replication, and are therefore hard to recognize or attack. Also, the formation of capsules puts parasites out of reach for the host's immune defences. Changing the LPS surface signature by bacteria such as *Neisseria meningitidis*, *Haemophilus influenzae*, and *Streptococcus pneumoniae* (a Gram-positive that lacks LPS) evades, in particular, the early response by complement. Perhaps, for this reason, these bacteria can early and rapidly multiply, and so become dangerous pathogens (Salyers and Whitt 2002).

8.2.2 Active interference

With active evasion, parasites produce specific molecules that interfere with recognition, block or modulate the host's regulatory networks, distort essential cellular functions such as cell motility or apoptosis, and impede the actions of effectors. Often, the parasite also induces the host to produce the manipulatory molecules instead. This must be

Table 8.1 Passive evasion of immune defences.

Type of evasion[1]	Observation	Source
Invading immune-privileged tissue (A)	Eye flukes (*Diplostoma spathaceum*) residing in eye. Myxozoa (*Myoxbolus cerebralis*) in brain. Parasitoids hide in privileged sites not patrolled by immune system, e.g. insect fat body.	2, 14, 16
Removing, modifying recognition tags (A)	Viruses scavenge or camouflage tags that give away their presence (e.g. vaccinia virus).	15, 17, 23
	Bacteria degrade secretory immunoglobulins (e.g. *Haemophilus influenza*), modify or shield their PAMPs (*Salmonella*). Modification of surface LPS molecules (e.g. *Haemophilus influenza*, *Neisseria meningitidis*).	6
	Leishmania exposed to antibodies shed antigenic surface components.	4
Molecular mimicry (A)	Trematodes: Molecular mimicry and masking to avoid recognition in snail hosts.	20
Antigenic variation (C)	Changing the antigenic surface during an infection. Known from African trypanosomes, *Plasmodium*, *Babesia*, *Giardia*, *Paramaecia*, *Tetrahymena*, nematodes. Also in bacteria.	3, 12, 18, 21
Novel types (C)	Recombination among strains during an infection produces novel types. Leads to 'escape mutants', e.g. during HIV infections. Also in *Plasmodium*.	5, 8
Quiescence (U)	Parasite is inactive and no longer produces signals. Effective against antibiotics that target cell division activities. Latency in viruses where production of viral proteins strongly downregulated (herpes simplex virus).	11 7.
Capsule formation (U)	Bacteria form polysaccharide capsules (also polypeptides, or protein–carbohydrate mixtures). Escape complement activation, phagocytosis.	13
Biofilms (U)	Bacteria (e.g. *Klebsiella*, *Staphylococcus*, and many others) can form biofilms, which are very inert against host defences, or disinfectants. Can cause serious problems with medical implants.	1, 9, 10, 19, 22

[1] Categories are: Avoidance of recognition (A), changing identity (C), and becoming unavailable (U).

Sources: [1] Arciola. 2018. *Nat Rev Microbiol*, 16: 397. [2] Bhopale. 2003. *Comp Immunol Microbiol Infect Dis*, 26: 213. [3] Blaxter. 1992. *Parasitol Today*, 8: 243. [4] Bloom. 1979. *Nature*, 279: 21. [5] Gomes. 2016. *Front Microbiol*, 7: Article 1617. [6] Hornef. 2002. *Nat Immunol*, 3: 1033. [7] Kapadia. 2002. *Immunity*, 17: 143. [8] Kent. 2005. *Trends Microbiol*, 13: 243. [9] Koprivnjak. 2011. *Cell Mol Life Sci*, 68: 2243. [10] Le. 2018. *Front Microbiol*, 9: Article 359. [11] Lewis. 2007. *Nat Rev Microbiol*, 5: 48. [12] Sacks. 2002. *Nat Immunol*, 3: 1041. [13] Salyers. 2002. *Bacterial pathogenesis*, 2nd ed. ASM Press. [14] Schmidt. 2001. *Bioessays*, 23.4: 344. [15] Seet. 2003. *Annu Rev Immunol*, 21: 377. [16] Sitjà-Bobadilla. 2008. *Fish Shellfish Immunol*, 25: 358. [17] Tortorella. 2000. *Annu Rev Immunol*, 18: 861. [18] Turner. 2002. *Parasitology*, 125: S17. [19] Van Acker. 2014. *Trends Microbiol*, 22: 326. [20] Van der Knaap. 1990. *Parasitol Today*, 6: 175. [21] van der Woude. 2004. *Clin Microbiol Rev*, 17: 581. [22] Vuotto. 2017. *J Appl Microbiol*, 123: 1003. [23] Yewdell. 2002. *Curr Opin Microbiol*, 5: 414.

the case for viruses but also occurs in most other parasites. For example, *Toxoplasma* secretes proteins that activate host kinases that, in turn, affect interferon regulation. *Leishmania*, in its chronic stage, shows antigenic molecules that activate macrophage proteins, which in turn deactivates specific signalling pathways (Mahanta et al. 2018). *Listeria* secretes (virulence) factors that induce the host to recruit actin, which helps to form a coat, protecting the bacterial evader (Bah and Vergne 2017). These mechanisms change the genomic or proteomic function in the host, e.g. by altering gene expression (Adamo 2013; Biron and Loxdale 2013). Even memory formation can be affected. Counterintuitively, several protozoans (e.g. *Plasmodium*, *Leishmania*, *Toxoplasma*) produce homologs of proinflammatory macrophage inhibitory factor (MIF). However, this

prolongs the survival of macrophages, where the parasites reside and inhibit immunological memory (Ghosh et al. 2019). There is a mind-boggling diversity of active evasion mechanisms in all parasite groups—from viruses that induce Ig-superfamily proteins for modulation at all levels (Farré et al. 2017), to secreted modulatory proteins of helminths (Robinson et al. 2013; Maizels et al. 2018). This diversity is much too broad to do justice here (Table 8.2). Collectively, active interventions can change the shape of the disease space, allowing pathogens to follow a trajectory more favourable to their success (Box 8.3).

Especially in bacteria, the modulatory molecules are deployed in different ways—acting at short range or more distantly. For example, bacterial adhesins and invasins are typically membrane-bound proteins acting at short range, by contact with host cells

Box 8.3 Manipulation and evasion in disease space

Depending on the mechanisms, the parasite load within a host can be very low when evasion and manipulation start to take effect. Alternatively, manipulation may only be effective when the parasite has replicated to high numbers (in microparasites) or has grown to large body size (for helminths). High numbers or large size may be needed to produce a sufficient level of manipulative effectors. So far, these quantitative effects on manipulation are not known well (but see section 9.1.2). Regardless of the mechanism, evasion and manipulation change the infection trajectory through disease space in various ways. Note that this refers to the manipulation of different elements like resistance, tolerance, and transmission.

In the idealized rendering of Figure 1, bold manipulation changes the path of the infection through disease space. The path becomes more direct and leads towards the favoured region from the parasite's point of view. In this case, the manipulation mechanisms will affect the immune defences massively and perhaps even change their nature (e.g. preventing the adaptive response). Evasion, by contrast, is shown here as resulting in localized, small changes of the defences (e.g. somewhat reducing the adaptive response or lowering the titre of antibodies). Evasion nudges the infection trajectory, such that the parasite takes another route to its favoured region, but may not wholly reach it. As mentioned in the main text, manipulation and evasion are related phenomena. The distinction made here is for illustrative purposes only, serving to guide thoughts on how such strategies may alter the disease space and the resulting infection trajectory.

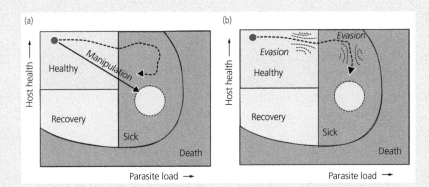

Box 8.3 Figure 1 Manipulating parasites in disease space. In this sketch, (a) host defences normally force the infection on a tortuous path (broken line) that will not reach the preferred zone (blue area) where the parasite reaches its highest fitness, e.g. an area in disease space associated with allowing high transmission rates. With parasite manipulation (solid line), host defences are altered, such that the parasite may take a more direct path and reach its preferred region. (b) Evasion avoids host defences or reduces their effect, but may only slightly change the infection path. Here, evasion is symbolized by two regions with dotted lines. The infection path (broken line) is similar to the 'normal path', but may lead the parasite into or close to the preferred zone. There are alternative renderings of the effect of manipulation or evasion.

(e.g. M-protein in *Streptococcus pyogenes*). Secreted proteins can also act at short range by taking effect near or on the bacterial surface, and in the immediate surroundings of the parasite (EndoS protein, *S. pyogenes*). However, many bacteria inject their modulatory proteins by specialized secretion systems directly into the host cell. This method, too, necessitates close contact with the host cell.

Bacteria have evolved several different secretion systems that allow the transport of proteins from the cell interior across the membrane to the outside

(Tseng et al. 2009; Galán et al. 2014) (Figure 8.1). The molecular details differ among these systems, and the secreted proteins follow different nomenclatures. The so-called type III and type IV systems are of particular interest in the pathogenic bacteria. Type III systems are known from Gram-negative bacteria and are responsible for transporting molecules such as toxins to the bacterial surface and into a host cell. These systems become activated upon contact with host cells (Galán et al. 2014) ('contact-dependent secretion'). For example, *Shigella* injects several

Table 8.2 Examples of active immune evasion.

Evasion mechanisms, targets, organisms	Source[1]
(a) Recognition:	
Viruses: Sequester, modify, relocate, degrade, cleave intracellular receptors, e.g. RIG. Produce decoy receptors to divert recognition. MHC decoy molecules to interfere with antigenic presentation (poxvirus, herpesvirus). Disabling MHC class I molecules by interference (cytomegalovirus).	9, 11, 19, 38, 60
Bacteria: Degrade secretory immunoglobulins (*Haemophilus*), modify or shield their PAMPs (*Salmonella*). Modulate co-stimulatory molecules. Interfere with antigen presentation. Interference with receptor crosstalk. Blocking of neutrophil receptors with CHIPS.	24, 25, 27
Protozoa, helminths: Presentation of competing ligands to divert recognition (*Plasmodium*, schistosomes, nematodes). PfEMPs proteins in *Plasmodium*.	21
Use of antigenic variation in many taxa: African trypanosomes, *Plasmodium*, *Babesia*, *Giardia*, *Paramaecium*, *Tetrahymena*, nematodes.	57
(b) Regulatory networks:	
Viruses: Interfere with interferon signalling; modulate TNF by producing homologs; block receptor activation in natural killer cells (HIV, HTLV); regulate MHC class I expression (HCMV HIV, SIV, MCMV). Vaccinia virus (VCAV): suppressing innate responses, inflammation, etc. Parasite protein A49 blocks NF-κB activations. Homologs block the Imd pathway in insects. Target adaptor proteins (MAVS, TRIF, STING), kinases (IKKε, TBK1), transcription factors (NF-κB, STAT) (hepatitis C virus, adenovirus, papillomavirus, KSHV, Lassa; Ebola, HIV). Co-opt negative regulatory pathways. Interfere with all complement pathways.	12, 14, 22, 23, 28, 29, 33, 35, 40, 43–45, 53, 55, 61
Bacteria: Interfere with proinflammatory cytokine secretion and signalling (*Mycobacterium*, *Bacillus*, *Yersinia*, *E. coli*), and with antigen presentation by DCs (*Mycobacterium*, *Chlamydia*); disrupt signalling for effector function of T- and B-cells (*Yersinia*, *Hyalobacterium*, *Neisseria*); prevent cell fusion by retaining crucial host signals. Interfere with all complement pathways. Block ubiquitination. Cleave Igs by own proteases.	10, 18, 48, 63, 64
Leishmania: Manipulates host cytokines and so disables T-cells; prevents induction of IL-12 in macrophages selectively but leaves other proinflammatory cytokine pathways intact; induces IL-10 to avoid clearance; inhibits cytolytic pathway of the complement cascade by proteolytic degradation of host proteins or by active release of signalling compounds. Block recruitment of immune cells to site of infection.	34, 37
Trypanosoma: Suppresses in DCs the induction of IL-12, TNF-α. Components of the Variable Surface Glycoproteins (CVSG) are immunomodulatory for complement. Capture of regulatory signals from bloodstream.	41, 43, 49, 51
Toxoplasma: Affects host gene transcriptions, dysregulation of signalling pathways; affects cytokines, cell migration, inflammation, apoptosis.	30

(c) Effector mechanisms:

Viruses: Interfere with macrophages, apoptosis, B-cells. Escape cytotoxic T-cells. Inhibit translation of effector proteins. Block effectors of complement, e.g. MAC. Prevent antibody production.	15, 58
Bacteria: Degrade antimicrobial peptides or reduce efficacy (*Staphylococcus*). Prevent being phagocytosed, interfere with macrophages (Yops in *Yersinia*), neutrophils. Induction of host cell apoptosis (*Salmonella*, *Shigella*); disorganize host cytoskeleton (*Yersinia*); escape from or impair vacuoles/lysosomes (*Listeria*, *Yersinia*); evade toxicity from immune system-produced reactive intermediates by catabolizing these molecules (*Mycobacterium tuberculosis*). Disorganize host cytoskeleton (*Listeria*, *Yersinia*). Disrupt function of T- and B-cells (*Yersinia*, *Hyalobacterium*, *Neisseria*). Prevent binding of MAC (*Streptococcus*, *Staphylococcus*). Escape trapping by neutrophils (Pneumococci).	3, 6, 18, 38, 39, 47
Hookworms, *Trypanosoma*, *Entamoeba* subvert neutrophils, trapping by extracellular nets, effects of NOx.	5, 8
Leishmania evades lysis by preventing insertion of host attack complex into its membrane with extended surface molecules and cleavage of crucial host proteins.	
Malaria-infected red blood cells form rosettes that are hard to attack. Cytoadherence to host endothelia, auto-agglutination to prevent attack and clearance (e.g. *Plasmodium falciparum*, *Babesia bovis*).	2, 21
Trematodes: Schistosomes deploy serine proteases to disrupt host protein production.	16

[1] References covering several topics: 1, 4, 7, 13, 17, 20, 26, 31, 32, 36, 42, 46, 50, 52, 54, 56, 59, 62.

Sources: [1] Agrawal. 2017. *Front Microbiol*, 8: article 1117. [2] Allred. 1995. *Parasitol Today*, 11: 100. [3] Bah. 2017. *Front Immunol*, 8: Article 1483. [4] Beachboard. 2016. *Curr Opin Microbiol*, 32: 113. [5] Begum. 2015. *Front Microbiol*, 6: Article 1394. [6] Beiter. 2006. *Curr Biol*, 16: 401. [7] Blaxter. 1992. *Parasitol Today*, 8: 243. [8] Bouchery. 2020. *Cell Host & Microbe*, 27: 277. [9] Bowie. 2008. *Nat Rev Immunol*, 8: 911. [10] Brubaker. 2003. *Infect Immun*, 71: 3673. [11] Chan. 2016. *Nat Rev Microbiol*, 14: 360. [12] Chatterjee. 2002. *Science*, 298: 1432. [13] Christiaansen. 2015. *Curr Opin Immunol*, 36: 54. [14] Coccia. 2015. *Semin Immunol*, 27: 85. [15] Deng. 2017. *Proc Natl Acad Sci USA*, 114: E4251. [16] Dvorak. 2018. *Int J Parasitol*, 48: 333. [17] Epperson. 2012. *Immunol Rev*, 250: 199. [18] Flynn. 2003. *Curr Opin Immunol*, 15: 450. [19] Gack. 2014. *J Virol*, 88: 5213. [20] Ghosh. 2019. *Front Immunol*, 10: Article 1995. [21] Gomes. 2016. *Front Microbiol*, 7: Article 1617. [22] Guidotti. 2001. *Annu Rev Immunol*, 19: 65. [23] Hahn. 2003. *Curr Opin Immunol*, 15: 443. [24] Hajishengallis. 2011. *Nat Rev Immunol*, 11: 187. [25] Hmama. 2015. *Immunol Rev*, 264: 220. [26] Hornet. 2002. *Nat Immunol*, 3: 1033. [27] Khan. 2012. *PLoS Path*, 8: e1002676. [28] Lalani. 2000. *Immunol Today*, 21: 100. [29] Lamiable. 2016. *Proc Natl Acad Sci USA*, 113: 698. [30] Lima. 2019. *Front Cell Inf Microbiol*, 9: Article 103. [31] Locksley. 1997. *Parasitology*, 115: S 5. [32] Loukas. 2000. *Parasitol Today*, 16: 333. [33] Ma. 2016. *Cell Host & Microbe*, 19: 150. [34] Mahanta. 2018. *Front Immunol*, 9: Article 296. [35] Mansur. 2013. *PLoS Path*, 9: e1003183. [36] Marques. 2016. *Curr Opin Microbiol*, 32: 71. [37] Martinez-Lopez. 2018. *Front Microbiol*, 9: Article 883. [38] Mercer. 2013. *Trends Microbiol*, 21: 380. [39] Mitchell. 2003. *Nat Rev Microbiol*, 1: 219. [40] Neufeldt. 2018. *Nat Rev Microbiol*, 16: 125. [41] Norris. 1991. *J Immunol*, 147: 2240. [42] Nothelfer. 2015. *Nat Rev Microbiol*, 13: 173. [43] Onyilagha. 2019. *Front Immunol*, 10: Article 2738. [44] Orange. 2002. *Nat Immunol*, 3: 1006. [45] Orzalli. 2014. *Annu Rev Microbiol*, 68: 477. [46] Pieters. 2001. *Curr Opin Immunol*, 13: 37. [47] Pietrocola. 2017. *Front Cell Inf Microbiol*, 7: [48] Portnoy. 2005. *Curr Opin Immunol*, 17: 1. [49] Ramirez-Toloza. 2017. *Front Microbiol*, 8: Article 1667. [50] Sacks. 2002. *Nat Immunol*, 3: 1041. [51] Saeij. 2002. *Parasitology*, 124: 77. [52] Seet. 2003. *Annu Rev Immunol*, 21: 377. [53] Shao. 2019. *Front Microbiol*, 10: Article 532. [54] Stijlemans. 2016. *Front Immunol*, 7: [55] Taylor. 2013. *Immunology*, 138: 190. [56] Tortorella. 2000. *Annu Rev Immunol*, 18: 861. [57] Turner. 2002. *Parasitology*, 125: S17. [58] Uebelhoer. 2008. *PLoS Path*, 4: e1000143. [59] Underhill. 2002. *Annu Rev Immunol*, 20: 825. [60] Unterholzner. 2019. *Immunology*, 156: 217. [61] Weinheimer. 2015. *PLoS Path*, 11: e1004711. [62] Yewdell. 2002. *Curr Opin Microbiol*, 5: 414. [63] Young. 2002. *Nat Immunol*, 3: 1026. [64] Zhou. 2015. *Cell Microbiol*, 17: 26.

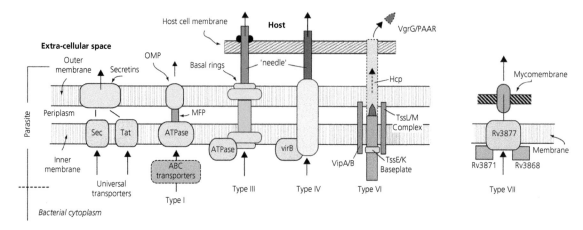

Figure 8.1 Secretion systems of bacteria. Gram-negative bacteria have an inner and an outer cell membrane, separated by the periplasm. To shuffle molecules from the inside (cytoplasm) to the outside, several secretion systems exist. *Universal transporters* are found in Gram-positive and Gram-negative bacteria. Sec (general secretory pathway) and Tat (two-arginine pathway) transport molecules into the periplasm from where they are transported to the outside, e.g. by secretins. *Type I secretion systems* consist of three major components: ABC transporters, membrane-fusion protein (MFP), and the outer membrane protein (OMP). The system provides a transmembrane channel to excrete bacterial proteins (examples of excreted molecules are: virulence factors such as metalloproteases in several plant pathogenic bacteria). *Type III secretion systems* deliver bacterial proteins across the bacterial membranes and across the host membrane into the host cytosol. This 'injectosome' consists of basal rings, spanning bacterial membranes and periplasm, a translocation pore that is inserted into the host membrane (symbolized by the black dots) that tips a needle (e.g. in *Yersinia*), a filament (*Salmonella*), or a pilus (*Pseudomonas*); the injectosome penetrates the host cell membrane (example of excreted molecules: Yop proteins of *Yersinia*). The systems are energized by the associated ATPases (incl. virB). *Type IV secretion systems* transport nucleic acids in addition to proteins; the virB protein is common to all bacteria that have this system. Type IV is found in Gram-positive bacteria with cell envelopes, too (example of excreted molecules: pertussis toxin of *Bordetella pertussis*). *Type VI secretion systems* are contact-dependent systems, possessing bacteriophage-like subunits that act like a loaded spring. The tube is formed by Hcp (haemolysin co-regulated protein), with a cap (VgrG/PAAR proteins). Six proteins (Tss) provide base plates and cross-membrane complexes; VipA/B is a tubular structure, similar to phage tail sheaths. The system can deliver effectors into a wide range of eukaryotic and prokaryotic cell types (example of effector: evpP in *Edwardsiella tarda*). Drawing adapted from Ho et al. (2014), with permission from Elsevier. *Type VII secretion system*: in some Gram-positive bacteria (especially the mycobacteria), the cell wall is rich in lipids (the mycomembrane). This secretion system is specialized to cross these particular membranes using conserved proteins B to E (green; example of excreted molecules: virulence factor ESX-1 in *Mycobacterium*); sketched after Abdallah et al. (2007) and Houben et al. (2014). Adapted from Tseng et al. (2009), under CCBY 2.0.

bacterial proteins (invasion plasmid antigens; Ipa's) into the cells of the host gut epithelium. Many genes of the type III machinery are similar to genes involved in the flagellar export proteins. Hence, they probably evolved from the ancestral system that is responsible for flagellum assembly (Mescas and Strauss 1996; Abby and Rocha 2012). Some pathogenic bacteria use the type IV secretion systems to transport virulence factors across the membrane (Burns 2003). The type IV system is ancestrally related to those permitting bacterial conjugation (Schröder and Dehio 2005; Grohmann et al. 2018). Type VI is a contact-dependent injector. It shares similarities with tail-associated protein complexes of bacteriophages (Silverman et al. 2012). Note that these types themselves each con-

tain extensive and quite diverse families of such secretion systems.

Some parasites—the viruses in particular—have 'captured' genes from their hosts and integrated them into their genome to produce molecules that disarm host immunity (Howell 1985). These genes code for molecules that are close to the original host molecules ('molecular mimicry'), regulating, for example, the host immune response, such as receptors and interleukins (Farré et al. 2017). By gene capture, original host cytokines are turned into 'virokines' that can interfere with the cytokine network of the host (Kotwal and Moss 1988; Lamiable et al. 2016). Capture also co-opts and deploys negative regulators of cytokines (Taylor and Mossman 2013). Host genes originally coding for receptors can also become captured and turned into decoy receptors that very effectively subvert the host's immune response (Upton et al. 1991; Struzik et al. 2015). Gene capture can even lead to the mimicry of the binding surfaces of the host, such that the host's own signals bind to a parasite decoy instead (Guven-Maiorov et al. 2016). Molecular mimicry, due to captured or independently evolved genes, is taxonomically widespread in viruses, protozoans, and bacteria.

8.2.3 Functional targets of immune evasion

An infecting parasite must overcome a series of successive immune responses by the host. In the case of vertebrate hosts, for example, the subsequent steps in the defence cascade have prompted the evolution of corresponding mechanisms of immune evasion by the parasite (Figure 8.2). Each of the primary defence compartments—recognition, the signalling network, and the effectors—have become targeted.

8.2.3.1 Escape recognition

Passive evasion and active interference impede recognition in a variety of ways. This includes 'molecular mimicry', or the presentation of competing ligands to misguide recognition. Viruses modify antigenic peptides, produce decoys, or otherwise disable MHC molecules. Viruses also sabotage intracellular receptors in many different ways. Coronaviruses are RNA viruses that have caused several pandemics, such as SARS (during 2000–2003), MERS (2013–2014), and the more recent Covid-19 (2020–2021). They can evade sensing by coding for a non-structural protein, nsp15, that degrades RNA signatures (Kindler and Thiel 2014; Deng et al. 2017); besides, nsp1 suppresses the translation of host proteins (Schubert et al. 2020).

8.2.3.2 Evasion of early responses

Many parasites have evolved mechanisms to evade complement. Viruses, for example, target key

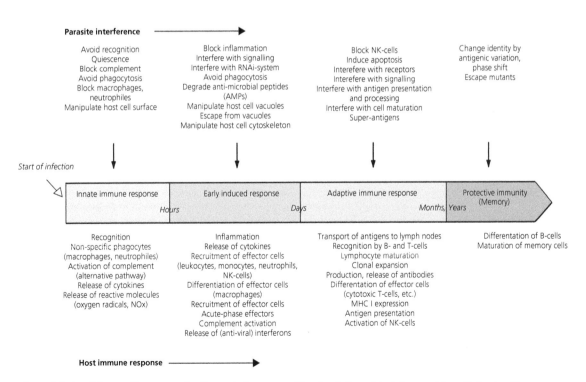

Figure 8.2 Host defence and immune evasion. The graph shows some of the steps in the vertebrate immune defence (bottom) and the corresponding interference by parasites (top). The infection starts on the left and is cleared or tolerated towards the right. Adapted from Schmid-Hempel (2008), with permission from Elsevier.

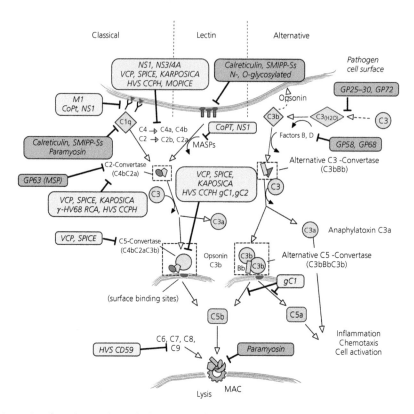

Figure 8.3 Parasite evasion of complement. The graph shows the complement pathways as in Figure 4.3, with an overlay of interference molecules (in *bold*) used by viruses (enclosed in light blue cartouches, after Agrawal et al. 2017), or by protozoa and helminths (light red cartouches, after Shao et al. 2019). Blocking symbols point to the targeted elements of the cascade. *Abbreviations*: KAPOSICA (Kaposi's sarcoma-associated herpesvirus inhibitor of complement activation), NS1 (non-structural protein 1 of flaviviruses), NS3/4A (non-structural protein3/4A of hepacivirus), gC1and gC2 (glycoprotein C of HSV-1 and HSV-2), CoPt (human astrovirus coat protein), M1 (M1 protein of INFLV), SMIPP-S (scabies mite proteolytically inactive serine protease paralog), MSP (major surface protein). Further definitions, see cited literature.

components, such as C3 convertases, C5b formation, or the membrane attack complex (MAC) by a variety of means (Figure 8.3). In hepatitis C, specific viral products bind to complement and disable T-cells (Hahn 2003). Bacteria interfere with the MAC by using self-expressed or host-derived proteases to prevent its binding to the parasite membrane (Pietrocola et al. 2017). Neutrophils or macrophages are essential for an early and generalized response, and these are targets of evasion, too (Urban et al. 2006). *Bordetella pertussis* uses its type III system to inject the effector molecule, BopN, into host cells. This stimulates anti-inflammatory interleukin-10 (IL-10) and blocks neutrophil recruitment (Kobayashi et al. 2018). *Mycobacterium tuberculosis* infects and persists in macrophages. It can subvert the cell's defence mechanisms. For this purpose, *M. tuberculosis* expresses

molecules that have a broad range of effects. For example, the molecules impede phagocytosis and autophagy, suppress the apoptosis of infected host cells, disrupt the inflammasome, alter the production of reactive oxygen and NOx, impede antigen presentation and MHC II expression, and distort vesicle trafficking (Hmama et al. 2015; Liu et al. 2017).

8.2.3.3 Manipulate the signalling network

Parasites interfere with signalling networks in many ways (Figure 8.4). Furthermore, targeting signalling seems to be a major evasion strategy more generally. For example, IL-10 is a prominent natural negative regulator of inflammation, T-cell responses, and the maturation of dendritic cells (DCs). Due to its many functions, IL-10 is a significant target of viruses. Viruses either stimulate its

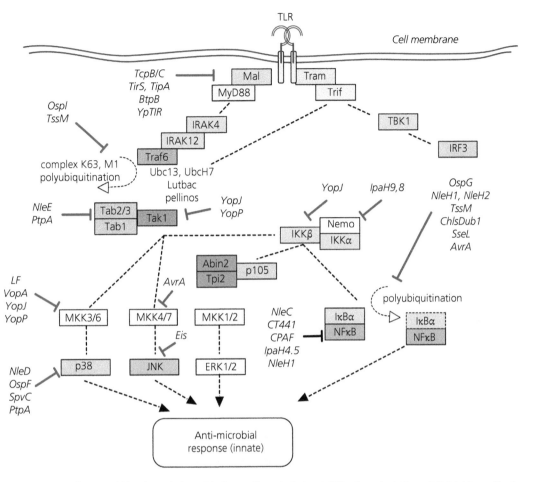

Figure 8.4 Blocking of antimicrobial pathways by bacterial effectors. Shown is the innate TLR pathway (as in Figure 4.6). It is triggered by the transmembrane domains of the TLR and the recruitment of adaptor proteins (e.g. MyD88). Boxes and dashed line indicate the signalling steps (*Left:* MAPK path; *Right:* NF-κB path). The cascades result in the transcription of cytokines and antimicrobial compounds. Bacterial effectors interfere with signalling at various steps, as indicated with the red blocking symbols; effector molecule abbreviations shown in italics. Further details or abbreviations, see source. Adapted from McGuire and Arthur (2015), under CCBY.

production (e.g. HIV, hepatitis B and C virus) or code for their own, viral orthologs (herpesvirus, poxvirus, human cytomegalovirus) (Christiaansen et al. 2015). Furthermore, viruses interfere with proper early signalling by targeting adaptor proteins and kinases. For instance, some DNA viruses (adenovirus E1A, HPV18, KSHV) code for proteins that interfere with adaptors needed for viral DNAsensing, notably STING (Orzalli and Knipe 2014; Ma and Damania 2016; Unterholzner and Almine 2019). Others sabotage the receptor family of TNF (a cytokine) by producing homologs of signals (Seet et al. 2003; Beachboard and Horner 2016).

Finally, transcription factors as the last step in the signalling cascade are targeted by viruses, too, e.g. by degrading NF-κB. Hence, almost any aspect of the chemokine signalling system has become targeted by viruses.

Bacteria, too, interfere with the signalling network (Portnoy 2005), such as downstream of the detection by Toll-like receptors (McGuire and Arthur 2015; Phongsisay 2016) (Figure 8.4), as do virtually all other parasite groups (Alto and Orth 2012). *Plasmodium, Entamoeba, Toxoplasma,* or *Leishmania,* for example, produce homologs of the proinflammatory, multifunctional MIF cytokine

(Ghosh et al. 2019), and also manipulate several other cytokines (Mahanta et al. 2018). These manipulations target the innate arm of the immune system. Nevertheless, the adaptive arm of jawed vertebrates is also attacked by bacteria and other pathogens. For example, *Francisella tularensis* suppresses the release of cytokines to prevent the maturation of DCs, which would be necessary for stimulating T-cells and a full response (Bosio and Dow 2005). MHC class II expression is similarly attacked by various bacteria. Moreover, viruses subvert the host's protein degradation or trafficking pathways to block MHC class I molecules. An infected cell can therefore not signal its status by the external presentation of the MHC–peptide complex (Tortorella et al. 2000; Hewitt 2003). Similarly, viruses, bacteria, or protozoa subvert the functioning of B-cells by activating B-cell suppressors, either by diluting the response through secretion of non-specific antibodies, or by reducing antibody affinity, or by the diversion of cell activity (Nothelfer et al. 2015). As mentioned in section 5.6.5, the robustness of the immune response against sabotage by parasites is under strong selection, and perhaps most important for the architecture of signalling networks. Not least, a signalling network misguided by parasite manipulation can also lead to severe damage to the host, i.e. to immunopathology. The massive release of cytokines (the 'cytokine storm') is one example where damage is caused by misregulation.

8.2.3.4 Avoid being killed by effectors

When the infection is recognized, the immune system deploys cellular defences (e.g. killer cells, phagocytes) and humoral effector molecules (e.g. AMPs, reactive oxygen species (ROS), metalloproteases). On the other hand, parasites have evolved a range of mechanisms to evade or neutralize these effectors. For example, *Staphylococcus aureus* evades neutrophils by secreting interfering molecules (McGuinness et al. 2016). Highly toxic radicals (ROS, NOx) are catabolized and degraded by *M. tuberculosis* (Flynn and Chan 2003). Bacteria evade the effects of AMPs in a variety of ways. This can happen by their sequestration or by repulsion from the bacterial surface. *Staphylococcus*, *Salmonella*, and *Bacillus* modify surface components needed for

attachment, alter electrical membrane charge (e.g. *Staphylococcus*), change membrane permeability, or degrade membranes through secretion of proteases; they also interfere with the cell's efflux pumps (Joo et al. 2016). Evasion correlates with progression to disease and virulence, e.g. in *Typhimurium* (Hornef et al. 2002), *Entamoeba* (Begum et al. 2015), or *Trypanosoma* (Stijlemans et al. 2016). Last but not least, biofilm formation by bacteria is a highly effective way of evading the action of immune effectors or medical interventions (Van Acker et al. 2014; Arciola et al. 2018).

Overall, the different parasite groups show a remarkable degree of parallel evolution in their immune defence evasion mechanisms. For instance, trypanosomes and fungi use equivalent signals to target the host; they also deliver modulating factors in similar ways (Haldar et al. 2006). Unrelated bacterial pathogens mimic the same host proteases (Sikora et al. 2005) and target the same elements (e.g. the Rho protein) of the regulatory cascade (such as the Kruppel-like transcription factors, KLF) to block inflammation or phagocytosis (O'Grady et al. 2007). Parasites as wide apart as viruses, bacteria, protozoa, or helminths have evolutionarily converged on targeting the same signalling components, often those with pleiotropic effects, to enhance their success.

8.2.3.5 Manipulation of auxiliary mechanisms

Parasites need not manipulate the immune system itself to ensure success. For instance, defence often involves the transport of the parasite into specialized cell vacuoles (lysosomes, phagosomes), where the invader is degraded (see Figure 4.5). Bacteria have evolved several ways to block this transport. Should they nevertheless end up there, bacterial pore-forming proteins are released to escape from the lysozyme into the cytoplasm, where the bacteria can survive and spread further, as with lysins in *Listeria monocytogenes* (Portnoy et al. 2002). Many bacterial parasites also manipulate the internal organization of cell vacuoles to disrupt host defences; for example, *M. tuberculosis* arrests the development of the phagosome. Interference with the maturation of the cell vacuole and blocking the associated destruction mechanisms are widespread (e.g. *Legionella*, *Coxiella*, or *Chlamydia*; Underhill and

Ozinsky 2002; Young et al. 2002). Hampering the proper trafficking of molecules is another strategy to evade host immunity. For example, *Salmonella* has evolved ways that prevent the delivery of (toxic) oxidases to its vacuole (Underhill and Ozinsky 2002). Obligate intracellular bacteria, such as *Chlamydia*, *Coxiella*, or *Rickettsia*, have evolved a bewildering variety of mechanisms to alter the functioning of the host cell cytoskeleton by secreting effector proteins into the cell cytosol. Some effectors mimic host proteins to manipulate within-cell signalling, to affect actin and the microtubules of the cytoskeleton. These changes aid pathogen dissemination within the host body (Colonne et al. 2016).

Similarly, protozoans like *Leishmania* and *Toxoplasma* cause long-lasting infections as they modify the host vacuole membrane by the production of homologs of regulatory proteins; as such, they prolong cell life and time for their development (Sacks and Sher 2002). *Plasmodium* is known to produce and deliver hundreds of proteins into the surrounding host cell cytoplasm. These function to remodel its host cell (the erythrocyte) by manipulating the cytoskeleton. The reshaping allows the uptake of nutrients and removal of waste products. *Plasmodium*'s virulence factor, PfEMP1, changes the physical properties of the host cell membrane to enhance cell adhesion to blood vessels such as to avoid being flushed out. All apicomplexan parasites (e.g. *Plasmodium*, *Toxoplasma*, or *Theileria*) share conserved pathways to export manipulative proteins. Some have evolved various mechanisms to manipulate the host cell cytoskeleton (Cardoso et al. 2016; de Koning-Ward et al. 2016).

Typically, host metabolism changes upon infection by a parasite. Some of these changes are an unavoidable consequence of infection, but some may be the result of active manipulation. Any defence is costly in terms of energy and time; metabolic processes are therefore limiting for a response, e.g. the glucose uptake for T-cell activation (Jacobs et al. 2008). Mitochondria are the key for energy metabolism, but also for the release of ROS and other vital functions, and thus a suitable target for parasite manipulation (Lobet et al. 2015). Intracellular bacteria have evolved several mechanisms to affect mitochondria. Examples include inducing host cell enzymes that reduce the production of cytotoxic ROS and NOx. Further manipulations stimulate the production of more metabolites for parasite consumption or changing the lipid metabolism, which alters signalling pathways and causes metabolic shifts. Several other processes exist (Eisenreich et al. 2013; Lobet et al. 2015). *Plasmodium* or *Trypanosoma* reduce blood glucose levels and interfere with the insulin pathways. The subverted host metabolism thereby weakens the defences (Freyberg and Harvill 2017).

8.2.3.6 Microbiota as a target

The microbiota is an essential component of the host's defences (see section 4.8), and parasites have evolved mechanisms to interfere with this function, too. An example is the elimination of microbial competitors from the host gut. For this, *Salmonella enterica* injects selective bactericidal effectors with its type VI secretion system, or stimulates host immune responses, using its type III system to induce inflammation. The latter also affects *S. enterica*, but it subsequently can use the cleared room more efficiently than its competitor to recolonize the gut; in evolutionary biology, this is a strategy of 'spite'. Besides, *S. enterica* also induces more specific responses, e.g. by stimulating host cytokine expression (IL-22) for the production of AMPs, which kill *E. coli* but not *S. enterica* (Anderson and Kendall 2017).

Helminths are no less effective for manipulating the microbiota. The intestinal nematode *Trichuris muris* alters the mouse microbiota after infection to keep out a second, competing infection by nematodes. Such manipulation also facilitates long-term infections for the parasite. A long residence allows for more growth and reproduction (White et al. 2018a); the trematode *Echinostoma caproni* presumably does the same (Cortes et al. 2018). Viruses typically enter the host via epithelial and mucosal surfaces and therefore immediately encounter the specific microbiotas associated with each of those. Nevertheless, some viruses (e.g. human and murine norovirus; Baldridge et al. 2015) even require bacteria for a successful cell entry, likely because membrane binding on bacteria increases virion stability and prevents immediate clearing (Kuss et al. 2011). Several such supporting effects, e.g. of bacterial flagellin, are discussed and could prove to be essential for viral infections (Dominguez-Diaz et al. 2019).

Finally, the microbiota affects the host diet. Therefore, inducing a carving for a diet that may disadvantage potential microbial competitors, or causing a motivational state (hunger, satiation) that leads to the search for certain foods, may help a parasite to establish and multiply. Parasites could directly act on targets such as taste receptors (Alcock et al. 2014), but also indirectly via a change in the microbiota. Currently, these processes are poorly known (Biron et al. 2015). When manipulating the host via the microbiota, parasites may take advantage of the fact that the microbiota itself can be a manipulator of its host (see section 4.8). The microbiota, for example, secretes neuroactive compounds (Biron et al. 2015), affects host feeding preferences, and influences psychological status (Ezenwa et al. 2012; Alcock et al. 2014). Hence, parasites could tap into these processes for their benefit.

8.3 Manipulation of the host phenotype

Beyond the evasion of the immune system, parasites are known to affect their host's phenotype, i.e. host behaviour, morphology, or host life history. Natural selection will favour mechanisms that induce such changes when they increase components of the parasite's fitness—the life span of the infection and the rate of transmission to the next host (Figure 8.5).

8.3.1 Extending infection life span (parasite survival)

A longer duration of infection allows for more extended development, growth (e.g. worms that grow in body size), or multiplication (e.g. bacteria populations), and for more opportunities of transmission. Manipulation by parasites can achieve this goal in various ways.

8.3.1.1 Fecundity reduction

Full or partial castration (fecundity reduction) curtails the host's investment into reproduction and can lead to reallocating resources from host survival to parasite development instead (Baudoin 1975). The crustacean *Sacculina*, parasitizing shore crabs (*Carcinus maenas*; see Figure 3.4), is a spectacular example of host castration. Among other things, infected crabs move to deeper water and start fanning the parasite eggs (as they would typically do with their brood), thus ensuring favourable development (Rasmussen 1959). Snails acting as intermediate hosts for trematodes (e.g. *Schistosoma*) eventually become castrated, which frees resources for the parasite. In this case, the host (e.g. *Biomphalaria*) accelerates reproduction to compensate for some of the expected loss of fecundity. 'Fecundity compensation' is a general strategy of host defence (see section 15.1). Not surprisingly, castration has different effects on males and females, with a higher capacity for parasite growth often found in females (Duneau et al. 2012).

8.3.1.2 Gigantism

Such growth to a large body size reflects a strategy rather than a simple tissue effect, as observed with human elephantiasis caused by the filarial worm, *Wuchereria bancrofti*. It belongs to the various manipulations of the host life history for the benefit of the parasite (Table 8.3). With gigantism, resources are strategically reallocated beyond a simple saving effect, yet these resources become invested in host body growth (Cressler et al. 2014). The rerouting can be induced by a parasite and takes resources away—mostly from reproduction—towards host maintenance and survival (Baudoin 1975; Hurd 2003). Whereas an expected outcome of infection is a shorter host lifespan, the resource reallocation can instead prolong the lifespan of hosts and the infection. Even without gigantism, longer host life spans are found, for example, in beetles (*Tenebrio*) infected by cestodes (*Hymenolepis*) (Hurd et al. 2001). However, in many cases, no difference is found between infected and healthy individuals, e.g. for tsetse flies infected by trypanosomes (Maudlin et al. 1998). The parasite could nevertheless have increased host life span, because without reallocation of resources it could have died earlier (Lefèvre and Thomas 2008).

Gigantism could also cause hosts to store resources that a parasite can consume at a later

(a) Life span of infection

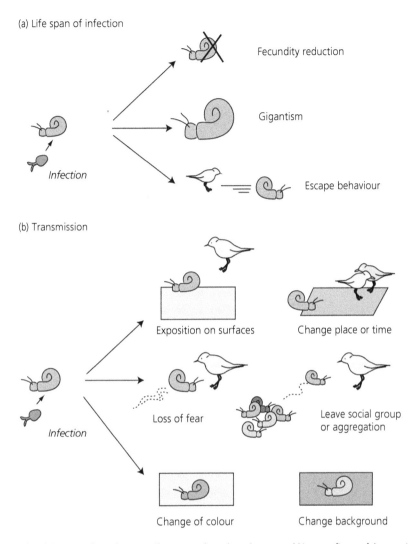

Fecundity reduction

Gigantism

Infection

Escape behaviour

(b) Transmission

Exposition on surfaces   Change place or time

Loss of fear   Leave social group or aggregation

Infection

Change of colour   Change background

Figure 8.5 Parasite-induced changes in host phenotype. The cartoon shows how changes could increase fitness of the parasite. In this hypothetical example, a snail intermediate host is infected by a manipulating parasite (a trematode, red) that needs to become transmitted to a final host (a bird). (a) Manipulations to extend the lifetime of the infection include: reduce host fecundity and re-route resources into host growth and maintenance—gigantism (often associated with fecundity reduction or castration) increases survival and might reduce predation risk; manipulate host escape behaviour to avoid times and places of high predation risk. (b) Manipulations to increase transmission include: parasites change host behaviour such that host moves to surfaces or to times and places where the risk of predation by the next host is higher; losing fear of predators, or getting closer to predators, e.g. due to increased foraging activity; leaving a group of individuals that offers protection; changing to conspicuous colour, or increased preference for the 'wrong' background.

stage; a strategy of 'temporal storage' (Ebert et al. 2004). The benefit of such a strategy depends on how parasite growth compares to the growth of the host body (Figure 8.6). Such complications make it difficult to test the real effect and its consequences. Experimentally, the manipulation by the parasite could be eliminated and compared to infections where the parasite has an intact manipulatory repertoire. Infections by castrating parasites or those with terminal transmission seem to be involved in many cases of gigantism (Baudoin 1975; Sousa 1983; Minchella 1985; Blaser and Schmid-Hempel 2005).

Table 8.3 Parasite groups and parasite-induced changes of host life history.

Parasite	Host	Observation	Possible function for parasite	Source
Fungi: *Balansia cyperi* *Epichloe glyceriae*	Grasses: *Cyperus virens* *Glyceria striata*	Infected plants grow up to four times the size of uninfected plants. Parasite induces vivipary in the host plant.	Fungus spread by this kind of vegetative reproduction.	3, 16
Bacteria: *Pasteuria ramosa*	Water flea *Daphnia magna*	Castrating parasite. Infected hosts grow larger. Infected hosts also show fecundity compensation.	Maximum parasite spore production at intermediate host life span.	6, 10
Protozoa: *Plasmodium*	Mosquito (*Anopheles*)	Reduction in fecundity of the vector host. Vitellogenesis disrupted, with frequent resorption of follicles.	More resources for parasite.	8
Cestodes: *Hymenolepis diminuta* *Spirometra mansonoides* *Schistocephalus solidus*	Beetle (*Tribolium confusum, Tenebrio molitor*) Many rodents Stickleback (*Gasterosteus aculeatus*) Whitefish (*Coregonus laveratus*)	In beetles: fewer matings. Vitellogenesis downregulated by parasite-produced molecule. In ovary, egg development disrupted by host-derived inhibitor of juvenile hormone. Gigantism in rodents. Infected fish grow faster and are in better condition.	More resources for parasite.	1, 9, 13–15, 17, 18
Trematodes: *Schistosoma* *Trichobilharzia ocellata* *Diplostoma phoxini*	Snail (*Biomphalaria glabrata, Lymnaea stagnalis, Littorina saxatilis*, and many others) Minnow (*Phoxinus phoxinus*)	Infection leads to castration and gigantism. Host neuronal peptide released; inhibits production of gonadotropic hormones, stimulates various growth hormones.	Prolongs host life by reallocating host resources from reproduction. Shell shape change allows more metacercaria to be housed.	2, 4, 5, 7, 11–14

Sources: [1] Arnott. 2000. *Proc R Soc Lond B*, 267: 657. [2] Ballabeni. 1995. *Funct Ecol*, 9: 887. [3] Clay. 1986. *Can J Bot*, 64: 2984. [4] de Jong-Brink. 2001. *Parasitology*, 35: 177. [5] de Jong-Brink. 1997. In: Beckage, editor. *Parasites and pathogens: Effects on host hormones and behavior*. Chapman and Hall, 338 pp. [6] Ebert. 2004. *Am Nat*, 164, Suppl.: S19. [7] Hordijk. 1992. *Neurosci Lett*, 136: 139. [8] Hurd. 2003. *Annu Rev Entomol*, 48: 141. [9] Hurd. 2001. *Proc R Soc Lond B*, 268: 1749. [10] Jensen. 2006. *PLoS Biol*, 4: e197. [11] Krist. 2000. *J Parasitol*, 86: 262. [12] McCarthy. 2004. *Parasitology*, 128: 7. [13] Minchella. 1985. *Parasitology*, 90: 205. [14] Mouritsen. 1994. *J Exp Mar Biol Ecol*, 9: 887. [15] Pai. 2003. *J Parasitol*, 89: 516. [16] Pan. 2002. *Oikos*, 98: 37. [17] Phares. 1996. *Int J Parasitol*, 26: 575. [18] Pulkkinen. 1999. *J Fish Biol*, 55: 115.

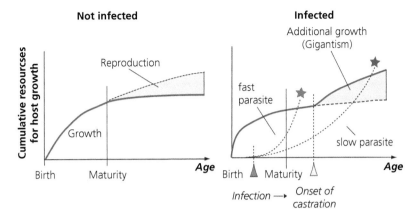

Figure 8.6 Parasite strategies of gigantism and castration. Shown are the cumulative resources that the host allocates to growth under normal conditions (left). When not infected at maturity, some resources are allocated to reproduction (green area), which reduces further growth. *Right*: When hosts are infected, and gigantism and castration sets in, resources that would normally be used for reproduction become invested in further growth (yellow area). This is also an extra supply for the parasite. However, the effect on the host only appears some time after infection. A fast-developing parasite, therefore, might not be able to use this extra supply because it has transmitted earlier (blue asterisk). A slowly developing parasite (red asterisk), however, can use this extra supply for its final growth and in preparation for transmission ('temporary storage hypothesis'). The heavy red lines indicate resources allocated to growth. The dotted lines indicate parasite growth and multiplication within the host. Adapted from Ebert et al. (2004), with permission from University of Chicago Press.

8.3.1.3 Changes of the social context

In social insects (ants, bees, wasps, and termites), infected individuals often desert their colony. For example, social wasps (*Polistes dominulus*) parasitized by Strepsiptera (*Xenos vesparum*) stop working, become inactive and eventually leave the nest. This behaviour presumably favours the completion of the parasite's life cycle because inactive workers are aggressively challenged by their nestmates, which increases the risk of host death and thus the destruction of the parasite (Hughes et al. 2004a). In some cases, nest desertion is a defence strategy of the host; for example, when bumblebee workers infected by parasitic flies stay outside the nest, especially at night, which seems to retard or stop the parasite's development (Müller and Schmid-Hempel 1993).

8.3.2 Manipulation of the host phenotype to increase transmission

Parasite-induced changes in behaviour to increase transmission are manifold (Figure 8.5). They are observed for every transmission mode that parasites use—from direct contact between hosts to vector-borne transmission (Table 8.4).

8.3.2.1 Transmission site

Many parasites rely on durable propagules to wait for an encounter with a next host. Examples include spores of *Bacillus anthracis* or *Clostridium* (causing severe human diseases, i.e. anthrax, tetanus by *C. tetani*, botulism, *C. botulinum*, or gangrene, *C. perfringens*) that withstand very adverse conditions and remain infective for years or decades. Also, manipulation of hosts can ensure the deposition of spores in suitable places. Nematodes, such as *Sphaerularia bombi*, actively manipulate the behaviour of their hosts (bumblebee queens) so that larval stages become deposited in overwintering sites of the next generation queens (Lundberg and Svensson 1975; Colgan et al. 2019). Mermithids (Nematoda) and hairworms (Nematomorpha) manipulate their hosts to move into water, where the offspring worms emerge and seek out new hosts (Herbison et al. 2019) (Table 8.4).

Being in the right place is of the essence for parasite transmission, and animal movement can ensure this (Binning et al. 2017). Parasites that depend on transmission by dispersal in the air have evolved various ways to force their hosts to climb to exposed locations from where parasite propagules can spread. For instance, the fungus *Ophiocordyceps*

Table 8.4 Parasite-induced changes in host behaviour in the major transmission modes.

Parasite	Host	Observation	Transmission and possible benefits	Source
(a) Direct transmission:				
Viruses: Rabies virus (RABV) Borna disease virus (BDV) Tick-borne encephalitis virus (TBE) Feline immunodeficiency virus (FIV)	Mammals Mouse (*Mus musculus*) Rats (*Rattus norvegicus*) Cats (*Felis sylvestris catus*)	Approach other animals more often and biting more frequent (increased aggressiveness). BDV infects neurons in limbic cortical areas and olfactory bulb. TBE: Increased testosterone. FIV: More frequent behaviour that increases wounding. Inflammation of CNS.	Direct contact and transfer via saliva of the virus to next host.	10, 11, 14, 27, 30, 33–35
Protozoans: Ciliate, *Lambronella clarki* *Toxoplasma gondii*	Mosquito (vector) (*Aedes sierrensis*) Mice, rats	Infected mosquitoes return to water and show false oviposition behaviour. Toxoplasma: Increased aggression and dominance. More active social exploration, less defensive behaviour.	Dissemination of ciliate parasite that can infect new mosquito larvae in pond. More contacts with hosts. Effect on dopamine levels unclear.	1, 8, 9, 14, 16, 36, 41, 42
Fungi: *Phycomycetes Entomophthora Ophiocordyceps* Microsporidia	Black flies (Simuliidae) Ants (*Formica* spp.) Shrimp	False oviposition behaviour. Infected ants stay at grass top. Feminization of shrimp behaviour and sex ratio distortion.	Parasites reside in ovaries and get transmitted. Spores can disperse more widely from elevated position. Fungal gene expression yields candidate genes for manipulation. Parasites transmitted via females.	5–7, 19, 22, 39, 44
Acantocephala *Polymporphus laevis*	Amphipod (*Gammarus pulex*)	Infected intermediate host approach predators more often, based on olfactory cues.	More likely to be eaten by final host, a fish.	2
Nematodes: Mermithidae *Gasteromermis*	Arthropods Mayfly (*Baetis bicaudatus*)	Host moves into water. Feminization of males: infected males join females upstream and show oviposition behaviour.	Spread of parasite to other hosts in water.	21, 38, 40
Cestodes	Stickleback (*Gasterosteus aculeatus*)	Infected fish change diet and competitive ability (among other things). Parasite candidate genes identified that mimic host proteins and affect cell signalling and development.	More contacts when infective.	12
Trematodes: *Gyrodactylus bullatarudi*, *G. turnbulli* Eye fluke (*Tylodelphis clavata*)	Guppies (*Poecilia reticulata*) Perch (*Perca fluviatilis*)	Infected individuals become sluggish and attract other guppies. Females infected by parasite are less selective in mate choice. Fluke infection impairs foraging efficiency; higher vulnerability towards predators.	Parasite eggs can spread to neighbouring fish.	20, 29, 32
Parasitoids: Conopid flies	Bumblebee (*Bombus terrestris*)	Bumblebee hosts dig themselves into soil before being killed by parasite.	Protects developing parasite from cold (during winter) and from hyperparasitoids.	28

(b) Intermediate hosts:

Protozoans: *Toxoplasma gondii*	Mouse, rat	Infected mice are less fearful. Infected rats do not avoid cat odours.	25
Acanthocephala *Polymorphus* spp.	Gammarids	Changed behaviour, photo- and geotaxis, disturbance, risk taking.	13, 25
Cestodes: *Schistocephalus solidus* *Hymenolepis diminuta*	Stickleback (*Gasterosteus aculeatus*) Beetle (*Tribolium* spp.)	Sticklebacks approach predatory fish. Beetles more often close to surface.	24, 43
Trematodes: *Euhaplorchis californiensis* *Microphallus papillorobustus* *Schistosoma mansoni* *Gynaecotyla adunca* *Dicrocoelium lanceolatum*	Killifish (*Fundulus parvipinnis*) Snails: *Biomphalaria glabrata, Bulinus truncatus, Potamopyrgus antipodarium* Amphipod: *Corophium volutator* Ants	Various effects: infected fish closer to surface; snails change feeding times; amphipods more often on mud flats. Infected fish surface more often and show conspicuous jerking movements.	4, 17, 18, 23, 31, 37

(c) Vectors:

Bacterium *Yersinia pestis*	Fleas	Bacteria in large numbers block feeding apparatus of flea. As a consequence, the flea more often regurgitates its blood meal. Bacteria more readily transferred from flea to vertebrate host.	26
Protozoans: *Plasmodium gallinaceum* *Leishmania major*	Mosquito (*Aedes aegypti*) Sandfly (*Phlebotomus doboscqi*)	Feeding behaviour changed so that more hosts (mice, humans) are attacked. Higher transmission rate is likely.	3, 15

Sources: [1] Arnott. 1990. *Ann Trop Med Parasitol*, 84: 149. [2] Baldauf. 2006. *Int J Parasitol*, 37: 61. [3] Beach. 1985. *Am J Trop Med Hyg*, 34: 278. [4] Carney. 1969. *Am Midl Nat*, 82: 605. [5] de Bekker. 2014. *Integr Comp Biol*, 54: 166. [6] de Bekker. 2015. *BMC Genomics*, 16: 620. [7] Dunn. 1993. *J Invertebr Pathol*, 61: 248. [8] Egerter. 1989. *J Med Entomol*, 26: 46. [9] Egerter. 1986. *Proc Natl Acad Sci USA*, 83: 7335. [10] Fromont. 1997. *Can J Zool*, 75: 1994. [11] Gardner. 1995. *Curr Top Microbiol Immunol*, 202: 135. [12] Hebert. 2015. *Parasites & Vectors*, 8: 225. [13] Jacquin. 2014. *PLoS ONE*, 9: e101684. [14] Klein. 2003. *Physiol Behav*, 79: 441. [15] Koella. 2002. *Behav Ecol*, 13: 816. [16] Kristensson. 2002. *Curr Top Microbiol Immunol*, 265: 227. [17] Lafferty. 1996. *Ecology*, 77: 1390. [18] Levri. 1999. *Behav Ecol*, 10: 234. [19] Loos-Frank. 1976. *Z Parasitenkd*, 49: 281. [20] Lopez. 1999. *Anim Behav*, 57: 1129. [21] Maeyma. 1994. *Sociobiology*, 24: 115. [22] Marikovsky. 1962. *Insectes Soc*, 9: 173. [23] McCurdy. 2000. *Can J Zool*, 78: 606. [24] Milinski. 1990. In: Barnard, eds. *Parasitism and host behaviour*. Taylor & Francis. [25] Moore. 1984. *Am Nat*, 123: 572. [26] Moore. 2002. *Parasites and the behavior of animals*. Oxford University Press. [27] Moshkin. 2002. *Psychoneuroendocrinology*, 27: 603. [28] Müller. 1994. *Anim Behav*, 48: 961. [29] Muñoz. 2019. *Parasit Res*, 118: 2531. [30] Rupprecht. 2002. *Lancet Infect Dis*, 2: 327. [31] Schneider. 1971. *Naturwissenschaften*, 58: 327. [32] Scott. 1985. In: Rollinson, eds. *Ecology and genetics of host-parasite interactions*. Academic Press. [33] Solbrig. 1995. *Curr Top Microbiol Immunol*, 190: 93. [34] Solbrig. 1996. *Curr Top Microbiol Immunol*, 222: 332. [36] Stibbs. 1985. *Ann Trop Med Parasitol*, 79: 153. [37] Swartz. 2015. *J Exp Biol*, 218: 3962. [38] Thomas. 2002. *J Evol Biol*, 15: 356. [39] Undeen. 1977. *J Invertebr Pathol*, 56: 150. [40] Vance. 1996. *Proc R Soc Lond B*, 263: 907. [41] Wang. 2015. *Infect Immun*, 83: 1039. [42] Webster. 1994. *Parasitology*, 109: 37. [43] Yan. 1994. *Am Nat*, 143: 830. [44] Yeboah. 1984. *J Invertebr Pathol*, 43: 363.

causes its ant hosts to climb to the top of the vegetation and fix themselves by biting into leaves. The subsequent mandible muscular atrophy maintains this state. The fungus then grows out of the host body and forms fruiting bodies to disseminate the spores by wind (de Bekker et al. 2014). Some fungi additionally start to 'tie' the ant down to the leaf surface by hyphae that grow out of the host's body (Loos-Frank and Zimmermann 1976; Roy et al. 2006; Hughes et al. 2011).

8.3.2.2 Transmission to a next host

The behaviour of intermediate hosts that need to be consumed by the final host for the completion of the parasite's life cycle seems very often manipulated by parasites. Behavioural changes that favour are associated with infection by a wide range of parasites (Table 8.4). A spectacular case is digenean trematodes (*Dicrocoelium*) that infect intermediate hosts (ants, snails). In these hosts, the metacercariae encyst in the host's nervous system and await predation by the final host. One of the metacercariae will leave the common site and migrate to the protocerebral ganglia to encyst there (the 'brain worm'). This metacercarium seems able to affect the host's behaviour such that the host becomes more prone to move towards the light instead of moving towards the dark as usual (Moore 2002; Mehlhorn 2015; Hughes and Libersat 2018).

Changing the risk taking of the current host is an effective strategy to become transmitted to a next host in cases where predation is essential. For example, sticklebacks infected with cestodes (*Schistocephalus solidus*) venture closer to the water surface and are also less intimidated by stimuli that indicate a predator's presence. The predator is a next host of the cestode. This change of behaviour does not happen for infections by microsporidians, which spread by the indiscriminate release of spores (Milinski 1985, 1990; Ness and Foster 1999). A well-studied case is *Toxoplasma gondii*, where small mammals serve as intermediate, and cats as final hosts. Infection changes several behaviours, notably the risk-taking behaviour of rats and mice. Whereas mice typically avoid cat odour, infected individuals are attracted to it and therefore are more likely to become prey (Berdoy et al. 2000; Kaushik et al. 2014). Inflammation of the brain induced by an

appropriate number of cysts seems to be the mechanism that leads to such loss of fear (Boillat et al. 2020).

Transmission requires some kind of spatial proximity to a next host. However, the successful transmission also requires a temporal overlap in the activities of the involved parties. Indeed, all organisms show certain daily, monthly, or annual rhythms. Parasites, therefore, have evolved to utilize a favourable time window for transmission (Tinsley 1990; Moore 2002). For example, *Schistosoma mansoni* sheds cercariae into the water from the intermediate host, a snail. Interestingly, in areas where humans are the principal host, most cercariae are shed from the snails over midday. However, where rats—a nocturnal species—are the primary host, the peak of shedding occurs later in the afternoon (Théron 1984). Crossing experiments show that this difference has a genetic component (Théron and Combes 1988).

8.3.2.3 Transmission by vectors

Many parasites are dispersed between hosts by vectors, mainly by blood-sucking arthropods such as mosquitoes, lice, and ticks. For the vector, taking up a blood meal serves to develop eggs (hence, only females are useful vectors); vector feeding behaviour thus becomes a significant determinant of parasite transmission success. Malaria parasites (*Plasmodium*) vectored by mosquitoes again are a case in point. To increase the chances of transmission, infected hosts should be more attractive. This is, indeed, observed in many blood-sucking insects (O'Shea et al. 2002; Lacroix et al. 2005). For example, a factor (HMBPP) produced by *Plasmodium falciparum* affects human red blood cells such that more attractants (CO_2, aldehydes, monoterpenes) are released (Emami et al. 2017); changes in the skin microbiota may additionally change attractiveness to mosquitoes (Busula et al. 2017). In mice, host attractiveness furthermore varies with the developmental stage of the parasite within its host, being high when mice are highly infectious (De Moraes et al. 2014).

At the same time, effects on the vector itself occur. *Plasmodium* activates the insulin-signalling pathway in the mosquito midgut, which, among other things, leads to a changed sensitivity of odorant receptors. As a result, mosquitoes are more likely to bite hosts carrying infective stages of the parasite. However, the same change results from a challenge with inactivated bacteria and is therefore not a specific benefit for malaria transmission but could simply be an unavoidable consequence of resource allocation—even though the parasite is likely to benefit (Cator et al. 2015; Stanczyk et al. 2017).

Vectored parasites also often cause their host to take up smaller blood meals, e.g. by *Culex* mosquitoes, vectoring avian malaria (Cornet et al. 2019). Manipulation of feeding behaviour also occurs in many other vectors, such as sand flies, ticks, or lice. It seems a common strategy of parasites to manipulate blood-sucking arthropods (Table 8.4) (Moore 2002). The size of blood meals is also affected by blood characteristics. In infected hosts (malaria, dengue, or trypanosomes), the number of erythrocytes (a side effect of parasite-induced anaemia) and platelets in the bloodstream characteristically is reduced. This incidentally associates with reduced blood viscosity and thus with a faster uptake of the blood meal (Rossignol et al. 1985; Taylor and Hurd 2001). Similarly, dilatation of blood vessels facilitates the uptake of a blood meal (Moloo et al. 2000) and might represent yet another way in which parasites can increase their transmission success to a vector.

8.3.3 Change of host morphology

8.3.3.1 Colouration and odour

Parasitic infections often change the externally visible host morphology, e.g. body size with gigantism (see section 8.3.1.2 above), or sexual morphology, or colouration (Table 8.5). In some spectacular cases, the host seems wholly taken over by the parasite in both behaviour and colouration. These hosts have been described as 'zombies' (Weinersmith and Faulkes 2014). Examples are carpenter ants seemingly guided by their fungal parasites to elevated positions in the vegetation (de Bekker et al. 2015), or land snails infected by the digenean *Leucochloridium paradoxum*. In this classic case, the snail's eyestalks start to pulsate in all colours (Loos-Frank and Grencis 2017). Note that such parasite-induced changes in colouration may make a male host more

conspicuous, yet not in terms of becoming more attractive to females (where the effect of parasitism is typically the opposite; see Chapter 6), but to make the manipulated hosts more attractive for predators.

A change in host body colour often also contrasts with the typical background of the habitat where the hosts live (Figure 8.5). This is likely to attract a predator that is the final host. For example, cestode-infected workers of the ant *Leptothorax nylanderi* not only become more sluggish in their behaviour but also develop a yellow colouration, which makes them highly visible among their brownish nestmates (Trabalon et al. 2000). The final host of the cestode is a bird that typically feasts on ants, such as the green woodpecker. A conspicuous contrast between host and background also occurs when an infected host shows an altered preference for the background. *Armadillidium vulgare* infected by an acantocephalan parasite prefer to stay on a lighter substrate and so become more visible than on their regular, dark background substrate (Moore 1983).

In the natural context, other changes might be even more critical. In the example of *L. nylanderi*, cestode-infected workers also differ in the composition of their cuticular hydrocarbons from uninfected nestmates (Trabalon et al. 2000). These hydrocarbons are used as chemical signals to interact with nestmates. Hence, such changes are not immediately visible to the human eye but are relevant for transmission to a next host. Examples include changes in pheromones that attract birds to insects (Saavedra and Amo 2018), in the UV spectrum of light, which is visible for many birds (notably birds of prey, Lind et al. 2013; Ödeen and Håstad 2013), or a change in body temperature that can guide heat-sensing predators such as snakes to their prey.

8.3.3.2 Morphology and feminization

Changes in host morphology also include deformation in limbs that impede locomotion. Decreasing metabolic capacities similarly decrease locomotory activities and dispersal distance of infected hosts. Such changes may be a beneficial side effect or an active strategy of the parasite to keep the transmission in a neighbourhood (Binning et al. 2017).

A particular case is the feminization of infected hosts. In some cases, a genetic male becomes

Table 8.5 Parasite-induced changes of host morphology.

Parasite	Host	Observation	Possible function for parasite[1]	Source
Bacteria Serratia marcescens	Cricket (Acheta domesticus)	Earlier reproduction by females infected by bacteria.	Fecundity compensation to limit effect of later castration or early death.	1
Myxozoa: Tetracapsuloides byrosalmonae	Colonial bryozoan (Fredericella sultana)	Reduced growth of colony, but increased size of modules (gigantism).	More resources for parasite.	5
Protozoa: Plasmodium	Mosquito (Aedes aegypti, and many others)	Reduction of fecundity of vector.	Favours vector survival and frees resources for parasite development.	6, 7
Fungus: Entomophthora muscae	House fly (Musca domestica)	Abdomen extended with white spots of fungal conidia on outside.	Attracts males to approach dead fly, presumably mistaken for a fecund female.	13
Acantocephala: Polymorphus spp.	Gammarid	Orange colouration by carotenoids.	Experimental tests show no effect of colouration on increased predation by final host.	9
Nematodes: Heterorhabditis bacteriophora	Beetle (Pterostichus madidus)	Change in host colour and odour (post-mortem).	Aposematism and odour repels predators on host and protects parasite.	4, 8
Cestodes: various spp., Schistocephalus solidus	Ant (Leptothorax nylanderi, and many other species) Stickleback (Gasterosteus aculeatus)	Infected ants have yellow instead of normal brown colouration (cestode infects larval stage). Colouration change with visible spots.	More conspicuous for predator.	12
Trematodes: Leucochloridium, Neoleucochloridium, Schistosoma solidus, Curtuteria, Acanthoparyphium	Snails Stickleback (Gasterosteus aculeatus) Mussels (Austrovenus stutchburyi)	Sporocysts of parasite are brightly coloured. Snail tentacles swollen, colourful and pulsating. In fish: larger pectoral fins, lack of pigment on integument. Parasite migrates to foot of mussels, replaces muscle tissue and causes foot to become smaller.	Attracts attention of final host, a bird, a predatory fish, a shore bird.	2, 10, 11, 14–16
Crustaceans: Cymothoa exigua	Fish (snapper)	Parasite attaches to tongue, grows and eventually replaces tongue with its own body.	Parasite can easily secure its own share of food. 'Tongue' can be used otherwise normally by fish.	3

[1] Function has rarely been tested in the same system and therefore often remains conjectural.

Sources: [1] Adamo. 1999. Anim Behav, 57: 117. [2] Brønseth. 1997. Can J Zool, 75: 589. [3] Brusca. 1983. Copeia, 3: 813. [4] Fenton. 2011. Anim Behav, 81: 417. [5] Hartikainen. 2013. Parasitology, 140: 1403. [6] Hurd. 2003. Annu Rev Entomol, 48: 141. [7] Hurd. 1995. Parasitol Today, 11: 411. [8] Jones. 2015. Behav Ecol, 27: 645. [9] Kaldonski. 2008. Proc R Soc Lond B, 276: 169. [10] Lefèvre. 2008. Infect. Genet Evol, 8: 504. [11] Lewis. 1974. J Parasitol, 60: 251. [12] LoBue. 1993. Am Nat, 142: 725. [13] Møller. 1993. Behav Ecol Sociobiol, 33: 403. [14] Mouritsen. 2002. Parasitology, 124: 521. [15] Ness. 1999. Oikos, 85: 127. [16] Thomas. 1998. Parasitology, 116: 431.

converted into a genetic female, for example in insects with complementary ('haplo-diploid') sex determination (Kageyama et al. 2012). A genetic male can also be converted into a phenotypic, functional female (e.g. crustacean hosts of Microsporidia; Jahnke et al. 2013), or express female secondary organs (e.g. *Sacculina*; see Figure 3.4). Feminization by endogenous *Wolbachia* is widespread among insects, and well investigated in various contexts (Zug and Hammerstein 2015). Typically, feminization leads to increased transmission of the parasite via the female route, i.e. vertical transmission. The mechanisms involved in feminization are now increasingly better understood (Kristensen et al. 2012; Ma et al. 2014), even though many riddles remain.

8.3.4 Affecting transmission routes

Some parasites use transmission routes that ensure transfer into a next host with high certainty. Vertical transmission is such a safe route since the next host is the offspring of the current host. Transmission via breast milk (transmammary transmission), as in nematodes (Casanova et al. 2006; Jin et al. 2008), or transovarial transmission from a mother into her eggs are obvious cases. Transovarial transmission is used by many groups of parasites, such as viruses (e.g. dengue, Zika virus; Thangamani et al. 2016), bacteria (Perlman et al. 2006; Duron et al. 2008), or microsporidia (Smith and Dunn 1991; Dunn et al. 2001). Because of its high value as a safe route, parasites manipulate vertical transmission in many ways (Box 8.4).

Box 8.4 Manipulation of vertical transmission

Vertical transmission happens within a host line, e.g. from parent to offspring (see Figure 9.1). Several host reproductive traits can be exploited by the parasite to increase the chances of successful transmission.

Manipulating a female host: When parasites transmit transovarially, the chances of success are higher due to:

1. *An increase in host fecundity*: More eggs are produced and can carry the parasite. Increased fecundity likely comes at the cost of future female survival.
2. *Elimination of competing females that do not carry the parasite*: Incompatibility factors, transferred by infected males, lead to abortion of eggs if the sperm-receiving female is not infected ('cytoplasmic incompatibility'). This strategy is followed by *Wolbachia* and other bacteria (Duron et al. 2008). A de-ubiquitylating enzyme is responsible for this effect in *Drosophila* hosts (Beckmann et al. 2017).
3. *Induction of (thelytokous) parthenogenesis*: This increases the fraction of purely female-derived infected hosts in the population. Examples are *Wolbachia*, *Rickettsia*, and other bacteria (Duron et al. 2008), which induce parthenogenesis in hosts. Complementary sex determination systems, such as those present in Hymenoptera, facilitate this strategy. With this system, males typically are haploid and develop from unfertilized eggs, and females develop from fertilized (diploid) eggs. The parasite prevents chromosomal separation during early development of the unfertilized egg, thus turning the male into a genetically diploid female (Huigens and Stouthamer 2003).

Manipulating a male host: For transovarial transmission, the parasite is in the wrong sex. Transmission can nevertheless be ensured by turning its male host into a female mimic.

4. *Feminization of males*: Feminization of the host by parasites is quite widespread, e.g. by *Wolbachia* (Hurst 1993) or microsporidia, and primarily in crustacean hosts (Dunn et al. 2001). In this process, genetic males are turned into morphological, and sometimes functional females (Dunn et al. 1993; Bouchon et al. 2008). For this, the parasite suppresses the proper functioning of the male androgenic gland during development (Rodgers-Gray et al. 2004). Male behaviour, too, can become feminized. In these cases, the parasite induces female traits in males, such as oviposition behaviour. For example, mayfly males, infected by nematodes, show 'oviposition' behaviour and thereby deposit the nematode larvae in upstream waters (Vance 1996). In several host–parasite systems, the formation of intersex morphs (i.e. partly male and partly female) is observed, perhaps, resulting from incomplete feminization (Kelly et al. 2006).
5. *Eliminating males*: The male egg or embryo is killed, which frees resources to produce female offspring instead. For vertical transmission of some microsporidia (*Thelohania*, *Ambylospora*) from infected mothers, developing males are killed. For horizontal transmission, the host larva is killed, and spores are released into the water (Dunn et al. 2001).
6. *Distorting the primary sex ratio*: A female-biased offspring sex ratio benefits the parasite by an increased frequency of next-generation hosts of the 'right' sex. Sex ratio distortion is generated by various mechanisms (Hatcher and Dunn 1995; Dunn et al. 2001).

Sexual transmission can combine horizontal (to another individual or lineage in the population) with vertical transmission. For example, transmission via seminal fluids allows the parasite to pass from the male to the female line and to offspring, as seen in microsporidia (Peng et al. 2016). In general, however, an infected individual is avoided (de Roode and Lefèvre 2012; Kavaliers and Choleris 2018) and becomes less attractive, or rejected as a mate. For instance, male mice recognize and avoid infected females by odours in the urine (Kavaliers et al. 2004). Nevertheless, parasites can suppress, for instance, the inflammatory response and thereby keep the host as an attractive mate, e.g. in sexually transmitted iridovirus in crickets (Adamo 2014). Sexually transmitted parasites often produce few symptoms in their hosts, which aids in their spread (Knell and Webberley 2004; Antonovics et al. 2011).

A parasite can also induce movement and dispersal behaviour (Boulinier et al. 2001). Host movements can serve both as a defence—by leaving or avoiding areas with increased infection risks—and as a transmission strategy, when the parasite manipulates dispersal (Binning et al. 2017). For example, microsporidia cause migratory locusts to resume their solitary behaviour and to disperse. This happens by antagonistic action on the host's native gut microbiota that is responsible for the production of an aggregation pheromone (Shi et al. 2014).

8.3.5 Affecting social behaviour

Group and socially living animals offer many opportunities for parasites to get transmitted (Schmid-Hempel 2017). However, the benefit of manipulation depends on transmission mode. Directly transmitted parasites gain from proximity among (genetically related) host individuals. However, vectored parasites, or those in need of a final host, benefit from isolating the current host from its social context, e.g. making it more vulnerable to predation than when sheltered by the social group. For example, proximity among hosts in shoaling fish favours transmission (Richards et al. 2010). Fish infected by mobile ectoparasites tend to shoal together more often. This should increase the chances of an ectoparasite to reach a next host

nearby (Barber 2000). On the other hand, shoals reduce the risk of predation for each host individual (the 'selfish herd' effect; Hamilton 1971; Barber 2000), and decreases the chances of the parasite transmitting to a predatory fish. Parasitized host fish often show impaired swimming abilities that make them unable to participate in shoals fully; instead, for example, they swim in risky peripheral positions (Barber 2000; Ward et al. 2005; Seppälä et al. 2008a). Parasite manipulation can even enhance the risk-taking behaviour of other, uninfected members of a group (as in sticklebacks; Demandt et al. 2018).

A change in aggressive behaviour also seems to be commonly associated with infection by directly transmitted parasites. Examples are viruses such as rabies or hantavirus, that need to get in contact with the next host, and that can enter a new host via wounds. These are inflicted during aggressive encounters between the current and the next host (Klein 2003). Effects of social living extend to further behaviours that, for instance, change contacts among individuals and increase or decrease transmission success of a parasite (Behringer et al. 2018). Section 15.1.3 discusses parasitism and group living further.

8.3.6 Affecting the neuronal system

Effects of parasites on the host's neuronal system are the reason for many of the changed behaviours mentioned above. The respective targets include neuronal, endocrine, neuromodulatory, or immunomodulatory systems (Adamo 2013; Lafferty and Shaw 2013). 'Neuroparasitology' labels the study of parasite-induced neuronal, cognitive, psychological, or behavioural changes (Libersat et al. 2018). Modern technologies offer many opportunities to study this field (Hughes 2013). This toolbox includes methods that are also useful in other contexts. Examples include comparative genomics, transcriptomics, proteomics, metabolomics, as well as experimental approaches, e.g. reverse and forward genetics with CRISPR–Cas gene-editing technologies (see Box 4.2).

As mentioned before, a simple way to affect host behaviour is to induce inflammation in the neuronal tissue, e.g. the central nervous system (CNS), as observed for cerebral parasites (Klein 2003). *Toxoplasma* releases compounds toxic to neurons and can lead to neuropsychiatric diseases (Henriquez et al. 2009). Neuroinflammation generally leads to disruption and irregular behaviour, which may or may not be advantageous for the parasite. The underlying processes may be more complex, however. For example, the transmission of rabies virus is from the saliva of infected animals to a next host when bitten (Rupprecht et al. 2002). Infection is indeed associated with the onset of aggressive behaviour and loss of fear by infected hosts, e.g. foxes, which would favour transmission when induced by the virus (Niezgoda et al. 2002). However, the virus usually lodges in parts of the CNS that are not directly involved in aggressive behaviour. Moreover, two clinical syndromes of rabies exist—the mild form (dumb, paralytic) and the furious form (encephalitic). Only the latter is associated with aggressive behaviour. Hosts of the furious form furthermore mount an intact immune response. Hence, an intact immune response may eventually lead to aggressive behaviour rather than any direct action of the parasite (Thomas et al. 2005). Infection by rabies also activates the hypothalamic–pituitary axis, which increases the levels of adrenalin and neurotransmitters that are typical for eliciting aggressive behaviours (Hemachudha et al. 2002). Similar processes might be present in trematodes that affect the behaviour of their intermediate snail hosts (*Lymnaea stagnalis*) (de Jong-Brink et al. 2001). Such kinds of host manipulation have evolved in many different parasite groups, and the mechanisms are often convergent.

Parasites lodging in or near the CNS may have a variety of additional mechanisms to change host behaviour more precisely. For example, endoparasitic fungi grow hyphae, and trematodes dispatch metacercariae, that become settled near the central parts of the neuronal system (Moore et al. 2005). Borna disease virus that infects sheep and horses renders hosts more aggressive. This seems to be due to the infection site (the limbic and cortical brain regions), where the virus modulates neurons and dopaminergic receptors (Klein 2003). Parasites can furthermore directly infect neurons, glia cells, or endothelial tissues to induce behavioural changes.

The intrinsic link between the immune and neuronal system is a major axis of how parasites can manipulate host behaviour with neuropharmacological interference, such as with biogenic amines (e.g. dopamine, octopamine, serotonin) that are key modulators of neuronal activities (Adamo 2013). In gammarids, neuroinflammation in response to helminth infections affects the biochemical cascades, leading to dysregulation by serotonin. This becomes visible as behavioural changes, such as altered phototaxis, geotaxis, or olfactory preferences (Helluy 2013). Similarly, wasp-derived venom acts as a neuromodulator that paralyses a cockroach host and makes it unable to walk in a coordinated way—except when the wasp 'leads' the victim back to its nest. Serotonin, octopamine, noradrenalin, and the neurotransmitter GABA, are involved in such parasite-induced changes of host behaviour.

Close spatial proximity of the parasite with the neuronal system is thereby not always necessary to take control. For example, mermithid worms or parasitic wasps can control their host's behaviour from the outside, or when residing somewhere else in the body; e.g. the larva of the parasitic wasp *Hymenoepimecis argyraphaga* develops in the body cavity but manipulates its host spider to spin a modified web that protects the developing wasp. The host is then finally killed (Eberhard 2000). Even changes in life history, e.g. by fecundity reduction, might be due to neuroactive products (de Jong-Brink et al. 1997). For example, intermediate snail hosts infected by the schistosome *Trichobilharzia ocellata* cease to produce eggs under the influence of schistosomin. This parasite-secreted protein acts as a neuromodulator. Schistosomin reduces the responsiveness of specific cells in the snail's ganglia (the caudodorsal cells) that normally release signals to induce ovulation and egg-laying (Hordijk et al. 1992). The same infection also leads to an upregulation of other neuromodulators.

8.4 Strategies of manipulation

8.4.1 Common tactics

The kinds of manipulations, i.e. the tactics used to achieve a strategic goal (such as to increase transmission rate) vary among parasite groups. In a large-scale study of 55 parasite genera (mostly helminths) (Lafferty and Shaw 2013), trematodes and cestodes mostly affected host activity or combined behavioural changes with changes in host microhabitat choice. In particular, nematodes relied on either host activity or habitat choice, whereas Acantocephala affected host microhabitat choice. These parasites also vary in the location where they lodge. Most parasite groups are found equally often in muscles, in the CNS, in various other tissues, and in the body cavity. Acantocephala, by contrast, almost exclusively use the body cavity. Another meta-analysis across 76 studies confirmed that effects on host performance, such as endurance, speed, manoeuvrability, or efficiency of movement, depended on the kind of tissue that the parasite infects, notably the connective tissue. Only host age was similarly significant (McElroy and de Buron 2014).

Some of the differences in the kind of manipulations or target tissues are determined by the repertoire of molecular and biochemical mechanisms that a given parasite can deploy, i.e. by phylogenetic constraints. However, much remains to be studied on how a particular tactic has evolved. For example, manipulation mechanisms must work more generally, across different hosts, when the parasite's host range is broad, but can be specialized and perhaps more targeted when the host range is narrow (see section 7.2).

8.4.2 What manipulation effort?

The choice of the best manipulation effort depends on the costs of manipulation. The costs are not only under the control of the parasite but also determined by the host. Efficient host defences, for example, would necessitate higher investments by the parasite to succeed in manipulation. Costs of manipulation are assumed to be physiological, e.g. a higher demand of energy (Thomas et al. 2005; Poulin 2010), or by loss of opportunities when ending up in a dead-end host (Seppälä et al. 2008b). Empirical evidence for manipulation costs is scarce, however. Trade-offs of manipulation with other traits, e.g. parasite fecundity, are known (Maure et al. 2011), but costs remain difficult to assess (Hafer-Hahmann 2019). The particular mechanism

should also affect its costs. Molecular mimicry, for example, can be expensive, because signalling molecules are usually rapidly metabolized and hence must continuously be produced at a cost (Adamo 2013). At the same time, there is a risk of overshooting that can lead to immunopathology. Molecular mimicry is thus subject to several trade-offs that vary with the host–parasite system (Hurford and Day 2013).

In some cases, costs and benefits are not in the same individual parasite. For example, the 'brain worms' of trematodes migrate to the host brain, from where they affect host behaviour and thus carry a physiological cost of manipulation. They will, however, not develop into infectious stages and, therefore, not become transmitted. Instead, the manipulation benefits the infective metacercariae present elsewhere in the same host. These become transmitted but have not paid a cost for manipulating. Similarly, in the example of the trematode stages lodging in the foot of a mussel, they destroy or reduce its size. The host mussels can no longer bury into the substrate and are more likely to be eaten by birds and fish, the final hosts of the trematodes. However, the affected part of the foot is usually not consumed by the predator, and therefore no benefit is realized for the manipulating individual (Poulin et al. 2005). Such strategies can only evolve if the manipulating and benefiting individuals are closely related or clonal copies, e.g. by kin selection.

Simple cost–benefit considerations suggest that manipulation effort to increase the life span of the host (and thus infection duration) should be higher when host longevity is otherwise short. Similarly, manipulation effort should increase when, otherwise, transmission without manipulation is rare (Poulin 1994). Theoretical models also show that manipulating the host to disperse is advantageous to the parasite, but the details depend on the spatial dynamics of host and parasite (Lion et al. 2006). The parasite might also be in a position to increase the defence costs for the host. At some level, it then becomes cheaper for the host to tolerate a certain amount of manipulation rather than to fight it at high defence costs. This parasite strategy is reminiscent of methods of extortion payments in organized crime and is therefore known as the 'mafia strategy'. As with the mafia, co-operation of the host with the parasite (i.e. no defence) per se is not beneficial. However, the parasite has evolved mechanisms so that if hosts fail to co-operate, this will make matters worse. In other words, the parasite steps up its virulence (the damage done to the host) if the host actively defends itself against manipulation. For example, parasites could induce inflammatory cytokines in a vigorously defended but not in a weakly defended host. Plastic defences in hosts may promote this strategy (Chakra et al. 2014).

Any infection has a life history that unfolds over time. A parasite can thus 'decide' over manipulation effort at different stages during the duration of the infection. As will be discussed in the context of 'virulence' (see Chapter 13), the timing of manipulation has many ramifications on what effects appear on the host and what fitness advantages will result for the parasite. We should expect to see that manipulations are particularly common in the early stages, that is by avoiding recognition or during the early stages of host signalling.

Finally, the activity of organisms always follows some kind of periodicity, be it diurnal, tidal, lunar, or seasonal. Host immune defences or parasite virulence, for example, follow a circadian cycle in mice, fish, insects, or plants (Westwood et al. 2019; Orozco-Solis and Aguilar-Arnal 2020), and in many cases annual cycles are also observed (Martinez-Bakker and Helm 2015). In the example of malaria transmission, periodicity is important (Prior et al. 2020). The gametocytes become infective at night when the host blood cells burst, thus achieving higher transmission success (Mideo et al. 2013b; Schneider et al. 2018; Westwood et al. 2019). Parasites—and hosts in defence—can use such patterns to their advantage. Effects of parasites on host rhythms are quite common but, in most cases, it is not clear to what extent targeted manipulation is the reason for such changes (Martinez-Bakker and Helm 2015; Carvalho Cabral et al. 2019). For instance, climbing and biting behaviour in ants infected by trematodes (*Dicrocoelium*) or fungi (*Ophiocordyceps*) peak at certain times of day, presumably making consumption by the next host more likely. Studies of gene expression suggest that this may be a result of manipulation (de Bekker et al. 2014). Interfering with the host's molecular

clock is also likely for viruses (Mazzoccoli et al. 2020), but the topic is still far from understood.

8.4.3 Multiple infections

In the wild, most hosts carry more than one infection—either infection by different strains of the same parasite, or infections by different parasite species. Co-infecting parasites will influence each other, and the combined effect of manipulation is probably more than just the sum of manipulations by the single parasites. Moreover, the different parasites may not have the same interests (Hafer 2016) when, for example, infecting an intermediate host (Lafferty et al. 2000) (Figure 8.7). When two co-infecting parasites are both capable of manipulating a host but each has a different final host or uses different transmission routes, there is a conflict over the optimal manipulation strategy. Parasites might then avoid infecting an already infected host, as seen in egg-laying by parasitoid females. Another solution is to eliminate the competitor. For example, two species of cestodes, *Hymenolepis diminuta* (the final host is a rat) and *Raillietina cesticillus* (the final host is a chicken), infect the same intermediate host, the flour beetle, *Tribolium*. Both cestodes manipulate the behaviour of the beetle to increase transmission to their final host, but the respective manipulations differ. In this case, *R. cesticillus* prevents the successful establishment of an infection by *H. diminuta*, and so eliminates its competitor from the same host (Gordon and Whitfield 1985). Conflicts over manipulation should also emerge when one parasite uses the horizontal and the other the vertical transmission route. Little or no conflict is expected when different parasites manipulate the host in very non-specific ways that can serve to achieve different goals (Cézilly et al. 2014). Furthermore, two co-infecting parasites might also evolve towards a kind of 'division of labour' when they share a similar final host.

A non-manipulating parasite might benefit from the manipulation by co-infecting parasites, thus saving the costs of manipulation yet benefiting from the effects (Thomas et al. 1998; Thomas et al. 2005). Such 'hitch-hiking' could evolve when two parasites are routinely associated. Alternatively, it could result from a chance co-infection, where one

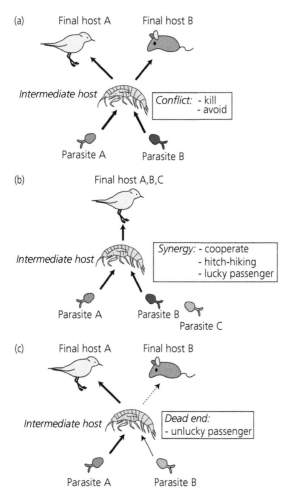

Figure 8.7 Co-infection by manipulating parasites. The examples illustrate the infection by a manipulating (red) and non-manipulating (green) parasite of the same intermediate host. (a) The final hosts are different for the two manipulating parasites. The conflict might be solved by killing the competitor, or by avoidance of hosts already infected by the competitor. (b) The final host is the same. The two manipulators might cooperate; a non-manipulator might benefit by hitch-hiking, or when it co-infects by chance ('lucky passenger'). (c) The final hosts are different and the manipulator wins (solid arrow); the competitor is transferred to the wrong final host ('unlucky passenger').

the amphipods to be closer to the water surface where the final host—a bird, which *M. subdolum* can also use—can eat them. For this benefit, the cercariae of *M. subdolum* seek out hosts already infected by the other, manipulating parasites. In particular, cercariae of *M. subdolum* move to the surface and so are more likely to infect an amphipod that is already infected by the other, manipulating parasite (Thomas et al. 1997).

Hence, different outcomes of manipulation in co-infections are possible—including dominance by the interests of one party, a compromise, suppression of one party by the other ('sabotage'; Hafer and Milinski 2015a), or complete suppression of both parties' effects (Hafer 2016). With complete dominance by one party, the inferior parasite might simply be unlucky and become transferred to the wrong host. Nevertheless, if it happens to survive in this new host, this could pave the way to the evolution of complex life cycles. The mechanisms of manipulation remain essential, in any case. For instance, an unavoidable side effect of infection (that nevertheless serves the parasite) can add up in multiple infections. This happens even though it may only be advantageous for one of the parasites (e.g. when the transmission is possible) but not the other (e.g. when the parasite still develops). Such a situation is observed in sequentially infected cestodes in sticklebacks (Hafer and Milinski 2015b).

Furthermore, co-operation among co-infecting manipulating parasites evolves with the degree of relatedness among them, and unrelated groups of parasites should typically be less co-operative. Finally, the overall effect of manipulation depends on the average effort invested in manipulation by the group of co-infecting parasites. How much an individual parasite should invest in the group effort, in turn, depends on the strategy of all other co-infecting parasites, accounting for the possibility of 'cheaters'. Game-theoretical analyses can define the evolutionarily stable strategy of manipulation effort that is expected to evolve. Some such models suggest that manipulation should be zero when group size is below a certain threshold number. Above this threshold, the per capita effort should generally decrease (while total group effort increases) (Brown 1999) (Figure 8.8).

parasite becomes the 'lucky passenger' (Figure 8.7). A possible example is the case of two co-infecting trematodes, *Maritrema subdolum* (which does not manipulate) and *Microphallus papillorobustus* (which does manipulate). Both are using amphipods as their intermediate host. *M. papillorobustus* induces

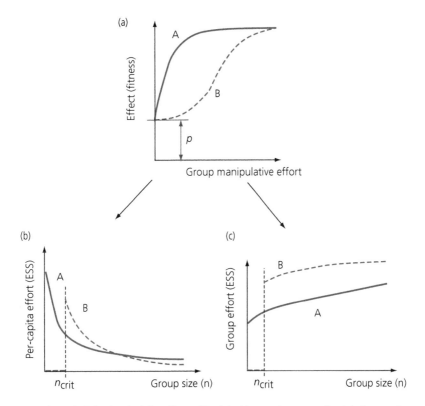

Figure 8.8 ESS investment in manipulation (manipulation effort; red lines). In this example, a group of co-infecting parasites cooperates to manipulate their host. (a) The relationship of the total manipulation effort of the group and group fitness. In case A, the fitness function saturates; in case B, there is a logistic relationship. The value p is parasite fitness without manipulation. (b) The ESS effort per capita based on the two fitness functions (A, B) in relation to the number of co-infecting parasites (group size). Below a minimum group size, n_{crit}, there should be no manipulation in case B. For case A, manipulation is always advantageous. (c) The ESS combined group effort in relation to group size for the two fitness functions. Adapted from Brown (1999), with permission from The Royal Society.

Important points

- All parasites manipulate their hosts to modify immune defences, host behaviour, host life history, and many other host functions for their benefit.
- Passive evasion includes strategies such as molecular mimicry, changing surface identity, and quiescence. Active evasion involves the production of molecules that actively interfere with the host immune defences, and other host functions such as metabolism and the neuronal system.
- Immune evasion targets all major steps of the immune response; that is, recognition (e.g. blocking opsonization), signalling (manipulating cytokines), effector systems (inducing apoptosis of phagocytes, destroying antimicrobial peptides), and the defence function of the microbiota.

- Manipulation increases the duration of the infection, or the rate and mode of transmission, by affecting host behaviour, colouration, body size, social behaviour, activity patterns, and many other traits. Vertically transmitted parasites affect sexual and reproductive characteristics of their hosts.
- Manipulation also has costs, but these are often difficult to measure. The effort that parasites should invest in manipulation and which kind of manipulation to choose varies with several factors, including the defence level of the host. Furthermore, the manipulation strategies of co-infecting parasites can conflict. Possible outcomes of this conflict include avoidance of already infected hosts or the elimination of competitors. Non-manipulating parasites might exploit the efforts of manipulating parasites.

CHAPTER 9

Transmission, infection, and pathogenesis

A parasite that has encountered a host needs to infect and establish next. Only then will it be possible to grow or multiply in numbers, and to gain transmission to a new host eventually. During these steps, the infection leads to pathogenesis; that is, the parasite inflicts damage to its host.

9.1 Transmission

Parasites use a bewildering variety of ways to get from one host to the next, a process known as 'transmission'. For clarity, here the term transmission 'mode' refers to the method a parasite uses to pass from one host to the next, e.g. by droplets in the air or by transport in water (Figure 9.1). The transmission mode is a physical characteristic and results from the traits of a parasite (and its hosts). By contrast, transmission 'route' is the actual path taken by the parasite that, in a given situation, results from the mode. Transmission route is a path that begins at the port of exit from the current host and ends where the parasite enters the next host. For instance, an oral-to-oral route starts with the shedding of infective cells through coughing by the current host. It ends with their inhalation by a new host (oral-to-oral), regardless of the actual mode (e.g. via droplets or via smear infections by direct contact) (Antonovics et al. 2017). The site where host entry takes place is often referred to as the 'infection route' (i.e. the last part of the transmission route). In practice, the infection route mainly refers to the nature of the host entry process, e.g. by inhalation, ingestion, or through the skin. Mode and route of transmission are important to define the epidemiological characteristics of the parasite and the possible ecological and evolutionary outcomes.

9.1.1 Exit points from the host

For transmission, a parasite must use any of the possible exit points:

1. *Faeces* are one of the most common exit points. In humans, some dangerous pathogens are shed in faeces, e.g. *Clostridium tetani* (causing tetanus or 'lockjaw', also shed in cattle or horse dung), or *Vibrio cholerae*, which also induces diarrhoea and thus increases the rate of shedding at the expense of the host. The epidemic potential of faecal exit also depends on how long the microbes can persist in the environment. For instance, *C. tetani* can form spores under adverse conditions; these are extremely hardy and survive many years or even decades in the soil. The 'faecal–oral' route (exit via faeces, entry via mouth) is particularly crucial for such pathogens. This route highlights the enormous importance of sewage disposal and clean-water supply for a healthy life in cities.

2. *Exit via saliva*: Several pathogens infect the salivary glands and are subsequently present in saliva, through which they exit the host. Transmission is by biting, or—in humans—during talking or kissing. Examples are mumps virus, Epstein–Barr virus, or cytomegalovirus, present in salivary glands.

3. *Exit through the respiratory tract*: Coughing transports microbes from the lower airways and throat to the mouth opening, from where they can exit. Similarly, sneezing expels large numbers of droplets into the air. In both cases, microbial pathogens can

Evolutionary Parasitology: The Integrated Study of Infections, Immunology, Ecology, and Genetics. Second Edition.
Paul Schmid-Hempel, Oxford University Press. © Paul Schmid-Hempel 2021.
DOI: 10.1093/oso/9780198832140.003.0009

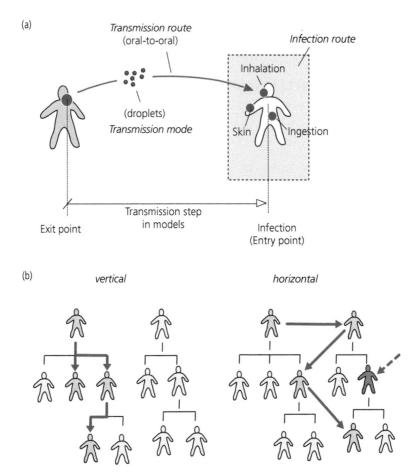

Figure 9.1 Transmission. (a) Transmission mode refers to the physical method of getting from one host to the next, e.g. by droplets through air. Transmission route is the path from the exit point from the current host (e.g. through the mouth when coughing) to the entry point in the next host (e.g. through the mouth when inhaling droplets). The example would apply, for example, to a common cold (rhinovirus) or influenza virus. The last part of the transmission route (grey box) is referred to as the infection route, i.e. the path of entry into the host by, for example, inhalation, ingestion, or through the skin. In epidemiological models such as the SIR model (see Chapter 11), the rate of transmission, i.e. how frequently new hosts become infected upon a contact, typically refers to the completed step from exit to successful infection of the next host. (b) The two major contexts of transmission are vertical, i.e. from parent to offspring within the same host line (left panel), or horizontal (right panel), which covers all other cases. In each case, the context is determined by the physical method of transmission, such as transovarial (vertical), or by direct contact, e.g. hand-to-hand (horizontal). Most parasites can use both modes. Infected individuals shown in red.

disperse with aerosols (particles suspended in air). Large particles will drop to the bottom within a few metres, but tiny particles (a few microns in diameter) can stay in the air for very long times and disperse over long distances. Although some microbes are thereby quickly inactivated by UV light or dry air, with the help of aerial currents hardy pathogens can spread over long distances. For example, during the foot-and-mouth disease outbreak in 1981, winds carried the virus across the sea from France to England (Sørensen et al. 2000), although it remains unclear whether any infections resulted.

4. *Exit through the urogenital tract*: Some specialized pathogens, including leptospiral bacteria, and some viruses can exit via this point. A particular case is sexual transmission, where mucosal contact is essential. The agents of the major human venereal diseases use this exit, e.g. gonorrhoea (*Neisseria gonorrhoeae*), syphilis (*Treponema pallidum*), *Chlamydia*, genital herpes (herpes simplex virus, HSV type 2),

and HIV. Polyomavirus in mice, which infects the kidney, is an example from animals.

5. *Exit via blood*: This is a central exit point for vectored pathogens, e.g. malaria and the mosquitoes. In public health, exit and transfer via contaminated needles is a challenge for the prevention of diseases such as AIDS and hepatitis.

6. *Shedding from the skin*: A human sheds an estimated 10^8 skin scales per day, of which a sizeable fraction carry microbes. Transfer to the next host occurs by inhalation or direct contact. For example, *Staphylococcus aureus* causes severe infections in humans and resides on skin and mucosa. Farmers are at risk through contact with the infected skin of an udder when milking cows. Strain USA300 of *S. aureus* has risen to prominence. It is now endemic in many parts of the world but has emerged as a skin disease, e.g. among (American) football players in Pennsylvania, a sport where bruises and lesions are commonplace (Tenover and Goering 2009).

7. *Exit via milk in mammals*: Milk contains microbes and is an exit route for pathogens such as cytomegalovirus, or rickettsias. Several other exit points exist but are not as commonly used. For example, exit via semen is relatively rare, although it is important for cases such as HIV.

9.1.2 Entry points

Almost all exit points are also entry points. This includes entry through the skin, e.g. through occasional lesions and minor damages. More extensive wounds from biting also play a role, for example, for rabies virus transferred via the saliva of an infected mammal. Vectors, such as mosquitoes, can inject a pathogen directly into the bloodstream. Entry via the respiratory tract is essential for airborne pathogens, e.g. the common cold, influenza, or the recent SARS-CoV-2. Humans inhale an estimated 7–10 m^3 of air per day via breathing and the resulting exchange of air (Stetzenbach 2009). In this volume, an estimated 10^4 to 10^6 bacterial cells per m^3 of air are present. Not surprisingly, therefore, the microbiome in the respiratory airways is diverse (Huffnagle et al. 2017; Kumpitsch et al. 2019). The gastrointestinal tract is a central entry point for microbes that can cope with low pH, bile, or proteases. Furthermore, the gut continually moves and changes shape; hence, parasites

using this entry must be able to attach to the gut wall, and often use adhesion molecules for this purpose. Vital entry points are the 'Peyer's patches'—patches of lymphoid tissue just below the gut epithelial cells, which consist of M-cells (microfold cells). M-cells specialize in the uptake (endocytosis) of antigens, which helps to initiate mucosal immune responses, but which also makes these patches an entry point for viruses and many bacteria.

9.1.3 Horizontal vs vertical transmission

For many considerations, the most critical distinction is between the context of 'vertical' transmission (parent to offspring), and that of 'horizontal' transmission (all other cases; Figure 9.1). The same parasite can use different modes and routes of transmission, resulting in a different context. Indeed, most parasites use vertical and horizontal transmission at the same time, or at different stages (Table 9.1). For example, HIV can spread vertically to offspring or horizontally by sexual and non-sexual direct contacts. Similarly, *Toxoplasma gondii* uses cats (including house cats) as its main hosts. It sheds as oocytes in cat faeces that can (horizontally) infect most other warm-blooded hosts, including sheep, mice, or humans. Nevertheless, *T. gondii* is also (vertically) transmitted to offspring as a congenital (i.e. acquired before, or at, birth), or neonate (i.e. to newborns after birth) infection (Rejmanek et al. 2010). In fact, parasites with strict vertical transmission share the fate of their host lines. Uninfected host lines would outcompete these if the parasite were to impose a cost of infection. Exclusively vertically transmitted parasites are therefore unlikely to persist indefinitely. Hence, 'mixed-mode transmission', i.e. using both contexts, is likely to be very common (Ebert 2013).

The lack of knowledge about modes, routes, and the transmission context for most parasites is a serious hindrance to a better understanding of parasite ecology and evolution (Antonovics et al. 2017). Such knowledge is furthermore essential to assess public health measures or occupational health risks in factories or laboratories (Sewell 1995). Genetic markers can help to clarify some of these points. For example, during the severe outbreak of foot-and-mouth disease in Britain in 2001, the sequenced

Table 9.1 Classification of transmission contexts, with examples of human diseases.[1]

Transmission mode		Transmission routes (examples)	Infection routes (examples)	Diseases (causative agent)
Vertical		Transovarial, transplacental, postnatal (breast feeding), seedborne (plants)	Ingestion, mucosa	AIDS (HIV), rubella (RuV), cytomegalovirus (CMV), Lyme disease (Borrelia)
Horizontal	Sexual	Genital–genital, flower–flower (plants)	Ingestion, skin lesion, mucosa	AIDS (HIV), syphilis (Treponema)
	Non-sexual: Direct contact	Hand–hand, mouth–mouth, faecal–oral route, skin–skin	Ingestion, inhalation, skin lesion, mucosa	AIDS (HIV), chickenpox (ZVZ), measles (MeV), hepatitis A & B (HAV, HBV), influenza, typhoid (Salmonella), plague (Yersinia)
	Non-sexual: airborne — By droplets, aerosol	Mouth–mouth, nose–nose	Inhalation, skin lesion	
	Non-sexual: indirect — Environment	Soil–mouth, soil–skin	Ingestion, skin lesion, cutaneous penetration, mucosa	Anthrax (B. anthracis), fungi
	Animals	Mouth–mouth, mouth–skin, faecal–oral route		Rabies (rabies virus), toxoplasmosis (Toxoplasma), Q fever (Coxiella)
	Waterborne	Water–mouth, water–skin	Ingestion, skin lesion, cutaneous penetration, mucosa	Hepatitis A (HAV), typhoid (Salmonella), cholera (Vibrio), river blindness (filaria), bilharzia (Schistosoma)
	Foodborne	Food–mouth	Ingestion	Listeriosis (Listeria), shigellosis (Shigella)
	Materials (fomites)	Clothing–skin, needle–bloodstream, doorknob–hand	Ingestion, skin lesion, cutaneous penetration	AIDS (HIV), hepatitis A (HAV), plague (Yersinia)
	Vector-borne	Vector–bloodstream, vector deposits–oral	Cutaneous penetration, ingestion	Yellow fever (YFV), Malaria (Plasmodium), Lyme disease (Borrelia)

[1] Adapted from: Antonovics. 2017. *Philos Trans R Soc Lond B*, 372: 20160083, under CC BY 4.0.

viral isolates demonstrated transmission by direct contact at cattle markets, by the transport of animals, and by aerial spread (Cottam et al. 2006). This knowledge helped the decision about how to slow down the epidemic.

Furthermore, whereas host genetic markers (e.g. mtDNA) and parasite genotypes are completely linked with maternal transmission, deviations from this interspecific linkage disequilibrium can provide an estimate of the amount of vertical vs horizontal transmission (Wade 2007; Brandvain et al. 2011). Phylogenetic distances of parasites and their hosts can also provide clues to the prevailing transmission pattern. Direct observation or experimentation are classical tools to identify transmission modes and routes. For example, tests can investigate how a new and probably unknown virus is transmitted. A case in point is the sudden outbreak of Chikungunya virus (CHIKV) in South Asia during 2005–2006 (Dubrulle et al. 2009). Such experiments can at least provide evidence for the pathogen's potential to spread.

9.1.4 The evolution of transmission

Transmission mode, by definition (Figure 9.1), is something physical and results from organismal traits. It can therefore evolve, whereas transmission route results from a given mode and therefore puts selection on the various traits associated with it. Transmission modes can be very different, even among closely related parasites, suggesting genetic variation in the underlying organismal traits. The transmission context can also be experimentally selected. For example, the ability to transmit horizontally is almost completely lost when the bacterium *Holospora undulata* (infecting *Paramaecium*) is cultured under conditions that promote vertical transmission (Dusi et al. 2015). In influenza virus, few nucleotide changes in the genomic sequence can lead to a different transmission mode, i.e. variants transmitted either by direct contact or as airborne particles (Luk et al. 2015). This is of obvious concern for public health, as viruses might rapidly evolve to become airborne, e.g. to transmit from an animal reservoir to the human population (Taubenberger and Kash 2010; Schrauwen and Fouchier 2014). As an aside, the genetic basis for transmission mode and its evolution are not

congruent with the basis for pathogenicity ('virulence'; see Chapter 13).

Theoretically, the evolution of transmission can be analysed based on the trade-offs in the associated benefits and costs. Examples are the evolution of polymorphism, i.e. mixed vertical and horizontal transmission (Ferdy and Godelle 2005), and the evolution of vector transmission (Gandon 2004). At the same time, theoretical studies typically assume that transmission is under the control of the parasite, which is not generally warranted. Actual cases of evolutionary changes in transmission are known, too. For example, free-living bacteria can evolve towards symbionts or parasites, with horizontal transmission as the ancestral state and vertical transmission as a likely irreversible endpoint (Sachs et al. 2011). On the other hand, among the *Rickettsia*, some vertically transmitted insect symbionts have evolved to become pathogens of vertebrates that are transmitted horizontally by vectors (Perlman et al. 2006). Vector transmission is particularly relevant for the spread of many serious human diseases, such as malaria, sleeping sickness, dengue, and yellow fever. Similar to the evolution of complex life cycles (see section 3.4.4) this transmission mode could evolve by 'upward incorporation', e.g. by acquiring the ability to survive in insects (or arthropods more generally) that blood-feed on the habitual vertebrate host, and which become the new vector. Also, 'downward incorporation' may evolve when an insect pathogen evolves the ability to produce transmissible infections in an otherwise dead-end vertebrate host, as seems the case in some viruses. Alternatively, a pathogen able to survive in insects and vertebrates could undergo a cross-species transfer, and evolve to use the insect as the vector (Antonovics et al. 2017). By contrast, sexual transmission is not likely to cross host-species boundaries. In fact, sexual transmission has evolved in many ways, but is typically a phylogenetically derived trait, often repeatedly emerging within the same groups (Antonovics et al. 2011).

Typically, 'emerging infectious diseases' result from host shifts; that is, when the parasite becomes able to transmit from an animal reservoir to humans (leading to zoonotic diseases; see section 15.6). Furthermore, transmission can evolve towards a mode that makes it easier to spread in social groups, such as in humans. For example, the influenza virus H5N1 ('bird flu') is a high-pathogenicity type with a high case fatality rate. It circulates in waterfowl, where it spreads along the faecal–oral route. If it evolved to transmit by airborne particles, H5N1 would become a severe threat for human public health. Viral traits that allow binding to receptors in the human airways (Shinya et al. 2006) are of particular interest. In H5N1, probably only five amino acid substitutions might be needed, of which two are already common in the viral populations (Russell et al. 2012). SARS-CoV-2 is a case where this has happened. The virus acquired the ability, likely in a bridge host such as pangolins, to better bind to human AEC2 receptors in the upper respiratory airways. In doing so, it became efficiently transmitted among humans via airborne particles (Zhang and Holmes 2020). Hence, the evolution of transmission mode relates to host usage by parasites (Longdon et al. 2014; Antonovics et al. 2017).

9.2 Variation in infection outcome

The outcome of infection can vary between different host individuals and among parasites, even if all other conditions are kept constant. For instance, isogenic *Drosophila melanogaster* (i.e. hosts bred to be genetically identical) of the same age, with a controlled microbiota, and reared and kept in the same environment were infected with the same, identical pathogens (Providencia *rettgeri*). Nevertheless, individually variable outcomes resulted (Duneau et al. 2017a). Likewise, the outcome of the infection varies with the parasite, e.g. with different bacterial species (Figure 9.2). Host and parasite genotypes matter for the outcome (see Chapter 10)—yet, there are several additional factors which relate to the infection process itself. Such variable outcomes of infections have stimulated discussions as to whether a disease, e.g. AIDS, can be cured by nudging the infection dynamics into a region of the disease space that leads to recovery or long-term control rather than to deteriorating health (Conway and Perelson 2015).

In addition to infection success, the pathogenic effects also vary, even among pathogens that have a very similar ecology. For example, the effects of infection by human rhinovirus (HRV, Picornaviridae)

and influenza virus type A (IVA, Orthomyxoviridae) are quite different. Although they belong to different groups, both are small ssRNA viruses, are transmitted by aerosols or direct contact, and primarily infect the upper airways. Infection by HRV has mild symptoms, with a sore, running, and stuffy nose that subsides 10–12 days post-infection. Up to 40 per cent of infections are asymptomatic altogether (Heikkinen and Jarvinen 2003; Kirchberger et al. 2007). With an infection by IVA, symptoms appear relatively suddenly a few hours to days post-infection, and include a dry throat, headache, muscle pains, fatigue, and high fever, which last for 7–14 days. But in contrast to HRV, the influenza virus can also cause severe pathological symptoms that can be life-threatening. Sometimes, pneumonia (a lung infection associated with oedema) develops, and the patients may virtually drown in their body fluids. The 1918 influenza strain (labelled 'H1N1', type A) caused the big pandemic at the end of the First World War (the 'Spanish flu'). It was one of the deadliest viral infections in human history, with an estimated 50 million deaths. So, why do these two viruses with matching lifestyles cause such different pathologies? This question has many answers. On the one hand, from the evolutionary point of view, differences in the effect on the host will also have different consequences for the success and further transmission of the parasite and are selected for or against accordingly (see Chapter 13). On the other, from the point of view of the underlying mechanisms, the biochemical and molecular processes eventually yield the effect on the host. How these effects come about is the study of 'pathogenesis'. These processes can be related to disease space (Box 9.1).

9.3 Infection

9.3.1 Infective dose

Macroparasites, such as helminths or parasitoids, typically infect and establish as single individuals. However, bacteria, viruses, or protozoa infect with an 'infective dose'; that is, many single pathogens are needed for the infection to occur. In practice, a dose is administered or comes naturally as an 'inoculum' with a given size. There is a remarkable

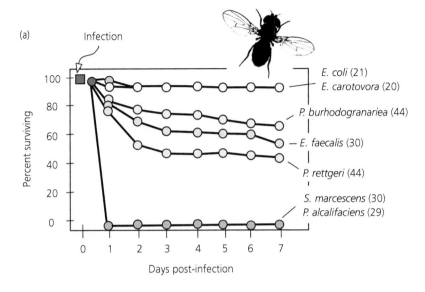

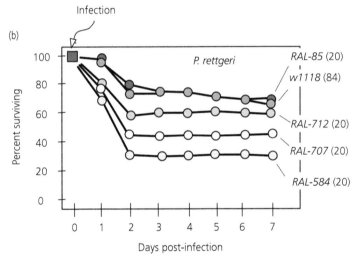

Figure 9.2 Infection outcome varies. The graphs show cohort survival of fruit flies, *Drosophila melanogaster*, when experimentally infected by bacteria. (a) In this experiment, males of fly line Canton S were infected with an inoculum (*c*.30 000 cells) of different bacterial species. The result varied according to species, with *Providencia alcalifaciens* and *Serratia marcescens* being lethal for all, while *E. coli* and *E. carotovora* did not reduce survival. Other bacterial species were intermediate in their effects. (b) Males of isogenic lines (from the Drosophila Genome Resource Panel, plus white mutant, *w*1118) of the same age, with a controlled microbiota, and reared in a common environment, were infected with the same inoculum of the bacterium *Providencia rettgeri*. Nevertheless, the outcome varied significantly, as survival varied among isogenic lines (labels at right) (Cox test: $c^2 = 19.23$, $df = 4$, $p < 0.001$). Figures in parentheses are numbers of replicates. Silhouettes from phylopic.org. Adapted from Duneau et al. (2017a), under CC BY.

variation in the infective dose needed to start an infection (Table 9.2). Some parasites can start an infection with a dose of only a few cells, for example, enteropathogenic bacteria (*Shigella* spp.) or protozoans (*Cryptosporidium parvus*; Englehardt and Swartout

2004) in humans. Similar low doses exist for viral infections in mice (Gammaherpes; Tibbetts et al. 2003), and cattle (foot-and-mouth disease; Schijven et al. 2005). Other parasites, e.g. *Campylobacter*, *Staphylococcus*, or *Salmonella* (Sewell 1995), are only

Box 9.1 Infection in disease space

Infection happens after a parasite or pathogen is transmitted to a new host. Once arrived, the parasite takes a particular route of infection to enter the host body (or to settle on a surface). Depending on the infection route, a given infection, therefore, may end up at a specific spot in the disease space to start the infection; for example, at point A or B shown in Figure 1. At point B, the further course of infection would take it much more rapidly to the domain in disease space where the host becomes sick, whereas this is farther away when the infection starts at point A. In both cases, the infection happens in the primary organ, such as in the peripheral airways or the gut. Most parasites, however, subsequently enter their 'preferred' site, the secondary organ: for example, the lung or the liver. The process of migrating towards this site is also known as 'tissue tropism'. After this migration, the infection may start at different points of the disease space that characterizes the respective secondary organ. For example, the secondary organ may be refractory to infection and immediately recover, which results in low fitness of the parasite. Nevertheless, the 'preferred' secondary organ (no. 3 in Figure 1) usually provides maximum fitness for the parasite. Selection would thus lead to efficient mechanisms of tissue tropism that take the parasite from the primary organ of infection to this preferred secondary organ (Figure 1).

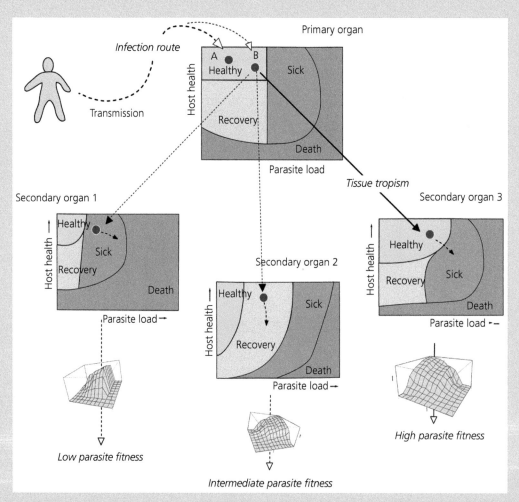

Box 9.1 Figure 1 Infection and disease space. Parasites transmit and enter via a given infection route into the primary organ, e.g. to site A or B in disease space. From there, typically, within-host migration occurs to a (secondary) target organ (tissue tropism) that ideally ensures a high fitness (right-hand route; green fitness surface). Depending on the characteristics of the disease space within each organ, these secondary cases start at different points of that space. If several secondary organs become infected, the overall parasite effect on the host is a composite of the effects of infections in different parts of the body.

The success of infection also depends on the dose, i.e. on how many cells of the parasite will be reaching the host. There are marked differences in the dose required to start an infection, depending on factors such as infection route, host status, host and parasite genotype, and the infection strategy used by the parasite itself. In general, small doses are less successful, and the host's early defence response may rapidly clear infections. Large doses have the advantage of numbers with many invading cells. These cannot quickly be cleared in the early stage. Large doses are often also diverse, containing different parasite genotypes that may take different routes through disease space. Such diversity can make a defence against large doses difficult (Figure 2).

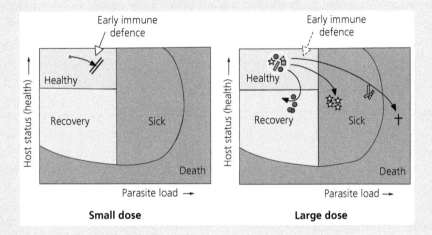

Box 9.1 Figure 2 Dose and disease space. Small doses (left panel) may be successfully cleared by an early immune response. Large doses (right panel) can override such early responses by numbers. Moreover, large doses may contain many different pathogen types (symbols) that can potentially take different routes through disease space. An orchestrated immune response against diverse infections is known to be typically less effective. Populations of some pathogen variants may reach the sick and death zones.

infective in high doses. Huge differences even exist between closely related pathogens, e.g. for *Escherichia coli* with type EHEC (ten cells) vs type EPEC (10^8–10^{10} cells; Table 9.2).

The estimate of an infective dose is based on the dose–response curve (Figure 9.3). The 'response' is defined by the effect of interest, such as infection status (infected or not), infection intensity (the number of parasites in the host), or the number of lesions on a plant leaf (the symptoms). Because the effect of interest varies according to the question under scrutiny, there are many definitions for dose, based on different responses (Box 9.2). A few are most commonly used, such as ID50, the dose needed to infect 50 per cent of the hosts. Similarly, dose per se can be defined in different ways, depending on the question asked; e.g. the amount (mg) of toxin injected, the volume of polluted water consumed, and so forth. Most commonly, it means the number of parasitic cells (or a measure of virus density) administered to the host. Most parasites can also use more than one route of infection, and this matters for infective dose (Figure 9.3).

In the food industry, or for public health management, a set of tools collectively known as 'quantitative microbial risk assessment' (QMRA) is used. A quantity of interest is the likelihood that a consumer, or an average person in an area, becomes sick (Haas et al. 2014; Battersby 2017). This assessment is again based on dose–response curves, i.e. the risk of illness given that an infection has happened. In addition, the fraction of the population exposed and susceptible to the agent of interest is taken into account (Box 9.3). Hence, the dose–response curve takes centre stage; the curve results from how the infection process unfolds.

Table 9.2 Minimum infective doses for various human bacterial pathogens.

Bacterium	Infection route	Disease, clinical symptoms [Fatality rate][1]	Dose (pathogen cells)	Source
Bacillus anthracis	Skin	Anthrax disease [20%]	$8 \cdot 10^3 \dots 50 \cdot 10^3$ ($\approx 10^3$ by inhalation)	3, 5, 7, 16, 15, 17
Bacillus cereus	Ingestion	Food poisoning [0%]	10^6	5, 7, 9, 16
Campylobacter jejuni	Ingestion	Campylobacteriosis (enteritis, gastroenteritis) [0.4%]	$10^2 \dots 10^5$ (50% infection) (FDA: 400–500) by ingestion	3, 5, 8, 16, 15–16, 18
Clostridium perfringens	Ingestion	Foodborne illness. Diarrhoea, necrosis of intestines [0.1%]	10^5 $>10^8$	5, 7, 9, 16
Coxiella burnetii	Inhalation	Q fever [3%]	≈ 10 by inhalation	1, 7, 10, 18
E. coli, enterohaemorrhagic (EHEC)	Ingestion	Diarrhoea (common serotype is O157:H7) [0.3%]	≈ 10	3, 5, 7, 8, 14, 14
E. coli, enterotoxic (ETEC)	Ingestion	Gastroenteritis [0.02%]	$10^8 \dots 10^{10}$	5, 7, 13
E. coli, enteropathogenic (EPEC)	Ingestion	Children's diarrhoea [0.02%]	Adults: $10^8 \dots 10^{10}$ Children: Very low	5, 7– 9, 13
Listeria monocytogenes	Ingestion	Septicaemia, meningitis, encephalitis, uterine infections [30%]	$\approx 10^3$	3, 5, 15, 16
Mycobacterium tuberculosis	Inhalation	Tuberculosis [0.02%]	≤ 10 by inhalation	2, 5, 7, 15, 16, 18
Salmonella paratyphi	Ingestion	Paratyphoid fever, enteric fever [4%]	$>10^3$, perhaps lower	4, 5, 7, 18 16
Salmonella typhi (enterica, typhimurium)	Ingestion	Gastroenteritis (salmonellosis) [4%]	10^5 (50% infection), perhaps lower	3, 4, 7, 9, 11, 18
Shigella flexneri	Ingestion	Bacillary dysentery Flexner Shigellosis [0.1%]	≈ 100	3, 5, 7, 8, 11, 15, 16, 18
Staphylococcus aureus	Ingestion	Food poisoning. Nausea, abdominal pain [0.02%]	10^5 per gram of food	5, 12, 16
Streptococcus A (pyogenes)	Inhalation	Sore throat, scarlet fever, septicaemia [19%]	10^3, perhaps lower	43, 4, 5
Vibrio cholerae (serotypes O139, O1)	Ingestion	Cholera [0.9%]	$10^2 \dots 10^{11}$; 10^8–10^9 by ingestion (FDA: $\approx 10^6$)	3, 5, 7–9, 15, 16
Yersinia enterocolitica	Ingestion	Gastroenteritis, diarrhoea, fever, abdominal pain [0.5%]	10^6	3, 5, 7, 8, 16
Yersinia pestis	Skin	The plague [90%]	10	1, 3, 6

[1] Fatality rate after Leggett. 2012. *PLoS Path*, 8: e1002512.

Sources: [1] Azad. 2007. *Clin Infect Dis*, 45: S52. [2] Cain. 2007. *Am J Respir Crit Care Med*, 175: 75. [3] Center for Disease Control and Prevention. 2009. *Alphabetical index of parasitic diseases*. Available: http://www.cdc.gov/parasites/az. [4] Edelman. 1986. *Rev Infect Dis*, 8: 329. [5] Food and Drug Administration. 2003. *The bad bug book*. [6] Hacker. 2003. *Science*, 301: 790. [7] Pathogen safety data sheets. 2003. Available: http://www.phac-aspc.gc.ca/lab-bio/res/psds-ftss/index-eng.php. [8] Kothary. 2001. *J Food Saf*, 21: 49. [9] Leggett. 2012. *PLoS Path*, 8: e1002512. [10] Levine. 1983. *Microb Rev*, 47: 510. [11] Medscape. 2009. Infectious Disease Articles. Available: http://emedicine.medscape.com/infectious_diseases.[12] Mims. 1987. *The pathogenesis of infectious disease*. Academic Press. [13] National Institute of Allergy and Infectious Disease. 2011. *Health and research topics A–Z*. [14] Microbial Pathogens data sheets. 2009. Available: http://www.foodsafety.govt.nz/elibrary/industry/Staphylococcus_Aureus-Science_Research.pdf. [15] Rasko. 2008. *J Bacteriol*, 190: 6881. [16] Rivera. 2010. *Proc Natl Acad Sci USA*, 107: 19002. [17] Salyers. 2002. *Bacterial pathogenesis*, 2nd ed. ASM Press. [18] Sewell. 1995. *Clin Microbiol Rev*, 8: 389. [19] Teunis. 1999. *Risk Anal*, 19: 1251. [20] The Institute of Food Technologists. 2011. Science reports. [21] Todar's online textbook of bacteriology. 2009. Available: http://www.textbookofbacteriology.net/index.html. [22] World Health Organization. 2009. Publications and fact sheets. Available: http://www.who.int/research.

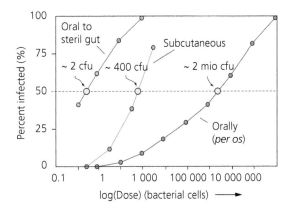

Figure 9.3 The dose–response curve. Shown is the percentage of mice infected with *Salmonella enteriditis* three weeks post-infection, in relation to dose (log scale). The number of bacteria was measured as the number of colony-forming units (cfu's) that were present in faecal samples. The curves differ according to the mode of administering the inoculum (the infection route); orally to a mice with a sterilized gut, or a control mouse, and subcutaneously. The ID50 (dose that leads to 50 per cent infected hosts) is shown as a yellow dot. Adapted from Johnson (2003), with permission from SAGE Publications.

Box 9.2 Definitions of dose

A dose is typically the number of infective parasites administered to a host in a defined way, e.g. parasites fed *per os*, or hosts exposed to infective spores.

Various infective doses are defined as follows:

- ID50 (the 50 per cent-infective dose): Dose needed to infect 50 per cent of the tested hosts.
- LD50 (the 50 per cent-lethal dose): Dose needed to kill 50 per cent of the tested hosts.
- LD10, LD99, etc.: Dose needed to kill 10 per cent, 99 per cent, etc. of hosts.
- LD50/30, LD50/60: Dose needed to kill 50 per cent of the tested hosts during a period of up to 30 (or 60) days. Mostly used for studies of radiation effects.
- LDLo: Lowest published dose needed to infect a host.
- EID50 (the egg 50 per cent-infective dose): Dose needed to infect 50 per cent of the tested eggs. Used when testing the infectivity of viruses on fertilized chicken eggs (e.g. to produce vaccines).
- ELD50 (the egg 50 per cent-lethal dose): Dose needed to kill 50 per cent of the tested eggs.
- TCID50 (the tissue culture 50 per cent-infectious dose): Dose needed to produce a pathogenic effect in 50 per cent of the cultures inoculated.

Usually, many host individuals are needed to test for a lethal dose. Fewer test individuals are necessary with the 'fixed-dose procedure' (usually used for tests when infected orally). For this test, a fixed number of host individuals are tested, and the lethal dose is extrapolated from the results.

In medical practice, several further measures are used. These are not necessarily referring to parasitic infections but often refer to drug treatment, or to protection from radiation, toxins, and poisons.

- Booster dose: Dose of an active immunizing agent, usually smaller than the initial dose, which is needed to maintain immunity.
- Effective dose (ED): Dose of a drug that produces the desired effect.
- Fractionated dose: Fraction of the total quantity of a drug needed for effect, and which is given at intervals.
- Maximum dose (safe dose): The largest dose that can safely be given.
- Median curative dose (CD50): Dose removing the infection symptoms in 50 per cent of subjects.
- Median lethal dose: Same as LD50.
- Minimum lethal dose (MLD, minimum fatal dose): The smallest dose that is capable of killing a host. This depends on body mass.
- Median effective dose (ED50): Dose producing an effect in 50 per cent of subjects.
- Median immunizing dose: Vaccine or antigen dose to generate immunity in 50 per cent of subjects.
- Median toxic dose (TD50): Dose producing the toxic effect in 50 per cent of subjects.
- Minimum dose (threshold dose): Smallest dose producing the effect.
- Priming dose: Dose needed to establish an effect (in drug treatments).
- Tolerance dose: The largest dose that can be administered without effect.

Box 9.3 Quantitative Microbial Risk Assessment (QMRA)

Quantitative microbial risk assessment seeks to quantitatively predict the level of harm that is expected for an average person and a given pathogen threat. The standard approach follows several steps, each one based on quantitative observational or experimental data as well as modelling to generate probability distributions of outcomes. Among other inputs, the dose–response curves play an important role, i.e. the probability of infection when an average person is exposed to a specific dose of the pathogen of interest (Teunis and Havelaar 2000) (Figure 1).

therefore routinely used in the context of food contamination or water safety management, e.g. dangers for beaches or safe drinking water regulations (Schijven et al. 2005; World Health Organization 2016). The analysis is often also separated by transmission routes, such as risks associated with transmission through animals, fomites (contaminated materials), food, drinking water, wastewater, and other possible sources of infection risks. Together, QMRA can lead to recommendations for engineering, e.g. for the threat of pneumonia by *Legionella* infections in water systems, such as whirlpools,

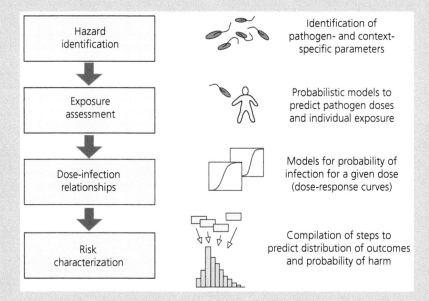

Box 9.3 Figure 1 Flow chart of QMRA. In a typical QMRA, (1) the kind of pathogen, together with its biological characteristics and the context of the analysed risk, is identified (Hazard identification). (2) Predictions are made regarding the pathogen's occurrence and abundance, and the likelihood of exposure for the average host individual (Exposure assessment). In subsequent steps, (3) dose–response curves are combined with previous steps, and (4) the expected harm is calculated (Risk characterization). Adapted from Brouwer et al. (2018), with permission from Springer Nature.

QMRA is particularly relevant to predict the risks of infection and harm for low doses of a pathogen, which is a typical situation for many everyday public health problems or in the food industry—in contrast to the situation where, for example, an epidemic runs through a population. QMRA is

spas, or showers (Buse et al. 2012). QMRA can serve many goals, such as treatment and vaccination strategies, and also aids in understanding disease outbreaks by tracing the risks and exposure patterns, e.g. salmonellosis in beef burgers or hepatitis among drug users (Brouwer et al. 2018).

9.3.2 Generalized models of infection

Generalized models of infection make no specific assumptions about the mechanisms that lead to an infection, and consider the infection process as happening instantaneously (i.e. a point event). In the

simplest case, every single parasite has the same probability, p, to infect a host—regardless of how this is happening. With all hosts equally susceptible and all pathogens fully infectious—that is, $p = 1$—every single parasite propagule would infect with

Box 9.4 Formalizing infectious dose in general models

For any practical consideration dose, D, is a random variable because it is rarely possible to administer the same number of parasites to a given host. Hence, the number of parasites in a given inoculum is varying around a mean number, D. Because parasites occur in discrete numbers, mean and variance of parasite numbers in a dose are approximated with a Poisson distribution. More precisely, $P(k, D)$ is the probability that any given dose contains k parasites, provided the average number is D. This Poisson-distributed probability formally is:

$$P(k,D) = \frac{D^k e^{-D}}{k!} \qquad (1)$$

With this distribution, the fraction of hosts not receiving any parasite at all, Q, is $Q = e^{-D}$, i.e. where $k = 0$. Therefore, the fraction of hosts receiving at least one parasite is given by:

$$R = 1 - Q = 1 - e^{-D} \qquad (2)$$

Assuming every parasite is fully infectious, the infection starts when at least one parasite ($k \geq 1$) arrives. Consequently, R is also the fraction of hosts responding with an infection when a dose, D, is used. At the point where $R = Q = 0.5$, by definition, dose is ED_{50} (the effective dose for 50 per cent infection; see Box 9.2). Using eq. (2), this corresponds to a dose $D = 0.69$ parasites (recall that D is a mathematical mean). Consideration of probabilities thus allows prediction of the expected dose, D, for a successful infection under explicit assumptions, as made above. Among other things, such general models are of great interest in predicting the probability of infection for low doses, in the context of QMRA (see Box 9.3), and when addressing common public health concerns. Unfortunately, low doses are often outside the range of actual measurements. The extrapolation of the model into this low-dose zone is fraught with difficulties. The necessary conditions therefore need to be carefully evaluated in any given case (Ngo et al. 2018).

(a) *With the independent-action model*, there is a probability, p, per parasite of starting an infection independent of dose, D. If all hosts are homogeneous and respond to the parasite in the same way, then p assumes a constant value. (If hosts are heterogeneous, then p is a random variable with a given distribution; for example, with a distribution similar to Figure 9.4b.) With homogeneous hosts, the probability of not becoming infected (P_0), given D, is:

$$P_0 = (1 - p)^D \qquad (3)$$

Because p is generally small, while D is large, one can approximate the fraction of host not becoming infected

(i.e. not responding, setting $Q = P_0$), and the fraction becoming infected (responding), R, with:

$$Q = (1 - p)^D \approx e^{-Dp}; R = 1 - Q \approx 1 - e^{-Dp} \qquad (4)$$

similar to eq. (2) above. Using the results of eq. (2) for the 50 per cent-effective dose, $ED_{50} = 0.69$, we find that

$$R = 1 - e^{-0.69p} \qquad (5)$$

When hosts are heterogeneous, p varies. In this case, the fraction of hosts not becoming infected has to be summed up over the distribution of p-values in the population. These values are given by the distribution of values over all host–parasite interactions. This distribution is similar to the distribution of IDEs, as shown in Figure 9.4b. If p is continuous, then:

$$Q = 1 - \int_1^0 e^{-Dp} f(p) dp \qquad (6)$$

where $f(p)$ is the probability distribution for the different values of p in a population of hosts ($0 \leq p \leq 1$). The fraction becoming infected is $R = 1 - Q$.

(b) *'Birth-death process' models* assume that a population of parasites starts growing in a host with births (at rate λ) and deaths (at rate μ). The parasite population will eventually grow to a population size that approaches the infective threshold, C. After reaching C, the infection is established. C can be reached when $\lambda > \mu$. If $\lambda < \mu$, the infection never occurs. When $\lambda = \mu$, the parasite population fluctuates, and infection or failure can result. The chances of having $\lambda > \mu$ increase with dose, D. If the value of D is large, the growth of the parasite population can be approximated by the standard exponential growth of classical population biology as:

$$D_t = D_0 e^{(\lambda - \mu)t} \qquad (7)$$

where D_0 is the infecting dose at the start of the infection process and D_t is the size of the parasite population at time t. With a birth–death process, the fraction of non-infected host subjects should decrease with dose, as:

$$Q = e^{-pD} \qquad (8)$$

With birth–death processes, the establishment of infection occurs with some time delay. This time can be estimated depending on how the model is formulated. In the simplest case, the time delay to infection decreases with $1/log(D)$ above a given threshold, i.e. above the individually effective dose. Additional complications arise when the values of λ and μ are themselves not constant but change with either the duration (time dependence) or the size (number dependence) of the infection (Ercolani 1984).

certainty. The infective dose is $D = 1$; i.e. one patho-gen cell is enough. However, a first complication arises because D is a random number in a given situation; in practice, therefore, the actual number of parasites varies around a mean of D. Box 9.4 shows how dose, D, can be analysed under explicit assumptions for specific infection models.

9.3.2.1 Independent action hypothesis (IAH)

In this scenario, infection results from the inde-pendently acting parasites in an inoculum. This model has been proposed early in the history of the subject (Halvorson 1935; Druett 1952). IAH models are characterized by the probability, p, per parasite of causing an infection, independent of dose. Furthermore, p is considered to be a biological char-acteristic of the parasite and the host it encounters. The infection results from a process of parasite births (at rate λ) and deaths (rate μ) as the single parasites are entering and colonizing the host (Box 9.4). Such 'birth–death' processes are useful for practical purposes, e.g. when important variables cannot be measured. The infecting population grows if $\lambda > \mu$, and the probability of infection is $P_{inf} = 1 - \lambda/\mu$ (Ercolani 1984). With birth–death pro-cesses, infection occurs with some time delay, which can be estimated from the model.

A somewhat analogous situation exists for mod-elling the effects of antibiotics; i.e. how many mol-ecules need to attach to a bacterial cell in order to kill it (Hedges 1966)? For an infection process, one might assume that a single parasite can start an infection ('single-hit models'; Teunis and Havelaar 2000; Heldt et al. 2015). At the same time, an inocu-lum would consist of a mixture of either fully infec-tious or non-infectious parasites. Hence, the capacity of an inoculum to cause infection varies with its size, D, and the fraction of competent para-sites, which depends on what determines the infec-tion process.

9.3.2.2 Individual effective dose (threshold models)

In this model, infection occurs when the inoculum contains more than the 'individual effective dose' (iED). iED is the number of parasites, n, needed for an infection. In particular, numbers below iED

cause no infection, and numbers at least or above iED cause full infection. Hence, the probability of infection is a step function with $p = 0$ for $n <$ iED and $p = 1$ for $n \geq$ iED. However, in practice, the actual number of parasites in a given dose varies randomly; therefore, the dose–response curve becomes sigmoidal (Figure 9.4a). Alternatively, the iED can vary among hosts. In this case, the dose–response curve also departs from a step function and becomes more gradual, given by the collection of steps at varying levels of iED corresponding to each host (Figure 9.4b). Hence, although the indi-vidual effective dose model is essentially a thresh-old model, in practice a gradual dose–response curve is observed, since individual subjects vary in a population. This makes such a model exceedingly difficult to test.

9.3.2.3 Host heterogeneity models (HHS)

Taking into account host heterogeneity in suscepti-bility is an extension of the individual effective dose models above (Regoes et al. 2003). The approach originated from frailty models used in survival analyses (Halloran et al. 1996; Ben-Ami et al. 2010). In the models, the dose–response curves (i.e. the susceptibilities) vary according to the combination of host and parasite types. The overall infection characteristic emerges from the ensemble of cases, i.e. across a host x parasite infection matrix, where the distribution of susceptibilities can be defined arbitrarily. As with all generalized models, there are no specific assumptions about why such variability exists. For example, hosts may have already encoun-tered the parasite and are now partially immunized and thus less susceptible. In practice, the compilation of such data is not easy, as different studies use dif-ferent doses, sensitivities of the tests, administration routes, or varying sample sizes. If infection prob-ability is determined for multiple inoculum doses, HHS models can be fitted and the variance in sus-ceptibility can be estimated (Ben-Ami et al. 2008; Gomes et al. 2014).

9.3.2.4 Within-inoculum interaction models

In these scenarios, there are interactions among pathogens in an inoculum, and among different strains of a given pathogen in particular. Interactions among co-infecting different species of pathogens

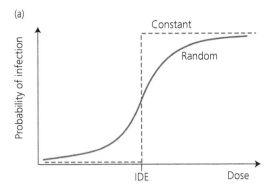

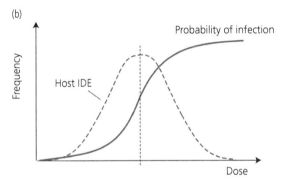

Figure 9.4 Individual effective dose (iED) model. (a) All hosts are homogeneous and have the same iED. As a result, the infection succeeds if the dose is at least or above the iED, but fails if the dose is below iED; the dose–response curve is a step function (dotted line). In practice, inoculum size (the administered dose) deviates randomly from the mean, and, hence, the observed curve is sigmoidal (solid line). (b) Hosts vary in their iEDs, as shown by the distribution indicated by the dotted line. The observed dose–response curve (solid line) is a gradual function reflecting the variation in individual iEDs in the population.

follow the same principles. The co-infecting pathogens act antagonistically or synergistically by choice of model parameters (Regoes et al. 2003) (Box 9.4). Synergistically acting pathogens will reduce the dose required to infect, whereas antagonism will increase the dose. With no interaction, the models collapse to the independent-action scenario above.

9.3.2.5 Sequential models

In many cases, the exposure to a pathogen repeatedly occurs over a limited time window. Sequential models assume that a certain number of sequential exposures is needed for infection. After each exposure, some invaded pathogens are cleared before the next pathogens arrive. Thereby, the host's defensive

capacity may erode and become exhausted. These sequential invasion events are therefore not independent of one another, as the effect of the next one depends on the previous ones. Success becomes more likely as the host approaches a critical state where the infection can establish. This scenario is sometimes also called a 'cumulative dose' model. It gives rise to a sigmoidal dose–infection curve that can be fitted to data. Examples include infections by *Cryptosporidium* and viruses in people (Pujol et al. 2009), or *Serratia marcescens* in wax moth larvae (Anttila et al. 2016). With sequential models, a (simple) explicit mechanism is added to a generalized model; that is, pathogens are cleared by the host immune system.

Because of their nature, generalized models can be statistically fitted to most cases and are helpful if a crude dose–infection curve is needed, without regard to the actual process. This is often the case in QMRA. In the event, a range of different mathematical functions is available. They describe similar scenarios for the infection process but can generate very different predictions for values outside the fitted measurements. Predictions for doses beyond the observed or tested values are the most valuable ones (Box 9.3).

9.3.3 Process-based models

Process-based models explain the observed dose–infection relationship with an explicit mechanism. Often, such models look at different aspects of the infection process and are therefore not mutually exclusive.

9.3.3.1 The lottery model

This scenario has initially focused on the colonization of a gut by the members of a microbiota. For pathogens, the lottery model assumes stochastic processes and a bottleneck for infection and colonization. Bottlenecks can result from physically limited regions used by invading pathogens, e.g. the Peyer's patches in the vertebrate gut. For gut infections, the inoculum passes through with a certain speed. Due to the resulting constraints of space (suitable patches) and time (speed), the infection becomes a stochastic process. Therefore, only a random subset of all available pathogens in the

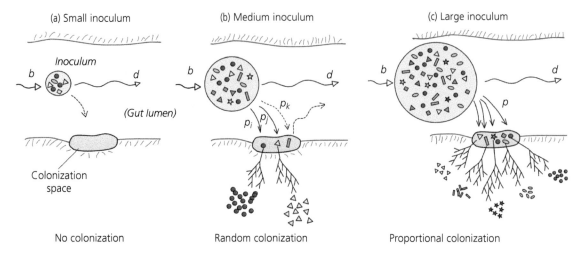

Figure 9.5 The lottery model of infection. Infection in the gut is a stochastic process governed by the number of pathogens in the inoculum (dose; the mix of different types shown by various symbols), the size of the 'colonization space' (yellow patches), and the rates at which the pathogens arrive in the gut (*b*; a birth rate) and disappear by being either destroyed or cleared (*d*; death rate). As the inoculum drifts past certain regions of the gut where invasion is possible (the colonization space, e.g. Peyer's patches in the vertebrate gut), each pathogen has an independent probability of infecting (shown as p_i, p_j, or p_k for three different pathogen types). Because the size of the colonization space is physically limited, it can only allow a limited number of pathogens to settle there, generating a bottleneck for the invading pathogen population. Combined with the different abilities of pathogens to actually infect, this limitation leads to a near-random draw from all available pathogens in the inoculum. This can result in either (a) no infection, when, by chance, none of the pathogens settles from a small inoculum, to (b) random infections, generating among-host variation in outcome, or (c) to a 'proportional infection' with a large inoculum, such that all types of pathogens are represented. Sketched after descriptions in Obadia et al. (2017).

inoculum has the chance to infect, and this subset varies from case to case. The lottery model therefore leads to variable outcomes for the dose–infection relationships (Figure 9.5). Some experimental infection studies use parasite cells that are tagged. When these represent only a fraction of the inoculum, the infection dynamics becomes stochastic, as assumed in the lottery models—that is, the number of tagged cells that are successful varies stochastically. This experimental procedure improves the statistical power for estimating infection rates (Kaiser et al. 2013).

9.3.3.2 The manipulation hypothesis

Mechanisms that parasites use to manipulate the host might be particularly relevant for dose–response relationships (Schmid-Hempel and Frank 2007). For example, some parasites secrete modulator molecules into the environment to gain entrance into hosts ('distant action' method). The molecules need to be sufficiently numerous to have an effect, but at the same time they diffuse in the vicinity of

the pathogen before they reach their receptors on a host cell. Hence, parasites succeed if they are numerous enough to produce sufficient quantities of immune modulators. *V. cholerae*, for example, releases toxins and requires a high infective dose (Table 9.2). By contrast, parasites that depend on short-range effects, or even directly transfer modulatory proteins into host cells ('local action'), may not have this numerical constraint. Even a single pathogen cell could manage to invade and initiate an infection that subsequently spreads inside the host. Some *Shigella*, for example, enter the host cell upon direct contact and injection of manipulation proteins via a type III secretion system; the infective dose is very low (Table 9.2).

In this view, the infective dose depends on how parasites manipulate the host to gain entrance into the host, especially the spatial scale over which parasite-derived modulators are useful. High doses, i.e. the collective action of many parasites, are required with diffusible toxins (Rybicki et al. 2018). By contrast, low doses are sufficient with local

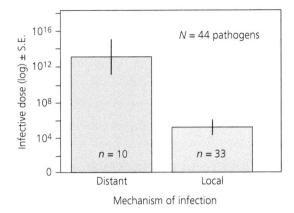

Figure 9.6 The manipulation hypothesis. Infective dose (the number of pathogen cells required to start an infection) is significantly higher for pathogens using a 'distant action' method (see text) as compared to those using a 'local action' method ($F_{1,40} = 25.79$, P < 0.0001). In total, $n = 43$ human pathogens were included in this analysis, including viruses ($k = 5$ species), bacteria ($k = 35$), fungi ($k = 1$), and protozoa ($k = 2$). Bars are means ±1 S.E. Adapted from Leggett et al. (2012), under CC BY.

delivery of modulators. This pattern is observed for the infective doses of 43 human pathogens (Leggett et al. 2012) (Figure 9.6). The distinction is reminiscent (but not equal) to the 'frontal attack' vs 'stealth' strategy made by parasitologists, based on the observation of how aggressively parasites attack the host and its immune system (Merrell and Falkow 2004).

9.3.3.3 Early infection dynamics

Infection success, for practical reasons, is evaluated sometime after exposure to an inoculum. Hence, between exposure and assessment, the infection could already have died out, for example, when the immune defence quickly manages to clear out the invaders. If so, this case would be counted as 'non-infected', and the outcome used to estimate infective dose (usually inflating its size). Modern techniques now allow tracking an incipient, early infection with appropriate markers (Grant et al. 2008) to study these early steps. For example, sequences of viral envelope genes revealed that most HIV infections (78 out of 102 cases) were by a single genotype, while the remainder was by two to five genotypes (Keele et al. 2008). As a more advanced method, sequence tag-based analysis of microbial populations (STAMP; Abel et al. 2015a)

uses high-throughput DNA sequencing in combination with extensive libraries of tags known for the pathogens. Such methods can help to estimate the inoculum size that has survived the infection barriers (Abel et al. 2015a, 2015b). The studies also elucidate how the subsequent infection dynamics depends on inoculum size (Hotson and Schneider 2015). So far, the methods have not often analysed infective doses in the first place.

9.4 Pathogenesis: The mechanisms of virulence

'Pathogenicity' is the capacity of a pathogen[1] to cause damage and to generate disease in its host. A broad range of biochemical, molecular, and physiological processes can cause damage to host cells and tissues, including side effects of the immune defences (immunopathology). They eventually appear as the pathogenic symptoms of an infection (Figure 9.7). Pathogenicity is sometimes used interchangeably with the term 'virulence'. The latter should be reserved for denoting the result of pathogenesis that is relevant for evolutionary ecology (see Chapter 13). In that context, virulence has been defined, among other things, as host mortality rate (May and Anderson 1983b). This clearly falls short of covering the full spectrum of possible outcomes of a parasitic infection. The quest for understanding the fitness-reducing effects of infection, therefore, touches on pathogenesis and virulence, but these are a set of rather diverse phenomena (Table 9.3). Importantly, though, the pathogenic mechanisms can generate fitness costs to hosts and advantages to parasites, and are therefore under selection.

9.4.1 Impairing host capacities

Parasitic infections lead to the loss of some function for the host, as parasites can lodge in places where they affect the capacity to hear, see, breathe, and so forth. For example, many trematodes specialize in attaching to the gills of fish (Valtonen et al. 1997;

---

[1] The word 'pathogen' is a composite from the Greek 'pathos' (suffering) and 'gennan' (to generate), whereas virulence stems from the Latin word 'virulentus', which is derived from 'virus' ('poison', i.e. 'full of poison').

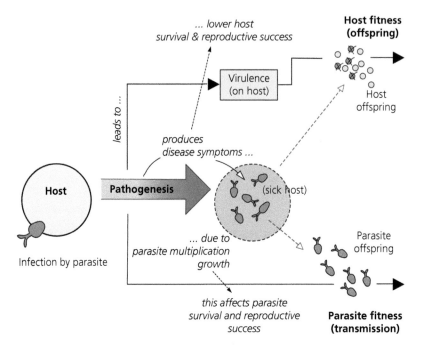

Figure 9.7 Pathogenesis and virulence. A healthy host individual becomes exposed to a parasite. The parasite infects, establishes, and invades tissues, where it starts to grow and multiply. These steps are typically associated with pathogenesis—a term summarizing processes based on many different molecular or physiological mechanisms that eventually damage the host. The selective consequences of pathogenesis on the host are macroscopically summarized as 'virulence'. Virulence is measured as a reduction in host fitness, i.e. in lower host survival and reproductive success, e.g. by a reduction in the number of surviving offspring. Pathogenesis also affects the chances of the parasite surviving inside the host and gaining transmission to a new host, i.e. parasite fitness. Note that the consequences for transmission will select for the level of pathogenesis, as far as it is caused by the parasite, as will be discussed for the evolution of virulence (see Chapter 13).

Morozinska-Gogol 2006). A heavy load of such external parasites will reduce water flow through the gills, damage its delicate structures, cause swelling, reduce oxygen uptake, and result in respiratory diseases (Reed et al. 2002). Apart from this direct pathological effect, such impairment also increases the chances that the fish becomes eaten by a predator. The impairment of vital functions is a generic principle that is characteristic of parasite-induced damage, even though the detailed effects vary among different systems and the consequences for host and parasites are quite variable.

9.4.2 Destruction of tissue

Some parasites destroy the host's tissue very directly. For example, females of ichneumonid wasps lay eggs into a host's body, typically the larvae of other insects. After hatching, the growing ichneumonid larva starts to consume the host's body fluids and

the host's tissues and internal organs. Eventually, the host dies and the parasite larva pupates and develops into the adult parasitoid. In such cases, host death is often essential for the completion of the parasite's life cycle ('obligate killers'; see section 13.10). Regardless, pathogenesis is primarily due to the destruction of tissues even though additional mechanisms contribute. For example, symbiotic viruses in many parasitoids help to evade and suppress the host's immune response (Schmidt et al. 2001).

Destruction of tissue as a generic element of pathogenesis is not restricted to parasitoids but also occurs with viruses or bacteria. Haemorrhagic viruses, such as Ebola (Baseler et al. 2017) or dengue (Screaton et al. 2015), destroy blood vessels, which leads to uncontrolled internal bleeding, an essential process of pathogenesis and a factor for host death (Mahany and Bray 2004; Bray 2006). Similarly, the primary site of replication for Rift Valley fever virus

Table 9.3 Mechanisms of pathogenesis.

Mechanism	Description	Pathogenic effect (examples)	Sources
Impairing capacities	Presence of parasite leads to loss of full functionality for important capacity. Associated tissue damage not a main effect. Many parasites induced behavioural changes that impair the host's capacity to function normally.	Phototactic behaviour that leads to host death by predators acting as next host of the parasite.	15, 22
Destruction of tissue	Parasite destroys critical or a large mass of tissue, which leads to failure of organs and eventual host death.	Parasitoids consume internal tissues or organs of the host. Haemorrhagic viruses cause necrosis and failure of vascular system.	6, 12
Virulence factors	Mainly described from bacteria and viruses, but also known from protozoa and fungi. This general category also includes toxins. Adhesins, invasins: Leukocidins of *S. aureus* are pore-forming toxins essential at the start of an infection. *C. diphtheriae* toxin (DT) in epithelia, causes thickening of membrane, assists colonization. *V. cholerae* toxin (CT): causes water loss, aids transmission. *Shigella* toxin (SHT, SHLT) aids transmission.	Various mechanisms, such as disruption of host cell cytoskeleton, cytokine signalling, neutralization of host defences. Often associated with severe necrosis of tissues or inflammatory processes. Pathogenic effect correlated with presence and expression of virulence factor. DT: necrosis, suffocation, nerve paralysis. CT: diarrhoea, shock, acidosis. SH, SHLT: causes diarrhoea, haemorrhage, colitis.	4, 10, 11, 16, 17, 21, 23, 24, 25
Toxins	Secreted proteins (exotoxins) or components of cell walls (endotoxins, enterotoxins) that allow pathogen to invade and spread within the host. Toxins have high biological activity and act like enzymes.	Many cytotoxins lead to apoptosis and tissue necrosis. Disruption of cytokine functioning causes fatal septic shock.	1, 5, 9, 14, 25
Proteases	Enzymes described from all major parasitic groups aid in breaking into tissues and across cell membranes.	Similar role as toxins. Proteases are antigenic and can induce inflammation and other severe pathogenic effects.	13
Response exhaustion (secondary infections)	Parasites deplete host immune response in various ways. Antigenic variation a major mechanism where pathogen persistently changes epitopes recognized by the host. Escape mutations produce new variants in an infecting population. Opportunistic infections damage the host.	Host forced to respond to continuously changing parasite antigenic surface. Eventually, defence breaks down, progression to disease follows. In many cases, weakening of the immune response allows secondary infections by other pathogens that lead to severe pathogenesis.	3, 4, 19, 20
Self-damage (immunopathology)	Parasite induces host immune response that causes damage to own tissue. (This general process applies to many of the above cases.)	Immune response with cytotoxic lymphocytes destroys own, uninfected tissue. Continuous destruction leads to pathogenesis.	2, 7, 8, 18

Sources: [1] Abrami. 2005. *Trends Microbiol*, 13: 72. [2] Bray. 2006. *Curr Opin Immunol*, 17: 399. [3] Deitsch. 2004. *Trends Parasitol*, 20: 562. [4] Dustin. 2006. *Annu Rev Immunol*, 25: 71. [5] Fukao. 2004. *Lancet Infect Dis*, 4: 166. [6] Godfray. 1994. *Parasitoids*. Princeton University Press. [7] Graham. 2005. *Annu Rev Ecol Syst*, 36: 373. [8] Guidotti. 2006. *Annu Rev Pathol Mech Dis*, 1: 23. [9] Lapaque. 2005. *Curr Opin Microbiol*, 8: 60. [10] Lee. 2014. *Int J Mol Sci*, 15: 18253. [11] Leo. 2015. *Int J Med Microbiol*, 305: 276. [12] Mahany. 2004. *Lancet Infect Dis*, 4: 487. [13] McKerrow. 2006. *Annu Rev Pathol*, 1: 497. [14] Moayeri. 2004. *Curr Opin Microbiol*, 7: 19. [15] Moore. 2002. *Parasites and the behavior of animals*. Oxford University Press. [16] Mühlenkamp. 2015. *Int J Med Microbiol*, 305: 252. [17] Nash. 2015. *Mim's pathogenesis of infectious diseases*. Elsevier, Academic Press. [18] Pawlotsky. 2004. *Trends Microbiol*, 12: 96. [19] Picker. 2006. *Curr Opin Immunol*, 18: 399. [20] Rall. 2003. *Annu Rev Microbiol*, 57: 343. [21] Rappleye. 2006. *Annu Rev Microbiol*, 60: 281. [22] Reed. 2002. *Monogenean parasites of fish (FA28)*. Fisheries and Aquatic Sciences Department Series. University of Florida, Institute of Food and Agricultural Sciences. [23] Rikihisa. 2015. *Annu Rev Microbiol*, 69: 283. [24] Spaan. 2017. *Nat Rev Microbiol*, 15: 435. [25] Wilson. 2007. *Postgrad Med J*, 78: 216.

(RVFV) is the liver. The affected cells undergo apoptosis, resulting in tissue destruction and severe pathologies. Experimentally infected mice (an animal model for RVFV) usually die three to six days post-infection with acute hepatitis and a delayed encephalitis (Smith et al. 2010). Host cell apoptosis is the natural defence that removes virus-infected cells (see Chapter 4). Pathogenic effects occur when this process overshoots or becomes modified by parasite modulators encoded by viral genes or by virus-induced expression of host genes. Many pathogenic bacteria manage to survive inside phagocytes and, after being engulfed, multiply and eventually destroy their cells and host tissues (Amulic et al. 2012). Similarly, streptococci proliferate in epithelial cells and destroy these tissues through the action of toxins (Berube and Wardenburg 2013). Hence, the actual causes of cell and tissue destruction are manifold and can involve toxins and other factors.

9.4.3 Virulence factors

'Virulence factor' is a generic term that refers to any factor (a molecule, a protein, a secretion system) critical for a pathogen's success, regardless of whether the factor itself causes pathogenic effects or not; yet, a successful pathogen typically generates such effects. Virulence factors govern the parasite's ability to enter into host cells, to persist, and to transmit from the host. In bacteria, these factors are often encoded by genetic elements on 'pathogenicity islands' (see section 10.3.2). The factors are classified by the functional role they play in the pathogen's life cycle:

9.4.3.1 Adhesion factors (adhesins)

Many bacteria must first attach to a surface of the host to successfully infect, such as to the skin, or the mucosa in the alimentary, respiratory, or urogenital tracts. Bacterial molecules, typically proteins or polysaccharides, mediate adhesion to cell surfaces and promote attachment ('adhesins'). Because this step is essential, an adhesion factor is considered a virulence factor whose absence renders the bacterium 'non-virulent'. For example, the capacity of different strains of *E. coli* to cause diarrhoea-related deaths (mainly in infants) is tightly linked to the capacity of the bacteria to attach to the cell lining of the gut (Kaper et al. 2004; Dean et al. 2005). Similarly, *Helicobacter pylori* use a range of different adhesins to attach to cell linings, such as neutrophil-activating protein (NAP, targeted to lipids on neutrophil surfaces), adherence-associated proteins (AlpA, AlpB), outer-membrane protein (HopZ), LacdiNAc-binding adhesin (LabA), or sialic acid-binding adhesin (SabA, for attachment to gastric epithelium). These factors also contribute to pathogenesis; for example, NAP induces macrophage inflammatory proteins that are associated with chronic gastritis (Kao et al. 2016). Yersinia adhesin A (YadA) is based on a type Vc secretion system and mediates the attachment to extracellular matrix components of the host. YadA is essential for the parasite's invasion success, as only tight adhesion allows for the subsequent injection of effectors into the host cell via a type III secretion system (Mühlenkamp et al. 2015).

9.4.3.2 Colonization factors

These molecules serve the bacterium to survive in a hostile host environment. In the example of *H. pylori*, within-bacterial cell urease activity provides acid resistance and is required to survive in the host's stomach; urease also has several other functions (Kao et al. 2016).

9.4.3.3 Invasion factors (Invasins)

Parasite success can depend on how well they enter hosts cells or spread through the extracellular space in a host's body, as for *Streptococcus aureus* or *Pseudomonas aeruginosa*. Specialized bacterial enzymes cleave proteins in connective tissue, or degrade the extracellular matrix to spread in this space. Furthermore, bacterial invasion factors induce host cells to endocytose the pathogen for transport into the cell interior, or facilitate tissue invasion that results in tissue necrosis. For example, *Salmonella* species possess a large variety of proteins secreted to penetrate and destroy the cells of the intestinal mucosa at specific locations (the M-cells in the Peyer's patches and the lymphoid follicles), which is an essential step in the pathogenesis of these infections (Jones and Falkow 1996). *Shigella* invasins (IpaA, IpaC) cause a rearrangement of the host's actin cytoskeleton such that the bacterium becomes internalized in the host cell. This strategy is common among many bacteria (Lee et al. 2014a). For some bacteria, residence inside a host cell is obligatory (e.g. *Chlamydia, Rickettsia, Mycobacterium leprae*). For others, cell entry can provide a means of being spread and transported to other host tissues but is not an essential step in the life cycle (Wilson et al. 2007).

9.4.3.4 Immune evasion factors

Many virulence factors are molecules that inhibit the host's immune defences (Table 9.3). For example, hosts phagocytose bacteria and then produce reactive oxygen radicals, change the pH, or mobilize degrading proteases to destroy the invader in the parasite-holding vacuoles. However, many bacteria can lodge and survive in phagocytic host cells. Immune evasion factors play a prominent role in allowing the parasites to overcome the defence mechanisms. These evasion factors thus allow the parasite to become pathogenic.

9.4.4 Toxins

Bacteria produce 'toxins' that are associated with damage to the host. A concise definition is problematic because toxins come in many forms and functions, and some are considered virulence factors. 'Exotoxins' are typically proteins released from the bacterial cells that diffuse into the surroundings. Exotoxins can thus take effect far from the site of bacterial growth. 'Endotoxins' are compounds whose effects are associated with the bacterial cells themselves. Endotoxins characteristically are components of the cell wall, which usually are not released into the surroundings. Endotoxins present within the cell can become released when the host cells lyse as a result of host defence or, perhaps, as a byproduct of antibiotic treatment that destroys the cells.

Typically, non-pathogenic bacterial strains produce far less, or no, toxins. In contrast, pathogenic strains produce ample quantities, usually of a toxin that is a significant determinant of the pathogenic effects on the host. Toxins have a high biological activity, even at low concentrations, but they are denatured by heat, acids, or proteolytic enzymes. Toxins thus act like enzymes and are more or less specific regarding the host functions they affect. Toxins are therefore classified as neurotoxins (acting

on the nervous system), leukocidins (acting on leukocytes), or haemolysins (acting on the blood). Some toxins are specifically cytotoxic, i.e. attack and destroy specific cells of the host (such as tetanus or botulinum toxins that attack nerve cells). Others are broadly cytotoxic for a variety of cells (e.g. toxins produced by staphylococci, such as *Clostridia*). Toxins (phospholipases) furthermore can cut the phospholipids from cell membranes, which leads to the loss of a stable host cell structure and to cell death and tissue damage. Some toxins play a significant role at the site of infection and invasion. Prime examples are pore-forming toxins. RTX toxins, for example, are a family of cytolysins found in Gram-negative bacteria. They can translocate bacterial molecules into the host cell by forming ion channels in host cell membranes. The mechanism thus also disrupts the ionic gradients and membrane potential, which leads to cell death. Similar effects come from α-haemolysin (HlyA) of *E. coli*, or adenylate cyclase toxin (Act) produced by *Bordetella pertussis* (Benz 2016). Pneumolysin (Ply, a cholesterol-dependent cytolysin) in *Streptococcus pneumoniae* is required for invasion of tissue, probably weakens host defences, but is also haemolytic, acts as a neurotoxin, and contributes to inflammation and cell death (Peraro and van der Goot 2016).

Some bacterial toxins are based on a two-component system (e.g. in *V. cholerae, Corynebacterium diphtheriae, B. pertussis*). In this case, one component (subunit A) generates the enzymatic (toxic) activity by a variety of mechanisms. Another component (subunit B) is responsible for the binding of the entire complex to a receptor on the host cell membrane (typically a glycoprotein) and its transport into the cell interior. Neither of the two subunits is active unless combined with the other part. The two separately synthesized subunits come together on the cell membrane, or when a single polypeptide with structure A/B is cleaved into the two subunits and so becomes activated. Cell entry of the entire complex can be direct when subunit B binds and opens a pore in the membrane. Alternatively, subunit B binds, and the entire AB complex is transported into the cell's interior by receptor-mediated endocytosis. In the process, the toxin complex becomes enclosed in a membrane-derived vesicle (the endosome) and is later activated (Wilson et al. 2007). The synthesis of toxins by the bacteria is tightly regulated and under genetic control. However, synthesis is also affected by the environmental conditions that the bacterium encounters. For instance, diphtheria toxin (DT) depends on the availability of iron, while cholera toxin (CT) depends on osmolarity and temperature (Beier and Gross 2006). The pathogenic effects of bacteria, therefore, depend on the action of toxins plus other virulence factors, in combination with a given environmental condition.

Toxins are proteins that become degraded and turn into a 'toxoid'. These remain antigenic, however. Toxins can also be artificially de-toxified by various reagents (e.g. formalin, iodine) to become a toxoid. As these remain antigenic yet non-toxic, toxoids are used for immunization against diseases caused by toxin-producing bacteria. Classical vaccinations based on this principle are those against diphtheria (caused by *Corynebacterium diphtheriae*) and tetanus (*Clostridium tetani*) (Table 9.4). In all, bacterial toxins are among the most potent toxic substances for mammals known in nature (Table 9.4). For example, botulinum and tetanus toxins are ten million times (a factor of 10^7) more potent than strychnine and even a hundred thousand times (a factor of 10^5) more toxic than the most potent snake venoms.

A more general serious pathogenic condition in bacterial infections is 'septic shock', which results from the combined action of cytokines and components of the complement as the host's immune system responds. The events are triggered with the recognition of pathogen-associated molecular patterns (PAMPs) by the innate immune system, e.g. LPS of Gram-negative bacteria (*E. coli*, *Pseudomonas aeruginosa*). Subsequently, lipid A (the toxic component of LPS, an endotoxin) stimulates the release of inflammation-triggering cytokines and activates the immune cascades. Gram-positive bacteria (e.g. *Staphylococcus aureus*, *S. epidermalis*) lack LPS but can induce a septic shock, too. This is triggered by peptidoglycan fragments and various PAMPs (Wilson et al. 2007; Hotchkiss et al. 2016; Cecconi et al. 2018).

9.4.5 Proteases

Proteases, by definition, are enzymes able to cut other proteins into pieces. Hosts use proteases to

deal with invading parasites, and parasites use them to manipulate hosts or to degrade their immune effectors. Proteases are, therefore, one of the critical elements of a host–parasite interaction that appear in very many functions and are produced by all major parasite groups (McKerrow et al. 2006). Among the different classes, cysteine proteases of protozoan and simple metazoan parasites mediate tissue invasion and inflammation (EhCP1, EhCP5 in *Entamoeba*) and degrade and mobilize nutrients. Cysteine proteases ensure immune evasion and allow macrophage infection (CPB in trypanosomes), or act to modify other proteins (*Leishmania*, *Plasmodium*), similar to the effect of toxins. For example, cruzain, a cystein protease produced by *Trypanosoma cruzi*, is a major proteolytic enzyme and affects different processes at different stages of the parasite's life cycle. Among other things, cruzain helps evade immune defences and remains highly antigenic during chronic infections. It is, therefore, a significant promoter of pathogenesis in Chagas disease (McKerrow et al. 2006; Doyle et al. 2011). Metalloproteases show a similar range of activities against coagulation for tissue invasion and nutrient degradation (McKerrow et al. 2006).

Serine proteases have manifold effects, too. They prevent coagulation (helminths), lead to the degradation of immunoglobulins (schistosomes), and are involved in processing other proteins (malaria). For instance, serine protease HtrA (high temperature requirement A) produced by *Campylobacter jejuni* is essential for cellular invasion and to cross the epithelial layer. At the same time, HtrA is crucial for the induction of cell apoptosis, tissue damage, and the associated immunopathology (Heimesaat et al. 2014). Similarly, *Pseudomonas aeruginosa*, an opportunistic infection, produces elastase that can degrade complement factors and destroy the junctions between epithelial cells. This facilitates the entry of the pathogen but also leads to tissue destruction and eventual haemorrhage, e.g. during lung infections (Gellatly and Hancock 2013). Therefore, proteases help the parasite to overcome host defences but, at the same time, induce inflammation and prepare for pathogenic conditions.

Table 9.4 Extracellular bacterial toxins and lethal doses.

Species	Toxin	Effect in host	Lethality[1, 2] (per kg of body mass)
Bacillus anthracis	Anthrax toxin (EF, LF)	Edema factor (EF) causes increased levels of cAMP in phagocytes and formation of permeable pores in membranes, leading to host cell lysis. Lethal factor (LF) induces cytokine release; is cytotoxic to host cells.	EF <200 μg LT <100 μg, <114 μg (rat)
Bordetella pertussis	Pertussis toxin (ptx)	Blocks inhibition of adenylate cyclase in host cells.	15 μg, 21 μg i.p.
Clostridium botulinum	Botulinum toxin	Inhibits neurotransmission at neuromuscular synapses; leads to flaccid paralysis.	0.4 ng i.p., 1–5 ng/kg ~1 ng (humans)
Clostridium difficile	Toxin A, B	Loss of actin polymerization; changes in cytoskeleton. Disrupts gut epithelium, loss of fluids, diarrhoea.	A: 25 ng (MLD$_{100}$) B: 50 ng
Clostridium perfringens	Perfringens entertoxin (CPB)	Activates adenylate cyclase leading to increased cAMP in epithelial cells.	81 μg i.v. ~310 ng
Clostridium tetani	Tetanus toxin	Inhibits neurotransmission at inhibitory synapses; leads to spastic paralysis.	1 ng (MLD)
Corynebacterium diphtheriae	Diphtheria toxin (dtx)	Ribolysation of ELF 2 (elongation factor 2), thus leading to inhibition of protein synthesis in host cells.	1.6 mg s.c. (MLD) <100 ng (humans)
Escherichia coli	*E. coli* LT, ST toxin	LT: Similar to cholera toxin. ST: Promotes secretion of water and electrolytes from intestinal epithelium.	250 μg i.v.
Pseudomonas	Pseudomonas exotoxin A (A/B), PE	Inhibits protein synthesis, resulting in cell death.	3 μg PE: ~200 ng
Shigella dysenteriae	Shiga toxin	Cleaves rRNA, thus inhibiting protein synthesis.	1.3 μg i.p., 450 ng i.v.
Staphylococcus aureus	Staphylococcus enterotoxins (SEB) Toxic shock syndrome toxin (TSST-1)	A family of molecules. Diverse and massive action against immune system, including lymphocytes, macrophages. Leading to emesis (vomiting). Interact with MHC class II molecules on antigen-presenting cells and beta chains on T-cell receptor. Leads to production of IL-1, TNF, and other lymphokines (which is used in diagnosis). TSST: Acts on vascular system, causing inflammation, fever, shock.	20 μg i.v. (monkey) SEB: 1.6 μg/g intra-nasal
Vibrio cholerae	Cholera toxin	Increasing level of cAMP, promoting secretion of fluids, electrolytes in intestinal epithelium; diarrhoea.	250 μg

[1] Taken from refs. 1, 2, 4–9.

[2] If not indicated otherwise, dose refers to LD$_{50}$ in mice. Abbreviations: intraperitoneally (i.p.), intravenously (i.v.), subcutaneously (s.c.), minimum lethal dose (MLD). See ref. 3.

Sources: [1] Fisher. 2006. *Infect Immun*, 74: 5200. [2] Gill. 1982. *Microbiol Rev*, 46: 86. [3] Hagiwara. 2001. *Vaccine*, 19: 1652. [4] Holmes. 2000. *J Infect Dis*, 181: S156. [5] Moayeri. 2003. *J Clin Invest*, 112: 670. [6] Montecucco. 2005. *Curr Opin Pharm*, 5: 274. [7] Savransky. 2003. *Toxicol Pathol*, 31: 373. [8] Tian. 2012. *Vaccine*, 30: 4249. [9] Weldon. 2011. *The FEBS Journal*, 278: 4683.

9.4.6 Pathogenesis via the microbiota

An intact microbiota is vital for the host's health. In particular, the resilience of the microbiota against disturbance maintains stability and provides 'colonization resistance' as a protection against disease (Sommer et al. 2017). Disease symptoms appear when the microbiota is altered by a parasitic infection (Vonaesch et al. 2018). For example, the relative abundance of two bacteria (*E. coli* and *Faecalibacterium prausnitzii*) in the human gut can be used as a marker for inflammatory bowel disease (Burgess et al. 2017).

The most dramatic case is 'dysbiosis', i.e. the severe change or even disintegration of the microbiota's composition and structure, for example when pathogens outgrow the microbial commensals. Dysbiosis is typically associated with inflammatory processes that can also lead to fatal sepsis

(Levy et al. 2017). In mice, *Salmonella enterica typhimurium* infections not only trigger inflammation of the mucosa but also increase lactate levels in the gut, thus allowing the pathogen to grow at the expense of the native microbiota (Gillis et al. 2018). This highlights the naturally existing intense competition among co-inhabiting gut bacteria for limited polysaccharides that is needed for their metabolism and replication. Many other pathogens follow a similar strategy. For instance, *Clostridium difficile* disrupts the microbiota to liberate sugars and also benefits from fermented products (succinate) that are produced by a disturbed microbiota. The changes also result in mucosal inflammation and diarrhoea, with pathogenic effects on the host (Hryckowian et al. 2017; Kho and Lai 2018). Many different mechanisms can thus be involved in the process of dysbiosis. These include mobilization of nutrients, sequestering of ions (e.g. iron), competition among microbes by way of interfering proteins, altering host immune responses, usurpation of host cellular respiration, and the stimulation of antimicrobial peptides by the pathogens that turn against the resident microbes. Members of the microbiota or their metabolites circulating in the gut can even induce the expression of virulence genes in the pathogen (Levy et al. 2017; Vonaesch et al. 2018).

An intact microbiota also affects the B- and T-cells (especially the regulatory T_{reg} cells) that reside in the gut mucosal layer of the adaptive immune system of higher vertebrates. This arrangement helps to suppress the immune response against harmless commensals and therefore maintains gut homeostasis and the integrity of the mucosal barrier. Dysbiosis—as a result of infection—can therefore cause autoimmune disorders by the improper activity of the T-cells—an immunopathological effect (Honda and Littman 2016). Hence, infection-related changes to the microbiota are an essential factor that leads to pathogenesis by various mechanisms.

9.4.7 Pathogenesis by co-infections

Pathological effects by an existing infection can result from the weakening of the immune system that allows another infection to become established

or to become activated. For example, infection by the measles virus only rarely (in about 1 per 100 000 infections) leads to an ultimately fatal, progressive neurodegenerative disease. Instead, the actual dangerous effects are most often caused by secondary infections (Rall 2003). Such infections are likely because the measles virus transiently suppresses almost the entire immune system (mainly via suppressing dendritic cells and macrophages), which predisposes the host to opportunistic infections. Similarly, in infections by the influenza virus, mortality is correlated with secondary infections by bacteria. Perhaps more than 95 per cent of the fatal cases observed during the 'Spanish flu' epidemic of 1918 were the result of a complication by bacterial pneumonia. Many processes add to this exacerbation. These include a decrease in tolerance to the infection, i.e. when tissue repair fails, as with a co-infection of influenza virus and *Legionella pneumophila* (Jamieson et al. 2013). Also, virus-induced changes in the host immune responses or damage to lung physiology and the epithelial barrier can be secondarily exploited by the bacteria (McCullers 2014). Bacterial, secreted compounds (and virulence factors) can promote viral growth and eventual pathogenesis, including the compounds that interfere with the host's immune defences. Bacterial proteases, e.g. from *Staphylococcus aureus*, can cleave the haemagglutinin from influenza virions, transforming the molecule to a functional state that mediates binding to host cells, which makes the virus more infectious. Co-infection with *S. aureus*, therefore, can amplify the viral infection and lead to pathogenic effects of influenza (Tashiro et al. 1987). Such secondary co-infection effects are common and make the occurrence and distribution of concomitant infections a prime concern for the epidemiology and pathogenicity of infections.

9.5 Immunopathology

Immune responses clear infections or limit their effects, but also control aberrant host cell populations. In the process, effectors are produced that damage or destroy cells and tissue, and inactivate invading parasites. Not surprisingly, such defences can also turn against the host itself. The pathological process and self-damage resulting from such a response are commonly referred to as 'immunopathology'. As discussed in Chapter 4, immunological tolerance limits the scope of a self-reaction by eliminating T-cells that too strongly bind to self-peptides. Insufficient elimination, or an immune response erroneously triggered by self, can nevertheless lead to autoimmune reactions and damage at any time. Similarly, the infection itself can unleash immunopathology (Table 9.5). Among other things, misdirected processes of killing the parasite, or the production of 'wrong' antigens, both of which can lead to the destruction of own cells and tissues, are involved. Immunopathology by self-reactivity also occurs in invertebrates (Sadd and Siva-Jothy 2006; Dhinaut et al. 2017; González-González and Wayne 2020). Immunopathology in invertebrates is not well studied so far.

The immune response thus carries the benefit of fighting an infection as well as having the potential cost of causing severe self-damage. Eventually, some intermediate level of cytokines, such as TNF-α, will probably balance the benefits of having a defence vs the risk of immunopathology (Graham et al. 2005). As discussed before (see section 4.3.4), the constitutive response can respond quickly but at the expense of a false alarm. Theoretical considerations suggest that correct identification is selected to become more reliable when the mortality hazard of infection is high and can be less reliable when the risk of immunopathology is increasing, especially when parasites have more benign effects (Metcalf et al. 2017). By contrast, induced defences should be more relevant when highly virulent parasites are abundant, even at the expense of immunopathology

Table 9.5 Immunopathology associated with parasitic infections.

Parasite	Immunopathology	Sources[1]
Influenza virus	Allergy	5
B3 cocksackie virus	Myocarditis, Type 1 diabetes mellitus (T1DM)	20
Herpes simplex virus	Keratitis (inflammation of the cornea in the eye)	22
Mouse adenovirus type I	Encephalomyelitis	17
Dengue virus	Haemorrhagic fever, liver damage	13, 16
Mycobacterium tuberculosis	Adjuvant arthritis Prolonged disease; lung damage not related to bacterial load	6, 10, 18
Borrelia burgdorferi	Lyme arthritis	2
Staphylococcus aureus	Allergy	7
Streptococcus pyogenes	Rheumatic fever	12
Helicobacter pylori	Autoimmune gastritis	1
Leishmania major	Skin lesions, liver damage	14
Plasmodium yoelii	Disease severity	19
Trypanosoma brucei	Damage to central nervous system	15
Trypanosoma cruzi	Damage to heart muscle	11
Nippostrongylus brasiliensis	Hypersensitivity of the respiratory tract	4
Brugia malayi	Hypersensitivity of the respiratory tract	9
Schistosoma spp.	Damage to liver and urinary tract	3, 21

[1] General information in ref. 8.

Sources: [1] Amedei. 2003. *J Exp Med*, 198: 1147. [2] Benoist. 2001. *Nat Immunol*, 2: 797. [3] Booth. 2004. *J Immunol*, 172: 1295. [4] Coyle. 1998. *Eur J Immunol*, 28: 2640. [5] Dahl. 2004. *Nat Immunol*, 5: 337. [6] Ehlers. 2001. *J Exp Med*, 194: 1847. [7] Ennis. 2004. *Clin Exp Allergy*, 34: 1488. [8] Graham. 2005. *Annu Rev Ecol Syst*, 36: 373. [9] Hall. 1998. *Infect Immun*, 66: 4425. [10] Hirsch. 1996. *Proc Natl Acad Sci USA*, 93: 193. [11] Holscher. 2000. *Infect Immun*, 68: 4075. [12] Kirvan. 2003. *Nat Med*, 9: 914. [13] Libraty. 2002. *J Infect Dis*, 185: 1213. [14] Louzir. 1998. *J Infect Dis*, 177: 1687. [15] Maclean. 2004. *Infect Immun*, 72: 7040. [16] Mongkolsapaya. 2003. *Nat Med*, 9: 921. [17] Moore. 2003. *J Virol*, 77: 10060. [18] Moudgil. 2001. *J Immunol*, 166: 4237. [19] Omer. 2003. *J Immunol*, 171: 5430. [20] Rose. 2000. *Cell Mol Life Sci*, 57: 542. [21] Wamachi. 2004. *J Infect Dis*, 190: 2020. [22] Zhao. 1998. *Science*, 279: 1344.

(Boots and Best 2018). Regardless of the accuracy of such models, immunopathology persists in populations due to the potential benefits of efficient pathogen clearance (Sorci et al. 2017).

9.5.1 Immunopathology associated with cytokines

Many factors and processes cause pathogenesis, and all can contribute to immunopathology. The misregulation of cytokines is a particularly important mechanism. Tumor necrosis factor, TNF-α, for example, recruits immune cells to the site of infection, activates resident immune cells in the infected tissues (Turner and Farber 2014), and induces direct killing of parasites by activation of cytotoxic T-cells. TNF-α also dilates blood vessels and makes them permeable such that immune cells can enter infected tissues from the bloodstream—with neutrophils being the first ones to arrive. Proinflammatory cytokines, such as TNF-α, are part of a regulated process where anti-inflammatory cytokines (e.g. IL-10) act as counterparts, such that a well-orchestrated response against an infection unfolds. Sometimes, however, this response runs out of control, and the massive release of cytokines—the 'cytokine storm'—results in severe tissue damage caused by massive cell apoptosis, overzealous macrophages, and misdirected cell repair mechanisms, that together lead to eventual organ failure. As cytokines make blood vessels more permeable, massive blood loss into the surrounding tissues occurs, too. This is followed by a dramatic drop of blood pressure (hypotension), a compensatory excessive heartbeat rate, and rapid breathing; these are signs of a clinical state of septic shock. In mice, the addition of IL-7 blocks apoptosis and restores normal functions. A fatal outcome can thereby be avoided (van der Poll et al. 2017).

Infections by IVA and other respiratory viruses provide examples of infection-induced immunopathology (Newton et al. 2016). The viral infection is detected by innate sensors such as TLRs or NOD-like receptors and leads to the release of cytokines and other factors in the epithelium. Then, resident macrophages are activated and neutrophils recruited. However, these cascades can easily overshoot, additionally driven by pathogen interference, causing uncontrolled inflammation that damages the tissue (Pechous 2017). Moreover, cytotoxic T-cells (CTLs) become activated by cytokines and start destroying lung tissue, too, where the accumulating debris from infected cells add to organ dysfunction. Cytotoxic T-cells are thus crucial players that destroy viruses but also contribute to immunopathology (Duan and Thomas 2016). In fact, besides the effects of bacterial co-infections, immunopathology is a significant cause of damage from influenza (McNab et al. 2015; Newton et al. 2016).

9.5.2 Immunopathology caused by immune evasion mechanisms

Parasite-derived immune evasion factors contribute to pathogenesis, as discussed above (section 9.4.3). This is likely to be an important class of mechanisms for self-damage under immunopathology (Table 9.6). Whereas the host's defence machinery is responsible for the actual damage under immunopathology, in perhaps most of the cases the host's defence machinery has been manipulated by the parasite. The pathogenesis might therefore be under the control of the parasite rather than the host. However, the distinction between self-damage as a result of parasite control (as an immediate or delayed result of manipulation) and damage under host control (a result of overresponding) is not easy to establish. Likely cases are the pathogenesis of *Bacillus anthracis*, where anthrax toxin, LT, promotes vascular leakage and causes massive damage. Similarly, modulatory botulinus exotoxins of *Clostridium botulinum* lead to severe muscle paralysis. In these cases, the parasite-derived compounds are also involved in host manipulation (Frank and Schmid-Hempel 2008). Not always, therefore, does the host have full control over its immune defence machinery and the ability to limit its detrimental effects (Table 9.6).

Table 9.6 Modulation of pathogenic effects by immune evasion mechanisms.

Parasite	Immune evasion mechanisms	Relation to pathogenesis	Source
Poxviruses	All poxviruses modulate the host's signalling network in some way; different groups use different mechanisms, e.g. targeting interferons, TNFs, interleukins, complement, and chemokines.	Change in chemokine signalling is of central importance to pathogenesis. In many cases, the respective genes have been acquired from the host, or homologues to host proteins are produced.	2, 17
Hepatitis C virus	Various mechanisms to disrupt innate immunity, delays organization of effective response. Virus also disables interferon production and response. Escape mutants accumulate in infecting population.	Progress to chronic infection strongly correlated with emergence of escape variants. Chronic infections can lead to complications, e.g. liver failure, cirrhosis, and carcinoma.	6, 8, 15
Cytomegalovirus	A variety of viral functions target the immune response, mainly the cellular response, e.g. leukocyte migration, apoptosis, and MHC function.	Failure to control virus; inflammation favours spread of virus.	12
Influenza virus	Virus produces anti-IFN protein, preventing activation of host transcription factors.	Immune evasion allows spread of virus and generates pathogenic effects.	19
Filial haemorrhagic viruses (Zaire Ebola)	Immune suppression by a variety of mechanisms that act in concert, e.g. preventing dendritic cells from activating T-cells, disrupting interferon response in macrophages and dendritic cells.	Systemic suppression of immune defence allows dissemination within host. Infection of cells can end with severe inflammation, necrosis, release of tissue factors that lead to intravascular coagulation and of cytokines and chemokines that induce vascular dysfunction and organ failure.	10
Brucella	Type IV-secretion system and unorthodox LPS ensure immune evasion, e.g. bacterium becoming resistant to macrophage attacks.	Pathogenic effects related to bacterial replication and persistence in host cells, which is directly dependent on capacity to evade host immune response.	9
Yersinia enterocolitica, Y. pseudotuberculosis	Invasion allows adhesion and invading host cells, regulated by the parasite's *RovA* gene.	Dissemination in tissue leads to potentially fatal pathogenesis.	7
Bacillus anthracis	Anthrax toxin (several components) impairs host immune system, e.g. via disruption of cytokine regulation.	Infection in lung: bacteria can multiply because immune response impaired. Leads to host suffocation and/or vascular collapse. Toxins also induce cytokine-independent shock late in an infection.	1
Pseudomonas syringae Salmonella enterica	*Pseudomonas*: Bacterial protein targets immunity-associated host-plant protein and destroys it by usurping host proteosome. Thereby host vesicle trafficking pathways and extracellular defence disabled. *Salmonella*: Also usurps host proteins to maintain a membrane-bound compartment that holds the parasite. In both cases, type III secretion systems involved.	Suppression of host immune response leads to spread of pathogen and potentially devastating pathogenic effects.	14
Neisseria, Streptococcus	Extracellular proteinases. Cleave host immunoglobulins (IgA1) at specific sites and inactive host immune defence.	Leads to infection and allows bacteria to multiply.	3
Pseudomonas syringae	Usurpation of host immune proteins. Destroys host immune protein and blocks vesicle trafficking.	Suppression of innate immune response allows building up of infection that destroys tissue of the plant host.	14
Bacteria, Protozoa	Microbial proteolytic enzymes (proteases) allow invasion of host tissue and immune evasion. Effect counteracted by host protease inhibitors.	Dissemination of pathogen leads to adverse effects, e.g. induction of inflammation and degradation of blood components.	11
Plasmodium falciparum	Parasite erythrocyte membrane proteins (PfEMP1) promote adhesion to endothelial receptors and immune evasion. PfEMP1 is based on family of antigenic variant genes (*var*).	Adhesion and immune evasion combine to generate pathogenic effects.	5

(Continued)

Table 9.6 Continued.

Parasite	Immune evasion mechanisms	Relation to pathogenesis	Source
Leishmania	A large diversity of lipophosphoglycans (LPGs) used to bind to macrophages, to inhibit macrophage signalling and host cytokine production.	Parasite effect depends on establishment, which is dependent on LPG.	18
Trypanosoma cruzi	Cruzain is the major parasite protease that facilitates host cell invasion and permits immune evasion.	Cruzain important for pathogenesis, e.g. by triggering autoimmune responses.	11
Trypanosoma brucei	Kinetoplastid endosomal system is crucial to evade host immunity; in particular, resistance to TLF and antigen recognition.	Infection and dissemination generates pathogenic effects.	13
Helminths	Release of proteases that inhibit host cystatins and serpins, and may function as potent anticoagulants.	Typically reduces inflammation normally caused by parasite. Reduces pathogenic effects.	11
Fungi	A fungal structure (the cryptococcal capsule) prevents phagocytosis of fungus.	Growing fungus causes pathogenic effects.	4
Dimorphic fungi	Proteins that facilitate attachment to host macrophages and downregulation of TNF. Antigenic glycopeptides on surface are actively removed by fungal proteases. In the extracellular space, fungal production of melanin provides protection against host-generated peroxides.	Most cause respiratory diseases. Tissue damage and fungal dissemination associated with alkalinization of host environment by fungal ureases.	16

Sources: [1] Abrami. 2005. *Trends Microbiol*, 13: 72. [2] Alcami. 2003. *Nat Rev Immunol*, 3: 36. [3] Armstrong. 2001. *Trends Immunol*, 22: 47. [4] Burgwyn Fuchs. 2006. *Curr Opin Microbiol*, 9: 346. [5] Crabb. 2006. *Curr Opin Microbiol*, 9: 365. [6] Dustin. 2006. *Annu Rev Immunol*, 25: 71. [7] Ellison. 2004. *Trends Microbiol*, 12: 296. [8] Guidotti. 2006. *Annu Rev Pathol Mech Dis*, 1: 23. [9] Lapaque. 2005. *Curr Opin Microbiol*, 8: 60. [10] Mahany. 2004. *Lancet Infect Dis*, 4: 487. [11] McKerrow. 2006. *Annu Rev Pathol*, 1: 497. [12] Mocarski Jr. 2002. *Trends Microbiol*, 10: 332. [13] Morgan. 2002. *Trends Parasitol*, 18: 491. [14] Nomura. 2006. *Science*, 313: 220. [15] Pawlotsky. 2004. *Trends Microbiol*, 12: 96. [16] Rappleye. 2006. *Annu Rev Microbiol*, 60: 281. [17] Seet. 2003. *Annu Rev Immunol*, 21: 377. [18] Turco. 2001. *Trends Parasitol*, 17: 223. [19] Yewdell. 2002. *Curr Opin Microbiol*, 5: 414.

Important points

- Transmission is a key process in the parasite life cycle. It follows a route (e.g. oral–oral) with a mode (e.g. by aerosols). The major contexts are vertical and horizontal transmission. All components of transmission can evolve.
- Infection success and the eventual outcome vary enormously among hosts and parasites due to several factors and processes. Macroparasites infect as individuals, but microparasites infect with a dose. Infective doses also vary widely; they are measured as dose–response curves.
- The infection process follows different scenarios, such as given by generalized independent action, individual effective dose, host heterogeneity, within-inoculum interaction, or sequential models. Process-based models for infection assume random processes (lottery), parasite manipulation, or focus on the early dynamics.
- Pathogenesis is the physiological process leading to damage to the host. Pathogenesis follows from impairment of host capacities, tissue destruction, the effects of virulence factors, toxins, and proteases, changes in the microbiota, or from co-infections. Even apparently similar parasites cause very different pathogenic effects.
- Immunopathology is caused by self-damage from the immune defences. The misregulation of cytokines and mechanisms of parasite immune manipulation are significant processes.

CHAPTER 10

Host–parasite genetics

10.1 Genetics and genomics of host–parasite interactions

10.1.1 The importance of genetics

Host–parasite interactions physically occur at physically close distances. The malaria parasite, *Plasmodium falciparum*, for example, resides inside human red blood cells. At this stage, *P. falciparum* erythrocyte protein 1 (PfEMP1) is expressed on the surface of the infected blood cells. Their primary function is to ensure the adherence of infected blood cells to the endothelium of small blood vessels. Adherence prevents the cell with the parasite from being flushed into the spleen. The PfEMP1 protein family is one of the best-studied surface protein families. Moreover, a single parasite lineage has up to 60 different variants of PfEMP1 in its repertoire. The details of adhesion and binding are now much better understood (Lau et al. 2015). At this close distance, one or a few residues (here, F656) of a PfEMP1 protein domain (here, CIDRα1) accomplish the binding (Figure 10.1). Different variants of PfEMP1 vary in the molecular sequences, but the general pattern is the same (Lau et al. 2015).

Whatever the exact nature of the binding proteins or the parasite epitopes recognized by the host immune system, their structure is encoded as a sequence in the host's genome. Even though the processes that lead from gene to protein are complex, variation in the gene sequence ultimately generates variation in the structure of binding sites or

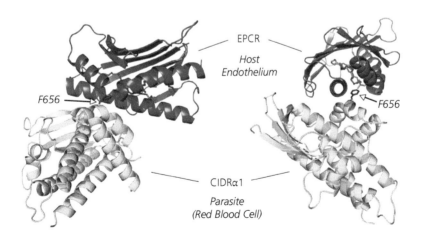

Figure 10.1 Host–parasite encounter at the molecular level. The three-dimensional protein model shows the binding between *P. falciparum* parasite-expressed protein PfEMP1 (domain CIDRα1, variant HB3var03, shown here; yellow) on the surface of red blood cells and the endothelial protein C receptor (EPCR; blue) on the inner walls of host blood vessels. The actual binding is by the small residue *F656* (dark red; arrows) protruding from the CIDRα1 domain. *F656* reaches into a hydrophobic groove of EPCR to bind. The two images show the same complex, rotated by 90° against one another. Adapted from Lau et al. (2015), under CC BY 3.0.

Evolutionary Parasitology: The Integrated Study of Infections, Immunology, Ecology, and Genetics. Second Edition.
Paul Schmid-Hempel, Oxford University Press. © Paul Schmid-Hempel 2021.
DOI: 10.1093/oso/9780198832140.003.0010

antigenic epitopes. The gene sequence thus affects binding properties and recognition. Gene sequences underlying the PfEMP1 proteins, or other surface proteins, are increasingly better characterized. Similarly, genetic sequences encoding parasite epitopes are accumulating in databases such as the Immune Epitope Database (IDEP) (Dhanda et al. 2019; Vita et al. 2019). The elements of the immune defence pathways, e.g. receptors, signalling molecules, and effector proteins, are similarly based on gene sequences encoded in the host's genome (Lazzaro and Schneider 2014). Genetics is, therefore, a key element of any host–parasite interaction. In 'classical' genetics, a gene and its alleles (the variants of a gene) are assigned to a 'locus', i.e. an idealized physical position on the genomic sequence. In reality, genes typically consist of several exons scattered over various places in the genome. When the gene becomes expressed, the single pieces are 'stitched' together to generate a complete transcript for the encoded protein. Nevertheless, the classical genetics approach allows us to readily study, for example, genetic variation among individuals in a host population and thus in their phenotypic defences—this variation is the raw material upon which selection can act and cause evolutionary change. The genetic underpinnings eventually also affect the shape and functioning of the disease space and, thus, how infections unfold (Box 10.1).

10.1.2 Genomics and host–parasite genetics

Genomics studies the structure, gene sequence, gene function, evolution, or the editing of genomes, and aims at including all genes of a genome. So far, the genomics of hosts is often known better than that of the parasites. This is due to their economic value (e.g. cattle, poultry), or their status as invasive or pest species (rabbits in Australia, aphids). Several tasks are essential for the study of host–parasite interactions and genetics.

10.1.2.1 Diagnostics

A recurrent problem in host–parasite studies is in identifying what kind of parasite is present. Public health management strategies, in particular, rely on diagnostic tools that can identify, for example, a virus that circulates in a human population. Modern technologies such as high-throughput sequencing and other molecular techniques like multiplex PCR have expanded the

scope of diagnostics and increased accuracy, speed, and cost-effectiveness (Gwinn et al. 2017).

10.1.2.2 Reading the genome

An essential part of genomics is to analyse the gene sequence itself. Advances in sequencing technology have been enormous (Box 10.2) and now allow for efficient screening of entire genomes, and even populations of genomes. Two elements are important: the sequencing technology per se (finding the sequence of nucleotides on a massive scale) and the mathematical and computational tools (collectively known as 'bioinformatics'). These tools allow the assembly of entire genomes from the large number (many millions of reads) of much smaller fragments that is the typical output from a sequencing instrument. Also, the respective algorithms carry out critical steps such as error checking, quality control to identify genes, and automated gene annotation by comparison with curated databases.

10.1.2.3 Association with a phenotype

The phenotype determines a host–parasite interaction, e.g. the level of host resistance or tolerance to infections. Therefore, the identification of the genes responsible for a given phenotype is a key task of genetics and genomics. This knowledge can, for example, inform breeders to improve their stock, or medicine to identify a possible target for drugs. With the *candidate-gene approach*, the search focuses on some chosen genes. They are chosen since, from previous information, these might be associated with the variation in the phenotype. Often, these genes are those involved in immune defences. The method thus allows focusing on good candidates quickly. Often, candidate-gene studies use a comparison of controls with a test group of individuals. A *whole-genome study*, by contrast, has no prior assumptions about which genes may be essential. A challenge, then, is to statistically filter out significant effects and judge which genes are of importance among the large numbers of possible effects.

Several different approaches can reveal the role of genes:

1. *Gene expression/transcriptome*: Which genes are up- or downregulated in their expression levels can reveal how an immune response unfolds genetically. Typically, treatment is compared with a

Box 10.1 Host–parasite interaction in disease space

A parasitic infection crosses the disease space along a trajectory that often leads out of the comfort (healthy) zone into sickness and eventual recovery. Sometimes, the trajectory also leads to the death zone. A large number of factors from both the host and parasite affect this outcome. Here, we can illustrate the outcome of host–parasite genotypic interaction by considering the trajectories in a comparison of two host lines and two parasite strains.

As sketched in Figure 1, genotypically different hosts are likely to have different shapes of the disease space. This results from genetic variants that encode different variants of proteins and other factors (e.g. regulatory RNAs) in the defence cascade. Some variants may respond earlier or regulate the immune response more effectively. As a result, the

boundaries where the infection trajectory, for example, enters the recovery or death zone are different. At the same time, infection trajectories depend on the parasite. Therefore, the infection trajectories of different host–parasite combinations can vary substantially. For instance, a parasite infecting one host line may cause a mild infection, whereas it can become fatal in another host. The underlying molecular and physiological mechanisms shape the zones in the disease space and vary among hosts. Therefore, even when an infection trajectory reaches the same point in disease space, it is likely to have activated along its path different mechanisms in the different hosts. This may help explain why different host (and parasite) combinations unfold so differently and seemingly unpredictably.

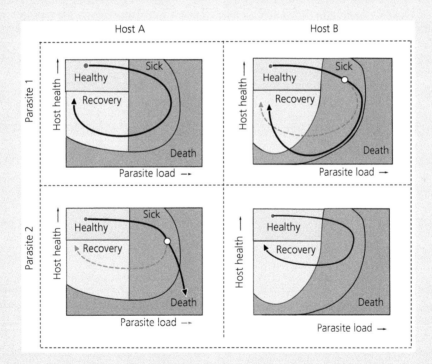

Box 10.1 Figure 1 Host–parasite genetics in disease space. Host lines A and B have different genotypes; the same is true for the two parasite strains 1 and 2. In this hypothetical example, Parasite 1 in host A follows a trajectory that returns to recovery after some time. The same parasite in host B becomes more dangerous, as its trajectory skims the death zone for a long time before returning to recovery. A particular step in the host defences, or in parasite actions, at the point indicated by the white circle may be responsible for this difference. Likewise, Parasite 2 in host B is a relatively mild infection that quickly leads back to recovery. The same parasite in host A, however, and due to an event at the point indicted by the white circle, takes a different course and can become fatal. Infection occurs at the red dot. The dashed lines represent the 'expected' trajectory from the other host and correspond to the parasite 'main effect'. The difference between the 'expected' and the realized curve reflects the host–parasite interaction term (see Box 10.4).

Box 10.2 Sequencing technologies

New sequencing technologies are among the key methodical advances for the study of evolutionary parasitology. These are therefore worthy of a short summary. The now-existing sequencing technologies, as listed below, exploit the parallel progress in bioinformatics that uses the primary information from an instrument (typically, very short DNA sequence fragments) to build much longer sequences ('contigs'), from which the genome is eventually assembled. No doubt, this development will continue, and the short overview given here is likely to be obsolete in the near future.

'Next-generation', 'second generation' sequencing:

1. *Next-generation sequencing* (Reuter et al. 2015; Slatko et al. 2018): Introduced in the mid-1990s, especially with the *Qiagen/Roche 454-Pyrosequencer*. DNA-template fragments from a sample are placed in pico-sized fluid chambers. In the process, light pulses show which nucleotides sequentially are incorporated into a growing DNA strand. This strand is complementary to the 400–700 bp long template (the sample sequence). Subsequently, different companies have developed a variety of instruments based on this principle.
2. *Illumina/Solexa*: The basic technique uses clonal amplification of thousands to millions of DNA fragments fixed on a glass slide, where each slide allows for millions of parallel amplifications. Progressive rounds of base incorporation sequence the template strand. Fluorescent imaging detects which base is added at which strand in a given round. The MiSeq, HiSeq, NextSeq 20000, or NovaSeq 6000 are different instruments, with different capacities and application ranges. Currently, the top-end NovaSeq 6000 produces 6000 Gb (gigabases) per output, which comes in at 20 billion reads of 250 bp in length.
3. *Thermo Fisher/Ion Torrent*: This also clonally amplifies DNA fragments. These are fixed on microbeads placed in microwells of a semiconductor chip. The template strands are sequenced by sequential rounds of flushing with the different nucleotides. Adding a nucleotide leads to a change in the pH value in the microwell, detectable by a voltage change. The change is indicative of which base

is added in a given round. Thermo Fisher's Ion 550 yields 40–50 Gb in 100–130 million reads, with read lengths of 200–1 500 bp.

In 'third-generation' sequencing machines (van Dijk et al. 2018), longer fragments in the order of dozens of kb in length are sequenced. At the same time, the focus is on sequencing single DNA molecules independently (in parallel arrangements):

4. *Pacific Biosciences (PacBio)* has built on the work of *nanofluidics* that pioneered single-molecule real-time sequencing (SMRT). In this technology, DNA-template fragments are prepared by adding adapters to the ends, such that a circular fragment is generated. The fragment can be read around the circle many times over (which increases accuracy). For this, a single-template DNA molecule is placed in a tiny chamber (of zeptolitre size) and flushed with labelled nucleotides. These are incorporated in the synthesis of a DNA strand as given by the template. Detection is by fluorescent imaging.

'Fourth-generation' sequencing:

5. *Oxford Nanopore Technologies* has pioneered a technology where a single DNA fragment or nucleotides pass through a tiny pore/channel. For this purpose, a molecular motor enzyme is added to one end of the template (and complement strand). The motor controls the translocation of the molecule through the nanopore. Depending on which base is passing through the hole, the electric current through the nanopore changes. The change allows for the reconstruction of the sequence. The process can be run massively, in parallel with many such pores. The MinION was the first such device that could even be hooked up to a computer as a USB stick. Currently, the top-end PromethION 48 (with 48 flow cells of 3000 nanopores each) produces up to 7000–9000 Gb per run. With this technology, the single reads are highly variable in size, but around 30–40 kb on average (N50 value). Some fragments can be up to 270 kb, and thus much longer than with all other technologies so far.

control group, e.g. infected vs non-infected individuals, and their respective gene expressions studied. Gene expression studies scrutinize the abundance of the mRNAs (the transcripts). The ensemble of all mRNAs is the 'transcriptome', active at a given time, and in a given cell and tissue (see below, section 10.4.3). Typically, hundreds or even thousands of genes are up- or downregulated when an animal becomes infected. The study of the proteins ('proteomics') recruited upon infection is the other tool and can reveal, for instance, how host defences are regulated (Chetouhi et al. 2015). Proteomics directly shows what proteins are circulating, and is, therefore, closer to the macroscopic phenotype than a transcriptome. However, proteomics is technically demanding due to the small quantities of proteins involved and their inherent instability. This makes it a delicate and expensive technology.

2. *Quantitative Trait Locus (QTL) mapping* is a method that yields statistical associations of a particular value of the phenotype with known markers on the genome. The markers can, for example, be microsatellite alleles or single-nucleotide polymorphisms (SNPs). SNPs are variants of nucleotides at the same position, i.e. SNPs are single-nucleotide alleles. They are present in high numbers throughout the genome and have become the most commonly used unit of genetic variation. SNPs can have functional consequences, e.g. by coding for different mRNA transcripts at crucial positions. As such, they become associated with a phenotype, e.g. of lower or higher resistance.

The data for QTL is a set of polymorphic genomic markers, i.e. loci that contain different alleles among the individuals of the studied groups, combined with the value of their phenotypes. Traditional QTL studies use parental lines that differ in these variables. The variation (similarities) of phenotypes is then compared between study groups, or among parents, offspring, and siblings. Statistical methods can relate the phenotypic variation to genetic variation in the markers. For the latter, the laws of quantitative genetics, e.g. inheritance of parental alleles, predict the contribution of genes. In particular, a part of the observed phenotypic variation is due to segregation of alleles as they are passed to offspring. With this knowledge, unknown 'genes' that are statistically contributing to the phenotype can retrospectively be mapped onto the genome. Given a large enough data set, QTL analysis can produce a genomic map that shows the approximate position of loci that affect, for example, resistance against a given parasite. The loci identified in this way are typically quite large regions of the genome and contain one or several genetic elements that significantly affect the phenotype.

3. *Association studies with limited gene sets*: A sample of individuals that vary in phenotype and genotype is screened. Variation can be natural or exists as a result of experimental crosses. Statistical processing yields 'odds ratios' ('relative risks') that describe the probability of a gene contributing to the phenotype. This identifies SNPs, haplotypes, indels, copy number variants, or simply extended regions of the genome, which may or may not contain genes with known functions.

4. *Genome-wide association study (GWAS)*: For this, the whole genomes of a set of individuals are sequenced (Bush and Moore 2012). The SNPs are identified and mapped onto the genome, which results in a characteristic plot (Figure 10.2). A large project of this kind is the Human Haplotype Map that lists all SNPs found in various human populations and their interdependencies (Altshuler et al. 2005). After correcting for many possible other sources of errors, the association of SNPs with the phenotype is derived by statistical methods, e.g. by generalized linear models or logistic regression models. In all these calculations, the effects of confounding variables, stratified sampling, or multiple testing need to be corrected. Again, the identified SNPs themselves could code for different proteins. Alternatively, the SNP is a tag ('tagSNP') that indicates a region of the genome associated with the phenotype under scrutiny. Genetically, a tagSNP is in 'linkage disequilibrium' with the actual genes that affect the phenotype; i.e. the tagSNP is closer to the causative genes than expected by chance. SNPs identified by these procedures are investigated further in follow-up studies.

5. *Loss-of-function tools*: Genes suspected to affect a phenotype can be neutralized to test for their role ('functional genomics'). Such 'loss of function' is achieved in several ways (Housden et al. 2017; Schuster et al. 2019a). Classical methods include chemically induced mutagenesis or insertion of transposons. More modern methods use RNAi (interference RNA) technology to produce a knock-out ('knock-down variants'). RNAi can block the expression of specific genes, as naturally occurs in virus-interference pathways (see Box 4.3). In recent years, CRISPR–Cas technology (see Box 4.2) has become a method of choice. It allows the targeting of specific genes with unprecedented precision. With CRISPRko (knock-out CRISPR) a short piece of DNA is degraded; the subsequently induced default repair mechanisms insert indels that lead to frameshifts and thus loss of function.

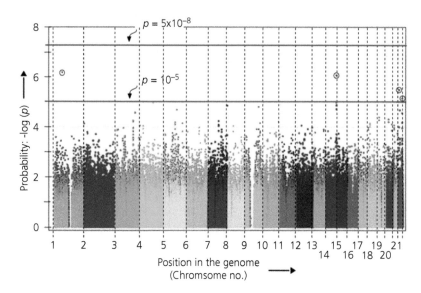

Figure 10.2 A 'Manhattan plot' from a GWAS. Shown is the map of SNPs (coloured dots) that may be associated with susceptibility to *Staphylococcus aureus* in humans (the phenotype). The x-axis shows the position of the SNPs in the entire genome, marked by the starting point of each of the 23 chromosomes. The y-axis shows the probability, *p* (given as -log), that a SNP at this position is associated with susceptibility to infection. Values higher than the upper red line are considered to be significant at the genome-wide level (the generally accepted criterion is $p_{crit} = 5 \; 10^{-8}$; Fadista et al. 2016). In this case, none of the SNPs reaches this threshold, but four loci (encircled) are above the $p_{crit} = 10^{-5}$ threshold; these are candidates for further investigation. Adapted from Ye et al. (2014), under CC BY.

With CRISPRi (interference CRISPR), gene transcription is inhibited, using the transcription start site (TSS) as a guide; Cas9 targets gene promoters or the regulatory elements near the TSS. Also, the activation of genes is feasible with CRISPRa (activation CRISPR).

10.1.2.4 Changing the genotype

A powerful way to investigate the genetic underpinnings of host–parasite interactions is to compare the performance of different genotypes. GWAS, for example, uses natural variation to find associations with the genotype. Natural variation contains many unknown differences. In contrast, experimental modifications change only one element and leave the rest of the genome intact; this allows for robust analyses. Essentially, modifications can result from any method that causes mutations. However, the classical tools, such as chemical mutagenesis or scheduled crossings, are either very laborious or quite unspecific. Current methods have refined the possibilities. For example, gene editing can modify functional DNA sequences. The various protocols

all have their advantages and disadvantages. As the technologies develop further, precision and reliability will further increase.

6. *Recombinant DNA-technology*: Here, two strands of DNA are recombined into one, thus creating a new genomic DNA sequence. For this purpose, the chosen DNA sequence is inserted into a 'cloning vector'—a phage, or a plasmid—and infected into host cells. In the process, the new DNA becomes inserted into the host cell's DNA and is eventually expressed as a recombined DNA (Roberts 2019).

7. *CRISPR base-editing* is a very recent technology. It allows a single nucleotide at a chosen position to be changed (Schuster et al. 2019a; Strich and Chertow 2019). Class II systems (see Box 4.2) are used because they are more straightforward, as they use only a single nuclease and have the defining Cas9 protein. The basic principle is to cleave the targeted DNA at a precise location; subsequently, the intrinsic DNA-repair mechanisms, particularly homology-directed repair, insert the new base as it reads from the desired template.

10.2 Genetics of host defence

The 'genetic architecture' of resistance (or tolerance) describes a landscape of genetic contributions to resistance against a parasite (the phenotype of interest). This refers to the number of genetic loci involved, their distribution in the genome, their mutual interactions and effects on resistance. Additionally, the interactions with the environment and their frequencies in the population are studied (Timpson et al. 2018). For example, the genetic architecture of resistance (measured as prevalence and infection intensity, respectively) of the water flea *Daphnia magna* to infections by the bacterium *Pasteuria ramosa* or the microsporidian, *Hamiltonella tvaerminensis*, was investigated with QTL studies (Routtu and Ebert 2015). Resistance to the bacteria was primarily based on a single QTL, explaining more than 60 per cent of the variance. This locus later became identified as a genetic region that codes for either preventing or permitting the attachment of the bacterium to the host oesophagus during the early infection stage. Additional QTLs with minor effects were also present, of which a further one (explaining 15 per cent of variance), located on the same linkage group (i.e. chromosome), affected disease severity in a quantitative manner. Some 23 additional loci, spread out over the genome, have gradual effects on host life history, including the probability of survival under infection (Hall et al. 2019). Resistance to *Hamiltonella* is based on multiple loci with epistatic effects that do not co-localize with those for bacterial resistance.

Similarly, GWAS identified dozens of SNPs (one-quarter located within coding regions) associated with resistance and tolerance of *Drosophila melanogaster* against bacterial infections (*Providencia rettgeri*; Howick and Lazzaro 2017). As far as the function of these genes is known, they belong to general categories, e.g. 'protein kinase activity' or 'stress response'. SNPs for tolerance are often located in genes affecting the 'negative regulation of immune responses'. These examples are quite typical for studies of genome-wide associations of genetic markers. Indeed, dozens of SNPs are usually found associated with a defence phenotype, many with—at first sight—unclear relationships to defence. In the fly study, RNAi knock-down confirmed the role of specific genes in tolerance to

infection, e.g. genes involved in cell repair, suppressing the expression of immune genes, or affecting autophagy. Together, tolerance to infection in *Drosophila* seems to have a genetic architecture with considerable diversity in the regulation of immune and stress responses (Howick and Lazzaro 2017).

For humans, too, GWAS is helpful for studying the genetic architecture of resistance and tolerance for infections. Examples are influenza, HIV, hepatitis B and C, herpes simplex, or norovirus. Among the identified genes, there are virus receptors, receptor-modifying enzymes, and many genes that code for proteins of innate and adaptive immunity (Kenney et al. 2017). Large-scale studies of human populations reveal a large amount of allelic variation in genes encoding for elements of immune defences, and which correlate with susceptibility. For example, the MHC-associated loci can account for more than 70 per cent of the viral load for hepatitis B virus (HBV) infections (Matzaraki et al. 2017). Similarly, SNP variants in human leukocyte antigen (HLA) alleles (MHC class I) correlate with HIV viral load (Ekenberg et al. 2019), and variation in HLA from the MHC class II region with susceptibility to *Streptococcus pyogenes* invasive group A, a rare but feared bacterial infection (Parks et al. 2020). In these examples, a single or a few identified genetic variants in a defined region of the genome (e.g. the MHC) contribute a large proportion to the observed outcomes of infection. However, in other cases, the genetic basis seems more complicated. For instance, resistance to *Mycobacterium tuberculosis* has a strong genetic basis. As of today, however, disappointingly few genetic markers or elements have been found that associate with the infection phenotype (Abel et al. 2018).

Genetic architecture of resistance, or any trait more generally, also refers to how the genes quantitatively affect the phenotype. On the one hand, genes (loci) can take effect independently of other genes in the genome; this is the 'main effect' in classical genetics. On the other, 'epistatic' effects emerge when a gene at one locus influences the effect of genes at another locus (Figure 10.3a). Epistasis (see section 10.5.1) can explain a further substantial amount of variance in resistance beyond the main effects. Moreover, the vast majority of the loci involved in epistasis are not identical to the main effect loci. However, the importance of epistasis

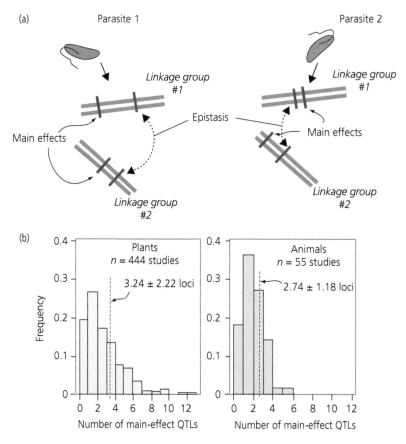

Figure 10.3 (a) Sketch of a genetic architecture for host resistance. Shown are two linkage groups (LG#1, LG#2, i.e. chromosomes) in a diploid host. Resistance to parasite 1 is due to a few main effect loci on different chromosomes (one each on the two LGs; red lines). Additional effects come from epistatic interactions between loci (arrows; blue lines). Loci involved in epistasis are not necessarily the same as the main effect loci; they also can reside on the same linkage group. The architecture often changes when a different parasite strain (parasite 2) infects the host (right panel). (b) The observed number of main effect loci. These are reported for QTL studies of resistance/susceptibility in plant (*left*) and animal hosts (right); the dotted line indicates the means (± S.E.). The difference in the mean between animals and plants is significant but may also reflect a methodical artefact. Adapted from Wilfert and Schmid-Hempel (2008), under CC BY 2.0.

varies among systems and cannot easily be generalized (Table 10.1). Finally, there is also the problem of 'missing heritability': the commonly used methods cannot reliably detect rare alleles (typically, with frequencies of less than 5 per cent) and small effects. This is a typical problem of GWAS. Even more sophisticated approaches (such as 'Genome-Wide Complex Trait Analysis', or GCTA) may not be able to resolve this limitation altogether (Krishna Kumar et al. 2016). Large sample sizes ('big data'), by contrast, alleviate the problem to some degree (Zuk et al. 2014; Kim et al. 2017).

Altogether, resistance and tolerance to infection do not show the uniform and straightforward genetic architecture that a naïve look at the immunological pathways would suggest (Lazzaro and Schneider 2014). Instead, resistance and tolerance can be based on a limited set of loci, often with a few identified genes having significant effects. Some of these loci are clustered on the same linkage group, but many are widely distributed over the entire genome. In rape (*Brassica*), for example, 57 per cent of receptor genes (NBS-LRR) were resistance genes, and 59 per cent of them were physically clustered (and more

Table 10.1 Epistasis in host–parasite interactions.

Study organisms	What was done	Finding	Source
HIV virus in cell cultures	For each viral infection, the sequence of the virus is known. Variation in sequences tested against performance of virus.	Most viral infections show epistasis, and predominantly of positive sign.	1
Arabidopsis thaliana and fungi, bacteria	Detailed molecular analysis of the genes involved in the defence signalling and effector pathway.	The gene of interest, OPC3, interacts with other genes to generate a successful response.	2
Daphnia magna	Six sets (different genetic backgrounds) of three increasingly inbred *D. magna* clones used to assess their relative fitness according to changes in frequency in a competition experiment against a tester clone. To examine whether an interaction between inbreeding and parasitism exists, each inbred clone tested with and without a microsporidium infection (*Octosporea bayeri*).	Logarithm of fitness decreases non-linearly across the three levels of inbreeding, indicating synergistic epistasis.	7
Drosophila melanogaster vs *Serratia marcescens*	*D. melanogaster* lines homozygous for genes of interest created and infected by the bacterial pathogen.	Substantial epistasis in immune defence detected between intracellular signalling and antimicrobial peptide genes.	5
Drosophila melanogaster vs viruses	Using advanced crosses of *Drosophila*, flies were infected. Genomic effects assessed by QTL method.	Resistance against Drosophila C virus primarily based on two loci, sigma virus on many, but only one epistatic effect.	3
Tribolium castaneum vs tapeworm (*Hymenolepis diminuta*)	QTLs identified in a mapping population tested for resistance against tapeworm infection.	A total of five major loci found. In one case, the major loci explained 29 per cent of variance in resistance; epistatic interactions among pairs of loci explained another 39 per cent.	8
Mouse vs Mouse cytomegalovirus (MCMV)	Experimental infection of mice and linkage analysis from F2 progeny from resistant and susceptible mouse lines.	Combination of alleles in the *Klra* cluster on chromosome 6 and in the MHC locus on chromosome 17 determines resistance against virus. The interplay is between a receptor and the MHC complex.	4
Humans vs Chagas disease (*Trypanosoma cruzi*)	Screen of candidate genes for resistance against Chagas disease.	Analysis of gene frequencies and expected equilibria suggests epistatic effects between genes at the MHC complex and those coding for the humoral immune systems (probably IL-10).	6

Sources: [1] Bonhoeffer. 2004. *Science*, 206: 1547. [2] Coego. 2005. *Plant Cell*, 17: 2123. [3] Cogni. 2016. *Mol Ecol*, 25: 5228. [4] Desrosiers. 2005. *Nat Genet*, 37: 593. [5] Lazzaro. 2004. *Science*, 303: 1873. [6] Moreno. 2004. *Tissue Antigens*, 64: 18. [7] Salathé. 2003. *J Evol Biol*, 16: 976. [8] Zhong. 2003. *Genetics*, 165: 1307.

polymorphic), with the remainder spread out over the genome (Alamery et al. 2018). However, the critical loci (and the genes belonging to those) are not necessarily those known to be part of the immune system. Many genes with an effect code for essential supportive functions, such as energy supply, or nutrient metabolism.

Despite these provisos, often only a limited number of loci are responsible for much of the host resistance or tolerance. The number of main effect loci is essential for how quickly a host population can adapt to its parasites. This number varies for different host–parasite systems but often

seems relatively low (Figure 10.3b). For example, in many insect hosts (such as mosquitoes, *Drosophila*, bees), a part of the resistance is based on only a few loci of major effect, in combination with partial dominance effects (Carton et al. 2005). Similarly, in plants, five loci have main effects on the resistance of maize (*Zea mays*) against *Gibberella zeae* (the asexual stage of the fungus *Fusarium* spp., causing stalk rot). Five main effect loci affect the resistance of barley to powdery mildew (the fungus *Erysiphe graminis*) (Heun 1987). There are probably four to nine major genes involved in the resistance of oat (*Avena sativa*) against mildew, and two to three in winter wheat (*Triticum aestivum*) (Jones 1986). Most remarkably, the genetic architecture is not invariable, but changes depending on which infection is occurring in which host (Wilfert and Schmid-Hempel 2008; Chapman et al. 2020). The loci showing an effect are therefore rarely the same. Therefore, each infection seems to 'shine its beam on the genome', but the involved loci that thereby light up are different each time. This variation reflects the diversity and complexity of immune defence regulation, where each host can 'choose' among several mechanisms of how to best respond to an infection, and each infection takes a different course through disease space (Box 10.1).

10.3 Parasite genetics

Parasite genetics focuses on genes affecting parasite infection, establishment, and parasite virulence. As discussed in Chapter 8, virulence factors are often genes coding for molecules that disarm and manipulate host defences. Chapter 13 discusses the evolution of virulence.

10.3.1 Viral genetics

With the development of powerful and affordable molecular technologies, virology, and especially the field of viral genetics, has entered a golden age. Viruses are tiny in size and, in fact, are just a bag of genes—either DNA or RNA sequences—packaged and wrapped in a hull. Viruses have evolved mechanisms to enter host cells and to exploit the host's molecular machinery to make more viruses. Astonishingly few genes, one or a few dozen, are necessary for this strategy. Viral genomes are simple in structure and can change extraordinarily quickly. Furthermore, viruses form large infecting populations that make it more likely that 'useful' mutations occur at least once, i.e. mutations that favour the viral infection and propagation. However, viruses are a very diverse group of pathogens, and their genomic structures and molecular biology vary considerably. Two examples may illustrate this diversity.

Influenza A virus (IAV; Orthomyxoviridae) is the principal aetiological agent of human influenza and has caused several major epidemics in the past (e.g.

the 'Spanish flu' in 1918–1919 and the 'Hong Kong flu' in 1968). It has a complex genomic organization, with a linear, segmented, negative-sense single RNA strand genome ('–ssRNA') of *c.*14 kb in length. That is, the genome is written with RNA and occurs as a single copy. The coding is in the reverse direction, which therefore has to be first converted into positive-sense RNA to produce the mRNA that is translated into proteins. Furthermore, eight viral ribonucleoprotein complexes (segments) contain the complete genetic information. These are packaged into a single virus and code for a total of 11 proteins. Hence, the 'genes'—in the sense of coding for a given protein—are contained in a segment, but some of the same segments can code for two proteins that result from reading the RNA in alternative ways (Figure 10.4a) (Louten 2016; Dadonaite et al. 2019).

The viral subtypes (such as 'H1N1' of the Spanish flu) refer to the proteins and their variants, i.e. haemagglutinin (HA, a hull protein) and neuraminidase (NA, a factor involved in virion release). Notably, the splitting of the genome into the eight separate segments facilitates the exchange and thus the reassortment of these pieces into new variants. This occurs when at least two different strains circulate in the same host, and the synthesized segments from different parental viruses reassemble into a new virion. Moreover, in this way, 'proven' pieces of genetic information, i.e. segments that have already been successful, can be reshuffled. The process generates a new subtype of the virus with different properties. Thereby, 'antigenic shift' can occur, meaning that the antigens presented by the new virus (its epitopes) are now different from the ones recognized by the host before (Louten 2016; McDonald et al. 2016b). This reshuffling is the primary reason why, each year, new subtypes of influenza virus can circulate in the human population, and why vaccines that worked against last year's flu may not be very effective against this year's suite of flu variants.

The causative agent of the recent Corona-pandemic, SARS-CoV-2 (2019-nCov), also belongs to the linear, single-stranded RNA viruses, but has its genetic information coded as positive-sense strands (+ssRNA) and is about 30 kb in length. Coronaviruses typically possess at least six open reading frames that code for about two dozen proteins (Chen et al. 2020). Despite being an RNA virus (which generally have

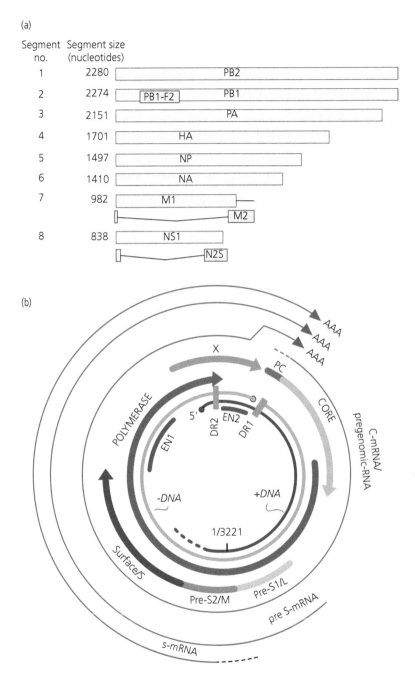

Figure 10.4 Viral genomes. (a) The RNA-genome of influenza A virus is linear and packaged into eight segments, together encoding for 11 proteins (PB2, PB1, etc.). The code for the small protein PB1-F2 results from an alternative reading frame in segment 2. The products M2 and N2S result from alternative splicing of products derived from segments 7 and 8. Adapted from Louten (2016), with permission from Elsevier. (b) The circular genome of hepatitis B virus (HBV). It contains a complete minus-sense DNA strand (–DNA; green), and an incomplete positive DNA strand (+DNA; blue). All major mRNA sequences (outer three thin red lines) end at a common position with the polyadenylation signal (AAA) located in the CORE open reading frame. Dashed lines in the 5′ ends of mRNAs symbolize staggered starting points for transcripts. All open reading frames run in a clockwise direction. The single-strand gap in the +DNA strand (grey bar, near DR1) is filled in the process by viral reverse transcriptase. DR1, DR2 are small (~11 bp) repeats important for viral DNA synthesis. EN1, EN2 are viral envelope proteins. The marker 1/3221 is a reference genomic location on the positive DNA strand. Other labels stand for further viral proteins. Adapted from Seeger and Mason (2015), with permission from Elsevier.

higher mutation rates than other viruses), SARS-CoV-2 seems relatively stable. Based on information from a previous coronavirus in humans (SARS-CoV-1), this is probably because re-assortment (as in influenza virus) seems to be less critical than the accumulation of substitutions within lineages (Consortium 2004; Holmes and Rambaut 2004; Zhao et al. 2004). Nevertheless, mutations do occur, and may affect crucial antigenic properties of the virus, such as the binding sites (Jia et al. 2020b; Korber et al. 2020).

HBV (Hepadnaviridae) illustrates another genetic organization. Here, DNA is the genetic material, and the genome is circular (with *c.*3.2 kb in length). The circle consists of a complete minus-sense strand (i.e. forming a full circle) and an incomplete plus-sense strand (i.e. covering only segments of the circle); this forms a partially double-stranded DNA type. In this genome, the overlapping sequences of the transcribed major mRNAs are all ending at a common polyadenylation signal in the primary open reading frame (Figure 10.4b) (Seeger and Mason 2015). After infecting, the viral genome becomes converted into a closed circular DNA that serves as the template of the viral mRNAs. In contrast to the influenza virus, but similar to SARS-CoV, a significant driver of antigenic escape in HBV is point mutations in the genomic sequences. These can lead to new antigenic variants that are not yet known to the host's immune system ('escape mutants'). HBV causes long-term, chronic infections. This provides sufficient time to accumulate mutations that make escape mutants more likely to appear (Sanjuán and Domingo-Calap 2016). In all, this is a less efficient way to produce novelty compared to the re-assortment of functioning segments seen in influenza viruses.

10.3.2 Genetics of pathogenic bacteria

In combination with other genes, virulence factors of bacteria are essential determinants of pathogenesis and infection success. For many, their genetic basis is known in considerable detail. In bacteria, the responsible genes are often found in blocks called 'pathogenicity islands'.

10.3.2.1 Pathogenicity islands

These are parts of the genome that harbour DNA sequences typical for mobile genes, i.e. genes that can be transferred horizontally among bacteria. In particular, the respective DNA sequences are indicative of transposases—enzymes that facilitate the copying and transfer of genetic elements within the genome (Kjemtrup et al. 2000). Such islands were discovered first in *Escherichia coli* (Knapp et al. 1986), but have since been identified in a large number of pathogenic bacteria. Pathogenicity islands are an example of the more general phenomenon of 'genomic islands', i.e. genomic regions that transfer horizontally. Genomic islands often contain genes for essential bacterial functions such as nitrogen fixation, sucrose uptake, the biosynthesis of specific biochemical compounds (Hentschel and Hacker 2001; Dobrindt et al. 2004), as well as functions important in symbiotic relationships (Sullivan and Ronson 1998), but also functions in the context of infection. Genomic islands can also be physically located outside the bacterial chromosome, e.g. on plasmids or sequences derived from lysogenic phages. Horizontal mobility of genes allows bacterial populations to adapt rather quickly. It is an important process that shapes the evolution of microbes more generally, especially of pathogens (Dobrindt et al. 2004; Bellanger et al. 2013; Ilyas et al. 2017; Fillol-Salom et al. 2018).

Bacterial genomes are highly dynamic over short evolutionary time scales as genes are added or removed by several processes. Mutations thereby play a prominent and vital role. However, horizontal gene transfer (also called lateral gene transfer) seems to be particularly common in bacteria. Horizontal gene transfer occurs by several mechanisms; that is, via transposons, via phages, via transformation (by the uptake of plasmids or naked DNA), and via bacterial conjugation (and transfer of plasmids). Furthermore, the transfer of chromosome islands is also induced by helper phages ('phage-induced chromosome islands'; see 'PICIs and gene-transfer agents' below). Some genes can change places within the genome by transposition to another location ('jumping genes') (Figure 10.5).

Due to their origin, the genomic sequences of pathogenicity islands differ in several distinct ways from the residual background genome of the bacterium. For example, the GC content of their DNA sequence differs from the background and shows conserved features, also visible by differences in codon usage. Typically, pathogenicity islands insert after a tRNA gene and are flanked by small directly repeated sequences (nearly or entirely identical sequences that repeat themselves with the same reading direction). Furthermore, they carry func-

tional genes or pseudogenes that code for (gene) mobility factors (Figure 10.6) (Hacker and Kaper 2000), and are some 10–100 kb or up to 200 kb in size. Still smaller pieces of DNA that code for virulence elements are termed 'pathogenicity islets'. These

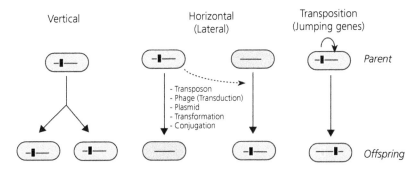

Figure 10.5 Movement of genes. The gene of interest is symbolized by the black bar. *Left*: Standard Mendelian inheritance; this corresponds to vertical transmission of genetic information. Middle: Horizontal (lateral) transfer; genes are transferred from one line (or species) to another, via a transposon, phage, plasmid, or bacterial transformation and conjugation. With conjugation, two parents contribute to a given offspring (biparental reproduction, sex). *Right*: Genes can also move to a new location within the genome (transposition) with the help of a transposon ('jumping genes'). The movement of genes can be inferred from comparing the characteristic DNA sequences around the site of an inserted gene.

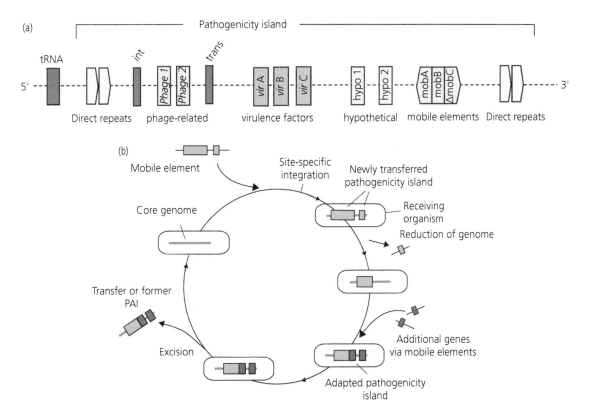

Figure 10.6 Continued.

(c)

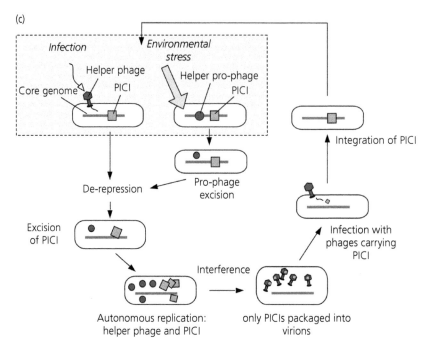

Figure 10.6 Pathogenicity islands. (a) Schematic structure of a pathogenicity island. It starts after a tRNA (red block), and contains factors for integration and excision (*int, trans*), virulence factor genes (*virA, virB*, etc.), phage-related genes (*Phage 1, Phage 2*), or inserted sequences (*mobA, mobB*, etc.). Integration is often followed by the subsequent inactivation and degradation of mobility genes (Δ*mobC*, through accumulation of mutations), because selection no longer maintains their function. The elements are flanked by direct repeats (yellow) at both ends, which are essential for the cycle, as they are recognized by integrases and factors contributing to excision. Adapted from Che et al. (2014), under CC BY 3.0. (b) The life cycle of a pathogenicity island starts with the horizontal transfer of a mobile element and its site-specific integration into the genome of the receiving organism (oval rectangle). Subsequently, the acquired genetic information is reduced, as not all elements have a useful function anymore (e.g. the mobility genes), or become redundant. Additional genes may become integrated via other mobile elements, leading to an overall adapted pathogenicity island. Eventually, the newly assembled island may again become excised, and is horizontally transferred, as a mobile element, to another line. Environmental factors (e.g. pH, temperature, oxidative stress) affect the rate of excisions. (c) The cycle of a 'phage-induced chromosomal island' (a pathogenicity island, PICI; orange block in core genome, grey bar) starts when a 'helper phage' (red) infects the bacterium, or when a prophage (the integrated genomic sequence of a phage; red circle) becomes activated by environmental stressors. Subsequently, the PICI becomes de-repressed and leaves the bacterial genome (excision). Phage and PICI replicate autonomously (red circles, orange blocks), but the island interferes such that the phage is not packaged efficiently. The PICI is contained in the virions that leave the host cell (by lysis) and which can infect a new host bacterium. Eventually, the PICI integrates itself into the background core genome and is ready for a next cycle. Adapted from Penadés and Christie (2015), with permission from Annual Reviews, Inc.

features, together with the now-available wealth of bacterial genomic sequences, allow searching for pathogenicity islands with computer algorithms (Che et al. 2014; Lu and Leong 2016). Several public databases list pathogenicity islands for many different bacteria; for example, DGI (Database of Genomic Islands), GI-POP (Lee et al. 2013a), and many others. Overall, however, pathogenicity islands are not homogeneous in their sequence but are rather mosaic-like structures. Evolutionarily speaking, they are relatively unstable over time and have a life cycle of their own (Figure 10.6b).

Functionally, the pathogenicity islands code for elements that affect bacterial virulence or infectivity, such as adhesion factors, mechanisms for host cell entry, but also for secretion systems, for the acquisition of crucial metabolites (such as for iron uptake), and for the production of toxins (e.g. enterotoxin) (Table 10.2). Genes on islands also code for products that interfere with host immune defence or provide antibiotic resistance. Because pathogenicity islands give their carrier bacteria an advantage, they tend to spread in populations. The evolution and radiation of bacterial pathogens is often a history of the

Table 10.2 Bacterial virulence factors coded by genes on pathogenicity islands (PAIs).[1]

Factor category[2]	Description	Examples of factors coded on PAIs[3]
Adhesins	Mediate capacity of bacteria to attach to host cell receptors.	*E. coli* (UPEC): P-fimbriae coded by specific PAIs. Can bind to receptor molecules on host uroepithelial cells. *Vibrio cholerae*: Toxin-co-regulated pilus coded by the Vibrio pathogenicity island, containing genes *tcp* and *acf* (accessory colonization factor cluster), the latter responsible for chemotaxis.
Invasion, effectors, modulins	Factors that allow invasion of host cells; induce apoptosis or inflammation.	*Salmonella* has several PAIs containing factors that facilitate invasion of epithelial cells (located on island SPI-1), induction of apoptosis in macrophages (SPI-1), proliferation within host cells (SPI-2), survival in macrophages (SPI-3, SPI-4), affecting inflammation (SPI-5). *Yersinia*: Yersinia virulon encodes a type III secretion system and effector proteins (Yops) that, for example, inhibit phagocytosis (YopH), disrupt host cell cytoskeleton (YopE), or have cytotoxic effects (YopT).
Toxins	Toxins are a broad class of molecules with various effects, e.g. formation of pores in host cells, proteases, and many other modulatory functions.	*E. coli* (UPEC): Haemolysins (HlyA) transported by type I secretion system (all located on PAI), can lyse host red blood cells. Cytotoxic necrosis factor 1 modifies a GTPase (RhoA protein). *Shigella*: Shiga toxin (encoded by *stx*) cleaves specifically 28S RNA of eukaryotes. *Listeria*: *prf* virulence gene cluster with phospholipases.
Type III secretion systems	Typically forms a channel through inner and outer membrane. Contact dependent, so that outer components fuse with host cell membrane to form a syringe injecting protein directly into host cell. Structure consisting of more than 20 different proteins. This system is typical for pathogenic species.	Most bacteria have only type II system encoded on an island, e.g. *Erwinia*, *Xanthomonas*, *Shigella*. *Salmonella*: Has a complete type III system on both SPI-1 and SPI-2 islands. *Yersinia enterocolitica*: Type III system is encoded with an island on plasmid as well as on chromosome.
Type IV secretion systems	System with a large number of proteins forming pilus-like structure that typically directly injects proteins into host cell. Can transfer large molecules.	*Legionella pneumophila*: Gene cluster *dot/icm* essential for virulence, with sequence similarity to virB proteins. Also described from *Helicobacter pylori*: Translocates CagA protein to host that induces growth changes in host cells. *Bordetella pertussis*: Transport of pertussis toxin.
Other secretion systems	Type I: A simple, one-step secretion system that spans inner and outer membrane.	*E. coli*: Found on pO157 plasmid of the highly infectious strain O157:H7 (EPEC, enteropathogenic *E. coli*). *B. pertussis*: Exports adenlyate cyclase. *Pasteurella haemolytica*: Exports leucotoxins.
	Type II: A two-step system dependent on *sec* (first step). Apparatus located in outer membrane (secretins). Exports degrading enzymes, sometimes also toxins.	*Klebsiella oxytoca*: Secretes pullulanase that hydrolyses sugars. *Pseudomonas aeruginosa*: Secretes toxin A, phospholipase C. *Vibrio cholerae*: Secretes cholera toxin, chitinase, proteinase.
Iron uptake systems	Iron is essential for bacterial growth and acquired by the production of siderophores or expression of iron receptors.	*Yersinia*: Yersiniabactin (siderophore system) encoded by HPI. *Shigella flexneri*: Aerobactin (siderophore system) encoded by *aer* (*iut*) locus on PAI-2 (SHI-2).

[1] Modified after Hacker. 2000. *Annu Rev Microbiol*, 54: 641.

[2] Secretion by type II, IV, and V systems additionally require the general secretory pathway, *sec*. The type III and IV (and sometimes type V) systems are typically associated with pathogenicity islands.

[3] Italics indicate names of genes.

pathogenicity islands. For example, the closest relatives of the human-pathogenic *Vibrio cholerae*, the aetiological agent of cholera, live in aquatic environments. Over time, *V. cholerae* has, in several steps, acquired its virulence genes by horizontal transfer. In particular, its primary virulence genes are in several genomic regions, which have been acquired recently via phages or some other as yet unknown horizontal transfer process. Each acquisition of new elements has been correlated with the outbreak

of a major cholera epidemic (Faruque and Mekalanos 2003) (Figure 10.7).

A similar case is *Salmonella*: foodborne, enteric pathogens of cold- and warm-blooded animals (Cotter and DiRita 2000; Nieto et al. 2016; Ilyas et al. 2017). *Salmonella* diverged from *E. coli* around 120–160 million years ago, and *S. enterica* split from *S. bongori* shortly thereafter (Desai et al. 2013). *S. enterica* has a broad host range. It originally included cold-blooded organisms such as fish,

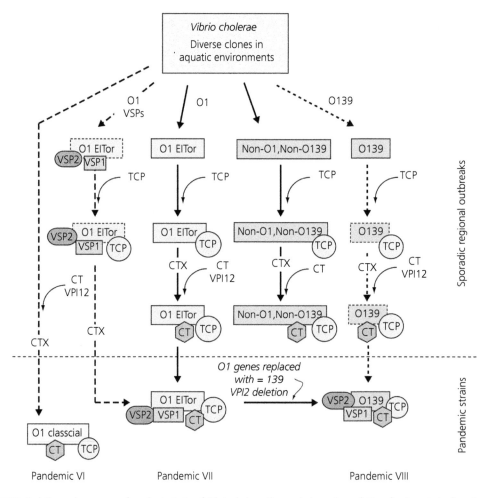

Figure 10.7 Evolution and emergence of pandemic strains of *Vibrio cholerae*. The graph shows the evolution of various strains from the ancestral *V. cholerae* in aquatic environments. In particular, the gain and loss of virulence factors is shown. Boxes contain names of the strains; boxes with solid lines are observed strains, boxes with dotted lines are conjectured. Solid arrows symbolize horizontal gene transfer of pathogenicity islands, broken lines are likely, yet hypothetical, transfers. The strains in the bottom row have caused subsequent cholera pandemics (pandemic VI to VIII). Acronyms: CTX, cholera toxin phage; CT, cholera toxin; O1 and O139, various O antigen gene clusters; TCP, toxin co-regulated pilus gene cluster, also referred to as VPI; VPI, Vibrio pathogenicity island; VPI2, the island that includes the nanH gene cluster; VSP, Vibrio seventh pandemic island; VSP1 and VSP2, the islands found in seventh pandemic El Tor strains and eighth pandemic O139 strain. Adapted from Faruque and Mekalanos (2003), with permission from Elsevier.

amphibia, or reptiles. Later, warm-blooded animals, e.g. humans, cattle, or poultry, were added. Currently, more than 2600 serovars or 'species' in the *S. enterica* complex have been described (Worley et al. 2018). The evolution of this clade is very complex due to many horizontal transfers from various lineages, such that different gene families vary in their phylogeny (Desai et al. 2013; Timme et al. 2013).

Again, the evolution of *S. enterica* is closely associated with the acquisition of pathogenicity islands (Timme et al. 2013; Langridge et al. 2015). *Salmonella* first acquired genes, by horizontal gene transfer on a pathogenicity island (SPI-1), that allowed the colonization of host intestines. The genes on SPI-1 encode a type III secretion system, plus various factors exported via this system (Cotter and DiRita 2000). In the next evolutionary step, *S. enterica* acquired a second pathogenicity island (SPI-2) that also encodes a type III secretion system. It facilitated the survival of the bacterium within the host macrophages, and thus extended the bacterium's capacity to cause long-lasting systemic infections. SPI-1 and SPI-1 are now characteristic for members of *S. enterica* (Jacobsen et al. 2011). At least 15 other islands are known; for example, SPI-4 is involved in adhesion to epithelial cells and SPI-6 is essential for invasion and acting against colonization resistance (Ilyas et al. 2017). The regulation of these virulence genes affects the ability of the bacterium to invade, persist, and multiply in its host, and is thus crucial for the parasite's success. Interestingly, the host specificity of serovars is loosely associated with the formation of pseudogenes (i.e. the degradation of functional genes in the islands; Figure 10.6), and such degradation seems to occur after the split of lineages (Langridge et al. 2015).

10.3.2.2 PICIs and gene-transfer agents

Besides the 'classical' (integrative–conjugative) transferable elements, additional processes of horizontal gene transfer exist (Novick and Ram 2016). For example, 'phage-induced chromosomal islands' (PICIs) were found to be widespread in bacteria (Penadés and Christie 2015; Fillol-Salom et al. 2018). Here, the pathogenicity island is an active element. It parasitizes the life cycle of a temperate 'helper phage' to become transmitted. When a helper phage infects the same cell, or when an already integrated phage becomes activated by environmental stress,

the pathogenicity island genome excises and replicates using its genes. It then becomes packaged into phage virions to reach a next host cell. In the process, the pathogenicity island interferes with the replication of the helper phage (Penadés and Christie 2015). In yet another horizontal transfer process, phage-related chromosomal sequences encode for the production of small infectious phage-like particles that can infect a next host ('gene-transfer agents'). In this way they transfer genes related to the pathogenicity and general success of their bacterial carriers (Novick and Ram 2016). The PICIs are probably derived from ancient prophages, similar to the classical pathogenicity islands. The gene-transfer agents likely evolved from conjugative plasmids (Novick and Ram 2016).

10.4 Genetic variation

Genetic and genotypic variation—also known as genetic polymorphism—is an essential characteristic of living systems, and a prerequisite for selection to cause evolution. One of the most ubiquitous selection factors is parasitism. Therefore, the role of genetic variation for susceptibility to infection and disease is an important topic. In essence, the existence of genetic polymorphism is also the basis for the association studies discussed above.

10.4.1 Individual genetic polymorphism

An impressively large number of studies show that variants of genes, as, for instance, described by SNPs, are associated with variation in the resistance, tolerance, or severity of disease (see Table 10.3). At the same time, such genetic variation (polymorphism, genetic diversity) exists not only among individuals in a population but also within individuals. For example, diploid individuals inherit two different alleles from their parents or have different allele combinations in gene complexes, such as the MHC locus.

Genetic diversity within individuals can be expressed by the fraction of loci that are heterozygous, i.e. that have two different alleles. Concerning parasitism, evidence for a heterozygous advantage exists. A well-known case is the genetics of sickle cell anaemia in the context of malaria infections. In humans, the locus coding for the Hbβ chain of the

Table 10.3 Genetic markers (SNPs) associated with susceptibility to infections in humans.

Variants	Pathogen and effects	Remarks	Source
SNPs in HLA-DP gene screened.	Several SNP polymorphisms affect susceptibility to HBV.	Loci analysed by PCR in 3036 cases and 1342 control individuals.	8
Polymorphism at locus CCR5 gene (chemokine receptor 5), especially a 35 bp deletion (Δ32) in the coding region.	Homozygotes for CCR5 (Δ32/Δ32) have reduced susceptibility to HIV-1, especially in Caucasian ethnicities.	A meta-analysis with 4786 cases and 6283 control individuals.	4
SNPs in various TLRs (TLR1, TLR6).	SNPs associate with susceptibility to several bacterial and fungal infections, malaria, and leprosy.	Literature review.	5
Allelic variants in TLR2 genes at SNP positions G2258 and T597.	Allele 2258G and the G/G genotype show decreased susceptibility to pulmonary tuberculosis. Polymorphism T597C shows no association. (*Mycobacterium tuberculosis*).	A meta-analysis with 1301 cases and 1217 control individuals.	7
TLR8 alleles at position 129.	SNP allele 129C associated with more pulmonary tuberculosis in male patients. (*Mycobacterium tuberculosis*) in Asian populations.	Monocytes with the allele show higher phagocytosis than wild type.	3
SNPs screened with genome-wide scans. Positions within HLA Class II genes scrutinized.	Several SNP polymorphisms affect susceptibility to *Staphylococcus aureus* infections. Polymorphisms are near coding regions for HLA genes, and HLA-DRB1*04 serotype especially affected.	Health cohort study with 4701 cases and 45 344 matched controls. More than 25 million SNPs screened also with the 1000 Genomes reference panel.	2
Humans have at least 34 different blood groups, and hundreds of antigens.	Different blood groups differ as receptors, co-receptors for various pathogens. Blood-group-specific antigens can affect immune-signal transduction, cellular uptake, and adhesion, and modify immune responses. Also interacts with microbiota.	A large literature survey.	1
Humans: Genome-wide scans for polymorphisms. Often detected in HLA genes.	Associations with 23 common infections found, including chickenpox, cold sores, common colds, mumps, HBV, scarlet fever, bacterial meningitis, etc. Suspected genes have roles in immunity and embryonic development.	A total of 23 genome-wide association studies in over 200 000 individuals of European ancestry.	6

Sources: [1] Cooling. 2015. *Clinical Microbioly Reviews*, 28: 801. [2] DeLorenze. 2015. *J Infect Dis*, 213: 816. [3] Lai. 2016. *Tuberculosis*, 98: 125. [4] Ni. 2018. *Open Medicine*, 13: 467. [5] Skevaki. 2015. *Clin Exp Immunol*, 180: 165. [6] Tian. 2017. *Nat Comm*, 8: 599. [7] Wang. 2013. *PLoS One*, 8: e75090. [8] Zhang. 2013. *J Virol*, 87: 12176.

haemoglobin protein has alleles *A* and *S*. Individuals heterozygous at this locus (labelled *HbAS*) are protected against malaria, the wild type (*HbAA*) is highly susceptible to infection, and for those who are homozygous (*HbSS*) it is a lethal condition (due to insufficient oxygen transport). Hence, concerning malaria, the heterozygotes (*AS*) have an advantage. Individuals heterozygous at the Fy-antigen locus (Duffy blood group) also resist malaria; the polymorphic locus is the amino acid at position 42 of the gene sequence (Cooling 2015).

Many studies across a range of host taxa show advantages for heterozygosity. For instance, the burden of endoparasitic gregarines (Apicomplexa, Protozoa) increases with the individual degree of homozygosity in damselflies (based on AFLP typing) (Kaunisto et al. 2013). Similarly, heterozygous fare better in racoons and their endoparasites (Ruiz-López et al. 2012), mongoose vs protozoans and helminths (Mitchell et al. 2017), guppies vs cestodes (Smallbone et al. 2016), salmon vs viruses and protozoa (Arkush et al. 2002), house finches vs their mycoplasms (Hawley et al. 2005), Soay and bighorn sheep vs their nematodes (O'Brien and Evermann 1988; Paterson et al. 1998; Luikart et al. 2008), and dairy cattle infected by tuberculosis (Tsairidou et al. 2018). Hence, individual heterozygosity often correlates negatively with parasite load or disease severity (Figure 10.8).

The MHC locus in jawed vertebrates is an example of a gene complex, as it consists of a series of tightly linked genes. It codes for recognition sites

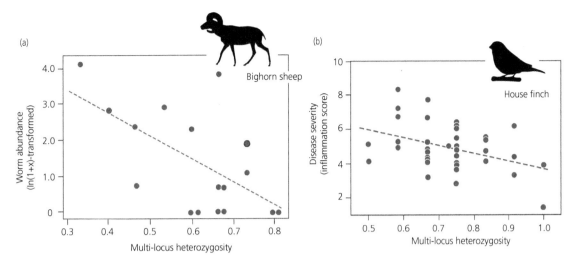

Figure 10.8 Host heterozygosity and parasitism. (a) Worm burden (the nematode *Protostrongylus* spp.) in bighorn sheep (*Ovis canadensis*) declines with heterozygosity (dotted line, $r^2 = 0.34$). The effect is especially strong when only loci linked to disease-related genes are considered. Adapted from Luikart et al. (2008), with permission from The Royal Society. (b) Disease severity (inflammation score) in house finches (*Carpodacus mexicanus*) infected with bacteria decreases with an increase in individual heterozygosity. Silhouettes from phylopic.org. Adapted from Hawley et al. (2005), with permission from The Royal Society.

that can bind to parasite-derived peptides. We can therefore expect that a high degree of heterozygosity within the MHC locus will lead to efficient defence against many different parasites and, hence, to a generally low parasite load. This is true in some cases (e.g. Soay sheep; Paterson et al. 1998) but not in others (Atlantic salmon; Lohm et al. 2002). In fact, in populations of sticklebacks, an intermediate number of MHC alleles, rather than a maximum number, associates with the lowest parasite load (Figure 10.9). Presumably, high MHC diversity lowers parasite load but increases the risk of having molecules that are too strongly self-reactive. In humans, genetic fine-mapping with SNPs suggests that heterozygosity at the MHC class I locus associates with slower progression to AIDS; heterozygosity in MHC class II associates with higher rates of clearance of HBV (Matzaraki et al. 2017). Similarly, SNP heterozygotes at the *S180L* polymorphism, within a coding region for TIRAP (also known as adaptor protein MAL) in the TLR cascade (see Figure 4.6), are better protected against severe malaria (Panda et al. 2016), tuberculosis, and several other infections (Ní Cheallaigh et al. 2016).

However, problems of study design, classification of phenotypes, or perhaps even SNP-typing methods can plague these conclusions. For example,

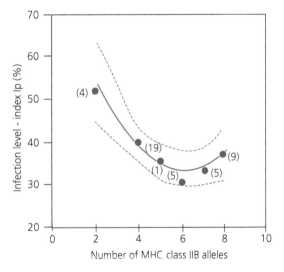

Figure 10.9 The relationship of parasite load with the number of MHC IIB alleles found in individual sticklebacks (*Gasterosteus aculeatus*). Each dot is the mean infection level for a group of sticklebacks (pooled per allelic diversity class) experimentally exposed to tapeworms (*Schistocephalus solidus*). Infection level was measured by the index Ip, as the ratio of tapeworm body mass to host body mass (as percentages). Small numbers in parentheses are sample sizes. The solid line is a polynomial regression with 95 per cent C.I. (broken lines). An intermediate number of MHC alleles is associated with the lowest level of infection. Adapted from Kurtz et al. (2004), with permission from The Royal Society.

whether or not the *S180L* polymorphism associates with susceptibility to malaria, bacterial infections, or vaccine failure is not fully clarified, as different studies use different methodologies and results differ between humans and the mouse model (Netea et al. 2012; Saranathan et al. 2020). In many cases, SNP the decisive factor is not heterozygosity per se, but the kind of allele present at an SNP (i.e. whether nucleotide G, A, C, or T is present). For humans, single SNP alleles define a high-risk or a low-risk category for infections by highly pathogenic bacteria (DeLorenze et al. 2015), and for many other common infectious diseases (Tian et al. 2017).

The heterozygous advantage is not universal, either. For example, humans homozygous in the CCR5/Δ32 locus are more, rather than less, resistant against infections by HIV-1 (Ni et al. 2018), even though heterozygotes progress more slowly to AIDS than those not having the deletion (Δ32) (Chapman and Hill 2012). Individuals that are homozygous for mannose-binding lectin genes better resist meningococcal infections (Emonts et al. 2003). Also, homozygosity in the KIR/HLA genotype protects better against hepatitis C virus (Jost and Altfeld 2013). Many further examples of a disadvantage of heterozygosity in various loci exist, e.g. for the HLA-DP gene and the risk of chronic HBV infection (Zhang et al. 2013), of bacterial infections of the urinary tract, or of infections by respiratory viruses (Skevaki et al. 2015).

Social insects illustrate a particular kind of advantage to genetic polymorphism. Colonies of bees, wasps, and ants are headed by a mother queen (sometimes several queens). Her daughters, the individual workers, tightly work together for the fitness of their mother and thereby become exposed to infectious diseases. In this context, workers can be considered individual genetic entities that together form the 'genotype' of the colony. This view is inaccurate in several ways (e.g. there are conflicts of interests among workers), but it serves to make the point. Polyandry (multiple mating by the queen) or polygyny (the presence of several queens) leads to genetically more variable workers being present in a colony. Such increased genetic diversity of the colony workers is associated with lower levels of parasitism (Baer and Schmid-Hempel 1999; Schmid-Hempel and Crozier 1999).

10.4.2 Genetic variation in populations

If genetic diversity is an advantage in the face of parasitism, genetically more variable populations should be less parasitized. In a study of the Galapagos Hawk (*Buteo galapagoensis*), for example, the average heterozygosity of each population, as expected, correlated positively with island size. Above and beyond this island effect, however, birds in less diverse populations had higher abundances of parasitic lice. They also had a lower average and less variable titre of antibodies when compared to more variable populations on larger islands (Whiteman et al. 2005). Antibody levels correlate with the abundance of certain louse parasites, whose lifestyle brings them in direct contact with the host's immune system. Several other studies show that genetically more variable populations typically contain fewer parasites or have lower parasite loads, even though this pattern is not universal (Table 10.4). In these comparisons, the direction of causation is often unclear. More heterozygous populations may have fewer parasites because it is more difficult for the parasites to cope with genetically diverse hosts. In contrast, more variable host populations may result from selection by more variable parasites.

The relationship between genetic diversity and parasitism is of interest in agriculture. Monocultures, i.e. genetically uniform fields and plantations, bear a considerable risk of being devastated by an incoming pathogen (Reiss and Drinkwater 2018). The relationship is also of interest for conservation biology. Are genetically less variable populations more susceptible to disease, which could lead to the final extinction of an endangered species? Empirical evidence suggests that small populations, having lost most of their genetic variability, indeed are prone to disease-induced population crashes, or even extinction. Section 15.1.5 discusses some of these parasite-induced crashes. The actual processes are complex, however. Typically, other factors drive populations to low numbers first. Also, a reservoir often must be present from which a parasite can transmit to the endangered population (de Castro and Bolker 2005). Parasites then act as the last push to a disastrous outcome.

Table 10.4 Parasitism and genetic diversity in populations.

Host species	What was done	Finding	Source
Plants	Comparing 182 species within five families (Pinaceae, Myrtaceae, Fabaceae, Poaceae, Asteraceae) for outcrossing rate and the number of fungal pathogen species.	Outcrossing rate and number of fungal species correlated significantly across the five families. Also, a meta-analysis from phylogenetically independent contrasts showed that the number of species of fungal pathogens is positively and significantly correlated with outcrossing rate within these families.	2
Bacteria, insects, water fleas, fish, frogs, birds	Meta-analysis of 23 studies with 32 host–parasite associations. Parasites include viruses, bacteria, protozoa, fungi.	Across all studies, low host population genetic diversity results in higher parasite success. Effect size is higher for field than for lab studies. Parasite host range has no effect.	4
Drosophila melanogaster	Lines of different inbreeding degrees tested against toxins from *B. thuringiensis* and infection by live *S. marcescens*.	More damage with increasing degree of inbreeding.	8
Galapagos hawk (*Buteo galapagoensis*)	Comparing island populations with different degrees of heterozygosity.	Low heterozygosity correlates with increased infection levels and decreased titres of constitutively expressed antibodies.	10
Deer mouse (*Peromyscus maniculatus*)	Enzyme allelic diversity measured in nine populations in Michigan and compared to prevalence of the nematode, *Capillaria hepatica*.	Heterozygosity correlates negatively with parasite prevalence, and also when controlled for population density. Density alone does not correlate with parasite prevalence.	5
Red deer (*Cervus elaphus*)	Measuring population-wide indices of heterozygosity (F_{IS}), based on multilocus genotypes.	Average ability of populations to control progression of tuberculosis (disease severity) increases with heterozygosity.	6
Termite (*Zootermopsis angusticollis*)	Experimental infection of worker groups with bacteria or fungal spores and with different levels of inbreeding.	More inbred groups show higher mortality rates and higher microbial loads.	3
Bumblebee (*Bombus terrestris*)	Experimental variation of genetic diversity among workers by artificial insemination of colony queen to generate high- (sperm from diverse males) and low- (sperm from similar males) diversity colonies. Colonies exposed to field conditions.	More diverse colonies have fewer parasites and higher reproductive success.	1
Honeybee (*Apis mellifera*)	Honeybee queens artificially inseminated to generate high- (sperm from diverse males) and low- (sperm from similar males) diversity colonies.	Genetically diverse colonies have lower variance of infected workers.	9
Ants	Comparing 119 ant species from different families for degree of genetic diversity among workers within colonies and number of recorded parasite species for each host species.	Negative correlation between genetic diversity and number of parasite species. Effect persists when considering phylogenetically independent contrasts.	7

Sources: [1] Baer. 1999. *Nature*, 397: 151. [2] Busch. 2004. *Evolution*, 58: 2584. [3] Calleri. 2006. *Proc R Soc Lond B*, 273: 2633. [4] Ekroth. 2019. *Proc R Soc Lond B*, 286: 20191811. [5] Meagher. 1999. *Evolution*, 53: 1318. [6] Queirós. 2016. *Infect, Genet Evol*, 43: 203. [7] Schmid-Hempel. 1999. *Philos Trans R Soc Lond B*, 354: 507. [8] Spielman. 2004. *Conserv Genet*, 5: 439. [9] Tarpy. 2003. *Proc R Soc Lond B*, 270: 99. [10] Whiteman. 2005. *Proc R Soc Lond B*, 273: 797.

10.4.3 Gene expression

10.4.3.1 Expression profile and transcriptome

Transcribed genes are present in a cell in the form of mRNAs, or non-coding RNAs (ncRNA; not translated into proteins such as rRNAs or regulatory microRNAs). As mentioned before in the context of association studies (see section 10.1.2.3), either a limited set (the 'expression profile') or the entire suite of all mRNAs (the 'transcriptome') can be used. Technically, gene expression profiles and transcriptome analyses are based on RNA sequencing (RNA-seq). However, RNA is a much less stable and tractable molecule than DNA. Although direct sequencing of RNA is possible, in general, RNA is enzymatically reverse-transcribed into the corresponding cDNA (coding DNA), which is then sequenced (Hrdlickova et al. 2017). In some cases, gene expression is resolved to a very fine level of single-cell transcriptomics (Merkling and Lambrechts 2020). Eventually, bioinformatics can identify those genes that are significantly up- or downregulated and narrow the data down to a manageable set.

Gene expression is variable among hosts, among host populations, among tissues, over time, and with changes in the host status, e.g. when infected (Oleksiak et al. 2002; Gibson 2003). Note that differences in expression reflect differences in the genetic sequences of regulatory elements. Gene expression itself is therefore heritable, too (Gibson 2003; Gibson and Weir 2005). Regulatory genes are mapped on the genome, for example, as 'expression quantitative trait loci' (eQTL), which define their locations (Gibson and Weir 2005; Rockman and Kruglyak 2006; Fairfax and Knight 2014). Such regulatory genes can be physically close to the target genes, known as *cis*-regulation, e.g. by a promoter sequence immediately upstream of the gene. Alternatively, *trans*-regulation is by regulatory genes somewhere else in the genome, distant from the target gene. Furthermore, the expression can be epigenetically modified, e.g. by environmentally induced DNA methylation in the sequence of a given gene (Suzuki and Dird 2008). In fact, a hallmark of eukaryotic gene regulation is post-translational modification. In the process, a primary transcript becomes modified secondarily before being used as a template for a protein, e.g. by phosphorylation to change the regulation of the innate TLR cascade (Liu et al. 2016b). Many of these processes are relevant for hosts and eukaryotic parasites such as protozoa (Doerig et al. 2014).

When a host becomes infected, typically several dozens or even hundreds of genes are differentially up- or downregulated. Typically, only a fraction of those have known functions to begin with, or are part of the immune defences. These nevertheless are valuable candidates for further study, e.g. with the candidate-gene approach. In *Drosophila* infected by various bacteria, a detailed study of 329 such candidate genes not only revealed differences in the regulation of immune-related genes. In addition, the expression of genes involved in pathogen recognition were good predictors for the resulting bacterial load during an infection (Sackton et al. 2010). Infections by *Enterococcus faecalis* in

Drosophila similarly showed a single SNP polymorphism in the vicinity of several genes associated with survival; this set could be traced to fewer, known genes (Chapman et al. 2020).

Infection is also a challenging process for parasites. In particular, the synchronized regulation of parasite genes is crucial for success. *Plasmodium*, for example, has evolved a piece of expression machinery that regularly changes the variant of PfEMP1 expressed at the surface of the blood cell where the parasite lodges. The parasite escapes by changing this epitope (a process known as antigenic variation; see also section 10.4.3.3). The variants of PfEMP1 are encoded by the multicopy gene family, *var*, having *c.*60 versions. At any one time, only one of those is expressed in a process known as 'mutually exclusive expression'. Several layers of regulation are thereby effective. It includes modification of histones, as well as cis-located modifiers and RNA transcripts that manage silencing or activation of a particular gene variant (Deitsch and Dzikowski 2017; Bunnik et al. 2018).

Gene expression also affects other essential processes, such as the level of quorum sensing and resistance against antimicrobial drugs (e.g. *Pseudomonas*; Cornforth et al. 2018); part of this regulation happens at the level of post-transcriptional modification (Grenga et al. 2017). Furthermore, expression cascades regulate the periodicity in pathogen effects or abundance to synchronize its life cycle with the diurnal rhythm of vectors (as in malaria or *Trypanosoma*; Prior et al. 2020). The same parasite may also have different host species, and therefore must regulate its genes in different ways, as bird malaria (*P. homocircumflexum*) does when infecting different bird species (Garcia-Longoria et al. 2020).

Expression differences alone can lead to biologically different pathogens. For instance, the four recognized subspecies of *Bordetella* (*B. pertussis'* two forms of *B. parapertussis*—*ov* and *hu*; and *B. bronchiseptica*) all infect the respiratory tract of mammals. For this, *Bordetella* secretes many factors that function in cell adherence or cytopathology, or induce host inflammatory cytokines that lead to severe damage. However, all of these forms of *Bordetella* use the same or similar virulence factors as are regulated by a nearly identical regulatory system (Cotter and DiRita 2000; Beier and Gross 2006). In particular, the BvgAS system is a two-component regulatory system of signal-transducing proteins and affects the secretion of virulence factors, cell motility, metabolism, surface molecules, electron transport, or toxin production—depending on whether the system is active (Bvg⁺ phase) or inactive (Bvg⁻ phase). Accordingly, the pathogen shows phases with different phenotypes, and these phases differ among the subspecies.

All subspecies express adherence factors during the initial infection process. However, pertussis toxin (Ptx) is expressed only by *B. pertussis*, and a specific type III secretion system exists only in *B. parapertussis*$_{ov}$ and *B. bronchiseptica*. The immunomodulatory effects of this secretion are associated with chronic infections. Similarly, all subspecies secrete TCT (tracheal toxin), but the respective amounts differ, which affects the virulence of the different bacteria. Hence, differences in expression lead to the establishment of either acute or chronic infections and different degrees of virulence, or even to asymptomatic infections (Cotter and DiRita 2000). Furthermore, *B. bronchiseptica* is able to survive in the external environment, which is characteristic of the Bvg⁻ phase phenotype. By contrast, *B. pertussis* and *B. parapertussis*$_{hu}$ are transmitted directly and via aerosol droplets that quickly reach a new host. Correspondingly, their Bvg⁻ phases differ substantially from the pattern in *B. bronchiseptica*. The BvgAS system is, therefore, able to sense whether the bacterium is within or outside the respiratory tract. It will then regulate the genes accordingly. In the *Bordetella* complex, the same or similar sets of genes have been acquired through pathogenicity islands. However, differential regulation, rather than a different set of genes, make it into different pathogens.

The evolution by variation in expression is based on somewhat different mechanisms and has a different dynamics as compared to evolution by changes in the gene sequence itself. Host populations may adapt faster and diversify more quickly with a system based on expression variation (Wittkopp et al. 2004; Carroll 2005), as the range of responses to infections is greatly extended even with the same 'hard-wired' genetic material (e.g. in *Bordetella*). Furthermore, individual hosts can plastically respond to particular infections, as illustrated in Figure 10.10. For instance, different colonies of the bumblebee *Bombus terrestris*, infected by different strains of the trypanosome *Crithidia bombi*, show different profiles of

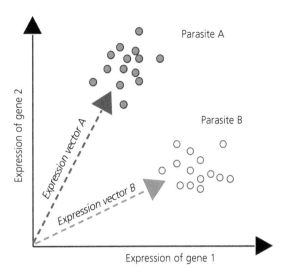

Figure 10.10 Variation in gene expression. In this hypothetical example, the red dots indicate the combined expression levels for gene no. 1 (x-axis) and gene no. 2 (y-axis) when a given host population is infected by parasite A; each dot should show a different host individual. The blue dots are the respective gene expression levels when parasite B infects. The 'expression vectors' indicate a particular combination of gene expressions. They differ according to infection and visualize the gene expression profile.

expression for the various antimicrobial peptides (AMPs; Barribeau et al. 2014). These are known to affect parasite loads (i.e. parasitaemia, Deshwal and Mallon 2014), and the host may express appropriate 'cocktails' for each infection (Marxer et al. 2016).

10.4.3.2 Copy number variation

Many genes occur in several copies in the genome. Differences in the copy numbers are an essential element of expression strategies that have evolved in different environments. For example, AMPs are often coded by genes that have variable copy numbers. A high copy number maximizes the response by allowing the fast expression of more products, e.g. for large titres of defensins (Machado and Ottolini 2015). Estimates for humans suggest that genetic variation in the immune system accounts for as much as 20–40 per cent of phenotypic variation in the response of individuals. At the same time, copy number variation probably exists for more than 15 000 loci in the human genome, many of which are known to affect infection success, disease severity, and immune responses (Brunham and Hayden 2013; Liston et al. 2016; Saitou and Gokcumen 2020).

10.4.3.3 Phase variation and antigenic variation

These processes refer to parasite strategies. 'Phase variation' denotes a frequent, stochastic, and reversible switching of gene expression pattern by rapidly switching some genes on and off. Phase variation is known, in particular, from bacteria. For example, *Neisseria meningitidis*, the cause of human bacterial meningitis, has more than 100 genes that can undergo phase variation, with an average of 47 phase-variable genes per bacterial genome. These genes include virulence factors responsible for bacteriocins or surface proteins, but also genes that affect iron acquisition (Caugant and Brynildsrud 2020). Phase variation is a widespread and influential adaptive phenomenon in bacteria. The mechanisms include frameshifts concerning their proximity to promoters, but also DNA methylation (van der Woude 2017; Phillips et al. 2019; Sánchez-Romero and Casadesús 2020).

'Antigenic variation' refers to the change of surface molecules that act as epitopes recognized by the host immune system. The prime example is the regular changes of the malaria parasite or of the surface coating in African trypanosomes (based on variants of PfEMP1) (Romero-Meza and Mugnier 2020) (see section 12.3.2). Antigenic variation is based on the regularly changing expression of different genes (in some cases, from an 'archive') that code for surface molecules (sugars and proteins).

10.4.4 Heritability of host and pathogen traits

Breeders of animals and plants have, since antiquity, known about the heritable basis of disease resistance, although the modern terms and insights have only been developed over approximately the last 100–150 years. Breeding is now as important as ever and drives much of the genomic research, as for maker-assisted breeding in many crops (Poland and Rutkoski 2016; Boutrot and Zipfel 2017). However, the identification of markers alone does not directly answer the question of what fraction of resistance is heritable in populations. For hosts, quantitative genetics (Box 10.3) is a classical and versatile tool to estimate the extent of heritability in traits such as resistance.

Heritability usually refers to narrow-sense heritability (h^2) and measures the fraction of phenotypic variance attributable to additive genetic variance (Box 10.3). This fraction is available for natural selection or for breeders to select, e.g. for improved resistance. Because heritability is a ratio relative to other quantities, the estimates of heritability vary among systems for many reasons. Examples are resistance of *Drosophila* to infection by ectoparasitic mites ($h^2 = 0.15$; Polak 2003), resistance of snails against trematode infections ($h^2 = 0.36$; Grosholz 1994), or plant resistance against fungal infections ($h^2 \leq 0.2$; Dieters et al. 1996). Freshwater fish parasitized by ectoparasitic copepods have an estimated

---

Box 10.3 Quantitative genetic effects

Quantitative genetics deals with quantitative traits (a phenotype) such as body size or host resistance. Such traits are affected by many loci and by the effects of the environment. With quantitative genetics, the total phenotypic variation among individuals in a population (V_P) is partitioned into contributions from different sources. In the simplest form, this partitioning is:

$$V_P = V_G + V_E + V_{GxE} \qquad (1)$$

Here, the total phenotypic variation is the sum of the variation among genotypes of individuals (genotypic variance, V_G), plus the variation among the environments where the individuals live or grew up (environmental variance, V_E). In addition, there is a variance contribution due to how the phenotypes of a given genotype are affected by the current environment (genotype x environment interaction variance, V_{GxE}). The genotypic variance can be further partitioned into components of interest:

$$V_G = V_A + V_D + V_I + \ldots + \left[\text{further terms}\right] \qquad (2)$$

Here, the total genotypic variance is the sum of variance that results from variance in the additive genetic effects (V_A), plus dominance effects (interaction between genes at the same locus, V_D), the interaction effects (between genes at different loci, V_I, i.e. epistasis), plus further terms (such as interactions between any of the foregoing). There are several statistical methods that allow these components to be extracted from actual data. Foremost among them is the analysis of variance (ANOVA), as first championed by R. A. Fisher (1890–1962). The methods can, for example, apply to data from defined crosses (e.g. between different selected lines), or from comparisons among relatives (for example, sib vs half-sib, or offspring vs parents) (Falconer 1989). With ANOVA, it is possible to determine statistical 'main effects', which here would refer to the additive genetic effect of single alleles. Besides, there are statistical 'interaction effects', which here might refer to the epistatic effects of combinations of alleles (Figure 1).

Using the same logic, it is also possible to reconstruct the phenotype from the single contributions. For this purpose, each allele (and combinations of alleles) is assigned a 'genotypic value' for the resulting phenotype. Hence, adding these values determines the genotypic effect on the phenotype. The environment will then add additional components or modify the expression of the genotypic value. From the genotype, the most significant genetic effects are the additive, dominance, and epistatic (interaction) terms. Figure 1 shows how these effects combine in the example of two loci (*Locus 1, Locus 2*), with the respective alleles (*A, B, C, D*). Additive effects are the independent contributions by each of the alleles. Dominance effects imply that one allele (e.g. allele *B*) overrides the effect of its partner allele on the homologous strand (e.g. allele *A*)—either entirely or partially. Epistatic (interaction) effects occur when the presence of alleles at other loci (e.g. allele *C* at *Locus 2*) affects the effect of a given allele at the locus under consideration (e.g. allele *A* at *Locus 1*). Figure 1 also shows how the phenotype is constructed from the single genetic effects for different genotypes (at a single locus).

Epistasis is notoriously difficult to analyse. A part of the problem results from how variance is partitioned with ANOVA. The method follows a hierarchical manner, such that epistatic effects are extracted later in the sequence of numerical evaluation and are of smaller values. Therefore, epistasis typically requires large sample sizes to become statistically significant (Templeton 2000). Epistasis at the population-wide level is also closely associated with genetic linkage. Population-wide linkage is a result of epistasis, which 'herds together' the respective alleles in the genotypes of the population and leads to nonrandom allele associations. Finally, epistasis is straightforward for two-locus, two-allele systems (such as in Figure 1). However, it becomes rapidly more difficult to study, quantitatively and conceptually, when several loci or several alleles are involved.

Partitioning the overall phenotypic variance into different components allows estimation of the contribution of different sources. Of note, heritability is defined as:

$$H^2 = \frac{V_G}{V_P},$$ broad-sense heritability (proportion of trait variance due to total genetic variation)

$$h^2 = \frac{V_A}{V_P},$$ narrow-sense heritability (proportion of trait variance due to additive genetic variation).

Because V_P is also affected by the environment, any measure of heritability depends on the environment. Broad-sense heritability (H^2) reflects all genetic components to the phenotype. Narrow-sense heritability (h^2) directly estimates what is responsible for parent–offspring resemblance, because components such as dominance or epistasis are typically broken up and rearranged between generations. Hence, narrow-sense heritability, based on additive genetic variance, yields the evolutionary response to selection.

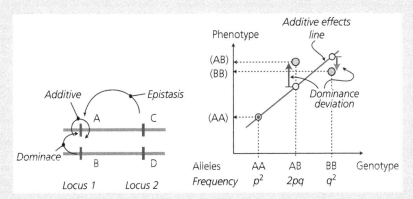

Box 10.3 Figure 1 Genetic effects on the phenotype. *Left*: Effects in a standard two-locus, two-allele system (A, B; C, D). Here, a phenotype (e.g. host resistance) is affected by genetic effects; the reference is allele A at locus 1 (red bar). The addition of the independent genetic effects from each allele at a locus yields the additive effect on the phenotype. When allele B suppresses the effect of allele A (either fully or partially), B exerts a dominance effect on the phenotype. If the presence of allele C can modulate the effect of allele A, then A and C have an interaction (epistatic) effect. *Right*: Graphical representation of how the phenotypic values (y-axis) emerge from the genotypes (x-axis). For simplicity, only one locus (with alleles A, B) is considered. The frequencies of the three possible genotypes (AA, AB, BB) are given as expected from the Hardy–Weinberg equilibrium (frequency of A: *p*; of B: *q* = 1 − *p*). The green dots are the observed phenotypes. If each additional allele B has the same incremental effect on the phenotype, the solid line will describe the phenotype belonging to each genotype (yellow circles; the additive effects on the phenotype). With dominance effects, the phenotypic values deviate by a certain amount from the additive line (blue arrows). Similarly, epistasis would lead to still further deviations from the values predicted by additive and dominance effects (not shown here).

heritability of $h^2 = 0.176$ for resistance and $h^2 = 0.188$ for infection tolerance (Mazé-Guilmo et al. 2014). The well-studied Soay sheep on St Kilda show heritable resistance to intestinal nematodes with $h^2 = 0.13$ (Hayward et al. 2014a). The economically important Pacific oyster has $h^2 = 0.49$–0.60 for resistance against oyster herpesvirus (OsHV-1). Elements of the immune defences are sometimes taken as a proxy for resistance, such as the enzymatic activity of the PO cascade in caterpillars with $h^2 = 69$, (Cotter and Wilson 2002), or $h^2 = 0.46$–0.54 for the PHA test in the common kestrel (Kim et al. 2013). Heritability to infectious disease has been studied in many food production systems. In these cases, the estimates are often based on pedigree relationship; they also vary considerably. For example, in the range of $h^2 = 0.01$–0.31 for resistance to white spot virus in shrimp production (Trang et al. 2019), $h^2 = 0.04$–0.11 for various infectious diseases in rabbits (Gunia et al. 2018), $h^2 = 0.11$–0.62 for resistance of Coho salmon (Barría et al. 2019), and $h^2 = 0.45$–0.62 for farmed rainbow trout against fish-infecting rickettsia (Yoshida et al. 2018).

The microbiota adds a twist to the story. Much of the genotypic variation in resistance may come from variation in the microbiota rather than from the host itself (Vorburger and Perlman 2018). Generally, this probably increases the amount of genotypic variation to the extent that the composition of the microbiota is itself varying with host genotype.

The estimate of heritability in pathogen characteristics, especially of infectious diseases, follows similar procedures but is based on traits that appear in the host as a result of the infection. Two major methodologies prevail (Mitov and Stadler 2018). On the one hand, 'resemblance estimators' use the similarity of parasite traits that emerge in individual hosts which are connected by transmission events, e.g. donor–recipient pairs. On the other, estimates of 'phylogenetic heritability' focus on parasite traits within hosts connected by a tree of phylogenetic descendence. These trees are reconstructed from parasite sequences, i.e. the closest phylogenetic pairs. However, pathogens such as bacteria or viruses are often clonal or near-clonal. Moreover, transmission causes bottlenecks such that only a small fraction of the pathogen's parental variation ends up in the next host. Finally, within-host evolution causes further deviations from

parental or transmitted types. Together, the fraction of genetic transmission is therefore highly variable and generally unknown, whereas for outbred, diploid organisms, this fraction is exactly 50 per cent. Hence, the biology of pathogen transmission typically violates the standard assumptions underlying the usual tools of quantitative genetics.

These deviations from classical quantitative genetics are implemented in improved methods. Examples are a modified analysis of variance (ANOVA) for close phylogenetic pairs (ANOVA-CPP) or the phylogenetic Ornstein–Uhlenbeck mixed model (POUMM) (Mitov and Stadler 2018). For example, using phylogenetic models, the heritability of set-point viral load of HIV (an essential trait for the disease progress to AIDS) was estimated to be 31 per cent (Blanquart et al. 2017). Similar estimates resulted when combined with data from transmission pairs. Also, the heritability of CD4[+] T-cell decline during HIV infections—a host trait variably induced by HIV genetics—was 17 per cent (Bertels et al. 2018).

10.5 Host–parasite genetic interactions

Populations genetics is an essential toolbox for the analysis of the ecological and evolutionary dynamics implicit in host–parasite genetics, i.e. to analyse the fate of genes in interacting and evolving host and parasite populations. For this purpose, the complexity of the underlying genetic architecture and gene expression has to be simplified and reduced. In this approach, a genome consists of separate loci—on the same or different chromosomes—that carry 'genes' for a given protein or function.

10.5.1 Epistasis

Epistasis is a fitness component that results from the combined effects of genes at different loci (Figure 10.3; Box 10.3). There are two different meanings of epistasis. 'Physiological epistasis' results from the fitness effects of specific allele combinations within an individual. Physiological epistasis is the immediate result of an interaction between an individual host and the parasites. 'Population-wide epistasis' results from the ensemble of individual physiological epistasis effects in a population. This reflects the overall fitness effect of the prevalent statistical association among alleles in the population (the population-wide linkage disequilibrium). Population-wide epistasis is particularly important in the context of host–parasite dynamics and co-evolution (see Chapter 14). In both cases, epistasis can be negative (a combination of alleles reduces fitness compared to their independent, additive effects) or positive (a combination of alleles increases fitness). Because of the practical difficulties of detecting epistasis empirically, its significance for host–parasite interactions is probably underestimated. However, there are many examples of physiological and population-wide epistasis (Table 10.1). In addition, there is an important link between population-wide epistasis, genetic linkage disequilibrium, and genetic recombination. This is because recombination

breaks genetic linkage disequilibrium and, therefore, changes the epistatic effects.

10.5.2 Models of genotypic interactions

A given parasite is not able to infect every host type within a population, nor every host species that it encounters. Likewise, a given host is not resistant to all possible parasites. Some of this variation is due to environmental factors or results from chance events. However, the host and parasite genotypes almost always matter. Cross-infecting host genotypes with parasite strains is a standard approach to studying these effects. The resulting host x parasite matrix of infections can be processed with ANOVA (Box 10.4).

Box 10.4 Cross-infection experiments

In a cross-infection experiment, several different host lines (Figure 1, five lines: A…E) are each experimentally exposed to several different parasite strains (Figure 1, six strains: nos 1…6). The outcome is observed. Parasite strains can be different clones or different isolates (an isolate is a parasite sample extracted from a given infected individual or a sub-population of hosts). The experiment yields a host x parasite matrix of infections that can be scrutinized with an ANOVA type of analysis. In the example, the entries to the cells of the host x parasite matrix are infection intensities in a given tissue (for example, cells per µl of tissue suspension). In the ANOVA, different host lines correspond to different levels of

the factor 'Host' (here, there are five levels) and, correspondingly, different parasite strains correspond to different levels of the factor 'Parasite' (six levels). If the outcome of the host–parasite encounter depends on varying the host lines, there will be a significant main effect of factor 'Host'. A host main effect describes the effect of differences among host lines, irrespective of the infecting parasite strain; it thus characterizes, for example, whether some host lines are generally more resistant than others. Similarly, when the outcome depends on the different infecting parasite strains and is independent of the host lines, a significant main effect for factor 'Parasite' would emerge. A parasite main effect tests

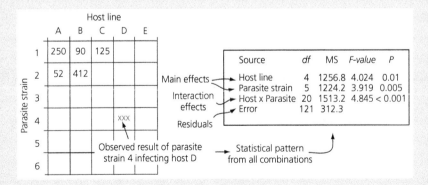

Box 10.4 Figure 1 The cross-infection scheme for a hypothetical example. *Left:* the table contains the observations of the outcomes for particular host and parasite combinations; entries indicate a fitness measure of interest; for example, infection intensity observed in a host (cells/µl). *Right:* the data are analysed with analysis of variance (ANOVA). The table shows the result with the source of variation, degrees of freedom (*df*), mean square values (MS), *F*-values, and the significance level (*P*) of the observed *F*-value. The analysis returns the main and interaction effects; the error term is the residual variance within a given cell of the host x parasite matrix that remains unexplained.

continued

Box 10.4 *Continued*

whether some parasite strains are generally more infectious than others. Finally, if the outcome depends on the particular combination of host line and parasite strain, the analysis will return a significant 'Host x Parasite'-interaction effect. Depending on the particular data, a standard ANOVA might be replaced by more sophisticated but equivalent tests.

An actual cross-infection experiment was done with the host *Daphnia magna*, a freshwater cladoceran, and its bacterial parasite, *Pasteuria ramosa*. *Daphnia* is a cyclic parthenogenetic species—it reproduces clonally during one part of the life cycle (in the summer months) and sexually in another (at the end of the season). In the experiment, host clonal lines and parasite isolates were from the same location and, therefore, the outcome of the infection should reflect the variation and interactions within the local population. As the experiment showed, each host–parasite combination produced

a slightly different result. These differences resulted from two of the three possibilities mentioned above. For example, host clones varied significantly in their propensity to become infected (the host main effect). Host clone B, for instance, was susceptible to almost all parasite isolates, while host clone E was mostly resistant. Variation also existed among the parasite isolates, but there was no statistically significant parasite main effect. Finally, the infection outcome depended on the particular combination of host clone and bacterial isolate, which yielded a statistically significant interaction term (Carius et al. 2001). None of the host clones or parasite isolates were universally resistant or universally infectious, respectively. Because *Daphnia* are genetically pure lines (clones) and the parasites narrowly defined isolates (representing one or a few bacterial strains at most), an implicit conclusion is that host and parasite genotypes affect the outcome of the interaction.

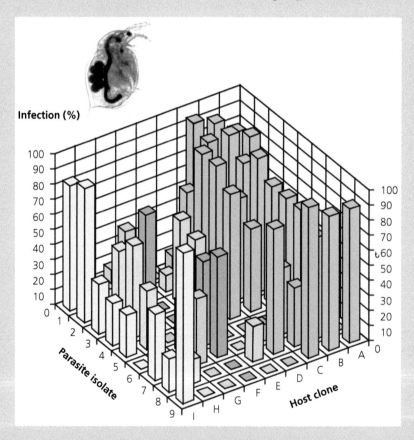

Box 10.4 Figure 2 Interaction between *Daphnia magna* and *Pasteuria ramosa*. The bars show the number of infected host individuals (infection in %), out of a sample of exposed hosts, and depending on parasite isolate (nos 1...9) and host clone (A...I). Bar colours symbolize host clone. Data analysed with a binary logistic procedure using mean deviance (with outcomes 'infected' vs 'non-infected'), which is similar to an ANOVA for continuous data. There is a significant main effect for host clones, no main effect for parasite isolates, but a significant host–parasite interaction term. Adapted from Carius et al. (2001), with permission from John Wiley and Sons.

If the outcome of the host–parasite interaction depends on 'who-is-infecting-whom', ANOVA should show a statistically significant host–parasite interaction effect. Host–parasite genotype x genotype effects are quite general and have been reported for many different systems, from microbes to plants, and from animals to humans (Table 10.5). Because of the overriding importance of genotypic interactions, a range of theoretical models trace the possible outcomes and explore the evolutionary dynamics of host–parasite interactions (Box 10.5). Of those, two models are prominent.

Table 10.5 Genetic interactions in different systems and at different levels.

Species	What was done	Finding	Prominent interaction type	Source
Flax (*Linum* sp.) vs flax rust (*Melampsora lini*)	Identification and sequencing of 13 host resistant alleles. Constructing recombinant plants with these.	Alleles providing resistance are precisely determined. Recombinants generate novel specificities against rust.	Gene-for-gene.	6
Flax (*Linum marginale*) vs rust fungus (*Melamspora lini*)	Cross-infection experiments with 67 wild flax plants and six isolates of rust.	Infection success can be classified into ten resistance types with one genotype entirely resistant.	Gene-for-gene.	2
Plants (*Glycine canescens*) vs fungus, soybean rust (*Phakopsora pachyrhizi*)	Nine races of the fungus tested against hosts from two populations. Crosses to determine genetic nature of resistance.	Variation within each host population. Effect based on single dominant gene, with an estimated 10–12 alleles.	Gene-for-gene.	1
Sugarcane (*Saccharum* sp.) vs fungus (*Bipolaris sacchari*)	Host clones of two sugarcane varieties experimentally inoculated with pathogen isolates.	Pairs of corresponding genes identified that associate with defence reaction and pathogenicity.	Gene-for-gene.	8
Wheat vs fungus (*Mycosphaerella graminicola*)	Statistical analysis of six data sets with 80 pathogen isolates and 47 host cultivars. Measures were presence or absence of necrosis and formation of fungal pycnidia.	Considerable genetic variance found for host and parasite. Around 25 per cent of variance explained by specific interaction.	Gene-for-gene, but also quantitative.	9, 10
Snail (*Biomphalaria glabrata*) vs trematodes (*Schistosoma mansoni*)	Strains of snails selected for resistance or susceptibility for two parasite strains.	Resistance against selected strain is heritable and does not affect resistance to other strain.	Quantitative.	15, 16
Snail (*Potamopyrgus antipodarum*) vs trematodes (*Microphallus* sp.)	Experimental infection of snail clonal lines with parasite isolates. Observing effect in juveniles and adults.	Effect of snail clone, life history stage, and condition. Rare snail clones generally less infected. Condition did not alter this rank order of susceptibilities.	Quantitative.	4
Moth larvae attacked by parasitoid (*Microplitis* sp.)	Experimental injection of calyx fluid from parasitoid into hosts and monitoring success of parasite larva.	Infection success depends on symbiotic virus in calyx fluid, which provides host-specific protection for the developing parasite.	Quantitative, among species.	7
Drosophila melanogaster vs parasitoids (*Leptopilina boulardi*)	Experimental infections of resistant and susceptible *Drosophila* strains with virulent and avirulent parasitoid strains.	Success of cellular immune response against parasitoid depends on host and parasite genotype.	Quantitative.	3
Drosophila melanogaster vs bacteria (*Serratia marcescens*)	Experimental infection of different host (homozygous) lines and study of polymorphism for candidate genes associated with immune responses.	Genetic variation among host lines correlates with infection success.	Quantitative.	12
Bumble bee (*Bombus terrestris*) vs trypanosome (*Crithidia bombi*)	Different parasite clones infected and passaged among different host colonies.	Infection success depends on the specific combination of parasite clone and host colony.	Quantitative.	14

(Continued)

Table 10.5 Continued.

Species	What was done	Finding	Prominent interaction type	Source
Atlantic salmon (*Salmo salar*) vs *Aeromonas salmonicida*	Juveniles infected with bacterium and survival recorded.	Survival depended on specific alleles present at MHC locus.	Qualitative, depending on allele.	13
Pig (*Sus domesticus*) vs bacteria (*E. coli*)	Experimental infection of pig lines with parasite.	Resistance depends on combination; pattern due to few genes.	Similar to gene-for-gene.	5
Rodents vs bacteria (*Bartonella*)	Experimental inoculation of mice and rates with bacteria.	Bacteraemia only occurs when inoculated with pathogen from same or closely related species.	Quantitative, among species.	11

Sources: [1] Burdon. 1987. *Diseases and plant population biology.* Cambridge University Press. [2] Burdon. 1991. *Evolution,* 45: 205. [3] Carton. 2001. *Immunogenetics,* 52: 157. [4] Dybdahl. 2004. *J Evol Biol,* 17: 967. [5] Edfors-Lilja. 1991. In: Owen, eds. *Breeding for resistance in farm animals.* CABI Publishing. [6] Ellis. 1999. *Plant Cell,* 11: 495. [7] Kadash. 2003. *J Insect Physiol,* 49: 473. [8] Kang. 1987. *Plant Dis,* 71: 450. [9] Kema. 1996. *Phytopathology,* 86: 213. [10] Kema. 1996. *Phytopathology,* 86: 200. [11] Kosoy. 2000. *Comp Immunol Microbiol Infect Dis,* 23: 221. [12] Lazzaro. 2004. *Science,* 303: 1873. [13] Lohm. 2002. *Proc R Soc Lond B,* 269: 2029. [14] Schmid-Hempel. 1999. *Evolution,* 53: 426. [15] Webster. 1998. *Proc R Soc Lond B,* 266: 391. [16] Webster. 1998. *Evolution,* 52: 1627.

Box 10.5 Genetic interaction models

There are several idealized models for how the genotypes of host and parasites interact. They are illustrated for the standard two-locus, two-alleles model (Figure 1).

Gene-for-gene (GFG): In its standard form, this model classifies the outcome of infection as resistant or susceptible. Resistance occurs when at least one parasite 'avirulence' gene is matched by a host 'resistance' gene. Parasites have some kind of elicitor that the host can recognize to prevent infection. Gene-for-gene interactions are asymmetric because a 'universally' infective/virulent parasite type (*AB*) and a universally susceptible host type (*ab*) exist.

Inverse gene-for-gene (IGFG): This model assumes that the parasite needs to recognize the host to become infective. Hosts can become resistant by losing receptors that the parasite can recognize. Thus, resistance occurs when hosts have a resistance gene or—if they have a susceptible gene—the parasites happen to have an avirulence gene at the same locus. Some of the systems that traditionally are analysed as GFG could be better modelled as IGFG genetics (Fenton et al. 2009) (Figure 1b). The IGFG is a mirror version of the classical GFG scenario but does not show the same dynamical behaviour. In this case, the parasite can only infect if it recognizes a receptor in the host. The task of achieving infection, therefore, is with the parasite. Therefore, there are no universally resistant hosts or universally infective parasites (Fenton et al. 2009). Moreover, in simulations of this model, co-evolutionary changes can occur in bouts; that is, with periods of stasis interrupted by rapid changes in the frequency of parasite virulence and host resistance alleles.

Matching-alleles model: This model suggests that self- vs non-self recognition acts like a lock-and-key system. Hence, infection is not possible unless the parasite possesses all or at least some alleles that 'match' those of the host (Figure 1c). The *general matching-alleles model* is an extension of the matching-allele models; the selection co-efficient for the host depends on how many loci are matched by the parasite, as shown in Figure 1c.

Linkage-based model (Nee model): In this model, the linkage between alleles at the two loci is essential. A parasite 'coupling genotype' (*ab*, or *AB*) infects when the host also has a coupling type (*ab*, or *AB*), but it cannot infect when the host has a repulsion type (*aB*, or *Ab*); and vice versa for parasite repulsion types. The model is similar to a matching type because the parasite is successful if it matches the linkage type, whereas the host is successful if it does not match the linkage type (Figure 1d).

(a) Gene-for-Gene

Host genotype

	ab	aB	Ab	AB
ab	0	x	x	xx
aB	0	0	x	xx
Ab	0	x	0	x
AB	0	0	0	0

Parasite genotype

(b) Inverse Gene-for-Gene

Host genotype

	ab	aB	Ab	AB
ab	xx	xx	xx	xx
aB	x	xx	x	xx
Ab	x	x	xx	xx
AB	0	x	x	xx

Parasite genotype

(c) Matching alleles

Host genotype

	ab	aB	Ab	AB
ab	0	x	x	xx
aB	x	0	xx	x
Ab	x	0	x	x
AB	0	x	x	0

Parasite genotype

(d) Linkage based (Nee)

Host genotype

	ab	aB	Ab	AB
ab	1	0	0	1
aB	0	1	1	0
Ab	0	1	1	0
AB	1	0	0	1

Parasite genotype

Box 10.5 Figure 1 Interaction models for two loci (*A,B*) with two alleles each (*A,a* and *B,b*). The tables show the degree of resistance for any combination of host and parasite genotype, coded as no resistance (0), resistance by one allele (x), resistance by two alleles (xx). The phenotypic effects are ordered as 0 < x < xx. The models are: (a) Gene-for-gene: The parasite has avirulence (*a,b*) and virulence alleles (*A,B*). The host has susceptibility (*a,b*) and resistance alleles (*A,B*). Resistance occurs when the host recognizes the avirulence signal (allele) from the parasite (i.e. the host has the corresponding resistance receptor allele). (b) Inverse gene-for-gene. Alleles are the same as in gene-for-gene. However, infection occurs when the parasite recognizes the host. Resistance occurs when the host has a corresponding resistance allele, or when parasite has an avirulence allele and the host a susceptible allele. A host resistance allele means the absence of a receptor that the parasite can recognize. (c) Matching alleles: Resistance occurs when parasite alleles correspond to those of the host. (d) Linkage-based interaction models for two loci (*A,B*) with two alleles each (*A,a* and *B,b*). The alleles *a* and *b* at the two loci can either be coupling types (*ab* or *AB*) or repulsion types (*aB* or *Ab*). Resistance occurs when host and parasite are of the same linkage and different types. Adapted from Salathé et al. (2008a), with permission from Elsevier.

10.5.2.1 Gene-for-gene interaction (GFG)

This model originally comes from the classical studies by Harold Flor on flax (*Linum marginale*) and its rust fungus (*Melampsora lini*) in Australia (Flor 1956). In plants, the outcome of an infection is often classified by the occurrence of lesions (yes/no) on leaves in a dichotomous scheme (susceptible/resistant). This is useful in practice but somewhat artificial, and will not distinguish between elements of infectivity, parasite multiplication within hosts, or degrees of virulence.

The GFG model assumes a genetically determined factor from the parasite (the elicitor, with a dominant 'avirulence' allele). A genetically determined factor recognizes the parasite factor from the host (the receptor; a dominant allele) such that only one combination of host and parasite factors results in resistance. If recognition by the plant occurs, the (non-specific) hypersensitive response unfolds in the tissue, and the pathogen attack is resisted. Infection is possible when the receptor does not match the elicitor (no recognition), or when the parasite does not produce an elicitor that can be recognized (Figure 10.11). GFG models are, therefore, asymmetric in their outcome. The model predicts a universally virulent parasite type (a 'super-parasite'),

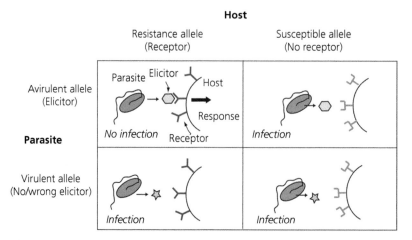

Figure 10.11 The gene-for-gene model as envisaged for plants. The avirulent allele in the parasite codes for an elicitor (green hexagon) recognized by a corresponding receptor in the plant (blue). With recognition (top-left panel) an immune response is elicited and resistance occurs. If the plant receptors cannot recognize an elicitor (all other panels), no response is triggered and infection can take place. Non-recognition can be due to absence of a receptor, a mismatch with the elicitor (top-right panel), or a new elicitor (lower panels; yellow asterisk).

and a universally susceptible host type (Figure 10.11; Box 10.5). However, such parasites and hosts are not known, even though some pathogens have an impressively wide host range (e.g. *Bacillus thuringiensis, Metarhizium anisopliae*). Therefore, gene-for-gene models are good approximations for some cases, but not a general model of host–parasite interactions.

The gene-for-gene model is a suitable description for some plant–pathogen interactions (Thrall et al. 2016). Plant genomes contain hundreds of receptor genes that recognize specific parasite components (Brown and Tellier 2011; Cui et al. 2015; Boutrot and Zipfel 2017). For example, resistance to *Zymoseptoria tritici*, a fungal disease of wheat, involves the plant *Stb6* gene. This gene encodes a protein (WAK-like) that recognizes the presence of a matching parasite effector protein and mounts a response; in this case, this is not a hypersensitive response (Saintenac et al. 2018). Of course, plant resistance is generally more complex than in this simple model and involves several loci, or quantitative resistance; that is, a gradual response rather than just resistance or susceptibility. Some insect–plant systems also show patterns very similar to GFG interactions, with resistance based on a few major genes. Examples include the Hessian fly on wheat (Aggarwal et al. 2014), gall

midges (Bentur et al. 2016), or the brown planthopper (*Nilaparvata lugens*) on rice (Kobayashi 2016). The resistance of *Daphnia* against a bacterial infection unfolds over several successive steps in the infection cascade. An early step (attachment to the oesophagus) shows qualitative resistance (all-or-nothing effect), which can assume the properties of a gene-for-gene interaction. In contrast, later steps are more quantitative (Hall et al. 2019).

10.5.2.2 Matching specificities (matching alleles)

This model assumes resistance to depend on whether genetically determined factors of the parasite match genetically determined factors on the side of the host, similar to a lock-and-key mechanism. A 'matching' between host and parasite alleles will result in infection or resistance, depending on how the model is formulated. Physical examples of such factors would be pattern-recognition proteins (PRPs) that bind to conserved epitopes, the pathogen-associated molecular patterns (see Chapter 4).

In many cases, a matching-allele model is a good description of the situation. For example, in *Daphnia magna*, a single locus (the PR locus) is polymorphic (has several alleles) and provides resistance against the bacterial pathogen *Pasteuria ramosa*. The interaction follows the model, and whole-genome sequencing, combined with transcriptomics, has identified the physical location of this host locus and the actual genes associated with it (Bento et al. 2017). Similarly, *Drosophila melanogaster* has two major genes that code for the ability of the host to either encapsulate the eggs of the endoparasitic wasp *Leptopilina boulardi* or eggs of *Asobara tabida*. Both genes are located on the second chromosome (Poirie et al. 2000). In turn, the ability of the wasps to escape the encapsulation depends mainly on the presence of different strains of endosymbiotic virus-like particles transferred into the host during egg laying. The virus-like particles show Mendelian inheritance and act as a single segregating gene, i.e. like alleles that either match or do not match the host alleles (Dupas et al. 1998). Together, host and parasitoids (their viruses in this case) act like a lock-and-key system to allow for infection or resistance.

Resistance can follow different model scenarios at different steps (Fenton et al. 2012). At the same time, the underlying molecular and immunological mechanisms can blur the distinction between models, even though basic properties may be present, e.g. 'matching infections' based on a correspondence of genetic factors on either side (Dybdahl et al. 2014). In most real situations, too, host and parasite genotypes do not interact strictly according to any of these models, and the effects are rarely symmetric. Nevertheless, the different interaction models matter, since they have different effects on the host–parasite co-evolution. For example, for whether and how genetic polymorphism is maintained in the host population (Engelstädter 2015; Thrall et al. 2016). The matching-allele model underlies much of the theory of host–parasite antagonistic co-evolution, especially in the context of the Red Queen scenario (Hamilton 1980) (see Chapter 14).

10.5.3 Role of the microbiota

The microbiota has its own 'genotype' or, rather, millions of genotypes present in the microbial symbionts. To some extent, the host genotype determines which symbionts can become part of the microbiota and, therefore, which genotypes are present. This is particularly the case for the 'core microbiome' (see section 4.8). This core could be regarded as an add-on to the host genotype. Its effects on the interaction would mostly be assigned to the host genotype. At the same time, a large part of the microbiota is not firmly associated with the host genotype (the 'transient microbiome'). It therefore has mostly independent effects on resistance or tolerance. Together, a tripartite interaction (parasite–host–microbiome) occurs, which shapes the ecological and evolutionary dynamics of host–parasite systems (King and Bonsall 2017).

10.6 Signatures of selection

Selection by parasitism will leave traces in the host (and, vice versa, in the parasite) genome, known as 'signatures of selection'. A recurrent theme in the search for such signatures is distinguishing the effects of selection from those of population history, including neutral genetic drift and mutation. Conveniently, population history leaves traces somewhat different from those of selection. This is because historical events, such as bottlenecks or migration episodes, affect the entire genome. By contrast, selection by parasites generally affects only a few loci. Several methods, therefore, exist to detect signatures of selection, especially also in the context of genome-wide scans (Box 10.6).

It is of fundamental interest to assess how selection changes the frequencies of different variants—or, genetically speaking, of different alleles. Selection can act 'positively' or 'negatively'. In this case, the selection is 'directional'. If positively selected, one variant is favoured consistently over others, such that a given allele increases in frequency over evolutionary time. It may eventually become 'fixed' in the population; i.e. it is the only allele left. The result is a change in the population mean. Selection on a trait can also be 'stabilizing'; that is, it favours existing variants around the population mean and removes deviating ones. Furthermore, selection can also be 'balancing', such that over time different variants are maintained by selection (i.e. a genetic polymorphism persists). In this case, the advantages and disadvantages of different variants can balance one another when in different combinations, e.g. with a heterozygous advantage, which protects them from extinction. Alternatively, 'disruptive' or 'diversifying' selection occurs when the two extremes in a distribution of traits are favoured.

Box 10.6 Signatures of selection

Phylogeny of haplotypes: Positive selection of a genomic region reduces nucleotide diversity and leads to fewer but more common haplotypes. This can be seen from phylogenetic haplotype networks.

Genetic divergence: Tajima's *D* test uses the observed nucleotide diversity, π, within a population or between species, within many segregating sites; i.e. looking at the number of positions in the sequence that have at least one different nucleotide across the sample. Under neutral evolution, the expected value of this diversity is $\theta = 4N\mu$, where N is the effective population size and μ is the (neutral) mutation rate. This expression stems from the assumption that every new mutation (a change in the nucleotide) happens at another locus in a so-called 'infinite site model'. In the practical test, the estimated value, π, is compared to the expected value, Θ, to calculate the test statistic, Tajima's *D*. If *D* is high, the difference between observed and expected (and thus the value of *D*) is too large for neutrality. Positive selection in this population is inferred.

A modification of this test uses phylogeny to give polarity to the character states, i.e. to define whether a nucleotide is ancestral or has been introduced by mutation (Fu and Li 1993). In the *HKA test (Hudson–Kreitman–Aguadé test)* (Hudson et al. 1987), a sample from *m* loci from different populations or species is considered. At each of these *m* loci, the genomic sequences are compared, and the differences in nucleotides at this position are counted. The number of loci that are different between a pair of species (or populations) is then compared to the average number of nucleotide differences between the two species (or sequences). The test statistic is calculated from these values. Positive selection is inferred when the differences are too large. The general assumptions of the HKA test include constant population sizes and no recombination within a locus, but recombination between loci.

Synonymous vs non-synonymous mutations: Not all changes in a nucleotide sequence of a coding region have the same effect. Due to the redundancy in the genetic code, some changes are synonymous (S), i.e. have no effect on the amino acid that is coded by a nucleotide triplet. This typically is the case for the third position in the triplet. By contrast, some changes lead to a different amino acid, i.e. are non-synonymous (N), as is typically the case for the first and second positions of a triplet. Because a non-synonymous change leads to a different amino acid and thus to a different protein, such a change can potentially be selected for or against. A synonymous change is typically neutral in terms of selection. Once the reading frame for a genomic sequence (defining the first, second, and third positions for each triplet) is known, this difference detects signals of selection from a genomic sequence.

The basic idea is that the neutral synonymous changes will reflect the effects of drift alone, while the non-synonymous changes additionally reflect the effects of selection. Hence, the comparison of changes at synonymous and non-synonymous sites reveals whether or not selection has happened. This test, therefore, is a convenient tool to study protein evolution. In practice, for each pairwise comparison, the number of non-synonymous changes per total possible non-synonymous site (d_N) and synonymous changes per total synonymous sites (d_S), is calculated to yield the respective ratio (d_N/d_S). Under neutral evolution, the two kinds of substitution should have the same (neutral) consequences and therefore evolve similarly (Hughes and Nei 1988). Hence, under neutral evolution we expect $d_N/d_S = 1$. If the sequences are under positive selection, non-synonymous should outweigh the synonymous changes and hence $d_N/d_S > 1$. The opposite is true for negative selection and, hence, $d_N/d_S < 1$. An extension of this logic compares the ratio of non-synonymous to synonymous mutations fixed between species to the ratio polymorphic within species through the McDonald–Kreitman test (MacDonald and Kreitman 1991).

Analysis of allele frequencies and linkage patterns: The distribution of allele frequencies at a locus (from common to rare alleles) reflects the kind of selection that has occurred. Negative selection, for example, removes disadvantageous alleles and leaves fewer, but more common, alleles in the population when compared to neutrality (Figure 1a). Allele frequencies also change when a selective sweep has occurred. In a selective sweep, an allele at a given locus (or a short genomic region) provides a fitness advantage and the frequency of the favoured allele increases rapidly in the population and goes to near or full fixation. A positively selected sweep increases the frequency of rare alleles beyond what is expected under neutrality (Figure 1a). At the same time, a locus that has undergone a selective sweep will show reduced nucleotide diversity. The sweep will also leave a trace on adjacent sites (the 'hitch-hiking effect', Maynard Smith and Haigh 1974) because neighbouring genomic sites are always somewhat linked to the locus under selection. The linkage patterns along the genome, measured as local linkage disequilibria (LD), can therefore be used as evidence for a signal of selection (Figure 1a,b). Under positive selection, there is local LD; under balancing selection, the local LD is reduced—provided selection had sufficient time to establish the polymorphism (if balancing selection is recent, LD might also transiently be increased).

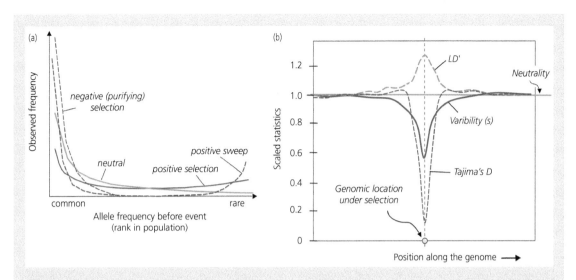

Box 10.6 Figure 1 (a) Frequency spectrum of alleles (from rare to common) and expected changes. Under neutral selection (no effects), the spectrum follows an expected distribution, as characterized by the green line. Before an event, alleles that will become selected are typically rare. Then, negative selection removes rare, disadvantageous alleles and leaves few common ones (dashed blue line). Positive selection favours rare alleles at the expense of others (solid red line). A selective sweep is an extreme case of this and exaggerates the increase in previously rare alleles (dashed red line). (b) Expected values of statistical measures (y-axis, scaled) in the vicinity of a location under positive selection (yellow dot), and in relation to distance (x-axis; position). Near the selected locus, the number of variable sites, s (solid red line), becomes reduced; the allele frequency spectrum, measured by Tajima's D (dashed red line), is skewed, as in (a). At the same time, there is a local increase in linkage disequilibrium (LD', dashed blue line) due to the hitch-hiking effect.

As a result, the population will lose the intermediate variants in favour of the two extremes, and the distribution becomes bimodal. In the longer run, a lineage can become split into two. Finally, an essential type of selection is 'negative-frequency-dependent selection', where rare types are favoured, and common types are selected against. All of these phenomena can be found in animal populations, as the following examples show.

10.6.1 Selection by parasites in animal populations

Genes in the immune system are often the fastest-evolving genes in host organisms, e.g. in mammals (Kosiol et al. 2008), birds (Ekblom et al. 2010; Sironi et al. 2015), water fleas (McTaggart et al. 2012), and in insects and nematodes (Palmer et al. 2018). In particular, genes involved in host defences have high rates of amino acid substitutions (Hurst and Smith 1999; Schlenke and Begun 2003). These cases also show signatures of positive selection, i.e. the

spread of favourable mutations, suggesting that selective pressure by parasites forces the hosts to adapt and drives the evolution of immune defences. Among the various components of the immune system, positive selection seems especially strong in the RNAi pathways involved in defence against viruses and mobile genetic elements. Across various insect lineages, for example, the coding regions for the antiviral pathways show high levels of sequence divergence—a result of positive selection acting in different directions (Palmer et al. 2018).

A temporal dimension can be added, for instance, with strong signals of positive selection in opsonins (thioester-containing proteins, or TEP) between *D. melanogaster* and *D. simulans*, two species that diverged *c.*2.5 million years ago (Jiggins and Kim 2006). The more detailed studies from species used in aquaculture or animal husbandry support the overall picture. Salmon, for example, show widespread positive selection on immune defence-related genes, such as for antiviral pathways or lymph node regulation (Zueva et al. 2018). Similar observations are made for cattle, bred for yield and disease resistance (Onzima et al. 2018), or for genes associated with resistance to intestinal nematodes in sheep and goats (Estrada-Reyes et al. 2019).

Overall, there are many examples of positive selection on host defence genes, as well as for parasite genes involved in evading host defences (Ford 2002). In other cases, diversifying (balancing) selection prevails. Examples are the immune genes of the MHC complex (representing adaptive immunity) and the TLR-receptor family (representing innate immunity) across different lineages of birds. In both cases, signatures of intermittent, periodic diversifying selection exist and are found at all levels (single codon sites, overall genes, within lineages). Generally, the effects are more pronounced in the MHC locus than otherwise (Antonides et al. 2019). In this context, non-passerine birds showed more robust signs of selection in MHC class II genes (defence against extracellular parasites). In contrast, passerines showed more substantial signs in the MHC class I genes (against intracellular parasites) (Minias et al. 2018), the reasons for this difference being unclear.

Balanced polymorphism seems quite common in immune systems. In fact, several evolutionary mechanisms result in balancing selection, including heterozygous advantage, negative frequency-dependent selection, or epistatic mutations and linkage that can 'shield' deleterious alleles from becoming extinct (Llaurens et al. 2017). However, balancing selection can sometimes be challenging to observe. For example, *Drosophila melanogaster* originated in Sub-Saharan Africa some 15 000 years ago (Li and Stephan 2006). This allows study of contemporary populations from the African site of origin together with the descendant populations in Europe or North America. Genome-wide scans found very few genes under balancing selection, and this was not due to the confounding effects of demography or population history (Croze et al. 2016, 2017). A possible reason is that much of the host–parasite dynamics is on a short scale and moves in various directions, such that it leaves fleeting signals in the host genome. Only with persistent balancing selection over more extended periods does a detectable signature emerge, as observed in ancient MHC polymorphisms that transcend species boundaries between humans and chimpanzees (Leffler et al. 2013) (see Figure 14.9). Subtle differences in the genomic sequence among the long-separated *Drosophila* populations from Africa, Europe, and North America indicate that a few AMPs, especially diptericin, show a pattern consistent with balancing selection against their genomic background (Chapman et al. 2019). AMPs are an essential class of effectors in the innate immune system (see section 4.4). They should be under selection by the parasites that have evolved means to escape their effects. Overall, however, AMPs are relatively well conserved (Rolff and Schmid-Hempel 2016).

10.6.2 Selection by parasites in human populations

Ever since modern humans evolved from their hominid ancestors, major infectious diseases have been present (Karlsson et al. 2014; Andam et al. 2016). Not surprisingly, therefore, signatures of selection are present in today's human genomes. For example, positive selection has favoured a mutation that disrupts the expression of Duffy antigens by *Plasmodium vivax* (causing malaria tertiana), and which has now become nearly fixed (i.e. is present to almost 100 per cent) in Sub-Saharan African populations (Kwiatkowski 2005). Infectious diseases, such as cholera, are particularly deadly, with premodern case mortality rates of more than 50 per cent (Harris et al. 2012). Some human populations are still exposed to cholera, such as in the Ganges river delta of Bangladesh. Genome-wide scans suggest that this population has undergone selection that adds to protection because clear signals of positive selection are found in immune genes involved in NF-κB signalling and in genes that regulate the ion flux from cells. The latter is particularly important, as water loss and the resulting diarrhoea are major symptoms of cholera and can have a fatal outcome (Lee et al. 2012). There are many hundreds of human protein-coding genomic loci that have been positively selected by diseases, and many more are 'non-coding' regulatory loci (Barreiro and Quintana-Murci 2010; Karlsson et al. 2014).

Balancing selection is also common in human populations. Well-known examples of human genetic polymorphism, already mentioned elsewhere, are the cases of sickle cell anaemia and MHC-allelic diversity. Also, the diversity of HLA (known as the MHC locus) correlates with pathogen diversity (Prugnolle et al. 2005). There is also the expected association between the binding properties of parasite-derived peptides (a primary function of the MHC locus) and the sequence diversity at the locus (Pierini and Lenz 2018).

10.6.3 Signatures of selection in parasites

Just as with the hosts, parasites are also under selection, and their genomes also show ample signatures of selection. There is also evidence for positive selection in parasite genes involved in evading host defences. For example, whereas human hosts are selected to suppress *Plasmodium*-antigen presentation on the surface of red blood cells, the pathogen shows selection on genes related to blood cell invasion and immune-response evasion (Mobegi et al. 2014; Shen et al. 2017). Even more tellingly, strong signals of selection in *Plasmodium* are associated with medical interventions. Genes providing resistance against commonly used drugs, such as chloroquine, thereby rise in frequency. This concerns the parasite's chloroquine resistance transporter (CTR), or the multidrug resistance region (MDR1) (Oyebola et al. 2017), plus several other genes related to parasite success (Oyebola et al. 2017).

10.7 Parasite population genetic structure

10.7.1 Determinants of structure

Host and parasite populations are genetically structured, and this structure keeps changing over time. In general, many ecological processes affect population genetic structure. These include dispersal, migration, population bottlenecks, microhabitat preference (including host specificity), mate choice, or general life history characteristics. Together these result in genetic consequences such as gene flow, gene drift, or assortative mating. Parasite populations, nevertheless, have a peculiar structure. They typically are composed of many subpopulations that reside inside the single hosts, which makes it more like a metapopulation overall. Among the many characteristics for structure, the 'effective population size' is an indicator for the extent of genetic variation in populations. This number indicates what size (number of individuals) a given population effectively has when 'judged' by the laws of population genetics. These laws determine, for instance, the response to selection and refer to a 'standard' population of diploid, outbred individuals. However, a highly inbred population genetically does not count as consisting of fully independent individuals. Instead, its population genetic properties are equivalent to a much smaller number of 'standard' outbred individuals. This lower number is the 'effective population size'. A given population, therefore, behaves in the standard way, like a population of effective size.

Viruses, for example, have enormous population sizes in terms of virus particles. However, population bottlenecks during transmission and purifying selection inside the host reduce the number of independent genotypes. Therefore, viruses often have surprisingly small effective population sizes, orders of magnitude lower than the number of viral particles counted in a population (Kouyos et al. 2006; Dolan et al. 2018; McCrone and Lauring 2018). Parasite population genetic structure, including the potential to respond to selection, is essential, though. Knowing these variables allows epidemics to be traced (Ciccozzi et al. 2019). Also, the success of medical interventions depends on the parasite population's evolutionary potential to respond to selection.

For the genetic population structure of many parasites, the host movement is a limiting factor, as they cannot disperse very far on their own. A comparative study on nematodes, for example, found that nematodes infecting very mobile hosts, such as ocean-going mammals, show virtually no population structure and spread wide and far. By contrast, those infecting hosts with localized populations and restricted movements, e.g. in small rodents, are highly structured and differentiated (Cole and Viney 2018). Strong geographical signals of clustering along with units such as cities, nations, and regions, are also evident in lineages of the human pathogen, *Staphylococcus aureus*. These signals correspond to the pattern of human movements (Andam et al. 2017). In fact, viral populations

(White et al. 2018b; Scherer et al. 2020) and those of many other parasites (Daversa et al. 2017) are also structured by host movements. Vector-borne transmission can modify this picture because bottleneck effects within a vector and selection for transmissibility affect the population genetic structure (Chisholm et al. 2019), although they do not fundamentally alter it. Vectors are typically effective at short-range distances. Their geographical distribution nevertheless remains important, similar to host specificity for transmission between—rather than within—host species. Some microbial pathogens can decouple themselves from host movements, e.g. *Bacillus anthracis* spores that can survive for decades in the soil. However, for most, host movements are critical for their dispersal.

10.7.2 Genetic exchange in parasites

The transfer of mobile genetic elements is an essential mechanism for genetic exchange among different pathogen lineages; for instance, in bacteria. As mentioned elsewhere, this fosters the exchange and spread of pathogenicity islands that carry virulence factors favouring successful infection, or elements that carry genes for antibiotic resistance. Genetic exchange can also occur when individuals of two different genotypes co-infect in the same host and 'mate' with one another. The rate at which this happens shapes the genetic structure of their populations, from purely clonal (with no exchange) to nearly or fully outbred (with obligatory exchange).

Some of the medically most important parasites are protozoa. The population genetic structure in parasites such as *Plasmodium falciparum*, *Trypanosoma cruzi*, *T. brucei*, and *Leishmania* across geographical areas and among patients is thus of considerable interest. The so-called 'clonal theory of parasitic protozoa' suggests that protozoan populations are basically clonal (Tibayrenc et al. 1990; Tibayrenc and Ayala 2002). This view, among other things, builds on the observation of genetic variation found in protozoan parasite populations. For example, the level of heterozygosity at single loci, and genetic linkage between loci, tests whether the observed values deviate from a Hardy–Weinberg equilibrium, which would result from random mating among individuals. Many protozoan populations

indeed show deviations from this equilibrium and have thus been considered clonal rather than outbred. However, our understandings of parasite 'strains', and 'clonal' population structures have made steady progress over the last decades (Heitman 2006; McKenzie et al. 2008). For instance, parasite populations typically represent isolates from individual hosts. These are therefore highly structured samples—not only in space and time but also concerning parasite genotypes in any given host—a result of host–parasite genotype–genotype interactions. A collection of parasite isolates is therefore unlikely to be in Hardy–Weinberg equilibrium (indicating random associations), even if panmixia (random mating between all individuals) were otherwise the normal condition. This problem is also known as the 'iceberg effect' because only the tips of the genetic structure of the entire parasite population are represented in the single hosts (Tibayrenc 1999).

On the other hand, the exchange of genetic material occurs in several protozoa (Gibson et al. 2017). Examples include *Trypanosoma brucei*, *T. cruzi* (Heitman 2006), *Leishmania major* (Akopyants et al. 2009), and the closely related bumblebee-infecting *Crithidia bombi* (Schmid-Hempel et al. 2011). Similarly, *Plasmodium* can undergo a sexual cycle with gametocytes that subsequently fuse to a zygote; this also can happen by self-fertilization (Hall et al. 2005). Besides, many other protozoan parasites (belonging to the Apicomplexa) have gametocysts and a sexual phase in their life cycle, e.g. *Cryptosporidium*, *Eimeria*, *Toxoplasma*, *Hepatozoon* (West et al. 2000). Sequencing projects of various parasites (e.g. *Giardia*) have furthermore turned up conserved genes known to be involved in meiosis, suggesting that many protozoans might be capable of cryptic sexual reproduction (Heitman 2006). Whether the underlying mechanism is always regular meiosis is often unclear, and alternative modes might exist (Gibson and Stevens 1999) (Figure 10.12). Nevertheless, not all protozoans are clonal/asexual by their very nature.

An occasional genetic exchange might routinely occur after more extended periods of clonal propagation—a form of 'epidemic clonality' (Maynard Smith et al. 1993). Depending on the relative frequencies of these events, the resulting population

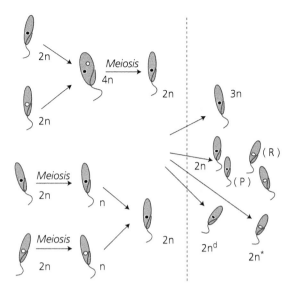

Figure 10.12 Two models of genetic exchange in trypanosomes. *Top row*: Fusion between two diploid cells (2n) yields a 4n cell (tetraploid), followed by meiosis and reduction to 2n daughter cells (diploid). *Bottom rows*: Meiosis occurs first, yielding haploid (n) cells, followed by fusion to 2n daughter cells. *Right*: Genetic analyses show that a number of products can emerge from these processes; in particular: a 3n-cell (triploid), parental (P) and recombinant (R) 2n-cells, and 2n cells of degenerate type that have either lost an allele ($2n^d$) or possess a novel allele ($2n^*$). Examples would be *Trypanosoma brucei* and *T. cruzi*. The filled and open symbols represent the nuclei. Adapted from Gibson and Stevens (1999), with permission from Elsevier.

structure might be anything from near-clonal to almost panmictic. The concept of 'strains' is useful to characterize variants of pathogens that differ in their properties and effects. Strains essentially refer to different pathogen genotypes. However, a given strain might thus only exist for a limited time, before being 'destroyed' by genetic exchange or being replaced by more successful variants on the epidemiological time scale. It might even be difficult to recover the same parasite strains more than once or twice, especially in areas of high transmission. In fact, a population structure is closer to clonal in low-transmission areas, as observed in the example of malaria (Razakandrainibe et al. 2005). Therefore, 'strains' are probably more like temporary collections of mutable, dynamic genetic entities that express certain antigenic or virulence characteristics. During some periods, strains might build up near-clonal populations, but these remain highly dynamic, with genetic exchange and mutations as essential processes of diversification. In viral populations, the concept of 'quasi-species' is one instance of such a structure (see section 12.3.1). Quasi-species are collections or swarms of slightly different variants that surround a 'master' genotype. Collectively, swarms can adaptively respond to selection much faster. Quasi-species is a controversial concept. Nevertheless, it is often useful for epidemiology and the prospects for medical interventions in viral infections (Dolan et al. 2018). It could also conceptually serve as an approximate tool for protozoan (Gibson et al. 2017) or other parasite populations.

Important points

- Genetics is a key for host–parasite interactions. Modern genomic methods cover diagnostics, sequencing, association studies, and the editing and change of genotypes. Genes responsible for host resistance or tolerance can thus be studied. SNPs are often used as markers for finding essential genes.
- The genetics of parasites varies among groups. Horizontally transferred pathogenicity islands are important in bacteria. Viruses can have linear or circular genomes. They also vary in mutation rates.
- Heterozygous individuals are often more resistant to infections. Similarly, more genetically variable populations often have lower parasite burdens. But the identity of alleles matters. Variable expression of defence-related genes adds further variation to the defence phenotype. This can generate biologically very different pathogens, even with very similar genomes.
- Host–parasite interactions vary with host and parasite genotypes. Main (additive) and epistatic effects are thereby important. Interaction models such as 'gene-for-gene' or 'matching alleles' explain host–parasite interactions from the underlying genetic interactions. Besides, the genes represented in the microbiota can additionally affect host–parasite interactions.

- Parasite-driven selection can take different effects, from positive-directional to balanced selection. Examples for all selection regimes exist. Directional selection seems common in antiviral defence pathways, whereas the MHC polymorphism shows signs for balanced selection. Likewise, signatures of selection are found in parasite genomes, especially also as a consequence of medical interventions.

- The genetic population structure of parasites is important for epidemics or the success of drug treatment. The structure is, among other things, strongly affected by genetic exchange among strains.

CHAPTER 11

Between-host dynamics (Epidemiology)

11.1 Epidemiology of infectious diseases

For evolutionary parasitology, epidemiology is the study of the ecological and evolutionary dynamics of a parasite population spreading in a population of hosts. It is, therefore, a branch of population biology that also takes into account rapid evolutionary change over ecological time scales. Epidemiology in this sense investigates how such a spread can be understood, modelled, and predicted, what factors and parameters would change the dynamics, and, in the context of public health policies, which interventions would reduce the further spread of a parasite. Furthermore, epidemiology is typically concerned with pathogens that cause infectious diseases in humans, livestock, and in wildlife (White et al. 2018c). This view differs from the approach taken in medicine, although it sometimes overlaps (Giesecke 2002). In medicine, epidemiology is the study of the distribution and determinants of health-related states or events in specified populations (according to a definition by the Centers for Disease Control and Prevention). It asks, for example, is leukaemia more common in the vicinity of high-voltage power lines, or is heart failure associated with smoking (Greenberg et al. 2004)?

Infectious diseases are among the most important causes of human suffering; these are also known as 'communicable diseases' in the public health domain. According to the 2018 data of the World Health Organization (WHO), 228 million people worldwide were infected with malaria (primarily in Africa), with an estimated toll of 405 000 deaths; two-thirds of the fatalities were among children (*WHO Malaria Report for 2019*). For the same year, other significant infectious diseases were tuberculosis, ten million new cases and 1.45 million dying (*WHO Global Tuberculosis Report for 2019*); HIV/AIDS (37.9 million infected, 0.8 million deaths), hepatitis (257 million living with chronic infections of hepatitis B, and 71 million with hepatitis C, with 1.4 million deaths; *WHO Progress Report on HIV, Viral Hepatitis, and Sexually Transmitted Infections 2019*). By comparison, the first 'severe acute respiratory syndrome' (SARS) outbreak in 2002–2003 led to 8100 cases, with 774 deaths. These low numbers were due to relatively low transmissibility of the virus and a rapid containment (Anderson et al. 2004). However, the recent outbreak of SARS-CoV-2/Covid-19 in 2020 again shows the enormous pandemic potential of a highly transmissible virus that gets out of control early on, causing morbidity and death for many and also inflicting enormous economic damage.

Outbreaks, 'epidemics' (meaning a more localized spread), and 'pandemics' (i.e. spreading to most parts of the world) of infectious diseases have been steady companions of human history and, by archaeological evidence, already occurred during prehistoric times (Andam et al. 2016) (Figure 11.1). The presence of influenza, for example, is dating back to antiquity. Today, three types of influenza viruses (types A, B, and C) circulate, with types A and B being the most prevalent and dangerous ones. Influenza virus has only 11 genes (!), all of which are known in sufficient detail. Nevertheless, this small number seems enough to cause significant trouble. One of the most famous pandemics, dating from spring to winter 1918 and known as the

Evolutionary Parasitology: The Integrated Study of Infections, Immunology, Ecology, and Genetics. Second Edition.
Paul Schmid-Hempel, Oxford University Press. © Paul Schmid-Hempel 2021.
DOI: 10.1093/oso/9780198832140.003.0011

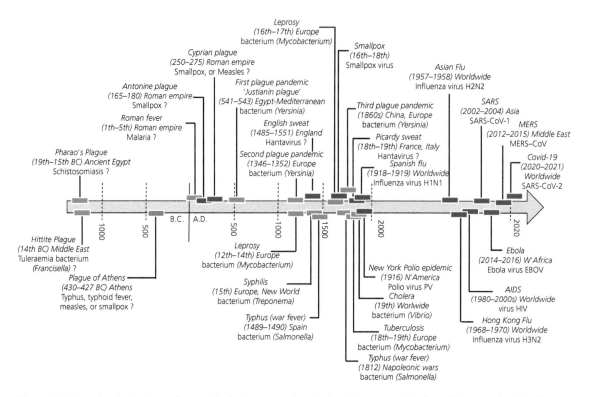

Figure 11.1 Examples of major human disease epidemics. Dates approximately placed. Question marks indicate likely disease but lack of firm knowledge; geographic locations refer to focal regions. Colours indicate viruses (red), bacteria (blue), and other (yellow). Adapted from Andam et al. (2016) with permission from Elsevier, with further information added.

'Spanish flu', was caused by influenza type H1N1. It killed somewhere between 20 and 40 million people—equivalent to the number of casualties during the First World War. Other major influenza outbreaks happened and spread from Hong Kong during spring 1957 (the 'Asian flu' pandemic, due to H2N2) (Fong 2017). H2N2 emerged from recombination of the still-circulating H1N1 strain with elements of avian flu; in the process, the new strain displaced the old H1N1. In 1968, a human H2N2 strain again re-assorted with avian influenza virus to produce the pandemic H3N2 'Hong Kong flu'. Again, the previously circulating strains were displaced. The serial replacement was undercut with the reappearance of H1N1 in 1977 in China, but this caused only mild symptoms. Such differences in the severity of the disease are typically due to either cross-protection gained from previous exposures, or, vice versa, the lack of resistance to new variants of H and N antigens (Fong 2017).

Hence, historical records and the analysis of viral serotypes and genotypes of influenza suggest that major pandemic strains interrupt the regular pattern of seasonal influenza at intervals of some 10 to 20 years. Furthermore, pandemic strains often emerge by recombination of human-adapted viruses with elements of other, notably bird- or swine-adapted, influenza, and they can be around for some time before becoming a problem. For instance, the 'bird flu' (H5N1) that appeared in 2004 had already been noticed in 1959. Up to 2006, it had caused 24 highly pathogenic outbreaks in birds (poultry), although each was restricted to limited geographic areas (Li et al. 2004a; Tiensin et al. 2005).

As cases of human infections accumulated during 2004–2005, migrating waterfowl, which are natural reservoirs of H5N1, were suspected of mediating the long-distance dispersal of the virus to Central Asia, Russia, and Europe. The study of bird migration, so far considered only an enjoyable pastime of bird lovers, suddenly looked like being crucial for human health. Also, the genomic sequence of viruses circulating in Asia during 2005 had undergone changes in several amino acids, which were feared to have affected transmissibility (Li et al. 2004a; WHO 2005). The virus was thus evolving and perhaps would become capable of direct human-to-human transmission. Not surprisingly, the fear in Europe and North America reached new heights. Trade restrictions were put in place by many countries. In Germany, army contingents were deployed to control access to outbreak areas; people had to leave and enter through locks with disinfectants. Panic buying of food items set in. Worries that there might not be enough medication or facemasks for everybody became prominent. Inevitably, religious fanatics claimed the epidemic to be a fulfilment of divine prophecies. Indeed, the response by the population, the authorities, and governments, was not very different from the SARS-CoV-2 pandemic of 2020–2021. Fortunately, during winter 2006–07, the principal dynamic of the H5N1 influenza epidemic subsided. H5N1 dropped out of the news and was forgotten. Science afterwards learned that dispersal by migrating waterbirds was more complicated than anticipated (Yin et al. 2017), and H5N1 case mortality rates were presumably lower than feared (Fong 2017). However, forgetting and underestimating potentially dangerous viruses is never a good public health strategy. A new avian influenza type, H7N9, has since emerged and caused human infections, notably in China (WHO 2015), while H5N1 continues to circulate among poultry.

11.2 Modelling infectious diseases

Infectious diseases, by the definition of the WHO, involve many different groups of parasites, including nematodes and other parasitic worms. However, the most prominent examples are viruses, bacteria, or protozoa. Epidemiology separates micro- from macroparasites not by size or taxon, but by how the host–parasite dynamics is best modelled. For 'macro'-parasites, the individual parasites can be tracked. For 'micro'-parasites, this is not feasible; instead, hosts are classified as infected or not infected (independent of how many viruses or cells the host carries). The study of infectious disease dynamics started with the Swiss mathematician Daniel Bernoulli. In 1766, he used a mathematical model to analyse the dynamics of a smallpox epidemic in Paris and to evaluate the effect of vaccination ('variolization' in his terms) (Box 11.1).

Box 11.1 Bernoulli's theory of smallpox

In 1766, Daniel Bernoulli (1700–1782) published a mathematical analysis of the epidemic behaviour of smallpox infections (*Variola major*) (Bernoulli 1766). During Bernoulli's lifetime in the eighteenth century, smallpox was endemic, with recurrent outbreaks in Europe. Furthermore, smallpox took a heavy toll, being responsible for perhaps one-tenth of all deaths at the time. According to Bernoulli's calculations, about three-quarters of all people must have had become infected at least once during their life. Nevertheless, his interest was aroused in particular by the observation that smallpox seemed to be a childhood disease. On average, children died from smallpox at the age of three to four years, whereas adults were more or less protected. From the available data, he calculated a case mortality rate of 12–14 per cent and noticed that this rate varied across countries and different epidemics. Bernoulli was already aware of the fact that a host that had survived infection by smallpox would become protected to some degree against a next infection. Although this was not the modern concept of immunization, the fact of protection after previous exposure was known.

Bernoulli developed his analysis in a very modern way, with numerical estimates of the degree of immune protection and of the case mortality rates at different ages. In this way, he essentially formulated a model that we would now recognize as belonging to the class of SIR models. From the model, Bernoulli estimated how many lives could be saved by vaccination with cowpox (which was already practised then). The calculations showed that an estimated 25 000 'useful lives' could be given to the society of France as a whole, by which criterion he understood that a person would reach the age of 17 and lead a working life. Bernoulli also applied the modern concept of sensitivity analyses

continued

Box 11.1 *Continued*

Box 11.1 Figure 1 Daniel Bernoulli (1700–1782). Portrait around 1750. ETH-Bibliothek, Zürich.

because he realized that the uncertainty about the exact value of model parameters is the most severe problem when applied to real cases. When presenting his analyses, just as in our days, Bernoulli had to deal with opponents to vaccination that feared the vaccine would cause an 'artificial smallpox'. However, Bernoulli's calculation showed that the number of fatalities would still be much lower in a population entirely, or even only partially, vaccinated, than if vaccinations were not carried out. Furthermore, with no vaccination, the natural infections would still be 32 times more common than from a possible breakthrough of the vaccine (i.e. cases resulting from the vaccine) (Bernoulli and Blower 2004).

Vaccination (from the Latin, 'vacca', the cow) against smallpox was gaining momentum after 1750–1770 in England and Germany. In 1796, Edward Jenner (1749–1823) finally demonstrated that inoculation with cowpox could protect against human smallpox. At this time, the nature of the causative agent, the smallpox virus, remained a profound mystery. The nature of viruses was discovered in the late nineteenth century, when Adolf Eduard Mayr (1843–1942) in 1879 first demonstrated the transmissibility of the tobacco mosaic disease in plants. Later, Dimitri Iwanowski (1864–1920) in 1892, and Martinus Beijerinck (1851–1931, who coined the term 'virus') in 1898 demonstrated that this occurred by particles that could not be removed by a porcelain bacterial filter and therefore must be smaller.

Current formulations of epidemiological models date back to those first suggested in the 1920s, e.g. the Kermack–McKendrick model (Kermack and McKendrick 1927), and have been continuously refined from the 1970s onwards (Dietz 1975; May and Anderson 1978, 1979; Anderson and May 1980, 1981; Diekmann et al. 1990). Eventually, the models were adapted to infectious diseases of humans (Anderson and May 1991).

11.2.1 The SIR model

The most widely used basic model of infectious disease dynamics is the 'SIR model'. It is a compartmental model and distinguishes three classes (compartments) of hosts. 'Susceptible' (S, i.e. individuals not yet infected, but that can be infected),

'infected' (I, those being currently infected), and 'resistant' (R, those that had been infected, cleared the parasite, and are now resistant to reinfection). The number of individuals in these classes at any one time can be calculated with the rates of transmission, clearance, mortalities, and so forth (Box 11.2). Hence, the course of an epidemic can be calculated and predicted when these parameters are known. Complication, such as evolutionary changes in the pathogens, in the course of an epidemic can be added by various refinements.

Within this general framework, the 'basic reproductive number', R_0, has arguably become the most important parameter to scrutinize an epidemic (Heesterbeek 2002). R_0 is defined as the expected number of newly infected hosts caused by an infected individual that enters a population consisting of

Box 11.2 The basic epidemiological model (SIR)

The SIR model (susceptible, infected, recovered) is the most widely used model in host–parasite epidemiology (Anderson and May 1981). It describes the dynamics of directly transmitted microparasites, where hosts can recover and remain protected (immune) for some time afterwards. In the model (Figure 1), the compartments S, I, and R are the number of susceptible, infected, and recovered hosts, respectively, at time t. Correspondingly, $N = S + I + R$ is the total population size at time t. From Figure 1, the change in numbers of S, I, and R per unit time (dt) is given by the set of differential equations:

$$\frac{dS}{dt} = b(S,I,R) - \mu S - \beta SI + qR$$

$$\frac{dI}{dt} = \beta SI - \mu I - \alpha I - \upsilon I = \beta SI - (\mu + \alpha + \upsilon)I$$

$$\frac{dR}{dt} = \upsilon I - \mu R - qR = \upsilon I - (\mu + q)R \qquad (1)$$

with terms, as explained in Figure 1. The parameter $\beta \cdot\cdot$ is the transmission rate, which indicates the probability that the parasite infects the next host upon encounter. The number of newly infected individuals is proportional to the product of S I. This assumes that infected and susceptible hosts meet at random and in proportion to their numbers; the assumption is known as the 'mass action principle', or 'homogeneous mixing'. With heterogeneous mixing,

susceptible and infected individuals meet according to stratifications by age, sex, behaviour, or spatial location. In eq. (1) transmission assumes density dependence, since new infections occur at a rate given by the numbers (densities) of susceptibles and infecteds ($\beta S I$); the term $\lambda = \beta \cdot I$ is known as 'force of infection'. Alternatively, new infections could arise in a frequency-dependent manner. In this case, the transmission term is changed to $\beta S (I/N)$, the latter term being the proportion of infecteds among all individuals, e.g. as applicable to sexually transmitted diseases. This simple model ignores some relevant biological situations; for example, populations with age structure or spatial heterogeneity. Such additional factors could be included, however.

The model of eq. (1) is simplified by observing that the total population size is often approximately constant if the parasite is not directly regulating population size. This happens when the parasite-induced mortality rate, α, is low. Then, with total population size $N = S + I + R$, and solving for the equilibrium:

$$\hat{S} = \frac{(\mu + \alpha + \upsilon)}{\beta}$$

$$\hat{I} = (\mu + q)\frac{\beta N - (\mu + \alpha + \upsilon)}{\beta(\mu + \upsilon + q)} \qquad (2)$$

Where $\hat{S}, \hat{I}$ are the endemic equilibrium values for the number of susceptible and infected individuals, respectively. The other equilibrium point would be $S = N$, i.e. no infection.

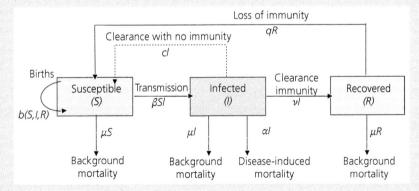

Box 11.2 Figure 1 Graphical representation of the SIR model. Susceptible individuals (S) are recruited into the population by births at rate b (S, I, R), which is a function of the number of susceptible, infected (I), and recovered (R) hosts. For any class, there is a background mortality rate, μ; for infecteds, an additional infection-induced mortality rate, α, applies. Newly infecteds arise at rate β S I (with transmission rate β) and leave the compartment I as they recover by clearing the parasite to become recovered (R) at rate υ. Recovereds re-enter the class of susceptibles at rate q, as they slowly lose immune protection. Alternatively, hosts may clear the infection without lasting immunity at rate c (broken line).

continued

Box 11.2 *Continued*

The SIR model unfolds as an epidemic. Eventually, the system converges to an 'endemic' state; that is, the equilibrium characterized by eq. (2), with a constant number, $\hat{I}$, of infected individuals.

The SIR model also shows when an epidemic can spread in the first place. With eq. (1), this is only the case when the number of infecteds increases; i.e. when $dI/dt > 0$, and therefore $\beta SI - (\mu + \alpha + \upsilon)I > 0$, which yields:

$$R = \frac{\beta S}{\mu + \alpha + \upsilon} > 1 \qquad (3a)$$

This requirement can change, e.g. with the addition of stochastic components. Of particular interest is the introduction of an infected individual into a population of susceptible individuals. In this case, $S \approx N$ and:

$$R_0 = \frac{\beta N}{\mu + \alpha + \upsilon} \qquad (3b)$$

R_0 is the so-called basic or net reproductive number of the infection. It describes the number of newly (secondary) infected hosts resulting from one (primary) infected individual that enters a wholly susceptible host population

(Diekmann et al. 1990). R_0 is thus approximately the maximum potential of a parasite to spread. The reproductive number, R_0, can also be interpreted as the product of the number of newly generated infections ($\beta \cdot N$) over the expected duration of the infection, which is $1/(\mu + \alpha + \upsilon)$. Therefore, R_0 refers to discrete 'generations', i.e. the time interval given by the duration of the infection. In other words, each parent is replaced by R_0 descendants after one generation, provided the epidemic unfolds in such discrete generational steps.

At the endemic equilibrium, $\hat{S}$, from eq. (2) with $N = S + I + R$, we find $R = \upsilon I/\mu$. For many infections, e.g. childhood diseases, we can furthermore assume that $\upsilon \gg \alpha$, and $\upsilon \gg \mu$. This yields approximately:

$$R_0 \approx 1 + \frac{\beta I}{\mu} \approx 1 + \frac{L}{A} \sim \frac{L}{A} \qquad (4)$$

where $L \approx 1/\mu$ is the average host life expectancy. During this time, susceptibles become infected at rate $\sim \beta I$, and therefore $1/(\beta I)$ is the approximate, average age, A, at infection in the host population. As intuitively expected, A generally is decreasing as R_0 increases.

---

susceptible individuals only. In public health studies, this first infected individual is the 'index case', i.e. the first identifiable individual that carries an infection into a given population. When all other individuals are susceptible, R_0 approximately describes the parasite's maximum rate of spread in the host population. In terms of population biology, R_0 is the number of (surviving) offspring per parent (the 'reproductive factor'), i.e. every parent is replaced by R_0 offspring in the next generation. The quantity $R_0 - 1$ is known as the per capita change, r, in classical population biology. If $R_0 = 1$, then $r = 0$ and the population size remains constant (see Box 15.1). Note, therefore, that R_0 is a 'per generation' measure (see the next paragraph). From the basic SIR model (Box 11.2), the standard equation for R_0 is:

$$R_0 = \frac{\beta N}{\mu + \alpha + \upsilon} \qquad (11.1)$$

where N is the host population size (variably the total size, or the number of susceptibles, S, at time zero). R_0, therefore, increases with transmission

rate, β, and decreases with parasite-induced mortality rate ('virulence', α), background mortality (μ), and the rate of clearance (υ). Eq. (11.1) provides a conceptual tool to understand how R_0 varies with a change in essential parameters. R_0 is often considered to measure the fitness of the parasite when analysing the evolution of virulence (see Chapter 13). However, the maximization of R_0 by natural selection is only valid under the simplifying assumptions made in the SIR model (Lion and Metz 2018). In this chapter, we restrict ourselves to the modelling of disease dynamics, which does not directly touch these issues.

R_0 typically is used to model a steadily unfolding epidemic, but the value of R_0 reflects the number of infecteds, per originally infected host (the 'parent'), after a defined time interval. This use, therefore, assumes that the epidemic follows a series of discrete steps; that is, it has 'non-overlapping generations' in terms of population biology. The generation interval lasts from the infection of the first host to the average time of infection of the next hosts. In medical practice, this interval typically is estimated

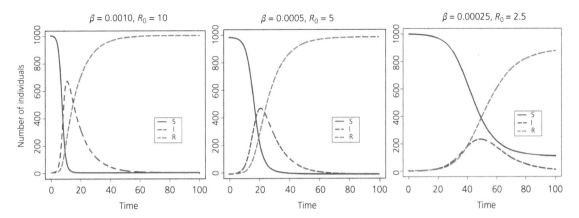

Figure 11.2 The time course of an infection according to a simple SIR model. The graph shows the number of infected (I, red), susceptible (S, blue), and resistant (R, green) individuals in a population of $N = 1000$ individuals. The initial population size includes one infected host entering at time zero (i.e. $S = 999$, $I = 1$, $R = 0$). Compared to the formulation in Box 11.2, the population size remains stable (parameters $b = 0$, $\mu = \alpha = 0$), the gain in immunity is $\nu = 0.1$, and immunity is permanent, i.e. no reinfection is possible ($q = 0$). The graphs (from left to right) show the effect of decreasing the transmission rate, β, and the resulting reproductive number, R_0, for the epidemic. With the model and parameters assumed here, R_0 is proportional to β. The exponential growth of infecteds with R_0 (dashed red lines) occurs at the beginning of an epidemic; this rate later decreases with a lack of susceptibles. Simulation run with R-script, courtesy of S. Bonhoeffer ETH Zürich.

from the time of onset of symptoms in the first host to the onset of symptoms in the next hosts. Regardless, the time interval is a 'generation time' for the infection, also known as the 'serial interval' in epidemiology.

Most epidemics, however, unfold with 'overlapping generations'; i.e. infections happen not only once in an interval, but continuously throughout the interval. In population biology terms, the number of infected hosts follows an exponential growth of the form $n_t = n_0 e^{rt}$ (see Box 15.1). Per one parental infected host, $n_0 = 1$, the number of newly infected hosts at the end of the generation interval ($t = 1$), therefore, is $n_1 = e^r = e^{(R_0-1)}$. In the example of SARS-CoV-2, a median generation time of five days, and a range of $R_0 = 2.5$–3.5 is estimated (Liu et al. 2020). With an uncontrolled $R_0 = 3$, we would have 7.39 instead of one infected host after one generation, i.e. after five days. This can be recalculated as an increase per day, which may be considered the 'instantaneous' growth rate (defined as per unit time). Note that, in real-time, a parasite with a low R_0 can nevertheless spread rapidly when its serial time is short, and a parasite with a high R_0 will still spread slowly if the serial time is long.

The basic reproductive number, R_0, is a characteristic of both the parasite and the host population it

enters. Hence, $R_0 = R_0(P \mid E)$; that is, R_0 is conditional on the environment (E) that the parasite (P) encounters. For example, when people pay attention to hygiene, the transmission rate, β, will decrease and the resulting R_0 will also decline—the (epidemiologically relevant) environment, E, has changed. Indeed, whereas the population-dynamic laws of an epidemic cannot be changed, the factors that affect E can. This is the basis of the control of an epidemic disease. Changes in E lead to a different dynamics that can be calculated and predicted, allowing, for example, for the evaluation of different control measures (Figure 11.2). Nevertheless, a part of R_0 remains a biological characteristic of the parasite and therefore is indicative of its epidemic potential (Table 11.1).

As an epidemic unfolds, more and more individuals become infected or have already returned to a resistant state by becoming immune (Box 11.2). Therefore, the pool of susceptible individuals becomes smaller with time. As a result, the epidemic slows down, and the value of R_0 decreases. The epidemics might eventually grind to a halt when not enough susceptible hosts are available. An epidemic can therefore only be sustained and eventually develop into an 'endemic' state (i.e. a state of the persistent presence of the parasite) when

Table 11.1 Basic reproductive rates, R_0, and critical vaccination coverage, p_{crit}, for various infectious diseases of humans.

Infection	Locality [Source]	Years	R_0 [1]	Coverage p_{crit} [2]
Malaria	Nigeria, hyper-endemic regions (*P. falciparum*)	1970s	80	99 per cent
	Nigeria (*P. malariae*)	1970s	16	
Measles	England and Wales	1950–1968	16–18	90–95 per cent
	Eastern Nigeria	1960–1968	16–17	
	Ghana	1960–1968	14–15	
	Cirencester, England	1947–1950	13–14	
	Ontario, Canada	1912–1913	11–12	
	Willesden, England	1912–1913	11–12	
	Kansas, USA	1918–1921	5–6	
	Various countries [6]	1838–2012	c.12–16 (median, but wide range)	
Pertussis	England and Wales	1944–1978	16–18	90–95 per cent
	Maryland, USA	1943	16–17	
	Ontario, Canada	1912–1913	10–11	
Rubella	Gambia	1976	15–16	82–87 per cent
	Poland	1970–1977	11–12	
	Czechoslovakia	1970–1977	8–9	
	England and Wales	1960–1970	6–7	
	West Germany	1970–1977	6–7	
Chicken pox	Baltimore, USA	1943	10–11	85–90 per cent
	England and Wales	1944–1968	10–12	
	New Jersey, USA	1912–1921	7–8	
	Maryland, USA	1913–1917	7–8	
HIV	Nairobi, Kenya (female prostitutes)	1981–1985	11–12	
	Nairobi, Kenya (heterosexuals)	1981–1985	10–11	
	England and Wales (male homosexuals)	1981–1985	2–5	
Polio	Netherlands	1960	6–7	82–87 per cent
	USA	1955	5–6	
Mumps	England and Wales	1960–1980	11–14	85–90 per cent
	Netherlands	1970–1980	11–14	
	Baltimore, USA	1943	7–8	
Diphtheria	New York, USA	1918–1919	4–5	82–87 per cent
	Maryland, USA	1908–1917	4–5	
Scarlet fever	New York, USA	1918–1919	5–6	82–87 per cent
	Maryland, USA	1908–1917	7–8	
Dengue	Colombia [11]	2010	1.5–2.7	
	Various countries [10]	1996–2017	4.74 (range 0.9–6.5)	
Zika	Various countries [10]	1996–2017	3.02 (range 0.1–9.4)	
	Florida [2]	2016	2.1–2.2 (median for peak times)	

Infection	Locality [Source]	Years	R_0 [1]	Coverage p_{crit} [2]
	El Salvador, Colombia, Suriname [13]	2015–2016	4–6	
Chikungunya	Various countries [10]	2014–2017	2.55 (range 0.46–6.5)	
Influenza A	H1N1 in Mexico [5]	2009	1.37 (1.26–1.42)	
MERS-CoV	Arabic peninsula [3]	2013–2015	0.6–0.7 (0.42–0.92) locally: 3.5–6.7, 2.0–2.8	
	South Korea [4]	2015	8.1	
SARS-CoV-1	China, Singapore [8, 9]	2002–2003	2–5, 2.2–3.6	
SARS-CoV-2	China [8, 9]	Jan/Feb 2020	2.4–4.2 (1.4–6.5)	
	City of Wuhan, China [16]	Jan/Feb 2020	before control: 4.71 after control: 0.76	
	Various countries, with low to high transmission [7]	March 2020	Low: 1.06–2.00 High: 4.52–6.64	
Ebola	The Congo [14]	2018–2019	0.9 (0–1.8)	
	West Africa, 'unsafe burials' [15]	2013–2016	2.58 (0.64–4.74)	
Smallpox	Globally			70–80 per cent

[1] From Ref. 1: Tables 4.1, 14.8, and [2] Table 5.1, with permission from Oxford University Press.
Sources: [1] Anderson. 1991. *Infectious diseases of humans*. Oxford University Press. [2] Boskova. 2018. *Virus Evol*, 4: [3] Breban. 2013. *The Lancet*, 382: 694. [4] Chang. 2017. *Biomed Eng Online*, 16: 79. [5] Chong. 2018. *Travel Med Infect Dis*, 23: 80. [6] Guerra. 2017. *Lancet Infect Dis*, 17: e420. [7] Kwok. 2020. *J Infect*, 80: e32. [8] Lipsitch. 2003. *Science*, 300: 1966. [9] Liu. 2020. *J Travel Med*, 27: 1. [10] Liu. 2020. *Environ Res*, 182: 109114. [11] Lizarralde-Bejarano. 2017. *Appl Math Model*, 43: 566. [12] Majumder. 2014. *PLoS currents*, 6: ecurrents.outbreaks. (PMC4322060). [13] Shutt. 2017. *Epidemics*, 21: 63. [14] Tariq. 2019. *Epidemics*, 26: 128. [15] Tiffany. 2017. *PLoS Negl Trop Dis*, 11: e0005491. [16] Zhao. 2020. *Int J Infect Dis*, 92: 214.

new, susceptible individuals are recruited to the population. This happens either by the addition of newborns, when infections are cleared with no immunity, when immune protection is lost with time, or when susceptible individuals immigrate.

Real situations require modifications of the basic SIR model in several ways. For example, the age structure in a host population can cause variation in transmission rates. The age structure can lower R_0 because some hosts will not transmit the disease. Similarly, a latency period to transmission will slow the process, depending on the survivorship curve of the host population. Sexually transmitted diseases affect the parameter β of Box 11.2, as transmission occurs only between sexes rather than among all individuals. This makes transmission frequency dependent, since meeting an infected individual of the opposite sex is proportional to their frequency in the respective population. Modifying the basic equations accordingly, R_0 becomes directly proportional to the number of sexual contacts.

The population of susceptible hosts can also grow on its own; for example, in proportion to their numbers, S. In the simplest case, the incremental increase of susceptibles is $dS/dt = bS$, where b is the per capita growth rate of the susceptible population (see Box 11.2). This modification can lead to periodic oscillations, especially when parasite-induced mortality is high. Further modifications of the basic model include multiple infections, cross-immunity, multiple transmission routes, or various kinds of heterogeneities. Together, a rich repertoire of dynamic behaviours for the host–parasite system can result, reaching from the standard dampened oscillations, to endemic states, periodic outbreaks, chaotic dynamics, travelling waves of infection in space, and various other outcomes.

The average age, A, at which an individual becomes infected is a useful parameter for the application of epidemiological theory in practice. From Box 11.2, we find that $R_0 \approx L/A$, where L is the life expectancy in a given population. Hence,

$A \approx L/R_0$, which shows that hosts are infected at younger ages when R_0 increases. This assumes that there is a single infection at age A, followed by lifelong immunity, and that the background and parasite-induced mortality rates are much lower than the rate of recovery from the infection. By and large, these conditions hold for most childhood diseases, for which A, therefore, is a useful indicator. Table 11.2 lists some numerical examples of A and shows that in highly endemic areas, children become infected at a very early age. The basic SIR model, and thus estimates of the relationship between R and A, obviously ignores a set of biologically reasonable conditions. Nevertheless, the estimate of A remains a good indicator of the dynamics of childhood diseases.

Epidemic modelling is vital for public health to track and control disease outbreaks in human populations, agriculture, or livestock, and to assess the impact of interventions. Note, however, that terminology may be quite different in various quarters and may not correspond to model terms. For example, in the public health domain, the 'case fatality rate' refers to the percentage of deaths per confirmed case. This corresponds to, but typically is

Table 11.2 Average age at infection for various childhood diseases and locations.[1]

Infection	Locality	Year	Average age (yr)
Measles	USA	1955–1958	5–6
	England, Wales	1948–1968	4–5
	Morocco	1963	2–3
	Ghana	1960–1968	2–3
	Senegal	1964	1–2
Rubella	Sweden	1965	9–10
	USA	1966–1968	9–10
	Poland	1970–1977	6–7
	The Gambia	1976	2–3
Chicken Pox	USA	1921–1928	6–8
Polio	USA	1955	12–17
Pertussis	England, Wales	1948–1968	4–5
Mumps	England, Wales	1975–1977	6–7

[1] From Anderson and May. 1991. *Infectious diseases of humans*. Oxford University Press, with permission.

not numerically the same as, the parasite-induced mortality rate in models. Furthermore, a 'case' means an infected individual in epidemiology. However, in medical terminology, a 'case' is an individual that has clinical symptoms or was tested for infection status with markers. A medical case is therefore necessarily appearing at a later stage because symptoms (or enough pathogens to be detected by a test) only appear a time after infection. This time is known as the 'incubation period' in medical terminology, but as a 'latency time' in modelling. Such delays are important, as they change the dynamics of a system.

For some infectious diseases, transmission can also be 'asymptomatic', i.e. a host can be infectious for others before symptoms appear (as for SARS-CoV-2), or remains without clinical symptoms for the entire duration of the infection, yet transmits the disease further. The infected host would, therefore, not be a 'case' in the clinical sense and could go unnoticed. This leads to unreliable databases for the development of predictive models. However, the testing for infection status can lead to unreliable data, too. For example, if a test has 97 per cent sensitivity (i.e. recognizes an infection correctly in 97 per cent of cases; see Box 12.1), and a specificity of 95 per cent (i.e. recognizes the absence of infection correctly in 95 per cent of cases), the numerical error can nevertheless be large. This is especially true for the early stages of an epidemic. Consider a situation with 50 000 true infecteds in a sample of one million individuals. Then by the test's sensitivity, c.30 000 individuals are tested positive but are not infected in reality, which is almost as many as genuinely infected. Likewise, 2500 cases of true infecteds would go unnoticed.

Finally, the SIR model is a conceptual model that is particularly useful for simulating an epidemic and capturing some essential features of disease dynamics. The basic reproductive number, R_0, is conceptually related to other parameters used in population biology. For instance, it is related to the basic (zero) term for the rate of increase of populations that results from mathematically factoring the age-dependent rate of producing newborns. Alternatively, similarity exists with the expected lifetime number of offspring in life history theory; this corresponds to the dominant eigenvalue of the growth matrix that maps the population from one generation to the next. Because R_0 is measured at the levels of infected hosts, it is strictly speaking a reproductive rate that pertains to a metapopulation of parasites across host individuals. Furthermore, transmission rate, or parasite-induced virulence, α, may vary with the age of the infection, while host age likely affects variables such as clearance rate. Therefore, R_0 describes some kind of 'lifetime infection pressure' caused by an infected individual on the population of all susceptibles.

11.2.2 Thresholds and vaccination

With the SIR model, an epidemic can only spread and grow if $R_0 > 1$; that is, an infected individual must infect more than one new host. With $R_0 < 1$, the epidemic will eventually die out. Hence, public health measures must aim at reducing R_0 to a value below one. The nominal threshold value of $R_0 = 1$

needs to be modified when the effects of stochasticity or heterogeneity become essential. But some kind of threshold characteristic pertains to the growth of any population, and to any model that is similar in structure.

In practice, to force $R_0 < 1$, interventions by hygiene, social distancing, or the wearing of masks are the classical tools of containing an epidemic. $R_0 < 1$ can also be reached by reducing the pool of susceptible individuals, S. For example, during the foot-and-mouth disease epidemic in the UK, the culling of cattle, pig, sheep, goats, and other livestock successfully helped to stop the spread of the virus (Kao 2002). Reducing S is also feasible by vaccination, which makes hosts unavailable for infection, too. A practical difficulty in all of these endeavours is to estimate the value of R_0 in an ongoing epidemic. This is necessary, however, to assess the situation and to evaluate the success of control measures and interventions (Box 11.3).

Box 11.3 Calculating R_0

By definition, the basic reproductive number, R_0, is the number of secondary infections per primary infection in a population of all susceptibles. This quantity is generally not equivalent to the parasite's fitness but has important information for the public health sector, e.g. highlighting when the spread of infection needs to be monitored. How to estimate R_0 is therefore of considerable practical relevance. Whereas the definition is clear, the task is far from trivial.

Common difficulties include the proper identification of cases (infections), delayed reporting, incomplete databases, undiscovered cases, uncertainties about infection times, and the length of the serial interval. The 'uncontrolled' R_0 is a measure of the pathogen's epidemic maximum potential in a given population, observed in a situation where the disease spreads without interventions. In many cases, however, changes in human behaviour, e.g. exerting more caution, will have already affected R_0 before public health measures become implemented. The estimates of R_0 after control measures are of prime interest for evaluating the effect of such interventions. Estimates of R_0 are typically associated with large confidence intervals, in particular at the beginning of an epidemic when numbers are low. Furthermore, R_0 is a retrospective estimate.

Direct estimates: When transmission routes, the number of infecteds, susceptibles, or contact rates, age at infection, or the serial time interval, are known, or can readily be estimated, the epidemiological equations allow for a 'direct' estimate of R_0. An example is the 'classical' estimate from survival functions (Heffernan et al. 2005). When $F(\tau)$ is the probability function that a newly infected individual remains infective for a time, τ, and b is the rate at which an infected host produces new infections, then $R_0 = \int_0^\infty bF(\tau)$. Similarly, the 'next-generation method' assumes discrete steps and uses an operator that maps the current numbers of infecteds or susceptibles into the next generation. Typically, this operator is a 'next-generation matrix', as used in the study of population dynamics. The approach is quite robust and can cope with a large number of complications such as age or spatial structure, or various host classes. Because the matrix converts the current population at time t to time $t + 1$, it also contains information on how each host compartment changes from one time step to the next. R_0 can be derived from the operator, e.g. the dominant eigenvalue of the matrix. Estimates can also result from the intrinsic growth rate for the population of infecteds, combined with the length of the infectious period, τ, although this results in

continued

Box 11.3 *Continued*

large uncertainties. Direct estimates also exploit the numbers at or near an endemic equilibrium. For example, the average age at infection, or the equilibrium fraction of infecteds (with the 'final size equation'; see Box 11.2), directly relate to R_0. Obviously, these latter methods are of little use for estimates during an ongoing outbreak.

Fitting models: SIR models predict the time course of an epidemic. If a fitting model is found and its parameters can be estimated, a value for R_0 can be derived. This is an attractive and often used method. However, modellers have to make numerous decisions and assumptions, e.g. what compartments are included, whether deterministic or stochastic processes are assumed, etc. More complex models include factors such as population mixing, stratification, or migration. Accordingly, more input parameters have to be estimated, which increase the uncertainties of the estimate. Other kinds of models can be based on inferred transmission networks, with a guess for the connection pattern and the underlying parameters that best fit the observations. Fitting the data to find values for the parameters is a crucial step. It requires appropriate statistical procedures, e.g. fitting time series to observations, or using Bayesian approaches to improve estimates from prior distributions. The fitting procedures must account for the effects of the many shortcomings in the database, e.g. missing reports on cases (reporting rate) or low reliability of reports. Nevertheless, this kind of approach has been used for many situations, such as for measles (Guerra et al. 2017), influenza (Chong et al. 2018; Nikbakht et al. 2019), Zika (Shutt et al. 2017), or the SARS-CoV-2 outbreak (You et al. 2020; Zhao and Chen 2020).

Phylodynamic estimates: This approach differs from the previous ones, as it reconstructs a phylogeny of parasite sequences (commonly, viral diseases) with the tools of phylodynamics (see Box 11.6). From the branching pattern of the tree, the ecological dynamics (transmission chain) is reconstructed. Together with the dates of sample collections, plus additional constraints and assumptions, R_0 can be estimated. Also, a time scale can be added, based on the molecular clock; a 'generation clock' reflects the accumulation of the diversity (of sequences) in the tree (and thus ignores the time within the host). The number of mutations between any two sequences can estimate the 'coalescent time', i.e. the generation that contains the last common ancestor of two sequences. Combined with the probability distribution of the serial time ('generation time') and the time between infection and sampling of this infection, this yields an estimate for R_0. Alternatively, a birth–death process is assumed, where a 'birth' means a transmission that generates a new infection, and 'deaths' are individuals becoming non-infectious, e.g. due to clearance or host death. For this procedure, the sampling of a sequence is also death; the sampled individual immediately becomes non-infectious. The method assumes that one individual starts an epidemic. This creates a transmission tree starting at a time, t_0, in the past. This tree will be sampled with some probability at subsequent time points; this shows up by sequence similarities. From this data, the most likely transmission tree and its underlying parameters, such as transmission rate (β), 'death' rate (d), or sampling rate (f), are calculated (with an algorithm based on Bayesian statistics). In the simplest case, an estimate would be $R_0 = \beta / (d + f)$ (Stadler et al. 2011).

Because of the intrinsic threshold characteristic of an epidemic (with $R_0 = 1$), there is a critical minimum fraction of the population, p_{crit}, that needs to become unavailable to stop the further spread. Based on eq. (11.1) and the threshold condition, $R_0 = 1$, we thus require that the number of susceptibles, S, should become lower than S_{crit}:

$$S_{crit} = \frac{\mu + \alpha + \upsilon}{\beta} \qquad (11.2)$$

With eq. (11.2), a minimum population size, S_{crit}, is in turn needed to sustain an epidemic infection in the first place. This suggests that a disease might

not sustain itself in small populations, e.g. on islands or in cities, if most of the interactions are among the residents. The model of a closed island or city population is not entirely correct, of course, but observations show the expected relationship with population size (Figure 11.3). In these cases, the recurrent disappearance of the infection is also a process of stochastic fade-out not covered by the basic SIR model. Instead, probabilities associated with epidemic events (infection, clearance) and a probability for a fade-out, associated with a given threshold size, are needed to model the situation.

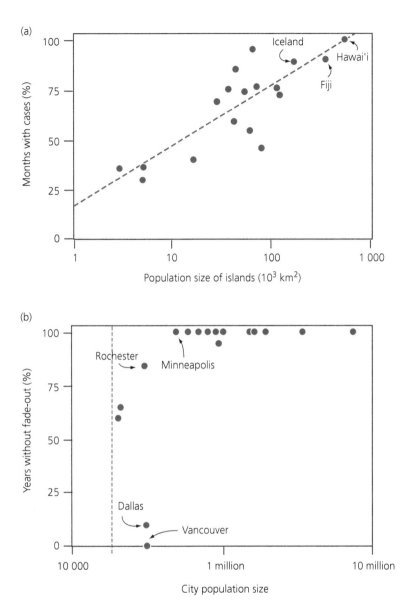

Figure 11.3 Population size threshold for epidemic measles. (a) The larger the population of an island, the more often an epidemic occurs. Recurrent epidemics can be caused by immigration of infected individuals from outside. However, a threshold population size of approximately 500 000 seems required to maintain measles permanently. (b) A permanent presence of measles in American cities does not occur below a minimum population size of approximately 200 000 people. The dotted line is the critical minimum community size for endemic maintenance. The data refer to the period from 1921 to 1940. Adapted from Nokes (1992), with permission from John Wiley and Sons.

With S_{crit} and total population size of N, the fraction of individuals that should become unavailable to stop an epidemic is $p_{crit} = 1 - S_{crit}/N$. Using eq. (11.2):

$$p_{crit} = 1 - \frac{\mu + \alpha + \upsilon}{\beta N} = 1 - \frac{1}{R_0} \qquad (11.3)$$

Hence, p_{crit} depends on R_0, such that infectious diseases with a large R_0 require a high percentage of the population to become unavailable (see Table 11.1). This proportion can be reached with culling (as in cattle), by measures of isolation, by vaccination, or with the natural course of an infection, where enough individuals have become infected, provided they get immunized. When p_{crit} is reached, the population becomes protected by 'herd immunity'. Immunized individuals lower the 'force of infection' (technically, the product of βI) in the population and so add to the protection of others.

The values for p_{crit} (as listed in Table 11.1) are based on the approximation by the basic SIR model. Additional factors, notably heterogeneity in transmission (VanderWaal and Ezenwa 2016), can increase (or sometimes decrease) the required threshold value. Nevertheless, it is clear that diseases with high values of R_0 are hard to eradicate completely, since a critical vaccination coverage close to 100 per cent is needed; that is, almost everybody needs to be vaccinated. Such high coverage puts high demands on infrastructure and logistics. Moreover, recurrent bouts of vaccination scepticism in developed countries (e.g. towards measles vaccination) and developing countries (e.g. against polio vaccination in Nigeria; Jegede 2007) have also added their share to the difficulties of reaching this goal.

The concept of critical vaccination coverage, p_{crit}, thus illustrates a classic example of the conflict between individual and group interests. A given parent, for example, might value the risk of vaccination for the child higher than the risk of disease and refuse. In this case, the benefits through herd immunity are still there, but the possible cost of vaccination is with others. The risk perception is also flawed: many childhood diseases have considerable fatality rates and very often cause lasting disabilities. Hence, the consequences of becoming infected are generally much worse than the statistically extremely rare side effects of vaccination. This disparity in risk perception is particularly strong when vaccination programmes have been so successful as to lower the individual risk of infection for everybody drastically. This vanishing awareness is known as the 'prevention paradox'. Vaccination is one of science's most significant achievements and saves an estimated several millions of lives every year by a combination of individual protection and reducing disease burdens and the force of infection in the population at large. Smallpox and brucellosis are two examples where vaccination programmes

have achieved complete eradication. With the Global Polio Eradication Initiative, poliovirus, which is currently endemic only in two Asian countries, is the next target.

11.2.3 Stochastic epidemiology

Random (stochastic) processes can affect the course of an epidemic in many ways. For example, a host may become infected only with a given probability that obeys a probability distribution calculated, for example, as a function of the force of infection. Whether or not an individual host actually becomes infected is like making a single draw from a pool of black or white balls, with their relative numbers representing the basic probability of becoming infected or not. If the number of draws is small (i.e. few individuals are present), considerable stochasticity occurs, since it is possible to accidentally draw a series of one colour before the alternative comes up, although the basic proportion of balls has remained the same. Stochasticity is particularly relevant at the beginning of an epidemic, or when control measures lower the number of infecteds. In either case, the epidemic may therefore stochastically fade out. However, an infectious disease can also be maintained even with low numbers of infecteds. The minimum population size required for an infectious disease to persist in an endemic state, where it is unlikely to fade out even without reintroduction from outside, is known as the 'endemic threshold', or 'critical community size' (Munro et al. 2020) (see Figure 11.3b for measles; Bartlett 1957).

Stochastic models use discrete rather than continuous numbers for S, I, or R. Also, the rates characterizing the dynamics, e.g. transmission rate, β, are replaced by probabilities of the corresponding events happening. Stochastic models allow an infection to die out, which is never the case in deterministic models such as that of Box 11.2 (where the number of infecteds converges to zero only in infinite time). Stochastic analyses show patterns that are much closer to real epidemics (Figure 11.4) as compared to simple, deterministic approaches. Note that the underlying dynamics is still the same; the only difference is the stochastic nature of the involved processes.

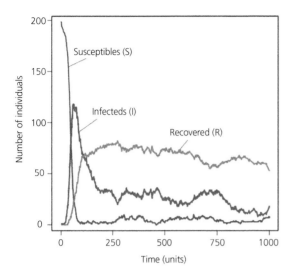

Figure 11.4 Example of a stochastic SIR dynamics. The simulation starts with $S = 199$ and $I = 1$ (population size $N = 200$) Parameters are birth rate (b) = 10^{-5}, transmission rate $\beta = 1.5\ 10^{-4}$, acquisition of immunity $\nu = 0.01$, loss of immunity $q = 0.005$, background mortality $\mu = 0$, parasite-induced mortality $\alpha = 0.005$ (see Box 11.2 for variables). The estimated reproductive number is $R_0 = 11.76$. Simulation run with R-script, courtesy of S. Bonhoeffer ETH Zürich.

11.2.4 Network analysis of epidemics

Individuals are never equal in their behaviour or resistance, and there is never a perfect mixing of infecteds and susceptibles as assumed in the SIR model. Ideally, therefore, individuals and their fates are followed throughout an epidemic. For this, 'individual-based' models increasingly are used for a wide range of systems and questions. However, as is often the case, the terminologies and diversity of model assumptions remain confusing (Willem et al. 2017). For example, 'agent-based models' treat individuals as 'autonomous agents' that behave and respond, and whose infection status is followed over time. Similarly, 'network models' can keep track of individuals (or defined groups) and their mutual connections over time.

All of these approaches have some drawbacks. For example, such models are computationally very demanding and therefore cannot deal with numbers of traced subjects that are too large. Furthermore, parameter choice and model parametrization, i.e. defining the variables and assigning numerical values to them, remain the most challenging parts for a

fit to reality. It is also never obvious whether the chosen model and the outcome is an excellent approximation to reality. In this situation, to make models more complex is usually a wrong choice, since it multiplies the difficulties of understanding what the model does, rather than clarifying the situation.

Network theory, too, has its terminology. Network theory uses graphs that show how objects, such as individuals, are connected by 'links' that reflect, for example, a pathogen transmission event between individuals A and B. These objects are the 'nodes' of the network; in epidemiology, links are the 'contacts' or 'edges'. Within a network, two individuals (nodes) are connected by a 'path', i.e. a series of contacts (links) along which one individual can be reached from the other, thereby traversing several individuals in between. The 'degree' of connectivity of an individual is the number of immediate contacts with the neighbouring individuals, i.e. those that can be reached in a single step. Sometimes, nodes in one part of a network connect well among each other but disconnected as a group from other such parts. In this case, the network consists of two or more separate 'connected components'. Furthermore, connections within a network form the 'adjacency matrix' that lists which individual has immediate contact with which other individuals. Hence, with n individuals, there are $n\ (n-1)$ possible contacts (links), but only a subset exists in the real world; a sparse matrix results when contacts are rare and restricted to specific subsets of individuals. Sometimes, only pairwise contacts between individuals matter; these are collected in a so-called 'incident list'. The kinds of connections within a network are crucial for the spread of disease.

Frequently, human contact networks that are important for epidemiology are 'scale-free'. In these kinds of networks, the distribution of connections per individual follows a power law, i.e. the probability, p, of an individual having k contacts is $p(k) \sim k^{-c}$, with c characterizing the distribution. In scale-free networks, individuals with very many contacts are more frequent than expected if contacts were purely at random (Figure 11.5). Such individuals can become 'superspreaders' in an epidemic. A retrospective analysis of the latest Ebola outbreak in West Africa (Guinea, Liberia, Sierra Leone) during

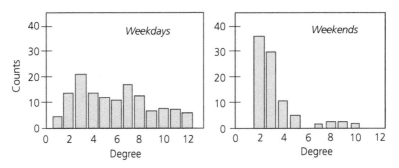

Figure 11.5 Degree distribution in networks. The degree is the number of contacts (edges) that a given individual (a node) has in a network. Data refer to social networks of Chinese college students in 2011, either during weekdays (left panel) or during weekends (right panel). During weekends the distribution follows a power law, characteristic for a scale-free network. Individuals at the far-right end represent 'superspreaders', i.e. those individuals that have a high degree (many contacts). The study was made in the context of a spread of respiratory diseases. Adapted from Huang et al. (2016), under CCBY 4.0.

2013–2016, found that superspreading was a key feature. Whereas it concerned only 20 per cent of the cases, it accounted for 61–73 per cent of the new infections (International Ebola Response et al. 2016). Superspreading was also a key driver for the 2002–2003 SARS and the 2014 'Middle East respiratory syndrome' (MERS) outbreaks (Lau et al. 2017).

A scale-free (power law) network does not always emerge, of course. Alternatively, an epidemic might have a Poisson distribution for the contact degrees or follow a 'small-world', or perhaps a structured, model. The latter corresponds to urban life, where households are epidemiologically separated from one another but connected by schools or workplaces. The spread of influenza among college students in China, for example, required a remarkably clustered network model because this reflects their hierarchical and ordered lifestyle as compared to their US counterparts. With an ideal free model, the epidemic would have been overestimated (Huang et al. 2016).

In network analysis, the status of each individual can be assigned to the SIR categories (susceptible, infected, recovered). However, the disease can only spread between individuals linked to one another. The spread of a parasite in this network also corresponds to a tree whose root is the first infection introduced into a susceptible population (the index case). Adding up the probabilities of transmission for each step in this tree yields the expected reproductive number, R_0. With $R_0 > 1$, the epidemic

should continue. However, encounters between hosts may not always lead to transmission, and the infection accidentally can die out despite $R_0 > 1$. Likewise, an epidemic can spread despite an average $R_0 < 1$ if the parasite hits a superspreader that causes a cluster transmission event (as in the case of MERS; Kucharski and Althaus 2015). Hence, the population-wide estimate of R_0 can be misleading, as it averages out the critical local and inter-individual variation that may fuel an epidemic much more than expected. Finally, network analyses can account for individual host characteristics, which relates to individually varying disease spaces (Box 11.4).

Network analyses can identify the most efficient control strategies. For the 2014–2016 Ebola outbreak in West Africa, network models showed that travel restrictions were most useful to stop the epidemic, even if compliance was only 40–50 per cent (Wong et al. 2016). However, the existing transmission network typically becomes known only retrospectively. Instead, proxies for transmission are used. For directly transmitted diseases (and, to some extent, also for vectored diseases), this amounts to finding the network of social contacts among individuals, as in the examples of Ebola (International Ebola Response et al. 2016) or influenza (Figure 11.6) (Huang et al. 2016).

Social networks can be of any form. For example, across 47 species of mammals, reptiles, fish, or social insects, solitary species showed the highest

Box 11.4 Epidemics and disease space

The spread of an infectious disease during an epidemic needs infection chains, i.e. transmission of the infection from one host to the next. Necessarily, transmission connects the disease spaces of the individual hosts along this chain—from the point in disease space where the infection enters to where it leaves the host. While in the host, the parasite population follows the infection trajectory through disease space.

Individual hosts vary in their physiology, condition, age, and so forth. Therefore, the disease space of different hosts

is different, even for the same parasite, and the respective infection trajectories will follow a different course (Figure 1). At some point along the infection trajectory, the parasite starts to become transmitted to a next host, e.g. when transmission stages form, or when the parasite has reached a suitable tissue or location.

Therefore, along the infection trajectory, there is a 'transmission window'. Transmission windows vary among hosts in terms of location, extension, and suitability (e.g. how many propagules are transmitted per time interval).

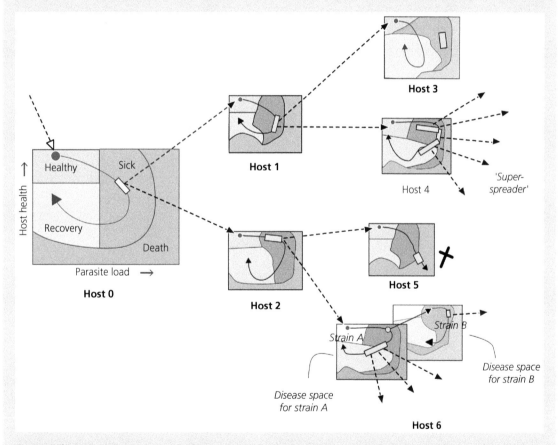

Box 11.4 Figure 1 Disease spaces in an epidemic. In this cartoon, the disease spaces vary among hosts. Accordingly, the infection trajectory for a given parasite is different in each host (black lines with arrow). At some point, there is a 'transmission window' (blue rectangles), during which the infecting parasite population can transmit to a next host. In host 5, this window is only reached as the host dies; in host 3, the window is in a region of the disease space that the parasite cannot reach, e.g. being suppressed or cleared by the host; no further transmission happens. Host 4 allows for long and rich transmission windows, which makes this host a (physiologically based) 'superspreader'. In host 6, a within-host mutation occurs (yellow dot) such that a new strain (strain B) descends from its ancestor (strain A) as the infection progresses. The two strains A and B experience the same host through different disease spaces, and therefore also reach different transmission windows. Transmission events are characterized by broken lines and connect the disease spaces of sequentially infected hosts. Host 0 could be the 'index case'; that is, the first identified host of an epidemic. Strictly speaking, $R_0 = 2$ in this case, as host 0 infects two new hosts.

continued

Box 11.4 *Continued*

Sometimes, the window is short and sparsely populated by the infection, and, hence, the probability of further transmission is low. The window may also not appear before the infection trajectory crosses into the host death zone. The window also may be located outside of the infection trajectory that the parasite can reach in this host (e.g. the parasite is cleared before being able to transmit). In both cases, no further transmission results. Alternatively, the transmission window may last a long time and be used intensely: this

would make this individual host a 'superspreader'. For relatively rapidly evolving parasites, e.g. viruses or bacteria, new variants (parasite 'strains') can also evolve as the parasite population follows its trajectory through disease space. A new variant will then experience the disease space differently from its parental type. A variant follows a slightly different infection trajectory that may or may not result in more transmission. With more transmission, the parental strain likely is replaced by the new strain as the epidemic unfolds.

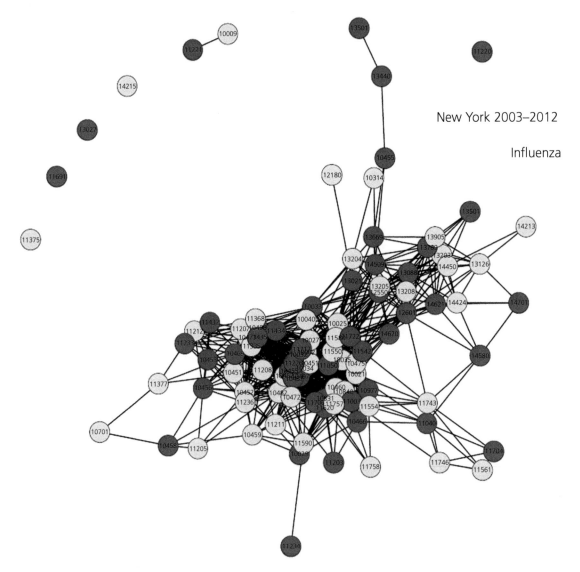

Figure 11.6 Transmission network of influenza in New York. Shown are the hospitalized cases between 2003 and 2012, and their mutual connections (contacts). This network shows 100 nodes and has a power law type distribution of the contacts (degrees). The highest observed degree was 86 in the Bronx (New York City). Each circle (a node) represents a small geographical area defined by its zip code (small letters) and having at least 70 patients. Each line (an 'edge' in the network) represents contacts with high correlation from the contact matrix (cut-off, $r \geq 0.9$). Nodes without lines have many patients but are isolated with respect to this network. Reproduced from Ljubic et al. (2019), with permission from Elsevier.

variance in the number of social contacts, and gregarious species were most fragmented in their social networks (Sah et al. 2018). The structure of social networks can, therefore, prevent the spread of disease to the most valuable reproductive individuals in highly social groups (Schmid-Hempel and Schmid-Hempel 1993). Furthermore, social networks plastically adapt to a disease threat (Rushmore et al. 2017; Stroeymeyt et al. 2018). Essentially the same process of changing a network structure happens with the classical epidemiological tools, such as quarantine, the installation of a 'cordon sanitaire' (i.e. the isolation of a location), restriction of human travels, locking down social and business life (as in the Covid-19 pandemic), or banning the shipping of animals between regions or farms (Marquetoux et al. 2016).

11.2.5 Spatial heterogeneity

Spatial heterogeneity can be at any scale and results, for instance, from social heterogeneity— where groups of individuals interact with one another in a local neighbourhood—geographical barriers, or microclimatic conditions affecting the survival of spores. With spatial heterogeneity, events have a spatial component, and this shapes the transmission networks. In the case of the Ebola outbreak, new infections occurred in a spatially restricted manner, with a median distance of only *c.*2.5 km between the source and new case (Lau et al. 2017). A spatially explicit metapopulation model also better traced the 2009 spread of H1N1-type influenza over 20 weeks in South Korea and helped to develop intervention strategies (Lee et al. 2018). A large number of other examples demonstrate the importance of spatial heterogeneity and spatial dynamics for the spread of diseases in humans, livestock, and wildlife, such as the spread of feline immunodeficiency virus (FIV) in mountain lions (*Puma concolor*) (Biek et al. 2006).

11.2.6 The epidemic as an invasion process

With the start of an epidemic, a parasite enters a host population and, at this point, the parasite's fitness during such an invasion process matters. In cases where the invading parasite may have to

compete with other, already-resident parasite variants and must cope with the hosts' status, analyses using the evolutionarily stable strategy (ESS) concept may be useful (Lion and Metz 2018). The so-called 'epidemic attractor' characterizes the situation. This point defines where the epidemic eventually will gravitate to, characterized with the 'endemic equilibrium', $(\hat{S}, \hat{I}, \hat{R})$, i.e. the equilibrium numbers of susceptible, infected, and recovered host individuals. Hence, the invading parasite's maximum R_0 that it can achieve in a field of resident parasite strategies is of interest to find the invading parasite's best actions. These considerations are more of a concern for the discussion on the evolution of virulence (Chapter 13).

11.3 Endemic diseases and periodic outbreaks

Infections are 'endemic' when they are sustained in a population for more extended periods, without the need to be reintroduced from outside. An endemic state can result from various histories. For example, a previous wave of infection may not have died out entirely after the epidemic has run its course. It is subsequently maintained, often at low levels, in dynamic equilibrium for extended periods. Alternatively, an endemic infection may suddenly generate the conditions for an epidemic outbreak caused by several processes, e.g. demographic stochasticity, seasonal variation, immigration of susceptibles, loss of immunity, or inherent time lags. When the epidemic thus returns to the endemic state, further recurrent outbreaks may follow because infected hosts remain a source. Hence, in the case of periodic outbreaks, repeated epidemic waves keep interrupting the longer-term endemic state.

Measles (rubeola) in the prevaccination period shows such a pattern of periodic outbreaks (Figure 11.7). It is a typical childhood disease caused by the measles virus. The virus transmits via droplets and infects the respiratory tract; it is highly contagious, with an estimated 90 per cent of exposed people contracting the infection (Rall 2003). The incubation period is 10–12 days, when the typical Koplik's spots appear—small, irregular spots in the mouth, named after the American paediatrician, Henry Koplik (1858–1927)—accompanied by flu-like

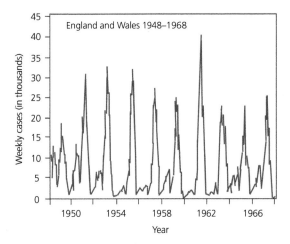

Figure 11.7 The dynamics of measles before the age of vaccination. Shown are the numbers of clinical cases reported weekly for England and Wales in the period from 1948 to 1968. Reproduced from Anderson and May (1991), with permission from Oxford University Press.

symptoms and, later, skin rashes. Complications with different degrees of severity (e.g. pneumonia, middle ear infections) occur in up to 30 per cent of the cases. Seizure and deafness are estimated to hit one out of 1000 cases. In one out of 100 000 cases, severe infections of the central nervous system develop, even months or years after the infection, associated with inflammation of brain tissue, which can lead to progressive motor impairment and eventual death.

In prevaccination times, measles caused an estimated two to three million deaths annually and left a large number of children with permanent disabilities. In 2018, the toll was still *c.*140 000 deaths, primarily in low-income countries with inadequate health systems. Periodic outbreaks in prevaccination times (Figure 11.7) resulted from the periodic exhaustion of susceptible hosts, e.g. by long-lasting immunity (which makes it a 'childhood' disease). In turn, numbers became sufficiently replenished by births and immigration after some time, such that a new wave could start. Such patterns can statistically be partitioned into the underlying constituent periodic waves. For prevaccination measles (Figure 11.7), the spectrum has a primary wave every two years and a minor wave every year. Other classical childhood diseases show similar prevaccination patterns. For example, pertussis

(whooping cough) has a seasonal cycle combined with a dominant three-year cycle. Similarly, mumps seems driven by a two- to three-year cycle (Anderson and May 1991). Such annual waves could result from simple seasonal effects, whereas periods of two to three years might reflect the periodic recruitment of children to school classes.

In 1963, a vaccine was introduced in the United States. Subsequently, cases dropped by 98 per cent. Measles has no known animal reservoirs and thus could be eradicated by vaccination programmes. Moreover, the vaccine is among the safest and most effective, providing 97 per cent protection with an extremely low incidence of vaccine-related serious complications (Paules et al. 2019). However, vaccination must reach 90–95 per cent of the population to establish herd immunity (see Table 11.1). The vaccination situation seemed on the right track around the turn of the millennium, with a declining number of cases and limited outbreaks. Logistical problems, e.g. the need for keeping vaccines in a cool environment for storage and transport, but also scepticism against vaccination, have slowed down progress since about 2010 (Patel et al. 2016; Paules et al. 2019). Periodic outbreaks as in earlier decades now re-emerge even in industrialized and wealthy countries.

11.4 Epidemiology of vectored diseases

Many important human infectious diseases are transmitted by vectors, especially by bloodsucking insects such as mosquitoes, flies, ticks, or bugs, and affect hundreds of millions of people worldwide. Numerous examples exist, e.g. malaria, yellow fever, dengue, Ross River fever, West Nile virus, leishmaniasis, Chagas disease, sleeping sickness, Lyme disease, encephalitis, and many others. The dynamics of vectored diseases can be traced with the SIR model by adding compartments for the uninfected and infected vectors, and by adjusting the transmission rates and other variables accordingly (Dye and Williams 1995). Such modelling was pioneered by Ronald Ross (1857–1932) and later extended by George Macdonald (1903–1967) into the Ross–Macdonald model (Macdonald 1957). For example, insecticide-treated bed nets are an essential tool to reduce the transmission of malaria.

Modelling helped to predict that weak repellents allowed for a lower coverage to achieve control; this is because strong repellents divert hungry mosquitoes to feed on other, non-protected hosts instead, which sustains the infections (Birget and Koella 2015). Furthermore, vectors (an insect, a tick) are short-lived compared to the final host of the parasite (a vertebrate animal). In a model, the vector populations can therefore be set to a numerical equilibrium while the processes in the host population unfold. Such considerations simplify the analysis and allow approximate solutions to a complex problem. Incidentally, with these simplifications, the dynamics of a vector-transmitted infectious disease can thereby come close to that of directly transmitted parasites.

Biologically, vectors often are also intermediate hosts where vital steps for the parasite's life cycle happen. These steps can be important for the dynamics of an epidemic. For example, the time *Plasmodium* needs for its development and to become infective (the latency period) is comparable to the life span of the vector. Hence, few mosquitoes usually live long enough to become infected and to become infective themselves. In hyperendemic areas, however, most vectors become infected at an early age and become infective, too. The mosquito biting rate is also affected by the vector and host conditions. Mosquitoes are twice as frequently attracted to humans infected by malaria (Lacroix et al. 2005), and malaria-infected mosquitoes prefer human odours (Smallegange et al. 2013) (see also section 8.3). Individual-based models, as mentioned elsewhere, have also been successfully used for vector-transmitted pathogens and the evaluation of control scenarios (Smith et al. 2018b).

For many human infectious diseases, vector control is, in fact, the best or only option for control. An example is dengue, a tropical and subtropical disease caused by the dengue virus (DENV). The virus is vectored by mosquitoes, primarily *Aedes (Stegomyia) aegypti*. Over the last two decades, dengue has increased eightfold and has spread globally. According to the WHO, there are approximately four million new infections per year (as of 2019) and 100–400 million people affected worldwide. There are four variants (serotypes) of DENV, and it is possible to become infected by all. Symptoms vary from virtually unnoticeable to flu-like symptoms, but complications can become serious or even lethal with severe, haemorrhagic dengue, which affects *c*.500 000 people per year (Salles et al. 2018). Because there is as yet no generally preventive vaccine (as of 2019, one vaccine is licensed for limited use) and no effective therapy, vector control is the only remaining option.

Aedes aegypti, which, for historical reasons is known as the 'yellow fever mosquito', is an excellent example of how diseases—yellow fever, or dengue—can follow in the wake of their vector's expansion. *A. aegypti* breeds in small, human-made water bodies such as small containers or used tyres that become filled in the rain. With the ongoing urbanization and much poverty in the less developed countries, such opportunities have become more common over the last decades and have fostered the spread of dengue.

Current knowledge suggests that *Aedes* originated in Africa. Probably about 500 years ago, the lineage leading to the primary vector of today became associated with human settlements, most likely in West Africa, as from the sixteenth century merchant ships from Europe would call at West African ports to pick up Africans (also recruited, for example, from Angola in the Portuguese domains) for the slave trade in the New World, but also to refresh their water supplies for the long crossing. At the time, yellow fever was present in West Africa; the records for dengue are less clear, also because many cases are asymptomatic. The ships then carried the virus and their vectors across the Atlantic. This was due to the water supplies as well as infected people. Given the life cycles, the infection, in fact, had to be sustained by ongoing transmission onboard the ships during passages of two to four months. The slave trade, therefore, also established the virus and viable populations of *A. aegypti* in the Americas. The first outbreak of yellow fever occurred in 1648 in Havana and the Yucatan. Epidemics during 1635 in Martinique and Guadeloupe, and 1699 in Panama, could have been dengue, but this remains debatable. The 1780 outbreak in Philadelphia was dengue, though (Brathwaite Dick et al. 2012). During the eighteenth century, the increasing trade brought the vector back across the Atlantic and into the Mediterranean basin.

As the Suez Canal opened in 1869, *Aedes* could rapidly spread along the shipping routes to Asia, Australia, and the Pacific region (Powell et al. 2018). Along with the vector, urban dengue appeared in Asia and reached Australia in 1897. Following the failure of the global *Aedes* eradication programme, the number of dengue outbreaks started to increase after 2000 (Brathwaite Dick et al. 2012). Today, vector control remains the best option to control dengue.

Modern technologies of vector control can include 'gene drive'. With this technique, CRISPR–Cas editing (see Box 4.2) is used to insert, for instance, a fertility-distorting modification into the genome of mosquitoes. The modification is further combined with a genetic driving element such that it copies itself into offspring instead of the original version (Scudellari 2019). In laboratory populations, the modifier spreads successfully and can crash *Anopheles gambiae* populations within 7–11 generations (Kyrou et al. 2018). More ambitious gene drives can, for example, deliver antiparasite effectors to the vector (Carballar-Lejarazú et al. 2020). The application to real-world vector control, e.g. a field release of genetically engineered mosquitoes, has political, legal, and ethical issues (Adelman et al. 2017; Kolopack and Lavery 2017).

11.5 Epidemiology of macroparasites

In epidemiology, macroparasitic infections are analysed by tracking individual life histories. Macroparasites, e.g. cestodes or trematodes, are large in body size and replicate relatively slowly. They are also not numerous and typically reside for prolonged periods inside their hosts. Hence, the per capita growth, body size, and fecundity become useful quantities. However, note that microparasites, e.g. bacteria, can be counted as individuals, too, and their sizes or rates of cell division can be measured by modern techniques. Moreover, the concept of a 'population virus load', i.e. summing up viral numbers over hosts, is used to predict transmission, e.g. for HIV infections (Montaner et al. 2010; Tanser et al. 2017). Nevertheless, in many cases, such approaches can be difficult or impractical.

11.5.1 Distribution of macroparasites among hosts

A prominent feature is the distribution of the number of parasites per host, also known as 'parasite load' or 'parasite burden'. This corresponds to infection intensity for microparasites. Parasite loads are important, since they typically correlate with an increased risk of host death or loss of host fecundity

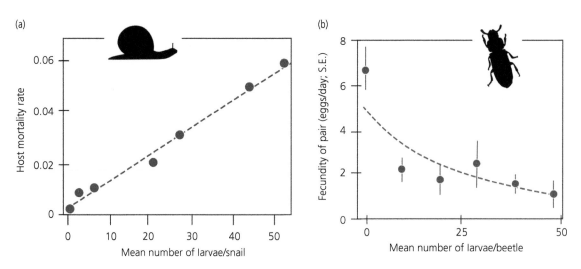

Figure 11.8 Effect of macroparasite load on host fitness. (a) A high number of nematode larvae (*Elaphostrongylus rangiferi*) increases mortality rate of the intermediate host, the snail *Arianta arbustorum*. (b) A high load of larvae of the cestode *Hymenolepis diminuta* in the flour beetle *Tribolium confusum* reduces fecundity. Silhouettes from phylopic.org. Adapted from Dobson et al. (1992), with permission from John Wiley and Sons.

(Figure 11.8). Among hosts, macroparasites rarely are evenly distributed, but rather are 'aggregated'. Some hosts have more, whereas others have fewer parasites than expected by 'chance', i.e. than expected from a Poisson distribution that characterizes independent events. With aggregation, parasites are 'overdispersed'. Statistically, such a pattern follows a negative-binomial distribution, which is characterized by a mean and an aggregation parameter, k. Empirically, many macroparasites show negative-binomial aggregations with $k = 0.1$ to $k = 10$, where low values indicate more aggregated distributions. Several other useful indices for aggregation are in use that focus, for example, on abundance-rank relationships, such as Poulin's D (McVinish and Lester 2020).

An aggregated distribution can result from various underlying processes, although these are still unknown for most cases. The sampling process itself may introduce artefacts, e.g. by limited sample sizes, and which allow for certain distributions but not others (Wilber et al. 2017). Nevertheless, the biological processes are most important and of prime interest (Johnson and Wilber 2017). For example, the hosts themselves might aggregate at certain sites, e.g. to search for food. Hosts can also vary in their susceptibility to infection, either by genotype, body condition, or differences in acquired immune protection. Furthermore, males typically carry heavier parasite loads than females; hence, a sex ratio bias in the sampling procedure would change the result, too. Additional effects can result from co-infections (Morrill et al. 2017).

In a large number of different systems, an increase in parasite load with host age is observed. This can result when the rate of new infections exceeds clearance or host mortality, and immune protection against subsequent infections is weak. Alternatively, the infection rate can decrease with age, e.g. due to changed behaviour, or by immune protection. In these cases, maximum parasite loads are observed at intermediate age classes, as for the number of ticks (*Ixodes ricinus*) per chick in a red grouse population (Anderson and May 1991; Hudson and Dobson 1995; Quinnell et al. 2004). Similarly, a population of young hosts often shows a more or less random distribution of parasites, whereas in older age classes an aggregated pattern emerges; this can result from differences in individual host development.

11.5.2 Epidemiological dynamics of macroparasites

Red grouse (*Lagopus lagopus scoticus*) infected by the nematode *Trichostrongylus tenuis* is an illustrative example of a macroparasite–host interaction (Hudson 1986). This case is dominated by the effects of the nematodes on host survival and fecundity. In particular, an increasing parasite load reduces fecundity, such that the population of host birds tends to oscillate over time (Hudson et al. 1992a). However, macroparasite–host systems show a range of dynamic behaviours. This includes a scenario of relatively stable host numbers over time, fluctuating populations (in terms of parasite burden per host), or even host population crashes as the infections take hold (Hudson and Dobson 1995). At least some form of host population regulation must be in place to explain such stable or fluctuating host population sizes. Regulation, in turn, implies density-dependent processes (see Box 15.1; section 15.1.4). For example, increasing host densities may lead to higher parasite loads that subsequently reduce host fecundity or increase host mortality rate (Figure 11.8). Among other things, such effects depend strongly on the degree of aggregation. Similar to the SIR model for infectious diseases, the epidemiology of macroparasites can be analysed with compartmental models (Box 11.5).

Modelling the epidemics of macroparasites again helps to devise control and vaccination strategies. For wildlife populations, a prime concern is management for conservation, or to prevent the spillover of infections to livestock (White et al. 2018c). The results of modelling are implemented in interventions, e.g to design vaccination or culling schedules (McCallum 2016). Disease control is also a significant aim for human macroparasites. For example, an extension from the model described in Box 11.5 was used to estimate the effects of vaccines against schistosomiasis (caused by the trematode *Schistosoma*). The parameters of interest were the differences in vaccine efficacy (low or high protection), the different vaccination schemes (only cohorts of infants, or mass vaccination), and what type of effect a vaccine has on the parasite, i.e. preventing establishment, reducing survival, or reducing transmission. From the model, the first two factors proved to be particularly important, with the duration of the protection

Box 11.5 Epidemiology of macroparasites

Models for the epidemiology of macroparasites keep track of different parasite classes; for example, juvenile and adult stages. Figure 1 shows a simple scheme that might apply for nematode infections of birds and mammals (Hudson and Dobson 1989). A set of differential equations can describe the changes in the number of hosts (H), adult (P), and free-living (W) parasites. With the symbols of Figure 1, but ignoring the juvenile stage for simplicity,

$$\frac{dH}{dt} = (b - \mu_H)H - \alpha P$$

$$\frac{dP}{dt} = \beta WH - (\mu_2 + \mu_H + \alpha)P - \alpha\left(\frac{P^2}{H}\right)\frac{(k+1)}{K}$$

$$\frac{dW}{dt} = \Phi P - \mu_0 W - \beta WH \qquad (1)$$

As before, β, is the transmission rate, but this time this is from the free-living stages to the hosts. For dP/dt, the parameter, k, characterizes the aggregation of parasites among hosts; the last term describes the average expected individual load, with its effect on host mortality beyond the linear effect. From these equations, the basic reproductive number of the parasite is:

$$R_0 = \frac{\beta \Phi H}{(\mu_2 + \mu_H + \alpha)(\mu_0 + \beta H)} \qquad (2)$$

In eq. (2), the number of newly generated parasites in the hosts (via parasite fecundity, $\beta\phi$) is weighted against the expected stay of the parasite inside the host (i.e. the average infection lifetime, $1/(\mu_2 + \mu_H + \alpha)$, and by the average lifetime of free-living stages, $1/(\mu_0 + \beta H)$). In the case of the nematode *Trichostrongylus tenuis* infecting red grouse, $R_0 = 5$ to $R_0 = 10$ (Dobson et al. 1992), thus reaching similar values as infectious diseases of humans (Table 11.1). It is possible to estimate the minimum number of hosts, H_{crit}, that allow an epidemic to unfold with the boundary condition $R_0 \geq 1$. Any treatment to eradicate the parasite would likewise require measures to reduce the number of susceptible hosts, H, below this threshold. For this model, equilibrium values for H, P, and W have been derived mathematically (Anderson and May 1978). The respective equations suggest that the equilibrium host number is directly proportional to the degree of parasite aggregation (as given by $(k + 1)/k$) and inversely correlated with parasite fecundity (ϕ) and transmission rate (β).

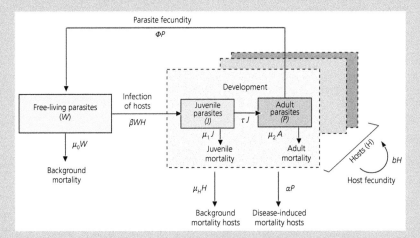

Box 11.5 Figure 1 A model of host–macroparasite dynamics. The compartments describe the number of juvenile (J) and adult (P) parasites, and of hosts (H). There is an obligatory free-living stage of the parasites (W), which, for example, could represent larvae, spores, or durable stages. Background host mortality rate is μ_H, parasite-induced host mortality is α. The parasites die at rates μ_2, for example, when being cleared by the host's immune system or from other causes. Developmental time to adulthood is indicated by τ. Transmission from the free-living stage to the host is given by the infection rate, β. A parasite-induced reduction of host fecundity could be added to the model, too. The model reflects the dynamics of nematodes (*Trichostrongylus tenuis*) infecting birds (red grouse; Dobson et al. (1992)).

being decisive, requiring a level of protection for five to ten years (Stylianou et al. 2017).

11.6 Population dynamics of host–parasitoid systems

Parasitoids are an essential group of parasites more generally. They grow as larvae inside or attached externally to a host, and typically kill their host as an obligate part of their life cycle. Therefore, the ecology and population dynamics of parasitoids are somewhere in between those of predators and infectious parasites (Hassell 2000). Observations from the field, as well as from experiments in the laboratory, often show fluctuations in the number of hosts over time, or host populations that are drastically reduced in size by the effects of parasitoids (Hassell and Godfray 1992; Godfray 1994).

The classical Nicholson–Bailey model predicts the dynamics of a host–parasitoid system (Nicholson and Bailey 1935). It assumes discrete generations and random encounters in large populations. The formulation is straightforward: the host population, H, basically increases from generation t to $t + 1$ with a given multiplication factor, R. At the same time, the number of hosts that survive to reproduce declines over time with the attack rate, a, of the parasitoids because infected hosts are eventually killed (Abram et al. 2019). Hence, $H_{t+1} = R\,H_t\,e^{-aP_t}$, where P_t is the number of parasitoids at time t. Parasitoid infections can furthermore alter host behaviours and reduce host fecundity, which is not considered here. Likewise, the increase in the number of parasitoids is proportional to the number of hosts that become infected, such that $P_{t+1} = c\,H_t(1 - e^{-aP_t})$; c describes the rate of conversion of hosts into new parasitoids. This simple model generates an equilibrium that is non-stable and shows oscillations with ever-increasing amplitudes in population size. Model extensions of the same general type produce a wide variety of dynamic behaviours, including stability, bifurcation, or chaotic outcomes, which require sophisticated mathematical treatment.

The basic Nicholson–Bailey model ignores biologically more realistic scenarios. For example, in nature, parasitoid attacks are typically aggregated. The parasitoids also interfere with each other's searches, and they compete for access to the same hosts. Such interference would affect the attack rate and make it density dependent. Host–parasitoid interactions can also vary in space and, for example, unfold in a metapopulation. Further complications arise as individual hosts vary in their susceptibility to parasitism, or by differences in microhabitats. Despite these challenges, the dynamics of host–parasitoid systems can be modelled successfully, as for moths and their parasitoid enemies in Northern Europe (Mutanen et al. 2020). This modelling is important because parasitoids are agents of biological control for many pest species. Moths and other insects attack agricultural produce and therefore need to be controlled. An example is the highly effective control of cassava mealybugs by parasitoid wasps (Neuenschwander et al. 1989).

11.7 Molecular epidemiology

The microevolution of many pathogens, such as bacteria or viruses, is very fast and happens within the same time scale as the unfolding of the ecological dynamics of an epidemic. This microevolution of 'measurably evolving' parasite populations (Drummond et al. 2003) involves processes such as mutation, recombination, selection, (population) genetic drift, and horizontal gene transfer common enough to generate a divergence of genetic sequences during the epidemic. Therefore, the genetic variation among isolates reveals something about the epidemic itself. For this, the evolutionary processes are aligned with the epidemiological processes by considering how an infecting population produces new variants that subsequently become differentially eliminated by natural selection (e.g. by host defences), and that becomes transmitted to a next host.

Co-infections are particularly relevant. In RNA viruses such as HIV, one virion can contain two different RNA strands. As the viral genome replicates in an infected host cell, the enzyme responsible for copying the sequence will first start reading the genetic information from one strand, but at some point jumps to read onwards from the other strand. It may subsequently jump back again to the first strand. This mechanism is quite different from the meiotic recombination by chromosome crossovers in eukaryotes. However, the result is the same—a new genetic sequence emerges with elements from two parental sequences; this is a mutation in the general sense. After some time, the infecting population will therefore have accumulated many new variants. When the parasite population persists and replicates within the host, the new variants become exposed to selection by the host's immune system. Eventually, a subset of the variants will survive and transmit to a new host; the parasite population, therefore, passes through a bottleneck. Furthermore, the transmitted subsample will now bear the respective signatures of mutation and selection.

'Molecular epidemiology' is the discipline that tracks an epidemic with genetic markers or entire genomic sequences. The sampling of pathogen isolates from many host individuals at a given time ('isochronic' samples), or over several time points during the epidemic ('heterochronic'), yields the primary data. This data is an array of genomic sequences, e.g. from virus isolates. The array carries information about the isolates' historical relationships and the genetic distance and microevolutionary steps connecting them. Together with the observation of the abundance and distribution of isolates, it thus becomes possible to reconstruct the past population-dynamic processes that generated these patterns, e.g. to reconstruct the transmission chains. There is, of course, only one real course of the epidemic, whereas the reconstruction is a statistical process. Therefore, any inferred scenario can only identify the most likely model for the course of a given epidemic. This blending of concepts from epidemiology (i.e. infectious disease dynamics) with concepts for reconstructing parasite phylogenies with the help of genetic sequences (i.e. molecular epidemiology) is also known as 'phylodynamics' (Grenfell et al. 2004) (Box 11.6).

Box 11.6 Phylodynamics

For microparasites, such as bacteria or viruses, the time scale of an epidemic is similar to the time scale of their evolutionary change. Such pathogens are 'measurably evolving' (Drummond et al. 2003). Modern genomic methods allow collecting a large number of viral sequences from different hosts and at different time points along the course of an infection. Computational-statistical tools can then yield a phylogeny from these sequences that connects viral types (and their hosts) in a tree, time-calibrated with a molecular clock—a method known as 'phylodynamics' (Grenfell et al. 2004) (Figures 1 and 2).

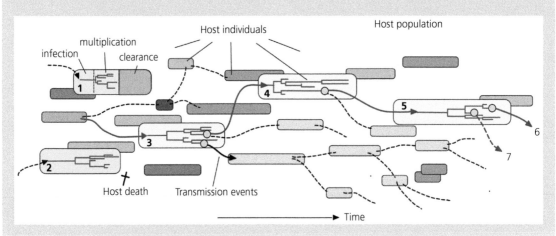

Box 11.6 Figure 1 The principle of phylodynamics. During an epidemic, individual hosts (rectangles) become infected at different times. Within a host, the parasite population (e.g. a viral infection) evolves, i.e. new variants appear and spread (characterized by the inset phylogenetic trees; red). From the within-host trees, only some variants (blue dots) transmit to a next host (solid lines). In the cartoon, host 3 infects hosts 4, 5, and then 6 and 7 in sequence as within-host evolution continues (light yellow rectangles); another variant from host 3 starts a transmission chain through other hosts (dark yellow rectangles). Transmission events can either be known (solid lines) or unknown (dashed lines) to the observer. With an explicit compartmental model, a host is infected or recovered; an infection either transmits or is cleared (as in host 1). Host 2 dies and takes the infection with it. Phylodynamics reconstructs these patterns statistically, e.g. by calculating the likelihood of the blue transmission chain, based on the observed genetic variation within and among-host infections.

A simplified procedure used in phylodynamics is the 'skyline plot' (Pybus et al. 2000). In the corresponding graph, the 'skyline' is a stepwise curve with ups and downs (like buildings in a city), which characterizes the (non-parametric) estimate of the number of infecteds (i.e. the heights of the buildings) at various time points during the epidemic. When the epidemic is spreading, the skyline is overall rising in height. The calculations are based on effective population sizes and the corresponding time to coalescence for lineages of genetic variants. In small populations, only a short time is necessary to go back to where two lines coalesce, i.e. where they split from their most recent common ancestor. This approach does not assume any underlying epidemiological mechanisms but simply reconstructs the pattern of past events in a phylogeny. A maximum-likelihood approach judges whether a hypothesized phylogenetic tree is a good explanation of the observed sequences (i.e. the pathogen variants sampled from the population), and finds the best fit.

Alternatively, with Bayesian approaches, 'posterior' values for the model and tree parameters are calculated, given 'prior' conditions. A 'prior tree' is a plausible tree with its respective parameters. Algorithms such as Markov Chain Monte Carlo simulations allow sequential improvement of the estimated trees together with their parameters (the posterior values), given the priors and the observed data.

In defining a prior tree, the coalescent approach (Volz 2012), or some kind of demography derived from a birth–death process (see Box 11.3), can generate the estimates (Stadler 2010). For example, the observed distribution of pathogen genetic variants yields a probability density for different possible birth–death processes, and hence estimates which priors are the most likely ones. The algorithm ideally converges to the best estimate for the true tree of this epidemic.

More comprehensive approaches include explicit epidemiological processes, such as compartmental models like the SIR model, or stochastic processes. A range of algorithms, using different assumptions and models for the epidemiological process, have been developed so far (Leventhal et al. 2013; Kühnert et al. 2014; du Plessis and Stadler 2015; Saulnier et al. 2017; Vaughan et al. 2019). Besides, viral sequences for reconstructing epidemiological phylogenies are available from open-access databases, such as *Nextstrain*.

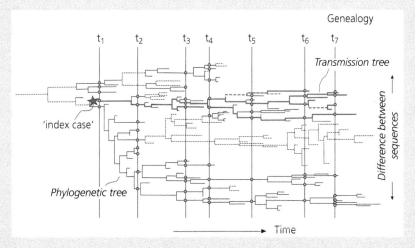

Box 11.6 Figure 2 Reconstruction in phylodynamics. When hosts and their infections are sampled at different time points ($t_1, ..., t_7$), the genomic sequences of variants are observed (sometimes only the final point, t_7, may be known). A true but unknown 'genealogy' connects these sequences. This is the true phylogenetic tree connecting all pathogen variants that are spreading in this host population. But only a part of the genealogy is sampled (green dots). Many sequences and their connections remain unknown (dashed lines); these may have become extinct, are not sampled or reported, etc. From the remainder (green dots), a best-fitting phylogenetic tree is reconstructed. This 'transmission tree' is one instance of all possible transmission chains that could be true. It will connect the sampled or inferred individuals (red lines). The likelihood that such a hypothesized transmission tree results from an explicit epidemic process, such as the SIR dynamics, is calculated. By stepwise improvements of the underlying parameters, the algorithm ideally converges to a most likely transmission tree, which yields the best description of the spread of a pathogen in a host population. In phylodynamics, the best 'transmission tree' (red tree) is considered to be the genealogy. The sketch is based on material from du Plessis and Stadler (2015) and Vaughan et al. (2019).

It estimates the underlying epidemiological parameters and transmission patterns, and analyses population structures or pathogen evolution. Together with the increase in sequencing capacities and computing power, with more and more sophisticated mathematical-statistical algorithms, the study of an epidemic from genomic data alone has grown into a potent tool.

Molecular epidemiology is particularly useful to investigate a disease outbreak where usually little is known about the pathogen (Grubaugh et al. 2019). Typically, a new infectious disease becomes noticed when a patient develops clinical symptoms that cannot be assigned to any known cause. At this point, genomic screens can identify what may be the agent, e.g. identify a new virus that has just entered the human population. The genomic screening of more cases, and possible sources and zoonotic reservoirs (see section 15.5) would subsequently reveal the likely origin and transmission routes. For example, the emergence and outbreak of several coronaviruses (Box 11.7) involved molecular epidemiology and phylodynamics. Adding information on geographic, economic, demographic, and travel patterns will additionally complete the picture and provide insights on risk

---

Box 11.7 Coronavirus outbreaks

With the recent, major Covid-19 pandemic, the Coronaviridae (CoV) have received public attention. It is a large family of enveloped, positive, single-stranded RNA viruses. CoV are the largest known RNA viruses, with genome sizes ranging from 25 to 32 kb, and a physical virion diameter of 118–136 nm. Virions are spherical and show protruding spikes that extend 16–21 nm from the envelope, which earned them their name of 'corona' (a crown). Spikes are formed by the spike glycoprotein (S) (Payne 2017). At least 2500 different genetic variants of CoV are known and grouped into 40 species (when to call them a 'species' depends on the phylogenetic position and genetic divergence from other taxa;

Coronaviridae Study Group 2020). Among the two subfamilies, Torovirinae are known from gastrointestinal infections of mammals and, rarely, of humans. The second subfamily, Coronavirinae, are widespread among mammals (notably, the Alpha and Beta genera), typically causing mild enteric or respiratory infections. Bats, in particular, appear to have a long co-evolutionary history with coronaviruses. Many species harbour their specific viruses. Bats are a prime animal reservoir that can lead to zoonoses (Figure 1).

Until recently, coronaviruses were considered dangerous for animals only, e.g. chicken, pigs, or domestic cats, in which they cause severe and deadly enteric or respiratory

Box 11.7 Figure 1 Worldwide distribution of bat species known to harbour coronaviruses (Coronaviridae). Shading reflects the number of bat species in a given area (with darker grey corresponding to more species). The viral data is based on the DBatVir database (Chen et al. 2014), and bat distribution data from the IUCN. Reproduced from Hayman (2016), with permission from Annual Reviews, Inc.

infections. For humans, a first CoV isolate was reported in 1965 and was associated with a common cold. By 2019, half a dozen different strains were known to infect humans, and typically were responsible for largely unproblematic common colds (Su et al. 2016). The 2002–2004 outbreak of 'severe acute respiratory syndrome' (SARS) changed the picture. Soon after this first SARS pandemic, an outbreak of another, novel coronavirus (MERS-CoV) occurred in 2014, causing the 'Middle East respiratory syndrome' (MERS). This pandemic started in Saudi Arabia, with the first case in 2012, and continued into Asia, most notably into South Korea (Figure 2). Both viruses are still present and have repeatedly caused local infection events. However, at the end of 2019, the most severe global pandemic since the 'Spanish flu' in 1918 hit a largely unprepared world. Named SARS-CoV-2, again it was a novel coronavirus, causing 'Coronavirus-disease 2019' (Covid-19) (Coronaviridae Study Group 2020).

The SARS pandemic of 2002–2003: The first case (the index case) of the new virus in humans, now known as SARS-CoV-1, was registered in the Guangdong province, Southern China, in November 2002. Over the next nine months, it had spread globally to 26 countries or regions (Figure 2). Some 8300 cases were reported at the time, with 774 deaths; this amounted to a case fatality rate of 9–10 per cent (Su et al. 2016; Payne 2017).

The reconstruction of the early spread of SARS-CoV-1 suggested that it emerged from a jump to humans from

masked palm civets (*Paguma larvata*) in a live animal market ('bushmeat'; see section 15.5). However, palm civets were only a bridge host that connected the background animal reservoir in bats to the human population. Unspecific symptoms of SARS have an incubation period of 2–11 days, and severe breathing difficulties appear 7–14 days after infection. Transmission is primarily by droplets from sneezing or coughing. The serial interval was estimated to be 8–12 days, resulting in a $R_0 = 2.2$–3.6 in a situation of uncontrolled spread (Lipsitch et al. 2003).

The SARS-CoV-2 pandemic (Covid-19): The first signs of this latest pandemic were noticed at the end of 2019 in Wuhan, China. However, SARS-CoV-2 is not a descendant of SARS-CoV-1, and the closest relative (RaTG13) at the whole-genomic level is found in bats. However, the sequence similarity in the critical receptor binding domain of SARS-CoV-2 is not close to RaTG13. At the same time, the genetically closest bat viruses exist at a considerable geographic distance away from the first outbreak point. It is likely, therefore, that bats are the background reservoir for SARS-CoV-2. Its emergence involved other hosts, similar to the history of SARS-CoV-1 (via civets) or MERS (via camels). The novel virus probably also emerged from contact with infected animals in a 'wet market', this time likely in Wuhan, China. However, the contact may have been via contaminated surfaces, and the virus could even have spread widely in China before it broke out (Zhang and Holmes 2020). Some CoVs

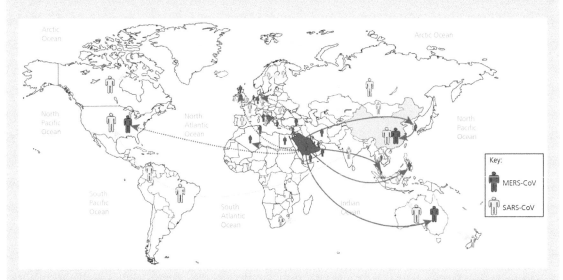

Box 11.7 Figure 2 The SARS-CoV-1 and MERS pandemics. SARS emerged in China in 2002, MERS in Saudi Arabia in 2014. The MERS pandemic 2014–2016 affected over 1600 people in 27 countries, and the virus is still circulating. It resulted in a high case fatality rate of over 30% (Peeri et al. 2020). Reproduced from Su et al. (2016), with permission from Elsevier.

continued

Box 11.7 *Continued*

---

found in pangolins, animals illegally imported for meat into the province, show close similarity in the crucial receptor binding domains, but they seem not similar enough to be a direct ancestor of SARS-CoV-2 (Zhang et al. 2020). This novel virus may therefore have evolved in an as yet unknown animal reservoir, perhaps in pangolins, or may have acquired the binding properties during a cryptic spread in humans before the registered outbreak. Also, emergence by recombination with other, closely related CoVs remains a possibility (Andersen et al. 2020; Zhang and Holmes 2020).

This new virus has a high, but not ideal, binding affinity to the human epithelial receptor *ACE2*. It is spread primar-

ily by droplets and aerosols and can be transmitted further already a few days post-infection. Around 90 per cent of all presymptomatic infections occur around four days (Ashcroft et al. 2020) before the symptoms appear; symptoms occur 5–14 days post-infection (Feng et al. 2020; Xie and Chen 2020). Asymptomatic transmission may account for nearly half of all infections. The serial time is probably around 3–5 days (Ganyani et al. 2020; Nishiura et al. 2020), and the uncontrolled $R_0 = 2.5$–3.5 (Liu et al. 2020). The average case fatality rate likely is be around 1–3 per cent (Petrosillo et al. 2020; Yuen et al. 2020).

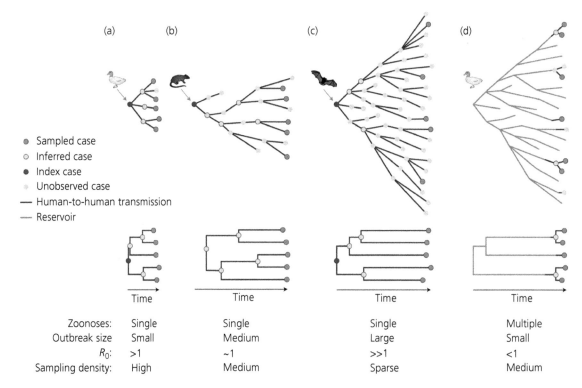

Figure 11.9 Virus outbreaks and the corresponding phylogenetic trees. The scenarios (a) to (d) differ in the number of cross-species transmission events from an animal reservoir (zoonoses), the size of the outbreak (the subsequent epidemic), the dynamics of the spread (R_0), and the coverage in sampling. In the scenarios, only some cases were sampled and confirmed, some had to be inferred, and others were not observed (see legend). In the phylogenies, the kind of outbreak leaves different traces. (a) A recent outbreak shows a short tree. (b) A medium-sized outbreak has a deeper tree, with internal nodes dispersed. (c) A large outbreak with high R_0 has nodes near the root of the tree, also suggesting that only a few cases are sampled. (d) In a tree from recurrent limited outbreaks, such as resulting from repeated spill-overs but sparse within human population spread (no single index case identifiable), the diversity of viruses is contained within the tree and cases of between-human transmission are closely related. The animal reservoirs are arbitrary and for illustration only. Reproduced from Grubaugh et al. (2019), with permission from Springer Nature.

factors for infection and its spread, or suggest possible countermeasures (Grubaugh et al. 2019). A phylodynamic analysis results in a tree containing information about the spread of a disease, similar to network analysis (Figure 11.9).

Molecular epidemiology also allows reconstruction of how a parasite has spread at larger temporal and spatial scales. An example is the global pandemic of the HIV type 1 virus. Its early spread could be reconstructed using Bayesian statistics, combined with historical records, and with geographical and economic information (Figure 11.10) (Faria et al. 2014). The reconstruction placed the origin of the major pandemic HIV-1 in Kinshasa (today, the Democratic Republic of Congo) around 1920. Initially, the virus had descended from simian immunodeficiency viruses circulating in different monkey species, which had recombined into a new variant in populations of chimpanzees of Southern Cameroon. This variant subsequently crossed the species boundary from chimpanzees to humans due

to hunting and consumption of primates, whereby HIV-1 groups M and N originated from distinct chimpanzee communities (Heuverswyn et al. 2007; Sharp and Hahn 2011). It also crossed from chimpanzees to western lowland gorillas, from where it later emerged as the epidemic HIV-1 group O (D'Arc et al. 2015). The type M from Southeastern Cameroon gave rise to the significant pandemic. Human activities transported it along the rivers of the Congo basin to the capitals of Kinshasa and Brazzaville. By the late 1930s, it had already spread further to the interior of the southern Congo basin, along roads and major rivers. The virus reached other regions in the Congo around 1950. Around 1960, further subgroups of HIV-1 diverged from one another and showed different pandemic behaviours. In the 1960s, one variant (HIV-1 group M, subtype B) had reached the Caribbean and, subsequently, must have been circulating undetected in North America for over a decade before the 'first case' was reported in 1981 (Tebit and Arts 2011; Kirchner 2019).

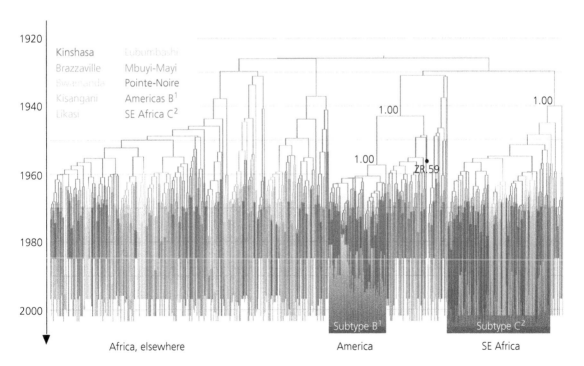

Figure 11.10 Phylogeny of the major pandemic HIV-1. The reconstructed tree shows the most likely descent of the virus (posterior probabilities given at major nodes) after type M had reached Kinshasa in around 1920. For details, see reference. Subtype B colonized the Americas. The sample ZR.59 is the earliest HIV-1 probe, sampled in 1959 in Kinshasa. For the reconstruction, Bayesian statistics with the MCMC algorithm (Markov chain Monte Carlo method) was used. Adapted from Faria et al. (2014), with permission from Springer Nature.

Molecular epidemiology thus provides rich information to disentangle the history of major pandemics with retrospective analyses. Such studies could guide informed prevention and management strategies for future pandemics—if political will and commitment existed.

11.8 Immunoepidemiology

The term 'immunoepidemiology' was first used in the late 1960s in connection with tracing different strains of malaria. The concept was expanded in the early 1990s, especially for studies of helminth infections of humans (Hellriegel 2001). Immunoepidemiology attempts to infer which immune protection elements may be essential given the course of an epidemic. Alternatively, it analyses the effect of immunity for infectious disease dynamics.

11.8.1 Effects of immunity on disease dynamics

The strength of individual immune responses, the degree of cross-immunity, or the extent of herd immunity are obvious factors that affect parasite survival or fecundity and thus play a critical role in the course of an epidemic. At the same time, immune protection in a population, or for any individual, is rarely complete. This is true regardless of whether the defence is by the B- and T-cell based acquired immunity of the higher vertebrates, or by an innate immune memory ('immune priming') as, for instance, in insects or crustaceans (see section 4.7). Compartmental epidemic models can account

for much more fine-grained levels of immunity, including age-dependent protection ('age-stratified immunity'), as successfully done for modelling the 2009 influenza epidemic in Hong Kong (Yuan et al. 2017).

Immune-stratified models are an example of accounting for heterogeneities that are used widely, e.g. for infections in wildlife populations, veterinary, or medical studies (Hayward 2013). Such models furthermore confirm that immune priming and memory can have profound effects on disease dynamics, depending on factors such as the strength of the protective effect or the costs of immunity in terms of fecundity losses. The modelled host populations can thereby exhibit a range of different dynamic behaviours, e.g. converge to an endemic equilibrium, oscillate with or without dampening, or show destabilization (Tidbury et al. 2012). Selection resulting from host–parasite interactions can also change the genetic composition of host and parasite populations—even as the epidemic unfolds—depending on the variation in immune responses and protection. Studying the effect of investment into immune defence and how this might vary among host genotypes in a population is, therefore, necessary (Koella and Boete 2003).

Co-infections by different strains (or different parasites) are especially interesting, not only for the possible effects on disease severity (e.g. in malaria, Sondo et al. 2019) but also for the dynamics of infection. The outcome is not a simple function of the infecting strains but depends on competitive processes and the parasite's differential effects of the

Table 11.3 Expected effects on transmission of the secondary infection (parasite success). A primary infection provides immune protection against the primary infection, and via cross-response to a secondary infection. Parasites can also suppress the host immune system. The actual outcomes depend on the relative strengths of these factors. The table shows idealized outcomes from continuum of effects.[1]

		Does primary infection generate suppression of immune system?	
		Yes	No
Secondary infection controlled by same mechanism as primary?	Yes	Transmission decreases slightly. Same (fitting) mechanism reduces success but host also suppressed.	Transmission decreases. Same (fitting) mechanism reduces success but host not suppressed.
	No	Transmission increases. Secondary (not fitting) mechanism combined with suppression greatly increases success.	Transmission increases slightly. Secondary (not fitting) mechanism increases success.

[1] Adapted from: Graham. 2007. *Trends Parasitol*, 23: 28, with permission from Elsevier.

host's immune system. In reality, co-infections usually have one parasite strain arriving first, and a further strain second. Hence, there is an immune response against the primary infection and, subsequently, a cross-reaction to the secondary infection. The combination of these responses will determine the outcome. In particular, this also concerns the success of a second infection after the immune system was exposed to a previous one (Table 11.3). Theory shows that cross-immune responses to different parasite strains have important implications for the dynamics of an infectious disease. For example, with co-infections, the reproductive

numbers, R_0, are modulated by the degree of cross-immunity. As a result, either one or both strains might become extinct even after an initial period of a seemingly standard host–parasite dynamics (Restif and Grenfell 2006).

Regardless of the details of these processes, immune responses alter the host–parasite dynamics and the dynamics, in turn, shapes immune defences. Together, this leaves traces in the genetic structure of parasite populations (Lythgoe 2002; Grenfell et al. 2004). Measles viruses, for example, show a phylogenetic tree with several groups separated by deep branches (Figure 11.11a). This pattern is

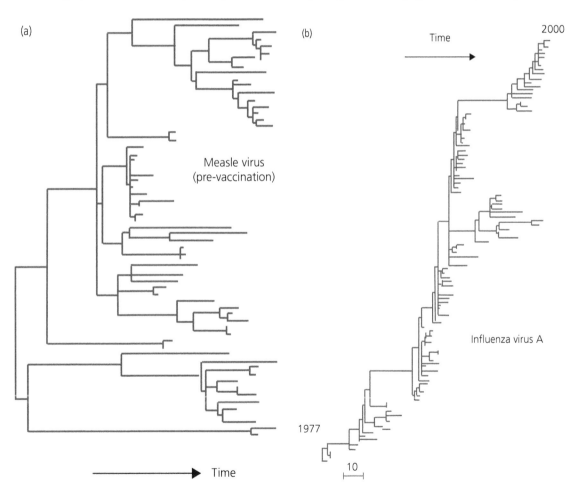

Figure 11.11 Phylodynamics of viruses. Shown are the phylogenetic trees for viral variants. Each tip of a branch indicates a different virus type (as given by the genetic sequence). (a) Measles in England before vaccination, based on the nucleocapsid gene (63 sequences). Adapted from Grenfell et al. (2004), with permission from the American Association for the Advancement of Science. (b) Influenza virus A, subtype H1, between the years 1977 to 2000 shows the phenomenon of antigenic shift (104 sequences). The bar indicates the number of nucleotide substitutions. Adapted from Ferguson et al. (2003), with permission from Springer Nature.

indicative of their evolution being driven by epidemiological processes under spatial heterogeneity. Thereby, relatively short infections and long immune protection drive pronounced epidemiological cycles, as susceptible hosts are locally quickly exhausted, yet a large variety of strains is overall maintained. For influenza virus, by contrast, at any one time, the diversity of parasite variants is limited, but the phylogenetic trace rapidly moves along a primary direction (Figure 11.11b). This pattern reflects strong selection by the host's immune system, combined with antigenic shifts in the virus population. In particular, transient cross-immunity limits persistent strain diversity within hosts and favours the continuous establishment of novel strains against which the hosts have not yet become protected (Ferguson et al. 2003). Influenza A virus provides an excellent example of a further application of immunoepidemiology. This is to predict the further course of the spread of viral variants. Such predictions are essential in the preparation of control measures, e.g. for the choice of a vaccine. Because the production of influenza A vaccine takes several months, it is necessary to predict as best as possible which variants will be most prevalent in the next flu season (Łuksza and Lässig 2014) (Figure 11.12).

11.8.2 Inferences from disease dynamics

This approach to immunoepidemiology is closer to the meaning of epidemiology in medicine: it analyses the observed epidemiological pattern of disease to infer which factors might favour a spread, or which factors could be responsible for making a host susceptible to infection (Krause et al. 2019). This approach essentially was used for smallpox by Bernoulli (Box 11.1), and again for schistosomiasis in the Congo during the 1930s (Fisher 1934). In both cases, the age-dependence of infection was a prime question, and its probable cause was inferred from epidemiological data. As mentioned elsewhere, age-specific infection curves are, in fact, commonly found (Bourke et al. 2011). They can exhibit a 'peak shift', i.e. maximum infection intensity shifts to younger age classes when transmission intensity (the force of infection) becomes high. For example, experimental infections of mice, as well as

infections by *Schistosoma* in humans (Woolhouse 1998; Woolhouse and Hagan 1999), show this pattern. The shift can result from acquired immunity by previous exposure to the same or similar strains, as the respective herd immunity builds up in certain age classes.

The capacity to identify factors that put hosts at risk again is especially valuable during the outbreak of a new disease, when it is usually not known whether immunity develops at all. Epidemic patterns also reveal the heterogeneity in infection risk and the acquisition of immunity across different groups in a population. In West Africa, for example, ethnic groups show differences in susceptibility to malaria, despite living in the same places and under the same exposure conditions. Based on inferences from epidemiological studies, the reasons for such differences became more transparent. From the results, it was clear that differences in the regulation and polymorphisms of immune-associated genes provided protection and were responsible for the observed differences (Arama et al. 2015).

Combining models for the acquisition of immunity in host individuals with those for the spread of diseases allows for a better understanding of the formation and duration of immune protection at the epidemic scale. In the case of influenza, the history of previous exposures to the different virus strains, and what part of the virus the host's antibodies target, emerge as important factors (Ranjeva et al. 2019). In all of these cases, the respective studies use observational data from various sources. This includes cross-sectional or longitudinal studies (as for Ross River fever; Tuncer et al. 2016), screening of genotypic or serotype variation of the parasites, the study of specific immune protection, possible cross-immunity (e.g. as for dengue and Zika viruses, Andrade and Harris 2018), or experimentation to combine immunological with epidemiological analyses.

11.8.3 Immunological markers in epidemiology

The merging of immunological methods and epidemiological tools can also serve to monitor the spread of an infectious disease with improved accuracy. Within this framework, the appearance of clinical symptoms is modelled as a function of

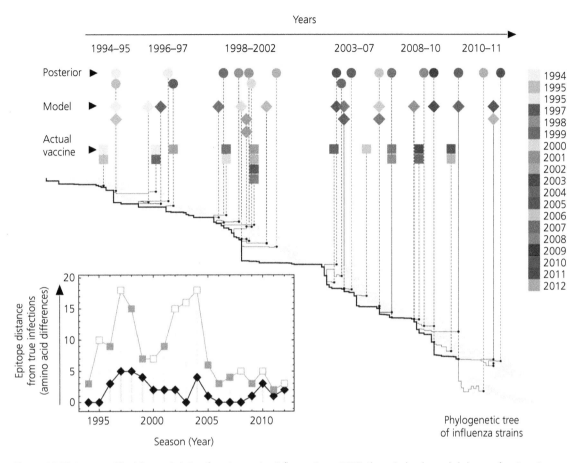

Figure 11.12 Immunoepidemiology and choice of vaccines against influenza A type H3N2. Shown is the observed phylogeny of strain variants from 1994 to 2012 (see colour scale), indicating the turnover and replacement of viral sequences from year to year. These changes reflect the adaptations in the antibody-binding epitopes of the virus (the haemagglutinin surface protein, H3) under selection by human defences, incl. vaccination (see Figure 11.11b). Immunoepidemiological models predict the fitness of the virus based on its epitope variant and, therefore, how likely it is to occur in the next flu season. The model-predicted epitopes for the next flu season inform which vaccine should be used. The symbols indicate the posterior, true identities of the viral strains (circles), the predicted epitopes (i.e. diamonds), and the actually deployed vaccine (squares) over the years. The goodness of the match between predicted epitopes (black line with diamonds) and actually deployed vaccine epitopes (grey line with squares) is shown in the inset, given as the distance in number of amino acids. The baseline refers to the true 'average' epitope sequences in the respective season. The binding of the model-predicted vaccines would have been closer to the circulating viruses than the vaccines that were actually used. Adapted from Łuksza and Lässig (2014), with permission of Springer Nature.

disease progression. This allows prediction of the time course of parasite load and, importantly, the time course of the immune response, too. For example, models of within-host disease dynamics (see Chapter 12) predict, with considerable accuracy, that influenza A virus reaches a maximum viral titre two days after infection. The immune response, expressed as the level of interferons, then peaks one day later, whereas the peak adaptive response, indicated with the titres of CD8+ T-cells or IgGs, appears

two to three weeks post-infection (Smith and Perelson 2011). These responses can therefore serve as 'immunological markers' for the time status of an infection in a given individual (Remoue et al. 2006), which would be difficult to estimate otherwise.

In a further step, the monitoring of the immunological status of individuals is input into epidemiological models, and this converges with the approach described above (section 11.8.1). These models can assess and predict the further course of an epidemic in a population, as done, for example, for influenza A infections in Pennsylvania (Lukens et al. 2014) (see also Figure 11.12). Hence, with this approach, the 'immunological phenotype', or the assessment of clinical symptoms (the 'disease phenotype'), serves to model the further epidemic at the population's scale. Together, the approach remains computationally demanding without using simplifying assumptions. But it can successively lead to a more accurate parameterization of the models.

Besides, the same approach can give better estimates for the effect of medical interventions and control measures to contain an epidemic. The effect of insecticide-treated bed nets to control the spread of malaria, for example, involved a survey of the immune responses in the human population. In this case, the IgG titres positively associated with the intensity of exposure to mosquito bites and the likelihood of infection. These titres were significantly reduced where treated bed nets were introduced, and this coincided with a drop in malaria infections (Drame et al. 2010). Thus, such immunological markers can serve as readily accessible tools to gain insight into the success of control measures. The method has been used in various contexts, e.g. for malaria in different regions (Wong et al. 2014; Sagna et al. 2017; Varela et al. 2020), or for other vectored diseases such as leishmaniasis (Gidwani et al. 2011). The approach furthermore allows the degree of exposure of individuals to potential infection risks to be retrospectively measured. For example, this applies to travellers that visit an area of high endemicity (Orlandi-Pradines et al. 2006). As a field, 'seroepidemiology' uses the host serotypes as biomarkers to trace an epidemic or to monitor the effect of vaccination programmes (Cutts and Hanson 2016).

Important points

- Epidemiology, in the sense used here, is the study of the population dynamics of host–parasite systems. Infectious disease dynamics is often analysed by the SIR model that can be extended in various ways. The reproductive number, R_0 (the number of secondary infections resulting from a primary infection in a fully susceptible host population), is an important characteristic.
- Epidemics show some kind of threshold dynamics because (in the simplest case) $R_0 > 1$ is required for the spread of disease. Vaccination programmes should thus reduce this value to below a critical value. Deviations result from spatial heterogeneities or stochastic processes.
- Individual-based models track the status of individual hosts during an epidemic. Network models are often used but are computationally demanding. Epidemics can also be analysed as an invasion process with evolutionarily stable strategies. Periodic outbreaks from an endemic state characterize some infectious diseases, e.g. childhood diseases in prevaccination times.

- Vectored diseases (e.g. the Ross–Macdonald model) or host–parasitoid systems (Nicholson–Bailey model) are analysed with models similar to SIR. For macroparasites, the distribution of parasite loads among hosts is typically aggregated, which is essential for the dynamics of these systems.
- Molecular epidemiology traces infectious disease dynamics based on parasite genomic sequences and reconstructs transmission chains by phylogenetic inferences. It is a powerful tool for ongoing epidemics or zoonotic outbreaks where little is known about a pathogen.
- Immunoepidemiology studies the effects of immune defence and protection, e.g. by memory or cross-immunity, on the disease dynamics. Likewise, relevant immune defence elements or other protective factors are inferred from the observed epidemic. Finally, immunological makers serve to trace and predict the further course of an epidemic.

CHAPTER 12

Within-host dynamics and evolution

The within-host dynamics of infection starts when a parasite population has successfully entered a host and established itself at a first site. Eventually, the parasite has to reach specific tissues or organs for its final development, except those that persist and multiply in the tissues reached first, such as pathogens of inner epithelia or in the skin. The infection life cycle of a parasite or a pathogen population ends with clearance or by host death and has several phases (Figure 12.1). After a comparatively short infection step, the 'primary phase' of varying duration follows, where the parasite, at the infection site, must overcome the first host defences to establish and to start migrating to its target organ and tissue. Once in this secondary site, the 'secondary phase' involves growth and multiplication of the parasite population. In biomedical practice, the primary phase extends over a pre-set phase (e.g. for HIV infections, usually three months), which includes the secondary phase as defined here, and during which the parasite has already reached the target organs. Regardless, transmission forms (transmission stages) will eventually develop and are released over some time, or in a burst, followed by parasite clearance or host death. These phases are not always clearly separated and have different names in different concepts. The associated selection regimes and the evolutionary consequences for parasite and host, furthermore, depend on both the within-host and the between-host (see Chapter 11) components of a parasite's life cycle.

12.1 Primary phase of infection

The primary phase of an infection is difficult to study because small numbers of parasites and local processes are involved. The primary phase of an infection unfolds—for the most part—at the site of infection (the primary site), which is in turn determined by the typical infection route of the pathogen. At the primary site, the host's early defences are crucial and should prevent the pathogen from multiplying, developing, or migrating to its preferred location. For the parasite, in turn, the prime

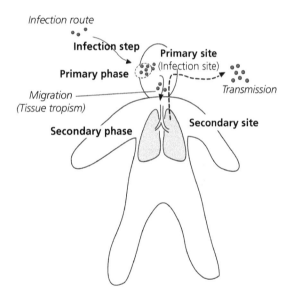

Figure 12.1 Simplified scheme of the phases of an infection. After the first encounter with a host, infection occurs along a given infection route (e.g. through the peripheral airways) in the infection step. Subsequently, the parasite typically establishes itself at or near the site of infection (e.g. the nose) during a primary phase. This is followed by migration to a target organ and tissue (the secondary site, e.g. the lung). The ability of a parasite to infect a specific organ and tissue is also known as tissue tropism. Some parasites are broadly tropic, i.e. can infect a wide range of organs, whereas others are narrow in their tropism. The infection cycle ends with transmission and/or when the parasite is cleared by the host. The sketch could refer to an infection of the respiratory pathways.

Evolutionary Parasitology: The Integrated Study of Infections, Immunology, Ecology, and Genetics. Second Edition.
Paul Schmid-Hempel, Oxford University Press. © Paul Schmid-Hempel 2021.
DOI: 10.1093/oso/9780198832140.003.0012

necessity is to escape the early, local defences, to develop, to overcome migration barriers, and thus to escape into other tissues—if a secondary tissue is part of the life cycle.

The significance of early events is demonstrated by *Drosophila melanogaster* infected with a dose of the bacterial pathogen *Providencia rettgeri* (Figure 12.2). Within the host, the bacterial population shows an approximately logistic growth. However, the overall population dynamics of the infection can be classified by the outcome. With host death as the outcome, population growth has continued to a lethal pathogen load. If hosts survive, population growth has stopped at some time point and returned to a lower load. Moreover, when host defences are disabled experimentally, e.g. using mutants deficient in the Imd pathway, bacterial populations always grow to lethal loads. This finding suggests that host defences are the prime mechanisms that control infections in this case. Therefore, the time a host needs to take control of infection is predictive for the eventual outcome. Time of control and other

characteristics thereby vary among host genotypes and pathogen species (Hotson and Schneider 2015; Duneau et al. 2017a).

Together, the dynamics of such an infection can be summarized in the model of Figure 12.2b. Here, the early phase is followed by a 'control phase', during which the host responds and starts to control the infection. If the pathogen population has already reached a sufficiently large size at this critical time—considered to be a 'tipping point'—it can outgrow the host defences. In this case, the population will reach a lethal load in the 'terminal phase', where host death is the almost inevitable outcome. Alternatively, suppose population load is lower than a critical level at the tipping point. In that case, host defences can control the infection and either clear the parasites or contain them at a lower level (the 'set point load'). Containment may lead to a chronic infection (the 'chronic phase'). This model is straightforward but fits observations in the *Drosophila* system and captures the crucial race between the host and the growing pathogen population.

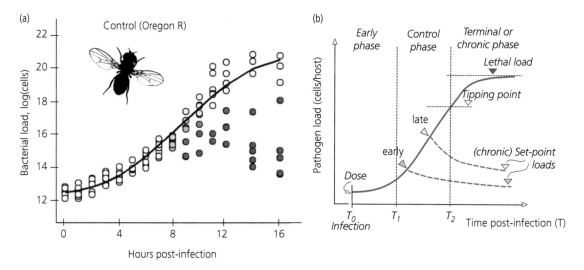

Figure 12.2 Dynamics of infection in *Drosophila*. (a) Pathogen load (bacteria per fly) over time after infection (*D. melanogaster* strain Oregon R, infected with *Providencia rettgeri*). Each dot represents the bacterial load in a single fly. The solid line is a commonly used growth model (standard Baranyi growth) fitted to the dots. The red dots indicate cases where the host was able to control the infection, with the intensity of red representing the probability of control. (b) Model of infection dynamics. After infection (at time T_0) with a given dose, pathogen load increases as long as the host's defences are not yet activated or effective ('early phase'). In a subsequent time window ('control phase', starting at T_1), host defences—activated early or late during this phase—can reduce pathogen load. If not cleared, the pathogen population is controlled at a 'set point' size that would be characteristic for a persistent, chronic infection (the 'chronic phase'). If the defence is too late or ineffective, the pathogen population size reaches a critical size ('tipping point') at a critical time, T_2, and escapes the host's control. It grows further to a lethal load (in the 'terminal phase') that eventually kills the host. Time and pathogen load points naturally vary among host types and individuals, and among parasite species or variants, or among ecological conditions. Adapted from Duneau et al. (2017a), under CC BY.

During the early primary phase, hosts must reliably distinguish between infection and non-infection. However, as signalling theory demonstrates, there are intrinsic limits to how accurate this can be (Box 12.1). In the case of an infection, the signal is a stimulus generated by, and within, the defence system. Any such signal follows a probability distribution for stimulus strength because it depends on many factors, e.g. on the number of pathogen epitopes, secondary signals, binding properties, host status, and environmental conditions. Inevitably, therefore, a signal for infection sometimes may be weaker than a signal generated when no infection is present. In other words, the distribution of signal strengths for 'infection' overlaps the distribution for 'non-infection' (Box 12.1). Regardless of how infection signals are generated, at some point the defence system must 'decide' whether it means infection or not. Hence, in a simple scenario, a threshold is implemented that converts the continuous distribution of signal strengths into a dichotomous outcome of 'yes' or 'no'. As Box 12.1 shows, the threshold for this yes/no decision determines the probabilities that the system responds correctly or incorrectly, and which errors occur. A correct decision adds to host fitness. A false-positive

decision, at the very least, wastes time, nutrients, or energy, but in the worst case can cause severe immunopathology.

The threshold level has the earliest consequences in the life history of the host–parasite interaction. Whereas the effects of immunopathology on the host appear relatively late, the costs of erroneous detection appear quickly when the parasite population outgrows the host defences. Correspondingly, life history theory predicts that early events are under stronger selection than later ones. Hence, threshold levels should be under particularly strong selection to maximize the chances to detect and control infections, while minimizing the chances for incorrect decisions. Simple models confirm such expectations (Metcalf et al. 2017). For example, sensitivity (the ability to correctly identify true infections; Box 12.1) should increase when parasites are more abundant or more virulent. However, the theory also predicts some less obvious patterns. For instance, the threshold should increase (increasing specificity, i.e. the ability to sense the absence of infection) when host life expectancy is higher. This seems counterintuitive. But too many 'false alarms' in a long life would increase the expected cost of immunopathology, relative to the gains of defence.

Box 12.1 Signalling theory and infection

Signalling theory deals with analysing signals by a receiver ('signal detection'), among other topics. Concerning an infection, the receiver is the host, especially its immune system. The system produces signals for infection. These include the abundance of activated transmembrane receptor compounds (indicating the binding to parasite-derived ligands), the abundance of the CpG-dinucleotide motifs in DNA or RNA accessible to toll-like receptors, and the products of several other processes. For example, the change in the composition of the microbiota can also indicate the entry of a pathogen (see Chapter 4). From these signals, the receiver—the host's immune system—has to 'decide' whether an infection has happened or not.

The situation would be simple if all signals were clear and unambiguous. However, all signals, including those generated by the immune defence system, follow a probability distribution of signal strength, such as illustrated in Figure 1.

In particular, signals are also present when no infection happens—a receptor may have bound to itself or to a meaningless ligand. Signals may also result from a background 'noise' inherent in any dynamic system, and so forth. Vice versa, an infection may produce a weak signal for similar reasons, not least because parasites often sabotage recognition (see Chapter 8). Because the immune system has to activate or not activate the defence cascade (i.e. in a simple scenario with a dichotomous outcome), there is a threshold for signal strength above which an infection is considered real.

As Figure 1 shows, changing the threshold for this yes/no decision changes the probabilities that the system responds correctly or fails in different ways. As the threshold is lowered, it becomes more likely that a real infection is detected. However, it also becomes more likely that
continued

Box 12.1 *Continued*

non-infection triggers the response. Two parameters charac-terize the signal detection system. 'Sensitivity' is the prob-ability that an infection is correctly recognized ('true positives'); 'specificity' is the probability that the absence of infection is correctly recognized ('true negatives'; Figure 1). With probability distributions of signals, it is never possible to simultaneously maximize sensitivity and specificity.

Sensitivity and specificity depend on how the receiver works, i.e. on the biochemical-molecular mechanisms of the defence system. A detection system may not always produce a 'yes' or 'no' signal but instead, it may generate a graded response. However, the fundamental detection problem remains, since the strength of the graded response, too, must somehow corres-pond to the detected signal.

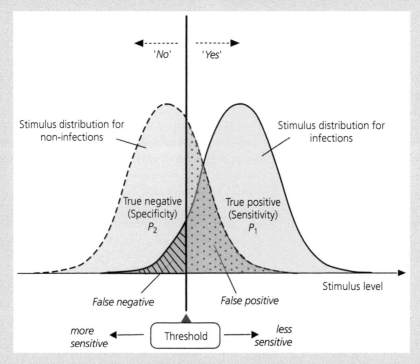

Box 12.1 Figure 1 Detection of infection. If a host becomes infected, its immune system generates a stimulus of varying strength (x-axis; light red, solid line). If the host is not infected, the system generates a stimulus at varying levels, too (light green, dashed line). The two probability distributions (y-axis) overlap. The decision to activate the defences (yes/no) is given by a threshold. Therefore, four outcomes are possible: (1) A true infection is correctly recognized (sensitivity of the system, with probability P_1, i.e. the area under the solid curve to the right of the threshold; light red). (2) A true absence of infection is correctly recognized as non-infection (specificity of the system, with probability P_2, i.e. the area under the dashed curve to the left of the threshold; light green). (3) A true infection is not recognized (hatched red area; 'false negatives', with probability $1 - P_1$). (4) A non-infection is erroneously recognized as infection (dotted green area; 'false positives', with probability $1 - P_2$). Shifting the threshold value to the right makes the system less sensitive but more specific, and vice versa when the threshold is lowered. A PCR test for SARS-CoV-2, for example, has a sensitivity of $P_1 \approx 0.98$, and a specificity of between $P_2 = 0.78$ to $P_2 \approx 1.0$; these values are typically difficult to measure (La Marca et al. 2020; Stites and Wilen 2020).

Thresholds should also change with host age—the lowest in the age classes are exposed to the high-est infection risk—and plastically change through-out all other life stages. Such a 'perfect' matching hardly is observed, and the realized solutions will vary with ecological conditions and physiological constraints.

However, nature may also have followed a differ-ent route to deal with the constraints set by signal-ling theory. Given the selective premium on being quick rather than accurate, the detection system should err on the positive side, i.e. have a low threshold that signals an infection more often than would actually be the case. However, at the same time, a slower error correction system may be put in place that can halt an unnecessarily triggered immune response (Frank and Schmid-Hempel 2019). Little is known about these alternatives. Nevertheless, as discussed in section 4.3.4, immune responses are regulated in various ways. In particular, while recognition may readily activate the defence cascades, all cascades have an impressive array of negative regulators that can halt it if needed (see Figure 4.6). Hence, the intrinsic constraints set by signal detection, combined with the need for an early response, may be linked to the evolution of negative immune regulators. Moreover, as negative immune regulators themselves must have become points of attack by the parasites (parasites can stop a response by activating or mimicking negative regulation; see Chapter 8), we should expect an evolutionary diversification in the negative regulators themselves.

12.2 Within-host dynamics and evolution of parasites

Hosts cannot change evolutionarily during the duration of infection, even though they can continuously fine-tune their responses by variation in gene expression or by somatic changes in their defence molecules. By contrast, an infecting parasite population can evolve. An example is an emergence of 'escape variants', first noticed in HIV infection in the 1990s and prominent in RNA viruses (Presloid and Novella 2015). Several approaches help to analyse these processes.

12.2.1 Target cell-limited models

A first approach is to focus on the target cells that a parasite usually infects, e.g. cells of the lung epithelia or the cells of the immune system itself. In these cases, infected cells produce new viruses that add to the population of freely circulating viruses in the host. Target cell-limited models (Box 12.2) are used

---

Box 12.2 Target cell-limited models

Target cell-limited models focus on the dynamics of host cells that a parasite infects, and which are assumed to be limited in numbers (de Boer and Perelson 1998). Within the host, the parasite has a free stage. Depending on the kind of parasite, other stages can be considered, too. Target cell-limited models first were applied to HIV, and viral infections more generally (Nowak and May 2000; Hadjichrysanthou et al. 2016; Ciupe and Heffernan 2017).

Consider a viral infection, where free viruses (V) circulate in the host that can infect their target cells (T). Infections convert them to infected cells (I). From infected cells, new free viruses emerge at a rate, k. This scenario yields the basic model of viral dynamics inside a host (Figure 1). In analogy to the SIR models (see Box 11.2), these models are also called TIV models ('Target cell-Infected-free Virus'). Formally, the dynamics over time (t) is:

$$\frac{dT}{dt} = bS - \mu T - rTV$$

$$\frac{dI}{dt} = rTV - \delta I$$

$$\frac{dV}{dt} = kI - cV - rTV - \beta V \tag{1}$$

where new target cells are proliferated ('born') at a rate, b, from some stock, S, and disappear with a background death rate, μ. Similarly, target cells turn into infected cells with an infection rate, r, and disappear at a rate, δ, which is a composite of the background death rate and the additional intrinsic cell death caused by the infection (e.g. clearance by cytotoxic T-lymphocytes). Free virus (called a 'virion') is usually cleared by the immune system, too, at a given rate, c. The pool of free viruses is also the source of virions transmitted to a next host at rate β.

Similar to the SIR dynamics, R_0, is the basic multiplication rate of infected cells when the virus enters the hosts. At the beginning, $I = V = 0$, and, ignoring between-host transmission ($\beta = 0$), this gives $R_0 = \frac{rT_0}{c + rT_0}\frac{k}{\delta}$. However, the analysis of such models is complex; for example, R_0 may not even be a good criterion for viral success (Ciupe and Heffernan 2017).

continued

Box 12.2 *Continued*

Furthermore, the models cannot easily explain what controls virus replication, especially during the early, acute phases of infections. At these stages, the misattribution of factors that control virus replication is a problem (Kouyos et al. 2010).

The 'loss' of viruses due to infection of target cells (i.e. term, rTV, in eq. (1)) is often ignored for simplicity, and only clearance is considered. This assumes that the dynamics of viruses is much faster than that of cells, such that a quasi-stationary state quickly is reached, where $V \approx kI/c$. In the simpler model, the infection spreads with $R_0 > 1$, i.e. the number of infected cells (I) increases and, eventually, may

level out at an endemic equilibrium. An infection will produce a higher endemic equilibrium, and the control of the process will be more difficult with an increase in R_0.

A drug treatment might operate to reduce the rate of new infections of cells, r, or facilitate the shedding of cells, including infected ones, i.e. increase death rates, μ and δ. A reduction of r leads to an increase in the number of target cells, T, which in turn, and paradoxically, can increase the number of newly infected cells under eq. (1). Hence, any treatment that reduces this transmission rate must be very effective to suppress the number of infected cells below a given threshold.

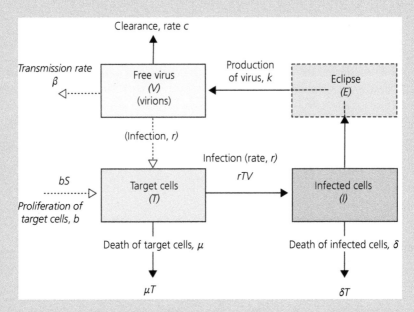

Box 12.2 Figure 1 Target cell-limited model for viral within-host dynamics. The basic model is extended with an 'eclipse' compartment (green, dashed box) where infected cells do not yet produce new viruses. These cannot be observed in the viral load data. Dashed lines indicate processes typically ignored in most practical models. Parameters described in text.

to analyse the dynamics of such infections. They are similar in structure as the SIR model of epidemiology (see Box 11.2), but do not explicitly consider the transmission step. They are, therefore, most appropriate for long-lasting, chronic infections where transmission events are infrequent or happen very late compared to the duration of the infection.

Target cell-limited models, as described in Box 12.2, have often been used for viral infections

such as HIV (Phillips 1996) (Table 12.1), but also for bacteria, e.g. *Mycobacterium tuberculosis* (Kirschner et al. 2017), or *Francisella* (tularaemia, Gillard et al. 2014). Target cell-limited models are used to generate testable predictions, and to assess how the infection dynamics or the viral load would change when a treatment is applied. When used in this way, models provide an estimation procedure that results in predictions to describe the course of

Table 12.1 Studies of viral infections with target cell-limited models.

Virus	Study goal	Findings	Remarks	Source
Influenza A virus (IAV) in humans	Model fitted to data to estimate parameters.	Numerical estimates for various infection characteristics, such as cell and virion life, time to virus production, R_0.	Viral load can have two peaks, likely as an effect of interferon response.	1
Dengue virus (DENV) in humans	Minimal model to accurately describe primary and secondary infections with DENV.	Dynamics of primary infection (innate response; IgM) seems differently controlled than in secondary infection (clearance of infected T-cells, enhancement of virus infectivity; IgG).	Typically, secondary infections can lead to complications and fatal outcomes.	3, 4
West Nile virus (WNV) in mice	Parameter estimation for model, then apply for dynamics of humoral immune response.	Numerical estimates for R_0, burst size, virion production, etc. Data often insufficient, but important values for viral spread relatively accurate. Knock-out mice show importance of humoral response.	WNV causes encephalitis in humans. Circulates in birds and small mammals. Vector-borne.	2
IAV in humans	Estimating key parameters of infection dynamics.	Numerical estimates for various parameters such as R_0. Most important are accurate data on cell and free virus lifetimes.	Models, evaluated with data from q2 patients.	5
IAV in humans	What are sources of error for parameter estimation in models?	Poor experimental data cause erroneous parameter estimates, also asynchrony in data points.	Various measurement errors assumed and tested.	8
IAV in humans, combined with *Streptococcus* co-infection	What controls infection, and how do co-infecting parasites benefit or interfere?	Many immune responses altered by co-infection. Exacerbation a main cause of disease severity. Non-linear dynamics important but difficult to model. Models help to find time scales.	A review of models. Co-infection can enhance severity of viral infections. Also reviews virus–virus co-infections.	9
Simian immunodeficiency virus (SIV) in rhesus macaques	Source of virus (which tissue).	Depletion of CD4+ T-cells in lamina propria (mucosa) from rectum associated (a few days before) with viral peak load.	SIV as model for HIV. Kinetics of CD4+ T-cells studied.	7
SIV in monkeys	What limits virus replication.	Limitation of target cells (CD4+ T-cells) alone not sufficient to control early virus replication. Inclusion of CD8+ improves model predictions.	Target cell-limited models fitted to data from sooty mangabeys and rhesus macaques.	6

Sources: [1] Baccam. 2006. *J Virol*, 80: 7590. [2] Banerjee. 2016. *J R Soc Interface*, 13: 20160130. [3] Ben-Shachar. 2015. *J R Soc Interface*, 12: 20140886. [4] Clapham. 2016. *PLoS Comp Biol*, 12: e1004951. [5] Hadjichrysanthou. 2016. *J R Soc Interface*, 13: 20160289. [6] Kouyos. 2010. *PLoS Comp Biol*, 6: e1000901. [7] Lay. 2009. *J Virol*, 83: 7517. [8] Nguyen. 2016. *PLoS One*, 11: e0167568. [9] Smith. 2018. *Immunol Rev*, 285: 97.

infection. Appropriate models fit the data on virus loads reasonably well and can predict other measures, too; for example, the titres of immune effectors. However, these models cannot easily explain the control of virus replication without more explicit reference to immune defences, especially in the early phases of viral growth. Besides, the quality of the available data is a significant limitation for the accuracy of parameter estimation and the model predictions. For instance, there can be too few data points, asynchronous sampling, measurement errors, or the essential variables may not have been considered in the first place, e.g. target cell numbers

and relevant immune response (Hadjichrysanthou et al. 2016; Nguyen et al. 2016).

Influenza A virus (IAV) infections in patients from a clinical study illustrate an application of target cell-limited models. A range of infection-related variables was defined, e.g. peak viral load or infection duration, which can be observed and measured (Figure 12.3a). The model, furthermore, is simplified by several assumptions, e.g. that there is no latent phase (infected cells produce new viruses immediately at infection). In this case, a quasi-stationary situation results with the number of free viruses, $V \approx kI/c$, and $R_0 = krT_0/c\delta$ (Hadjichrysanthou

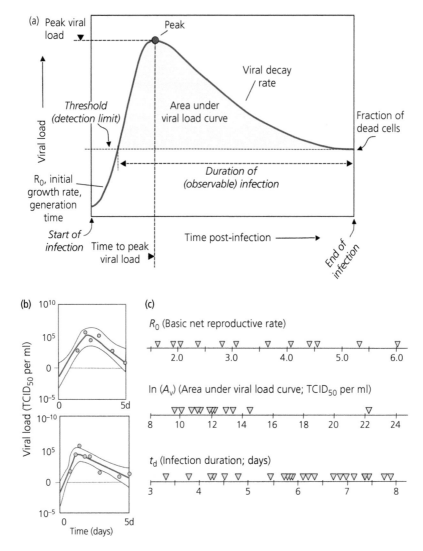

Figure 12.3 Application of a target cell-limited model to influenza A infections. (a) Typical course of within-host viral load over time. Some infection-related measures used for estimation of model parameters are indicated. Note that viruses are only observed above a detection threshold. (b) Fitting the model prediction for viral load (red line) to observed data from two individual patients (blue dots) who have received placebo treatments. The blueish area indicates the range of predictions from 10 000 random draws from the posterior distribution of estimated model parameters. (c) Numerical estimates of infection-related measures for individual patients (triangles). Individual hosts vary considerably in their characteristics. Adapted from Hadjichrysanthou et al. (2016), under CC BY 4.0.

et al. 2016) (see Box 12.2 for terms). The estimates from each patient then predict its viral load over time quite well, as shown in Figure 12.3b. Notably, too, the estimated infection-related variables show considerable variation among patients (Figure 12.3c).

Target cell-limited models furthermore are useful to assess the effects of treatment or the key immune

defence mechanisms. For example, models identified innate immune responses to be most important in a primary dengue virus (DENV) infection. In contrast, the elimination of infected cells by cytotoxic T-lymphocytes was most critical for a secondary (next) infection (Ben-Shachar and Koelle 2015). Similarly, protecting cells from becoming infected in the first place is most critical for mice experimentally infected with West Nile virus (WVN) (Banerjee et al. 2016). These models are thus valuable tools for many questions, even though their actual implementation and parameterization often are somewhat tricky.

12.2.2 Dynamics in disease space

In this book, the concept of the disease space is a guide to visualize the various topics. No explicit mechanisms, such as the action of cytotoxic T-cells or the activation of humoral components, are assumed. In disease space, host health status changes with parasite load (Box 2.2) and yields information about host resistance or tolerance. As the sketch in Figure 12.4a shows, host health may not be compromised for some time after infection, but eventually declines, while parasite load continues

to increase. At some point, host defences take effect, and the infection trajectory leads the host to a path of recovery, to parasite clearance or the control of infection, and to the restoration of the healthy status. Real infections show the corresponding infection trajectories in disease space (Figure 12.4b).

One can consider the trajectories to be the traces of a dynamic system through state space when it is 'disturbed' by an infection. In general, a host system is considered 'resilient' against infection when the trajectory returns to its original state after perturbation. At the same time, the return trajectory follows a path that is different from the one when the infection started. A large and well-developed theoretical toolbox exists to analyse such dynamic systems (Brauer et al. 2019), which can address levels of resilience, stable points, or what domains a trajectory crosses during the lifetime of an infection.

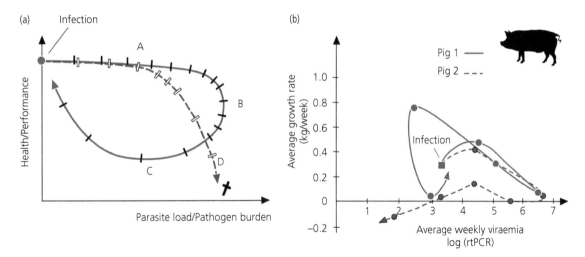

Figure 12.4 Infections in disease space. (a) The hypothetical course of an infection is plotted in a two-dimensional space with parasite load or pathogen burden on the x-axis, and some measure of host health on the y-axis. Infection occurs at a point where host health is best and parasite load yet zero (red dot). The trajectories of the infection through this space in two different hosts (red solid, blue dashed line) are shown here. The trajectories are further specified by the speed in which they traverse the disease space by adding equidistant time marks (bars). After infection, parasite load typically increases but host health is not yet much affected; only after a while, in region A of the space, does host health start to decline. The red host manages to reduce parasite load in region B, for example through an efficient immune response. Note that in region B—as illustrated by this solid curve—host health declines even though parasite load also declines; this is indicative of the negative effects of immune responses on host health. Also note that the solid host passes slowly through region B, but then rapidly recovers in region C (as indicated by the time marks). At point A, the trajectory of the dashed host starts to deviate from the solid host, and its health is rapidly declining as parasite load slowly but persistently increases. In region D, this host has lost control over the infection and is doomed. The solid host is said to be more 'resilient'. (b) Example of a disease space from animal husbandry. In this case, pigs were followed after their infection with porcine reproductive and respiratory syndrome virus (PRRSV). Viral load as well as weight gain or loss (a health measure) is monitored. Both pigs manage to reduce viral load, but pig 1 recovers to normal growth, whereas pig 2 eventually loses body mass. Adapted from Doeschl-Wilson et al. (2012), under CC BY.

Several parameters characterize the host health status. Hence, the infection trajectory is a trace through a multidimensional space. For example, in a study with mice experimentally infected by malaria (*Plasmodium chabaudi*), seven different measures were monitored. These included host body mass, the density of red blood cells, the titres of B-cells, NK-cells, granulocytes, reticulocytes, and body temperature (Torres et al. 2016). This high dimensionality can be reduced by statistical tools, such that a more straightforward, two-dimensional plot results. For this purpose, combinations of two variables were identified that would be most informative for the question under scrutiny. Generally, these are those two variables that result in trajectories enclosing the largest area, i.e. lines that curve through disease space far and wide, rather than just going back and forth in a straight line (which is hardly informative). These pairs must not necessarily include parasite load as one of their variables (as in Figure 12.4). Instead, these covariates can be any combination with parasite load and act as a covariate of, say, the dynamics of titres of red blood cells relative to NK-cells or reticulocytes. Such combinations were most informative in the case of mice and malaria.

Can the eventual fate of a host be inferred from its trajectory through disease space? This indeed seems the case. In the above example of malaria-infected mice, individuals that recovered followed a tight loop in disease space. In contrast, trajectories of non-resilient hosts showed a much wider loop that carries a high risk of entering the zone of danger, which may lead to host death. Monitoring the host status at a critical point in disease space can thus be predictive for the trajectory and the future fate of the host (Box 12.3). These methods would, for example, benefit patients admitted to hospitals where predictions about their likely fate are of the essence.

12.2.3 Strategies of within-host growth

For macroparasites (tapeworms, parasitoid larvae), within-host growth is growth in body size. For microparasites, this means the growth of population size. In brief, two strategies for the parasite are as follows.

1. *Continue to grow elsewhere*: Available resources in the current host limit growth. However, the large body size is an advantage, particularly for macroparasites. For example, a large tapeworm has more proglottids (body segments) that can produce eggs. If the resources of the current host are exhausted, one possibility is to continue growth elsewhere; that is, to transfer to a next host that offers more resources. This strategy can lead to the evolution of complex life cycles where the parasite needs to cycle through two or more different kinds of host, as discussed in section 3.4.
2. *Timing of damage to the host*: By definition, a fitness advantage to the parasite is associated with a fitness loss for the host. The time when such parasite-induced adverse effects appear and the nature of this damage do not only matter for the host but have consequences for the parasite, too. Too much damage, too early, can turn the host into an unfavourable environment for the parasite or even kill both. In contrast, too little damage, too late, will leave host resources unexploited. How the variation in timing and magnitude of parasite-induced damage affects parasite success, and how these patterns evolve, is discussed in Chapter 13.

12.2.4 Modelling immune responses

Apart from simple responses, the response dynamics of an immune system is complex and, therefore, hard to grasp intuitively. Luckily, the progress in various mathematical approaches for modelling immune systems can now contribute to clarify many problems (Eftimie et al. 2016).

12.2.4.1 Computational immunology

With this approach, numerical, computer-based calculations are used to understand the essential elements of immune defences (Chakraborty 2017; Perelson and Ribeiro 2018). Examples are the mechanisms of activation of T- and B-cells, especially the binding kinetics of T- and B-cell receptors to either MHC ligands or antigens. Similarly, the migration, lifetimes, birth and death of various immune cells, or the immune responses to a variety of infections such as HIV, malaria, or tuberculosis, are scrutinized.

Box 12.3 Predictions for infections from disease space

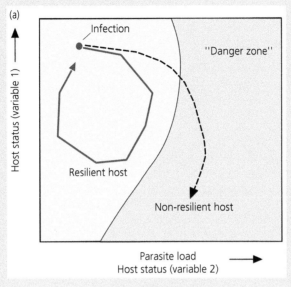

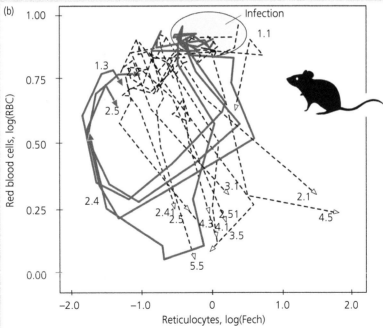

Box 12.3 Figure 1 Resilient and non-resilient hosts. (a) The trajectory of non-resilient hosts (dashed black line) describes a wider loop in disease space compared to resilient hosts (solid red line). Non-resilient hosts are therefore more likely to enter the 'danger zone', i.e. an area in disease space where recovery to a healthy state becomes difficult or impossible. Note that disease space could also be defined by different variables for the host status. (b) Example of infection trajectories of experimental mice infected by malaria (*P. chabaudi*) in the yellow region of the space. Mice that survived ('resilient'; solid red lines, $n = 4$) showed a tighter loop through disease space than those that eventually died from the infection ('non-resilient'; dashed black lines, $n = 11$). Axes refer to counts of red blood cells and a measure for reticulocyte activity (expression of *Fech*); both axes log-transformed. The infection was followed over 25 days; small numbers refer to host identity. Silhouettes from phylopic.org. Based on data in Torres et al. (2016), under CC BY 4.0.

continued

Box 12.3 *Continued*

The trajectory through disease space maps the course of an infection within a host. A trajectory starts with infection and ends with clearance or host death. However, the time point of infection is typically unknown for animals in the wild, or for patients admitted to a hospital. Instead, the patients in a hospital represent 'cross-sectional' data because individual patients are at different stages of their infection when they arrive. So, how could a trajectory, based on observations made at unknown times after infection, be correctly placed in a defined disease space?

One workaround to this problem is to assume that an individual host (e.g. a patient) would not randomly appear in disease space but be closest to other hosts that follow a similar trajectory (Torres et al. 2016). After measuring each patient's status in disease space (i.e. health status vs parasite load), connecting the nearest neighbours will, therefore, group together patients that follow a similar path. Moreover, a topological map can identify a network of connections between close neighbours in disease space, where the different regions of the network correspond to different health states of the host (healthy, in discomfort, sick). Various combinations of measures (e.g. NK-cells, red blood cell counts) can be relevant for separating these domains. Such networks are not identical to trajectories through disease space but bear obvious similarities, such that they become interpretable along the same lines. Typically, such analyses are done in multidimensional space because more than two variables for host health status usually are observed. However, for better visualization, the multidimensional data is reduced by statistical procedures to two dimensions.

Interestingly, trajectories of hosts that recover ('resilient hosts') look different from those that eventually succumb to the disease ('non-resilient'). In particular, trajectories of resilient hosts tend to follow a tight loop in disease space, whereas non-resilient hosts describe a much wider loop. The latter carries a high risk of entering the zone of danger, which is hard to leave again and may thus lead to host death (Figure 1a). Therefore, monitoring the host status at a critical point in disease space can be predictive for the future fate of the host. For mice experimentally infected with malaria, the count of red blood cells (RBC) on day eight after infection turned out to be predictive for survival or death of the host. Non-resilient mice become anaemic earlier than the resilient ones. On the other hand, the shape of the infection trajectory itself is a predictive tool when, for example, observing the infection trajectory in the RBC vs reticulocytes count space (Torres et al. 2016) (Figure 1b).

The loop shape of the trajectory (tight or wide loop) is better analysed with a polar diagram. With this, host status, at a given time, is plotted as a point with a radius (d) away from the centre, and at an angle (Φ) away from the start (Figure 2a). As the infection unfolds in non-resilient hosts, the trajectory in polar space moves away from the centre in a widening loop: hence, r increases with an increasing Φ. These polar coordinates are informative primarily in a specific range of values but then can predict the outcome of infection (Figure 2b).

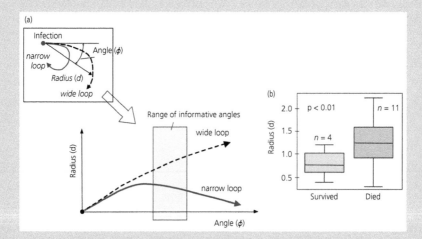

Box 12.3 Figure 2 Predicting host fate from infection trajectories. (a) Data from a (cartesian) disease space (inset top left) are replotted in a polar diagram, with radius (d) and angle from origin of infection (Φ). As the infection takes its path through disease space, the angle grows over time. The corresponding radius increases more rapidly for wide loops (dashed black line) than for tight loops (solid red lines). In a domain of the disease space, characterized by a certain range of angles from the origin (yellow zone), the difference between the two types of loops will be most informative. (b) In the example of malaria-infected mice, the difference in the radius within this zone predicts whether the individual will survive or die. Boxplot with sample sizes indicated; significance value from ANOVA test. Adapted from Torres et al. (2016), under CCBY 4.0.

Computational immunology can also predict the specificity of an immune response, e.g. which HLA alleles would recognize which peptides (Lundegaard et al. 2008). This is exploited for various practical applications, such as drug and vaccine development (Oli et al. 2020) (see Figure 11.12). Besides relatively simple models that can easily be matched with data, the modelling of complex pathways and immune networks contributes to a better understanding of immune defences.

Affinity maturation (see section 4.6.3) is another example where computational methods have been helpful. In this process, cell populations with antibodies (directed against specific antigens) experience selection and evolve. In particular, B-cells undergo somatic mutations in their development that generate an array of cells with different binding specificities. Those with increased binding affinities are positively selected. Computational models suggested that the process must be more complicated, however (Kepler and Perelson 1993). In particular, several rounds of mutation and selection with circulation through germinal centres were predicted, a theoretical expectation that eventually was confirmed empirically (Victora and Nussenzweig 2012; Tas et al. 2016).

12.2.4.2 Systems immunology

This approach treats the immune system as a complex dynamic system. In general, it also exploits the massive amounts of data ('big data') that become available from high-throughput genomic sequencing, metabolomics, proteomics, or from mass cytometry, and the progress in bioinformatics (Davis et al. 2017). The new technologies make it possible to monitor in parallel, for example, the bulk of the immune cell types, together with their status and activities (Qiu and al. 2011). Together with gene-editing techniques, e.g. the CRISPR–Cas9 system, the effect of genetic changes on immune defences can be observed with precision, too.

Systems immunology has addressed various topics, e.g. vaccination, by investigating the mechanisms of immunization against malaria (Kazmin et al. 2017). In this case, the titres of IgGs, polyfunctional CD4[+] T-cells, and the cytokines IL-2, TNF, and IFN-γ were identified as the major players for the protective effect. Similarly, mass cytometry has been instrumental in demonstrating the array and types of different functional variants or specificities of T-cells recruited for an immune response and in tracking them throughout an infection (Warren et al. 2011; Newell et al. 2013). However, systems immunology has mostly been descriptive so far. Connecting these tools with host or parasite strategies will be necessary for a more predictive perspective, based on the fundamental principles of evolutionary parasitology.

12.3 Within-host evolution

Within-host evolution is particularly important for microparasites such as bacteria or viruses, where infections grow as populations. As the cells and virions replicate, mutations accumulate, recombination occurs, or mobile genetic elements are transferred. Therefore, the infecting population changes its composition. To understand the within-host dynamic of infection, therefore, requires taking into account the evolutionary processes and studying the consequences for the host and the transmission of the parasite (Figure 12.5). This is not only a scientifically interesting problem but also one of high practical value. For instance, infecting populations can evolve variants resistant to treatment, or initially asymptomatic infections can evolve to become severe with life-threatening consequences (Young et al. 2012, 2017a).

New variants of a pathogen emerging within a host can have an advantage. They may be able to colonize a new niche (e.g. another tissue, another site in the gut), or to replicate faster. However, inevitably, there are also disadvantages because of trade-offs among different components of fitness. Furthermore, as the parasite adapts to its current host, it can lose the ability to transmit or to infect a different type of host efficiently. Hence, there is a danger of short-sighted evolution within the current host, which will not increase fitness over the entire parasite life cycle (Levin and Bull 1994; Martínez 2014). It is essential, though, to recall that although resistance must initially appear by mutation, for most microbial pathogens the primary path to acquire resistance within a host is by horizontally transferred genes from co-infecting strains.

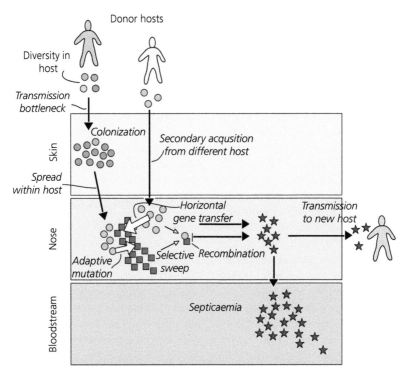

Figure 12.5 Within-host evolution of pathogenic bacteria. Infection starts with transmission from a donor host (top left, red) and can include a genotypically diverse primary infection (coloured dots). Because of the infection and colonization bottleneck, only a subset of these genotypes can establish (red and green strains). The infection subsequently spreads from the primary infection site (e.g. on skin) to a secondary site (e.g. the nose), where the bacterial population multiplies and grows. In this process, new mutations arise (red squares) that are better adapted to the within-host conditions. Such adaptive mutations can rapidly fix in the infecting population by a selective sweep. Here, a second infection from a different donor host (top-right, blue) happens, which establishes a different genotype (blue dots). Co-infections can lead to recombination or exchange of genes between strains by horizontal transfer that yield a novel type (red asterisks). The novel type is transmitted to a next host. It may also enter the bloodstream, multiply to high numbers, and cause a life-threatening septicaemia. The scenario fits the biology of *Staphylococcus aureus*. Adapted from Didelot et al. (2016), with permission from Springer Nature.

12.3.1 Evolutionary processes in infecting populations

The evolutionary forces that lead to changes in infecting populations are the same as for any other organisms. In short, evolution requires extant (genetic) variation and selection that acts on this variation. The relative importance of these 'standard forces' may vary among infecting populations, but the basic processes remain the same. It may be a truism, but it is probably fair to say that for most parasites, especially for bacteria and viruses, the evolutionary processes that lead to new variants and genotypic diversification of populations happen within a host (or a vector). During transmission, in contrast, selection among existing variants is the dominant process. This may be particularly obvious for pathogens having long-lasting spores or durable stages outside a host, which survive at different rates but where no replication occurs.

12.3.1.1 Processes of diversification

Mutations of single sites (point mutations of amino acids, changing the protein) occur randomly and at a given rate. In bacterial populations, estimates are 10^{-10} to 10^{-9}, and up to 10^{-6}, per base and replication (Ochman and Wilson 1987; Didelot et al. 2016). This value is smaller than in most viruses (where estimates range from 10^{-3} to 10^{-8} per base and replication, Duffy et al. 2008). However, bacteria have larger genomes than viruses and, therefore, the total number of mutations that appear in the entire genome becomes quite similar to viruses. Some genetic loci may be more prone to mutation than others; they show hypermutation. If mutations affect loci for gene expression, the phenotype of the pathogen may evolve to quickly change back and forth as the infection proceeds (known as 'phase variation'; see section 10.4.3). The functional genetic diversity can also become reduced, as mutational degradation can corrupt genes, leading to pseudogenes.

Recombination among different co-infecting genotypes is an additional source of evolutionary change. With orderly, homologous recombination, offspring receive genetic contributions from either parental strain. With non-homologous recombination, the ends of the double-strand breaks (that occur during meiosis) are joined with non-matching sequences. Hence, the offspring can be of a completely novel genotype, not resembling any of the parental ones. In both cases, the new genotype is likely unknown to the host's immune defences and thus has an advantage. Recombination can speed up evolutionary change by orders of magnitude, especially when one of the parental strains already contains 'successful' genetic material that has entered with horizontal gene transfer by mobile genetic elements, such as plasmids and transposons, or by vectoring through bacteriophage (Didelot et al. 2012). Nevertheless, any mutant will rise in frequency only when favoured by selection. As discussed in section 10.6, selection can be purifying, selecting for a narrow range of best-adapted types, or selecting for a diversity of variants that may persist in different niches inside a host. For populations immediately after the infection step, a numerical bottleneck is probably the most common situation; genetic drift, therefore, may be a non-negligible process. If so, the infecting population changes randomly, and selection may be a comparatively weak force. As a result, the likelihood of population extinction is high, and the infection may die out before it takes hold and can adapt further.

12.3.1.2 Evolution of bacteria

Horizontal gene transfer is a hallmark for the evolution of bacterial populations. In *Salmonella*, for example, essential genes such as type III secretion systems are acquired in this way. Because complete and functional genes are transferred as a whole, this kind of evolution is typically much faster than when dominated by point mutations. Nevertheless, the evolution of pathogenic variants without horizontal gene transfer can be fast, too, and, in particular, when regulatory genes are affected. Likewise, trade-offs among different components of fitness can limit adaptation. In *Salmonella*, the type III secretion system is costly, as variants not expressing this system replicate faster within a host but are not fit to invade the gut mucosa. In this case, genotypes with a functional set of secretion systems pave the ground for the co-infecting, non-expressing types.

Among the various pathogens, there is, therefore, a diversity of processes that dominate the generation of variation and within-host evolution. For example, *S. enterica* serovar typhi has hundreds of genes with remote or no orthologues elsewhere, an essential pathogenic island (SPI-7) involved in immune envasion, as well as genes for toxin production that are associated with disease symptoms. All of these were acquired by horizontal transfer and have experienced subsequent change by mutations. In *S. typhimurium*, by contrast, evolution to new host-adapted variants seems to have happened beyond gene transfer (Tanner and Kingsley 2018). Distinct, but characteristic patterns of bacterial within-host evolution also emerge from studies of *Pseudomonas aeruginosa*. In a Danish study, the whole-genome sequencing of 474 isolates showed that dozens of genes evolved convergently at the molecular level (Marvig et al. 2015). The changes found in these screens must have appeared by mutation events first and were subsequently positively selected to become frequent enough as to be detected. Besides, the succession of these mutations (affecting regulatory networks) is well ordered, suggesting that the change in regulation is a prime process during the adaptation to hosts. Any mutation that occurs upstream in the regulation cascade, in fact, sets the stage for the selective value of downstream mutations. The selective persistence of different mutations in a regulatory network is, therefore, contingent on each mutation and will require concerted changes to keep a cascade functional and make it more efficient.

12.3.1.3 Evolution of viruses

The mutation rate is relevant for within-host viral evolution. Across viruses, mutation rate correlates negatively with genome size in viruses (Figure 12.6) (this is also the case for prokaryotes). The RTdRp system (RNA-dependent RNA polymerase) of RNA viruses is particularly prone to making errors during copying of a strand, which results in some of the highest mutation rates (per site, per generation) known for all organisms (Duffy et al. 2008; Sanjuán et al. 2010; Sanjuán and Domingo-Calap 2016).

A single infection by a virus can leave the enormous number of 10^{15} descendants within a host (Dolan et al. 2018). However, for the evolutionary process, the (genetically) effective population size rather than its absolute number is crucial (see section 10.7.1). In viruses (and many other microparasites) the effective size is much smaller than the numerical population size. Moreover, the effective size is approximately the harmonic mean of numerical population sizes when it changes over time. Its value is thus heavily affected by population size bottlenecks (Bull et al. 2011), which can result from various processes as the pathogen population infects and adapts to its host (Pennings et al. 2014). Bottlenecks also occur during transmission when only a few variants succeed (even though they may occur in large numbers).

Similarly, strong selective sweeps, favouring variants that initially are present at low numbers, can pull along other genes by linkage ('hitch-hiking'), but the population keeps its small effective size. In such cases, the effects of genetic drift may also be strong and will push the genetic profile of an infecting population in different, somewhat random directions. Moreover, when deleterious mutations are accumulating in populations, the rate of adaptation of the population will slow down. This process is known as 'Muller's ratchet'. It may eventually lead to a mutational meltdown, i.e. a collapse of performance in the environment because of too many malfunctioning genes. Similarly, when several lineages are present that are adapted to the current host, they inevitably compete for the same resources ('clonal interference', Pandit and de Boer 2014), again slowing down the rate of evolution.

Viral genomes are packed densely with coding regions that also frequently overlap along the genome. Most mutations, therefore, are likely to be deleterious or lethal (Sanjuán 2010; Acevedo et al. 2014); high mutation rates should not be advantageous. Viral populations can evolve mechanisms that limit the phenotypic effects of deleterious mutations, a phenomenon known as 'robustness' (de Visser et al. 2003). Moreover, the mutation rate itself evolves, e.g. in the RNA viruses that encode their

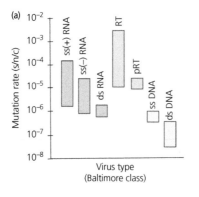
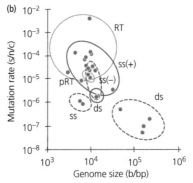

Figure 12.6 Mutation rates in viruses. (a) Rates found in different types of viruses—RNA (red), DNA (blue), and retroviruses (green). (b) Rates in relation to genome size. Larger viruses tend to have smaller mutation rates, mainly due to DNA viruses. Mutation rate is substitutions per nucleotide per infection of a cell (s/n/c). Baltimore classes of viruses are as follows: ss(+)RNA, single-stranded (positive sense) RNA viruses; ss(-) RNA, single-stranded (negative sense) RNA viruses; dsRNA, double-stranded RNA viruses.; RT, reverse-transcribing (retro-) viruses; pRT, para-retroviruses; ssDNA, single-stranded DNA viruses; dsDNA, double-stranded DNA viruses. Reproduced from Sanjuán and Domingo-Calap (2016), under CC BY 4.0.

replication machinery (in contrast to small DNA viruses that depend on the host cell) (Duffy 2018). It seems plausible, therefore, that mutation rate, generating new variants and thus increasing the speed of adaptation, is balanced against the fidelity of replication that preserves already adapted genotypes. However, high mutation rates in organisms, such as RNA viruses, are probably not an adaptation per se, but a consequence of selection on other traits. Various episodes of viral life history, including transmission, are such selective factors. For example, fast but accurate replication is advantageous when external factors limit infection duration. Replication can be slow and sloppy when a set viral load will eventually kill the host and terminate the infection (Regoes et al. 2013). Hence, certain conditions select for high recombination rates, and viruses have to cope with them (Duffy 2018). For this, viruses have secondarily evolved mutational robustness.

Many viruses show recombination upon replication. In viruses, different mechanisms from those of meiosis in eukaryotes exist. During the replication of RNA viruses, for example, the RdRp system generates a copy of a template (the genome), which subsequently associates with another template (a process called 'copy choice'). If the association is at the same site (locus), the recombination is homologous and viable new variants frequently emerge. If the association is with different sites, heterologous recombination occurs and defective RNA strands are likely to emerge, but also occasional functional, novel viruses may result. Such heterologous recombination can also occur between different viral 'species' or distant lineages, which may have severe consequences for pathology. For RNA viruses with segmented genomes, the recombination typically produces progeny that inherit a mixture of the parental segments. This is somewhat similar to the result of meiotic recombination in eukaryotes, where maternal and paternal chromosomes form novel combinations in the gametes. A most prominent example is IAV, whose proteins are located on eight genomic segments. As a result, the antigenic profile of IAV is re-assorted frequently and between distant lineages, e.g. between variants circulating in pigs, birds, and humans.

With these evolutionary processes, RNA viruses, in particular, can rapidly diverge into an array of slightly varying genotypes, typically covering the genetic neighbourhood of the ancestral type. Eventually, these populations consist of large clouds of sequences surrounding an ancestral master sequence known as a 'quasi-species' (Eigen 1996). This swarm-like behaviour of viral populations could lead to rapid further adaptation or consequences for pathogenesis (Vignuzzi et al. 2006; Trimpert and Osterrieder 2019). For instance, RNAi's are important antiviral defences and inhibit viral protein synthesis by targeting specific mRNAs. However, in almost all cases, viral mutations that resist this mechanism are known (Presloid and Novella 2015). Especially RNA viruses are feared to evolve such resistant escape mutants because of their high mutation rates and a possible quasi-species diversity. Such escape occurs against vaccines or drugs and is a long-standing problem, e.g. control of hepatitis B viruses (HBV) (Perazzo et al. 2015). Nevertheless, the relevance of the quasi-species concept for real infections remains controversial. It may merely describe a population in mutation-selection balance (Holmes 2010), and the associated genetic diversity may not be relevant in the first place (Fitzsimmons et al. 2018).

Particular elements stand out for the evolution of viruses, as compared to other organisms. For example, within-host evolution often happens in parallel in the different tissues colonized by a virus. In such cases, the different viral populations adapt to tissue-specific defence mechanisms. An example is poliovirus experimentally infected in mice. The populations in spleen, kidney, and liver diverge, and divergence also occurs among different individual mice (Xiao et al. 2017). The simultaneous presence of viral populations in different tissues is also a case of spatial dynamics within a host. Different methods can assess such compartmentalization, e.g. with an analysis of genomic distances among variants (e.g. with the so-called 'Hamming' distance). Alternatively, a spatial structure can be analysed with tree-based methods, where all variants are classified by a phylogenetic tree that, ideally, will reveal the history of compartmentalization of the entire infection into different tissues (Bons and Regoes 2018). The dynamics of such spatially structured systems adds further layers of complexity to within-host viral evolution (Gallagher et al. 2018).

12.3.2 Antigenic variation

With antigenic variation, an individual parasite repeatedly changes its antigenic surface during its stay inside the host. With a continuously changing antigenic surface, it becomes more costly for the host to keep track of the identity of the parasite, to the extent that long-lasting infections become possible, e.g. when the host defences cannot catch up with the rate of change. In protozoans, for example, this results from changes in gene expression and is not, therefore, a genuine process of within-host evolution. In viruses, antigenic variation is genetic change and thus an evolutionary process.

The African trypanosomes, e.g. *Trypanosoma brucei*, are the classic examples for antigenic variation (Horn 2014; Matthews et al. 2015). *T. brucei* activates this system when it reaches the salivary gland of its

vector, the tsetse fly (*Glossina* sp.), ready to become transmitted to a mammal. The system becomes inactivated when the parasite leaves the mammalian host and returns to its vector. The parasite's surface is densely covered by variant surface glycoproteins (VSG), which together obscure other surface components that could be recognized, too. The turnover of this coat is associated with cell division (approximately every six hours) where the new variants are expressed (Figure 12.7). Because the VSG surface is highly immunogenic, variants recognized by the host become eliminated. Novel surfaces not recognized rapidly enough survive and can rise in frequency. However, antigen switching seems not merely driven by the response of hosts (Matthews et al. 2015). A single infecting trypanosome expresses at least 100 distinct antigenic surfaces during an infection (Capbern et al. 1977), but likely many

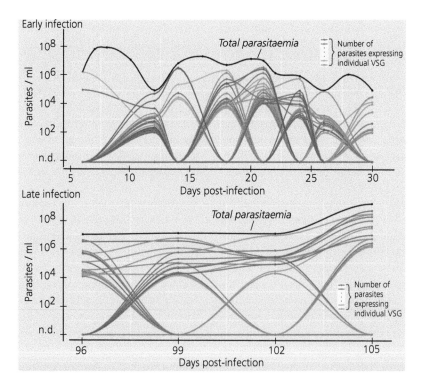

Figure 12.7 Antigenic variation in *Trypanosoma brucei*. The graphs show the number of parasite cells (titre in cells/ml) during the early (top panel, 5–30 days after infection) and late (bottom panel, 96–105 days after infection) period. The top line (black) is the total number of cells/ml (parasitaemia; scale on left, n.d. = not detectable). The coloured lines refer to the various cell populations that express different variable surface genes (VSG). The turnover of variants occurs over a matter of days. VSGs expressed during the early period do not reappear later in the infection. Reproduced from Mugnier et al. (2015), with permission from the American Association for the Advancement of Science.

more. Well over 2000 VSG variants are laid down in the genome (Cross et al. 2014; Horn 2014), of which the parasite expresses only one at a time.

The details of how VSGs are encoded and expressed are quite sophisticated. They involve unusual processes such as DNA recombination and gene rearrangement (similar to antibody formation; Aresta-Branco et al. 2019), trans-splicing, and poly-cistronic transcription (Horn 2014; Onyilagha and Uzonna 2019). The 15 to 25 VSG-expression sites in the genome regulate the processes, of which only one site is active at any one time. Whereas a given infection shows highly diverse VSG expression, frequently the same types appear in different infections. Furthermore, pre-existing immunity by earlier exposure of hosts can also reduce the success of antigenic variation in any real-life situation (Mugnier et al. 2015). As a whole, of course, the VSG repertoire evolves in a population of parasites. Comparative studies have revealed how these repertoires have evolved in trypanosomes and are reconstructed as a 'surface phylome' (Jackson et al. 2013; Onyilagha and Uzonna 2019).

Antigenic variation is, in fact, a more general phenomenon and not restricted to trypanosomes. As mentioned elsewhere, it also exists in *Plasmodium falciparum* (Haemosporidia, Apicomplexa) (Sacks and Sher 2002; Turner 2002; Guizetti and Scherf 2013; Gomes et al. 2016). In this case, single members of the parasite's *var* family of genes (with *c*.60 members, coding for PfEMP1) are expressed and transported to the surface of the red blood cell, where the parasite (the merozoite stage) resides. Sometimes, only a single member is expressed in a process called 'monoallelic expression' or 'allelic exclusion'. The development of the parasite inside the red blood cell takes around 48 hours. The *var* genes are active during the first 10–14 h of this stage, then silenced, but reactivated again in the next erythrocytic cycle. Many of the underlying processes that lead to these expression profiles are still unknown.

Antigenic variation also exists in bacteria, such as in *Borrelia burgdorferi* that causes Lyme disease (Verhey et al. 2019). *B. burgdorferi* is a chronic and systemic infection of mammals that persists for years. During this time, different antigenic variants of a *vlsE* lipoprotein ('variable protein-like sequence, Expressed') are expressed at the surface of cells. The underlying mechanisms are quite impressive. A gene encodes the lipoprotein on a plasmid (called 'lp28-1'); furthermore, a gene cassette of 15 silent (non-expressed), partly homologous sequences is located nearby. Repeated events of gene conversion then transfer elements from the silent cassette into the expressed variable region. A new variant is thereby generated, while keeping some structurally important regions intact. The *vlsE*-antigenic variation system allows the bacterium to present new variants to the host continuously. A related system is known from *Neisseria*, which also shows segmental gene conversion (Verhey et al. 2019). Finally, virus populations also show antigenic variation; for example, as mutant clouds evolving within the quasi-species collective (Domingo and Perales 2019), as diversity limited by a replication speed vs fidelity trade-off (Lauring 2020), or by antigenic drift under selection (e.g. influenza virus, Xue et al. 2018; HIV, norovirus, Debbink et al. 2014). These processes underlie within-host evolution and are likely subject to strong purifying selection (leading to dominant types).

12.3.3 Antibiotic resistance

Antibiotics are biochemical compounds that can kill microorganisms and slow the growth of their populations. For example, antivirals serve for the medical treatment of specific infections, e.g. influenza (Hsu et al. 2012) or can act more broadly (Vigant et al. 2015). Similarly, antifungals are important for medicine, or for the food industry to fight fungal infections (Campoy and Adrio 2017; Fuentefria et al. 2018). Microorganisms produce antibiotics, which likely are natural means to remove or outpace competitors, notably by bacteria and fungi. Nevertheless, all classes of antibiotics share the same problem of encountering increased resistance. Note that the term 'antibiotic resistance' (generally known by the acronym 'ABR') alludes to resistance against antibiotics with a focus on pathogens that are of medical importance; often, this is also called 'drug resistance'. The term 'antimicrobial resistance' (AMR) covers the same phenomena but has a broader focus. Here, we focus on antibiotic resistance against pathogenic bacteria as the best-known cases.

The beneficial effects of fungi against infections have been known since antiquity. In the late nineteenth century, the inhibiting effect of mould (such as *Penicillium*) on bacterial growth was noticed. However, only with the (re-)discovery of penicillin in 1928 by Sir Alexander Fleming (1881–1951) did the modern story of antibiotic use in medicine and agriculture begin. The treatment of wounds in the Second World War accelerated this development until the industrial-scale production finally was mastered. Over the years, derivatives of the first antibiotics and many new classes of hopeful antibiotics were discovered (Smith et al. 2015; Aminov 2016).

However, microorganisms, notably bacteria, but also fungi, viruses (Anderson 2005), and any other parasite, for that matter, quickly evolved resistance against these 'wonder drugs'. Resistant strains appeared within a few years of the introduction of a new drug (Smith et al. 2015) (Figure 12.8). For example, sulfonamide-resistant strains of *Streptococcus pyogenes* had already emerged by the 1930s. The penicillin-resistant *Staphylococcus aureus* appeared in the 1940s, soon after this new drug was introduced. Multidrug resistance (i.e. simultaneous resistance

against many drugs) was first observed in enteric bacteria in the late 1950s (Levy and Marshall 2004), and its current spread is a significant concern. In Europe and the United States alone, an estimated 50 000 people die every year due to antimicrobial-resistant infections (Blair et al. 2015; Aminov 2016). Hence, looking at these numbers, a crisis is looming (Rossolini et al. 2014). The search for new and better drugs is, therefore, pertinent. In the so-called 'golden age', from the 1940s to the 1990s (Peterson and Kaur 2018), new drugs were developed by modifications of existing ones, but this is no longer sufficient.

In general, antibiotics work more efficiently when the rate of bacterial replication, i.e. the rate of cell division, is high. In fact, the molecular mechanisms of resistance involve the inhibition of bacterial cell wall synthesis, inhibition of protein synthesis, inhibition of RNA or DNA synthesis, competition with bacterial biochemical pathways, and interference with other essential bacterial functions (Blair et al. 2015; Peterson and Kaur 2018) (Table 12.2).

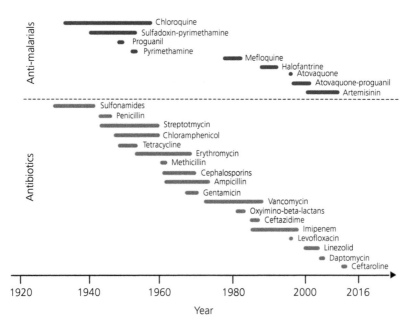

Figure 12.8 Time to emergence of resistance against antimalarial drugs (top) and antibiotics (bottom). Start of each bar marks the time of introduction of a substance; ends mark first observation of resistance. In all cases, substantial benefits for public health continued after first resistance, although not for an unlimited time and use. Resistance against vaccines also evolves but occurs much more rarely by comparison. Scale of years at bottom. Adapted from Kennedy and Read (2017), under CCBY 4.0.

Table 12.2 Some major groups of antibiotics and their mechanisms of action.[1]

Mechanism of action	Antibiotic families
Inhibition of cell wall synthesis	Penicillins, cephalosporins, carbapenems, daptomycin, monobactams, glycopeptides.
Inhibition of cell protein synthesis	Tetracyclines, aminoglycosides, oxazolidinones, streptogramins, ketolides, macrolides, lincosamides.
Inhibition of DNA synthesis	Fluoroquinones.
Competitive inhibition of folic acid synthesis	Sulfonamides, trimethoprim.
Inhibition of RNA synthesis	Rifampin.
Other mechanisms	Metronidazole.

[1] Adapted from: Levy. 2004. *Nat Med*, 10 (suppl.): S122, with permission from Springer Nature.

Bacteria, in turn, have evolved various mechanisms to resist the effects of antibiotics (Box 12.4). Some bacteria may have 'intrinsic resistance' due to their functional characteristics; for example, they lack the targets that the antibiotics need to act on. The antibiotic daptomycin, for example, works against Gram-positive but not against Gram-negative bacteria. Such drugs cannot cross the outer membrane of Gram-negative bacteria and, therefore, do not reach their targets (Blair et al. 2015). Otherwise, three major mechanisms provide antibiotic resistance: bacteria either reduce the concentration of drugs within the cell, modify the drug's target such that it becomes unavailable, or directly modify the drug, e.g. by its degradation (Box 12.4).

The genes that provide antibiotic resistance can emerge either by mutation, by gene transfer during conjugation of bacteria (a sex pilus allows transfer of plasmids), with the transfer of mobile genetic elements ('transduction'), or when acquired from other bacteria via uptake of 'free' DNA ('transformation'). For example, penicillin resistance entered *Streptococcus pneumonia* from naturally resistant *S. viridans* by transformation. These are the elements of horizontal gene transfer already discussed in section 10.3.2. They play an essential role in the rapid spread of antibiotic resistance in bacterial populations. Antibiotic resistance genes are found in most, if not all, bacteria explored so far (D'Costa et al. 2006; Forsberg et al. 2014). The emergence of antibiotic resistance is a natural phenomenon and has been around for millions of years (Aminov 2009).

However, several antibiotic resistance mechanisms are not based on the acquisition of mutations

or horizontal gene transfer. Such 'non-inherited resistance' occurs when bacteria become temporarily refractory to the action of antibiotics. Because antibiotics typically kill bacteria as they divide or when they are metabolically active, such risky activities are downregulated for protection. Therefore, non-inherited resistance is associated with a (plastic) change in the phenotype rather than with changes in the genotype. Different mechanisms are known. Bacteria, for example, can form biofilms, which are dense aggregates of bacterial cells embedded in a matrix and attached to a surface (Flemming and Wingender 2010). Biofilm formation is a strategy of cooperation (West et al. 2006; Leggett et al. 2014) (see section 13.9.2). Biofilms are particularly refractory to most antibiotics, and thus pose a serious medical problem (Stewart 2002; Hall and Mah 2017). The protective effect itself can result from the polysaccharide matrix, through which antibiotics cannot diffuse. Also, inside biofilms bacteria generally have low activities and are, therefore, less susceptible. Besides, refractory cells might accumulate. Some other possible mechanisms include a change in physiological state or the induction of stress responses, which also protects against antibiotics.

The rate at which bacteria become killed by antibiotics often declines with time, such that a small fraction of the population survives. This phenomenon is known as bacterial 'persistence', 'adaptive resistance', or 'phenotypic tolerance' (Kussell et al. 2005; Levin and Rozen 2006). Persistence can result when a non-dividing fraction of the population activates an 'emergency system' that temporarily arrests further cell divisions. The presence of antibiotics

Box 12.4 Mechanisms of antibiotic resistance in bacteria

Following Blair et al. (2015), a convenient grouping of anti-biotic resistance mechanisms in microorganisms, and in the case of bacteria, is as follows.

1. *Reduction of drug concentration within the cell*: This is achieved by reducing the influx and/or increasing the efflux of the drug from the cell. Gram-negative bacteria have a second outer membrane that acts as a permeability barrier, which intrinsically reduces drug uptake. Reductions in permeability also result from modification, replacement, or downregulation of outer-membrane porin channels. Examples are Enterobacteriaceae (*E. coli*), or *Pseudomonas*; drug treatments select for these changes. Bacteria increase the efflux of antibiotics by increased

expression of the respective transporters in the cell membrane ('efflux pumps'). This type of resistance mechanism is quite common, as many different mutations can affect regulation. Transport systems can be quite complex (Greene et al. 2018), and some transporters can be specific in what they accept. Many pumps use a wide range of similar drugs and are, therefore, involved in multidrug resistance phenomena, a well-studied example being the RND (resistance nodulation division) family of efflux pumps (Piddock 2006). A disquieting development is that multidrug pumps have become mobile with time, i.e. the respective genes are located on plasmids that can readily jump to other bacteria.

2. *Modification of the target for the antibiotic*: The targets of antibiotics involve ribosomes, metabolic enzymes,

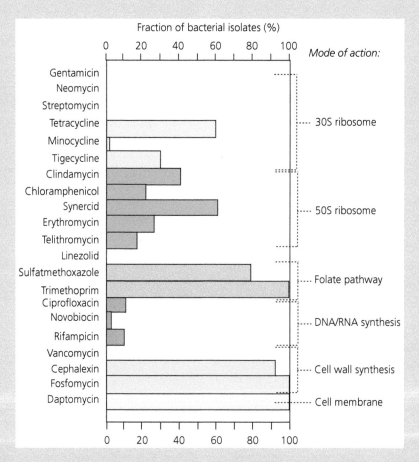

Box 12.4 Figure 1 Natural resistance of bacteria. The graph shows the percentage of bacteria (x-axis) with resistance against different antibiotics (y-axis), and classes of their action, in a sample of 480 isolates from soil. Adapted from D'Costa et al. (2006), with permission from the American Association for the Advancement of Science.

the DNA replication machinery, or cell wall components. A large number of resistance mechanisms exist that make these targets unavailable. The antibiotic linezolide, for example, targets the 23S rRNA ribosome in Gram-positive bacteria. Multiple copies of the same gene encode this subunit; resistance appears when one of these copies is mutated (Billal et al. 2011).

Other mechanisms include the acquisition of genes homologous to the original target, but which no longer bind to an antibiotic, e.g. the *SCCmex* element in methicillin-resistant *Staphylococcus aureus* (Shore et al. 2011). The modification of targets also occurs by methylation, e.g. for ribosomes that are targeted by aminoglycoside antibiotics. Furthermore, the drug–target complex resulting from binding to an antibiotic can secondarily become broken up. This restores the normal function of the targeted molecule, e.g. the *qnr*-resistance genes against quinolone (Blair et al. 2015).

3. *Direct modification of antibiotics*: Inactivation by hydrolysis is a major mechanism that confers resistance. The classic example is the bacterial penicillinase that inactivates penicillin. Penicillinase was discovered very soon after this antibiotic became used (Abraham and Chain 1940). Since then, thousands of bacterial enzymes have become known that degrade or modify antibiotics of many classes.

The search for derivatives of prime antibiotics is one strategy to circumvent the problem of degradation. Unfortunately, many of the underlying genes rapidly spread by plasmids, or even by the insertion of sequences from soil bacteria (D'Andrea et al. 2013). Another way to inactivate an antibiotic is the transfer of chemical groups such as acyl, phosphate, nucleotidyl, or ribitoyl groups by the bacteria, which leads to structural obstacles for binding. Antibiotics, where the molecule is extensive and involves many exposed and affine regions (notably the aminoglycosides), are particularly sensitive to the addition of chemical groups. Resistant bacteria effectively block such antibiotics.

More than one genetic modification in the bacteria can, therefore, produce the same type of resistance, and a given antibiotic can encounter several resistance mechanisms at the same time. Redundancy and the broad repertoire of bacterial mechanisms favour the emergence of multidrug resistance. Resistance mechanisms, furthermore, derive from general functions of the bacterial cell. In natural bacterial communities, therefore, antibiotic activity is widespread and acts against a wide range of antibiotic classes, although the exact functions are not always known (Figure 1).

activates this stop (Miller et al. 2004). The phenotypic switch itself has a genetic basis and, consequently, there is a variation for persistence in a population (Balaban et al. 2004). Antibiotics are also particularly effective when bacterial populations grow exponentially, but much less so when the population has reached the stationary phase (Tuomanen et al. 1986). The antibiotic treatment, therefore, becomes less effective when the infection is already more advanced, i.e. when the population has reached the refractory phase (Levin and Rozen 2006). This kind of non-inherited resistance thus depends on the phase of the population growth.

Today, current human activities massively affect the prevalence of antibiotic resistance. In particular, the everyday use of antibiotics in hospitals, in agriculture, or by the general public is directly linked to the emergence and spread of antibiotic resistance in bacterial populations (Figure 12.9). Such overuse and misuse of antibiotics, as in some instances of meat and fish production, accelerates the problem even for human populations distant from the locations of use. Additional factors contribute to the worldwide spread of antibiotic resistance through, for example, careless instances of transmission between patients in hospitals, travel patterns of humans, co-resistance to other drugs, and so forth (Holmes et al. 2016; Marston et al. 2016). Against this background, there is substantial concern about the evolution of multidrug-resistant strains in several dangerous bacteria, e.g. *M. tuberculosis* or *S. aureus*. These pathogens circulate in hospitals. Multidrug-resistant strains have also emerged in pathogens that can generate classical epidemics, e.g. *Vibrio cholerae* and *Salmonella enteritidis* (Levy and Marshall 2004). Some bacteria, such as *Burkholderia cepacia* and *Pseudomonas fluorescens*, have even evolved to utilize antibiotics (penicillin) as their prime source of carbon and nitrogen (Hamilton-Miller 2004). Such subsistence on antibiotics exists in many other bacterial groups, too

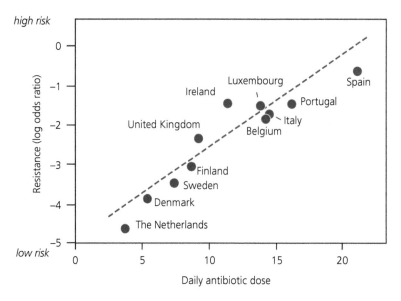

Figure 12.9 Evolution of antibiotic resistance. The graph shows the relationship between penicillin usage (x-axis: daily dose of lactam-type antibiotics) and risk of penicillin resistance (y-axis), expressed as the natural logarithm of the odds for resistance in any given isolate. Data refer to pneumococci infections in different European countries in 2005. Adapted from Livermore and Pearson (2007), with permission from Elsevier.

(Dantas et al. 2008). The increasing spill of medical antibiotics into the environment will, furthermore, alter the selective environment for natural populations of microorganisms (Martínez 2008).

Because so many promoting factors are human-made, intelligent strategies to manage antibiotic resistance seem possible and are urgently needed (Abel et al. 2015a). An ultimate goal would be to find an 'evolution-proof' drug, i.e. a drug against which microbes cannot evolve resistance, or at least one for which this would take a very long time (Read et al. 2009; Bell and MacLean 2017) (see section 14.2.1). Also, the consequences of competition and within-host dynamics could be harnessed to suppress the emergence of resistant strains. For example, limiting resources required more intensively by resistant lines will slow down the evolution of antibiotic resistance, when these lines have to compete with a sensitive strain (Wale et al. 2017).

12.3.4 Evolutionary perspectives of antibiotic resistance

In nature, the production of antibiotics should be an advantage against other, competing microorganisms

that are susceptible to a given compound. Likewise, antibiotic resistance provides protection against foreign antibiotics and own compounds ('self-resistance'; Peterson and Kaur 2018). Nevertheless, the evolutionary biology of antibiotic resistance in nature remains puzzling, since the concentration of antibiotics in soil may often be too low to be effective (Andersson and Hughes 2014). Antibiotic compounds can also be involved primarily in processes such as 'quorum sensing' rather than defence (Aminov 2009; Andersson and Hughes 2014). Similarly, good evidence for an advantage of the production of antibiotics for the microorganism is scarce, too (Andersson and Hughes 2014).

Whatever the natural advantages of evolving antibiotic resistance, it certainly provides a fitness advantage under selection by drug treatments. At the same time, it entails a fitness cost when the drug is absent, e.g. lower growth rates or loss of competitiveness. Empirical evidence shows that these costs are highly variable, with values ranging from losing well over half of the fitness in a drug-free environment, to little or no measurable costs at all. The costs also depend on drug class and, in no small degree, on how resistance genetically is implemented

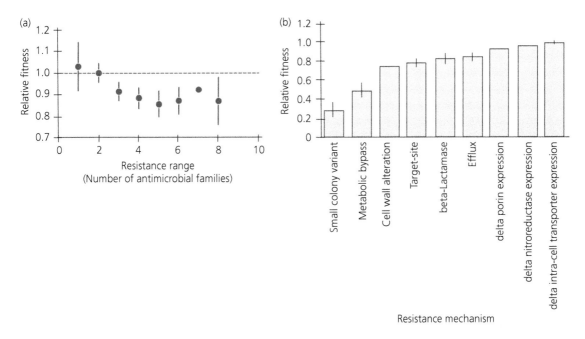

Figure 12.10 Costs of antibiotic resistance. (a) The fitness cost of acquiring a plasmid increases in proportion to the resistance range of a plasmid; defined as the number of antimicrobial families to which phenotypic resistance is gained when acquiring the plasmid. (b) Fitness cost in relation to the mechanism of resistance. Reproduced from Vogwill and MacLean (2014), under CC BY.

(Melnyk et al. 2014; Vogwill and MacLean 2014). For example, the acquisition of resistance via plasmids carries a lower cost than when acquired by a mutation in the existing genotype. Furthermore, plasmids carrying resistance to many families of antibiotics impose higher fitness costs, regardless of their size, than those carrying only a few resistance genes (Figure 12.10).

In many cases, the costs of antibiotic resistance in bacterial populations are astonishingly small or non-measurable (Melnyk et al. 2014). Bacteria can secondarily acquire compensatory mutations that reduce costs without much loss of resistance (Schrag et al. 1997; Andersson and Levin 1999; Andersson and Hughes 2010). For example, strains of *Salmonella typhimurium* experimentally selected for resistance against streptomycin (based on a mutation in the *rpsL* gene) and against fusidic acid (mutation of the *fusA* gene) (Björkman et al. 2000) pay a cost and grow more slowly in media as well as in test mice. The reason is that the same mutations also affect the rate of protein synthesis needed for multiplication (*rpsL* encodes ribosomal proteins and *fusA* encodes

elongation factor). However, when these resistant lines subsequently are selected for faster growth, resistance persists, but novel mutations accumulate that increase growth rates. By molecular analysis, a total of 53 different such compensatory mutations go with resistance against fusidic acid, and a total of 24 go with streptomycin resistance. The majority of those do not affect the already acquired resistance (Björkman et al. 2000).

Entire spectra of mutations, in fact, can reverse bacterial fitness losses due to the acquisition of antibiotic resistance. They differ according to the environment, e.g. whether the bacteria grow in medium or mice. Compensatory mechanisms, furthermore, can evolve very fast on both the bacterial chromosome and the plasmids. Sometimes, mutations may also be pre-existing, e.g. were induced with a former resistance acquisition and persisted when the resistance plasmid was subsequently lost (Vogwill and MacLean 2014).

The spread of antibiotic resistance is studied intensively in theory and with experiments (Legros and Bonhoeffer 2016; Lehtinen et al. 2017; Birkegård et al. 2018). Together, it appears that the relative accumulation of novelty determines the emergence and persistence of antibiotic resistance, i.e. how frequently mutations occur, the associated (pleiotropic) fitness costs, and the acquisition of compensatory mutations (Levin et al. 2000). Migration acts in similar ways to mutation, since it introduces new genetic variants that can spread in the resident population. For example, resistance was evolving faster, and the fitness costs of resistance declined more rapidly when migration into a population of *Pseudomonas aeruginosa* experimentally increased. The adaptation was also faster with only one antibiotic, or when two different kinds were alternated ('cycling therapy'), rather than given simultaneously ('bitherapy') (Perron et al. 2007). How resistance evolves and wanes with time is of obvious relevance to medical practice (Hall 2004; Andersson 2006); for example, medical practice and hygiene change the migration rate of genes among units and patients.

12.4 Multiple infections

Multiple infections of the same host by several strains of the same parasite are also labelled as 'mixed-genotype infections' or 'mixed infections', and are common (Table 12.3). Similarly, the simultaneous infection by different species of parasites is frequent. Of course, the genotypic heterogeneity within a host also increases by within-host evolution of the infecting parasite population, as discussed in section 12.3.1. 'Co-infection' is a simultaneous infection by different parasite strains and thus touches on their continued co-existence. In this case, competition among strains is comparatively weak but likely persists over the lifetime of an infection. By contrast, with 'super-infection', strains do not co-exist. In this case, strains replace one another throughout infection, and a process of competitive exclusion prevails during the short intervals of joint host occupancy by the different strains. Whether different strains co-infect or super-infect within a host is not always easy to show, and often difficult to trace over time. Newer technologies such as strain-specific real-time PCR, SNPs, or metagenomics analyses, allow the tracing of single strains as they multiply and grow within a host.

Regardless of the process, heterogeneity of the parasite population within a host sets the stage for competition or cooperation between parasite variants. The competition will be particularly intensive when there are many strains within the host, when these strains are similar in relevant characteristics, when the infection lasts a long time, or when it takes a long time from infection to eventual transmission. On the other hand, cooperation among strains and parasites can change the outcome of the infection considerably. Multiple infections, therefore, have a range of consequences on hosts and parasites (Balmer and Tanner 2011; Bordes and Morand 2011; Li and Zhou 2013; Kada and Lion 2015; Sofonea et al. 2017).

12.4.1 Competition within the host

Competition between co-infecting parasites is widespread and essential. Co-infecting parasites affect one another almost by default, as host resources are typically limited. Furthermore, an existing infection can prevent a second infection from establishing (e.g. in viruses, Hart and Cloyd 1990). Competition primarily is for host resources, either when co-infecting strains passively exploit the host and some variants can extract more resources, or when they actively interfere with the host and with each other (Hart and Cloyd 1990; Riley and Gordon 1999; West and Buckling 2003). Furthermore, competition is mediated by differential responses of the host's immune system (Cobey and Lipsitch 2013) (see immunoepidemiology, section 11.8). Competitive process, furthermore, can provide new windows to therapy. For example, a drug-resistant strain could be numerically contained through competition with a more sensitive one; this procedure could then prolong the time for treatment (Hansen et al. 2017).

For *Plasmodium* infections, competition is suggested because infection intensity does not necessarily increase with the number of co-infecting strains (Read and Taylor 2001; Sondo et al. 2019). Eventually, too, the more successful strains will dominate the infection. Multiple infection can, therefore, lead to the persistence of more virulent strains when they are more competitive (de Roode et al. 2005b). Clinically noticeable cases of human

Table 12.3 Multiple infections in natural situations.

Host	Parasite	Observation	Source
Fungi	Viruses	Multiple infections common in fungal hosts.	3
Plants (cotton) in Asia	Begomovirus	Very frequent, multiple infections by several strains and different viruses vectored by whiteflies.	5
Plants	Viruses	Typically heterogeneous viral populations in plants, either by mutation or multiple infections. However, diversity of infections typically dominated by a few common variants.	2
Insects	*Wolbachia* (bacterium)	Frequent occurrence of multiple infections in many host taxa.	9, 16
Bombus terrestris in Central Europe	*Crithidia bombi* (trypanosome)	Half of all spring colonies infected by three to six strains. High diversity of strains among all hosts.	14
Domestic fowl (chicken, turkey)	Oncogenic viruses (Marek's disease virus, avian leukosis virus, etc.)	Among infected hosts, 24–25 per cent had multiple infections in chicken and turkey, respectively, through different strains and different viruses.	1
Lions in the Serengeti	Feline immunodeficiency virus (FIV)	Multiple infections with two to three strains in 43 per cent of all FIV-positive individuals.	15
Mammals (sheep, mice, humans); fish	Various	Many examples of multiple infections, and of within-host interactions by parasites.	8
Humans (Southeast Asia, Caribbean)	Dengue virus	Of 292 samples, 5.5 per cent with two or more strains.	6
Humans (adults in Brazil)	TT virus	Six of eight patients analysed in detail were multiply infected by two to seven strains. High diversity of strains among all hosts.	7
Humans in USA	*Borrelia burgdorferi* (bacterium)	Two out of 16 investigated patients with symptomatic Lyme disease infected by two strains.	10
Humans in high endemic areas	*Plasmodium falciparum*	Many hosts multiply infected; less frequently observed in chronic infections.	11
Humans in Africa	*Plasmodium*, helminths (*Ascaris, Trichuris*)	Concurrent infections by all three parasites common.	13
Humans	Various	Review of a total of 2009 published studies of co-infection in humans. Co-infections generally have serious health effects; most often include bacteria.	4
Amphibians (frogs, newts)	Trematodes, chytrid fungus	More than 2000 individuals at 90 study sites in California. Five amphibian species. Multiple infections found at 92 per cent of sites and in 80 per cent of hosts.	12

Sources: [1] Davidson. 1999. *Acta Virol*, 43: 136. [2] Garcia-Arenal. 2001. *Annu Rev Phytopathol*, 39: 157. [3] Ghabrial. 1980. *Annu Rev Phytopathol*, 18: 441. [4] Griffiths. 2011. *J Infect Dis*, 63: 200. [5] Harrison. 1999. *Annu Rev Phytopathol*, 37: 369. [6] Lorono-Pino. 1999. *Am J Trop Med Hyg*, 61: 725. [7] Niel. 2000. *J Clin Microbiol*, 38: 1926. [8] Pedersen. 2007. *Trends Ecol Evol*, 22: 133. [9] Reuter. 2003. *Mol Biol Evol*, 20: 748. [10] Seinost. 1999. *Arch Dermatol*, 135: 1329. [11] Smith. 1999. *Parasitologia*, 41: 247. [12] Stutz. 2018. *Methods Ecol Evol*, 9: 1109. [13] Thsikuka. 1996. *Ann Trop Med Parasitol*, 90: 277. [14] Tognazzo. 2012. *PLoS One*, 7: e49137. [15] Troyer. 2004. *J Virol*, 78: 3777. [16] Werren. 1997. *Annu Rev Entomol*, 42: 587.

malaria, for example, can be associated with fewer and presumably more rapidly replicating clones of the parasite, when compared to asymptomatic patients (Mercereau-Puijalon 1996; Smith et al. 1999). Experimental tests with malaria in mice also suggest competition, with an outcome set by several factors (de Roode et al. 2005a). The initially more common clone in the inoculum usually becomes dominant afterwards. However, the further transmission to mosquitoes (the vectors) does not necessarily favour the majority strain circulating in the vertebrate host at the time of biting. Instead, the sexual gametocytes of the minority strain occur at higher frequencies in mosquitoes than expected from the blood of experimental mice (Taylor et al. 1997). This shift is caused by strain-specific

Table 12.4 Iron uptake in pathogenic microorganisms.

Bacterium	Disease	Uptake	Source
Enterobacteriaceae (*E. coli, Salmonella, Shigella, Klebsiella, Yersinia*)	Pathogenic strains cause urinary tract infections, septicaemia, wound infections.	Siderophores: aerobactin, enterobactin, salmochelin, yersiniabactin.	8
Yersinia pestis	Plague.	Siderophore: yersiniabactin, can remove iron from host transferrin and lactoferrin.	5
Legionella pneumophila	Legionnaires' disease.	Via FeoB system (transporter): inner-membrane protein. Required for intracellular infections and virulence. Via siderophore: legiobactin, a non-protein iron chelator.	4
Mycobacterium tuberculosis	Tuberculosis.	Via siderophore system: carboxymycobactin (released extracellularly), mycobactin (anchored in outer membrane). Two uptake pathways for haem from host, followed by haem degradation and release of iron.	3
Pseudomonas aeruginosa	Pneumonia, septicaemia, urinary infections.	Siderophores: mainly pyoverdine, pyochelin. Haem: direct uptake (receptor and transporter); haemophore-mediated system (extracellular binding protein, then interacts with haemophore receptors).	2, 6, 8
Staphylococcus spp.	Gastrointestinal, urinary tract infections, pneumonia, endocarditis, etc.	Siderophores: staphyloferrin A, B (endogenous); hydroxamate, catecholate, carboxylate (xenosiderophores, produced by others). Haem uptake: Isd system with receptors to bind haemoglobin, which is shuttled inside.	2, 7
Neisseria meningitidis	Meningitis.	Transferrin (receptors TbpA, TbpB) and lactoferrin (LbpA, LbpB) uptake systems. Release of iron and transport into periplasm.	2
Fungi: *Candida albicans, C. glabrata Cryptococcus neoformans Aspergillus fumigatus*	Mucosal infections, meningitis in HIV patients, aspergillosis.	Siderophores, transporter system: Arn1, Sit1 (*Candida*); Sit1 (*Cryptococcus*); Sit1, Sit2, MirB (*Aspergillus*). Haem, haemoglobin acquisition: CFEM proteins, haemophore (*Candida*); ESRT complex, haemophore (*Cryptococcus*), none (*Aspergillus*).	1

Source: [1] Bairwa. 2017. *Metallomics*, 9: 215. [2] Cassat. 2013. *Cell Host & Microbe*, 13: 509. [3] Chao. 2019. *Chem Rev*, 119: 1193. [4] Cianciotto. 2015. *Future Microbiology*, 10: 841. [5] Fetherston. 2010. *Infect Immun*, 78: 2045. [6] Marathe. 2015. *Int J Bioassays*, 4: 3667. [7] Sheldon. 2015. *FEMS Microbiol Rev*, 39: 592. [8] Wilson. 2016. *Trends Mol Med*, 22: 1077.

host immune responses that favour (up to 20-fold in the experiments) the initially rare clone as it develops into the transmission stage (the gametocytes). Success in the competition also depends on the differential effects of drugs when hosts are treated. A generally substantial effect, however, seems to be the order of infection, which typically gives an earlier strain an inherent advantage over a strain arriving later (de Roode et al. 2005a).

For microorganisms, iron is one of the most precious resources. Mammal hosts, in turn, store or transport iron intracellularly in ferritin complexes or various cell types, e.g. in enterocytes, macrophages, or hepatocytes. Mammals regulate iron availability by cell-associated transporters, hormones, and a range of extracellular iron carriers, e.g. transferrin or lactoferrin (Sheldon and Heinrichs 2015). Microorganisms, notably bacteria and fungi, have, therefore, evolved various means to extract iron from their hosts (Table 12.4). Siderophores, for example, are secreted, iron-scavenging molecules that help to extract this critical resource from the host. Chemically, siderophores are chelators that can form a bond with metal ions, such as Fe^{2+}, and fixate these in a complex. Pathogenic microbes also acquire iron by 'stealing' it from haem, an iron-porphyrin complex of the haemoglobin. Capturing host transporters (transferrin or lactoferrin proteins) with specific receptors, or garnering iron directly with specialized iron transporters are other mechanisms that ensure iron acquisition for pathogens (Cassat and Skaar 2013). Hosts, however, are not defenceless and withhold iron, or actively sequester it from their pathogens. For instance, hosts have evolved siderophore-binding proteins, the 'siderocalins', that act against microbial siderophores. Their structure allows them to bind to a wide variety of different siderophores, but the binding properties seem to be determined by the most frequent pathogens (Sia et al. 2013).

Siderophores are also a subject of competition among co-infecting strains. When released by microorganisms, these molecules are freely circulating and can be utilized by all co-infecting strains. Siderophores are a shared resource that benefits all (a 'public good'). However, the costs of production might not equally be shared among all strains (West and Buckling 2003). In particular, strains not producing siderophores (the 'cheaters') do not bear a cost and could thus outcompete the producers. Nevertheless, when relatedness among co-infecting strains is high, this is less likely to have consequences, as shown for *Pseudomonas aeruginosa* (Griffin et al. 2004). The respective costs and benefits of siderophore production, therefore, depends on the composition of the infecting population and can be experimentally tested (Kümmerli et al. 2015).

Within-host competition can occur by direct interference, too. Bacteria have evolved a wide range of interference strategies, including reciprocal and pre-emptive interference, spite, or suicidal attacks (Granato et al. 2019). A large number of bacteria produce biologically active compounds that have antimicrobial properties, called 'bacteriocins', to interfere with or kill nearby bacteria (Riley and Wertz 2002). The distinction from 'true' antibiotics is one of degree, as bacteriocins typically work against the same or closely related bacterial species, whereas antibiotics have a much wider spectrum. Furthermore, bacteriocins are produced during the bacterial growth phase and are usually small molecules that are easily degraded, whereas antibiotics often are more persistent secondary metabolites.

Bacteriocins are known from many groups, such as lactic acid bacteria (Zacharof and Lovitt 2012; Collins et al. 2017), *Pseudomonas* (Parret and De Mot 2002), or pneumococci (e.g. *Streptococcus pneumoniae*) (Bogaardt et al. 2015; Javan et al. 2018). Because bacteriocins are effective against own and closely related taxa, there is a danger of succumbing to one's own products. But the bacterial genomes also contain genes for self-protecting 'immune proteins'. In the genomes of pneumococci, for example, there are several bacteriocin cassettes, which code not only for several different bacteriocins but also for several putative immune genes (Bogaardt et al. 2015).

Bacteriocins are deployed by diffusion from the bacterial cell, by shedding of membrane vesicles, by direct injection with the contact-dependent type VI secretion system (see Figure 8.1) (Chassaing and Cascales 2018), but also by lysis and the release of the contents by the producing cell. Such 'suicidal' behaviour is known for *E. coli*, *P. aeruginosa*, and *S. enterica* (Michel-Briand and Baysse 2002; Cascales et al. 2007). It can evolve when it benefits co-infecting non-lysing variants of the same clone (Mavridou et al. 2018). Due to their effectiveness against a narrowly defined range of bacteria, and their relatively simple chemistry, bacteriocins have become the focus of promising applications, notably in the food industry, e.g. for processing (fermentation) and as food preservatives (to keep unwanted bacteria away). Furthermore, their use now extends to human health practices in the form of probiotics to regulate the gut microbiome, as a novel class of antibiotics, and as delivery molecules that can shuttle a specific compound to the site of action, e.g. in cancer therapy (Balciunas et al. 2013; Chikindas et al. 2018).

12.4.2 Cooperation within hosts

Microbes engage in cooperation for their mutual benefit when infecting and exploiting a host, as is known, for example, for viruses and their helper viruses (Turner and Chao 1999). In general, close relatedness among co-infecting strains favours higher investment in the cooperative exploitation of the common good (the host resources). This, in turn, can lead to increased virulence of the infecting population (Buckling and Brockhurst 2008). However, benefits and costs of cooperation are dependent on the environment; therefore, the outcome of the interactions or the resulting virulence is not always the same (Zhang and Rainey 2013; Kümmerli and Ross-Gillespie 2014).

Cooperation among parasites can involve a range of mechanisms. Examples include the production of digestive enzymes to mobilize host resources or the joint production of adhesive polymers to form biofilms. Examples are dental plaque or biofilms on implants that prevent healing and their integration into the body. A further important process for cooperation is 'quorum sensing' in bacterial populations (Box 12.5). This is also known from pathogenic fungi such as *Candida albicans* (Albuquerque and Casadevall 2012).

Box 12.5 Quorum sensing in bacteria

Quorum sensing is best-known from bacteria. Quorum sensing starts when bacteria release substances into the environment in concentrations that are too low to be effective. However, when many cells of a local aggregate contribute, the concentrations rise to a level where bacterial genes become activated (Waters and Bassler 2005). The phenomenon was discovered with bioluminescence in marine bacteria (*Vibrio*). Quorum sensing utilizes a wide range of different molecules (Figure 1).

Notably, the production of a signal depends on environmental conditions. In many cases, quorum sensing regulates the production of signals that can benefit others in a group, e.g. the directed activation of appropriate enzymes to gain access to specific food sources. However, quorum signals are freely diffusing in a neighbourhood and are, therefore, public goods. Hence, the signal benefits any bacterial cell, regardless of whether it has produced it or not. For example, *Pseudomonas aeruginosa* relies on quorum sensing to

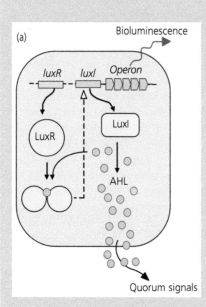

Box 12.5 Figure 1 Quorum sensing. (a) The classical quorum sensing pathway of *Vibrio fischeri*. Shown is the bacterial cell, with the *luxR* and *lux1* genomic loci. The latter activates a synthetase (LuxI) that produces the signal AHL (N-acyl homoserine lactones; compound no.1; blue dots). AHL is sensed by a receptor (LuxR) that is encoded by *luxR*. The activation of LuxR leads to expression of the LuxI-associated operon, which induces bioluminescence. The AHL signal is used for quorum sensing. (b) The enormous range of chemical compounds used by various Gram-negative and -positive bacteria for quorum signalling. Compound no. 8 contains amino acids in its structure (bases indicated by lettered circles). Adapted from Whiteley et al. (2017), with permission from Springer Nature.

activate extracellular proteases to mobilize milk proteins. Populations of mutants that do not produce the signal, and thus have no quorum sensing, cannot exploit this source. They will survive as 'cheaters' and persist at low frequencies in normal populations (Diggle et al. 2007). Various mechanisms exist with which signal-producing lines can exclude or limit the presence of cheaters to keep the benefits within collaborating bacterial lines (Majerczyk et al. 2016; Nadell et al. 2016; Popat et al. 2017). The relatedness among collaborating bacteria in a group—their kin structure—thereby plays a critical role (Strassmann et al. 2011).

Quorum sensing is important for several phenomena (Whiteley et al. 2017). For example, it is involved in the formation and degradation of biofilms. Quorum sensing in the gut microbiota—where it extends to inter-species

signalling—can affect the composition of the microbial community. There is also a link between quorum sensing and virulence of bacterial pathogens (Antunes et al. 2010). For example, *P. aeruginosa* causes pulmonary infections in mice only with lines that do communicate. In fact, quorum sensing is necessary for the production of important, extracellular virulence factors (Pearson et al. 2000). In turn, quorum sensing can be suppressed by the degradation of the signals, a process called 'quorum quenching' (Dong et al. 2001). Quenching occurs with a change in environmental conditions, e.g. in the pH value, or by the activity of enzymes produced by other microbes or by the host. Such findings offer prospects for new therapies of diseases; for example, quorum sensing inhibitors for *P. aeruginosa* block the sensing receptors and thus prevent biofilm formation and virulence expression (O'Loughlin et al. 2013).

12.5 Microbiota within the host

The microbiota, particularly the gut microbiome, is an important part of the host's defence system, as discussed in section 4.8 (see Table 4.7 for mechanisms of defence). Treatment by antibiotics damages not only an infectious pathogen but also the defences ensured by the resident microbiota, a problem that could be ameliorated by more specific therapeutics (Behrens et al. 2017). Furthermore, individuals differ in their microbiomes (Franzosa et al. 2015). The composition and functional profile of the microbiota is continually shaped by external factors such as diet (David et al. 2014; Phillips et al. 2017; Inamine et al. 2018), it is correlated with the season (Maurice et al. 2015), and it is subject to effects of the host's physiology and immune defences (Freilich et al. 2011; Bunker and Bendelac 2018), and by microbe–microbe interactions (McNally and Brown 2015). In particular, the strategies and mechanisms of microbial cooperation and interference also affect the structure and functioning of the microbiota. The microorganisms compete among themselves for access to space and resources in the gut. For the microbiota, an infecting parasite is 'just another competitor' for the same resources (Chassaing and Cascales 2018; Granato et al. 2019).

Changes in the microbiota affect the host health status and lead to pathogenic effects (see section

9.4.6). In fact, the infection by a pathogen, for example, changes the gut microbiome, as observed in a wide range of organisms such as lemurs (Aivelo and Norberg 2017), mice and rodents (Houlden et al. 2015; Kreisinger et al. 2015), sea birds (Newbold et al. 2016), or humans (Lee et al. 2014b; Kay et al. 2015), although with some exceptions (Cooper et al. 2013; Baxter et al. 2015). Mice experimentally infected by *Citrobacter rodentium* that causes inflammatory colitis show a detailed picture of these changes (Belzer et al. 2014) (Figure 12.11). The infection leaves an early trace in the presence of *Mucispirillum*, which normally is found in the mucus layer of the colon. Adherence of *Citrobacter* to the epithelium in the same region leads to a disruption of this association and prepares the way to inflammation. Later, when the parasite is cleared and inflammation subsides, the tissue is repaired. This shows up as a signal with the presence of *Clostridium,* for example. Hence, a complex pattern of changes in the gut microbiome is associated with the trace of the infection. The most noticeable effects occur with an infection-induced dysbiosis of the microbiome, i.e. a massive change in its composition that can seriously undermine the host's health status (see section 9.4.6).

If commensal microbes are essential for host defences, they should be under selection and evolve towards defensive capacities. Experiments support

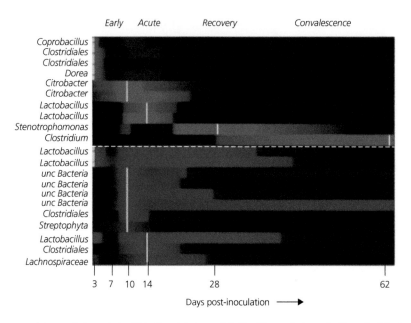

Figure 12.11 Time map of changes in the gut microbiota (ileum) during an infection. The graph shows the time after infection of mice with the bacterium, *Citrobacter rodentium* (x-axis in days; phase of infection at top). The colours indicate the change in abundance of bacterial taxa in the gut microbiome relative to uninfected mice; red indicates increase, blue decrease with infection, yellow bars at time of maximal change. Bacterial taxa given on the left. Adapted from Belzer et al. (2014), under CC BY.

this expectation. For example, experimental evolution of the mildly pathogenic gut-inhabiting bacterium, *Enterococcus faecalis*, in its host (the nematode *Caenorhabditis elegans*) and in the presence of a severe pathogen (*Staphylococcus aureus*) produces this result. In this case, *E. faecalis* evolves towards a protective microbe that defends the host against its more virulent rival (Figure 12.12). With closer inspection, this happens by the production of superoxidants, a potent killer of microbes (King et al. 2016). *E. faecalis* also exploits the siderophores produced by *S. aureus* for its own purposes. As a response, *S. aureus* evolves to produce fewer siderophores and becomes less virulent (Ford et al. 2016).

In theory, the situation could evolve towards the host trading the extra cost of harbouring defensive microbes against the benefits they provide for defence. In the end, this co-adaptation with defensive microbes, driven by their competition with the invader, might even render the corresponding host immune defences dysfunctional (King and Bonsall 2017). Such 'out-sourcing' of defences to commensals is known from experimental evolution in *Drosophila*. In this case, the commensal *Wolbachia* protects

against viral infections (Drosophila C virus). In the presence of *Wolbachia*, an essential resistance allele of *Drosophila* decreases in frequency, indicating that the host's defences are increasingly taken over by the commensal (Martínez et al. 2016). For similar reasons, some groups, like the pea aphid (*Acyrthosiphon pisum*), may have lost a canonical immune pathway whose function is replaced by a defensive symbiont (Gerardo et al. 2010).

12.6 Within- vs between-host episodes

For a parasite, a full life cycle consists of both the within- and the between-host episodes. Parasites are only successful when both of these parts are completed. Therefore, any trait that parasites evolve is shaped by both. Likewise, host defences, defensive behaviours, or life histories are shaped by both challenges, too. Empirical studies look at the entire parasite life cycle, although typically only one episode is investigated.

Experiments, for example, need to extract parasites from a host, prepare them for an inoculum, and reinfect a dose into a next host. This procedure

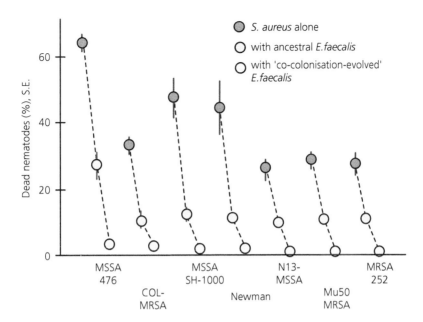

Figure 12.12 Evolution of host protection by a bacterium. Shown is the effect of various bacteria on the mortality of the host, the nematode *C. elegans* (scale at left). Several nematode lines (as indicated at bottom) were experimentally infected. *E. faecalis* is normally a mild pathogen ('ancestral', yellow dots), whereas *S. aureus* is highly pathogenic (red dots). *E. faecalis* experimentally co-evolved with *S. aureus* become protective for the host. When co-infected with *S. aureus*, the host has a low mortality (green dots). Reproduced from King et al. (2016), under CC BY 4.0.

corresponds to the between-host episode, even though the experimental procedure is different from a natural transmission process. This is obvious in serial infection experiments (Ebert 1998), where parasites are repeatedly administered to a next host in a controlled way. The experimental step, therefore, bypasses the natural transmission episode that would select for survival in the environment in vectors or intermediate hosts. In addition, it also bypasses selection for the capacity to enter a next host. These selective elements can be added to an experiment, but any protocol will have to define what elements of the between- or within-host episode are mimicked and in which way.

Empirical studies show how different regions of a parasite's genome may evolve, depending on whether within- or between-host selection is prevalent. In hepatitis C virus (HCV), for example, the E1/E2 region is essential for expression of antigenic patterns. It evolves exceptionally quickly under within-host evolutionary processes (Gray et al. 2011). Similarly, the HIV genome evolves rapidly in the region containing envelope genes, which indicates a higher impact of within- than of between-

host evolution (Alizon and Fraser 2013). Within-host processes furthermore affect the timing of transmission from the host, the representation of transmitted strains from a multiply infected host, or the order in which strains transmit further. Such differences will have consequences for the between-host dynamics (Mideo et al. 2008). In some cases, the transmission is by sexual stages, and the sex ratio (males relative to females) becomes relevant, too. Some species of *Plasmodium* produce more males than expected, given the mating chances when gametocytes are rare. This pattern is another instance where the details of the within-host processes are crucial to understanding the larger-scale dynamics (Greischar et al. 2016b).

Modelling often focuses on either the within- or between-host dynamics. For the latter, models such as the SIR model in epidemiology (see Box 11.2) are used. Some of the within-host models have been mentioned above, e.g. target cell-limited models (Box 12.2). Simple integration of both processes can deal with these two episodes without any more specifications. Theoretical efforts that more explicitly combine the within- and between-host episodes are

Table 12.5 Multiscale models.

Host	Parasite	Model scope	Finding	Method	Source
Mammals (incl. humans), birds	*Toxoplasma gondii*	Dynamics of the system, equilibria, reproductive number (R_0).	Stable and unstable equilibria can be found for subsystems. Pathogen remains present even when between-host $R_0 < 1$. Not matched with data.	Model separated into a 'slow' system (epidemiology between-host, plus environment) and a 'fast' system (within-host). Linked by analysing perturbations to a system. Deterministic, differential equations.	3
Humans	HIV	Emergence of drug resistance in population.	Higher drug treatment levels lower prevalence of resistant strains, but only up to a point.	Within-host dynamics of resistant vs susceptible strains generates time-dependent transmission rates. Deterministic, differential equations.	5
Humans	HIV	Evolution of HIV strains in relation to host cell (CD4+ T-cells) reservoirs.	Strains of moderate virulence evolve with large reservoirs because this slows down the within-host evolutionary processes.	Transmission between hosts made dependent on the representation of strains in actively infected CD4+ cells. Deterministic, differential equations.	2
Birds, mammals, arthropods	Arboviruses (arthropod-borne RNA viruses)	Does a slow or fast within-host replication strategy maximize transmission in population?	Slow strategy leads to higher rates of persistence in hosts and vector populations.	Slow strategy is low infection intensity of long duration; fast is high intensity, short duration. SIR model.	1
Humans	Guinea worm (*Dracunculus medinensis*)	What is the most efficient way to eradicate Guinea worm disease from the human population?	Most efficient method for eradication is elimination of intermediate host (copepods). Also, killing of female worms within human hosts.	Model considers three organisms: human host, copepod intermediate host, protozoan parasite. Reproductive number used as criterion for eradication ($R_0 < 1$). Model tracks infections of different compartments. Deterministic, differential equations.	4

Source: [1] Althouse Benjamin. 2015. *Philos Trans R Soc Lond B*, 370: 20140299. [2] Doekes. 2017. *PLoS Comp Biol*, 13: e1005228. [3] Feng. 2012. *Math Biosci*, 241: 49. [4] Netshikweta. 2017. *Comput Math Methods Med*, 2017: 29. [5] Saenz. 2013. *Epidemics*, 5: 34.

known as 'mixed', 'nested', 'embedded', or 'multi-scale' models. The latter nomenclature puts a focus on the fact that parasites move across different scales, i.e. from within-host to between-host and also to different spatial scales.

Around 200 studies on this topic have appeared over the last decades, with an apparent increase over the last ten years (Childs et al. 2019). In those models, most often viral infections of humans are analysed, with all other host–parasite systems underrepresented (see examples in Table 12.5). These attempts also span a range of different modelling techniques, such as deterministic models that use systems of ordinary differential equations and stochastic models where parameters are randomized during computation, but also individual-based models that focus on modelling the fate of individuals (or 'agents') that interact with each other or with a simulated environment (Murillo et al. 2013; Handel and Rohani 2015; Dorratoltaj et al. 2017; Willem et al. 2017; Garira 2018).

Often, the models for the within- and between-host dynamics are constructed on their own and then connected secondarily. For example, a connection is made by taking the predicted parasite load from a within-host model as the starting point to a between-host model for transmission. However, multiscale models that explicitly combine within- and between-host dynamics are required when the processes act both ways; that is, when the between-host epidemiological dynamics provides feedback on the within-host dynamics and vice versa.

For example, mixed models show that the within-host competition between strains of *Plasmodium* that are either resistant or sensitive to antibiotic treatments directly affects the spread of resistance in the population as a whole. The within-host dynamics thereby depends on the costs of drug resistance, but the chances of transmission for resistant strains depend on how many vectors are present. For example, in epidemiological situations characteristic of high-transmission areas (i.e. with many vectors) resistance does not spread fast (Legros and Bonhoeffer 2016). Another example concerns IAV in birds from where it can spill over to humans. The virus is transmitted via the environment, e.g. by droplets and contaminated surfaces, where it can survive for various lengths of time. However, persistence is temperature dependent. Some strains are better adapted to within-host conditions (around 35–40 °C), whereas others do well

under between-host environmental conditions (5–20 °C). The combined dynamics of the virus under both conditions will add up to its overall fitness. Combining both scales suggests that viruses remaining inside hosts for a relatively long time should adapt to warm host conditions. On the other hand, if the between-host transmission is frequent, adaptation to cold temperatures is beneficial. Whereas this may not be surprising as such, modelling would allow finding exact criteria for when one or the other strategy is best (Handel et al. 2013). In all, however, the empirical verification of mixed models with observed data remains an open field of many opportunities (Alizon and van Baalen 2008; Handel and Rohani 2015).

Important points

- Within-host processes start with a primary phase (entry and establishment) at the primary site (the infection site), from where parasites typically migrate to their target tissues (the secondary site) and grow or multiply in numbers. Rapid, early defences are essential to stop an infection, but sensitive detection cannot be without errors.
- Within-host infection dynamics can be analysed in various ways. Target cell-limited models look at the dynamics of infected cells. The disease space approach looks at infection trajectories in relation to infection outcome. Parasites, in turn, adopt different strategies of within-host growth and multiplication.
- Immune responses shape within-host infections and are studied with computational immunology or systems immunology. The microbiota is a further part of the defence system that affects within-host dynamics in various ways.
- Infecting populations of pathogens evolve within hosts. This can lead to genetic diversification of populations.

Evolution of pathogenicity islands in bacteria or escape mutants in viruses are essential elements. While inside the host, parasites also show antigenic variation to escape recognition.
- The evolution of antibiotic resistance is a particular concern and affects all kinds of drugs. Bacteria, for example, have evolved several mechanisms to become resistant. Resistance spreads very fast, also due to horizontal gene transfer. Compensatory mutations reduce the costs of antibiotic resistance in bacteria.
- Multiple within-host infections by different genotypes are common. Competition among strains can be direct, e.g. via bacteriocins, or indirect by competition over resources. For microorganisms, iron is a key resource; bacterial siderophores sequester iron. Co-infecting strains also cooperate, e.g. in biofilms or by quorum sensing.
- The within- and between-host episodes in the parasite life cycle affect one another in important ways. Multiscale (mixed) models combine the respective analyses using different approaches and methods.

CHAPTER 13

Virulence evolution

13.1 The meaning of virulence

A parasitic infection reduces the host's fitness and, typically, causes some organic damage. Damage is what commonly is thought to result from the parasite's 'virulence'. Chapter 9 discusses the pathogenic mechanisms that eventually cause virulence. Notoriously, the term has many different meanings, and its use has not been remarkably consistent. For example, in plant-pathogen studies, the term 'virulence' means a successful infection with parasite-induced lesions on leaves. In animal–parasite systems, 'virulence' is a measure of parasite-inflicted host damage defined in various ways. Regardless of these discrepancies, it is evident that some parasites cause severe damage and can be lethal, whereas others have only mild effects. Why do these differences exist? A large body of work has dealt with the question of how parasite 'virulence' should evolve under different scenarios and what explains this variation (Frank 1996; Day 2002b; Alizon et al. 2009; Bull and Lauring 2014; Cressler et al. 2016).

In this chapter, virulence has a function; that is, it is an adaptive trait that leads to an increase in the parasite's fitness. Moreover, 'virulence' is a generic term for the parasite-induced reduction in host fitness. Therefore, its consequences on the host measure it. Note that virulence is not a trait of the parasite alone, but results from the interaction with the host, and combines with effects of the environment. Hence, virulence is more like a statistical parasite main effect (see Box 10.4) that explains some of the variation in the outcome of the host–parasite interaction. With this meaning, it reflects the part of virulence that is under the 'control' of the parasite and, thus, can be an adaptive trait. Moreover, operational measures for virulence exist for any specific case, and include reduction in host fecundity, increase in host mortality, reduction in host body condition, or degree of tissue damage (Read 1994; Alizon and Michalakis 2015) (Table 13.1).

13.2 Virulence as a non- or mal-adaptive phenomenon

In some situations, virulence adds little or nothing to parasite fitness and is therefore not a trait selected for its consequences. For example, virulence can result from unavoidable side effects, or from 'mistakes' during the infection that are detrimental to both host and parasite. Also, when an infection spills over from a reservoir, virulence on the new host likely is mismatched. Virulence can also not evolve when the necessary genetic variation is missing, or when its selective effects are neutral. In such cases, virulence is not an adapted trait of the parasite.

13.2.1 Virulence as a side effect

With a 'side effect', the consequences for the host are unselected or unavoidable. They can be caused by the mechanisms that the parasite has otherwise evolved to extract resources. Similarly, the host defence system may have evolved to respond to infections with immunopathology (Graham et al. 2005; Best et al. 2012). In these cases, the host is under selection by the pathogenic effects, but the parasite might not be.

In the case of very recent host–parasite associations, there was not enough time to cause adaptive changes either. Such novel associations can even

Evolutionary Parasitology: The Integrated Study of Infections, Immunology, Ecology, and Genetics. Second Edition.
Paul Schmid-Hempel, Oxford University Press. © Paul Schmid-Hempel 2021.
DOI: 10.1093/oso/9780198832140.003.0013

Table 13.1 Measures of virulence used in different studies.

Virulence measure	Organisms	Scope of study	Source
Increase in host mortality.	Trematodes (*Rhipidocotye* spp.) in mussels (*Anodonta piscinalis*).	Condition-dependent effect of parasite.	11
Host longevity.	*Ophryocystis elektroscirrha* (protozoa, neogregarine) in monarch butterflies (*Danaus plexippus*).	Relationship of virulence to parasite fitness and genotypic variation in rate of parasite replication within host.	7
Host mortality rate in theoretical model Proportion of lysed host cells in experiment.	λ-bacteriophage (virus) in *E. coli* (bacteria).	Theoretical model combined with experiment to explore effects of spatial structure and transmission mode on parasite fitness.	5
Reduction in colony founding success (given by egg laying and offspring production).	Trypanosome (*Crithidia bombi*) in bumblebee (*Bombus terrestris*).	Host life history stage-specific effect of parasite.	6
Castration of host.	Several organisms.	Consequences of host castration and induction of host gigantism on virulence evolution.	8
Time delay to castration of host. But a broader definition used, too.	Bacteria (*Pasteuria ramosa*) in water fleas (*Daphnia magna*).	Effect of host age on parasite success.	10
Probability of developing into adult mosquito.	Microsporidia (*Vavraia culicis*) in mosquito (*Aedes aegypti*), parasitizing larvae.	Effect of food availability on virulence.	4
Degree of inflammation (myocarditis, i.e. cardiac inflammation).	Coxsackie virus in mice.	Effect of nutrition on virulence and on viral populations.	3
Strength of host immune response, host cell mortality in cultures.	Influenza virus.	Virulence of Spanish flu virus.	12
Delay to shedding of virus from host.	Marek's disease virus (MDV) in chicken.	Effect of mass vaccination on outbreaks.	1
Degree of anaemia in host.	Malaria (*Plasmodium chabaudi*) in mice.	Tolerance patterns. Effect of immunodeficiency of virulence.	2, 13
Eye score, i.e. a measure for swelling around and damage to eye due to the infection.	*Mycobacterium gallicum* in song birds.		9

Source: [1] Atkins. 2013. *Epidemics*, 5: 208. [2] Barclay. 2014. *Am Nat*, 184, Supplement: S47. [3] Beck. 2004. *Trends Microbiol*, 12: 417. [4] Bedhomme. 2004. *Proc R Soc Lond B*, 271: 739. [5] Bernrguber. 2015. *PLoS Path*, 11: e1004810. [6] Brown. 2003. *J Anim Ecol*, 72: 994. [7] de Roode. 2008. *Proc Natl Acad Sci USA*, 105: 7489. [8] Ebert. 2004. *Am Nat*, 164, Suppl.: S19. [9] Hawley. 2013. *PLoS Biol*, 11: e1001570. [10] Izhar. 2015. *J Anim Ecol*, 84: 1018. [11] Jokela. 2005. *Oikos*, 108: 156. [12] Kash. 2006. *Nature*, 443: 578. [13] Råberg. 2007. *Science*, 318: 812.

cause severe virulence effects that mean a dead-end for the parasite. Examples are accidental infections of humans by fox tapeworms (echinococcosis) or rabies virus, both of which have fatal outcomes for hosts if untreated, but do not allow for parasite transmission. Virulence side effects could also be maintained evolutionarily, because the responsible parasite genes have more important fitness benefits in other contexts (pleiotropic effects), or when the fitness trade-offs are generated by unrelated processes (Alizon and Michalakis 2015). For example, bacterial adhesins of *Escherichia coli* are essential to colonize the host from the gut but might cause severe inflammation in other organs. Similarly, adhesins help bacteria to attach to epithelia but damage the host's urinary tract. They provoke a host response that facilitates their clearance (Levin and Svanborg Eden 1990).

13.2.2 Short-sighted evolution

Short-term consequences can powerfully drive the evolution of parasite traits. At the same time, these traits may not be favourable for the longer-term components of fitness (Levin and Bull 1994; Lythgoe et al. 2013). 'Short-sighted' variants often emerge from the infecting population within a host, e.g. escape mutants, rather than being maintained in the population at large (Martínez 2014). Possible examples of 'short-sighted' evolutionary processes are human poliomyelitis, cerebral malaria, or bacterial meningitis. The latter is associated with several bacteria (*Haemophilus influenzae*, *Neisseria meningitidis*, *Streptococcus pneumoniae*) that sometimes colonize the cerebrospinal fluid instead of the usual respiratory tract. Severe damage to the central nervous system results. However, not all strains can expand into this unusual niche, which points to a genetic basis for the difference; they thus can evolve. Notably, the central nervous system is a dead-end for these parasites, as no transmission occurs from there. Similarly, poliovirus is transmitted via the oral–faecal route. However, the virus can also enter the lymphatic system and disseminate to different tissues, including the central nervous system. Such colonization causes severe damage, with the severe symptoms of poliomyelitis. However, the virus cannot return to the gut to become transmitted via its regular route. It is therefore locked in a dead-end.

Such aberrant and virulent variants nevertheless could be maintained in the population at low frequencies, e.g. by recurrent mutation. Also, the odd chance of transmission may be sufficient to counterbalance the adverse effects of colonizing the 'wrong' tissue. Alternatively, a rapidly multiplying parasite that has a selective advantage by producing many transmission stages in its normal tissue may also be more likely (by sheer numbers) to accidentally invade the wrong tissue. In this case, virulence is a side effect of rapid multiplication that is maintained for its other advantages. In any case, short-sighted evolution (and pleiotropic effects) can have severe consequences for human health and medical treatment (Hansen et al. 2017).

13.2.3 Virulence as a negligible effect for the parasite

The consequences of virulence depend on many factors and, therefore, can become overshadowed by other effects. Indeed, virulence effects vary with host genotype, age and nutritional status, the specific properties of the interaction, infective dose, environmental conditions, and co-infecting parasites (Thomas and Blanford 2003; Ebert and Bull 2008). The virulence consequences for the parasite's fitness could, therefore, be strong in some situations, but weak in others. On average, virulence effects may therefore be lost in the background noise and have, on average, negligible impact. For instance, the bacterial parasite *Pasteuria ramosa* sterilizes its host, the water flea *Daphnia magna*, at an ambient temperature of 20°–25 °C but has almost no effect at 10°–15°C (Mitchell et al. 2005). Environmental effects might be the best candidates to render virulence, as a parasite trait, relatively unimportant (Mitchell et al. 2005; Myers and Cory 2016; Mahmud et al. 2017).

13.2.4 Avirulence theory

Newly introduced parasites can be highly virulent for their novel hosts, as in the example of some zoonoses in humans, such as Ebola, Lassa fever, or bird flu. Likewise, infections caused by parasites that have a long co-evolutionary history with their hosts often produce less dramatic symptoms. These observations have led to the (wrong) generalization that parasites always evolve to become less virulent as they co-evolve with their hosts (Allison 1982), a hypothesis known as 'avirulence theory'. The avirulence theory assumes that virulence is an evolved and adapted trait. It overlaps with the myth that well-adapted parasites do not harm their host, because otherwise, they would also deprive themselves of new hosts. The imagined evolutionary process would lead to a 'balance of interests' by which the parasite is not too virulent and the host not too defensive. This old idea has been most prominently reiterated by one of the fathers of modern immunology, Frank Macfarlane Burnet (1899–1985) (Burnet and White 1972) and still exists in more modern debates (Zinkernagel and Hengarnter 2001). Until the 1980s, this view had been so dominant that it became known as the 'classical wisdom' (Anderson and May 1982).

However, the proposed evolutionary mechanism is unlikely to work. It would, for example, require that a parasite restrains its virulence in favour of the long-term survival of the host and the parasite population. Avirulence theory thus illustrates the classical 'tragedy of the commons' (Hardin 1968), where the prudent use of a shared resource may not be of the same interest to all parties. Consider a situation where two strains compete against each other in the same host. In this case, the more rapidly multiplying strain will quickly outcompete its more prudent counterpart and eventually dominate the infecting population. It is likely that the quicker strain will also cause more damage to the host (see section 13.4.3). Hence, the more virulent strain would gain more fitness. Furthermore, from all we know about the evolutionary process, the short-term effects of selection (that favour the quicker strain) are much more potent than effects that appear in the distant future (preserving the host population by a less virulent strain).

Therefore, the avirulence theory would have to overcome formidable obstacles in the evolutionary process in order to work. It is, therefore, not considered a likely evolutionary mechanism and is not part of the modern evolutionary theory of virulence. Also, the empirical evidence is mixed at best, with many instances of parasite virulence becoming intermediate or increasing, rather than decreasing with time (Allison 1982; Toft and Karter 1990; Read 1994) (see section 13.7). Moreover, an ascertainment bias in the data is likely. Unspectacular cases of invading parasites that cause only mild or no symptoms are rarely noticed.

13.3 Virulence as an evolved trait

Parasite virulence changes over time and evolves. A classical case is myxoma virus, introduced in Australia in 1950 from collections in Europe to control the burgeoning rabbit populations (Fenner and Ratcliffe 1965). The virus proved highly virulent for the rabbits (i.e. causing high mortality) but soon evolved towards lower grades of virulence in the years that followed. It has remained at an intermediate level of virulence for a long time since. Over the years, virus isolates were extracted from Australian rabbits and tested against standard breeds of rabbits in the laboratory. Likewise, wild rabbits were inoculated with defined, standard laboratory strains of the virus. In this way, it was possible to separate changes resulting from the evolution of the parasite (virulence) from changes resulting from the evolution of the host (resistance). Both variables changed during the co-evolutionary interactions—the population of viruses became less virulent, and the populations of Australian rabbits more resistant. This co-evolutionary process continues in Australia as well as in Europe and can be traced by changes in the viral genome (Kerr et al. 2017a, 2017b).

The 'evolutionary theory of virulence' suggests that a certain level of virulence evolves because it provides a fitness advantage for the individual parasite or, more precisely, for this strategy that is represented by its carriers. This concept developed in the early 1980s (Anderson and May 1982; Bremermann and Pickering 1983; Ewald 1983). The success of the parasite results from the entire life cycle, from infection and multiplication within the host, to transmission and infection of a new host. This can be illustrated by putting host disease spaces in a chain (Box 13.1). However, which measure should be used for parasite success (i.e. its fitness) is intensely debated (Lion and Metz 2018).

Besides a definition for parasite fitness, various trade-off relationships are assumed. The trade-off most commonly considered is between the benefits of high within-host parasite replication—favouring increased transmission to the next host—and the associated reduced life span of the infection, known as the 'virulence–transmission' trade-off. Most often, virulence is equated with host mortality rate. However, the term covers many different meanings. In practice, virulence can refer to case mortality rate (parasite-induced host death once infected), the expected life span post-infection, or the lethal dose. These variables may not be identical with variables used in the model analyses (Day 2002b; Cressler et al. 2016). For example, case mortality is conditional on the host not dying of natural causes; it typically counts the cases of recovered hosts as infecteds. Hence, infection-induced mortality rate, α, i.e. the actual virulence of the parasite, is not correlating with case mortality unless recovery from infection co-varies with α. Similar discrepancies emerge for other measures.

Box 13.1 Virulence in disease space

The infection trajectory shows the path through disease space of a given host. Along the way, the infection causes processes of pathogenesis that become visible as damage to the host ('virulence'), e.g. severe inflammation or anaemia. In disease space, the pathogenic effects will push the host status into different domains. Hosts become sick or can recover.

The evolutionary concept of virulence counts the average infection trajectory in the host population, i.e. a trajectory of the parasite population through the successive host disease spaces. It thus uses population-level processes and statistical expectations. When rolled out, the average cycle starts with the infection of a focal (primary) host, followed by the infection trajectory of growth and multiplication within the host. Eventually, transmission (between hosts) occurs to the point of infection of the next host. The 'choice' of virulence by the parasite population will affect this average cycle and, thus, the expected population-level trajectory with its associated fitness. The best choice will maximize the expected fitness of the parasite.

Figure 1 illustrates the basic cycle and the principle of virulence evolution. The parasite resides within a host for a specific time. At some point (the 'transmission point'), it leaves the current host and transmits to a next host. While inside the host, virulence emerges, and the damaging effects can accumulate. If the parasite leaves early, virulence effects will be lower, but the potential for transmission is also lower. If the parasite leaves late in the infection path, transmission potential is higher (e.g. more propagules will have been produced until then). At the same time, the overall virulence effects are also higher (e.g. the host is more likely to have died by then). Selection will therefore push transmission to a point in disease space that maximizes the chances of infecting new hosts within a given time. In the simplest model, this maximizes R_0, the term characterizing parasite fitness in the standard scenario (see Box 11.2).

The situation can be sketched by a fitness landscape that assigns a parasite fitness value for every point in disease space (Figure 2). This landscape refers to the average cycle as a given parasite population repeatedly encounters hosts, and thus follows its average trajectory through the current population. This expected fitness is similar to the age-dependent residual reproductive value known from life history theory. The shape of the fitness landscape will depend on the details of the host–parasite interaction, on prevailing environmental conditions, the average structure of the population, and many other factors. It will be tough to actually measure.

However, the illustration in Figure 2 lets us immediately understand that parasites would probably follow an infection trajectory that leads 'uphill' and which should reach the highest fitness peak. An existing population that currently reaches

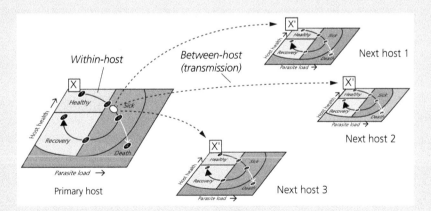

Box 13.1 Figure 1 The average parasite cycle. The evolutionary theory of virulence considers the (average) entire life cycle of a parasite from infection of the primary host (at point X) to infection of a next host (the secondary host, at points X'). Each host has its disease space (rectangles, see Box 2.2). The parasite gains transmission at some point (blue dot) along its path of infection through disease space (black line). Along this path, damage to the host occurs ('virulence'). At the same time, the point of transmission affects the chances of success. The combined within-host and between-host parts of the parasite's life cycle determine fitness. In the case shown here, the parasite infecting a primary host causes three secondary infections; in epidemiological terms, $R_0 = 3$.

continued

Box 13.1 *Continued*

point A in disease space would, therefore, evolve to reach point B. However, depending on the shape of the expected fitness landscape, this would mean evolution towards higher virulence (Figure 2a) or lower virulence (Figure 2b).

Note that 'virulence' in these evolutionary terms is not the damage done to the individual host, but the expected fitness loss of the average host in a population when infected. In practice, this may be measured by increased host mortality rate or reduced fecundity of the host population. With the evolutionary theory of virulence, the pathogenic effects themselves do not take centre stage. Instead, it is the parasite's fitness that results from these effects.

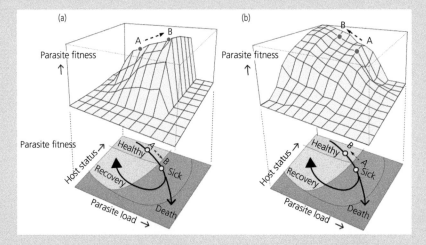

Box 13.1 Figure 2 Evolution of virulence. Every point in disease space is associated with a fitness value for the parasite, shown here as a fitness landscape above the disease space plane (rectangles). The fitness values reflect the average future expected fitness, resulting from transmission at this point. The black line in the disease space characterizes the infection trajectory of the parasite. The parasite population evolves by moving upwards in the fitness landscape (red dots). (a) Parasites will evolve towards higher virulence (moving the transmission point from A to B); it leaves later, making the hosts sicker. (b) Parasites will evolve towards lower virulence, leaving the host earlier and in a healthier state.

13.4 The standard evolutionary theory of virulence

13.4.1 The basic principle

The standard evolutionary theory of virulence makes use of the equations of SIR epidemics. The basic reproductive number, R_0, thereby is a helpful working definition for parasite fitness, while keeping in mind its restricted validity. In particular, from Box 11.2, we have $R_0 = \beta N/(\mu+\alpha+\nu)$, which is eq. (11.1) in the text. As before, R_0 depends on the number of susceptible hosts ($\approx N$, when the parasite enters a wholly susceptible population), transmission rate (β), background host mortality rate (μ),

parasite-induced mortality (α, virulence), and recovery rate from infection (ν). The theory suggests that R_0 should be maximized by selection.

As also discussed in Box 11.2, several assumptions underlie the SIR model. This basic approach will, therefore, not be appropriate for many real situations. Furthermore, there are many obstacles to measure and define the parameters used in this primary consideration (Metcalf et al. 2015). Box 13.2 describes some extensions of this basic scenario; multiple infections are discussed in section 13.9.

Eq. (11.1) shows that to maximize R_0, the parasite should evolve to minimize α. Hence, its virulence should decrease over time ($\alpha \rightarrow 0$). This is actually

Table 13.2 Trade-offs in the evolutionary theory of virulence.

Parasite/Host system[1]	Finding	Virulence measure	Source
E: Bacteriophage f1/*E. coli*	Virulence and phage production from host correlate positively.	Infection rate (density of infected hosts).	3, 15
O: *Pseudomonas syringae* (bacterium)/*Arabidopsis thaliana*	Higher symptom severity correlates with more bacteria present in leaf.	Leaf damage according to standardized symptom scale.	10
E: *Glugoides intestinalis* (microsporidia)/*Daphnia*	High spore loads increase host mortality.	Host mortality, host fecundity, host body growth.	6
E: *Ophryocystis* (gregarine)/*Danaus plexippus* (monarch butterfly)	Parasite virulence related to parasite multiplication rate (spore load). Also, spore load and transmission positively correlated. Parasite fitness maximum at intermediate spore load.	Hatching from pupa, mating success.	4
E: Avian influenza (H5N2)/Chicken	High-pathogenicity strains are more infectious and cause longer infections in next host than low-pathogenicity strains.	Infection success in next host.	18
E: *Glugoides intestinalis* (microsporidia)/*Daphnia*	Parasite strains forming more spores (higher host mortality) have higher probability of transmission.	Spore load that correlates with host mortality.	6
E: *Ophryocystis* (gregarine)/*Danaus plexippus* (monarch butterfly)	More new hosts infected with higher spore loads. High spore loads associated with later emergence and lower mating success.	Hatching from pupa, mating success.	4
O: *Plasmodium falciparum*/Humans	Expected transmission and virulence in different age classes positively correlated.	Morbidity: anaemia, body weight loss.	11
E: *Plasmodium gallinaceum*/Mice	Virulence and transmission correlate positively for host mortality but not for anaemia as measure.	Mortality, host condition (anaemia).	16
E: *Plasmodium chabaudi*/Mice	Virulence and transmission correlate positively with host anaemia as measure.	Host condition (anaemia).	7, 12–14
E: *Hyaloperonospora parasitica* (fungus)/*Arabidopsis thaliana*	Virulence and transmission correlate positively.	Host fecundity (seed production).	17
O: HIV-1/Human	Virulence and transmissibility correlate positively.	Early onset of AIDS symptoms.	8
E, O: Myxoma virus/Rabbit	Potential to transmit to mosquito increases with viral titre in primary skin lesions (where mosquitoes can take up blood).	Average survival time of host.	9
O: MDV/Chicken	Viral fitness maximized at intermediate virulence. Estimated as viruses shed in an infection (transmission potential) and estimate of R_0.	Virulence score (number of infecteds dying or developing clinical symptoms).	2
E: Cauliflower mosaic virus/Rape (*Brassica rapa*)	Transmission increases with virulence; probably saturating. Group with lower within-host accumulation showed more virulence.	Reduction in leaf area compared to healthy plants.	5
O: Mycoplasm (*Mycobacterium gallisepticum*)/North American house finch (*Haemorhous mexicanus*)	Virulence and transmission positively, but transmission and recovery negatively correlated. Variation among strains.	Average score of eye lesion over course of infection.	19
O: Myxoma virus/Rabbits	More virulent strains cleared more slowly (recovery).	Virulence grade (average host mortality rate, host survival time).	1

[1]Experimental (E), observational (O) study.

Sources: [1] Anderson. 1982. *Parasitology*, 85: 411. [2] Atkins. 2013. *Evolution*, 67: 851. [3] Bull. 1992. *Evolution*, 45: 875. [4] de Roode. 2008. *Proc Natl Acad Sci USA*, 105: 7489. [5] Doumayrou. 2013. *Evolution*, 67: 477. [6] Ebert. 1997. *Evolution*, 51: 1828. [7] Ferguson. 2003. *Evolution*, 57: 2792. [8] Fraser. 2007. *Proc Natl Acad Sci USA*, 104: 17441. [9] Kerr. 2012. *Antiviral Res*, 93: 387. [10] Kover. 2002. *Proc Natl Acad Sci USA*, 99: 11270. [11] Mackinnon. 2008. *Vaccine*, 265: C42. [12] Mackinnon. 1999. *Evolution*, 53: 689. [13] Mackinnon. 1999. *Proc R Soc Lond B*, 266: 741. [14] Mackinnon. 2003. *Parasitology*, 126: 103. [15] Messenger. 1999. *Proc R Soc Lond B*, 266: 397. [16] Paul. 2004. *BMC Evol Biol*, 4: 30. [17] Salvaudon. 2005. *Evolution*, 59: 2518. [18] van der Groot. 2003. *Epidemiol Infect*, 131: 1003. [19] Williams. 2014. *J Evol Biol*, 27: 1271.

Box 13.2 Extensions to the standard theory

There are a range of factors that can affect virulence evolution. These can be implemented or added to the basic considerations.

(A) Specification of defence mechanisms

In these cases, the variables of eq. (11.1) are specified more precisely, but the basic SIR model remains. Examples include:

Host tolerance: The discussion on virulence evolution has often been ignorant of whether hosts resist or tolerate the infection. Depending on assumptions, tolerance does not directly affect parasite fitness and, therefore, does not directly change expectations for virulence evolution (Best et al. 2009; Little et al. 2010; Vale et al. 2014).

Immunopathology: It might be necessary to separate the different processes that generate virulence. For example, extraction of host resources directly reduces host status and can damage tissue, but immune responses can also generate damaging side effects ('immunopathology'). These two components can be merged with the standard epidemiological equations (see Box 13.3). Pathogenic effects can, furthermore, result from the mechanisms that parasites use to manipulate their hosts or to evade, block, or circumvent its immune responses.

Immunodeficiency: Hosts can fail to respond, which is not the same as strategic tolerance. An immune-weak environment supports higher parasite densities. Therefore, virulent escape mutants are more likely to emerge in the parasite populations, and more intense competition favours more virulent strains (Barclay et al. 2014). Immunodeficiency would change the values of variables in standard eq. (11.1).

(B) Difference in ecology

The simple ecological scenario assumed in the SIR model may not be correct but can be improved by including essential deviations. Deviations include heterogeneity due to host differences or external ecological conditions, but also more complex transmission processes. Examples are:

Host condition: Hosts can vary in body condition, in access to resources, or immune responses. The differences can be persistent (in which case the evolution of virulence is affected by extant variation in host types), or transient and episodic. Episodic changes can result from weakened hosts

being more affected by infections. This is shown for trematodes infecting clams (Jokela et al. 2005), viruses in bees (Manley et al. 2017), or parasitoids infesting ladybirds (Maure et al. 2016). Transient host conditions have ramifications for the regulation of host population structure and size; it marginally adds to the random fluctuation in the host population. However, this is unlikely to change virulence evolution at large.

Environmental effects: Environmental effects can change the observed level of virulence. In particular, stressful conditions compromise the host's condition and, thus, change virulence. Environmental variables with documented effects are temperature (Blanford et al. 2003), pH level in water bodies (Blanford et al. 2003), anoxia (Jokela et al. 2005), or availability of nutrients (Maure et al. 2016). Both host condition and ecological effects cause changes in the mean or variance of model variables.

Vector transmission, reservoirs: These deviations from the standard scenario introduce additional compartments in the SIR model and are expanding eq. (11.1) with additional terms. A related approach is with mixed models that separate the calculations for the within-host and between-host episodes to generate an overall R_0.

(C) Failure of R_0 as a fitness measure

The basic scenario of the SIR model and the possible modification may not be correct altogether. This is notably the case when the ecological setting generates a 'feedback' for the success of the parasite, e.g. through complex, structured transmission processes. The quantity R_0 itself is not the appropriate parasite fitness measure, e.g. when combining within- and between-host dynamics (Greischar et al. 2016a; Greischar et al. 2016b). A more general solution is to analyse the dynamics of evolutionarily stable strategies (ESSs) (Lion and Metz 2018). Parasite fitness then becomes 'invasion fitness', i.e. the growth of a mutant parasite variant population that can invade a resident parasite population which had reached its epidemiological attractor (a region of the state space where the system dynamically remains). Formally, in this more general case, parasite fitness in the sense of eq. (11.1) is weighted with other terms that describe the ecological feedback in the entire system.

---

what avirulence theory suggests, although based on a very different argument. The evolutionary argument maximizes the current reproductive number of the parasite, instead of favouring restraint in

favour of long-term benefits. However, there is an important proviso: the variables in eq. (11.1) are not necessarily independent of one another. For example, the recovery (ν) or transmission rate (β) could be affected by the parasite-induced mortality rate (α), in which case avirulence ($\alpha \to 0$) no longer maximizes R_0. Instead, maximization depends on the trade-offs between these quantities. A large body of empirical studies and of theory has been exploring these (Table 13.2) (Bull 1994; Frank 1996; Alizon et al. 2009; Alizon and Michalakis 2015; Cressler et al. 2016).

13.4.2 The recovery–virulence trade-off

High levels of virulence can make it more difficult for the host to recover (Anderson and May 1982). For example, tissue repair is a limiting factor, or the parasite may already have multiplied to numbers that are difficult to clear. More virulent strains of myxoma virus indeed are cleared more slowly and cause more prolonged infections in their rabbit hosts (Figure 13.1a) (Anderson and May 1982). Therefore, in eq. (11.1), recovery rate, ν, should depend on virulence, α. Furthermore, with virulence that is too low, the host will recover too quickly, and the time over which the infection persists before clearance is too short for exploiting the parasite's full potential for multiplication and further transmission. Analogous consequences would be associated with virulence that is too high. In fact,

the myxoma virus achieves the maximum value of R_0 at some intermediate level of virulence. This level was also the most commonly observed at the time of the study (Figure 13.1b).

However, the myxoma story does not end there. Whereas the virus became less virulent over the earlier decades, isolates from the 1990s were again highly virulent. In laboratory rabbits, they induced a lethal immune collapse syndrome—a phenotype quite unlike the one induced by the ancestral viruses. Closer inspection showed that the viruses had evolved better capacities to suppress the host immune response (Kerr 2012; Kerr et al. 2017a). In other words, a completely new mechanism had evolved that allowed the virus to infect, multiply, and transmit successfully. Whether this will eventually again lead to a level of virulence that provides the maximum fitness for myxoma, as predicted by the evolutionary theory, remains to be seen.

This simple analysis ignores several factors. For example, the myxoma virus in Australia is also vectored by mosquitoes, a step ignored here. Similarly, the rabbit population size is substantially reduced by the virus, which is not assumed in the standard models, but which was the very goal of the control measure. Despite these shortcomings, the analysis shows how evolutionary changes in virulence,

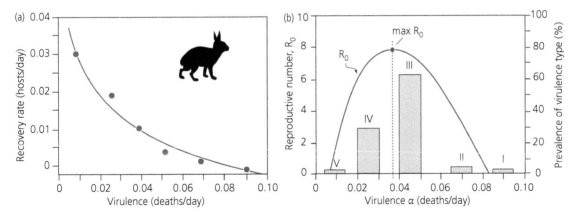

Figure 13.1 Recovery–virulence trade-off model for myxoma virus infecting rabbits in Australia. (a) Observed relationship between recovery rate, ν (y-axis, recovered hosts per day) and virulence, α (host deaths per day). The data were obtained from studying virus strains with different degrees of virulence for their rabbit hosts (dots). Adapted from Anderson and May (1992), with permission from Cambridge University Press.
(b) The calculated basic reproductive rate, R_0, as a function of virulence, α. The maximum of R_0 corresponds to an intermediate virulence (between categories III and IV). The other variables of the standard equation ((eq.) 11.1 in text) were estimated: $\mu = 0.011$ day^{-1}, $\beta = 0.1$ to 0.6 day^{-1}. The bars show the observed distribution of virulence types (I to V) one decade after introduction of the virus. The types close to the predicted maximum R_0 (red line) are the most common ones. Silhouettes from phylopic.org. Adapted from Mackinnon et al. (2008), under CC BY.

notably, the evolution towards lower virulence, as postulated with the avirulence theory, can be understood without referring to long-term effects or the benefit to entire populations. Of course, the analysis of eq. (11.1) would also predict an increase in virulence under the appropriate conditions. Despite the simplicity of this model, empirical studies also have provided some evidence that the recovery–virulence trade-off is an important mechanism to explain the observed variation in parasite virulence.

13.4.3 The transmission–virulence trade-off

Transmission rate, β, can depend on virulence, α, too. For example, a parasite variant that rapidly multiplies within a host generates a large number of transmission stages. At the same time, rapid within-host multiplication likely results from aggressive extraction of resources, which leads to damage and limits host life span. Empirically, this relationship exists for myxoma virus (Massad 1987), and data matching this scenario are known for several other cases (Table 13.2).

The density of viral particles in HIV—the viral load—in the peripheral blood circulation system is a useful measure. During the long asymptomatic phase that follows an initial infection, viral load fluctuates around a given value, the 'set point', which varies among patients and HIV variants. The study of patient cohorts shows that higher set-point viral loads correlate with shorter duration of the infection, with earlier onset of AIDS (the symptomatic stage of the infection; a measure of virulence), but also with increased transmission to new hosts (Figure 13.2a). Moreover, with the simple model, the predicted set-point viral load that leads to maximization of R_0 over the lifetime of the host is indeed close to observed values (Figure 13.2b) (Fraser et al. 2007). The finding suggests that HIV might have evolved to maximize its fitness in human hosts by adapting the set-point viral load. However, this conclusion is not yet entirely ascertained, given the difficulties of studying this infection in the first place.

The relationship of transmission with virulence also can be different, and (i) increase linearly with virulence, (ii) be an accelerating function, or (iii) be a decelerating function of virulence (Figure 13.3a). In the first two cases, the parasite is under selection to increase its virulence without bounds. However, suppose the transmission rate is a decelerating function of virulence. In that case, this is no longer true, and virulence should evolve towards an intermediate level to maximize the basic reproductive number of the parasite, R_0 (Figure 13.3b).

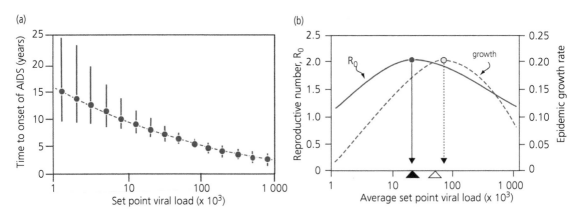

Figure 13.2 Virulence–transmission trade-off model for HIV in human populations. The data are estimated from cohort studies of patients in Zambia and the Netherlands (Amsterdam). (a) Observed relationship between the delay to the onset of AIDS (the clinical symptom) and the set-point viral load during the preceding asymptomatic phase. The points denote the best fit with 95 per cent confidence intervals; line drawn by eye. (b) Basic reproductive number, R_0, of the virus (left scale; solid red line) and the estimated growth rate of the viral population during an epidemic (right scale; dashed blue line) as estimated from a simple SIR model. Maximum values of R_0 are indicated by the arrows. The triangles at the bottom indicate the observed viral loads for Zambia (open symbol; referring to the calculations with growth rate) and Amsterdam (filled symbol; referring to direct calculations of R_0). Adapted from Fraser et al. (2007); copyright (2007) from the National Academy of Sciences USA.

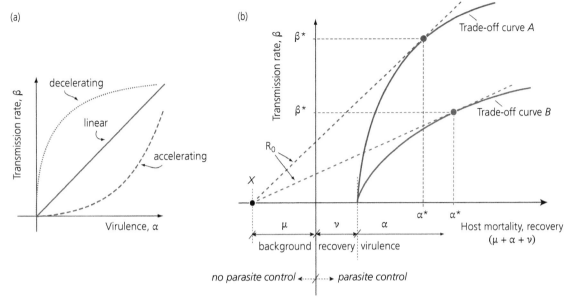

Figure 13.3 Transmission–virulence trade-off model. (a) Three possible relationships of virulence, α, and transmission rate, β. (b) With decelerating relationship, the optimum level of virulence, α^*, that maximizes the basic reproductive rate, R_0, can graphically be derived. The x-axis is extended to include all terms of the denominator of eq. (11.1), i.e. virulence α, background mortality μ, and recovery rate, ν (adding up to the rate at which hosts disappear from the pool of susceptibles). The slope of the broken line from point X represents the value of $R_0 = \beta/(\mu+\alpha+\nu)$, scaled with population size N. The maximum value of R_0 corresponds to the tangent from point X (indicating the background host mortality rate, μ, plus the loss due to recovery) to the trade-off curve. Optimal virulence, α^*, changes with the trade-off curves (A vs B). If the decelerating relationship were given by a generic function, such as $\beta = c\sqrt{\alpha}$ (shape constant, c), the optimal virulence is $\alpha^* = \mu + \nu$. Recovery and virulence are assumed to be under the control of the parasite, background mortality is not.

Box 13.3 Virulence evolution with immunopathology

Damage to the host can occur through parasite-induced effects and through the effects of the immune response itself, i.e. immunopathology (Graham et al. 2005; Sorci et al. 2017). The standard models of virulence evolution typically do not separate these effects. However, it is possible to expand the standard formulation eq. (11.1) for the basic reproductive rate, R_0, as follows (Day et al. 2007):

$$R_0 = \frac{\beta(z)}{\mu + \alpha(z,\upsilon) + \upsilon} \quad (1)$$

where z is a variable that reflects the level of exploitation of the host. It is assumed to be under the control of the parasite. Transmission rate, $\beta(z)$, shall decrease with exploitation level, z. Background mortality rate, μ, and recovery rate, ν,

remain as before. The parasite-induced mortality of the host, i.e. virulence $\alpha(z,\nu)$, is now a function exploitation level, z. The effects of immunopathology are included in this term as:

$$\alpha(z,\upsilon) = cz + f(z,\upsilon) \quad (2)$$

In this formulation, the effect of parasite exploitation (at level z) on virulence is scaled by a factor, c. The function $f(z,\nu)$ reflects the effects of immunopathology on virulence, which relates to clearance rate, ν, i.e. a proxy for the strength of the immune response that can cause damage. Combining eqs (1) and (2) and differentiating for the level of exploitation, z, yields the condition for the maximum basic reproductive rate, R_0^*, as:

continued

Box 13.3 *Continued*

$$\frac{d\beta(z)}{dz} = \left(c + \frac{\delta f(z,\upsilon)}{\delta z} \right)\left(\frac{\beta(z)}{\mu + \upsilon + cz + f(z,\upsilon)} \right) \quad (3)$$

This equation defines the optimal exploitation level, z^*, that the parasite should adopt to maximize R_0. The left-hand side reflects the changes in the benefits of the infection when increasing transmission, β, with increasing exploitation. With decelerating returns, we require $d\beta/dz > 0$ and $d^2\beta/dz^2 < 0$, i.e. an increase in exploitation will increase transmission but at a decreasing rate.

The right-hand side reflects the costs of the infection, i.e. the limitation of transmission, β, by the duration of the infection, which is set by host death, determined by μ, υ, plus the parasite-induced virulence, cz, and immunopathology, $f(z,\upsilon)$. Compared to the case with no immunopathology, the term $f(z,\upsilon)$ introduces additional factors that limit the duration of the infection. This selects for higher exploitation to compensate for the shorter duration of the infection. If immunopathology increases with z (i.e. when $\delta f/\delta z > 0$), parasite exploitation should be reduced (z^* decreases) to reduce the costs of killing the host prematurely. The opposite occurs (z^* increases) when exploitation has no effect on immunopathology (or even when the effect decreases).

Here, overall host mortality is the sum of exploitation-induced and immunopathology-induced damage (eq. (2)). Therefore, the virulence of the infection changes in a different way from in the standard model. The evolutionarily stable level of virulence will always be higher with additional effects of immunopathology, provided that at least some of the immunity-induced effect is independent of exploitation. Furthermore, if immunopathology increases with exploitation level, the combined effects of exploitation and immunopathology at point z^* may cancel each other out (under reasonable assumptions for the effects). This paradoxical outcome reflects that, on the one hand, z^* decreases as immunopathology increases. On the other, this also changes the relative contributions of exploitation-induced (cz) and immunopathology-induced $f(z,\upsilon)$ host mortality, even though overall infection-induced host mortality (virulence) might stay the same (Figure 1).

Of course, these patterns depend on the precise assumptions about how immunopathology arises. Regardless of the model's accuracy, the analysis illustrates how the different sources of virulence that evolve under different selective forces generate the observed host mortality, which can therefore change in unexpected ways.

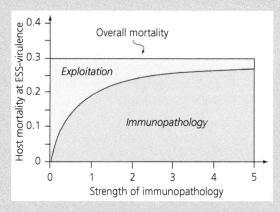

Box 13.3 Figure 1 Virulence under immunopathology. With a simple extension of the standard SIR model—where the parasite population has reached an evolutionarily stable level of host exploitation—the overall host mortality (y-axis) does not vary with the strength of the immunopathological effect (x-axis). However, the relative contributions of mortality due to exploitation or due to immunopathology vary with the strength of the immunopathological effect. Adapted from Day et al. (2007), with permission from The Royal Society.

The transmission–virulence trade-off hypothesis has developed into a major branch of the evolutionary theory of virulence. It has generated empirical results that are in line with the theoretical predictions, but also many results that do not agree with expectations. The study of such trade-offs remains difficult (Alizon et al. 2009; Alizon and Michalakis 2015).

The evolutionary theory is, therefore, not without its critics. Among other criticized points, the evidence for the necessary trade-offs is not as common as hoped for (Lipsitch and Moxon 1997; Weiss 2002). Also, there are additional factors that may limit parasite transmission, virulence measures may be unsuitable (Day 2002b), or the relationships of virulence, parasite replication, and transmission could be so complex as to defy any simple prediction (Bull and Lauring 2014; Greischar et al. 2016b).

Nevertheless, the evolutionary theory of virulence is a robust framework to study the changes in virulence expected to evolve. However, it primarily applies to cases where the basic structure of the trade-offs remains the same but defined, ecological conditions change, e.g. when opportunities for transmission vary (Frank and Schmid-Hempel 2008).

13.5 The ecology of virulence

The standard SIR model (eq. (11.1)) gives insights into which parameters and how their mutual dependencies affect the evolution of virulence to maximize parasite fitness, R_0. However, these primary considerations say nothing about what kind of ecological scenarios affect and change these parameters. Differences in transmission mode, host population structure and dynamics, or variation among individual hosts are the most prominent of these conditions.

13.5.1 Transmission mode

Parasites transmit horizontally directly to neighbouring hosts and by water or vectors to more distant ones. Other parasites primarily transmit vertically; that is, to offspring of the current host (Ebert 2013) (see Figure 9.1). This difference has consequences for the evolution of virulence. With horizontal transmission, the parasite does not carry the full cost of high virulence, because when the host dies, it can instead jump to a new host and continue its life cycle there (Ewald 1983). With vertical transmission, the parasite carries the full cost, since the number and condition of offspring (its next hosts) likely depend on the virulence effects on the current host.

Theoretical analyses support this intuitive insight and suggest that vertically transmitted parasites should generally evolve towards lower levels of virulence. It is, however, more appropriate to analyse this problem as meaning the evolution of transmission mode itself. With this approach, vertical transmission—and therefore lower virulence as the associated effect—evolves when new (susceptible) hosts are rare and current hosts have high fecundity. Horizontal transmission, and thus high virulence, evolves when new hosts are common (Lipsitch et al. 1995; Lipsitch et al. 1996; Berngruber et al. 2013; Ebert 2013; Berngruber et al. 2015).

In terms of the evolutionary theory of virulence, the virulence–transmission trade-off changes as the chances for transmission, given by the availability of new hosts, change. Strict vertical transmission, in particular, carries a considerable extinction risk, since the current host line can die out when a chance event—independent of the infection—kills the host (Lipsitch et al. 1996). Hence, vertically transmitted parasites are expected to evolve at least occasional horizontal transmission, too. With mixed horizontal and vertical transmission, however, the predictions are no longer simple. An increasing opportunity for horizontal transmission can, theoretically, lower the expected level of virulence. This happens because the increasing saturation with hosts having become horizontally infected (and thus no longer being available) gives more relative weight to vertical transmission, which, in turn, acts to reduce the overall virulence levels (Lipsitch et al. 1996; Day and Proulx 2004). Likewise, (almost) purely vertically transmitted parasites can persist and become virulent if, despite reducing host fecundity (i.e. fewer offspring that are new hosts), the parasite becomes more frequent in the next host generation, relative to competing variants. The bacterium *Wolbachia*, for example, is an intracellular parasite that can induce feminization or sex ratio distortion in its insect hosts. This lowers fitness for the host, i.e. causes high virulence for males, but also leads to higher representation via the female. *Wolbachia* is not exclusively transmitted vertically (Werren 1997; Bandi et al. 2001), however. The hosts of an essentially vertically transmitted parasite can also become 'protected' against other, competing, horizontally transmitted parasites (Haine 2008; Jones

Table 13.3 Transmission mode and virulence.

Scenario, *Main theme*[1]	Prediction, procedure, organisms[2]	Finding	Source
Mixed Vertical/Horizontal: Parasite transmits vertically and horizontally. (T)	Mixed transmission should reduce virulence.	Regardless of transmission mode, increased transmission leads to higher prevalence and, thus, to more vertical transmission.	4, 8
Environmental: Parasite propagules can survive in the environment, or disperse a long way in space. (T)	Verbal model, suggesting that long-lasting or dispersing parasites have lower virulence.	Parasites that leave hosts behind do not carry cost of virulence, as host mortality has weaker effect. Rough comparison among cases.	6
Environmental: Parasite produces long-lived spores for dispersal in time and space. (T)	Parasites that leave hosts behind do not carry cost of virulence, as host mortality has weaker effect. V: Host mortality rate. Host death as an obligate step to release spores considered as an option.	High virulence under broad range of conditions for dispersing or long-lived propagules. Spore strategy can also select for toxins that kill the host.	2, 3
Vertical/Horizontal: Changing experimental conditions to force either vertical or horizontal transmission. (Emp)	Vertical transmission should lower virulence. E: Manipulation of transmission mode for bacteriophage f1 on *E. coli*. V: Growth rate of infected bacteria.	Virulence decreases when experimentally transmitted vertically.	10
Vertical/Horizontal: Changing experimental conditions to force either vertical or horizontal transmission. (Emp)	Vertical transmission should lower virulence. E: Manipulation of transmission mode for barley stripe mosaic virus (BSMV) on barley (*Hordeum vulgare*). V: Reduction in lifetime seed production.	Virulence decreases when experimentally transmitted vertically. Virulence not correlated with virus concentration within host.	13
Vertical/Horizontal: Nematodes parasitize fig wasps when these are inside the fig. Number of co-inhabiting wasps varies and therefore opportunities for vertical transmission. (Emp)	Vertical transmission should lower virulence. O: Checking wasps in figs and their nematode parasites. Vertical transmission should reduce virulence. V: Host mortality.	Virulence decreases when only a few fig wasps inhabit a given fig and so force the nematode to transmit vertically. Comparison is among fig wasp species.	7
Vertical/Horizontal: Opportunities for vertical vs horizontal transmission changes as an epidemic unfolds. (Emp)	Vertical transmission should lower virulence. E: Experimental epidemic for bacteriophage λ on *E. coli*. During early epidemic, virulent variants should be favoured, as horizontal transmission is more common; situation reversed later in epidemic. V: Host mortality (lysis).	Virulent strain favoured early, but benign (latent) virus later in epidemic.	1
Vertical/Horizontal: Forcing transmission mode. (Emp)	Vertical transmission should lower virulence. E: Experimental, serial passage of cucumber mosaic virus (CMV) on cress (*Arabidopsis thaliana*) either strictly vertically or horizontally. V: Seed weight of infected hosts relative to control.	Vertical passaging reduces virulence, but horizontal passaging has no effect. In addition, hosts also increased resistance.	12
Vertical/Horizontal: Forcing transmission mode. (Emp)	Vertical transmission should lower virulence. E: Forcing host population growth of the host (*Paramaecium caudatum*) such that horizontal or vertical transmission of bacterial parasite (*Holospora undulata*) becomes more frequent.	High-growth conditions favour vertical transmission and lower virulence. In long-term experiment, parasites lose ability to transmit horizontally altogether.	5, 9
Environmental: Parasites with long-lasting, or far-travelling spores. (Emp)	Virulence should increase with survival time of spores outside host. O: Comparing properties of 16 human respirator pathogens. V: Mean per cent mortality of host.	Positive correlation of spore longevity and virulence.	14

Scenario, *Main theme*[1]	Prediction, procedure, organisms[2]	Finding	Source
Environmental: Virus propagule survival outside host cells as a result of history of co-evolution with different cell types. (Emp)	E: Evolving vesicular stomatitis virus in human HeLA or in normal cells. V: Percentage of dead host cells.	Viruses forced to survive outside host cells change in virulence. But increased virion survival associated with lower rather than higher virulence.	11

[1] Theoretical (T), empirical (Emp) study.   [2] Experimental (E), observational (O) study. V: measure of virulence.

Sources: [1] Berngruber. 2013. *PLoS Path*, 9: e1003209. [2] Bonhoeffer. 1996. *Proc R Soc Lond B*, 263: 715. [3] Day. 2002. *Ecol Lett*, 5: 471. [4] Day. 2004. *Am Nat*, 163: E40. [5] Dusi. 2015. *Evolution*, 69: 1069. [6] Ewald. 1983. *Annu Rev Ecol Syst*, 14: 465. [7] Herre. 1993. *Science*, 259: 1442. [8] Lipsitch. 1996. *Evolution*, 50: 1729. [9] Magalon. 2010. *Evokution*, 64: 2126. [10] Messenger. 1999. *Proc R Soc Lond B*, 266: 397. [11] Ogbunugafor. 2013. *Am Nat*, 181: 585. [12] Pagán. 2014. *PLoS Path*, 10: e1004293. [13] Stewart. 2005. *Evolution*, 59: 730. [14] Walther. 2004. *Biol Rev Camb Philos Soc*, 79: 849.

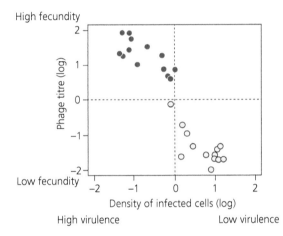

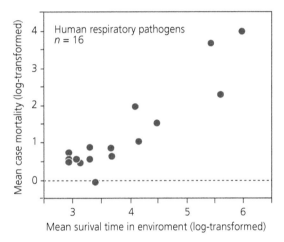

Figure 13.4 Vertical transmission lowers virulence. When experimental lines of bacteriophages f1 infecting *E. coli* are transmitted vertically (open, blue circles), virulence (x-axis) evolves to lower levels as compared to experimental lines that are primarily transmitted horizontally (filled, red circles). Virulence is measured as the initial replication capacity, i.e. the density of infected host cells. Parasite fecundity is the titre of free phage in the supernatant; it reflects replication rate within the host and transmission potential. In the experiment, high virulence was associated with high fecundity of the parasite. Adapted from Messenger et al. (1999), with permission from The Royal Society.

Figure 13.5 Virulence and environmental transmission. Virulence (y-axis: per cent case mortality in hosts) increases with the duration spent in the environment by the transmission stage (x-axis: spore survival in days) (Spearman's $r = 0.86$, $P = 0.0008$, $n = 16$ pathogens). Both axes transformed with $\log_{10}(1000\, x + 1)$. Adapted from Walther and Ewald (2004), with permission from Cambridge Philosophical Society.

et al. 2011). Experimental tests and observations in many different systems are generally consistent with the expectation of the theory and, especially, the expected lower virulence under vertical transmission (Table 13.3, Figure 13.4).

Horizontal transmission results from dispersal in space and time, too, e.g. by the production of long-lasting dispersal stages such as spores. The latter is also known as 'environmental transmission'. As argued above, this would select for higher virulence, according to a more exact scenario named the 'Curse of the Pharaoh' (Bonhoeffer et al. 1996), also sometimes called the 'sit-and-wait hypothesis' (Wang et al. 2017). This prediction has, by and large, been vindicated by theoretical models (Day 2002a; Caraco and Wang 2008; Brown et al. 2012) and by empirical evidence (Walther and Ewald 2004; Wang et al. 2017) (Figure 13.5). As with all general predictions, there are also results which contradict the expectations; for example, some virulent phages of *E. coli* have less, rather than more, persistence in the environment (De Paepe and Taddei 2006). Parasites evolve and adapt to the conditions outside the host, too, e.g. when predators on the host are present (Mikonranta et al. 2015), in which case virulence evolution might take a different course.

Vectors, in particular, are prime vehicles for horizontal transmission of many parasites. Unlike the case of environmental transmission, a predicted increase for virulence with vectors is only found in some cases (Ewald 1983; Ewald 1994), but is not a general fact—both from a theoretical stance (Day 2001) and from empirical studies (Froissart et al. 2010). So far, no simple prediction seems to apply to virulence evolution for vector-borne pathogens. Several reasons may be responsible. For example, vectors do not always serve as a purely passive vehicle (Lambrechts and Scott 2009) and parasites adapt to their vectors, too (Elliot et al. 2003).

13.5.2 Host population dynamics

The standard model includes a host background mortality rate, μ. When background mortality increases, infection duration decreases and parasites should, therefore, increase investment in the transmission step. Given a virulence–transmission trade-off, this would also select for higher virulence. Host mortality may not always limit transmission (Bull and Lauring 2014), but this general prediction is the most widely accepted one (Cressler et al. 2016)

and finds support in theory (May and Anderson 1983a; Ebert and Weisser 1997; Day et al. 2007; Shim and Galvani 2009) and empirical studies (Table 13.4). Beyond the simple scenario, things get more complicated, and predictions are less straightforward. For example, sources of mortality may not simply add up, but instead affect one another (Williams and Day 2001). Virulence also interacts with risk and mortality rate by predation (Choo et al. 2003; Morozov and Adamson 2011), and host immunopathology affects host survival in complex ways (Day et al. 2007).

In the standard model, host population size, N, is a given quantity. However, many new hosts may be born into the population independently of the parasite's action. This is the case when the population is expanding for other reasons. In this scenario, the parasite would have many new hosts to infect, even if it caused high mortality rates on the current host. Therefore, high levels of virulence would evolve despite its cost. In this case, the optimal virulence maximizing R_0 is $\alpha^* = \mu + \nu$ (Figure 13.3b); host birth rate and population size, N, do not matter. However, what happens if the parasites massively affect host population size? This can be due to a substantial reduction in host birth rate or by increasing host mortality. In these cases, not surprisingly, the predictions for the evolution of virulence change. Such scenarios are essential in the context of parasites invading a new population.

Table 13.4 Host population dynamics and virulence.

System	Scope, prediction, procedure[1]	Finding	Source
Glugoides intestinalis (microsporidia)/*Daphnia magna*	High host background mortality selects for higher virulence. E: Experimental removal of hosts to mimic higher host mortality rates. V: Host mortality.	Contrary to prediction, lower virulence with higher background mortality. Perhaps due to selection in multiple infections. But higher within-host growth correlates with higher virulence.	5
Nuclear Polyhedrosis Virus in gypsy moths (*Lymantria dispar*)	Timing of transmission (corresponding to host longevity) affects virulence. E: Virus lineages transmitted early or late. V: Host mortality (fraction dead host larvae after 9 d).	Early-transmitted viruses more virulent than late-transmitted. But late-transmitted viruses overall produced more propagules.	3
Bacterium (*Holospora undulata*) in ciliate *Paramaecium caudatum*	Shorter host life span selects for higher virulence. E: Serial passage of parasite when killing the host either after 11 d or 14 d. V: Host survival.	After 13 transfers (c.300 generations) parasites from early-killing treatment had shorter latency (time to production of infectious forms, associated with higher parasite loads) and higher virulence, but lower infectiousness.	6

System	Scope, prediction, procedure[1]	Finding	Source
Avian influenza virus (H5N1) in poultry (chicken)	Culling mimics increased host mortality and should increase virulence. Culling applied when epidemics occur. T, O: Model population dynamics and genetics of dominant allele for virulence. V: Host mortality.	Culling selects for higher virulence and increased transmissibility.	7
MDV in chicken	E, O: Is increased virulence of MDV over last 60 years due to vaccination and breeding practices. Estimating fitness parameters, etc. V: Virulence score (host mortality, clinical symptoms).	Vaccination prolongs host life span, which also extends infection duration for more virulent strains. But also, breeding for shorter life span in industry adds to more virulence.	1
Vesicular stomatitis virus in BHK-cell cultures	Host longevity should decrease virulence. E: Serial passage of virus in BHK (baby hamster kidney) cell cultures, with normal (after 24 h) or delayed (48 h) transmission events. V: Population growth, plaque size.	After 300 passages, equivalent to 120 generations, delayed transmission (longer host survival) associated with lower virulence. Also genetic changes observed.	8
Granulosis virus (PiGV)/*Plodia interpunctella* (moth)	Localized interactions select for lower virulence. E: Restricting movement of hosts by food distribution. V: Mortality induced per virion in solution.	Infectivity of virus reduced when hosts are localized.	2
Pleistophora intestinalis (microsporidia)/*Daphnia magna*	Evolved virulence should be locally adapted and higher on local hosts. E: Cross-infecting local and distant parasites. V: Host fecundity, survival.	Geographically more distant parasites cause less virulence and have lower spore production on a given host.	4

[1]Experimental (E), observational (O), or theoretical (T) study. V: measure of virulence.

Sources: [1] Atkins. 2013. *Evolution*, 67: 851. [2] Boots. 2007. *Science*, 315: 1284. [3] Cooper. 2002. *Proc R Soc Lond B*, 269: 1161. [4] Ebert. 1994. *Science*, 265: 1084. [5] Ebert. 1997. *Evolution*, 51: 1828. [6] Nidelet. 2009. *BMC Evol Biol*, 9: 65. [7] Shim. 2009. *PLoS ONE*, 4: e5503. [8] Wasik. 2015. *Evolution*, 69: 117.

13.6 Host population structure

Host populations are rarely uniform. Instead, individuals may vary in age or condition and by social groups, and different parts of the population may be separated from one another in space. Such within- and among-population variation affects the evolution of virulence.

13.6.1 Spatial structure

A spatial structure decreases the frequency of contacts between host individuals of different groups but increases the frequency of contacts among host individuals of the same group. At the same time, the pool of susceptibles becomes rapidly depleted within groups of limited size. Hence, the transmission rate becomes lower, and virulence could overall decrease; the benefits of higher between-group transmission often will not offset the within-group effect. A prediction of generally lower virulence with population structure is supported by a range of different theoretical models (Lipsitch et al. 1995; Boots et al. 2004; Lion and Boots 2010; Messinger and Ostling 2013), but dependent on the exact virulence–transmission curve. For example, with a saturating relationship (Figure 13.3a), virulence can increase or decrease depending on the degree of global transmission, with maximum virulence in a population of intermediate structure (Kamo et al. 2007).

The effect of spatial structure frequently is studied in bacteria–phage systems. In experimental metapopulations of *E. coli*, for instance, phage T4 evolved towards low virulence (defined by lysis rate) and lowered infectivity when migration between subpopulations was restricted, i.e. more structure was present. Likewise, the phages lysed earlier (higher virulence) but were less productive when the migration rate was high. This also caused local host populations to decline rapidly (Eshelman et al. 2010). Similar findings come from other studies (Kerr et al. 2006; Berngruber et al. 2015). The spatial structure also determines the distance over which transmission can take place. Long-distance transmission (to other populations) should select for higher virulence, compared to more localized transmission (Boots and Sasaki 1999; Haraguchi and Sasaki 2000). Indeed, granulosis virus (PiGV) became less virulent when the movements of its host, the larvae of the Indian meal moth (*Plodia interpunctella*), were restricted (Boots and Mealor 2007). More spatial separation—a higher degree of structure in host populations—co-localizes the benefits and costs of virulence, and generally leads to lower virulence. This has many practical ramifications, e.g. for agriculture or in hospitals.

13.6.2 Variation in host types

The host population can be heterogeneous because it consists of different host types. Such differences may be due to genotypic variation, differences in host condition or host behaviour, or because host individuals are subject to different microecological conditions. Such heterogeneity, too, affects virulence evolution (Zurita-Gutierrez and Lion 2015). Variation among and within local populations also prompts the question of what parasite strategy is best, i.e. whether to be a generalist or a specialist (Brown et al. 2012; Leggett et al. 2013b; Bruns et al. 2014).

In general, when hosts are sufficiently similar to allow for the evolution of a generalist exploitation strategy, the parasite population will evolve to intermediate levels of virulence. If hosts are too different in their characteristics, the gain associated with a given level of virulence in one host might not compensate for the costs associated with the same level of virulence in another host. As a result, the parasites may evolve to specialize on different host types and, therefore, a range of virulence levels can emerge (Regoes et al. 2000).

In many cases, the relevant heterogeneity is not spatial, but by the separation of relevant host properties. Polymorphism in virulence levels can result (de Roode et al. 2008). In the classical rust–flax system, for example, resistant host populations of flax (*Linum marginale*) harbour more virulent populations of the rust fungus (*Melampsora lini*) than susceptible host populations (Thrall and Burdon 2003). Whether ecological variation is directly associated with relevant differences among locally adapted hosts is unclear. However, ecologically driven host type variation can easily have repercussion for virulence (Tellier and Brown 2011; Mahmud et al. 2017). Of course, the current host condition will also affect the level of virulence. Such short-term conditions would normally not dramatically affect the evolution of virulence unless the environmental condition is a more permanent change, e.g. when water pools become eutrophic or dry out, and host anoxia, therefore, becomes a permanent challenge.

13.6.3 Social structure

Host populations also are structured by sociality and group living. In this case, spatial structure typically co-varies with genetic population structure. Most social groups consist of related individuals, e.g. family members or the worker castes in social insects, that separate themselves from other such social groups in space. The two elements have opposite consequences for virulence.

From the above, spatial structure generally selects for lower virulence. Transmission among similar hosts, by contrast, can select for higher virulence, as in the example of serial transmission (Ebert 1998). Hence, depending on the relative significance of the two processes, the parasite is expected to evolve towards higher or lower virulence (Schmid-Hempel 1998; Nunn and Altizer 2006). Specialized parasites can dwell in social groups particularly easily. They should become less virulent, as their lifestyle brings them closer to being symbionts (Hughes et al. 2008). Another way to look at social structure is to investigate the contact networks along which parasites can be transmitted. Models suggest that the structure of social networks affects virulence (Altizer et al. 2003). However, simple expectations do not readily capture the effect of host sociality on the evolution of parasite virulence, but this probably needs to be scrutinized for any given case.

13.7 Non-equilibrium virulence: Invasion and epidemics

The evolutionary theory of virulence implicitly assumes that hosts and parasites have reached some sort of equilibrium where short-term evolutionary processes no longer change the situation. In many cases, this assumption is not warranted: in particular, when parasites have recently invaded a new host population, either by transfer from some distant locality or an animal reservoir. If an invasion were very recent and most hosts are not yet infected, the starting conditions of the SIR process are formally applicable. However, the parasite might adapt quickly to this situation. The availability of hosts will then drive the subsequent evolution of its virulence and the chances of becoming transmitted, rather than this being led by the maximization of R_0 in a broader, static context (Bull and Ebert 2008).

For example, parasites can initially be too benign to maximize R_0, but 'catch up' later and become more virulent with time. Infectious haematopoietic necrosis virus in salmonid fish may be such a case. When the virus from the wild invaded the reduced host type diversity in fish farms (i.e. similar host genotypes), it increased its virulence (Kennedy et al. 2005). With invasion events, the evolution of virulence will depend on how fast the evolutionary process can track the expanding epidemic, which can lead to higher or lower virulence, depending on the conditions.

An invasion is also present at the start of a 'normal' epidemic when a new, susceptible host population becomes infected. Among many theoretical models, a consensus prediction is that virulence should be high at the beginning, where many new hosts are available and horizontal transmission, therefore, has more effect. As the epidemic progresses, susceptible hosts become rarer, and virulence should decline and converge to a long-term intermediate virulence that is given by the standard equations (Lenski and May 1994; Bull and Ebert 2008) (Figure 13.6).

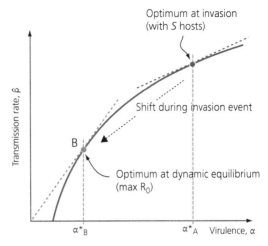

Figure 13.6 The trade-off model for an invading parasite. The solid red line shows the intrinsic relationship between transmission and virulence that the parasite can maximally achieve; here, transmission rate (β) is assumed to be decelerating as virulence (α) increases. When the parasite invades the host population, initially a large number of susceptibles, S, is available (hence, host population size $N \approx S$, at point A). The parasite settles for the highest possible transmission rate without regard for the resulting overall reproductive number R_0 *(blue case). When the epidemic has run its course, and has reached a (dynamic) equilibrium, the parasite evolves to maximize its reproductive number (R_0); this is given by the tangent from the origin (point B, green case; as in Figure 13.3b). Consequently, virulence of the parasite population shifts from α^*_A to α^*_B as the invasion process progresses. For simplicity, background mortality and recovery rate are ignored here. Adapted from Bull and Ebert (2008) with permission from John Wiley and Sons.

Empirical evidence for this general prediction is mixed (Table 13.5). Simulations based on actual data for several diseases (e.g. West Nile virus, myxomatosis virus) suggest that transiently high virulence at the beginning of an epidemic can indeed occur but then decreases (Bolker et al. 2009). Likewise, and depending on the exact conditions, the evolving virulence may be low in the advancing front of an epidemic but increases after that in the 'left-behind areas', where the disease becomes endemic (Hawley et al. 2013).

Under certain conditions, the basic SIR model predicts a general decrease in virulence. When the parasite enters a wholly susceptible population, selection favours virulent parasites that replicate quickly and transmit at high rates. As the parasite spreads, the density of available (susceptible) hosts declines to a level determined by parasite virulence. This decrease in transmission opportunities now

Table 13.5 Virulence in novel and established host–parasite associations.

System	General pattern after introduction	Observation	Remarks	Source
Phage (lc1857) on *E. coli*.	Evolves towards lower virulence.	Experimental evolution. Competition between latent and virulent phage mutant. Virulent mutant prevalent at beginning. Decline follows predictions from virulence–transmission trade-off.	Lysis prevents vertical and favours horizontal transmission.	2
Bacterial (*Flavobacterium columnare*) disease in fish (*Salmo*).	Evolves towards higher virulence.	Introduced to fish farms. Over 23 years, increase of host mortality and disease severity, especially since use of antibiotics after 1992 that controls mortality.	Bacterial strains differ in virulence. Virulent strains with advantage due to higher infectivity, better competitive abilities, and warmer summer temperatures.	7
Mycoplasma gallisepticum in house finches.	Evolves towards higher virulence.	Effect likely due to immune response that does not prevent reinfections but reduces associated host mortality and thus favours more virulent strains.	Mycoplasms emerged in the 1990s in free-living house finches.	4, 5
Nosema ceranae in bumblebees (*Bombus*).	Evolves towards higher virulence?	Spill-over from honeybees. Increased virulence in novel host, i.e. higher mortality, altered behaviour.	*Nosema* is a recent, emergent pathogen.	6
Needle blight (*Dothistroma septosporum*) on *Pinus radiata*.	Evolves towards lower virulence.	Fungus introduced in a single event in New Zealand in the 1960s. Virulence factor is dothistromin.	Did not acquire higher copper tolerance despite use of the corresponding fungicides.	3
Malaria (*Plasmodium falciparum*) in children (endemic areas of Africa).	Evolves towards intermediate levels.	Intermediate virulence.	Virulent for Europeans.	1
Smallpox, measles in humans. Zoonoses (rabies, Lassa virus, bubonic plague) in humans.	Evolves towards intermediate levels.	Highly virulent when introduced but much less so under long associations. Mild in animal reservoirs.	Properties of reservoir hosts often speculative. Often anecdotal reports only.	8
Trypanosomes (Chagas, *Leishmania*). Malaria (*Plasmodium*). Yellow fever virus in humans.	Evolves or remains at high level.	Highly virulent even after long co-evolutionary history.	Host resistance varies among populations.	8

Sources: [1] Allison. 1982. In: Anderson, eds. *Population biology of infectious diseases*. Springer. [2] Berngruber. 2013. *PLoS Path*, 9: e1003209. [3] Bradshaw. 2019. *Microorganisms*, 7: 420. [4] Fleming-Davies. 2018. *Science*, 359: 1030. [5] Gates. 2018. *J Evol Biol*, 31: 1704. [6] Graystock. 2013. *J Invertebr Pathol*, 114: 114. [7] Pulkkinen. 2010. *Proc R Soc Lond B*, 277: 593. [8] Toft. 1990. *Trends Ecol Evol*, 5: 326.

favours less virulent variants. As the epidemic unfolds further, this process continues and selects for decreased virulence—until a level of virulence is attained that maximizes the parasite's R_0 for the given functional relationships of virulence, transmission, and host density (Lenski and May 1994). The predictions from this scenario again converge with the classical avirulence theory, i.e. a decreasing virulence with longer duration of the host–parasite association (Stewart et al. 2005). In all, the expected change in virulence for invading parasites and emerging diseases is difficult to predict, and any outcome seems theoretically possible.

13.8 Within-host evolution and virulence

The evolutionary theory of virulence considers a full cycle from infection to transmission and the infection of a next host. Virulence will be determined by a combination of both between- and within-host evolution, although one of the two parts usually is more relevant in a given ecological context. Hence, within hosts, infecting populations of microparasites (notably viruses, bacteria, protozoa) evolve (see section 12.2) and can thereby change in virulence. The microbiota is a part of the host defence system, and its change throughout an infection also results in changes in virulence. Primarily, this is a change in the composition and functional structure of the microbiota. For example, infecting *Vibrio cholerae* uses the type VI secretion system to attack members of the host microbiota and to colonize the gut. This leads to changes in the level of virulence for the host (Zhao et al. 2018). Besides, there are evolutionary changes because the microbial populations that make up the microbiota evolve. As Figure 12.12 shows, microbes can evolve towards more protection of the host. How these evolutionary processes within the microbiota affect the within-host pathogen dynamics and virulence evolution remains a challenging topic.

13.8.1 Within-host replication and clearance of infection

The growth rate of the infecting parasite population results from its replication (growth) rate, r, and the rate at which the host's immune system clears parasites. Fast-growing parasites can outpace the response times of the immune system and rapidly reach a density that is lethal for the host (see Figure 12.2). The resulting infection duration would be short. In contrast, a parasite that continuously produces transmission stages could benefit from a longer duration. Accordingly, model calculations suggest that transmission success, and thus parasite fitness, is maximized at an intermediate growth rate, r^*. It allows the parasite to grow just below the critical host lethal density before being cleared by the immune system, while allowing for transmission (André et al. 2003). The example demonstrates that within-host replication strategies can be crucial for what level of virulence is observed. More realistic models include additional factors such as variation in the lethal density among hosts. The population-wide optimal growth rate, r^*, and the resulting parasite population size, then surpasses this limit in some hosts but not in others, contributing

to the observed variation in outcome (Alizon and van Baalen 2005).

13.8.2 Within-host evolution: Serial passage

Serial passage is an experimental method. With this protocol, parasite propagules are collected from one host and directly transferred to the next one. The natural transmission process to a next host is thereby bypassed (Box 13.4) (Ebert 1998). This protocol eliminates the costs of virulence for transmission. Serial passage, therefore, favours parasite strains rapidly growing within the host at the expense of other fitness components, such as the ability to transmit and survive in the environment—a case of 'short-sighted evolution'. The effect of short-sighted evolution is visible in *Plasmodium* serially passaged in mice. These increasingly lose the ability to form gametocytes (the transmission stages taken up by mosquitoes) and become more like a clonal line (Dearsly et al. 1990).

As expected from selection for higher within-host replication, and given the association with damage to the host, the virulence of parasites increases during serial passage. Examples include viruses passaged in different systems. e.g. nuclear polyhedrosis virus in moths (*Plutella xylostella*) (Kolodny-Hirsch and Van Beek 1997) or Newcastle disease virus (PPMV-1) extracted from pigeons but passaged through domestic chickens (Dortmans et al. 2011). Many other cases follow this rule (Rafaluk et al. 2015) (Figure 13.7). Exacerbation of virulence can be the experimental goal for parasites used in control measures, e.g. to eliminate an unwanted pest species. However, seemingly similar serial-passage experiments can produce contradictory results. The effects of a given transmission protocol can introduce non-intended elements of selection. An associated change in the duration of infection (and thus the potential to replicate) is one of the most common confounding factors. Sometimes, virulence also decreases during a serial passage, e.g. in microsporidian pathogens infecting water fleas (*D. magna*) (Ebert and Mangin 1997), or *Nosema whitei* infecting flour beetles (*Tribolium castaneum*) (Bérénos et al. 2010a).

Experimental evolution of virulence with serial passage also is used to develop therapeutic tools. Inherent trade-offs dictate that while serial passage

Box 13.4 Serial passage

With serial passage, a starting inoculum of the parasite is administered to a first host, where the infection develops. After a given time, the infecting parasite population is extracted, and a next inoculum prepared. This next inoculum is standardized in size and administered to a next host of the same kind, i.e. of the same species or the same breed. The process is repeated several times until a series of passages through the experimental host type has been completed (Figure 1). It is usually assumed or experimentally ascertained that the first inoculum contains some strain (genotype) diversity in order to foster such experimental evolution within a reasonable time (the arrival of useful mutations may take much longer).

Serial passage replaces the natural process of transmission with the experimental act of extracting and transferring the parasite from one host to the next. Sometimes, a more natural way of transmission may be mimicked by removing hosts at a regular schedule and replacing them with naïve ones. In this case, the transmission may fail, as hosts may 'die' too early. The chain of infections, therefore, collapses during the experiment.

Serial passage is a powerful tool to study how parasite characteristics change under experimental evolution (Ebert 1998).

Because of the limited (effective) population sizes in any such experiment, genetic drift and other uncontrollable sources of variation exist. Therefore, serial passages are run in several replicate lines to randomize these effects. Nevertheless, the major selective forces in serial passage experiments are competition among co-infecting parasite strains and adaptation to the experimental host type. Serial passage will select for those strains having higher numerical representation in the infecting population at the time of their extraction from the host. This favours strains rapidly growing in the experimental host. Therefore, a typical result of serial passage experiments is an adaptation of the parasite population to the type of host where it is passaged through, e.g. a higher virulence on this host.

Also, serial passage is an important tool for vaccine development. As the parasite population adapts to the experimental host, it typically loses virulence in the original host ('attenuation'). The attenuated viruses can be used as a live vaccine. Several major vaccines use this technology, such as Sabin's polio vaccine, Theiler's yellow fever vaccine, and the MMR (measles, mumps, rubella) vaccine administered to children. Louis Pasteur invented serial passage for vaccine production in the 1880s.

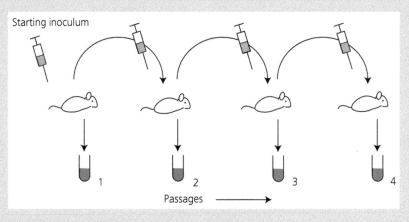

Box 13.4 Figure 1 Scheme of a serial-passage application.

increases virulence in an experimental host, typically the passaged parasite population loses virulence on the original host. Examples are morbilliviruses (Liu et al. 2016a) or plant pathogens (Meaden and Koskella 2017) (Figure 13.7). The parasite, therefore 'attenuates' for the original host.

This effect is exploited to develop so-called 'live-attenuated vaccines', an often-used method before the advent of molecular biotechnology. Attenuated pathogens elicit an immune response and immune memory but cannot cause damage. Nevertheless, if again serially passaged through an appropriate

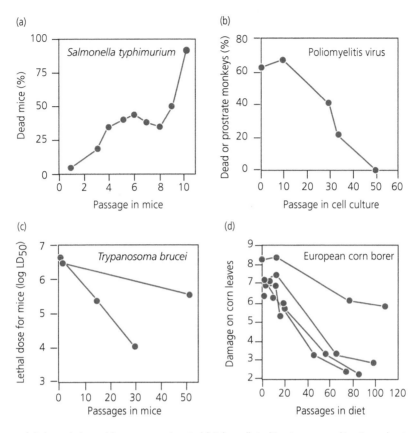

Figure 13.7 Change of virulence during serial passage experiments. (a) *Salmonella typhimurium* passaged in mice evolves towards higher virulence in mice. (b) Polio virus passaged in cell cultures loses its virulence on the original monkey hosts (*Cynomolgus*). (c) *Trypanosoma brucei* passaged in mice evolves towards higher virulence in mice, i.e. a lower LD_{50} dose is needed. (d) Corn borer (*Ostrinia nubilalis*) passaged on an artificial diet loses its virulence on the original host plant (corn). Adapted from Ebert (1998), with permission from the American Association for the Advancement of Science.

host, such attenuated strains can revert to virulence (Muskett et al. 1985). Together, these findings demonstrate that parasite virulence can evolve rapidly based on the within-host adaptation of parasites.

13.8.3 Within-host evolution and virulence in a population

Within-host evolution also has consequences on the evolution of virulence in the parasite population at large. For example, it can lead to the within-host emergence and spread of strains that cause severe infections (Young et al. 2017a). However, the effects depend on the details of a given host–parasite system. Unfortunately, there is much theory but little in the way of empirical study (Cressler et al. 2016).

Nested or multiscale models (Gilchrist and Sasaki 2002) (see section 12.6) cut across scales. They embed, for example, a model for within-host evolution within a between-host epidemiological model that describes the overall spread of the parasite in the host population (Mideo et al. 2008). The within-host part can make explicit reference to detailed processes, e.g. the dynamics of immune cells, or parasite numbers. The within-host dynamics yields the values of variables that matter for the between-host dynamics, e.g. the timing of transmission, the relative representation of different strains in the inoculum, or the order in which strains are transmitted and infect a next host. Such integration of scales can have unexpected consequences. For example, a virulence–transmission trade-off might emerge intrinsically from the dynamics itself, rather than being a result of mechanically coupled physiological processes (Alizon and van Baalen 2005). Also, a virtual transmission–recovery trade-off can emerge by travelling waves of a spreading epidemic (van Ballegooijen and Boerlijst 2004).

Sometimes, the integration of scales rests on decomposing the overall R_0 into a term for a within-host reproductive number, R_{w0}, considered to be a rapid process, and a term for a between-host reproductive number, R_{b0}, considered to reflect a slower, longer-term process. The two systems thus operate at different speeds. Theoretically, the coupled dynamics shows that the usual condition of $R_{b0} < 1$ no longer guarantees the elimination of infectious disease at large. For example, the coupled dynamics can 'rescue' the between-host disadvantage of virulence that is too high. At least in theory, this coupled dynamics can lead to multiple stable states for the virulence level, such that simple predictions may not always apply (Mideo et al. 2013a; Cen et al. 2014; Feng et al. 2015). For example, virulence can be zero in young, expanding epidemics.

Actual data on the course of an infection can parametrize a within-host model. For example, parameters such as the risk of infection, mutation rate to novel strains, or host mortality rate were numerically estimated for HIV infections and inserted into models (Lythgoe et al. 2013). For most cases, however, parametrization seems complicated. Alternatively, a generalized quantitative genetics framework can treat the transmission rate, the expressed virulence, or recovery/clearance rates as summary functional traits. This approach avoids the need for a mechanistic understanding and, therefore, the modelling of within-host processes and the respective quantification of parameters (Day et al. 2011). For rodent malaria, for instance, the experimental data were converted into life history traits of the parasite and their co-variances calculated. Together with epidemiological considerations, this generalized framework generated predictions about virulence and other disease characteristics (Mideo et al. 2011).

Regardless, the within- and between-host processes can select for opposing properties of the parasite, e.g. higher or lower virulence. Therefore, taking into account a multistep selection regime is

always illuminating. Unfortunately, no simple rules of thumb emerge, because the predictions vary with the conditions and assumptions.

13.9 Multiple infections and parasite interactions

As mentioned at various points, within-host competition among co-infecting strains is a crucial element that affects strategies of parasite replication, growth, and eventual virulence effects on the host (Cressler et al. 2016). Thereby, co-infecting parasites interact either antagonistically or cooperatively. These different kinds of interactions impact virulence evolution (Leggett et al. 2014). Empirically, such cooperative or competitive scenarios show various outcomes (Table 13.6).

In nature, multiple infections of hosts are common (see section 12.4, Table 12.3). Furthermore, a single primary infection can evolve into a genetically diverse infection over time within a host. Such diversification alone may change the level of virulence. However, the order of infection by different strains also is a simple mechanism that shapes virulence. For example, a low-virulence strain may infect first but could be displaced by a later arriving, more virulent strain, depending on the conditions. Also, the temporal spacing between infection events can affect the outcome of competition (Hargreaves et al. 1975; Taylor et al. 1997; Lipsitch et al. 2000). For instance, a competitively inferior but lower-virulence strain can establish an infection and multiply to become the dominant type before a more competitive strain of higher virulence arrives (Ben-Ami et al. 2008).

13.9.1 Virulence and competition among parasites

Observing or testing for the effect of multiple vs single infections has clarified that competition within hosts can have substantial effects on observed virulence. Different mechanisms for competition exist.

13.9.1.1 Resource competition

When co-infecting, parasite strains better at the extraction of resources from the host will grow faster

Table 13.6 Parasite interactions and virulence.

Host	Parasite	Multiple infection [type of interaction]	Observation	Source
In culture media	Bacterium (*Pseudomonas aeruginosa*)	Experimental evolution with wild type and deficient strains (i.e. not producing siderophores). [Cooperation for public good]	Deficient type outcompetes co-infecting wild type under iron limitation. Cooperative siderophore production evolves more likely under close relatedness. If competition is more local, relatedness becomes less important.	2
Biofilms in culture media	Bacterium (*Pseudomonas fluorescens*)	Formation of biofilms by specialized bacteria. Biofilms promote virulence, as they are resistant against immune response. [Cooperation for public good]	Biofilms made up of mixed genotypes (presumably specializing on different components of the biofilm) more resilient to invasion by non-cooperating bacteria.	1
Lytic bacteriophage	Bacterium (*Pseudomonas syringae*)	Infecting with different ratios of virus to host cells, thus affecting the relatedness among bacteriophages. [Cooperation among kin]	Virulence (lysis) reduced with low relatedness among viruses. Likely a result of cooperation and conflict over production of components needed to replicate. Phages produce replicative enzymes that can be used by all virions in host cell.	7
Moth (*Galleria monella*)	Bacterium (*Pseudomonas aeruginosa*)	Infection with siderophore-producing ('producers'), and siderophore-deficient strains ('cheaters'). [Cooperation for public good]	Deficient strains grow faster in mixed infections at the expense of producing strains. But mixture with deficient strains less virulent as a whole. Low relatedness more likely to include deficient strains. However, overall fitness of deficient and producing strains equivalent; no advantage for cheaters detected.	3
Moth (*Galleria monella*)	Bacteria (*Photorhabdus, Xenorhabdus*)	Infection of strains that either produce bacteriocins or are deficient. [Competition by direct interference]	Bacteriocin-producing strains outcompete deficient strains in mixed infection; virulence determined by producing strain. If both strains produce bacteriocins, both strains co-exist, and virulence is lower than with single infections.	5
Stickleback (*Gasterosteus aculeatus*)	Cestode (*Schistocephalus solidus*)	Double infection by related or unrelated strains (kinships of cestodes). [Cooperation among kin]	In male hosts: infection more likely when strains are related. Female hosts: infection more likely when strains are not related. Irrespective of sex: higher infection intensity when parasites are related. In sequential infections: later arriving parasite survived better, irrespective of relatedness.	4
Mice	Malaria (*Plasmodium chabaudi*)	Infection by one or two parasite clones in different ratios. [Competition]	In double infections virulence (loss of host body mass, anaemia) higher and more transmission stages produced, but effect not simple. Most likely, higher virulence induced by genetically more heterogeneous infections that are more difficult to clear.	6

Sources: [1] Brockhurst. 2006. *Curr Biol*, 16: 2030. [2] Griffin. 2004. *Nature*, 430: 1024. [3] Harrison. 2006. *BMC Biol*, 4: 21. [4] Jäger. 2006. *Evolution*, 60: 616. [5] Massey. 2004. *Proc R Soc Lond B*, 271: 785. [6] Taylor. 1998. *Evolution*, 52: 583. [7] Turner. 1999. *Nature*, 398: 441.

and generally outcompete slower variants. Therefore, multiple infections generally select for higher virulence under the extraction-associated damage. This expectation finds empirical support, even though the findings are quite heterogeneous (Table 13.6). For example, more virulent *Plasmodium chabaudi* (malaria) strains indeed are more competitive against less virulent strains and will eventually dominate the infection in mice (de Roode et al. 2005a; Bell et al. 2006) (Figure 13.8). A similar dominance of the most

virulent strain occurs in bacterial infections of water fleas, *Daphnia* (Ben-Ami and Routtu 2013).

However, most experimental tests use only one round of multiple infections vs one round of single infections. This protocol does not take account of the epidemiological processes in a natural population. When host background mortality removes many infected hosts, for instance, the force of infection and, thus, the potential for competition and increased virulence decreases; this can be relevant

(a)

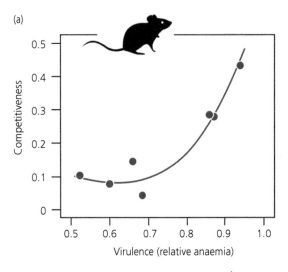

(b)

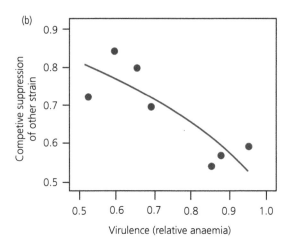

Figure 13.8 Virulence of malaria in experimentally infected mice. The virulence of *Plasmodium chabaudi* is measured as degree of anaemia in the mouse host, and as induced by the experimental strain relative to a standard strain. (a) Competitiveness against other strains increases with virulence; measured as proportion of cells of the focus strain among all parasite cells in mixed infections. (b) More virulent strains reduce the success of competing strains. Competitive suppression is the proportional reduction in infection intensity of the experimental strain compared to when it is infecting alone. Silhouettes from phylopic.org. Adapted from de Roode et al. (2005b); copyright (2005) from The National Academy of Sciences USA.

in wildlife management programmes, where culling removes the infecteds (Bolzoni and De Leo 2013). Furthermore, increased virulence under the test conditions could also reflect phenotypic plasticity of the parasites; for example, a higher virulence is expressed when another parasite is present.

13.9.1.2 Apparent competition

Competition between co-infecting parasites is also mediated by the host's immune system, a process known in ecology as 'apparent competition'. For example, hosts that mount a non-specific immune response can hit one competitor more than another. Similarly, cross-reactivity of the response can differentially affect competitors that are antigenically close (Cox 2001; Read and Taylor 2001).

For example, different strains of a trypanosome (*Crithidia*) infecting bees (*Bombus terrestris*) differ in their growth rates in media. Inside living hosts, however, the fastest strain does not necessarily outgrow the slower ones, a result that is due to (innate) host responses, and which adds to the 'filtering' of mixed infections before transmission (Ulrich et al. 2011; Ulrich and Schmid-Hempel 2012). In rodent malaria, a virulent clone competitively suppresses

an avirulent clone, when infected into an immunocompetent mouse. In immunodeficient mice, however, both clones are on more equal terms, suggesting that the host immune response affects the competition among strains (Råberg et al. 2006). Another study in the same system, using CD4[+] T-cell-depleted mice, did not find such an effect, however (Barclay et al. 2008).

Hence, the evidence remains mixed. Whether host immune responses cause less virulent strains generally to be outcompeted by more virulent ones still is debatable (Grech et al. 2008). At least, in theory, a previous infection can change the immunological responses for the next parasites with an overlapping antigenic profile. If so, cross-reactivity and immune memory at least could foster the diversity of infections and thus of virulence levels (Best and Hoyle 2013b). In the rodent malaria case, prior residency is also of the essence. Clones infecting first suffer less from competition than later infecting clones, almost irrespective of their virulence (de Roode et al. 2005a).

13.9.1.3 Interference competition

Interference competition refers to the direct, active interaction of competitors. For example, viruses can compete against each other by active mutual suppression that is also unrelated to the virulence effects exerted on the host (as in phages, Chao et al. 2000). Typically, in bacteria, interference involves the secretion of bacteriocins—antimicrobial toxins that can kill a competing bacterial infection (Riley and Wertz 2002) (see section 12.4.1).

Co-infecting strains also exert spite against one another, i.e. they damage the competitor at their own expense (Gardner et al. 2004; Hawlena et al. 2010). The concept is sometimes controversially discussed but can indeed evolve (West and Gardner 2010). Notably, spiteful interactions can affect the overall virulence of the host. For example, *Pseudomonas aeruginosa* releases bacteriocin by lysis, i.e. at a high cost to the producing bacterium itself. Bacteriocin-producing strains of *P. aeruginosa* infected in moth larvae (*Galleria mellonella*) produced most toxins when at intermediate frequencies relative to their non-producing competitors, i.e. when the advantage to kill competitors was highest. At the same time, this maximum investment into toxin production leads to lower investment into virulence-causing mechanisms, such as mobilization of resources for growth, and therefore to lowest virulence for the host at this point (Gardner et al. 2004; Inglis et al. 2009).

In all, the virulence of the infecting population can either decrease or increase, depending on the interference mechanism (Brown et al. 2009). Higher virulence is the primary expectation from theory. Phages, for instance, evolve to respond plastically to the presence of mixed infections and speed up the time to lysis, thus killing the host sooner (Leggett et al. 2013a). However, this may not be the typical pattern (Cressler et al. 2016). For example, when *Photorhabdus* and *Xenorhabdus* were infected into caterpillars, competition by bacteriocins did not exclude the less virulent strain, and multiple infections led to lower rather than higher virulence compared to single-strain infections (Massey et al. 2004). Similar reports in other systems, such as for *Xenorhabdus bovienii* (Bashey et al. 2012) *or Bacillus thuringiensis* (Garbutt et al. 2011), suggest that interference competition via bacteriocins leads to lower virulence in mixed infections—as also seen in the example of *P. aeruginosa*.

13.9.2 Cooperation among co-infecting parasites

Parasites also cooperate to exploit the host and gain transmission (see section 12.4.2). For example, all co-infecting strains can step up the exploitation of their host such that each one gains more resources than when not cooperating (Frank 1996). These scenarios require that parasites can plastically adapt interference or virulence levels to the current infecting population. Secreted molecules are essential for the plastically regulated cooperation among co-infecting microbes (Noguiera et al. 2009).

13.9.2.1 Kinship among parasites

The relatedness among co-infecting parasites is particularly crucial for the evolution of virulence. Relatedness is measured as the average coefficient of relatedness among different strains in the infecting population. As the number of co-infecting strains increases, their relatedness decreases. In general, virulence is predicted to evolve to lower levels with an increase in relatedness (Box 13.5) (Frank 1992). However, this prediction depends on the exact conditions and mechanisms (Figure 13.9). For example, it may result from more prudent exploitation of the host. Increasing cooperation by higher production of bacteriocins can, in turn, lead to higher virulence when relatedness increases, with virulence at a maximum for intermediate relatedness (Gardner et al. 2004).

Other factors also add to these effects. For example, increased antigenic diversity in mixed infections of unrelated *Plasmodium* strains suppresses immune defences more strongly than single infections. When the epidemiology of the infection in the host population additionally is taken into account, reduced virulence only is expected for linear trade-off relationships between virulence and transmission. Cooperation among parasites may collapse when the trade-off functions are different (Alizon and Lion 2011). Accordingly, the empirical support for kin effects is also heterogeneous.

13.9.2.2 Cooperative action

Parasites actively cooperate in many ways. For example, in some cases, viruses can only code for a subset of all necessary proteins needed to complete

Box 13.5 Kin selection and virulence

The standard equation for the SIR model, eq. (11.1), can be modified to allow for co-infection by strains that share some traits; for example, because they are related (Frank 1992, 1996). Suppose that the virulence of a parasite strain (genotype, g), its virulence phenotype, α_g, can be described by using genotypic values as follows: $\alpha_g = \alpha + dg$, where g is the genotypic value of this strain and d is the effect of this value on virulence; α is the population-average virulence. Correspondingly, the virulence in a mixed infection is $\alpha_g' = \alpha + dG$, where G is the average genotypic value of the mixed infection.

With a trade-off model, transmission and recovery rate depend on virulence (and thus the genotypes of the strains) as follows:

for transmission rate: $\beta(\alpha) = c\alpha^k$

for recovery rate: $\upsilon(\alpha) = \dfrac{d}{\alpha^m}$ (1)

where c, d, k, and m are shape parameters. Eq. (1) implies that transmission increases, but recovery rate decreases with increasing virulence, α.

The basic reproductive rate, R_0, assumed to be the parasite's fitness, now depends on its genotypic value. Each strain, therefore, differs in its basic reproductive rate, R_{0g}. When a population of parasite strains infects the host, the average, expected R_0 can be calculated using Price's covariance approach. The approach respects that the evolutionary consequences of virulence co-vary among the strains, depending on their (genetic) similarity. Furthermore, it takes into account how genotypes are associated with the basic reproductive rate; the details of the procedure are given in Frank (1992). With Price's procedure, the evolutionarily stable (ESS) levels of virulence, α^*, are:

(i) For a transmission–virulence trade-off only:

$$\alpha^* = \frac{k(\mu + d)}{(r - k)}, \text{ for } r > k$$

but $\alpha^* \rightarrow \infty$ otherwise. (2)

Here, r is the regression coefficient of relatedness ($-1 \leq r \leq 1$), i.e. the slope of the group genotypic value, G, on the individual genotypic value, g (as in kin selection theory). Eq. (2) suggests that α^* should decline with an increase in the relatedness among co-infecting strains (Figure 1). If transmission increases disproportionally with α, then $k > 1$ and $\alpha^* \rightarrow \infty$.

(ii) For a recovery–virulence trade-off only:

$$\alpha^* = \left(\frac{dm}{r}\right)^{\frac{1}{(1+m)}}$$ (3)

In this case, virulence also generally decreases as relatedness, r, among co-infecting strain increases. The ESS level, α^*, remains numerically stable over a wide range of conditions. These predictions assume that each genotype affects its transmission from within the group and thus benefits from its strategy. If transmission depended linearly on the effects of all genotypes, every strain would benefit, and the virulence would be as with a single infection.

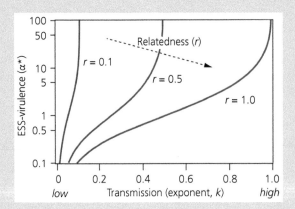

Box 13.5 Figure 1 Virulence, transmission, and kinship. Shown is the ESS level of virulence, α^* (y-axis), for co-infecting parasite strains related with a co-efficient of relatedness, r (x-axis). α^* depends on transmission rate, given by the shape parameter k (high value of k promotes transmission more strongly). Adapted from Frank (1992), with permission of the Royal Society.

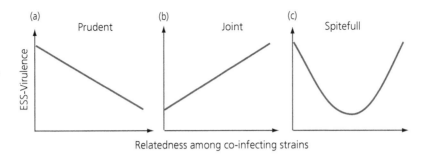

Figure 13.9 Virulence and kinship. Shown are the predicted relationships for the evolutionarily stable level of virulence (y-axis) as a function of the degree of relatedness among co-infecting strains (x-axis). Virulence is defined by host exploitation, i.e. by the growth rate of the parasites through extraction of resources. The graphs refer to: (a) 'Prudent' exploitation to increase transmission, i.e. each parasite exploits the host so as to maximize transmission over the lifetime of the infection. Low relatedness leads to exploiting hosts as quickly as possible, and thus to higher virulence. (b) Joint exploitation of the host, i.e. parasites generate a costly public good (such as iron-sequestering siderophores) that can be used by all. With low relatedness, fewer strains will invest in the public good but benefit from others producing the public good; hence, virulence will decrease with lower relatedness. (c) Spiteful behaviour of parasites, i.e. scramble competition among parasites (e.g. by production of bacteriocins to kill competitors). Spite is costly and reduces own growth, and thus the virulence for the host. Therefore, virulence is high (low spite) when relatedness is high, and when relatedness is low (when a spiteful line is rare and advantage of spite is therefore small). High levels of spite (low virulence) are preferred at intermediate levels of relatedness because many competing strains are affected, whereas damage to own line is still low. Adapted from Buckling and Brockhurst (2008), with permission from Springer Nature.

the replication within a host cell. As a result, a virus and its helper virus have to cooperate to be successful (Chao et al. 2000). As discussed in section 12.4.2, the production of secreted siderophores by bacteria is a classical case of active cooperation. When parasites cooperate in this way, virulence increases with kinship (Figure 13.9), since close relatedness favours higher investment into the exploitation of the common good (the host resources). Other mechanisms include the joint production of digestive enzymes to mobilize host resources, or the production of adhesive polymers to form biofilms. Because these different mechanisms are all public good situations, their primary effect on virulence is similar, too. Nevertheless, benefits (and costs) are dependent on the environment and, therefore, the outcome of cooperation on virulence is not always exactly the same (Zhang and Rainey 2013; Kümmerli and Ross-Gillespie 2014).

Finally, quorum sensing (see Box 12.5) allows parasite cells to respond to current conditions plastically. Quorum sensing is intertwined with the degree of relatedness among co-infecting strains (West et al. 2006; Diggle et al. 2007) and affects virulence. For example, the overall level of virulence in *Staphylococcus aureus* infecting wax moth larvae is negatively correlated with the fraction of

non-producing types ('cheaters' carrying the *agr* mutation) (Pollitt et al. 2014); similar results are found for *Pseudomonas aeruginosa* tested in mice (Rumbaugh et al. 2012). Plastic responses to current conditions are a powerful mechanism that affects parasite fitness and changes the expression of virulence (Mideo and Reece 2012).

13.10 Additional processes

13.10.1 Medical intervention and virulence

Within-host selection can have repercussions on the evolution of virulence, and within-host selection is a consequence of medical treatment. Treatment also affects the epidemiological parameters in the standard equation, eq. (11.1). For instance, treatment might increase the average recovery rate, v, of the host through faster clearance or by improving host body condition. We should, therefore, expect that the evolutionary response of a parasite population to treatment is to adapt virulence, α, to maximize its fitness under the new circumstances (e.g. R_0 -> max, but see provisos above). These effects are especially interesting in the context of vaccination (Gandon and Day 2007).

Vaccination protects the individual host and protects the population at large if a sufficient proportion is vaccinated (herd immunity). However, vaccines act in different ways and are never entirely effective. They occasionally allow parasites to transmit, which, therefore, can affect the further evolution of virulence. In fact, vaccines can either prevent infection, reduce growth inside the host, block transmission, or reduce damage to the host. Because all of these steps affect the parasite's fitness, the consequences for the parasite's adaptive response, and thus for virulence, depend on which of these elements is targeted by the vaccine (Gandon et al. 2001, 2003; Ganusov et al. 2006; Gandon and Day 2007) (Figure 13.10).

For example, when a vaccine blocks infection or transmission, the effect on virulence is negligible unless multiple infections occur. Some vaccines, especially novel therapeutics, reduce damage to the host. This works by targeting the parasite's virulence mechanisms, such as toxin production, or by improving host health with neutralizing virulence factors. However, when vaccines reduce disease severity for the host, the parasite no longer carries the full cost of its virulence; virulence, therefore, evolutionarily increases. When virulence factors were only neutralized or made ineffective, the parasite still carries the cost of their production; virulence will, therefore, evolutionarily decrease (Gandon et al. 2002).

The impact of vaccination on the further evolution of virulence is thus not straightforward and depends on the targets of the vaccine (Vale et al. 2014). Furthermore, the impact also depends on the efficacy of a vaccine, i.e. how 'leaky' it is in allowing virulent parasites to still transmit from vaccinated hosts. Leaky vaccines also introduce heterogeneity in the host population, with some hosts fully protected, while others are not safe, or still represent a source of further infections for others. Such heterogeneity can additionally affect the evolution of virulence (Williams and Day 2008). Imperfect vaccines can select for higher rather than lower virulence (Gandon et al. 2001, 2002). Medical intervention, therefore, has repercussions that are often counter-intuitive and can, in due course, make matters worse for susceptible or untreated individuals, even though the overall prevalence of the infection might decline.

These theoretical expectations have some empirical support. In mice treated with a vaccine that reduces within-host growth, the parasite (rodent malaria) indeed increased in virulence as expected from the competitive effects (Mackinnon and Read 2004; Mackinnon et al. 2008; Barclay et al. 2012). Diphtheria toxoid vaccine provides immunity against the toxin produced by *Corynebacterium diphtheriae* (the causing agent) but does not remove the cost of production. This persisting cost would explain the observation that the prevalence of toxin-producing (virulent) strains of *C. diphtheriae* and *B. pertussis* has been declining in countries with long-lasting programmes of anti-toxoid vaccines (Gandon et al. 2002; Ganusov et al. 2006). For Marek's disease virus (MDV) in farmed chickens, a vaccine introduced in the 1950s increases host survivorship and thus prolongs infection duration, even when infected by virulent strains. As expected, MDV has become more virulent over the last decades (Witter 1997). This change results from the vaccine and was more than would be expected from

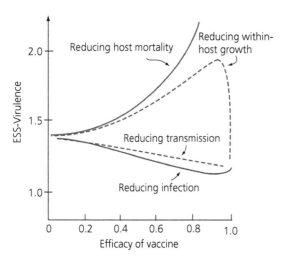

Figure 13.10 Virulence and vaccination. Shown is the predicted evolutionarily stable (ESS) level of virulence (y-axis) under different mode of actions of a vaccine, and its efficacy (perfect vaccines have an efficacy of one). When the vaccine acts to reduce parasite growth within the host, or when it reduces host mortality (red lines), virulence in the parasite population increases, especially for efficient vaccines. For vaccines that reduce the probability of infection, or of further transmission, virulence decreases (blue lines). In these scenarios, a vaccine has only a single effect, not multiple effects. Adapted from Gandon et al. (2001), with permission from Springer Nature.

shortening host life spans by way of breeding (Atkins et al. 2013). When chickens are protected with imperfect vaccines that prolong host survival but do not block infection, replication, or transmission, highly virulent strains of MDV that otherwise are too pathogenic to persist can survive and transmit. Vaccinated chickens are, therefore, a reservoir from which more virulent strains can spread (Read et al. 2015).

13.10.2 Castration and obligate killers

The vast majority of parasitoids, but also many microsporidia, kill their host as an essential condition for transmission. Pupation occurs, or spores are released, when or after the host is killed ('obligate killing'). Therefore, virulence is how long it takes until the host dies. It may be measured by what the expected residual host fitness would be if the host survived from the point of infection onwards and went on to die from another cause (Ebert and Weisser 1997). The strategy of obligate killers is to kill the host at a time when staying with the current host is no longer better than switching to a new host. Hence, the optimal killing time is when the marginal value for its further exploitation drops to what is expected in the population of hosts at large.

With logistic growth of the parasite within the host, the optimal killing time is when the instantaneous growth rate drops to the level of the background host mortality rate (Ebert and Weisser 1997). This time point is affected by various other parameters, e.g. the chances of successful transmission, and so forth. Empirically, the trade-offs assumed in these scenarios, especially between host life span (i.e. virulence) and residual parasite transmission, are sometimes elusive. For example, strains of SpexNPV (*Spodoptera exempta* nucleopolyhedrosis virus) that rapidly killed the host (*S. exempta*) indeed attained lower virus yields, setting the stage for competition with slower-killing but more productive strains (Redman et al. 2016). A robust trade-off exists for bacterial infections (*P. ramosa*) of water fleas (*D. magna*). However, the relationship is not simple and, notably, transmission can sometimes increase as virulence decreases (Ben-Ami 2017).

Parasite-induced castration is akin to premature killing. In both cases, host fitness tends to zero, i.e.

virulence is exceptionally high. The consequences for the evolution of virulence can be different, though (Abbate et al. 2015). Parasite-induced shortening of host life span directly translates into the parameters of eq. (11.1) via the parasite-induced mortality rate, α. Nevertheless, the reduction of fecundity has effects, primarily through the epidemiological feedback, because more resistant hosts are recruited into the local population. This also depends on spatial structure, fluctuations in host abundance, multiple infections, and variation in host quality (Abbate et al. 2015). Furthermore, castration frees resources for the parasite, but often does not directly affect the duration of the infection and thus transmission opportunities for the parasite. A reason could be that hosts reallocate their resources or express tolerance to infection, which would reduce host damage (Best et al. 2009). Interestingly, many parasites induce partial rather than full castration, perhaps due to these processes.

13.11 Virulence and life history of infection

13.11.1 The timing of benefits and costs

When a host becomes infected, the parasite population within the host begins its life history. As described in several other contexts, the parasites exploit the host by extracting resources and fight the immune response to avoid being cleared. The resources are converted into more parasites and eventually into transmission stages. At the same time, the parasite generates damage to the host (virulence). However, these effects need not occur at the same time. In particular, the effects of virulence (e.g. parasite-induced mortality) typically occur later than when the benefits from the parasite's actions appear, e.g. gains in transmission (Figure 13.11). This discrepancy yields the central argument: the cost of virulence could be paid later than the realization of benefits. This time lag is a significant characteristic of disease life history; analogous considerations would apply for the lag between clearance and transmission (Day 2003; Frank and Schmid-Hempel 2008).

According to general life history theory, earlier events have more selective weight than later ones.

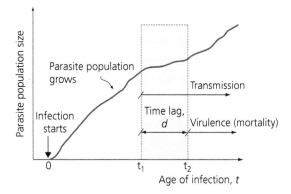

Figure 13.11 Life history of an infection. It starts at time $t = 0$, when the host becomes infected. The parasite population increases in numbers (or individual parasites grow) as the infection unfolds over time, t, which is the age of the infection (defined by the time since the infection happened). In this example, transmission starts at time t_1 post-infection, and the parasite-induced host mortality (virulence) takes effect starting at time t_2. There is a time lag, d, between onset of transmission and onset of mortality. Adapted from Day (2003), with permission from Elsevier.

Hence, when the onset of virulence (i.e. the costs of the parasite's actions) occurs later in the life of the infection than transmission (i.e. the benefits), virulence costs are discounted, relative to the benefits. Therefore, virulence can evolve towards higher levels. Likewise, with a short time lag, or when virulence effects emerge before transmission, virulence becomes very costly for the parasite and, hence, lower virulence should evolve. The time lag between such events is thus crucial for the evolution of virulence.

Nuclear polyhedrosis virus that infects the gipsy moth, *Lymantria dispar*, illustrates the case (Cooper et al. 2002). In serial-passage experiments, early-transmitted viruses, i.e. transmitted before virulence effects emerge, indeed evolved towards higher levels of virulence (host mortality) than the late-transmitted group. A similar result is reported for stomatitis virus serially transmitted in (hamster) cell cultures for more than 100 generations (Elena 2001). The timing of life history events, at the same time, calls for plasticity in the parasite's action, depending on current conditions. Plasticity in investment for within-host replication vs transmission considerably outperforms fixed strategies, and—true to life history theory—early allocation decisions have more effect (Greischar et al. 2016a).

The costs of virulence also become discounted when host background mortality is high, i.e. when

there is a high chance that the host will not live to experience the virulence effects (see also section 13.5.2). A similar effect occurs when the clearance rate is high. In both cases, higher levels of virulence should evolve. Moreover, multiple infections cause similar discounting, because a co-infecting strain might deplete the host's resources earlier, kill the host prematurely, or replace the founding strain. Finally, when a parasite population is rapidly increasing in size, early reproduction (i.e. rapid growth and transmission) should be favoured at the cost of a reduction in fitness later in life (i.e. when virulence occurs with a delay). The same number of parasite offspring produced early represent a larger share of the future parasite population. This asymmetry occurs in an early epidemic phase when the parasite population is rapidly expanding among its susceptible hosts (Lipsitch et al. 1995).

Genotypic variance for the expression of life history-specific traits of the parasite is a necessary precondition for such time lags and their associated consequences to evolve. Such variance against the life history background is known, for example, for rust fungus infecting oats (Bruns et al. 2012). For genetic analyses, the infection life history can also be split into different stages and the quantitative genetic variances estimated (Hall et al. 2017). In the same spirit, characteristics such as the pattern of transmission, host mortality, or recovery, can be considered as co-varying traits that shape a disease life history, embedded in the epidemiological (between-host) dynamics. Using such descriptive co-variation independently of underlying mechanisms might be a useful tool (the 'trait-value framework') for empirical studies; this has been mentioned before as a generalized quantitative genetics framework (Day et al. 2011; Mideo et al. 2011).

13.11.2 Sensitivity of parasite fitness

The standard evolutionary theory of virulence assumes a constant set of trade-off relationships. Correspondingly, the theory is well suited to analysing situations where the respective hosts and parasites are the same or closely related, and where only the environmental conditions change; for example, when comparing vertical vs horizontal transmission of the same parasite. The classical theory is difficult to apply when comparing different parasites strains and host types, or different species altogether (Frank and Schmid-Hempel 2008). In these cases, the basic assumptions, e.g. the same virulence–transmission curves, are unlikely to hold. It is, nevertheless, possible to expand the standard evolutionary theory by considering the relative sensitivity of parasite fitness to a change in the variables of the parasite's life history.

An illustration of this approach is in the context of immune-evasion mechanisms, by which a wide variety of parasite actions is meant that impede, block, or manipulate the host's response in order to infect, establish, multiply, and eventually become transmitted (see Chapter 8). Consider a parasite that can manipulate the host's immune response so as to enhance its transmission at a given time, but where the same mechanism also generates virulence effects somewhat later in the infection (Figure 13.12). With the life history argument, such an immune-evasion mechanism is favoured by selection despite its consequences for virulence, since the costs appear later than the benefits.

A parasite can also suppress the host's immune responses to avoid clearance at a cost for virulence. The time to clearance affects the parasite's fitness, since it determines how long the infection lasts and, therefore, for how long the parasite can grow, multiply, and transmit. At the same time, the timing of these events (transmission, clearance, virulence) determines the relative costs and benefits. The sensitivity of parasite fitness towards changing these parameter values is given by how much the parasite's fitness would change if, for example, the evasion mechanism was activated earlier or later, or if the associated virulence increased. Because no universal trade-off structure for all parasites or hosts is assumed, the sensitivity framework is more general than the standard evolutionary theory of virulence. For any real case, it would be a formidable task to quantify the sensitivities for various parameters.

The sensitivity framework nevertheless can provide some general insights. Significantly, it clarifies that avoiding clearance has different consequences for parasite fitness from enhancing transmission. For instance, when the parasite activates an evasion mechanism that reduces clearance at time, t, it extends the lifetime of the infection by an amount, x. Thus, the parasite is not only more likely to be alive at time $t+x$, but it is also more likely to continue replicating and to accumulate the benefits of transmission during the interval x (Figure 13.12a).

Such a 'carry-over effect' does not emerge when, instead, the evasion mechanism increases transmission. In this case, a manipulation at a time, t, only benefits the parasite at that time by enhancing transmission, and will not carry over into the future (Figure 13.12b). When both mechanisms generate the same virulence after the same time lag, d, both actions carry the same cost. However, the benefits of avoiding clearance are more considerable and,

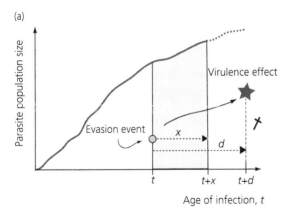
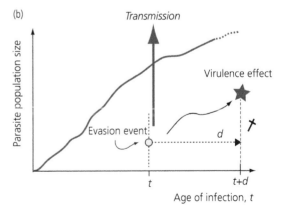

Figure 13.12 Timing of immune evasion events and parasite virulence. (a) An immune evasion event at time t (green dot) reduces clearance and thus prolongs the lifetime of the infection by an interval, x (green interval). The associated virulence appears at time $t + d$ (red asterisk). Avoidance of clearance increases growth and transmission during the entire interval x (green area). (b) An immune evasion (blue dot) increases transmission at time, t, and causes virulence at time $t + d$ (red asterisk). In this sketch, the transmission benefit of evasion (blue arrow) only occurs once, at time t. For simplicity, the time lag d is the same for both cases. Adapted from Schmid-Hempel (2008), with permission from Elsevier.

hence, this strategy is more likely than evasion to evolve to increase transmission. Likewise, matched for the benefits, an evasion mechanism that avoids clearance will be selected for, even if it has more severe consequences for virulence, as compared to a mechanism that increases transmission.

Therefore, the evolution of virulence, if generated by evasion mechanisms, should primarily be driven by mechanisms that avoid clearance, since the parasite will always gain disproportionally more from extending the duration of the infection (Frank and Schmid-Hempel 2008). Note that it will gain relatively little from preventing host death when death occurs near the time when the parasite would anyhow be cleared. Hence, the parasite will be selected to avoid clearance, but not necessarily to avoid host death. This difference is a difference in the sensitivity of parasite fitness towards changes in the life history of the infection.

Besides, the discounted costs of immune-evasion mechanisms will not only depend on the time lag between the activation of the mechanism and the onset of virulence effects but also how widely the virulence effects spread inside the host. For example, when a parasite manipulates the intracellular vesicle trafficking to avoid being cleared (as done by some trypanosomes, Chapter 8), the associated virulence effects might be immediate and localized. Suppose, instead, the parasite manipulates the cytokine network to avoid being cleared. In that case, the respective virulence effects might be delayed (as it takes a while to disrupt the cytokine signalling at large) but spread widely in the host.

According to the sensitivity framework, we expect evasion mechanisms that have delayed effects (discounted costs) and spread most widely in the host (costs likely to appear somewhere else than at the transmission site) to be associated with higher virulence. Classifying the virulence effects of immune-evasion mechanisms in a scheme of time lag to virulence vs the spread of the virulence effects would thus be essential to understanding the evolution of virulence from this perspective (Figure 13.13).

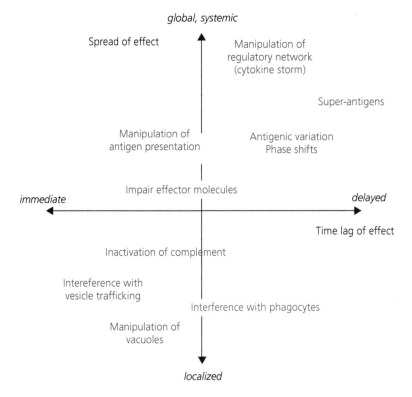

Figure 13.13 Immune evasion mechanisms and effects. Effects may occur long after the onset of the evasion mechanism (time lag of effect). The virulence effects can also spread very widely throughout the host body (global/systemic) or appear more locally (spread of effect). The placement of the mechanisms in this scheme is conjectural but shows how mechanisms can be linked to effects of relevance for infection life history and thus for host and parasite fitness. Adapted from Schmid-Hempel (2008), with permission from Elsevier.

Important points

- Virulence has different meanings but often means parasite-induced host mortality. Virulence results from the interaction of host and parasites, modulated by environmental conditions.
- Virulence can be a non-adaptive side effect of the infection, a result of short-sighted evolution within the host, or negligible. Alternatively, virulence can be an evolved trait that is adaptive for the parasite to increase its fitness.
- The evolutionary theory of virulence is a set of concepts and models that predicts the evolutionarily stable level of virulence based on trade-offs, e.g. between virulence and transmission. It assumes that parasites maximize their basic reproductive number, R_0. Under many conditions, parasites should evolve towards intermediate levels of virulence. However, avirulence or high virulence are also evolutionary outcomes.
- Vertically transmitted parasites should generally be less virulent than horizontally transmitted parasites, but the effects can be more complicated. Among other things, spatial structure (lower virulence) or higher background host mortality (higher virulence) is essential. Non-equilibrium virulence (e.g. during invasions) can take a different course.

- Within-host evolution of parasites affects virulence. In general, multiple infections selects for higher virulence, whereas increased relatedness among co-infecting strains sometimes can lower virulence. With serial passage elsewhere, parasite virulence can be attenuated for the original host, as used for vaccine development.
- Medical interventions, such as vaccination, affect the evolved level of virulence, depending on which component of parasite fitness is targeted. Under some conditions, medical intervention can increase parasite virulence even though the fraction of infected hosts decreases.
- The sensitivity framework is a more generalized theory of virulence evolution. It considers the life history of infection events, e.g. the timing of transmission, clearance, and appearance of virulence effects. A long time lag between early benefits (e.g. transmission) and later costs (e.g. host death by virulence) as well as spatial separation (i.e. systemic effects rather than local) select for higher virulence when matched for the same benefits. When immune evasion causes virulence, avoiding clearance should drive virulence evolution.

CHAPTER 14

Host–parasite co-evolution

Immune systems are among the most complex biological systems and have many functions. The characteristics of immune defences are shaped by several competing needs besides the defence against parasites, such as the control of aberrant cells and tissue regeneration (Saini et al. 2016). Nevertheless, parasites are ubiquitous and can generate strong selection on their host populations. Moreover, parasites and their hosts interact over many generations, and each population responds to selection imposed by the other. Hence, the essential properties of immune defence systems, and likewise the properties of parasites, result from host–parasite co-evolution. Because hosts typically are attacked by more than one parasite strain or species and parasites, in turn, infect more than one host line or species, the co-evolutionary processes can be very complicated.

Furthermore, it is often challenging to identify the most critical parasites that force host adaptations, and, vice versa, the essential hosts of a parasite. Two overlapping domains usually are considered. The evolutionary patterns and processes at the level of species or beyond (genera, families) are known as 'macroevolution'. Those happening within populations belong to 'microevolution'. Microevolution is the core process that, over time, eventually becomes visible as macroevolution.

14.1 Macroevolution

14.1.1 The adapted microbiota

The defensive microbes of social bees illustrate some of the macroevolutionary patterns (Figure 14.1). Among host species, the composition of the microbiota varies widely, and the dissimilarity among the microbiotas from different bee species increases with the phylogenetic distance between these hosts. At the same time, an 'ecological factor'—colony size (the number of workers), which is a generally important characteristic for social bees—also matters, as species with larger colonies have diverse microbiotas.

These patterns reflect the continuous gain and loss of microbes to and from the microbiota while a host line evolves, with a core microbiome more or less keeping its composition. Along the way, microbes can evolve to become defensive symbionts and part of the microbiota, especially when transmitted vertically, where spatial structure connects host and microbes, or where horizontal transmission co-varies with benefits (Shapiro and Turner 2014). Indeed, the microbiota has many characteristics favourable to its own evolution, and for co-evolution with the hosts. For example, the microbiota provides benefits for defences, and a core fraction transmits to offspring of the current host, therefore adding to heritable genetic variation for selection to act and evolution to happen. Benefits and the close association over time by vertical transmission also favour the evolution of close physiological-molecular interactions in intracellular or within-body niches (Vorburger and Perlman 2018).

The evolution of the microbiota shows the same patterns known for any organism. Hence, there is convergent evolution in different areas, or among different host clades (e.g. microbiota in cichlids of different African lakes; Baldo et al. 2017a), and they trace their hosts' phylogeography (as in lizards on islands; Baldo et al. 2017b). Whether microbiome composition primarily is driven by parasitism is not always clear. However, there is a correlation of composition with the presence or absence of parasites,

Evolutionary Parasitology: The Integrated Study of Infections, Immunology, Ecology, and Genetics. Second Edition.
Paul Schmid-Hempel, Oxford University Press. © Paul Schmid-Hempel 2021.
DOI: 10.1093/oso/9780198832140.003.0014

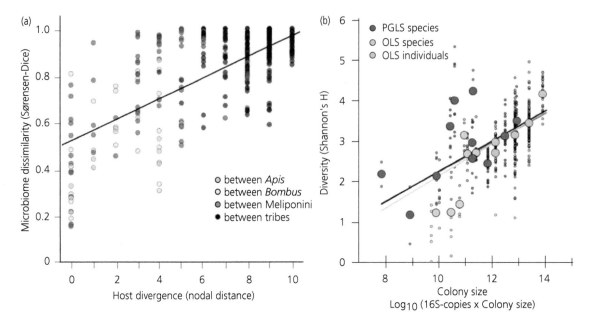

Figure 14.1 Adaptive and ecological patterns in microbiota of social bees. (a) Phylogenetic distance (nodal distance) correlates with dissimilarity in the composition of the microbiota ($r = 0.782$; Sørensen–Dice distance for presence/absence data). Distances are within a genus (*Apis*, *Bombus*), family (Meliponini), or between bee tribes (see legend). (b) Colony size of social bees (x-axis; number of bacterial 16S RNA copies weighed by number of workers) correlates with 'species' diversity of the microbiota (y-axis; Shannon–Wiener diversity index, H, for OTUs). Social bees with large colonies also offer a large gut volume for colonization by microbes. Correlations (see legend) were calculated with ordinary least squares (OLS; black line; $r^2 = 0.416$), and when corrected for phylogeny (phylogenetic generalized least squares, PGLS; grey line; $r^2 = 0.235$ for individuals). Adapted from Kwong et al. (2017), with permission from the American Association for the Advancement of Science.

e.g. among various insect host species and their parasitoids (Hafer and Vorburger 2019). In the example of *E. faecalis*, evolving to protect its host, *Caenorhabditis* (see Figure 12.12), the underlying (microevolutionary) process was in line with a fluctuating selection dynamics between the defensive microbe and the pathogen, *S. aureus* (Ford et al. 2017).

14.1.2 Co-speciation

A speciation event in the host population results from a reproductive barrier that cuts off the gene flow among subpopulations. In many instances, this will also separate the two parasite populations in the newly separated entities. Host separation sets the stage for the parasite populations to evolve themselves towards different species, each population thereby tracking its hosts. If so, each new

species of parasite will become more host specific. When such speciation events repeat themselves, and in turn dominate parasite evolution, a set of host species with their associated, specific parasite species will form. Such a pattern of strict co-speciation is known as the 'Fahrenholz rule'.

Alternatively, the parasites might still transmit with sufficient frequency between the two new host species, e.g. using a vector. In this case, the parasites are less likely to speciate themselves because selection to survive and reproduce in either host can still prevail. Nevertheless, many other evolutionary outcomes are possible (Figure 14.2). Because the events themselves can no longer be observed, evolutionary histories need to be identified from the extant macroevolutionary patterns of host–parasite associations. The methods of phylogenetic reconstruction (as also discussed in the context of reconstructing epidemics, see section 11.7) provide the

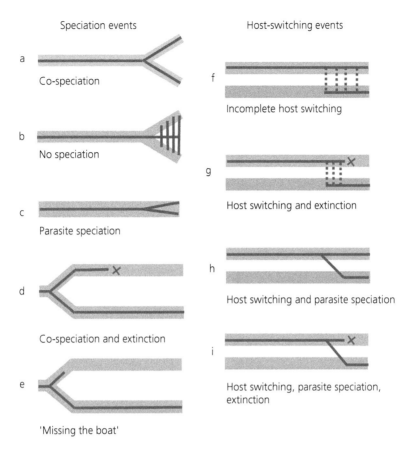

Speciation events

Host-switching events

a

Co-speciation

f

Incomplete host switching

b

No speciation

g

Host switching and extinction

c

Parasite speciation

h

Host switching and parasite speciation

d

Co-speciation and extinction

i

Host switching, parasite speciation, extinction

e

'Missing the boat'

Figure 14.2 Phylogenetic patterns of host–parasite associations. Hosts lineages are shown as grey lanes, parasite lineages as red lines. *Left panel*: Evolution with host speciation events. (a) Strict co-speciation (Fahrenholz rule). (b) Association, but parasites do not speciate. (c) Association with a host line, and speciation restricted to the parasite. (d) Co-speciation followed by extinction in one host. (e) Co-speciation (parasite evolves into a 'chrono-species' that is soon extinct), but parasite follows only one of the hosts ('Missing the boat'). *Right panel*: Evolution with host switching by parasites. (f) Incomplete switching, such that the parasite uses both hosts. (g) Host switching followed by extinction in the former host. (h) Host switching with subsequent speciation of the parasite. (i) Host switching with speciation of the parasite, followed by extinction in the former host. Reproduced from Johnson et al. (2003), with permission from Oxford University Press.

respective toolbox. Molecular clocks can add a time scale to the various events, such that a more concise picture emerges.

The study of host–parasite co-evolutionary scenarios and the reconstruction of their history nevertheless pose several technical challenges and some caveats (Dowling et al. 2003; Page 2003). For example, the real association of parasites and hosts might not be detected, due to sampling errors or sampling bias that tends to miss rare parasites or hosts. Parasites could also coincidentally infect a given host. Such 'straggling' can occur, for instance, when a predator preys on the regular host and the

parasite can subsequently survive in the predator; this does not necessarily indicate successful host switching. Furthermore, speciation of hosts and parasites can occur for many other reasons and may not be a co-speciation event. Hence, the exact timing of events in the host and the parasite line is a crucial element in identifying true co-speciation. A problem also arises when parasites are lost from host lineages. Such losses might, in particular, occur when a founder event with a small number of individuals promotes host speciation. Because of the small founding population, the parasites might be absent from the founders only by chance ('missing the boat'; Figure 14.2). The invasive bumblebee species *Bombus terrestris*, for example, has recently colonized the island of Tasmania with very few individuals (perhaps one or two inseminated queens) and has left behind most of its parasites (Allen et al. 2007).

Empirical studies of co-speciation show all possible kinds of co-divergence, parasite losses, and host shifts as the defining processes. This evidence paints a rather complex picture of host–parasite co-evolution (Table 14.1). For example, among 103 different studies covering many taxa and also involving many host–symbiont associations, only nine cases showed evidence of true co-speciation. In the remainder, there was a mismatch in the time of events, host shifts occurring instead of true co-speciation, or the data was simply inconclusive (de Vienne et al. 2013). Not surprisingly, perhaps, co-speciation was more common for hosts and their symbionts than for hosts and their parasites.

Ectoparasites, especially of rodents and birds, are among the best-studied examples of co-speciation. For example, sound data come from three species of pocket gophers (*Geomydoecus*) and their lice (seven species of *Cratogeomys*). These show co-phylogeny with a corresponding timing of splits in the host and the parasites (Light and Hafner 2007). In birds, cases of co-speciation with lice are numerous. For example, phylogenies are largely congruent for ground doves (Claravinae) and their body lice (Ichnocera, Sweet et al. 2017). Similarly, lice (*Pectinopygus*, 17 spp.) infesting birds have regularly co-speciated with their waterbird hosts (Pelecaniformes, 18 spp.), though not without exceptions (Hughes et al. 2007) (Figure 14.3). Frequent co-speciation is also likely for pinworms (48 spp. of Enterobiinae nematodes) that are parasites of African monkeys (36 spp., Hugot 1999), and for primate viruses that more or less co-speciated with their hosts for at least 30 million years (Switzer et al. 2005).

However, in all of these cases, parasites also often fail to speciate when their hosts speciate. For example, in another group of small mammals (43 spp. of heteromyid rodents) and their lice (of the genus *Fahrenholzia*) (Light and Hafner 2008), a loose co-divergence but not strict co-speciation is much more likely. Such processes generate cases where host and parasite phylogenies become partly incongruent,

as in some of the examples. If failure is frequent, the phylogenies diverge more strongly, e.g. in the case for chewing lice (*Columbicola* spp., Phthiraptera) on doves (Johnson et al. 2003). For lice of mammals, co-speciation on a broader time and taxonomic scale even appears to be uncommon (Taylor and Purvis 2003). Hence, co-speciation occurs, but it is neither the rule nor the only possible outcome.

14.1.3 Host switching

In many cases, the parasites must frequently have switched hosts during evolution (Huyse and Volckaert 2005) (Figure 14.4). Host switching can be more likely than co-speciation, for example, when the parasite is vectored. In this case, the behaviour of the vector determines the transmission within and between host species. Vector transmission can prevent local reproductive isolation and thus impedes co-speciation of parasites. For avian malaria (a total of 181 putative parasite species) in New World birds, host switching is a typical process. It happens at all taxonomic levels, i.e. from between species to between families and orders (Ricklefs et al. 2014). Frequent host switching also is inferred for clades of cichlid fish that undergo adaptive radiation, even though these events vary according to group and locality (Vanhove et al. 2016). From the examples in Table 14.1, host switching exists in any group of hosts and parasites and includes 'small' systems, such as viruses in bats (Streicker et al. 2010), as well as 'big' systems, such as digenean flatworms in whales (Fraija-Fernández et al. 2015).

Ecological barriers are one crucial determinant. For instance, host switching is more common when the hosts share the habitat (Clayton et al. 2004). Similar host body sizes are an essential correlate of success for ectoparasites on different hosts (Clayton et al. 2003), and so forth. Furthermore, predation can facilitate transfer to a new host (Whiteman et al. 2004), and transfer by passive transport will also remove the ecological separation between host species (Clayton and Johnson 2003).

Besides ecological barriers (compare the 'filters'; see Figure 7.4), what other factors can favour host switching? The question is related to what affects the host range and parasite specificity. An obvious

Table 14.1 Host–parasite macroevolution.[1]

Host	Parasite	Finding	Remarks	Source
Drosophila spp.	Nematodes (Tylenchidae; *Howardula*).	A few co-speciation events. Frequent host switching. Repeated host colonization in shared breeding sites.	Host feeds on mushrooms. Nematodes potentially have large host range. Based on molecular data.	13, 14
Fish (various families)	Copepods (26 spp., *Chondracanthus*).	Co-speciation common, host switching rare.	Co-speciation, especially occurring within fish orders.	11
Fish (8 spp., Paralichthydae)	25 virus isolates (causing lymphocystis).	Rare co-divergence, many duplications and sorting events.	Divergence of hosts and viruses seems entirely independent.	19
Frogs and toads (23 spp.)	Monogenean flatworms (26 spp.).	Many co-divergences, duplication, and extinction events, few host shifts.	Grand geological scale. Timing of events may not fit co-divergence.	1
Lice in several groups of birds: (a) 21 spp. of Passeriformes (b) 13 spp. of doves (c) Flamingos, ducks	(a) 15 spp. of lice (*Brucelia*). (b) Wing (*Columbicola*) and body lice (*Physconelloides*). (c) 43 genera of wing lice (*Anaticola*).	(a) Seven possible co-speciation events, but phylogenies not congruent. (b) Many co-speciation events inferred. (c) Co-speciation, host switching, loss events.	(a) *Brucelia* considered host specific. (b) Congruence of phylogenies likely, but other scenarios possible. (c) Timing unknown.	(a) 7, 8 (b) 2 (c) 9
Galapagos dove (*Zenaida galapagoensis*) Galapagos hawk (*Buteo galapagoensis*)	Chewing lice (Phthiraptera): *Columbicola*, *Physconelloides*.	Frequent straggling of lice from dove to hawk.	Differences in prevalence of lice on hawks driven by opportunities for switch.	18
Birds (29 spp., from different families)	Feather mites (26 spp., Avenzoariinae).	Co-speciation, but also some secondary extinctions (sorting), duplications, and a few host switches.	Host switches more common when hosts share habitat, but timing not clear.	3, 4
Sea birds (Procellaridae)	Lice (39 spp.).	Co-speciation and host switching common.	Some clades with high fidelity to hosts. Sequence divergences in host and parasites match.	10
Birds, 79 spp. from 20 families	Malaria: *Plasmodium*, *Haemoproteus* (68 lineages).	Co-speciation more common than expected, but also frequent switching and extinction. New hosts frequently colonized by different parasite lineages.	Parasites rather specific to host family; probably follows acquisition of new host.	16, 17
Small rodents (various subfamilies, 95 genomes)	Hantavirus (65 lineages).	Mostly host shifts, and some degree of co-divergence.	Timing of lineage splits overlap to some degree.	15
Vertebrates (birds, mammals)	72 sequenced viruses (Polyomaviridae).	Common co-divergence; many losses, duplication events, and few host switches.	Avian viruses have broader host ranges.	12
Humans, carnivores, ungulates	Cestodes (*Taenia*).	Repeated host switch from carnivores to humans. Switches pre-date domestication of animals.	Host switches associated with changes in human diets. *Taenia* uses ungulates as intermediate, and carnivores as definitive hosts. Humans can have both roles.	6

[1] Further information in Ref. 5.

Sources: [1] Badets. 2011. *Syst Biol*, 60: 762. [2] Clayton. 2003. *Evolution*, 57: 2335. [3] Dabert. 2003. *Acta Parasitologica*, 48: S185. [4] Dabert. 2001. *Mol Phylogen Evol*, 20: 124. [5] de Vienne. 2013. *New Phytol*, 198: 347. [6] Hoberg. 2001. *Proc R Soc Lond B*, 268: 781. [7] Johnson. 2002. *Biol J Linn Soc*, 77: 233. [8] Johnson. 2003. *Syst Biol*, 52: 37. [9] Johnson. 2006. *Biol Lett*, 2: 275. [10] Page. 2004. *Mol Phylogen Evol*, 30: 633. [11] Paterson. 1999. *Syst Parasitol*, 44: 79. [12] Pérez-Losada. 2006. *J Virol*, 80: 5663. [13] Perlman. 2003. *Evolution*, 57: 544. [14] Perlman. 2003. *Mol Ecol*, 12: 237. [15] Ramsden. 2008. *Mol Biol Evol*, 26: 143. [16] Ricklefs. 2002. *Proc R Soc Lond B*, 269: 885. [17] Ricklefs. 2004. *Syst Biol*, 53: 111. [18] Whiteman. 2004. *Int J Parasitol*, 34: 1113. [19] Yan. 2011. *Virus Genes*, 43: 358.

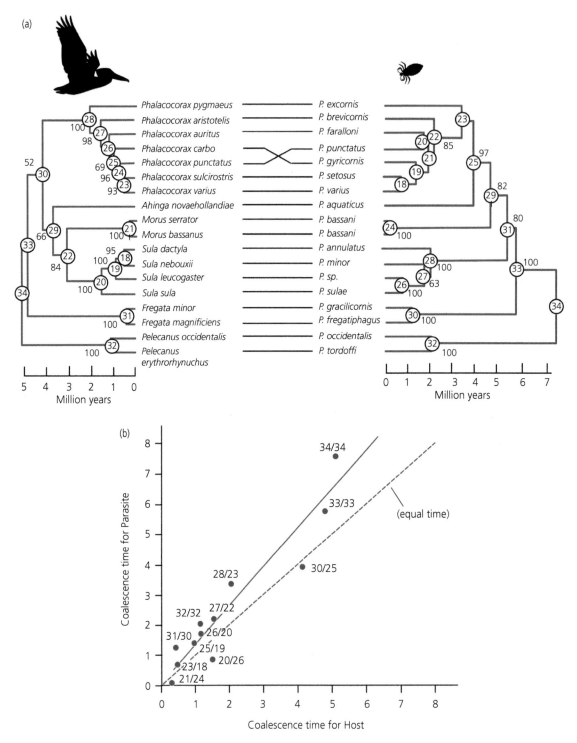

Figure 14.3 Co-speciation (Fahrenholz's rule). (a) Phylogenetic trees for hosts (18 spp. of Pelecaniform birds, left) and their lice (17 spp. of *Pectinopygus*, right). Trees were based on molecular data, constructed with maximum parsimony; small figures give percentage bootstrap support for the nodes (numbers in circles). (b) Estimate of the time since the corresponding splits of the evolutionary lines (coalescent time; in millions of years). The dots correspond to the phylogeny in (a), with number pairs listing the involved nodes. The overall regression line (solid line) is not different from unity (dashed line; same time for both parties), indicating that splitting events in host and parasite phylogenies are congruent in time and therefore represent co-speciation events. Silhouettes from phylopic.org. Adapted from Hughes et al. (2007), with permission from Oxford University Press.

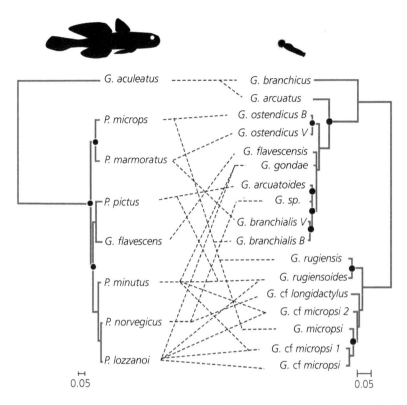

Figure 14.4 Frequent host switching. Phylogenies of host fish (gobies, left) and their trematode parasites (*Gyrodactylus*, right). Black dots refer to apparent co-speciation events, yet the timing of events is not congruent. Data were collected in Belgium and Italy; trees are based on 12S and 16S mtDNA, and on 18S rRNA in the parasites. Horizontal branch lengths represent magnitude of evolutionary change (host and parasite rates of 1 per cent and 5.5 per cent per million years). Dotted lines connect hosts with parasites. Silhouettes from phylopic.org. Adapted from Huyse and Volckaert (2005), with permission from Oxford University Press.

candidate is a phylogenetic distance between hosts. Close relatives of a current host are often more susceptible to the same parasite. Rabies viruses of bats, for example, seem constrained in this way, as both host shifts and cross-species transmission decline with increasing phylogenetic distance of their hosts (Streicker et al. 2010). A similar pattern exists for many other viruses of primates and humans, a finding corroborated by the observation that overlapping geographical ranges are necessary (Davies and Pedersen 2008).

The phylogenetic 'distance effect' exists in all major host and parasite groups, e.g. protozoans in New World monkeys (Waxman et al. 2014), plants and their fungi (Gilbert and Webb 2007), beetles and bacteria (Tinsley and Majerus 2007), fruit flies and viruses or nematodes (Perlman and Jaenike 2003;

Longdon et al. 2011); or in a study with 793 parasite species belonging to various groups and infecting a total of 64 mammalian carnivore species (Huang et al. 2014). However, the 'distance principle' is not universally true. Notably, many parasites have evolved to become host generalists and can switch among large host phylogenetic distances, being more driven by ecological than 'physiological' factors. Generalism occurs in a range of further parasites of apes and monkeys, where host distance does not explain the patterns well (Cooper et al. 2012a). Several specific reasons can be important for this. Viruses, for example, may readily evolve to use a different host receptor for cell entry and thus can enter a novel host as soon as ecological barriers are removed. This probably happened in the recent SARS-CoV-2 pandemic, where the virus evolved to enter via human ACE receptors in the upper respiratory tract as soon as it came in contact with the human population (or with a physiologically similar bridge host). Interestingly, parasites with an indirect life cycle (requiring intermediate hosts) may switch hosts as readily as those with a direct life cycle (Lymbery et al. 2014).

Ecology and phylogeny, therefore, are strong predictors of host switching. On the ecological side, there are overlapping host ranges, similar niches, or transmission via generalized vectors and ubiquitous phoresis that matter (Harbison and Clayton 2011). On the other, phylogenetically close distance facilitates host switches (e.g. for monogeneans infesting neotropical fish; Braga et al. 2015). These two main dimensions may also be important during different episodes. For example, when host species remain geographically isolated, co-speciation may be predominant, whereas, during periods of host geographical expansion and radiation, host switching would prevail (Hoberg and Brooks 2008). Theoretical models do underpin the importance of both phylogenetic distance (Foster 2019) and ecological constraints (Araujo et al. 2015).

Nevertheless, the tempo and mode of host–parasite co-evolution at the macroevolutionary scale remain a significant challenge to understand. Understanding factors that prevent or facilitate host switching is of high relevance, in particular in connection with the emergence of novel infectious human diseases from an animal reservoir (zoonoses; see section 15.5).

14.2 Microevolution

Microevolution happens in populations, based on the principles of Darwinian evolution by natural selection. The study of microevolution highlights how parasites can select for characteristics of their hosts and, vice versa, how hosts select for characteristics of the parasites. Many of these elements are discussed in other chapters. For example, the evolution of antibiotic resistance is an iconic case of host–parasite co-evolution that quickly happens within populations of pathogens (see section 12.3.3). At the level of within-host processes, microevolution connects the disease spaces of hosts and how they are utilized by the co-evolving parasites (Box 14.1).

Box 14.1 Co-evolution and disease space

Host–parasite co-evolution is—by definition—antagonistic. At the same time, each party lags behind the other, since selection can only proceed when the host (or parasite) population is exposed to a changed trait in the other party. Evolutionary change becomes visible in the offspring of successful parents, i.e. in the next generation. Selection also changes the disease space.

As illustrated in the hypothetical example of Figure 1, a parasitic infection initially may follow a given trajectory through the host disease space (1). In a first selective episode, the host population is selected to resist more or to have more tolerance to the infection. As a consequence, the changed defence mechanisms generate a new disease space for a parasitic infection; for example, it may reduce the length of the trajectory in the 'sick' compartment and force the parasite to the recovery area, where it is cleared. The infection, therefore, no longer reaches its zone of highest fitness (2). This zone is a (hypothetical) area in the disease space where the parasite population can, for example, produce many transmission stages.

Hence, in a further step (3), the parasite population adapts in turn and now follows a new infection trajectory that can reach the current zone of high fitness. This trajectory makes the host sick again and reduces its fitness. Therefore, in the subsequent episode (4), the host population adapts again and modifies its defences. This adaptation results in defence mechanisms that, for example, can nudge the 'sick' zone of the disease space somewhat away from the current infection trajectory. This modification increases host fitness but can also dislocate the zone of high fitness for the parasite.

In this hypothetical example, this high-fitness zone has now moved into a dangerous corner of the disease space. If the parasite population would subsequently be able to adapt to this new situation, it could generate a new infection trajectory into this danger area. As a macroscopic consequence, the parasites would have become more virulent for the host. At the same time, comparing situation (1) with situation (4), co-evolution over several episodes has led to new mechanisms for infection in the parasite, and defences in the host population. These new mechanisms become visible in changed infection trajectories and changed disease spaces; that is, both parties have evolved. Although trajectories and disease spaces kept changing, the macroscopic result, e.g. parasite virulence, may remain the same.

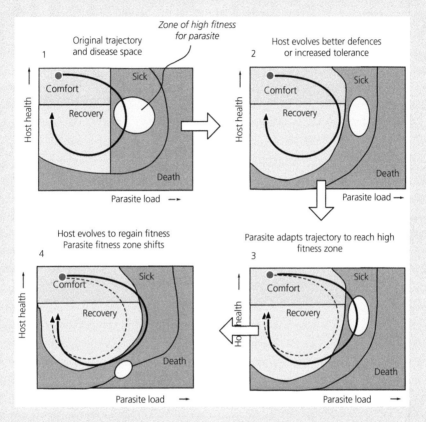

Box 14.1 Figure 1 Co-evolution and disease space. The panels (1) to (4) refer to successive selective episodes on host or parasite populations. Refer to text for description of the scenario. Solid line is the infection trajectory, with red dot as infection point; broken line is original trajectory from (1). Blue zones are where the parasite reaches high fitness.

14.2.1 Co-evolutionary scenarios

The dynamics of host–parasite co-evolution reflects how selection operates on the two parties. The following scenarios often are discussed.

14.2.1.1 Selective sweeps

Antagonistic co-evolution can proceed in a series of 'selective sweeps'. With a selective sweep, a particular allele (for example, coding for resistance) is under strong, directional selection. As a consequence, the allele rapidly rises in frequency and reaches fixation at some point, i.e. displaces all other variants; the population has become monomorphic.

Selective sweeps in the context of host–parasite evolution are well documented. They differ from 'normal' evolutionary changes in the speed and scope of the dynamics, but not by principle. For example, when host defences adapt to the parasites, favourable parasite alleles, in turn, may quickly rise in frequency and sweep through their populations. These alleles subsequently may persist in the population or go extinct again, before the next sweep sets. Hemagglutinin, one of the key proteins of human influenza A virus, illustrates the case. Host adaptation here means change by the naturally acquired host immune memory or vaccination that defends against a limited set of influenza strains. As a consequence, viruses with different hemagglutinin variants replace each other in rapid succession (Figure 14.5).

A series of selective sweeps in rapid succession could also merge with a polymorphism in the population; for example, when the previous sweep has not yet reached fixation at the time when the next sweep is underway, or in reciprocal sweeps at different speeds. Then, 'old' and 'new' alleles for host resistance are both present and can persist, as in *Drosophila* being selected to resist a virus (Wilfert and Jiggins 2012).

More generally, directional, positive, purifying selection (as needed for selective sweeps) on host or parasite genes is frequent. Directional selection, for instance, has led to the diversification of the members within gene families for host defence (Stotz et al. 2000; Ford 2002). Directional selection is also

suggested for RNAi genes that function in antiviral defences of *Drosophila* (Obbard et al. 2006).

The effect of selective sweeps might furthermore combine with, or be superimposed to, other processes. For example, the R-genes of plants are involved in recognizing pathogens. They evolve very fast, indicating strong positive selection. However, within-species polymorphism remains common; hence, selection does not merely occur in the form of successive selective sweeps (Bergelson et al. 2001). Similarly, the exceptionally high polymorphism of TEP1 (thioester-containing protein 1), which is a critical defence gene of the vector *Anopheles gambiae* against infections by *Plasmodium*, indicates balancing selection. However, a closer inspection suggests otherwise. Rather, TEP1 diverged first by selection and became secondarily polymorphic by gene conversion. Selective sweeps in TEP1 nevertheless still occur and have changed allele frequencies in natural populations recently (Obbard et al. 2008).

In all, immune genes show signs of selective sweeps more often than genes with other functions

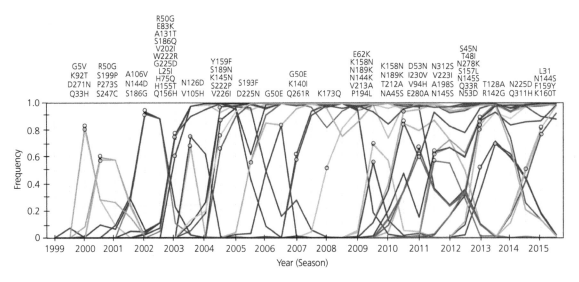

Figure 14.5 Selective sweeps. Shown are the frequencies of different variants of the seasonal H3N2 influenza A virus hemagglutinin over the years 1999 to 2015 (coloured lines). In the successive sweeps, some variants rise to fixation (i.e. from zero to one in frequency). Others rise to intermediate frequencies, are selected against, and become lost again from the population. The positions of amino acid changes within the genome that undergo selective sweeps are indicated by their acronyms above the panel. Their respective first emergences are indicated by the dots. Adapted from Klingen et al. (2018), under CC BY 4.0.

(Hurst and Smith 1999; Schlenke and Begun 2003). The HapMap project also suggests that many genomic regions in humans show signs of recent selective sweeps, in particular loci known to be associated with disease risks and host defences (Nielsen et al. 2005). Especially for small host population sizes, however, mutations that trigger a selective sweep might not appear frequently enough to outpace the rapidly evolving parasites. Alternatively, frequent immigration by gene flow could fuel the selective sweeps.

14.2.1.2 Arms races

In an arms race, both host and parasite are under directional selection. As a result, host resistance and parasite infectivity both tend to increase over time,

or the abundance and diversity of parasites select for an increase in the defence repertoire (Figure 14.6). Arms races can be based on a 'slow' process, where allele frequencies persistently and continuously change, or they could result from a series of selective sweeps in one or both parties. In other words, an arms race is a phenotypic view of evolutionary changes, whereas the underlying genetics can be of various kinds. Typically, an arms race is noticed when the phenotypic characters increase over time (Figure 14.7).

14.2.1.3 Antagonistic, time-lagged fluctuations (Red Queen)

This scenario has received much attention, since it combines several genetically essential elements. For

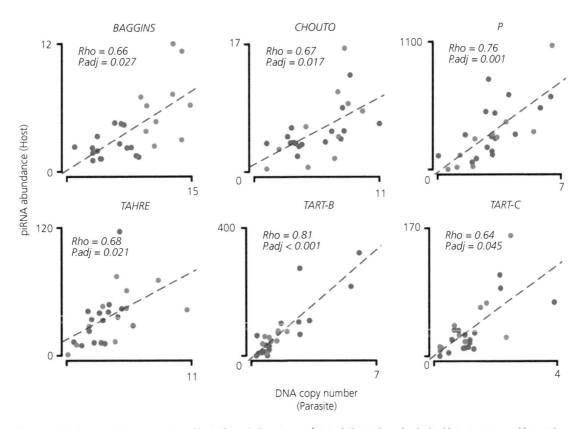

Figure 14.6 Arms race between parasite and host. Shown is the outcome of co-evolution at the molecular level between transposable genetic elements ('parasites', similar to viral challenges; names above each panel) and the defence machinery in *D. melanogaster*. The host's piwi-interacting piRNAs repress transposable elements. Across a worldwide collection of *Drosophila* strains, an increase in the copy numbers of transposable elements (x-axis, DNA copy number) is matched by an increasing abundance of piRNAs, as expected from an arms race scenario. Red dots are DGRP strains, green dots GDL strains of the host; Spearman's rho is given. Reproduced from Luo et al. (2020), under CC BY 4.0.

the process, some kind of 'matching' between host and parasite genotypes is assumed (see section 10.5). Hence, a given parasite type can only infect certain host types; likewise, host types are resistant to some but not all parasite types. Furthermore, the parasite population should adapt more rapidly to the prevalent genotypes in the host population than vice versa. As a consequence, the parasite population rapidly evolves and becomes fitter on the more common host types, while the frequency of host types changes only slowly during this episode.

This asymmetric process gives rare host types an advantage because the parasites have not adapted well to them. This pattern is known as 'negative frequency-dependent selection' for the host types. Over time, therefore, the common types decline in frequency, while the rare types become more frequent until they, in turn, become common enough for the parasites to adapt to them rapidly. Inevitably, though, there is a time lag in this co-evolutionary process, as parasites 'run behind' the changing host types. At any one time, the current parasite population reflects the composition of the host population to which it has adapted in the recent past. With such 'time-lagged, negative frequency-dependent selection', parasites perpetually track their host populations through genotype space, a process that is known as 'Red Queen' dynamics (Figure 14.7) (Box 14.2).

14.2.1.4 'Evolution-proof' strategies

Would it be possible to utilize the underlying principles of host–parasite co-evolution for better treatments or vaccination strategies? In essence, for this, parasite evolution, as induced by the use of antibiotics or insecticides, should be decoupled from host evolution. It is, of course, impossible to halt evolution completely. However, at least the speed of adaptation could be slowed down. If successful, such 'evolution-proof' strategies would allow the use of antibiotics or insecticides for much longer before resistance evolves. Several practices already take this route, e.g. the breeding of more resistant crops, or procedures to lower mutation rates of microbes. The latter reduces genetic variation and, thus, the response to selection. Further methods include the fragmentation of populations, such that stochasticity and (random) genetic drift overshadow selection. Also, the artificial selection for traits that compromise increased resistance could be considered, e.g. by staging competition (Wale et al. 2017). A range of less practicable measures have been suggested, too (Bull and Barrick 2017).

However, the modern techniques of genome editing now open up unprecedented possibilities to implement evolution-proof mechanisms. Examples include the genomic engineering of viral sequences such that naturally favoured codons are distorted while maintaining the genomic sequence (known as 'codon deoptimization'). This slows the speed of

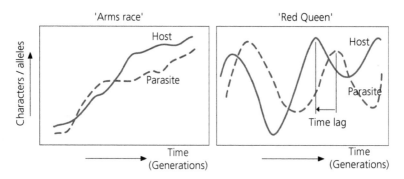

Figure 14.7 Co-evolutionary scenarios. *Left*: In an 'arms race' between host and parasite, both parties are under persistent directional selection in response to changes in the other party. As a result, each party changes in one direction; here, the characters increase over time, e.g. with increased resistance and infectivity. The process leads to loss of polymorphism; new variants only emerge by new mutations; arms races typically refer to phenotypes. *Right*: With a 'Red Queen' dynamics, time-lagged fluctuating antagonistic selection continuously changes direction in response to previous changes in the other party. In this scenario, previous variants (e.g. alleles) are 'recycled' after a while, such that they do not disappear but persist in the population indefinitely. The process maintains diversity and (genetic) polymorphism over time; the scenario typically is studied at the genotypic level.

Box 14.2 History of the Red Queen hypothesis

Species and higher taxa as a whole show survival curves with a constant risk of extinction over millions of years, rather than showing signs of ageing, i.e. increasing extinction rates with an increased life span of the clade (as individuals do). Leigh Van Valen (1973) used the metaphor 'Red Queen'[1] for his 'new evolutionary law' to explain this observation. According to his law, the constant rates of extinction are generated by the ongoing evolution among competing species, with a given species losing its 'race' and going extinct at an arbitrary time, i.e. when it happens to become outpaced by competitors and enemies. Graham Bell adopted this term for the specific ecological interaction between hosts and parasites, and also reduced it to microevolutionary processes (Bell 1982). This covers the idea of genotype frequencies changing through time. In a closed system (i.e. where no new genes appear), genotype frequencies would thus tend to oscillate in hosts and parasites. Both populations are 'running' in genotype space just to escape the other party and avoid extinction. True to the original metaphor, they remain in the same place, as both can persist. Note that population persistence is a consequence of selective processes at the level of individuals, not necessarily a selection for the purpose of population survival.

Relation to the evolution of sex. Probably the first reference to the possibility that the Red Queen processes based on host–parasite interaction could lead to selection for sexual reproduction came from William D. Hamilton's (1975) review of Michael Ghiselin's (1974) and George Williams's (1975) books on the evolution of sex. His review contained these quotations: 'it seems to me that we need environmental fluctuations around a trend line of change' and 'for the source of these we may look to

fluctuations and periodicities…generated by life itself' (Hamilton 1975, p. 366).

Hamilton clearly is referring to species interactions leading to oscillations in selection, but, as he later stated (Hamilton 2001), he was not thinking of host–parasite interactions at that time. Hamilton did later, however, become very focused on the relevance of parasites for generating periodicities in selection (Hamilton 1980; Hamilton et al. 1990). Remarkably enough, J. B. S. Haldane, 30 years before, was already thinking of parasites as a significant selective factor to maintain diversity in host populations (Haldane 1949). In his time, however, why sexual reproduction evolved and is maintained, was not yet connected to parasites.

John Jaenike (1978) presented the first explicit outlay of parasites as a selective agent for the selection of sexual reproduction. He borrowed heavily from the work of Bryan Clarke, who was among the first to recognize the possibility of genetic oscillations in species interactions (Clarke 1976). Jaenike realized that these kinds of genetical dynamics readily could be generated in host–parasite interactions. Moreover, they could lead to the build-up of linkage disequilibrium, which would negatively affect fitness in the near future. Sex (and recombination) reduces such linkage. Outcrossing, i.e. biparental contributions, could thus be favoured over uniparental forms of reproduction. Essentially, parasites create the kind of environment that John Maynard Smith (1978) had reasoned would select for recombination. This is an environment that alternates predictably, so that previously selected genotypes would be less fit in future generations and become selected against.

A first formal model eventually was developed by Hamilton (1980). He showed that the geometric mean fitness of a sexual population under selection by parasites is more than two-fold higher than that of an asexual population. The factor of two would make up for the 'two-fold cost of sex' as defined by Maynard Smith (1978). However, his analysis required highly virulent parasites, i.e. large selection coefficients on the host (May and Anderson 1983a; Howard and Lively 1994; West et al. 1999). Since then, a broad diversity of theoretical models has developed.

---

[1] In Lewis Carroll's novel 'Through the Looking Glass', Alice meets the Red Queen in a magic forest. To Alice's amazement, all the trees around her stay in the same place, even though she and the Red Queen run as fast as they can. So, the Red Queen explains 'see, in this place it takes all the running you can do, to keep in the same place'. This metaphor covers the idea that a population should evolve as fast as possible just to hold out against its co-evolving enemies and competitors.

---

translation and leads to attenuation of the virus; the altered viruses also only slowly revert to wild type (Coleman et al. 2008). Similarly, use can be made of the co-regulation of different genes, e.g. by operons in bacteria, which naturally shields against mutations that would change the regulation of the entire

set. Genetic engineering now allows genes of interest to be coupled to undesired ones in a co-regulated set. With the construction of bidirectional promoters, the set is to some degree self-protected against evolutionary changes by its complementary in reading direction (Bull et al. 2012; Yang et al. 2012).

However, even without reverting to sophisticated genomic technologies, the application of the principles of evolution can slow down adaptation in pathogens. For example, the suppressive effect of sugar feeding on parasite development within the fly can be exploited, or the repressive effects of the symbiont, *Wolbachia*; all of which does not directly select for the resistance of the pathogen itself and thus slows adaptation (Huijben and Paaijmans 2018).

An illuminating scenario is the targeting of insecticides against older mosquitoes. This slows the evolution of resistance, based on a rationale as follows (Read et al. 2009). After mosquitoes have taken up malaria parasites, it takes several gonotrophic cycles to become infective. With the high natural death rate, however, most mosquitoes do not live long enough for the parasite to complete these cycles and, therefore, do not become infective. Applying selection on those relatively few mosquitoes that live long enough to become infective is a desirable strategy. On the one hand, this will remove only a small fraction of the mosquito population; therefore, it reduces the force of selection and slows adaptation in the mosquitoes. On the other, most other mosquitoes will already have laid the vast majority of eggs in the population. Because these parental mosquitoes have not been under selection yet, they cannot transmit adaptations to the next generation. Therefore, this strategy slows down evolution by putting selection not on the bulk of the reproducing population, but only on a small minority that happens to be infective as vectors. For the practical implementation, insecticides that act late in a mosquito's life cycle are the choice. Existing insecticides deployed in lower concentrations can, for example, mostly kill older and weaker females (Glunt et al. 2011). Several other methods were suggested, too.

14.2.2 Parasites and maintenance of host diversity

The observation of extensive population-wide biochemical polymorphism prompted John B. S. Haldane in 1949 to suggest that antagonistic co-evolution with parasites maintains diversity in host populations. In J. B. S. Haldane's words: 'to put the matter rather figuratively, it is much easier for a mouse to get a set of genes which enable it to resist *Bacillus typhimurium* than a set which enable it to resist a cat' (Haldane 1949).

Whether or not this statement is true, the host–parasite interaction has a genetic underpinning, as discussed in Chapter 10. Plasticity in the expression of the genotype and environmental variation might blur the effect of direct genetic interactions, and so reduce the evolutionary effects of host–parasite specificity (Mitchell et al. 2005). Nevertheless, selection by parasites is a pervasive effect. Furthermore, if co-evolving parasites select for diversity in their host populations, the same reasoning will apply for the parasite population under selection by their hosts. The general problem of the maintenance of diversity unites, in a nutshell, many topics of evolutionary parasitology.

14.2.2.1 Host–parasite asymmetry

Because parasites typically have larger population sizes and shorter generation times, parasite populations should adapt to hosts faster than the host population to the parasites. This widely accepted argument is not universally applicable, however. Many parasites only become transmitted when their host dies, as is the case for parasitoids and other obligatory killers. Similarly, parasites, such as helminths or some protozoa (e.g. *Plasmodium*, trypanosomes), stay inside the host and produce chronic infections that are comparable in duration to the lifetime of the host, even though there is an evolutionary change of the parasite population within the host. In these cases, the generation times of host and parasite might not be very different.

At the same time, the numerical population size of parasites like bacteria or viruses can indeed be huge (and reach billions of cells or particles). However, the genetically 'effective population size', which determines the pace of adaptation (see section 10.7), usually is much smaller. Where hosts are infected by only one parasite, or only a single parasite strain survives to be transmitted, the effective population size of the parasite can be minimal, indeed. Host populations, too, can effectively be much smaller than the census size (i.e. the number of host individuals); for example, when a selective sweep reduces the genetic diversity and thus the genetically effective size. Hence, the divergence in the evolutionarily relevant population sizes between hosts and their parasites needs clarification for each case.

However, a basic, significant asymmetry persists: parasites—by definition—depend on their host but not vice versa. Therefore, selection on the parasite is usually stronger than selection on the host. This has been described as the 'life-dinner principle' (Dawkins and Krebs 1979). Note that this asymmetry sets a potential for evolutionary change, but it does not imply that parasites will always be ahead in the evolutionary race. For example, hosts might keep up in this struggle and escape their parasites by other means, such as by rapidly changing receptor identities.

14.2.2.2 Red Queen and host diversity

A remarkable property of the Red Queen dynamics is that it can maintain genetic polymorphism in the population over time. As just described, rare host genotypes are protected from extinction because selection against them is weak. Overall, a diversity of types is, therefore, maintained in the population.

Given the epistatic effects of host defence genes (see section 10.5.1), selection by parasites additionally generates linkage disequilibria. These fluctuate over time, with the rise and fall of different genotypes. With the Red Queen dynamics, the population does not converge to one, best genotype (for example, the one with lowest mutational load or the best combination of genes). Instead, it will evolve towards containing an array of different genotypes. Hence, a 'wild type' never exists—just temporarily favourable gene combinations (Hamilton et al. 1990). The fundamental Red Queen dynamics is also independent of a particular reproductive system. Host and parasite can either be clonal, sexual, or facultatively sexual.

For the Red Queen dynamics to take effect, however, some requirements must be met. For example, there should be some degree of specific interaction between host types and parasite strains. As an aside—in theory, at least, a mixture of frequency dependence and density dependence, i.e. ecological processes, may also generate a fluctuating selection in the absence of any specificity (Best et al. 2017). Nevertheless, a large body of evidence, discussed in Chapters 7 and 10, shows that the specific interactions

between hosts and parasites are indeed strongly affected by the respective genotypes (see Table 10.3; Box 10.4).

An advantage for rare host genotypes, implicit in the Red Queen scenario, has empirical support in some but not all cases (Table 14.2). Also, over-infection of common host genotypes is not consistently found either (Jarosz and Burdon 1991; Vernon et al. 1996; Little and Ebert 1999). However, the effects of rare or common types are lagged in time. Hence, the currently rare types could still be heavily infected if they had recently been common, and the currently common types could still be under-infected because they were rare one or two generations ago (Dybdahl and Lively 1995). What exactly is meant by 'lag' and 'rare vs common' is not straightforward to define in empirical studies, where the relevant time axis is not precisely known.

In 'time-shift experiments' these temporal cycles can be separated by infecting parasites from different time points to hosts of the same or different time points (Gaba and Ebert 2009). For example, spores of the bacterium *Pasteuria ramosa* and eggs of the host *Daphnia*, retrieved from pond sediments at different depths, represent temporal archives that can be reactivated and tested. The cycling of host and parasite types thus becomes visible (Decaestecker et al. 2007) (Figure 14.8); the results fit the Red Queen expectations. Similar findings of time-shift changes fit *Tribolium castaneum* infected by *Nosema whitei* (Bérénos et al. 2010a), co-evolution experiments of bacteria and phage (Buckling and Brockhurst 2008; Forde et al. 2008), and the nematode *Caenorhabditis elegans* and its bacterial infections (Papkou et al. 2019).

In the field, direct observation of cycling of genotype frequencies (clonal and sexual) in the New Zealand freshwater snail *Potamopyrgus antipodarum*, infected by trematodes, was observed. The time series show that the most common clonal genotypes in the snail population declined over a ten-year study period. Cross-infection experiments between allopatric and sympatric host–parasite combinations additionally suggest that this decline follows the adaptation of the trematodes to the prevailing host clones (Jokela et al. 2009). A similar observation comes from the flax–fungus system (Thrall et al. 2012). There is also more direct evidence for time-lagged

Table 14.2 Negative frequency-dependent selection in host–parasite systems.

Host	Parasite	Finding	Source
Bacteria (*Pseudomonas fluorescens*)	Phage SBW25Φ2	Experimental co-evolution. Largely driven by directional selection; hosts become resistant to more phage strains, and phage more infective to more host strains. Evidence for negative frequency-dependent selection weak, but more likely to explain field patterns of polymorphism.	5
Bacteria (*Pseudomonas aeruginosa*)	Different phages	Bacterial genes show negative frequency-dependent selection under low to moderate phage diversity; no such effect with high phage diversity.	3
Plant (*A. holbellaii*)	Rust fungi, herbivores	Rust fungi with evidence for local adaptation, over-infection of common host clones with time lag. Lower parasite infections in populations with higher levels of genetic diversity. Herbivore insects overshadow frequency dependence.	11
Barley (*Hordeum vulgare*)	Mildew fungus (*Erysiphe graminis*)	Newly introduced barley variants stay resistant initially when still rare. As variants increase in frequency, they become more heavily parasitized.	2
Wheat (*Triticum*)	Rust fungus (*Puccinia striiformis*)	Wheat genotypes planted at different frequencies. Fungus generates frequency-dependent selection; other (unknown) factors also favour rare genotypes. Disease-induced frequency dependence likely not sufficient to maintain genotypic polymorphism.	4
Water fleas (*Daphnia*)	Protozoa (*Caullerya*)	Parasite genotypes surveyed over four consecutive years. Common types decreased and rare types increased, corresponding to negative frequency-dependent selection.	7
Water flea (*Daphnia magna*)	Bacterium (*Pasteuria ramosa*)	Controlled infections in a panel of host clones from different ponds. The pattern of host and parasite types across localities consistent with negative frequency-dependent selection.	1
Freshwater snail (*Potamopyrgus antipodarum*)	Trematodes (*Microphallus* spp.)	Locally common genotypes of snails are more heavily infected; not the case for foreign genotypes, i.e. those not co-evolving with the tested parasites. In experimental co-evolution, initially common genotypes become more susceptible.	6, 8, 9
Freshwater guppy (*Poecilia reticulata*)	Trematode (*Gyrodactylus turnbulli*)	Novel MHC variant has an advantage.	10

Sources: [1] Andras. 2018. *Mol Ecol*, 27: 1371. [2] Barrett. 1988. *Philos Trans R Soc Lond B*, 319: 473. [3] Betts. 2018. *Science*, 360: 907. [4] Brunet. 2000. *Evolution*, 54: 406. [5] Buckling. 2002. *Proc R Soc Lond B*, 269: 931. [6] Dybdahl. 1998. *Evolution*, 52: 1057. [7] González-Tortuero. 2016. *Zoology*, 119: 314. [8] Jokela. 2009. *Am Nat*, 174 Suppl: S43. [9] Koskella. 2009. *Evolution*, 63: 2213. [10] Phillips. 2018. *Proc Natl Acad Sci USA*, 115: 1552. [11] Siemens. 2005. *Evol Ecol*, 19: 321.

frequency-dependence selection in several systems. Examples are again the New Zealand freshwater snails (Dybdahl and Lively 1995, 1998; Lively and Dybdahl 2000; Jokela et al. 2009; Koskella and Lively 2009), but also water fleas (Ebert 2008), plants (Antonovics and Ellstrand 1984; Barrett and Antonovics 1988; Brunet and Mundt 2000), or bacterial microcosms (Levin 1988).

Experimental co-evolution studies can address cycles of genetic frequencies with more precision. For example, allele frequencies in populations of the nematode *Caenorhabditis elegans* fluctuated more when co-evolving with *Bacillus thuringiensis* as compared to control populations (Schulte et al. 2010). However, a simple allelic model was not supported in this case; more complex changes at the whole-genome level occur, which differ between host and parasite (Papkou et al. 2019). Despite much supportive evidence, parasite-driven cycling in natural or experimental populations, therefore, remains hard to study (Little 2002; Brisson 2018). In the field, many additional factors take effect, not least the various environmental conditions (Oleksiak et al. 2002; Gibson and Weir 2005; Rockman and Kruglyak 2006). Furthermore, when hosts simultaneously co-evolve with many different parasites, selective sweeps may simply overshadow a possible Red Queen dynamic, the latter being more likely the case for single host–parasite pairings (Betts et al. 2018).

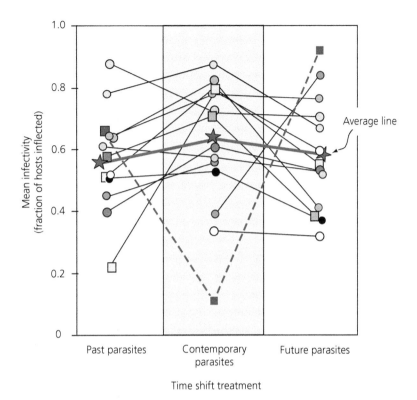

Figure 14.8 Time-shift experiment with co-evolving parasites. Bacterial spores (*Pasteuria ramosa*) and resting eggs of *Daphnia* were retrieved from different sediment depths at the bottom of natural pools. *Daphnia* raised from eggs were infected by reactivated spores. The graph shows host line vs parasite isolate combinations (coloured lines) that were contemporary, past, or future parasites of their *Daphnia* hosts. On average (thick red line, asterisks), the contemporary parasites were most infective on their contemporary hosts, but this adaptation quickly is lost for future parasites, with a time window of a few years. All except one isolate (dashed line) showed the same pattern. Sediment depth is a corollary of time (2 cm corresponding to 2–4 years of time). Adapted from Decaestecker et al. (2007), with permission from Springer Nature.

However, support for Haldane's initial suggestion that antagonistic co-evolution with parasites should maintain the genetic diversity of the host population exists. For example, clonal diversity in the freshwater snail *Melanoides* was positively related to the prevalence of various trematode parasites in a given population (Dagan et al. 2013). Similar relationships are known for water fleas (Cabalzar et al. 2019). Experimentally, *Tribolium*, co-evolving with the microsporidium *Nosema*, shows this effect, too (Bérénos et al. 2010b).

14.2.2.3 Trans-species polymorphism

In terms of population genetics, balancing selection maintains genetic diversity. Balancing selection, in turn, results from the temporal dynamics under the

Red Queen. The corresponding genetic signatures are found in genes involved in defence against parasites, such as immune genes of *Drosophila melanogaster* (Croze et al. 2016; Chapman et al. 2019), or in the TLR genes of free-living rodents (Kloch et al. 2018) (see also section 10.6). Balancing selection can operate over extended periods and even continue through speciation events. As a result, genetic polymorphisms will be shared by species that have otherwise diverged, even millions of years ago (this turns it into a macroevolutionary pattern).

Such 'trans-species polymorphism' is observed, for example, for allelic polymorphism in resistance genes of the cabbage plant *Arabidopsis thaliana* against infections by bacteria (*Pseudomonas*). This polymorphism has persisted for millions of years, and antagonistic co-evolutionary processes with fluctuating gene frequencies remains the best explanation (Stahl et al. 1999). A similar finding pertains to defence against fungal parasites (*Peronospora parasitica*) of *Arabidopsis* (Bittner-Eddy et al. 2000). Most conspicuously, however, trans-species polymorphisms are observed for genes of the MHC locus, as well as in a few other immune-related genes in jawed vertebrates (Key et al. 2014; Těšický and Vinkler 2015; Lighten et al. 2017). It is, in fact, a remarkable observation that such trans-species polymorphisms of immune-related genes closely connects the immune repertoire of humans to that of the great apes (Figure 14.9; Azevedo et al. 2015). Also, among guppies (*Poecilia* and relatives) living in Trinidad and other islands, species that diverged more than 20 million years ago share MHC types.

Genes for a range of other functions also show trans-species polymorphism; for example, genes that underlie fertilization systems based on compatibility alleles in plants (Dwyer et al. 1991), fungi (van Diepen et al. 2013), or in the blood groups of

primates (Ségurel et al. 2012). These examples show the generality of these processes. Different selection regimes may operate at the same time and even on the same genes. Some alleles are subject to a host–parasite arms race, whereas entire functional complexes may be driven by a Red Queen dynamics of balancing selection (Azevedo et al. 2015).

14.3 Parasites, recombination, and sex

From an evolutionary point of view, sex and recombination are numerically inefficient mechanisms of reproduction (Box 14.3), and the quest for understanding this contradiction has generated a large body of research.

14.3.1 Theoretical issues

A range of theoretical scenarios for the advantage of sex and recombination exists (Box 14.3). Parasites are particularly important in some of these scenarios (Dapper and Payseur 2017). A key observation

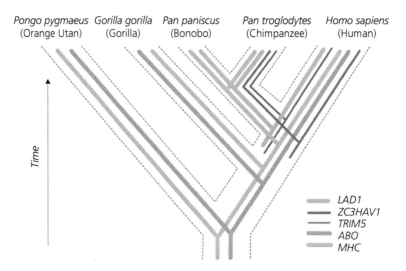

Figure 14.9 Trans-species polymorphisms. The evolutionary tree of humans and the great apes are decorated with known examples of trans-species polymorphisms in this clade. The various lines symbolize the shared repertoires of the same alleles in the different species; loci indicated in the legend: *LAD1* (ladinin-1; a filamentous protein associated with IgA and autoimmune diseases), ZC3HAV1 (a zinc-finger antiviral protein), TRIM5 (tri-partite motif-containing 5; antiviral protein), ABO (blood-group alleles), and MHC (multihistocompatibility complex). Adapted from Azevedo et al. (2015), under CC BY 4.0.

is that long-lived and large organisms generally also have low fecundity (few offspring per lifetime), and thus evolve at a plodding pace. Such organisms should thus be particularly in peril from the numerous, small, and rapidly adapting parasites. So, how can they even exist in the face of ubiquitous parasitism (Hamilton et al. 1990)?

The Red Queen hypothesis is one answer to this paradox. Parental lines that produce offspring with rare and novel genotypes will more likely escape the prevailing parasites. Rarity is what sexual reproduction and meiotic recombination generate (Box 14.3). For the intense and rapid fluctuations of the Red Queen to persist, parasites should be

sufficiently virulent (West et al. 1999; Gandon and Otto 2007). However, even with 'mild' parasites, the combined effects of infections by multiple parasitism might be severe enough. Effects of parasitism on host fecundity also selects for sex over asex in many cases (Lively 2006). Note that the causes of the evolution of sex may differ from the maintenance of sex. With the 'pluralist view' (West et al. 1999), a combination of mechanisms facilitates the latter. Overall, the conditions under which sex can evolve and is maintained are less restrictive than initially assumed, even though the actual mechanisms can be quite complex (Neiman et al. 2017).

Box 14.3 The masterpiece of nature: Sex and recombination

To explain the 'masterpiece of nature' (Bell 1982), a large number of hypotheses exist. Among those, time-lagged antagonistic host–parasite co-evolution—the Red Queen dynamics—is a prime contender.

Sexual reproduction: Sexual reproduction ('sex') in eukaryotes is symmetric and amphimictic. It leads to the mixing of genes from both parents to form progeny. Two parents are needed to produce a given offspring—it is numerically very inefficient. There is, therefore, a fundamental 'two-fold' cost of sex when compared to asexual reproduction, where only one parent is needed (Maynard Smith

1978) (Figure 1). As a part of this cost, during meiosis, the (diploid) stock of parental genes is halved to form the (haploid) gamete. The gametes subsequently fuse (in syngamy) to produce an offspring (the zygote). At the same time, recombination further alters the genotype (see below). Sexual reproduction, therefore, generates offspring whose genotypes not only contain just half of the parental genes, but which are also assembled in different ways from their parents and their siblings. This is counterintuitive, since the seemingly successful parental genotypes are thereby not preserved. This comes in addition to the numerical disadvantage.

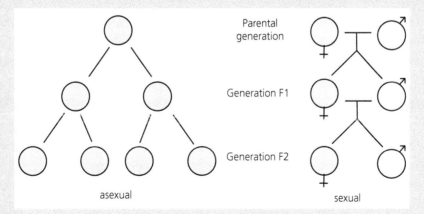

Box 14.3 Figure 1 Sexual reproduction has a two-fold cost. Measured by the number of offspring, a single asexual individual leaves four descendants in generation F2. A sexual pair only leaves two. In both cases, two offspring per generation are produced.

continued

Box 14.3 *Continued*

If sex is to evolve, phenotypic and genetic variation for this trait (i.e. sex) must be present in populations. What exact mechanism leads from asex to sex can be different in each case, and, hence, the genetic basis for sex varies as well. In systems such as the freshwater snail *Potamopyrgus*, or fish of the genus *Poecilia*, the segregation of chromosomes during meiosis occasionally fails in one parent, leading to asexual, triploid (3n) offspring as compared to the usual diploid (2n) offspring. Non-segregation might have a genetic basis or might be a simple error, but it continuously generates new clonal lines splitting off from the sexually reproducing population. This seems to be a common phenomenon in species where sexual and asexual lines co-exist in the same population (Jokela et al. 2009).

One key element of sexual reproduction—the contribution of genes from different individuals to form offspring—is already present in a simple form in prokaryotes, e.g. by the genetic exchange in the transformation of bacteria (Redfield 2001). Such 'asymmetric' sex requires only a small advantage, e.g. by the transfer of antibiotic resistance. Among prokaryotes, many different schemes combine various elements of sex, reproduction, or recombination in their life cycle, and that do not necessarily follow the simple Mendelian segregation schemes (Bell 1982). Similarly, chromosome segregation most likely was the initial step in the evolution of meiosis and, in combination with isogamy (equal size of 'male' and 'female' gametes), would not have included some of the additional costs of sex.

Meiotic recombination: During meiosis, pairs of 'homologous chromosomes'—maternal chromosomes that match the paternal ones—associate with one another. They subsequently undergo a DNA-strand crossing-over that leads to genetic recombination, and finally segregate into the daughter cells. The main genetic effect of meiotic recombination is to shuffle the genetic material from both parents of the reproducing individual into novel combinations, which are present in the gamete. Genetic recombination also results from segregation of the complete paternally and maternally derived chromosomes. The physiological processes of meiosis involve several different genes, some of them conserved across large evolutionary distances (Brooks 1988; Cohen et al. 2006; San Filippo et al. 2008).

Recombination does not change alleles and, hence, any offspring inherits single genes in standard diploid systems, with a probability of one half. However, a particular combination of parental genes is broken up by recombination with a probability that decreases with the distance between the concerned loci on the same chromosome. As an essential consequence, recombination also breaks an existing linkage disequilibrium in the population.

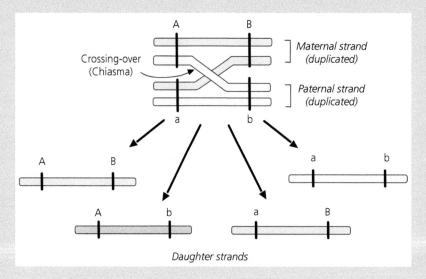

Daughter strands

Box 14.3 Figure 2 Genetics of recombination during meiosis. In the standard model, a diploid parental cell duplicates each strand (maternal, paternal), which subsequently pair with a homologous strand. A crossing-over (microscopically visible as a chiasma) can happen between two loci on the same chromosome, as shown here. After separation, four daughter strands emerge. Of those, two are of parental genotype (*AB* or *ab*), and two are of recombinant genotype (*Ab*, *aB*). Alleles *A*, *a* at locus 1, and *B*, *b* at locus 2 are for illustrative purposes.

Theories for the evolution of sex and recombination: A large number of theoretical models fall into three classes of ideas (Sharp and Otto 2016).

1. Common genotypes are currently less fit than the rare ones. The rare genotypes result from sexual reproduction and recombination. This fits the *Red Queen hypothesis*, suggesting that a rarity advantage results from time-lagged processes, i.e. over time. With *Spatial Heterogeneity* (where the dynamics unfold in space), there is a direct benefit in offspring fitness for rare types at the same time.

2. Intermediate genotypes are more fit than extreme genotypes (a case of negative epistasis). Sex breaks up (intermediate) genotypes and shuffles together beneficial alleles (yet also deleterious alleles) in the same genotype. This reduces mean fitness of offspring but increases genetic variance for fitness; 'good' genotypes can therefore spread more quickly. Hence, sex serves to improve the response to selection and long-term fitness of the population; a sex allele thereby spreads by hitch-hiking. *'Synergistic epistasis'* and *'mutational deterministic'* models (selection primarily removes deleterious alleles) implement this basic idea.

3. Beneficial and deleterious alleles are not present in all possible combinations within the genotypes, due to final population size. Selection will then either eliminate (for the poor genotypes) or fixate (for the good genotypes) the tails of the genotype distribution. Because alleles are always selected as part of the genotype, selection at a given locus also is generally reduced due to selection at other, linked loci (known as 'selective interference'). Sex can regenerate the genetic variance by breaking up associations, and thus increase variance for fitness to improve response to selection and long-term fitness (as in class (2)). This is implemented in the *'selective', 'clonal', and 'Hill–Robertson' interference* concepts where sex promotes mixing of alleles and can then hitch-hike with good allele combinations. *With 'Muller's ratchet'*, sex 'rescues' genotypes that have accumulated many deleterious alleles through the formation of new combinations.

Note that these hypotheses are not mutually exclusive. Consequently, the *'pluralistic view'* (West et al. 1999) suggests that either two (or more) different mechanisms may act together, or that the effects of another mechanism may enhance the effects of a core mechanism (such as the Red Queen). This view also proposes that the responsible mechanisms are not universally the same but can vary according to taxa or with a particular system. Concerning host–parasite co-evolution, the most frequently discussed and tested pluralistic approaches are the *Red Queen* combined with either *Muller's ratchet* or the *mutational deterministic* hypothesis (Hodgson and Otto 2012; Neiman et al. 2017).

Recombination is the 'little sister' of sexual reproduction, a process that became associated with the production of gametes in virtually all multicellular organisms. The physical sign of recombination is crossings-over (chiasmata) between paired chromosomes. In a wide variety of organisms, the average number of observed chiasmata is around 1.5 per chromosome (Otto and Lenormand 2002). Partly, this results from the requirement that at least one crossing-over per chromosome physically is needed for successful segregation (Dumont 2017). Stabilizing selection for a limited number of crossings-over per chromosome seems the most parsimonious explanation for this observation (Otto and Payseur 2019). Nevertheless, recombination rates differ among various groups, with fungi and microorganisms having exceptionally high rates (Stapley et al. 2017). Besides, the recombination rate is variable among individuals in a population, including humans (Coop and Przeworski 2007). Recombination rates, furthermore, are heritable and involve known genes (Johnston et al. 2016). Rates can experimentally be selected (Kerstes et al. 2012) and, therefore, can be subject to selection by parasites.

Theoretical studies clarified the critical conditions under which antagonistic co-evolution could select for recombination. A constant environment thereby serves as the benchmark, since under these conditions recombination disrupts well-adapted genotypes and is therefore generally selected against (the 'reduction principle') (Altenberg and Feldman 1987). Most theories assume a gene (a 'modifier') that affects the recombination rate between two functional loci involved in defence against parasites. The modifier can or cannot be linked to these loci. Recall that the primary genetic effect of recombination is to break existing linkage disequilibria among loci. Recall also that defence-associated fitness components result from the population-wide epistasis among functional loci.

At any given time, alleles involved in resistance against parasites could be locked in negative epistasis, i.e. having a combined fitness that is lower than the sum of their average fitness. In such cases, recombination frees these alleles and can allow for new and fitter combinations; a modifier would have an advantage. Antagonistic co-evolution generates fluctuations in the sign of epistasis and linkage over time. Hence, with host–parasite co-evolution such episodes of negative epistasis regularly occur. Against this background, host–parasite co-evolution can favour the spread of a recombination-enhancing modifier. The theoretical calculations show that this may happen over a wide range of conditions, provided the parasite population, in turn, is under selection and therefore diverse (Salathé et al. 2008b). Because of the time lags in the co-evolving host–parasite system, the fitness effects of such a modifier extend to different time scales—an immediate short-term effect in the next generation, a delayed short-term effect in the generations immediately following the next, and a long-term effect extending beyond the next few generations.

14.3.2 Empirical studies

Many empirical studies have tested whether recombination has an advantage. The approaches and, in particular, the studied host–parasite systems range widely, but advantages exist, although the details often make the stories complicated (Table 14.3). In general, an increase in recombination rate results from directional selection. The change often happens rapidly, in the order of five to ten generations (Kidwell 1972; Dewees 1975; Kerstes et al. 2012). Furthermore, populations or lines with recombination possess general fitness advantages over their non-recombining counterparts. With few exceptions (Bourguet et al. 2003), recombining populations also show a faster response to selection than those with reduced levels of

recombination, or with none at all (Charlesworth and Barton 1996; Korol 1999; Rice 2002; Rodell et al. 2004). Detailed genomic studies suggest that the underlying processes can be fairly complex, involving selective interference (i.e. suppressing selection on beneficial alleles by linkage with deleterious alleles nearby) in combination with frequency-dependent selection and selective sweeps (McDonald et al. 2016a).

Sex and recombination lead to genetic diversity among offspring. If this is an advantage against parasites, genetically diverse populations and heterozygous individuals should have fewer parasites (see Figure 10.8). Studies on wild populations are particularly relevant. Their collective results suggest that different factors favour sexual reproduction. For example, sex reduces the effects of deleterious mutations, ensures a higher rate of adaptive evolution, and asexual vs sexual lines benefit from the differentiation of their ecological niches. Hence, while there is no single mechanism that promotes sex in all systems or circumstances, the role of host–parasite co-evolution seems prevalent.

Currently, the best empirical field evidence again comes from the New Zealand freshwater snail *P. antipodarum* and its *Microphallus* trematodes. Snails are an intermediate host; the final hosts are birds and fish. A large pool of clonal snail lines co-exists with the sexual population in any one lake on South Island. In a study covering many years, the frequencies of the clonal lines cycled over time, together with time-lagged cycling of parasite infections (Dybdahl and Lively 1998; Jokela et al. 2009). Heavily infected clones declined in frequency in future generations, and rare clones enjoyed an advantage against their locally co-evolving parasites. Hence, the snail population seems to be tracked by its parasites. The rise and fall of clones, co-existing with the sexual types, changes under the effects of parasite pressure (Vergara et al. 2014). At the same time, sexual reproduction is more common when parasite pressure is high (Lively 1987; Jokela et al. 2009; Vergara et al. 2014), demonstrating an advantage of sexual reproduction with parasitism (Figure 14.10).

Table 14.3 Advantage of sexual reproduction and meiotic recombination in host–parasite systems.

System	Question, approach	Finding	Source
(a) Advantage for sexual reproduction			
E. coli, infected by plasmids (F-episome)	Compare fitness of parasitized and non-parasitized genotypes carrying known numbers of mutations.	No evidence for synergistic epistasis. However, the average effect of deleterious mutations was greater in parasitized than parasite-free genotypes. Sex perhaps maintained by combination of mutations and parasites.	4
Flax (*Linum marginale*) vs rust fungus (*Melamspora lini*)	Comparing inbred with outbred populations for infection.	Outbred populations show higher mean resistance to six pathogen isolates, and higher average number of pathogen lines to which hosts are resistant.	3
Water flea (*Daphnia magna*) vs bacteria (*Pasteuria ramosa*)	Relative fitness of asexually vs sexually produced offspring.	Tested in lab, but with field-derived hosts and parasite. Sexual offspring less infected; processes happening on a short time scale. Also, parasite population evolves rapidly. Genetic studies show strong linkage, and epistasis between resistance loci.	1, 15
Snail (*Potamopyrgus antipodarum*) vs castrating trematodes (*Microphallus* spp.)	Relationship (among populations) between the percentage of individuals infected by trematodes and the frequency of sexual individuals in the population. Competition of sexual vs asexual clones.	In many separate studies in the field, a positive relationship between prevalence of infection and frequency of sexual individuals. Sexual individuals generally less infected, but considerable variation over years. Long-term study suggests that sexual population maintained in co-existence with asexual lines, e.g. by episodes of selection in time or space.	8, 11, 12, 13, 19
Psychid moths	Comparing 49 populations of co-existing sexual and asexual moths for the prevalence of parasitoids.	For each of three different years, positive correlation between frequency of sexually reproducing females and frequency of individuals infected by hymenopteran parasitoids.	10
Nematode (*Caenorhabditis elegans*) vs bacteria (*Bacillus thuringiensis; Serratia marcescens*)	Experimental co-evolution. Advantage of outcrossing, of males, of sexual reproduction.	Although males can be less fit under parasitism, sexual reproduction provides more diversity and higher resistance in offspring. Red Queen dynamics constrains spread of self-fertilization, and favours biparental sex. No evidence for an increase in genetic re-assortment in the host. Co-evolved bacteria presumably exchange plasmids more frequently.	14, 16–18
(b) Advantage for recombination			
Protozoa (*Toxoplasma gondii*)	Experimentally testing the virulence of different strains.	Recombined strains from ancestral clones are more successful and virulent than original strains.	6, 7
Plants	Comparing recombination rates (frequencies of chiasmata) among populations and species.	Among species, frequency of chiasmata positively correlated with number of B-chromosomes. B-chromosome acts as a parasite for the rest of the genome.	2
Mosquito (*Aedes aegypti*) vs microsporidia (*Vavraia culicis*)	Do infected mosquito females have more recombinant offspring than uninfected ones.	Yes, infected females produce offspring with higher rates of recombination events. A case of condition-dependent response. i.e. plastic adjustment of recombination.	20
Beetle (*Tribolium castaneum*) vs microsporidium (*Nosema whitei*)	Experimental co-evolution. Does recombination rate change when parasites co-evolve?	Higher recombination rates when co-evolving with parasites, and higher resistance.	5, 9

Sources: [1] Auld. 2016. *Proc R Soc Lond B*, 283: 20162226. [2] Bell. 1990. *Parasitology*, 100: S19. [3] Burdon. 1999. *Evolution June*, 53: 704. [4] Cooper. 2005. *Proc R Soc Lond B*, 272: 311. [5] Fischer. 2005. *Biol Lett*, 1: 193. [6] Grigg. 2001. *Science*, 294: 161. [7] Grigg. 2003. *Microb Infect*, 5: 685. [8] Jokela. 2009. *Am Nat*, 174 Suppl: S43. [9] Kerstes. 2012. *BMC Evol Biol*, 12: 18. [10] Kumpulainen. 2004. *Evolution*, 58: 1511. [11] Lively. 1987. *Nature*, 328: 519. [12] Lively. 1992. *Evolution*, 46: 907. [13] Lively. 2002. *Evol Ecol Res*, 4: 219. [14] Masri. 2013. *Ecol Lett*, 16: 461. [15] Metzger. 2016. *Evolution*, 70: 480. [16] Morran. 2011. *Science*, 333: 216. [17] Schulte. 2010. *Proc Natl Acad Sci USA*, 91: 7184. [18] Slowinski. 2016. *Evolution*, 70: 2632. [19] Vergara. 2014. *Am Nat*, 184: S22. [20] Zilio. 2018. *PLoS ONE*, 13: e0203481.

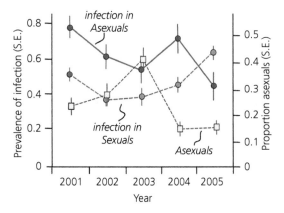

Figure 14.10 Maintenance of sexual reproduction in the snail *Potamopyrgus*. Across a number of lakes in New Zealand, the average prevalence of infection (y-axis, left-hand scale) by trematodes (*Microphallus* spp.) varies among years, but is generally lower in sexual (dashed line, open red circles) than in asexual (clonal) snails (solid line, full red circles). At the same time, sexuals and asexuals co-exist at varying proportions in the populations (dotted line, open blue squares; y-axis, right-hand scale). Note, for example, an increase in the proportion of asexuals in 2003 is followed by an increase in their infection in 2004. The data suggest that sex is maintained by a Red Queen-like mechanism, despite the costs of having to produce males. Adapted from Vergara et al. (2014), with permission from University of Chicago Press.

14.4 Local adaptation

Antagonistic co-evolution requires the close association of the evolving host and parasite populations. Eventually, this can become visible in differences among different populations or localities. Such 'local adaption' is an excellent opportunity to scrutinize the processes of host–parasite co-evolution. In general, the extent to which local adaption evolves depends on several factors, e.g. spatial heterogeneity or the strength of local selection vs the right balance of gene flow among localities (i.e. the migration rate of individuals). With limited gene flow, local selection can generate specialized adaptations. By contrast, with excessive gene flow, locally adapted genotypes are swamped by the incoming genes. At the same time, small local populations are subject to genetic drift and the random fixation of genotypes that are not necessarily locally adapted (Blanquart et al. 2013).

Host–parasite systems can deviate remarkably from this simple picture, because antagonistic

co-evolution also generates fluctuating selection. According to theoretical studies, intermediate levels of gene flow become decisive in such scenarios. The most rapidly evolving party—usually the parasites—is then expected to locally adapt, whereas the other party may not (Gandon and Michalakis 2002). However, a general 'co-evolutionary advantage' for parasites in a given system is challenging to estimate (Nuismer 2017). The simple observation that hosts have their parasites virtually anywhere and throughout all times vividly illustrates that neither party can easily outpace the other.

Allopatric parasites (those from a different location) might be less virulent or infective to local hosts when compared to sympatric parasites (those from the same location) (Ebert 1994). To study such local adaptation empirically, researchers have swapped hosts between different locations and let them become infected by the local parasites (a 'transplant experiment'). Alternatively, the parasites become swapped to infect the respective local hosts (a 'cross-infection experiment'). Both exchanges can also be done at the same time. Often, hosts and parasites from different locations are also kept together in a shared environment to randomize environmental conditions (a 'common garden experiment'). The fitness of either party, e.g. host resistance or parasite infectivity, is then measured under different local vs foreign combinations.

At this point, different estimates for local adaptation exist. For example, the 'home vs away' difference compares the average fitness of a host population in its native (home) location with the fitness when transplanted to another site (away). Alternatively, the 'local vs foreign' difference refers to the average fitness of a host population in its home location vs the average fitness of all other populations transplanted to its home patch (Kawecki and Ebert 2004). Whereas these alternatives are not describing fundamental differences in the concept of local adaptation, their numerical estimates reflect different biological processes and can thus lead to different conclusions.

Numerical estimates are additionally affected by the background environment, the quality of the local habitats, and variation in the overall 'quality' of the local populations. Together, this can lead to inconsistent results and even statistically blur any effect of local adaptation. At least in theory, repeating the combinations across many populations and localities can improve the situation (Blanquart et al. 2013). There are many limitations to fulfil all requirements in actual empirical studies, such that our insight into the generalities of local adaptation is steadily revised.

Despite the many difficulties, however, adaptation by either host or parasite is observed across a range of organisms, whereas joint adaptation by both sides is somewhat less common. In meta-analyses (Hoeksema and Forde 2008), or from a survey of 57 host–parasite systems (Greischar and Koskella 2007) (Table 14.4), local adaptation (parasites being more infective on their own host populations) typically was associated with parasites having higher migration rates (i.e. gene flow) than their host. Relative generation times played a minor role—as expected from theory (Gandon and Michalakis 2002). Contrary to theoretical expectations, however, virulence levels did not affect the occurrence of local adaptation; it also did not matter whether parasitism was obligatory or facultative. For sticklebacks and their parasites, the MHC genotypes correspond to the differences in the overall parasite load in different habitats and are independent of the background genetic differences (Eizaguirre et al. 2012a). Other mechanisms, such as the capacity to manipulate a host, may not show local adaptations, however (Franceschi et al. 2010; Hafer 2017).

Given their large population sizes and short generation times, microbial systems such as bacteria and their viral parasites (phages) are often used to study local adaptation. The conclusions from these systems converge with those gained from other systems (Greischar and Koskella 2007). At the same time, bacteria–phage systems allow investigations at the molecular level. For example, there is a limitation to local adaptation of phages when the bacterial CRISPR–Cas defence system stores an excessive diversity of templates (Morley et al. 2017). There are also many cases where no local adaptation is present, e.g. in great tits infected by malaria (Jenkins et al. 2015). Also, mismatches ('maladaptation'; i.e. parasites are less adapted to local hosts) are possible (Greischar and Koskella 2007). Mismatches can be experimentally generated, e.g. in the protozoan *Paramaecium*, infected by bacteria (Adiba et al. 2010).

A closer analysis, therefore, reveals a great deal of complexity in the underlying host–parasite dynamics that does not always yield simple predictions (Morand et al. 1996; Kaltz and Shykoff 1998). In elaborate geographical mosaics, determined by spatial variation in resource supply and other environmental conditions, the differences in gene flow, furthermore, are blurred by a diverse set of patterns, with local maladaptation as well as local adaptation (Nuismer 2006). Geographical mosaics are a significant driver for the divergence in host–parasite co-evolutionary trajectories (Hochberg and Holt 2002; Laine 2009).

Local adaptation is also relevant when some offspring which are better protected against parasites outcompete their siblings at the same site. This should be particularly relevant when dispersal distances from the natal site are short, and the likelihood of competing against siblings is high. Parents might therefore be selected to genotypically diversify their offspring, such that at least some of them are well protected by having a resistant genotype. This 'tangled bank' hypothesis predicts that sex and recombination should be favoured when dispersal distance is short and many offspring are likely to settle at the same place (Bell 1982). The tangled bank hypothesis, therefore, refers to spatial dynamics rather than a temporal dynamics, as implemented in the Red Queen scenario. In a study of plants, the frequency of recombination events indeed decreased as dispersal distance increased, consistent with the theory (Koella 1993).

Table 14.4 Patterns of local adaptation illustrating the range of taxa, conditions, and outcomes.[1]

Host	Parasite	Life cycle[2]	Migration[3]	Generation[4]	Virulence[5]	Obligate[6]	Transmission[7]	Infectivity[8]	Severity[8]	Main effect Host	Main effect Parasite
Plant hosts:											
Pinus sylvestris	Crumenulopsis soraria	Simple	P>H	P<H	W	Yes	H	N	-	Yes	No
Plantago lanceolata	Podosphaera plantaginis	Simple	P>H	P<H	F	Yes	H	LA	N	No	No
Amphicarpa bracteata	Synchytrium decipiens	Simple	P<H	P<H	M	Yes	H	LA	LA	-	-
Phaseolus vulgaris	Colletotrichum lindemuthianum	Simple	P>H	P<H	F	Yes	H/V	LA	LA	No	Yes
Pinus edulis	Matsucoccus acalyptus	Simple	P<H	P<H	W	Yes	H	N	-	No	Yes
Rhus glabra	Blepharia rhois	Simple	P>H	P<H	W	No	H	N	-	Yes	Yes
Borrichia frutescens	Asphondylia borrichiae	Simple	P>H	P<H	W	Yes	H	LA	LA	Yes	-
Agrostis capillaris	Rhinanthus serotinus	Simple	P<H	P=H	W	No	H	N	N	No	-
Animal hosts:											
Bulinus globosus	Schistosoma matthei	Complex	P>H	P<H	M	Yes	H	LA	N	No	Yes
Bombus terrestris	Crithidia bombi	Simple	P<H	P<H	W	Yes	H/V	N	N	-	No
Danaus plexippus	Ophryocystis elektroscirrha	Simple	P<H	P=H	W	Yes	V	-	N	Yes	No
Daphnia magna	Pasteuria ramosa	Simple	P>H	P>H	S	Yes	H	N	N	Yes	No
Ochlerotarsus sierrensis	Lambornella clarki	Simple	P<H	P<H	M	No	H	LA	N	Yes	Yes
Gammarus duebeni	Nosema granulosis	Simple	P<H	P=H	F	Yes	V	N	LA	-	-
Gasterosetus aculeatus	Diplostomum pseudopathaceum	Complex	P>H	P<H	M	Yes	H	-	MA	-	-
Parus major	Ceratophyllus galinae	Simple	P=H	P<H	N	No	V	-	N	-	-
Galliota galliota	Haemogregarine	Complex	P<H	P<H	W	Yes	H	MA	-	-	-

[1] From Greischar. 2007. *Ecol Lett*, 10: 418, with permission from John Wiley and Sons.

[2] Simple (one host) or complex (several hosts) life cycle.

[3] Relative migration rates of parasite (P) and host (H).

[4] Relative generation times of parasite (P) and host (H).

[5] Virulence, increased mortality (M), sterility (S), little or no effect (N), reduced fecundity (F) and fitness (W).

[6] Parasite obligate on this host.

[7] Transmission vertical (V) or horizontal (H).

[8] Infectivity (ability to infect) or infection intensity (severity). The observed pattern is either maladapted, i.e. foreign or away-parasites more infective/severe (MA), locally adapted, i.e. local or home-parasites more infective/severe (LA), or no difference observed (N). Main effects refer to statistical effects of either host or parasite.

Important points

- Macroevolution concerns the phylogenetic patterns and processes of host–parasite associations at various taxonomic levels and time scales. Co-speciation and host switching are two of many possible outcomes. Phylogenetic distance among host lines and ecological factors shape the co-evolutionary history.

- Microevolution unfolds within populations and eventually leads to macroevolutionary patterns. With selective sweeps, favourable alleles (e.g. for host resistance) rapidly rise in frequency and can become fixated. In arms races, host and parasite traits increase directionally over time. Co-evolutionary processes can be harnessed for developing 'evolution-proof' strategies for the use of antibiotics or insecticides.

- Parasites generally are under more substantial selection to infect than the host is to defend. With Red Queen dynamics, common genotypes are selected against by parasites, but rare genotypes have an advantage (negative frequency-dependent selection); host and parasite genotypes can fluctuate in frequency for long periods.

- Red Queen dynamics can maintain genetic diversity in the host population. Empirical evidence for a fitness advantage of rare host types, time-lagged effects, and continuous cycling comes from field and laboratory studies. However, it remains unknown how general these patterns are. Diversity can be maintained as trans-species polymorphisms in host defence genes.

- Sexual reproduction is numerically inefficient. However, with recombination, it generates rare, novel, and different genotypes in offspring. With antagonistic co-evolution favouring rare types, sex and recombination are maintained under parasitism. Empirical studies support this prediction.

- Antagonistic co-evolution results in local adaptation of host and parasite, depending on the migration rates among subpopulations and embedded in a geographical mosaic of varying ecological conditions. Empirically, local adaptation, defined and measured in various ways, has been found in many systems and overall is consistent with theoretical expectations.

CHAPTER 15

Ecology

Ecology seeks to explain the observed patterns of abundance and distribution of organisms. Important factors that determine these patterns are the interactions with the abiotic (physical, chemical) and with the biotic (other organisms) environment. Biotic interactions are known as mutualism, symbiosis, competition, predation, and parasitism. Within this context, the life history of an organism is a pattern of birth, survival, reproduction, and death that reflects the adaptive responses to the selective effects of such biotic interactions and from the determinants of the abiotic environment. Here, we focus on the perspective of hosts, but the parasite's view is often the mirror image.

15.1 Host ecology and life history

15.1.1 Host body size

Body size is a simple but essential trait for host ecology and host life history. The costs and benefits of small or large size are well studied, e.g. as a correlate of fecundity, mating success, or thermoregulation. Across organisms, body size correlates with other important traits, e.g. longevity, or with dispersal abilities, all of which affect the likelihood of acquiring infections (Cooper et al. 2012b; Blasco-Costa and Poulin 2013). Host body size, therefore, relates to parasitism in many ways. For instance, host body size often correlates positively with parasite abundance or parasite species richness. Furthermore, across all organisms, physiology, metabolism, and energetics typically correlate with body size. Because immune defences are embedded in these mechanisms, body size should affect the immune response as well, e.g. its effectiveness or costs.

Nevertheless, across a set of 236 vertebrate and invertebrate species, life span or body mass did not correlate directly with the costs of immune system activation—measured as a loss of body mass, growth rate, or fecundity. Instead, small but relatively long-lived species incurred the highest costs, while the effect of phylogenetic group was considerable (Brace et al. 2017). The example demonstrates that an important aspect of immune systems—the cost of defence—has a complicated relationship with host body size. This also reflects on the association of parasitism with host body size. Host body size often correlates with parasite abundance or richness (Kamiya et al. 2014a; Harnos et al. 2017) (see section 7.2.4), but there are also significant differences among host and parasite taxa and localities (Morand 2015). Parasite body size, in turn, generally increases with host body size; a pattern termed 'Harrison's rule'. Among 581 species of avian lice and their bird hosts, this pattern is observed, but exceptions abound (Harnos et al. 2017).

15.1.2 Host reproductive patterns

An individual's life history is the schedule of growth and development (the juvenile phase) to reach maturity, followed by one or several episodes of reproduction, senescence, and eventual death. Selection shapes this schedule such that fitness—survival with reproduction—is maximized over an organism's expected lifetime (Stearns 1992). At any one time, along with this schedule, a host survives or dies, and it can reproduce in the next episode. Which outcome is more likely differs between infected and healthy individuals. For example, when a host becomes infected by castrating parasites, the gradual consumption of the internal organs

Evolutionary Parasitology: The Integrated Study of Infections, Immunology, Ecology, and Genetics. Second Edition.
Paul Schmid-Hempel, Oxford University Press. © Paul Schmid-Hempel 2021.
DOI: 10.1093/oso/9780198832140.003.0015

leads to the loss of reproductive capacity and causes premature death.

Within the life history framework, however, the statistical expectations matter. This is because evolution happens in populations and, therefore, the probabilities of different outcomes matter. For example, when parasite pressure is high (e.g. a high force of infection), an infection becomes more likely. At a given point in time, the expected future fitness is, therefore, lower for any individual, even if a particular host is not infected at the time and perhaps never will be. Hosts can plastically change their individual life history to 'rescue' as much of their future fitness if future expectations are low. This occurs when really infected or in anticipation, e.g. when the respective cues for imminent infection are sensed. When hosts thus expect increased mortality and reduced fecundity, they should invest more in current reproduction. Such early reproduction comes at the price of overall reduced fecundity; it would nevertheless minimize the expected fitness loss due to parasitism. This strategy of 'fecundity compensation' (Figure 15.1) is discussed from the parasite perspective in section 8.3.1.

A classic example is a freshwater snail (*Biomphalaria glabrata*). Egg production peaks a few weeks after maturity and then slowly decreases. However, when snails are infected by the trematode *Schistosoma mansoni* around the time of maturity, egg production immediately starts in the first week post-infection and ceases prematurely. The snails thus advance their reproductive peak in time. With the advance, infected snails will have a lower total reproductive output than uninfected snails, but a higher one than had they followed the standard scheme of reproduction, since castration sets in before most eggs are produced (Minchella and Loverde 1981; Minchella 1985; Thornhill et al. 1986). When snails are only exposed to chemical cues of *S. mansoni* but not infected, they also reproduce earlier. The response is therefore anticipatory, in response to the presence of the parasite, and independent of the infection itself.

Similar examples of fecundity compensation are observed in several other cases (Table 15.1). In the above example, age at maturity does not change much. Nevertheless, a changing pattern of reproduction might also affect the age at maturity, i.e. the age when first reproduction occurs (Figure 15.1).

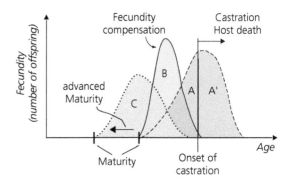

Figure 15.1 Life history change and parasitism. The dashed curve shows the normal pattern of reproductive activity over time. The area under this curve (A plus A'; red area) corresponds to the total number of offspring produced for the lifetime of the individual (i.e. host fitness) under normal conditions. When the host is infected and, therefore, becomes castrated or dies earlier at a certain time (vertical line; 'onset of castration'), the host loses fitness corresponding to the area A', leaving a residual fitness of A. Strategies to reduce this expected fitness loss are: (1) fecundity compensation, i.e. a higher reproductive effort early after maturity and before castration sets in (fitness B; green, solid line); or (2) advancing age at maturity and starting to reproduce at a smaller body size (fitness C; yellow, dotted line). In both cases, the fitness is less than for an uninfected host (A plus A'), but more than for an infected host (A) that keeps to its normal schedule.

A younger age at maturity comes with a cost to fecundity, too. Because individuals reproduce at a smaller body size, they have lower fecundity. There is, therefore, a trade-off between earlier reproduction, i.e. securing some reproduction before the parasite has damaged the host, and reduced fecundity at smaller body size. It is likely that this represents a so-called 'best of a bad job' strategy; that is, the result is less than would be optimal, but the best possible given the current conditions.

Even after the young are born, hosts might adaptively change their behaviour and their life history strategies. For example, when a nest of blue tits (*Parus caeruleus*) becomes infested by ectoparasites (hen fleas), the investment into brood care is increased. In this case, the male starts to feed the chicks more frequently. This reduces the parasite's effects (caused by blood loss) and increases the chicks' chances of fledging and becoming independent. However, with the increased effort, the male incurs a cost, as this compromises its survival over the next winter (Richner et al. 1995; Christe et al. 1996).

Table 15.1 Changes in host life history with possible defence function.

Host	Parasite	Observation	Possible function for host	Source
Cricket *Achaeta domesticus*	Bacterium (*Serratia marcescens*); Parasitoid fly (*Ormia ochracea*)	Increased egg production one day after infection by bacteria. No response to parasitoid infection (an obligate killer). No response to injection with Sephadex beads.	Fecundity compensation.	1
Fly *Drosophila melanogaster*	Mite (*Macrocheles subbadius*)	Infested males increase courtship behaviour in relation to infection intensity. Mite causes higher mortality.	Fecundity compensation by higher mating frequencies.	9
Snails *Biomphalaria glabrata* *Lymnaea stagnalis*	Trematode (*Schistosoma mansoni, Trichobilharzia ocellata*)	Infected snails increase reproductive output. Response occurs even when only exposed to water having contained parasite.	Fecundity compensation for impending castration by parasite.	8, 12
Lizard *Lacerta vivipara*	Blood parasite	Individual weight of offspring and total clutch mass increased when infected.	Fecundity compensation by increasing viability of young.	13
Bird *Parus major, P. caeruleus*	Hen flea (*Ceratophyllus gallinae*); Blow fly (*Protocalliphora* sp.)	Males increase rate of provisioning of young, thus increasing survival of young. Increased feeding rate when infected.	Fecundity compensation by increasing viability of already produced young.	3, 11
Amphipod *Corophium volutator*	Trematode (*Gynaecotyla adunca*)	Soon after infection, reproduction starts, and at a smaller size than normal.	Ensure reproduction before damage; fecundity compensation.	7
Amphipod *Gammarus insensibilis*	Trematode (*Microphallus papillorobustus*)	Infection has no effect on reproductive timing.	No effect.	10
Snail *Potamopyrgus antipodarum*	Trematode (*Microphallus* spp.)	Infected snails have smaller body size at maturity. When infected as juveniles, slower development and smaller body size.	Ensure reproduction before damage; fecundity compensation.	5, 6
Snail *Zeacumantus subcarinatus*	Trematode (*Maritrema novazealandensis*)	Infected snails develop faster and are smaller.	Reproductive assurance at lower fecundity.	4
Mosquito *Culex pipiens*	Microsporidium (*Vavraia culicis*)	Infected females mature at younger age and smaller body size; smaller body size correlates with lower fecundity.	Reproductive assurance at lower fecundity.	2

Sources: [1] Adamo. 1999. *Anim Behav*, 57: 117. [2] Agnew. 1999. *Proc R Soc Lond B*, 266: 947. [3] Christe. 1996. *Behav Ecol*, 7: 127. [4] Fredensborg. 2006. *J Anim Ecol*, 75: 44. [5] Jokela. 1995. *Evolution*, 49: 1268. [6] Krist. 1998. *Oecologia*, 116: 575. [7] McCurdy. 2001. *J Parasitol*, 87: 24. [8] Minchella. 1981. *Am Nat*, 118: 876. [9] Polak. 1998. *Proc R Soc Lond B*, 265: 2197. [10] Ponton. 2005. *Mar Ecol Prog Ser*, 299: 205. [11] Richner. 1999. *Oikos*, 86: 535. [12] Schallig. 1991. *Parasitology*, 102: 85. [13] Sorci. 1996. *Oikos*, 76: 121.

Infection or increased parasite presence in the habitat does not necessarily lead to a change in host life history, though. For example, juvenile snails infected by trematodes do not always show fecundity compensation later. Instead, the time of infection relative to the age at maturity is crucial for triggering a compensatory response (Ballabeni 1995; Gérard and Théron 1997; Krist and Lively 1998). Similarly, when crickets become infected by parasitoid flies (that eventually will kill them), no response towards earlier reproduction is observed (Adamo 1999). Infected hosts also often grow to an enormous size. This phenomenon—'gigantism'—has already been

discussed in section 8.3.1. As a cautionary note, it is not always sure that the observed changes in life history are an adaptive response by the host in the first place. These changes could also be under the control of the parasite and reflect strategies of host manipulation instead (see Chapter 8).

15.1.3 Host group living and sociality

Group living and sociality affects many characteristics of the host species' ecology and life history that matter for host–parasite interactions. In groups, a large number of hosts live in close spatial proximity and have many contacts with one another. Therefore, group living has various epidemiological consequences because transmission rates generally increase with the density of individuals (i.e. density-dependent transmission) and with the frequency of interactions between hosts. However, beyond host density in groups, sociality has an additional important dimension for parasitism. Social groups very often consist of closely related individuals, e.g. family groups of mammals or the colonies of social insects. Therefore, parasite transmission is among genetically similar hosts. This would additionally favour successful transmission and the within-group spread by an adapted parasite.

Several studies show that parasite prevalence often increases with social group size (Table 15.2). The effect of group size seems more pronounced for hosts that form huge colonies, as seen in many bird species (Rifkin et al. 2012). In mammals, prevalence and intensity increase with group size for contagious parasites that need close contact for transmission. The reverse pattern is observed for parasites that have mobile stages, capable of seeking out their hosts by themselves (Côté and Poulin 1995; Patterson and Ruckstuhl 2013). An increase in prevalence is typical, but it depends on many other factors. These include the host taxa, their mating systems, investment in immune defences, the particular parasite group, host specificity, and whether parasites transmit directly or indirectly (Altizer et al. 2003; Wilson 2003; Ezenwa 2004; Patterson and Ruckstuhl 2013; Schmid-Hempel 2017; Kołodziej-Sobocińska 2019).

Other measures, too, such as the species richness or species diversity of parasites, correlate with host social group size or host population density (Figure 15.2), but again not universally. In many cases, host group size is not, or is negatively, correlated with parasite diversity or with their specialization (Ezenwa et al. 2006b; Patterson and Ruckstuhl 2013). When social groups consist of genetically closely related individuals, and within-group transmission dominates the evolution of a parasite, this additionally corresponds to a kind of 'serial passage' situation (see Box 13.4). The scenario would, therefore, favour specialization on the social host group's genetic and physiological background.

Spatial proximity in a colony or nest and close genetic relatedness among hosts make colonies of social insects a rewarding target for parasites. Their societies are typically based on cooperation among close kin (i.e. genetically similar individuals). Furthermore, colonies can range in size from a few individuals (as in some wasps) to the millions (as in leafcutter ants). Physical interactions between individuals are frequent, e.g. when sharing food or caring for a brood, and the nest, therefore, provides a common ground for parasites to infect and spread (Schmid-Hempel 1998). Parasite loads, indeed, often increase with colony size. However, the opposite is also observed; that is, larger colonies have lower parasite burdens (Hughes et al. 2002).

Not only might the basic risk of contracting a parasite increase with group size but also the possibilities for defences. For example, larger colonies show more hygienic behaviour, as in ants (Hughes et al. 2008), and generally invest more into individual defences or can benefit from social immunity (Cremer et al. 2007) (see section 4.1.3). For example, the level of individual-level antimicrobial responses increases with the degree of sociality in bees (Stow et al. 2007), and fungus-exposed ant colonies use

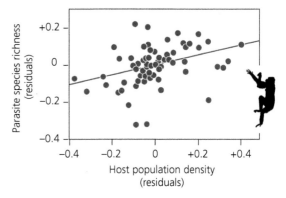

Figure 15.2 Effects of host density on parasite species richness in primates. Shown are the phylogenetically independent contrasts (residuals) for parasite richness (y-axis: number of parasite species reported from a host species) for free-ranging host populations, based on primate phylogeny, and corrected for the intensity of study. Parasites include viruses, bacteria, fungi, protozoa, helminths, protozoa, and arthropods. Host density (x-axis) is high in social groups, and the values shown are corrected for body mass (residuals). Across host species, parasite richness increases with host density. Silhouette from phylopic.org. Adapted from Nunn et al. (2003), with permission from University of Chicago Press.

Table 15.2 Host social group size and parasite infection.

Host	Parasite	Observation	Source
Guppies (*Poecilia*)	Cestodes (*Gyrodactylus*)	In experiments: grouped fish had higher peak loads and more persistent parasite loads than isolated fish. Effect smaller than scaled by group size.	16
Freshwater fish (60 spp.) in Canada	Contagious ectoparasites (copepods, monogeneans)	Mean number of parasite species (richness) does not correlate with group living.	14
Australian lizard (*Egernia stokesii*)	Apicomplexan blood parasites (*Hemolivia, Schellackia*)	No relation of prevalence to group size. Parasites transmitted by ticks. With *Hemolivia*, genetically closely related groups have higher prevalence.	9
Passerine birds (45 spp.)	Feather mites	Higher prevalence in group-living species.	13
Cliff swallows (*Hirundo pyrrhonota*)	Ectoparasites: fleas (*Ceratophyllus*), bugs (*Oeciacus*)	More ectoparasites in larger breeding colonies, and smaller nestlings with more bugs per individual. Experimental removal of bugs reduced effect on large colonies but not on small ones.	3
Barn swallows (*Hirundo rustica*)	Blow flies (*Protocalliphora hirundo*)	Parasite prevalence higher in larger colonies. Parasites are major source of nestling mortality.	15
Ground squirrel (*Xerus inauris*)	Ectoparasites (fleas, lice, ticks), endoparasites (helminths, protozoa)	No effect of group size. Males had more ectoparasites, and females more endoparasites.	10
Hoofed animals (96 spp.)	Various groups (601 spp.), incl. viruses, bacteria, protozoa, helminths, arthropods	Parasite species richness decreases with social group size, depending on the mating system (higher effect in monogamous species).	6
African bovids	Nematodes	Within species parasite prevalence varies with group size, and with host specificity of parasite. Territorial species more affected.	5
Grant's gazelle (*Nanger granti*)	Gastrointestinal nematodes	After helminths were removed by drugs, individuals in larger groups regain infections sooner.	7
Badgers (*Meles meles*)	Fleas (*Paraceras*)	Social group size, based on interconnected burrows, correlated with flea load. But group size within burrows (setts) not correlated.	11
Greater mouse-eared bat (*Myotis myotis*)	Wing mite (*Spinturix myoti*)	No relationship of prevalence and colony group size.	12
Grey wolves (*Canis lupus*)	Sarcoptic mange (mites; *Sarcoptes sabiei*)	Infection risk does not depend on group size, but wolves in groups survive better, likely because of social benefits with hunting and territory defence.	2
Monkeys in Amazonia	Malaria (*Plasmodium brasilianum*)	Infection prevalence increases with sleeping group size.	4
Monkeys (*Cercopithecus albigena, C. mitis, C. ascanius, Papio anubis*)	Intestinal protozoa	In *C. albigena* (species with largest sample size of groups) protozoan species richness increases with social group size. Otherwise similar with respect to group densities and land use.	8
Baboons (*Papio cynocepehalus*)	Helminths (e.g. *Trichuris*, strongyles)	Larger group sizes correlate with higher egg counts of *Trichuris*, but less for strongyles.	1

Sources: [1] Akinyi. 2019. *J Anim Ecol*, 88: 1029. [2] Almberg. 2015. *Ecol Lett*, 18: 660. [3] Brown. 1986. *Ecology*, 67: 1206. [4] Davies. 1991. *Funct Ecol*, 5: 655. [5] Ezenwa. 2004. *Behav Ecol*, 15: 446. [6] Ezenwa. 2006. *Oikos*, 115: 526. [7] Ezenwa. 2018. *Proc R Soc Lond B*, 285: 20182142. [8] Freeland. 1979. *Ecology*, 60: 719. [9] Godfrey. 2006. *Parasitol Res*, 99: 223. [10] Hillegass. 2008. *Behav Ecol*, 19: 1006. [11] Johnson. 2004. *Behav Ecol*, 15: 181. [12] Postawa. 2014. *Parasitol Res*, 113: 1803. [13] Poulin. 1991. *Condor*, 93: 418. [14] Poulin. 1991. *Oecologia*, 86: 390. [15] Shields. 1987. *Ecology*, 68: 1373. [16] Tadiri. 2016. *Parasitology*, 143: 523.

low-dose inoculations that protect the colony (Konrad et al. 2012). Similarly, large groups could share their defensive microbiota more effectively (Kaltenpoth and Engl 2014).

Furthermore, genetically more diverse colonies of social insects tend to have lower parasite burdens (Schmid-Hempel 1998; Tarpy 2003; Seeley and Tarpy 2007; Wilson-Rich et al. 2008; Desai and Currie 2015; Simone-Finstrom et al. 2016). Multiple mating by the mother, therefore, is another possibility to reduce infections. Although larger social insect colonies might generally be more prone to parasitic infections, the combination with size-dependent defence strategies could lead to a negative relationship between group size and parasite burden. In other groups of organisms, too, the parasite-induced costs of living in groups can be outweighed by the group-related benefits for avoidance and defence (Kappeler et al. 2015; Ezenwa et al. 2016). Such beneficial mechanisms include better resource acquisition, mutual help to remove ectoparasites (grooming behaviour) (Wilson et al. 2020), the learning of parasite avoidance strategies (Curtis 2014; Hart and Hart 2018), or social avoidance of infected group members (e.g. by odour in mandrills; Poirotte et al. 2017).

In all, parasitism is a potential cost to social groups and generally becomes more of a burden with larger and more tightly knit groups. At the same time, larger groups may be better able to tolerate infections (Ezenwa et al. 2016). The effects of parasitism, therefore, may become weak, and the costs of parasitism not visible, as long as environmental conditions remain favourable.

15.1.4 Regulation of host populations by parasites

Infection of hosts by parasites causes lower fecundity and higher mortality; many such examples have been mentioned in other chapters. These effects additionally depend on environmental conditions. Starvation, for example, reduces the capacity of geometrid moths to resist their parasitoids. Together with shortages of food, parasitism could thus cause the population crashes of these moths (Yang et al. 2007). Likewise, an infection can render hosts more susceptible to adverse environmental conditions or increased predation. The latter may also result from host manipulation to increase transmission to a final host (see Chapter 8).

However, effects such as the general weakening of the host, its inability to clear an infection, or a decline in sensory acuity are quite general for any physiological stress. In one study, for example, tadpoles (*Rana aurora*) infected by the yeast *Candida humicola* were less able to recognize odours emanating from predators (newts) than their uninfected counterparts; they suffered increased predation rates (Lefcort and Blaustein 1995). Other studies provide correlative evidence that lower immune defence capacities, or higher parasite burdens, relate to increased predation. Parasitic infections (by helminths and coccidia) also lower the escape capacity of hares (Alzaga et al. 2007), and make hosts more prone to predation in a variety of other examples, such as game birds (Hudson et al. 1992b), or moose that are shot by hunters or caught by wolves (Joly and Messier 2004).

By increasing mortality and reducing fecundity, parasites affect host population size and abundance, e.g. population density (the number of individuals in a given area). However, to regulate the size of populations, parasite effects should act in a density-dependent manner, such that the population is reduced when large, but can increase when small (Box 15.1). This is, therefore, not equivalent to demonstrating a negative effect on the host population per se, for instance in the context of controlling a pest species, or when managing crops or aquacultures. For regulation, parasite-induced mortality rate should increase (or fecundity decrease) as the host population increases in size (Table 15.3).

Density dependence results from two different processes that have been identified in ecology more generally: with a generalized 'functional response', functional parameters such as infection rate, or the per capita effect on the host, vary with host abundance. For example, when hosts are in denser populations, more intense competition for resources would reduce host condition, thereby lower defence levels, and generally increase rates of parasitism. Measures of immune defence are indeed condition dependent in taxa as different as birds (Alonso-Alvarez and Tella 2001; Ewenson et al. 2001) and insects (Lee et al. 2008b), with unfavourable environments associated with lower defence levels.

Box 15.1 Basic population ecology

The size of a population, i.e. the number of individuals, is an essential variable in ecology. Large populations, for example, can dominate the habitat they live in and consume most of the available resources, and are buffered against disturbances. Small populations run the risk of increased inbreeding and are more prone to extinction. Classical ecology suggests that population size is both determined and regulated (Figure 1).

Populations are 'determined' because they have a specific size that is set by various factors. For example, populations can reach a 'carrying capacity' (K) in their environment. K is determined by food availability, habitat productivity, competi-

tive interactions, parasite pressure, and so forth. Populations are 'regulated' because they decrease in size when too large and grow when too small. (What exactly 'too large' and 'too small' mean is beyond the scope of this section.) Regulation implies that the growth of the population (a decrease or increase) depends on the size of the population—population growth is said to be 'density dependent'.

The search for processes that regulate population size is a long-standing problem in ecology. The simple model of Figure 1 is hardly met in reality. For example, K is a theoretical limit that is difficult to identify. Nevertheless, the model serves as a useful metaphor to illustrate the principles of determining and regulating processes in populations.

Populations change their size by changes in birth (b) and mortality rates (μ). If the population is not 'closed', size also changes because individuals can immigrate or emigrate. For closed populations, the population grows when $b > \mu$, it shrinks when $\mu > b$. The (net) growth rate is $r = b - \mu$, where r is the per capita rate of increase (also known as the 'Malthusian parameter'). If b and μ stay constant over time, r will remain constant, too. With $b > \mu$, the population grows exponentially with rate, r, such that population size, n, at time t is $n_t = n_0 e^{rt}$, with n_0 the initial size. The population size, n_t, will eventually increase to infinity with time.

No natural population follows such a trajectory. Therefore, b and μ must be changing over time, e.g. in a density-dependent manner as sketched in Figure 1. This can happen, for example, with increased competition for resources, more aggressive territorial behaviour, or a shortage of nest sites, etc. These limitations lead to fewer births (b) and more deaths (μ). Concerning parasitism, increasing population size leads to higher host densities, which facilitates transmission and might favour more virulent parasites associated with even higher mortality rates.

With density dependence, r continuously decreases as the population grows, and converges to $r = 0$ when the population reaches its carrying capacity, K. The maximum value of r is, therefore, only reached at the beginning when the population grows at its maximum potential. Note that when using discrete time intervals, the growth rate, r, turns into a growth ratio per generation (the interval), which is equivalent to $R_0 - 1$ used in epidemiology (see Chapter 11).

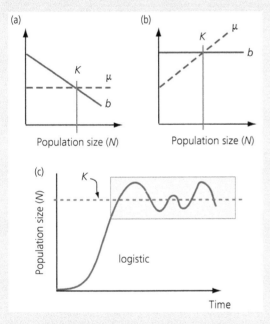

Box 15.1 Figure 1 Idealized population growth dynamics. (a) Birth rate, b, declines with population size. (b) Mortality rate, μ, increases with population size. In both cases, the net growth rate of the population declines to zero as N approaches the 'carrying capacity' in the environment (K). The level of K is a determining factor for population size. (c) The density dependence of b and μ is a regulating process that keeps a population around its carrying capacity, K (grey area). Initially, the population shows a logistic growth.

Table 15.3 Effects of parasites on host population and species.

Host	Parasite	Observation[1]	Source
Solitary bee *Osmia rufa*	Parasitoids, cleptoparasites	Inverse density dependence of rate of parasitism, i.e. strong effects of parasites only in smaller populations. Cannot regulate in density-dependent manner. 39 per cent of provisioned brood cells lost to parasites.	18
Polistes wasps	Parasitoids	Regulation associated with seasonal cycle of wasps. Numerical response important.	14
Leafhoppers (a plant pest)	Parasitoids	Rate of parasitism positively correlated with host density, at spatial scale level of fields.	16
Freshwater isopods *Austridotea*	Trematodes	(E) In mesocosm: Four 'sympatric' species. Habitat selection differentially affected by parasitism.	6
Intertidal snail *Zeacumantus subcarinatus*	Trematodes	Parasites reduce individual fecundity. With increasing parasite prevalence, host populations have higher mortality and lower density. Effect somewhat stronger at high population density.	5
African buffalo	*Mycobacterium bovis*	Infection reduces fecundity and increases mortality in prime-age bulls but compensates for other mortality factors in aged individuals. Regulation effect on population through density-dependent effects on prime-age groups.	8
Bridled goby *Coryphopterus glaucofraenum*	Copepod gill parasite (*Pharodes tortugensis*)	Negative effects of high density and refuge shortage more severe for infected hosts.	4
Caddisfly *Brachycentrus americanus*	Microsporidium	Density-dependent infection rate with time lag of one generation. Path analysis shows strong effect of parasites on density-dependent growth of host population. Study over 15 years.	9
Red grouse *Lagopus lagopus scotticus*	Nematode (*Trichostrongylus tenuis*)	Similar climatic condition across wide geographic areas can synchronize population cycles. Effect likely due to parasite transmission, i.e. parasites trigger epidemic outbreaks and force populations into synchrony. (E) Reduction of parasites leads to larger broods, higher population densities, and, partially, smaller population losses in autumn and spring. Combination of parasitism and territoriality (by male aggressiveness) explains population cycles.	2, 11, 12
Rock partridge *Alectoris graeca saxatilis*	Nematode	Cyclic populations have less aggregated parasites than non-cyclic host populations. According to theory, weak aggregation favours regulation.	13
Vole *Arvicola, Microtus, Myodes*	Nematode (*Trichuris arvicolae*)	Parametrized models show that parasite can regulate host populations, but not drive cycles.	3, 17
Mouse	Nematode (*Heligomosomoides polygyrus*)	(E) Parasite introduced into large enclosures with mice. Under conditions favourable for transmission (wet soil), host population size suppressed to a dramatically lower level (by a factor of 20) over a period of four months. Experimental removal of parasite (anti-helminthic drug) leads to an increase in the host population.	15
Reindeer *Rangifer tarandus Platyrhynchus*	Nematodes	Experimental infections and modelling show that nematodes affect host body condition and female fecundity in density-dependent ways and are thus likely to regulate host population sizes.	1, 19
Soay sheep	Nematode (*Teladorsagia circumcincta*)	(E) Removal of parasites from sheep by anti-helminthic drug increases winter survival, which is an important factor for population regulation.	7
Darwin's finches	*Philornis* fly	Simulations show effects of reduced infestation on population growth.	10

[1] E: experimental study.

Sources: [1] Albon. 2002. *Proc R Soc Lond B*, 269: 1625. [2] Cattadori. 2005. *Nature*, 433: 737. [3] Deter. 2008. *Eur J Wildl Res*, 54: 60. [4] Forrester. 2006. *Ecology*, 87: 1110. [5] Fredensborg. 2005. *Mar Ecol Prog Ser*, 290: 109. [6] Friesen. 2018. *Biol Lett*, 14: 20170671. [7] Gulland. 1993. *Proc R Soc Lond B*, 254: 7. [8] Jolles. 2006. *Am Nat*, 167: 745. [9] Kohler. 2001. *Ecology*, 82: 2294. [10] Koop. 2016. *J Appl Ecol*, 53: 511. [11] New. 2009. *Am Nat*, 174: 399. [12] Redpath. 2006. *Ecol Lett*, 9: 410. [13] Rizzoli. 1999. *Parasitologia*, 41: 561. [14] Rusina. 2013. *Entomol Rev*, 93: 271. [15] Scott. 1987. *Parasitology*, 95: 111. [16] Segoli. 2016. *Biol Control*, 92: 139. [17] Smith. 2008. *J Anim Ecol*, 77: 378. [18] Steffan-Dewenter. 2008. *Ecology*, 89: 1375. [19] Stien. 2002. *J Anim Ecol*, 71: 937.

Sometimes, measures of immunocompetence increase with host density, e.g. with 'density-dependent prophylaxis' in the desert locust. These upregulate phenoloxidase activity and resistance to fungi when in dense populations (Wilson et al. 2002). As a result, the infection rate would decrease with increasing host density. A functional response appropriate for density-dependent regulation could also result from a rapid evolution of the parasite population; for example, when parasites evolve to become more virulent as the host population increases in size.

With a 'numerical response', the number and density of parasites disproportionally vary with host population size, e.g. when parasites aggregate in high-density host patches. Mortality rate would thereby disproportionally increase in large or dense host populations and reduce further population growth. The numerical response envisaged here concerns the distribution of parasites among different (sub)populations, rather than the distribution and aggregation of parasites among host individuals within a population (see section 11.5). According to theory, regulation is more likely to occur when parasites are randomly distributed among individual hosts than when aggregated (May and Anderson 1978; Tompkins et al. 2001a).

Some evidence exists, therefore, that parasites can regulate host populations in size, and in the wild, although much remains to be studied (Table 15.3) (Myers 2018). A classic example is nematodes in free-ranging populations of the red grouse in Scotland; their removal by drug treatment leads to the disappearance of host population cycles (Hudson et al. 1998, 2002). The processes of parasite-induced host population regulation are complex and include a range of factors.

15.1.5 Host population decline and extinction

The history of biological pest control demonstrates that host populations are reduced in size by predators and parasites (Hajek and Eilenberg 2018), even though ecological conditions can mitigate against the effects (Tscharntke et al. 2016). Several cases of historical species extinctions are associated with parasites. An example is the extinction of endemic birds after bird malaria was introduced

in Hawaii (Warner 1968). Also, the spread of the African rinderpest at the end of the nineteenth century (Box 15.2) demonstrates severe declines of host populations. The charismatic Tasman tiger, the thylacine (*Thylacinus cynocephalus*), became extinct from a combination of excessive hunting and a viral disease (de Castro and Bolker 2005).

Headlines are prominent when iconic host species are affected. For example, the Tasman devil (*Sarcophilus harrisii*) is threatened by a contagious viral infection causing facial tumours (Bostanci 2005). A vaccine may help to protect these populations (Flies et al. 2020). Similarly, the populations of African wild dogs (*Lycaon pictus*), and lions (*Panthera leo*) in the Serengeti, were severely decimated by the spread of a canine distemper virus (CDV, Morbilivirus) in the 1994 epidemic. Vaccines against CDV may aid in conserving the African wild dog (Loots et al. 2017). CDV and rabies virus (RABV) also hit the highly endangered Ethiopian wolf (*Canis simensis*) (Gordon et al. 2015). Attempts to control RABV by vaccination of wolves and, in particular, the reservoir of domestic dogs, have been successful but require continued efforts (Haydon et al. 2006; Sillero-Zubiri et al. 2016; Marino et al. 2017).

Human populations are a reservoir, too, such as for polio and measles viruses that were transferred to endangered species including chimpanzees and mountain gorillas, respectively (Cleaveland et al. 2001a). Since 1993, and starting from the Eastern United States, a bacterial epidemic, *Mycoplasma gallisepticum*, has been spreading among songbirds, causing mycoplasmal conjunctivitis. The population abundance of the most affected house finches (*Haemorhous mexicanus*) decreased to some 40 per cent of the disease-free abundance (Dhondt et al. 2005). The bacterium has since expanded its range westwards, and is infecting a number of other bird species, too (Ley et al. 2016).

Entire groups can become threatened, too. 'White-nose disease' (caused by the fungus *Pseudogymnoascus destructans*) caused severe declines in at least six species of bats (Thogmartin et al. 2013); similarly, the West Nile virus (WNV) severely affected 23 out of 49 investigated bird species (George et al. 2015). One of the worst disease-induced extinction waves ever recorded in the wild is the current pandemic chytridiomycosis of amphibia, caused by fungal infections (chytrids, *Batrachochytrium* spp.) in combination with deteriorating ecological conditions, including global warming. In the late 1980s, it was first noted that many species that were abundant before were suddenly in severe decline, had come to the brink of extinction, or had already disappeared. Now, chytrid fungi infect over 700 amphibian species on all continents (Lips 2016).

A particularly difficult situation exists in Central and South America, where more than half of all described amphibian species live. For example,

about two-thirds of the *c.*110 New World tropical species of the genus *Atelopus* (the harlequin frogs) have vanished from Central America (Pounds et al. 2006). Only ten of the species are not threatened. At least 40 species have not been seen for years and are presumably extinct (La Marca et al. 2005; Lips 2016). Similar extinctions are observed in the Atlantic forests of Brazil (Figure 15.3). A conservative estimate is that the diseases have led to the decline of at least 500 amphibian species worldwide (Scheele et al. 2019).

Box 15.2 The African rinderpest epidemic

Rinderpest is a disease of cattle that has been known since antiquity. It is caused by the rinderpest virus (RPV), a morbillivirus related to measles and distemper virus. Distemper also affects cats, dogs, seals, and a variety of other carnivorous mammals. RPV presumably originated in Asia, from where it spread to Europe more than once, causing several epidemics, mainly in the eighteenth century. Rinderpest had, for example, also been prevalent in Egypt and was spreading southwards around the middle of the nineteenth century. However, the most severe epidemic occurred in the late nineteenth century in Africa.

The most likely account of this Great African Epidemic is as follows. By 1887, the Italian Army had invaded and occupied Ethiopia. To feed the soldiers, the authorities imported cattle from India to their camps. However, some of the cattle were infected and carried the rinderpest virus. Subsequently, the imported animals transferred the disease to other livestock. The virus also jumped into populations of wild animals. Within a decade, the epidemic had spread from Ethiopia through the interior of Africa to reach the Cape, where around 2.5 million cattle died (Mack 1970) (Figure 1).

In the course of the epidemic, an estimated 80–90 per cent of livestock perished. Besides, wild ungulates such as antelopes, buffaloes, and giraffes were heavily affected, too. As a result of the loss of large natural herbivores, the vegetation changed, and the landscape became covered with dense bush. This, in turn, allowed specific insect populations to flourish. Tsetse flies, in particular, increased in numbers and spread to settled areas. Tsetse flies are the classical vector of human sleeping sickness (caused by the protozoan *Trypanosoma brucei*), which therefore also increased in prevalence. As their livestock died, the Masai people of East Africa starved and were severely reduced in numbers. Together

with the subsequent smallpox epidemic, this resulted in large tracts of emptied land that likely favoured the European colonizers (Mack 1970). Elsewhere in Africa, too, many millions of people died due to loss of their livestock.

Box 15.2 Figure 1 The Great African rinderpest epizootic 1887–1897. The epidemic originated in 1887/1889 from infected, imported cattle at the Horn of Africa, and swept to Cape Town within ten years. The Zambezi river is several kilometres wide and provided a natural barrier. It halted the spread of the virus for about three years. Eventually, a herd of infected cattle was taken across the river and sold, which triggered the further, southward spread. Adapted from Mack (1970), with permission from Springer Nature.

At the time, the Cape Colony had one of the few functioning medical services in Africa, and Arnold Theiler (1867–1936), a Swiss-born veterinarian and father of Max Theiler (of yellow fever Nobel fame), was charged with fighting the disease by the British government. By first developing a culling strategy for cattle and later developing a vaccine, he succeeded in containing the epidemic by 1897.

In Europe, the last large rinderpest epidemic was recorded in Bulgaria, 1913, as a consequence of the war in the Balkans, and 1920 in Belgium. In Africa, the Pan-African Rinderpest Campaign (PARC), an organization active in 34 countries, was launched in 1986 by the OAU (Organisation of African Unity). Its goal was to eradicate the disease by information campaigns, vaccination programmes, and improved veterinary services. PARC coordinates information and expertise, maintains a vaccine bank, and finances national efforts. It is supported by the European Union (EU), the

United Nations Food and Agricultural Organization (FAO), the United States Agency for International Development (USAID), and many other funding agencies. The last outbreak of rinderpest in the world was reported in 2001 from Kenya. By 2009, rinderpest had basically disappeared. In a few African countries, the Arabian Peninsula, Russia, and a few Central and Southeast Asian countries, the final assertion is still pending. However, by the end of 2010, the FAO was confident that all known lineages of the viruses had disappeared. Therefore, the FAO Conference of June 2011 in Rome adopted the resolution that rinderpest is globally eradicated. A collection of strains is kept in the laboratory as a safety net to combat an unexpected reappearance of the disease. Besides smallpox (and, foreseeably, polio), rinderpest is the other great triumph of vaccination that has led to the eradication of a horrible disease. It needed the efforts of many, investment of money, and the insight of science.

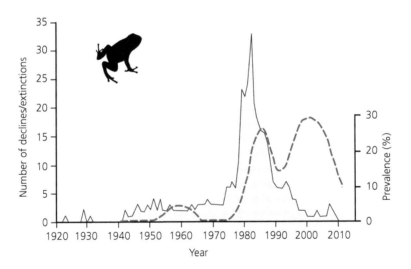

Figure 15.3 Extinction of amphibia in the Atlantic Forest of Brazil. Shown is the number of population declines and extinctions reported over 95 years (curve and green area, left axis). The dashed red line is the prevalence of tadpoles found infected with the fungus *Batrachochytrium dendrobatidis* (right axis). Rising infections associate spatio-temporally with species declines and extinctions; in later years, many species were already extinct, and populations may have become more resistant. Silhouette from phylopic.org. Adapted from Carvalho et al. (2017), with permission from The Royal Society.

Comparable reports for invertebrate populations are not as detailed but point in the same direction. For instance, land snails of the genus *Parula* are highly endangered, and some species have depended on breeding in captivity for their survival. A microsporidian infection in a captive population of *P. turgida*

led to the death of the last known individuals of this species in London Zoo in 1996 (Cunningham and Daszak 1998). Based on records for bumblebees (*Bombus* spp.) over the past decades, besides habitat change, parasitic diseases may also account for the dramatic decline of several native species of bumblebees in North America (Cameron et al. 2011, 2016). Similarly, the decline of honeybees has caught attention. Among other factors, infections by the *Varroa* mite and associated viruses (McMahon et al. 2016; Wilfert et al. 2016) inflict serious stress on these populations. Parasites are thus critical, but several factors together are responsible for bee decline (Meeus et al. 2018).

According to the Red List of the International Union for the Conservation of Nature (IUCN) in 2020, around one-quarter of all mammals are currently threatened with extinction. Among those, many populations are considered to be threatened by diseases (Figure 15.4). Viruses are the most frequently involved parasites. Moreover, directly transmitted, generalist parasites are more important than more specific parasites that are transmitted by vectors or intermediate hosts (Pedersen et al. 2007). Estimates based on the historically documented extinctions since the year 1500 suggest that infectious diseases seem involved in global extinctions in less than 4 per cent of all cases (affecting 833 species of plants and animals) and in less than 8 per cent of species

currently classified as critically endangered (2852 species) (Smith et al. 2006). However, these estimates are troubled by generally insufficient data, short observation periods, and the sheer diversity of taxa and systems (McCallum 2016). These estimates are likely to severely underestimate the real role of diseases.

There are general ecological and biogeographic factors that obviously worsen a situation. For example, amphibia particularly prone to decline by chytridiomycosis have a restricted geographical range (e.g. a narrow altitudinal range) and large body size, and live in aquatic habitats that facilitate frequent transmission by spores (Blooi et al. 2017; Scheele et al. 2019). Small ranges and large body size typically mean small population sizes and, thus, higher extinction risks more generally (Ducatez and Shine 2017; Chichorro et al. 2019). A spreading infectious disease threatens, moreover, populations and species that are in poor condition for other reasons.

Taken together, parasitism and infectious diseases often act as the ultimate push to extinction after

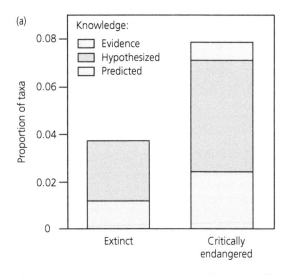

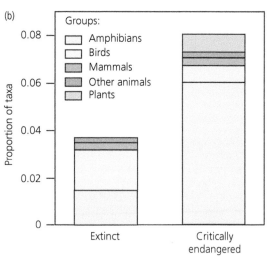

Figure 15.4 Species extinction due to parasites. (a) Proportion of known species extinctions facilitated by disease (after the year 1900) and critically endangered species threatened by disease according to the World Conservation Union Red List (IUCN 2004). Knowledge level—'Evidence': the proportion of taxa for which the threat is known to exist; 'Hypothesized': the proportion of taxa for which the threat is only a hypothesis, i.e. no evidence currently exists to support the claim, or ongoing research is attempting to discern the threat; 'Predicted': the proportion of taxa for which parasites are expected to become a threat in the future. (b) Proportion of known species extinctions facilitated by disease (after 1900), and critically endangered species, ordered by taxonomic group. 'Other animals' include invertebrates, fishes, and reptiles. Adapted from Smith et al. (2006), with permission from John Wiley and Sons.

other processes have already taken their course. Indeed, the risk of extinction by parasites rapidly increases as a species become more and more endangered anyhow, as can be seen when relating disease risk to status ranking in the Red List of the IUCN from 'least concern' to critically 'endangered' (Heard et al. 2013). The exact path to extinction and the contributing factors are also not always straight-forward and can take different routes. Nevertheless, parasites can push a species towards a critical path to start with and then also cause its final disappearance (Hilker et al. 2009). Likewise, parasites can go extinct as well, e.g. when host populations pass through a bottleneck (Hesse and Buckling 2016). Similarly, active management of threatened species, such as with the removal of feather lice during captivity breeding programmes for the highly endangered California condor, leads to parasite extinction (Koh et al. 2004).

15.2 Host ecological communities

Ecological communities consist of populations of different species in a given area. The role that parasites might play in structuring their host communities has been discussed for some time, and the subject was reviewed early on (Holmes 1982; Freeland 1983; Minchella and Scott 1991).

15.2.1 Parasite effects on host competition

Competition within and between species is an essential process that structures communities of organisms, i.e. what number and abundance of co-existing species occur. Parasitic infections alter host phenotypes and, therefore, affect the ecological profile of their hosts, e.g. habitat choice, diet, foraging activities, or predator avoidance. This changes the ecological interactions with other species, including levels of competition (Lefèvre et al. 2009). Modifying 'apparent competition' is probably the most critical effect that parasites have on the co-existence of their hosts, and thus on host community structure (Price et al. 1988; Holt and Lawton 1994). With apparent competition, hosts compete via the differential effects that parasites have on them. In a classical study, two species of the flour beetle *Tribolium* (*T. castaneum*, *T. confusum*) only co-existed when a parasite was present (a microsporidium). Otherwise, one species (*T. castaneum*) was driving the other to extinction, being the superior competitor in the absence of parasites (Park 1948). There are many examples of such parasite-mediated competition (Table 15.4). In most cases, the effects of apparent competition result from different susceptibilities of the host species.

15.2.2 Communities of hosts

In contrast to predators, parasitic infections do not immediately consume their hosts, such that there is a time lag in effects. Regardless, they are 'direct consumptive effects' by the parasite (i.e. utilizing the host as 'food'). For example, myxomatosis virus 'consumes' rabbits, and therefore competes with the predators, such as foxes or lynx, at the inter-kingdom level (Janzen 1977). 'Non-consumptive effects' include changes in life history and host avoidance behaviour, changed competitive behaviour (see above), modified dispersal patterns, and so forth, which can change host ecology (Lefèvre et al. 2009; Buck et al. 2018; Buck 2019). Hence, parasites affect the ecological interactions of their hosts in many and subtle ways, which has ripple effects throughout the host community (Hatcher and Dunn 2011). Theory can analyse how parasite-induced changes impact on the stability of host communities, e.g. by causing fluctuations in species abundance (Fenton and Rands 2006; Rogawa et al. 2018).

Empirical studies have covered many different groups of parasites and their hosts, suggesting that the role of parasitism in structuring host communities is rather general (Table 15.5). In particular, parasites that infect the dominant competitors in the community can play a similar role as the 'keystone predators' in classical ecology. For example, experimental removal of a generalist, parasitic vine (*Cuscuta salina*) in a salt marsh community of plants confirmed that the parasite suppressed the dominant competitor and had promoted species diversity (Grewell 2008).

These effects might generally be more likely for microparasitic infections because these tend to be less aggregated, a factor known to be more effective for host population regulation in the first place (Fenton and Brockhurst 2008). Recent metagenomics

Table 15.4 Parasite effects on host competition.

Host	Parasite	Observation[1]	Source
Plants *Lolium, Trifolium Festuca, Holcus*	Hemiparasitic plant (*Rhinanthus minor*)	(E) In mixed host plant settings, parasite effect on *Trifolium* weaker than on *Lolium*; the same finding when *Trifolium* planted with other plants.	7
Isopoda *Austridotea*	Trematodes	Segregation of microhabitats in two co-existing species driven by parasites. In particular, one species selects habitat when infected.	6
Crayfish *Oronectus rusticus*	Trematode (*Microphallus* sp.)	(E) Infection reduces feeding rate (at short-term scale) and increases predation risk. Infection can change aquatic community.	11
Beetles *Tribolium confusum T. castaneum*	Trematode (*Hymenolepis diminuata*)	(E) *T. castaneum* superior competitor. This advantage even increases when parasite is present. Effect perhaps due to changed feeding behaviour.	13
Ants *Pheidole dentata Solenopsis texana*	Parasitoid (phorid flies) (*Apocephalus*)	*Pheidole* changes foraging behaviour and avoids day times when flies are present. This gives inferior competitor (*Solenopsis*) an advantage.	4
Flies *Drosophila putrida D. falleni*	Nematode (*Howardula aoronymphium*)	(E) Competitive superiority of *D. putrida* is lower when parasite is present. This fits with observation that *D. putrida* is more abundant in the field where parasite is absent.	9
Flies *Drosophila mela-nogaster D. simulans*	Parasitoid (*Leptopilina boulardi*)	Parasitoid preferentially attacks *D. melanogaster* that is superior competitor. Co-existence of both species observed when parasite is present.	3, 5
Moths *Plodia interpunctella Ephestia kuehniella*	Parasitoid (*Venturia canescens*)	(E) *E. kuehniella* is eliminated by *P. interpunctella* when parasitoid is present but can persist when alone. Effect due to numerical response of the parasitoid.	1, 2
Frogs *Rana cascadae Hyla regilla*	Water mould (*Saprolegnia ferax*)	Outbreak of parasite affects larvae of the two frogs differentially and reduces *Rana* more than *Hyla*.	10
Lizards *Anolis*	Malaria (*Plasmodium azurophilum*)	*A. gingivinus* superior competitor against *A. wattsi*. However, reptile malaria affects *A. gingivinus* more strongly; this species is absent where malaria is prevalent.	12
Hares and rabbits	Intestinal helminth (*Graphidium strigosum*); myxoma virus	Helminths more prevalent in shady woodlands; infected hares become excluded in favour of rabbits. The rise and fall of hare and rabbit populations likely due to these effects.	8

[1] E: Primarily an experimental study.

Sources: [1] Bonsall. 1997. *Nature*, 388: 371. [2] Bonsall. 1998. *J Anim Ecol*, 67: 918. [3] Boulétreau. 1991. *Redia*, 84: 171. [4] Feener. 1981. *Science*, 214: 815. [5] Fleury. 2004. *Genetica*, 120: 181. [6] Friesen. 2018. *Biol Lett*, 14: 20170671. [7] Gibson. 1991. *Oecologia*, 86: 81. [8] Hudson. 1998. *Trends Ecol Evol*, 13: 387. [9] Jaenike. 1995. *Oikos*, 72: 235. [10] Kiesecker. 1999. *Ecology*, 80: 2442. [11] Reisinger. 2016. *Ecology*, 97: 1497. [12] Schall. 1992. *Oecologia*, 92: 58. [13] Yan. 1998. *Ecology*, 79: 1093.

studies (Zhang et al. 2018a) show that viral infections are prevalent in nature, including infections by several viruses in a single host. If common, viral infections, by their known consequences for interactions between individuals, are particularly significant factors for populations and entire communities of their hosts, too (French and Holmes 2020). Generalist pathogens pose the greatest threat in this context because, by definition, they can infect a wide range of hosts (Cleaveland et al. 2001b).

15.2.3 Food webs

Food webs are an alternative way of describing and analysing ecological communities. In food webs, species take the places of nodes in a network.

ECOLOGY 431

Table 15.5 Parasite effects on host communities.

Host	Parasite	Observation[1]	Source
Grasses, forbs	Parasitic plants (*Rhinanthus* spp.)	Parasite strongly reduces biomass of host plants; effects differ by functional host groups. In one study, presence of parasite had positive effect on species richness. Meta-analysis.	1
Plants in a coastal wetland	Holoparasitic vine (*Cuscuta salina*)	(E) Parasites primarily affect superior competitor, the salt marsh plantain (*Plantago maritima*). Experimental removal confirms that species diversity in the host community increases with parasite.	6
Plant zonation in marshland	Parasitic plant (*Cuscuta salina*)	Presence of parasite increases species diversity in community and changes zonation of plants along the marshland. Effect strongest when competition between hosts is highly asymmetrical.	3
Woodland in East Africa	Anthrax	Anthrax outbreaks drastically reduce impala populations. As a result, acacia trees can establish better, and shrub cover increases.	10
Plant communities	Various pathogens	Many examples of density-dependent and density-independent effects of parasites. Reversal of competition among plants if infected.	8
Invertebrates on coastal flats	Trematodes	The cockle *Austrovenus stutchburyi* is prevented from burrowing into sand when infected by trematodes. Cascading effects through the community, such that rate of parasitism correlates with species diversity in the overall community.	9
Freshwater crustaceans	Trematode (*Maritrema poulini*)	(E) Differential effects on host species changes relative abundance of species in the community.	5
Intertidal snails and algae	Trematodes	Infection reduces algal grazing by snails. This in turn changes algal communities and affects other dependent species.	11
Phytophagous insects	Parasitoids	Apparent competition generates 'enemy-free space' that allows co-existence of host species and structures communities.	7
Deer (Cervidae) in North America	Meningeal worm (*Paralophostrongylus tenuis*).	Only white-tailed deer (*Odacoileus virginatus*) is able to persist in areas that are heavily infested by worms.	2
African ungulates	Rinderpest virus.	Ungulate and other wild species (warthogs, etc.) differentially affected. Pan-African epidemic changed composition of ungulate communities.	4

[1] E: Primarily an experimental study.

Sources: [1] Ameloot. 2005. *Folia Geobot*, 40: 289. [2] Anderson. 1972. *J Wildl Dis*, 8: 304. [3] Callaway. 1998. *Oecologia*, 114: 100. [4] Dobson. 1995. In: Sinclair, eds. *Serengeti II: Research, management, and conservation of an ecosystem*. Chicago University Press. [5] Friesen. 2019. *Parasitology*, 147: 182. [6] Grewell. 2008. *Ecology*, 89: 1481. [7] Holt. 1993. *Am Nat*, 142: 623. [8] Mordecai. 2011. *Ecol Monogr*, 81: 429. [9] Mouritsen. 2010. *Mar Biol*, 157: 201. [10] Prins. 1993. *J Ecol*, 81: 305. [11] Wood. 2007. *Proc Natl Acad Sci USA*, 104: 9335.

Connections between these nodes indicate the trophic relationships (i.e. who consumes whom). The strength of the connection varies with the rates of consumption, e.g. the average transfer of biomass from a host species to its parasite. To investigate these links and their strengths is not an easy task.

Parasites themselves can represent a considerable fraction of the total biomass in a given system (Kuris et al. 2008). Furthermore, manipulative parasites affect food webs, too, as they change the phenotype of their host (Sato et al. 2019). A changed behaviour, in turn, can put the host into a habitat that it usually avoids, and so establish a new link in the food web. For instance, trematode-infected cockles no longer burrow themselves into the sand, so they become prey to new predators (Mouritsen and Poulin 2010). Similarly, crickets infected by nematomorph parasites enter water bodies such as streams where the parasites reproduce. Crickets thereby become a substantial new food supply for fish, with consequences for the stream communities (Sato et al. 2012).

Changed behaviour can also increase consumption rates within existing links of the food web.

An infection reduces escape behaviour and makes the prey more vulnerable to their usual predators (Lafferty et al. 2008). 'Predation' on parasites, i.e. the collateral consumption of parasites with the prey, is an important element of food web structure. In coastal marine ecosystems, for example, an estimated 4–14 predators exist for every parasite life cycle stage (Thieltges et al. 2013), and a majority of interactions in food webs are probably between parasites and their hosts (Lafferty et al. 2006). Parasitism can also stimulate and reroute the flow or mobilize nutrients through the food web (Kuris et al. 2008; Sánchez Barranco et al. 2020). The parasitic infection can furthermore decrease the consumption of food, as with reduced feeding by isopods infected with acantocephalan worms (Hernandez and Shukedo 2008). Bees infected by parasitoids (conopid flies) change their floral preferences, thus pollinating a different set of flowering plants from their healthy counterparts (Schmid-Hempel and Schmid-Hempel 1990; Schmid-Hempel and Stauffer 1998); comparable effects exist for protozoan infections of bee pollinators (Otterstatter et al. 2005).

Parasites in food webs, therefore, are essential elements and nodes, but their role has only been studied more recently (Lafferty et al. 2008; Jephcott et al. 2016; Michalska-Smith et al. 2018). Considerable uncertainty is still surrounding these issues. For example, should parasites be placed at the top of the food web and the biomass pyramid? If so, there could only be a limited sustainable mass of parasites in this ecosystem, simply because the 'space' at the top is somewhat limited. Nevertheless, adding parasites at the top is just one of several alternative placements that have different effects. Adding parasites changes connectance (i.e. how many species have ecological interactions with each other), nestedness, and the length of trophic chains in the food web. Parasites might thus render food webs, under certain conditions, more stable. On the other hand, food webs with parasites can be vulnerable when a key host is removed (Lafferty et al. 2006). Losses of many connections in the food web, and disproportional extinction of parasites, have carry-over effects on other host species in the community (Lafferty and Kuris 2009). Hence, parasites can affect food webs in many ways, and it remains a formidable challenge to investigate these effects in natural ecosystems (Jephcott et al. 2016).

15.2.4 Dilution effect

In many different systems, the community-wide host species diversity correlates negatively with the prevalence of pathogen infections in the individuals (Ostfeld and Keesing 2012; Civitello et al. 2015; Huang et al. 2017). For example, the prevalence of Sin Nombre virus (a hantavirus) decreases with increasing species diversity of small mammal hosts (Clay et al. 2009; Dizney and Ruedas 2009); the effect is supported by experimental manipulation of the host species' abundance in a field experiment (Suzán et al. 2009).In the Eastern United States, the incidence of WNV that is vectored by mosquitoes in the human population is lower in areas with high richness and diversity of bird species, even when a number of potentially confounding variables (climate, vector species, socioeconomic factors) are controlled for (Swaddle and Calos 2008; Allan et al. 2009) (Figure 15.5). Experimental tests support these observations. For example, the prevalence of infections by *Borrelia* in ticks, a significant risk factor for human Lyme disease, is reduced by the addition of further host species, e.g. squirrels, to an otherwise impoverished mouse host community (LoGiudice et al. 2003). Similarly, with more species of snails experimentally added, a reduced prevalence of *Schistosoma* in these intermediate hosts results (Johnson et al. 2009b).

Within food webs, parasites are embedded in a network of interactions. Consequently, for helminth parasites of freshwater fish, and fleas infesting small mammals, the degree of connectance in the network decreases as the number of species in the host–parasite interaction network increases. This also suggests that most of the possible connections are realized in host-species-poor communities, but not in host-species-rich ones (Poulin 2007a); a dilution effect would result.

There is nevertheless some controversy about how real dilution effects in the wild might be (Randolph and Dobson 2012; Salkeld et al. 2013; Wood and Lafferty 2013; Halsey 2019). However, an increasing number of studies in various contexts and systems support the negative relationship of

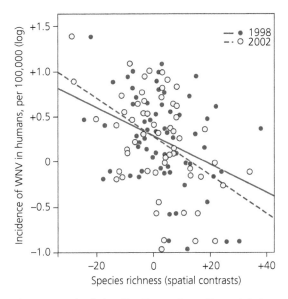

Figure 15.5 The dilution effect. The prevalence of human infections by West Nile Virus (WNV; y-axis) is lower in areas where there are species-rich communities of birds (x-axis), the natural reservoir of the virus. Shown is the reported incidence of WNV by county of the eastern United States, in relation to the contrast in bird species richness between different areas (corrected for spatial effects). Closed symbols (solid line, red) refer to data from 1998, open symbols for 2002 (dashed line, blue). Adapted from Swaddle and Calos (2008), under CC BY.

diversity with the presence of pathogens (Turney et al. 2014; Civitello et al. 2015; Johnson et al. 2015; Huang et al. 2017), even though different relationships exist in some cases (Wood et al. 2014). It, therefore, remains essential to identify possible mechanisms that can result in a dilution effect and to investigate whether such mechanisms exist in a given system.

A major reason for a negative relationship is a change in the fraction or abundance of 'competent' hosts, i.e. those that the parasites can readily infect. On the one hand, such competent hosts may be displaced by low-competence hosts through ecological competition or predation. On the other, the contact rate among competent hosts may decrease as more low-competence hosts are present with an increasing diversity of host species (Huang et al. 2017). Experimental changes in host species abundance can indeed alter infection patterns, as in the examples above. Similarly, when larger mammals were removed from large-scale field enclosures, i.e. host

species diversity was reduced, helminth infection levels increased in small rodents (that also increased in density; Weinstein et al. 2017). Experiments where the density of focal hosts is manipulated are also in line with such a mechanism (Civitello et al. 2015).

A summary measure of species diversity, such as the Shannon–Wiener index or species richness, obviously ignores other factors that can be important for the presence or absence of a dilution effect. Mechanisms such as behavioural changes in the competent hosts, in the vectors, changes in host densities, or the particular kind of host species being present or absent, can decouple variation in species diversity from the relevant traits that affect parasite transmission and prevalence patterns. Evidence shows that, indeed, the identity of the species that are in the set of species being present in an area is essential for a dilution effect; for example, in mammals (LoGiudice et al. 2008; Werden et al. 2014) or in frogs (Johnson et al. 2019). The dilution effect is significant enough to be studied, and it has an obvious bearing on conservation efforts (Young et al. 2013) and on the risk of emerging diseases for human populations (Keesing et al. 2010; Johnson et al. 2013, 2015). Infection by WNV is an example where the biodiversity of bird communities lowers the fraction of infective mosquitoes (Ezenwa et al. 2006a), and thus protects humans from infection (Figure 15.5). Similar effects on humans are observed for Lyme disease with the diversity of the rodent reservoir (Werden et al. 2014).

15.2.5 The value of parasites for hosts

Parasites may, in fact, not merely be detrimental to hosts. Instead, parasites are essential parts of host communities, ecosystems, and host species, too. For example, parasites can maintain genetic diversity in host populations, which allows for more rapid adaptation to changing environments (see section 14.2.2). Parasites can also protect against additional infections that may be worse (Ashby and King 2017). The continuous exposure could maintain resistance when species are reintroduced to their former area for conservation purposes (Almberg et al. 2012).

Similarly, host populations exposed to parasites may harbour more diverse defences that could resist novel pathogens. Hence, parasites are an essential component of host biology and can glue together species in a community and ecosystems. Their effects have evolved and co-evolved with their hosts and structured the populations and communities. A loss of parasites will, therefore, change the face of biodiversity. It is therefore not an absurd proposition to consider programmes for the conservation of parasites (Hudson et al. 2006; Dougherty et al. 2016; Carlson et al. 2020).

15.3 Parasite ecology

Parasites also form ecological communities. However, only hosts represent suitable environments for a parasite to grow and reproduce. Therefore, parasites typically live in separate patches connected by transmission events. In ecological terms, this is a metacommunity where local extinction (by clearance, or host death) is balanced by colonization events (through transmission and infection) (Mihaljevic et al. 2018). The metacommunity, in turn, contains the metapopulations of the different parasite species. A metapopulation is a set of subpopulations connected by migration. Although not often used in this book, the respective terms are different in parasitology. Parasites of the same species within a given host individual form the 'infrapopulation'. All infrapopulations within a host (i.e. the different parasite species) are the 'infracommunity'. Furthermore, the ensemble of all parasites found in the population of conspecific hosts is the 'component parasite community'. Finally, parasitologists define the 'compound parasite community' as the community of all parasites in a given ecosystem, i.e. the parasite species in all host species together. By implication, any ecological process will therefore cut across several hierarchical levels of this organization.

15.3.1 Geographical patterns

The presence or absence of parasites varies across geographical scales and regions. A variety of factors, such as habitat differences, climatic conditions, the availability of hosts, dispersal barriers, or anthropogenic influences, contribute to these patterns. The resulting variation is a subject of the macroecology

of parasitism and infectious diseases (Han et al. 2016; Stephens et al. 2016). Two topics stand out.

15.3.1.1 Relation to area size

One of the most robust findings in ecology is the positive correlation between area size and the number of different species living in this area, and the decrease in species numbers with increasing isolation of the area from a larger mainland. The underlying ecological concepts are known as 'island biogeography' (MacArthur and Wilson 1967).

For parasites, too, geographical patterns in the distribution of parasites and their hosts were noticed for some time, e.g. the association of parasite richness with geographic host range (Price et al. 1988). Similarly, parasite species number (richness) or species diversity often relates to area size (Morand 2015) and decreases in isolated communities, although effects may be weak (Jean et al. 2016). The situation in parasites is involved, however, because a concomitant variation in host species numbers blurs superficial relationships. Furthermore, host body size, niche utilization, diet, colonization ability, or migratory habits are also likely to co-vary with host geographic range. Hence, the increasing heterogeneity in host characteristics becomes more relevant as geographical ranges increase in size (Johnson et al. 2015; Dallas et al. 2018).

At any one locality, historical and regional processes affect the number and diversity of species being present. The local set of parasites can be thought of as a sample from the regional pool, either in strict proportion to the available parasites ('proportional sampling') or as a saturating function. With proportional sampling, the local species number is likely limited by regional processes and dispersal opportunities. With a saturating curve, processes such as local competition and frequent local extinctions are more critical and may limit the number of locally co-existing parasites. In both cases, the locally present parasites are fewer than what is present in the regional pool, because a locality does not have suitable hosts for all of the regional parasites. However, the difference in these curves indicates the relative strengths of local vs regional processes. Similar logic would apply to the relationship of the infracommunity with the component community of parasites (Poulin 1997). Implicit in these scenarios is the importance of dispersal (by transmission, by host movements) of parasites from one suitable location (a host, a locality) to another. In fact, geographically close parasite communities are generally more similar to one another than more distant ones (Poulin and Morand 1999).

Geographical distance is only a proxy for a variety of factors that do affect the exchange of parasites among hosts in different localities. Dispersal in different types of parasites, furthermore, is dominated by different processes (Guégan et al. 2005). For example, ectoparasites on fish are mobile and can disperse over more considerable distances. Probably for these reasons, the local ectoparasite community is not saturated, empty niches are common, and the local richness is proportional to what is available in a region (Morand et al. 1999; Norton et al. 2004). Intestinal parasites, on the other hand, have tighter requirements for successful transmission and establishment. They likely experience more intensive competition; accordingly, local communities are often observed to saturate (Kennedy and Guégan 1994), even after controlling for several confounding factors such as geographical distances (Calvete et al. 2004).

Finally, dispersal and establishment need time, in addition to the effects of a local saturation. This requirement becomes a dominant constraint when the host–parasite assemblage is recent. For example, comparing introduced and native hosts, the parasite richness in introduced hosts (e.g. in crustaceans, molluscs, fish, amphibians, reptiles, and mammals) is typically lower than that of native hosts (Torchin et al. 2003). Similarly, parasite burden can be less at the invasion front than in the core (Stuart et al. 2020). Therefore, introduced host species might not have had enough time to accumulate enough parasite species, and a saturating pattern is unlikely to be observed. Different dispersal abilities also affect the time needed to colonize the new hosts. For instance, in wolves reintroduced to their former area (Yellowstone National Park), the colonization by resident viruses was very rapid. In contrast, a mite-associated disease (sarcoptic mange) was considerably slower to spread (Almberg et al. 2012). How host communities subsequently become assembled affects the time course of geographical patterns of parasitism, too (Halliday et al. 2019).

15.3.1.2 Latitudinal gradients

Latitude, similar to elevation, is known to associate with the presence of species and parasite richness. In general, species richness is higher in lower latitudes (near the equator) than in higher latitudes (towards the polar regions). Such a pattern is known, for example, for fish parasites (Rohde and Heap 1998), for blood parasites in birds (Fecchio et al. 2019), and for many helminths. The presence of human diseases, too, follows a latitudinal gradient. This relationship persists even when confounding factors such as social and demographic differences among countries, or geographical-physical parameters, are controlled for (Guernier et al. 2004). The latitudinal patterns reflect the availability of host species (Figure 15.6a). Hence, this pushes the problem back to the reasons for latitudinal differences in species richness more generally (Gaston and Blackburn 2000). However, the latitudinal pattern has exceptions and is not monotonic (Morand 2015). Measured as parasite richness per host species (which reflects parasite pressure), the pattern deviates from the latitude–parasite richness relationship (Dallas et al. 2018) (Figure 15.6b).

Spatial variation in the presence of hosts, in ecological conditions, or historical legacies (e.g. by dispersal barriers) generates spatial variation in the presence of parasites, with the effects of latitude being one such example (Gaston and Blackburn 2000). As a result, immune responses also vary on a spatial scale. The study of spatio-geographic variation of immunity has emerged as the field of 'macroimmunology'. However, as with field studies more generally, studying the spatial patterns of immune responses is plagued by a variety of difficulties, such as methodical inconsistencies, lack of replication, biased representation of taxa, and weak statistical procedures (Becker et al. 2020).

Nevertheless, standardized immune responses indeed vary with latitude in a range of organisms such as birds (Hasselquist 2007), bats (Becker et al. 2019), and insects (moths from Finland compared to those from the Caucasus, Meister et al. 2017). The mechanisms may be similar to those caused by general physiological stress levels. In the neotropical vampire bat *Desmodus rotundus*, for example, the relative proportions of different immune cells from the blood (e.g. neutrophils or lymphocytes) differed

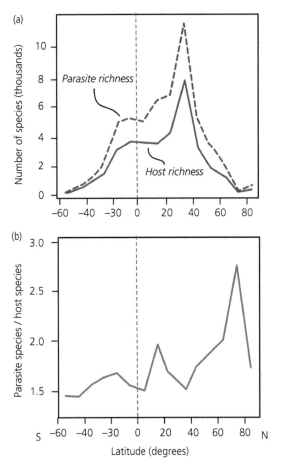

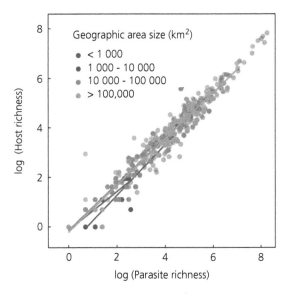

Figure 15.7 Parasite species richness. Host and parasite species richness are closely associated across different geographical scales (with country sizes ranging from less than 1000 km² to more than 100 000 km²). The data refer to the worldwide distribution of helminth parasites. Adapted from Dallas et al. (2018), with permission from John Wiley and Sons.

Figure 15.6 Latitudinal pattern of parasite diversity. (a) The number of host and parasite species (richness) peaks in mid-latitudes (x-axis) in the northern hemisphere, the two curves follow each other fairly closely. (b) When parasite richness per host species is calculated, the peak shifts towards higher northern latitudes. The data refer to helminths and is based on the largest available data base so far. Artefacts due to oversampling of iconic arctic species may nevertheless bias the result. Recall also that the bulk of landmasses is located in the northern hemisphere. Adapted from Dallas et al. (2018), with permission from John Wiley and Sons.

15.3.2 Parasite community richness and diversity

There is a generally positive relationship between species richness in a host community and the corresponding parasite community (the compound parasite community), across many systems and geographical scales (Figure 15.7) (Kamiya et al. 2014a; Johnson et al. 2016; Dallas et al. 2018). This observation has become known as the principle of 'host diversity begets parasite diversity' (Hechinger and Lafferty 2005). The pattern does not necessarily contradict the dilution effect discussed above, since a focal host species can still experience lower parasite loads in more diverse host communities. Similar to the effects of area size (see section 15.3.1.1), parasite diversity is limited by hosts being local 'resource islands', and by the dispersal opportunities among these islands (Poulin 2014).

Local parasite assemblages result from colonization–extinction dynamics in a regional setting. In this view, local parasite communities result from colonization from other localities. At the same time,

between populations in the core of the geographical distribution and those from the northern or southern limits of the range. At these edges, the populations live under more stressful conditions (Becker et al. 2019). Hosts might, therefore, choose a different trade-off between immune defences and reproductive effort, depending on habitat quality at the different locations; in this case, a latitudinal gradient results.

parasite species go extinct locally. Such a colonization–extinction dynamics in metapopulations leads to some parasite species (the 'core species' of metapopulation concepts) being common and present in most localities. Other species are rare and only present locally (the 'satellite species') (Hanski 1982). An analysis of 36 parasite communities of marine fish indeed showed a bimodal distribution of abundance, as suggested by the core–satellite scenario, and no evidence for interspecific competition among parasites (Morand et al. 2002).

Sometimes, too, parasite communities may contain 'nested sets' of species. Nested sets are a distinct set of parasite species, non-randomly assembled, and, in particular, present in species-poor communities. The same set is present in increasing numbers of species-rich communities, too. Hence, some parasite species occur in most communities, whereas some species only occur in species-rich ones (Poulin 2007a). Examples are fish ectoparasites on African cyprinids (Hugueny and Guégan 1997), parasites of tropical freshwater fish (Guégan et al. 1992), internal parasites of fish populations in Finnish lakes (Poulin and Valtonen 2001), and fleas on mammals (Guégan et al. 2005; Krasnov et al. 2005). However, in many parasite communities, there is no readily discernible structure; for example, for gastrointestinal helminths of mammals and birds (Poulin 1996).

Parasite communities can also be rather fleeting. In some well-studied fish communities, shorter episodes of more ordered community structures occur, but such patterns do not remain stable over time and space (Kennedy 2009). In some other cases, parasite communities are more persistent. For example, acanthocephalan parasite populations remained stable over 18 years in eels (Kennedy and Moriarty 2002), and for 32 years in roach living in a small lake in southern England (Kennedy et al. 2001). Such long-term data, therefore, suggest that parasite communities sometimes have a distinctive structure, but very often, too, they form stochastic species assemblages in their hosts that are never quite in equilibrium.

Parasite community structure also varies when the phylogenetic distance among hosts in a community becomes more extensive. Such phylogenetic overdispersion is expected to reduce transmission and thus lower parasite diversity (Parker et al. 2015b). This effect would run counter to the idea that more diverse host communities provide more diverse niches and thus have more diverse parasite communities, as mentioned above. However, it reiterates the point that a summary measure for host diversity may miss the essential characteristics of a community. For example, aquatic birds have richer helminth parasite communities than freshwater fish, and the presence of these birds, therefore, affects helminth communities in aquatic ecosystems. Possible reasons might be that birds are warm-blooded, have richer diets, and are more mobile than fish (Kennedy et al. 1986). Similarly, metabolic rate, longevity, and body sizes vary among host species. In mammalian hosts, for example, such physiological and life history parameters correlate with parasite species richness (Morand and Harvey 2000; Poulin 2007a). Hence, the identities of host species are just as relevant as their numbers.

Looking only at parasite species to identify regularities in communities might also not be the right choice, for other reasons. More consistent patterns could arise from using parasite biomass as the unit of consideration. For example, in a study of 35 species of marine fish, the biomass of all metazoan parasites (i.e. excluding bacteria or protozoa) was positively correlated with host biomass (explaining 79 per cent of total variance). The relationship, furthermore, scaled linearly with the processes that describe how biomass is produced and transferred between trophic levels, corrected for variation in body temperature (which affects the efficiency of metabolism) (Poulin 2007a). In other words, parasite biomass appears to represent a constant fraction of the biomass production of the host, obeying the same scaling rules as known from metabolism and biomass production more generally (van der Meer 2006). Additional factors and processes exist. For example, multiple infections of hosts are common. The resulting parasite–parasite interactions are important for the structuring of the community of parasites (Pedersen and Fenton 2006; Kennedy 2009). Such a persistent role of biotic interactions within parasite communities, furthermore, emerges from field observations and experimental demonstrations of (density-dependent) competition, e.g. between co-infecting nematode species (Ashworth and

Kennedy 1999; Fazio et al. 2008). Similar findings exist for acanthocephalans (Holmes et al. 1977) or cestodes in fish hosts (Kennedy 1996). The interaction between parasites could reflect differential effects of the host's immune system (the apparent competition), as suspected for co-existing fish and their eye flukes (Karvonen et al. 2009), or for gut parasites in mammals (Lello et al. 2004). As already discussed for the defensive role of microbiota (see section 4.8), defensive symbionts also mediate the occurrence and abundance of parasites (Hopkins et al. 2016, 2017).

The search for general laws in parasite ecology has nevertheless remained difficult. Compared to their non-parasitic counterparts, parasite communities appear to be more variable and fleeting. The ability to invade and utilize host resources seems essential, but the longevity of any resulting association might be short and change over time, with noticeable differences among parasite groups. The combination of these factors makes it challenging to infer general principles (Rohde 2005b).

15.4 Migration and invasion

15.4.1 Host migration

Organisms move in space, either as individuals or over generations. Seasonal movements covering more considerable distances typically are classified as 'migration'. Examples are shorebirds that move to their breeding grounds in the subarctic in summer and return to their quarters in Africa for the winter. Migration can also unfold over successive generations, as in the case of the monarch butterfly. This species moves from its winter quarters in Mexico to the north-eastern United States and Canada over four to five (short) generations, from where adults head back to Mexico.

Movements and migration are connected with parasitism. Migration can even be triggered by the presence of parasites in the first place, either as a result of host manipulation to facilitate parasite dispersal, as host escape strategies to leave heavily infested areas, or to recover from an infection ('migratory escape'). A range of different scenarios exists to understand if and when an infected host should move, migrate, or disperse, as compared to their uninfected counterparts (Binning et al. 2017). Migrating individuals also carry their infection with them. Hence, depending on the rather complex interplay between the adverse effects of the infection on the host's migratory performance and the properties of the infection itself (e.g. transmission mode, life cycle, incubation periods, infective doses), migrants transport an infection to new areas to different degrees (Hill and Runstadler 2016).

When the infection weakens individuals, they may not reach their breeding or wintering grounds. The infection is thereby weeded out ('migratory culling'). Infected monarch butterflies, for example, migrate over shorter distances than their non-infected counterparts. This reduces infection load in the long-distance migrating population but not in short-distance or sedentary populations (Altizer et al. 2015). Note that this reduction at the population level is a side effect of selection for individual migratory capacities or immune defences, rather than selection to 'purge' the population at large. Movement and migration can also reduce infection levels due to a reduction in transmission opportunities, or through upregulation of immune responses, e.g. in the context of density-dependent immune prophylaxis (Wilson et al. 2002).

As discussed before, the dynamics of an infection considerably changes with superspreading events (see section 11.2). Superspreading individuals may have an increased susceptibility to infection, increased capacities to transmit, higher contact rates, or more intensive contacts (Barron et al. 2015). Many of these characteristics change with migration. The respective physiological demands can, for instance, prolong infection duration and increase contacts along the migratory route, in migrating flocks, or the breeding grounds, and thus favour the spread of an infectious disease. Therefore, migration can potentially lead to superspreading; it is still unclear how often this occurs, though (Fritzsche McKay and Hoye 2016).

15.4.2 Host invasion

With dispersal, individuals permanently move away from their natal site into new areas. At the new site, organisms settle, reproduce, and form populations, which amounts to an 'invasion' event. The invasion of a species into a new area can strongly affect the local resident community and change its diversity. Invasions occur naturally, but human-made introductions play an increasingly important role; for instance, as a result of global trade (Seebens et al. 2016). The field of 'invasion biology' is concerned with several questions surrounding these issues, e.g. what are the characteristics of invasive species, which communities and ecosystems are particularly vulnerable to invasion, or what factors allow co-existence of invaders and residents (Gallien and Carboni 2017; Young et al. 2017b)? In general, invasion success depends strongly on the specific circumstances of the

introduction event: among other things, on the number of individuals that arrive or are released into a new location (Blackburn et al. 2015). Invasion by host species into another community, not surprisingly, affects host–parasite relationships in various ways (Figure 15.8).

15.4.2.1 Enemy release (parasite loss)

Dispersal can be associated with the loss of former parasites. The consequences are similar to the absence of competitors ('competitive release') on islands vs the saturated, species-rich mainland communities known from classical ecology.

Parasites can get lost for several reasons. With migratory culling, the weakened, infected individuals may not become part of the founding population

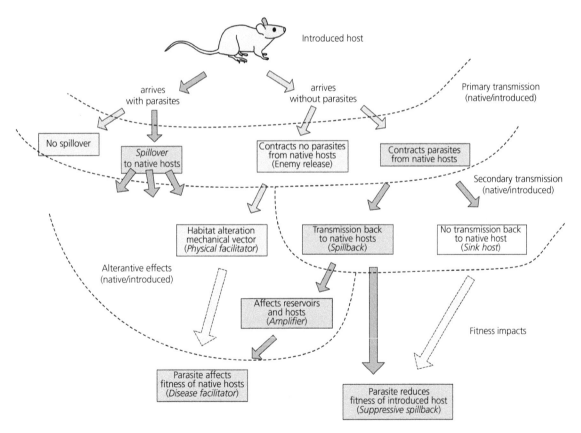

Figure 15.8 Many outcomes of a host invasion. The introduced host may or may not carry a parasite. If infected, it may or may not spill over to the native hosts. Even a host that arrives non-infected can change the resident host–parasite interactions. It can become infected by native parasites and transmit these further (and thus becomes a sink for local, resident parasites). Or it can transmit it back (spill-back) and thus change anything, from the force of infection to the competitive ability of native hosts. Adapted from Chalkowski et al. (2018), with permission from Elsevier.

in the new area. Also, a bottleneck effect results because invasions typically involve a small number of individuals that form the founding population. Hence, there is a good chance that the invading few hosts happen to be uninfected. These processes can also repeat themselves, such that over a series of stepping-stone founder populations, parasites are successively lost. For example, the invasive cichlid *Oreochromis* (tilapia) completely lost all its gill parasites from the native range in Mozambique during the introduction process of aquaculture into New Caledonia (Firmat et al. 2016).

Furthermore, parasites might reach a new area with the founder host population but get lost soon afterwards. The newly established host population could be too small to sustain an infection, the parasites may lack an appropriate intermediate host at the new location, or the stage of the invading host (e.g. seeds, spores, or juveniles) might not be suitable for the parasite. For instance, most ectoparasitic lice reached the new area. However, they failed to establish themselves after the arrival of their bird hosts in New Zealand, mostly because small population sizes reduced their transmission opportunities, or appropriate hosts could not establish (MacLeod et al. 2010).

In all, comparative studies support the notion that loss of parasites by invading species is the most common situation. Introduced plants and animals lose more than half of their parasites when invading a new area, as measured by the prevalence of infection and species richness (Mitchell and Power 2003; Torchin et al. 2003; Torchin and Lafferty 2009). Similar losses are reported for many other systems, too (Lymbery et al. 2010; Roche et al. 2010; Goedknegt et al. 2016).

Horizontally transmitted parasites should be lost more frequently than vertically transmitted ones, and rare parasite species more often than common ones. Similarly, the loss is not the same for all host groups; for example, freshwater host species seem to lose more parasites than either marine or terrestrial species. It often remains unclear, however, to what extent the release from parasites contributes to subsequent host invasion success (Torchin and Lafferty 2009). Many of the processes discussed to affect competition between hosts (see Table 15.4) are relevant for invasions, too.

15.4.2.2 Parasite spill-over

The invading host species can harbour parasites that become transmitted and subsequently spread within the native, resident community. The parasites may come from the invader's site of origin or be picked up en route (Plowright et al. 2017). Although spill-over is not the most common outcome of host invasions, this possibility has generated the most attention, given the potential new threats; it is also known as 'co-introduction' of host and parasite.

These concerns mainly refer to the cases where an introduced host species carries a parasite that jumps over to a native host species, where it has severe effects and can push it to extinction. The IUCN estimates that infective diseases are responsible for one-quarter of all disastrous invasions (Hatcher et al. 2012). Instructive examples are the extinction of half of the endemic bird fauna on Hawaii (Warner 1968) by malaria from introduced birds (and the later arrival of a competent vector). Similarly, parapoxvirus from introduced American grey squirrels (*Sciurus carolinensis*) decimated the native European red squirrel (*S. vulgaris*) populations in England (Tompkins et al. 2001b). A disastrous outcome, too, is the already mentioned worldwide decline of amphibians (Scheele et al. 2019) caused by chytrid fungi. It originated in Asia and spread by the commercial trade of these animals (O'Hanlon et al. 2018). Asymmetric effects on introduced vs native hosts are diverse in kind, but, indeed, seem to have facilitated invasion in several cases (Prenter et al. 2004), notably also in marine ecosystems (Torchin and Lafferty 2009).

Nevertheless, despite the spectacular cases with a disastrous outcome, considerable hurdles must be overcome for the successful spill-over of parasites. For instance, the introduced parasite must find other suitable hosts in the new range; hence, it must accomplish a host switch (see section 14.1.3) (Box 15.3). Furthermore, parasites with an indirect life cycle also need equivalent intermediate hosts, and vector-borne diseases need to find an appropriate vector. In the case of the lung fluke *Haematoloechus* (Trematoda), a parasite of bullfrogs introduced to Vancouver Island, several native species of dragonfly now serve as intermediate hosts.

Box 15.3 Spill-over and disease space

Disease emergence by spill-over connects disease spaces in different host species. In the reservoir community, parasites follow different infection trajectories through the disease space of their respective host species. The likelihood of a spill-over depends on the opportunities for transmission from these reservoir disease spaces to the target disease space. Numerically, this is probably more likely from abundant reservoir hosts. However, the target species must also be a competent host for the spilled-over parasite. This depends on its disease space and whether the infection trajectory in the new host allows the parasite to infect, establish, and further transmit in the population of new hosts with its existing mechanisms (Figure 1).

This initial similarity is essential because any possible adaptation by evolutionary change will take time. Hence,

similar disease spaces of reservoir and target greatly facilitate a spill-over.

A spill-over, therefore, may not necessarily connect the most common host species, but those connected by disease space similarities. From this primary spill-over, a secondary jump may occur to another species with a similar disease space in the target community. With time, evolutionary changes may expand the host range in the target community still further. Similar to the various 'filters' (see Figure 7.4) that determine the host range of a parasite, the probability of host switching and disease emergence by spill-over is given by ecological factors such as transmission opportunities, and by physiological factors, as captured in the disease space concept.

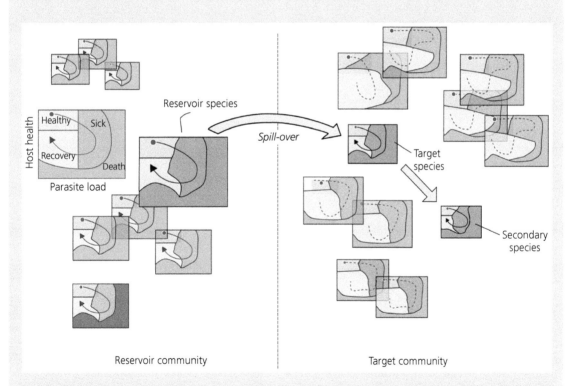

Box 15.3 Figure 1 Emergence by spill-over and disease spaces. In this cartoon, the rectangles symbolize the disease spaces of different host species in the reservoir (left) and target community (right). The size of the rectangles indicates species abundance. Within a disease space, parasites will follow an infection trajectory (lines with arrows). Unsuitable hosts for a focal parasite symbolized by dashed trajectories. Further explanations in text.

The probability of a spill-over therefore also varies among parasite groups. For example, in a set of 98 documented introductions, half of the introduced parasites were helminths, followed by arthropods and protozoa, whereas viruses accounted for only 9 per cent of the cases; fish were the most commonly reported hosts (Lymbery et al. 2014). In this set, no difference in host switching between parasites with and without a direct life cycle existed, however. Hence, it remains unclear whether the naïve expectation that spill-over should be more likely for directly transmitted parasites with simple, direct life cycles is valid. Occasionally, parasites also become introduced into a new area without their hosts; for example, as free-living stages of fish parasites in the ballast water of cargo ships (Chapman et al. 2012). Together, parasite spill-over is one of the fundamental processes that can facilitate disease emergence and zoonosis.

15.4.2.3 Parasite spill-back

Invading hosts can become infected, in turn, by parasites from their new locality (Kelly et al. 2009). With additional hosts becoming infected, this can also facilitate the spread of a resident parasite in the resident community. Spill-back can lead to the extinction of an invading host population (Faillace et al. 2017). Spill-back is less likely when the invading host can respond better to a new parasite. For example, birds introduced in New Zealand and other islands were more likely to establish (as measured by the number of necessary attempts) if the species was a habitat generalist. However, the species was also more likely to succeed if it showed a more robust response to a standardized immune challenge (conducted on nestlings), even when controlled for several additional factors, such as the migratory habit of the species, or body mass. The pattern was particularly strong for invading populations with many individuals, where the persistence of parasites in the invading population is more likely (Møller and Cassey 2004). Similarly, songbirds invading from areas with high pathogen diversity were more likely to succeed than those from low pathogen diversity areas, presumably because they were better buffered against the new, resident pathogens (O'Connor et al. 2018).

15.4.2.4 Facilitation

When an invading host population provides more opportunities for native parasites by spill-back events, this can also be considered a 'facilitation' for the native parasite community (Chalkowski et al. 2018). Facilitation occurs because the invaded hosts are a new reservoir or enlarge the pool of susceptible hosts (an amplification or dilution effect), or they are new vectors by themselves. Such sharing of parasites seems the case for the introduced and native birds of New Zealand (Howe et al. 2012) and Australia (Clark et al. 2015), where the same native blood parasites (Haemosporidia) are present, and the invaders likely served as amplifiers. Many vector species, such as the Asian tiger mosquito (*Aedes albopictus*) are highly invasive, which is also promoted by climate change (Rochlin et al. 2013). As a result, new pathogens can be spreading alongside the invasion of vectors (as with bird malaria in Hawaii). For human diseases, the spread of mosquitoes is of concern for yellow fever and dengue (see section 11.4), but also the Chikungunya fever virus (Vega-Rúa et al. 2014; Lounibos and Kramer 2016), or WNV (Sardelis et al. 2002).

15.5 Zoonoses and disease emergence

15.5.1 Reservoirs

Most parasites, notably infectious diseases, can infect several host species. This capacity paves the way for the emergence of new diseases in a host population. Novel pathogens are a severe problem for wildlife, e.g. the devastating chytridiomycosis in amphibia (O'Hanlon et al. 2018). However, the most obvious concern is the emergence of novel pathogens in human populations by a jump from an animal host.

To use the appropriate terminology (Haydon et al. 2002; Viana et al. 2014), a 'target' population is the population that is in danger of acquiring a novel pathogen, e.g. the human population of a given area. A 'reservoir' is where the infection is coming from. A reservoir can be a single population (the 'source population') or a more complex set of populations and species that together sustain the infection (the 'maintenance community'). Within this set, only a single or a few species may be relevant (the 'maintenance populations'), whereas others are involved but cannot maintain the parasite on their own (the 'non-maintenance populations'). Therefore, reservoirs are rather diverse in their structure. Furthermore, a reservoir is only functional when transmission from the sources to the target population is a reasonable option. The structure and dynamics of maintenance communities, as well as the potential for transmission, are strongly affected by man-made ecosystem changes and environmental factors (Gibb et al. 2020; Roberts and Heesterbeek 2020).

Pathogens, new to a naïve population, can, of course, be of any kind. However, novel bacteria and viruses are the most prominent examples. These are known to have the potential to be a source of severe local threats and pandemics. Furthermore, mammals are the most obvious candidates for a reservoir of emergent human diseases. Among those, the most species-rich orders typically also contain the most zoonotic species, i.e. species from which a disease has jumped to humans. This puts rodents and bats in focus. Carnivores are in focus because of the extraordinary diversity of zoonotic pathogens that they harbour (Han et al. 2016). Zoonotic viruses have repeatedly caused significant epidemics in humans; a particular focus on viruses seems, therefore, warranted.

Outbreaks of avian influenza in human populations have often emerged from domestic animals, e.g. from pigs (Bourret 2018), with wild and domestic birds as major reservoirs (Hurt et al. 2017). In contrast, haemorrhagic hantaviruses have a reservoir in small rodents, such as Sin Nombre virus (SNV) in deer mice of North America, or Puumela virus (PUUV), which circulates in bank vole populations of Northern Europe. Furthermore, for the roughly 50 hantaviruses species known, each seems associated with one or a few closely related reservoir host species (Forbes et al. 2018). Not surprisingly, the search for the reservoir can be a rather frustrating and challenging endeavour. For example, fruit bats have long been considered the reservoir for Ebola virus, yet the situation seems more complex (Leendertz et al. 2016). Similarly, contrary to earlier hypotheses, marsupials are competent reservoirs, but not the only, nor the major,

non-human host species for Ross River virus in Australia (Stephenson et al. 2018).

Bats (Chiroptera) rank prominently as reservoirs. However, they are surpassed by rodents as far as the number of known zoonoses goes, with the possible exception of tropical South America where bats are a highly diverse zoonotic group (Han et al. 2016) (Figure 15.9). Several thousand bat-associated viruses are known (Chen et al. 2014). Furthermore, viruses of all replication strategies and genomic structures use bats as their host. For example, more than 2500 sequence types ('species') of Rhabdoviridae have been identified. Among those, lyssaviruses are among the best studied in bats and include prominent examples, such as rabies virus (Banyard et al. 2011). Rabies and other lyssaviruses originated in bats and then crossed into other animal species and human populations (Banyard et al. 2014). The second most common viral group harboured by bats is the Coronaviridae, of which over 900 types ('species') exist worldwide (Hayman 2016). Also, the Paramyxoviridae are a large virus family that contain important pathogens such as measles, mumps, distemper, or Newcastle disease virus. These viruses and their relatives are circulating in bats in Africa, South America, Europe, or Southeast Asia (Drexler et al. 2012).

Bats as a group have several characteristics that may explain why they are such essential reservoir species (Han et al. 2015). For example, with more than 1200 species, the order Chiroptera is the most species-diverse among mammals, accounting for one-quarter of all mammalian species. Chiroptera also is an evolutionarily old clade. It originated 52 million years ago and thus has a long and rich history of co-evolution with viruses. Bats are also very long-lived for their body size. Being airborne, these hosts can cover long distances in their daily forays as well as during seasonal migration. Their physiology is adapted to dampen the effects of oxidative stress during flight, which may have a pleiotropic effect for the control of infections by the immune system (Brook and Dobson 2015). In fact, bats also seem to tolerate many viral infections and do not show symptoms. Finally, bats are typically social, and some caves can contain colonies with up to several million individuals. Large numbers and proximity, and, in some cases, also food sharing, facilitate

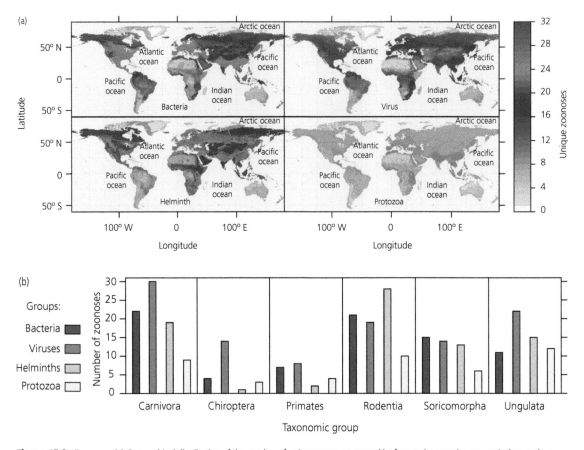

Figure 15.9 Zoonoses. (a) Geographical distribution of the number of unique zoonoses caused by four major parasite groups (colour scale on the right). (b) Histograms of zoonoses by four major parasite groups (see legend) which had spilled over from six host taxonomic groups. Carnivores and rodents stand out. Soricomorpha contain the primates, rodents, shrews, and moles. Adapted from Han et al. (2016), with permission from Elsevier.

virus transmission. Such spread in bat colonies can 'amplify' many emergent human pathogenic viruses (Drexler et al. 2011). Not all viruses or other parasites in wildlife will, of course, eventually jump into human populations. However, it would be imperative to know which pathogens have the potential to do so, and which reservoirs maintain these (Figure 15.9).

15.5.2 Emergence

When a parasite 'jumps' to a new host species, it 'emerges' as a new pathogen in this host. Because such host switching is quite common (see section 14.1.3), parasites also quite often must emerge in a new host. However, in the majority of cases, this probably either goes entirely unnoticed or has only mild effects. Although there is no firm evidence on this, only in a minority of such jumps does a devastating disease and a severe epidemic seem to result. Examples are again chytridiomycosis in amphibia, and the plague, yellow fever, AIDS, the Spanish flu, and Covid-19 in humans (French and Holmes 2020). Regardless, for any target population, a reservoir is only relevant when transmission also occurs. Hence, the process of emergence is of interest.

The likelihood of an emergence has at least two dimensions. On the one hand, there must be a sufficient contact rate to be likely to lead to transmission from the reservoir into the target population. On the other, once the pathogen has arrived in this new population, it must be able to spread, i.e. have

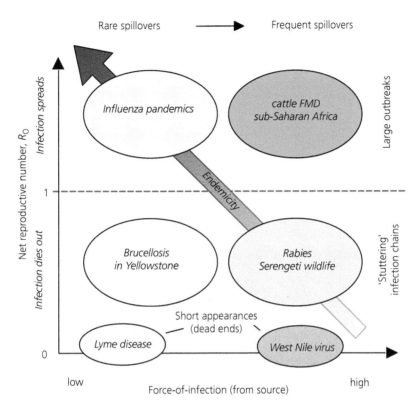

Figure 15.10 Disease emergence. Emergence depends on the force of infection from a source (e.g. a reservoir; x-axis), and the ability of the parasite to subsequently spread in the new host population (as represented by its reproductive number, R_0; y-axis). The various domains in this graph characterize different patterns of emergences (with examples given). For example, with low R_0, infections die out soon (a dead-end for the parasite) but may keep reappearing by reintroductions (an enzootic disease). On the other hand, with high R_0, large outbreaks may develop, almost regardless of how frequently introductions happen (given by the force of infection). Note that for the actual invasion step, criteria other than R_0 may be relevant. True endemicity becomes more likely towards the top left (green arrow). Adapted from Viana et al. (2014), under CCBY 3.0.

a reproductive number, $R_0 > 1$, or meet other relevant invasion criteria (see section 11.2.6) (Figure 15.10). Various ecological, evolutionary, physiological, and immunological factors affect these requirements. Transmission, for example, depends on contacts. Humans have frequent contacts with their domestic animals, e.g. pigs or domestic fowl, and these are indeed frequently involved in spill-overs and emergent diseases.

Foodborne transmission, too, offers many contacts and is a major concern. Many viruses use this path, often by the faecal–oral infection route; e.g. hepatitis E virus, enterovirus, and poliovirus. In particular, the handling, the preparation (butchering), and the consumption of animals from the wild, also known as 'bushmeat', carries a considerable risk of animal-to-human transmission (Kurpiers

et al. 2016). In Africa alone, estimates go as high as 3.4 million tons of bushmeat being consumed annually. Notable cases of this route are the emergence of HIV, monkeypox virus, rabies (via primates), Ebola, SARS-CoV-1 (from bat reservoirs, via civets), or MERS-CoV (via camels) (Kingsley 2018). The latter cases show that, often, an additional 'bridge host' is the connection between the background reservoir and the human population. Bridge hosts such as civets, or pangolins are bushmeat, and some of the bridge hosts typically are consumed as exclusive delicacies. Most likely, the recent Covid-19 pandemic (SARS-CoV-2) also emerged along this route (see Box 11.7).

Habitat alteration or destruction is a further factor that brings wildlife and its reservoirs in contact with human populations (Afelt et al. 2018). Thereby, new food resources can become available for wild-life and lure animals near human activities (e.g. recreation, agriculture) (Muehlenbein 2017; Kingsley 2018) and into settlements (e.g. with urban waste, shelter, nesting sites) (Becker et al. 2015). Indeed, wildlife species with known zoonotic pathogens occur with higher abundance and species richness in areas of intensive human use, notably in urban areas (Gibb et al. 2020) (Figure 15.11). Large-scale development projects can also change habitats for the worse. For example, a dam-building in Senegal changed river flow regimes, salinity, and pH, lead-ing to an increase in the populations of intermediate snail hosts. This favoured the spread of schistosomiasis (N'Goran et al. 1997; Southgate 1997).

Deforestations, such as those observed in Amazonia, favour breeding habitats for mosqui-toes, especially for the most competent vectors, which increases the overall number of bites and malaria transmission (Vittor et al. 2006). A similar effect of forest fragmentation benefits rodent popu-lations as reservoirs for Lyme disease in the United States (LoGiudice et al. 2003). Current agricultural practices more generally, e.g. increased land use, the use of antibiotics and fertilizers, and globalized food production, are estimated to be involved in 25 per cent to 50 per cent of all infectious diseases that have emerged since the 1940s (Rohr et al. 2019). High-yield agriculture typically also means dense stands of crops and lower genetic diversity of the planted crop, both factors favouring the spread of a parasite.

A vector-borne viral disease, such as yellow fever, dengue, Rift valley fever, or Zika virus, transmitted by mosquitoes, can become an emergent infection when the vectors expand their range or when the

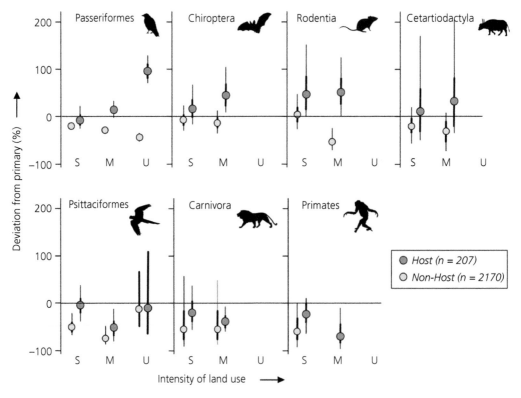

Figure 15.11 Zoonotic hosts and land use. Shown is the deviation in the abundance of zoonotic hosts (red dots; y-axis) and non-zoonotic hosts (green dots), relative to the base line of a minimally disturbed primary site (horizontal line, zero deviation). Zoonotic hosts are known to share pathogens with humans; seven taxonomic groups are shown here. The intensity of land use is grouped into three levels: secondary land (recovering from disturbance; S), managed land (pastures, cropland, plantations; M), and urban areas (U). Zoonotic hosts tend to be more abundant in more disturbed areas. Silhouettes from phylopic.org. Adapted from Gibb et al. (2020), with permission from Springer Nature.

pathogen can use an alternative vector (see section 11.4). Because an actively moving vector spreads the pathogen, such diseases can potentially spread far and wide, and also reach many different host species (Rückert and Ebel 2018; Huang et al. 2019).

The density of the target population that is within reach of transmission from the reservoir is essential, too. Urbanization, for instance, has increased dramatically in most parts of the world. An estimated two-thirds of the human population will be residing in cities by 2025 (United Nations Department of Economic and Social Affairs 2003). Urbanization generates more focal points of high host density and more intensified land use. Both factors increase the risk of disease emergence (Gordon et al. 2016; Eskew and Olival 2018). On the other side, it could also provoke broader defences by the adaptive immune system, as the background exposure is to a broader range of parasite types. Besides, the association of urbanization with more wealth, improved sanitation, and better public health management may counteract the increasing risks of transmission (Wood et al. 2017). Nevertheless, urbanization also alters the landscape in and surrounding the cities, which in turn increases contacts between parasites and wildlife, and between domestic animals and wildlife (Gibb et al. 2020). At the same time, a more intensive use might, for example, compromise immunocompetence in wildlife reservoirs by higher concentrations of pollutants (Bradley and Altizer 2006) and make infections more likely to persist.

The properties of the pathogen itself can increase or decrease the chances of coming in contact with the target population, too, and thus the likelihood of being a candidate for emergence (see also Figure 15.9). Parasites, such as bacteria, for example, can rapidly adapt by the transfer of pathogenicity islands, and viruses have high mutation rates and use conserved characteristics of the host for cell entry. Furthermore, bacteria and viruses often produce a large number of primary cases, e.g. when the new host is a predator that routinely consumes prey items of which the majority are heavily infected (Pulliam 2008). It appears that mammalian viruses, which replicate in the cytoplasm and have a phylogenetically broad host range, have properties that especially promote emergence (Olival et al. 2017). A sizeable geographical distribution further

increases contacts with a target population. Several other correlates of the potential for animal-to-human transmissibility have been identified; for example, viral richness in reservoirs and climatic factors (Brierley et al. 2016; Mollentze and Streicker 2020).

In the case of human emergent diseases, identifying whether and how much human-to-human transmission is possible is critical (Geoghegan et al. 2016). Many spill-overs collapse with a 'dead-end', because the parasite is not able to spread further in the target population. For example, the H5N1 and H7N9 subtypes of avian influenza frequently cross from poultry to humans but have so far not been able to establish persistent epidemics because human-to-human transmission seems very poor. The infection is 'enzootic', i.e. is endemic in animals, and appears by repeated events of transmission; for example, from live poultry markets to humans (Lam et al. 2015).

Many other viruses, instead, have achieved human-to-human transmission, as shown by genomic surveillance of virus variants during the 2014 Ebola outbreak in West Africa (Gire et al. 2014). In a comprehensive study of over 200 RNA and DNA viruses, transmissibility among humans was found in one-half of them (105 species), whereas the remainder intermittently entered by transient spill-overs. Moreover, a total of 69 viruses were vector-borne, and 25 caused chronic infections. In this set, the likelihood of human-to-human transmission is best predicted by the duration of the infection, with chronic infections being most likely. Furthermore, transmission becomes more likely with non-segmented viruses, if not vector-borne, or with less virulent variants, and—to some degree—whether the virus is enveloped (less likely) or not (Geoghegan et al. 2016). No difference existed between DNA and RNA viruses.

Obviously, these are not always good predictors. SARS-CoV-2, for example, is enveloped. Envelopes usually make viruses more vulnerable to environmental impacts, since their loss immediately exposes the viral genome. Low virulence seems a critical factor to breach the epidemic threshold of $R_0 > 1$, as it allows for more time for transmission and requires fewer susceptible hosts before the epidemic is slowed down. Different parasite groups each have their own correlates that make them more or less prone for emergence, but with an increasing database these factors will become better known. This will be extremely helpful to monitor and pre-actively interfere with possible spill-overs of catastrophic consequences.

15.5.3 Zoonotic human diseases

The World Health Organization (WHO) defines a 'zoonosis' as an infection or disease caused by a pathogen that is naturally transmissible from vertebrate animals to humans. The WHO estimates that over 60 per cent of all human infectious diseases are of zoonotic origin. Also, 75 per cent of all emerging diseases in the past decade were of zoonotic origin. In terms of the previously discussed processes, a zoonosis has a reservoir in animal populations, spilled over to humans, and has managed to establish and spread in the human population by human-to-human transmission.

Zoonoses in human populations have a long history, certainly dating back to prehistoric times and up to the latest Covid-19 pandemic. Furthermore, they include different parasite groups (Gordon et al. 2016; Spyrou et al. 2019) (Table 15.6). Zoonoses have also changed human history and caused massive economic losses. Such massive effects are not just a thing of the distant past but also happened within the last decades. Estimates for the avian flu (H5N1) epidemic starting in 2008 amount to a worldwide economic cost of $30 billion, and those for the first SARS-CoV-1 pandemic (2002–2003) amount to $30–50 billion (Zambon 2014). The recent SARS-CoV-2 surpasses these marks by far. Therefore, reason demands that we should know more about what kinds of parasites have a high risk of spill-over to humans, what kind of parasites are circulating in potential reservoirs, and set up programmes for monitoring possible outbreaks at the earliest stage. Initiatives such as 'One Health' attempt to cover these themes (Kelly et al. 2017). Nevertheless, the emergence of new diseases is a multifactorial process that remains difficult to study (Plowright et al. 2008).

Monitoring reservoirs requires collecting data on the presence and diversity of parasites circulating in reservoir populations, i.e. in wild animals, with a focus on mammalian species for the case of human zoonoses. Increasingly, such studies are done typically focusing on specific reservoir host groups and certain kinds of parasites (Han et al. 2016) (Figure 15.9). Examples are the worldwide surveys of bats for viral groups of particular significance, e.g. paramyxoviruses (measles, Newcastle disease virus; Drexler et al. 2012), lyssaviruses (rabies; Banyard et al. 2011) or flaviviruses (yellow fever, dengue, Zika; Pandit et al. 2018).

Other studies report data for several parasite groups at the same time, not all of which may be of concern for zoonoses. However, they add to a better understanding of parasite distribution and abundance, e.g. worldwide bat surveys (Hayman 2016), surveys of bats in Southeast Asia together with their viruses, helminths, and ectoparasites (Gay et al. 2014a), or mammals carrying viruses, bacteria, helminths, and protozoa in Central Europe (Duscher et al. 2015). Some studies focus on a particular disease, such as Ebola (Leendertz et al. 2016) or Ross River fever (Stephenson et al. 2018), and ask what the reservoirs could be?

Any such effort will bring us closer to an understanding of reservoirs. Not least, reservoirs are rarely static. Instead, reservoirs form a more complex, multispecies community of host species that sustain a dynamic parasite community which needs to be sampled and surveyed (Plowright et al. 2019). Given the potentially massive impact of zoonoses, the study and surveillance of reservoirs—as the essential background that breeds new threats—should probably elicit the same urgency as monitoring the trajectories of asteroids that can collide with Earth. Estimates suggest that there may be at least 320 000 different viruses (from a set of just nine virus families) in mammals alone. In fact, a worldwide effort to identify these would be quite affordable. If all viral species were surveyed and described, the price tag would be an estimated $6.3 billion, and only $1.4 billion if just 85 per cent of the estimated viral diversity were covered (Anthony et al. 2013). Even if these estimates are underestimated by a magnitude, this would still cost only a tiny fraction of any expected economic damage caused by a global pandemic.

In addition to reservoirs, the pathways by which a pathogen can enter a human population—the pathways of spill-over—need to be better understood

Table 15.6 Examples of more recent zoonoses that caused emergent diseases in humans.

Disease, agent[1]	Original host (reservoir)	Remarks	Source
BSE/vCJD (prions)	Cattle.	Outbreak 1994–1996 in the UK. Transmission by consumption of beef.	12.
Ebola virus (EBOV)	Multiple host species, incl. fruit bats and insectivorous bats. Further as yet unknown hosts are likely.	Fruit bats can be involved, but a general role is still unclear. Last major outbreak 2013–2015 in West Africa.	8
Influenza virus (H1N1)	Swine. Outbreak in humans also with H1N1 from birds, and seasonal flu viruses from humans.	Outbreak in 2009. Re-assortment of different swine virus lineages had occurred.	10
Bird flu (H5, H7subtypes)	Birds in poultry farms. Risk higher from outdoor stocks, i.e. turkeys, ducks. Probably, small rodents that are common in farms are also involved.	2007–2013 in the Netherlands. For outdoor stocks, risk decreases with distance from waterways and wild waterfowl populations.	1 11
Tick-borne encephalitis (virus, TBEV)	Deer a major reservoir for ticks.	Tick bites. Global warming may favour tick populations.	12.
Hantavirus pulmonary syndrome in North America	Small rodents.	Outbreak 1993 in Southwestern USA. Increased rainfall facilitated an ecological cascade, resulting in increased rodent reservoir populations, and subsequent transfer into human populations.	5, 6, 13
Hendra virus (HeV)	Bats.	Outbreak 1994 in Brisbane. Virus also infected horses.	12.
Nipah virus (NiV, NIPV)	Bats; pigs as bridge hosts.	Orchard planting around pig farms increased interactions of wild fruit bats with pigs and virus transfer; agricultural intensification allowed Nipah virus persistence within piggeries. Outbreak 1999 in Malaysia.	3
AIDS (HIV)	Primates, likely chimpanzees.	First emergence around 1920; first reported 1981. Ongoing pandemic.	4, 7
SARS coronavirus (SARS-CoV-1)	Bats, palm civets as bridge hosts.	Outbreak and pandemic 2002–2003.	12
Covid-19 (SARS-CoV-2)	Bats, pangolins?	Outbreak and severe pandemic 2019–2021.	2
Monkeypox virus (MPV)	West African rodents, prairie dogs as bridge hosts.	Outbreak 2003 in Midwest USA. Imported from Ghana with small rodents. Transfer to prairie dogs in a per animal vendor's facility.	12
E. coli, strain O157:H7	Cattle.	Outbreak 1982 in Northern USA. 1996 in Europe, Japan. Foodborne by consumption of beef.	12

[1] Bovine spongiform encephalopathy (BSE); variant Creutzfeldt–Jakob disease (vCJD); further source: 9.

Sources: [1] Bouwstra. 2017. *Emerging Infect Dis*, 23: 1510. [2] Coronaviridae Study Group of the International Committee on Taxonomy of. 2020. *Nat Microbiol*, 5: 536. [3] Daszak. 2005. In: Collinge, eds. *Disease ecology: community structure and pathogen dynamics*. Oxford University Press. [4] Eisinger. 2018. *Emerging Infect Dis*, 24: 413. [5] Glass. 2003. *Emerging Infect Dis*, 6: 238. [6] Hjelle. 2000. *J Infect Dis*, 181: 1569. [7] Kirchner. 2019. *Fundamentals of HIV medicine*. Oxford University Press. [8] Leenderz. 2016. *EcoHealth*, 13: 18. [9] Plowright. 2008. *Front Ecol Environ*, 6: 420. [10] Smith. 2009. *Nature*, 459: 1122. [11] Velkers. 2017. *Veterinary Quarterly*, 37: 182. [12] Woolhouse. 2005. *Trends Ecol Evol*, 20: 238. [13] Yates. 2002. *Bioscience*, 52: 989.

and scrutinized (Kuris 2012; Han et al. 2015; Allocati et al. 2016; Plowright et al. 2017). As mentioned above, this can concern bridge hosts, such as livestock (Hassell et al. 2017), but also those species that serve as 'bushmeat' (Kurpiers et al. 2016). Livestock, for example, could thereby serve as living sentinels for defined pathogens that are a danger for humans (Anderson et al. 2017; Filippitzi et al. 2017). Not the least, monitoring should also cover the distribution and possible range expansion of competent vectors for potentially dangerous pathogens.

Once a spill-over has happened, it will be essential to monitor the further spread of a pathogen. Given that it is not precisely known where and when a novel pathogen enters the human population, monitoring may start with the mapping of the distribution of human infectious diseases around the globe. So far, only a minority of known diseases are systematically surveyed. However, current capacities of data processing would allow integrating data from many different sources, such as official reports, social media, geographical information, climate, weather reports, and so forth. This information allows a much better picture of the actual spread of infection than was possible before (Hay et al. 2013). The big data approach partly was used for the Covid-19 pandemic (Jia et al. 2020a; Lin and Hou 2020; Wang et al. 2020).

15.6 Climate change and parasitism

Climate change, particularly global warming, is in progress and hard to stop. The outward signs, such as retreating glaciers, changes in precipitation regimes, or extended heat waves in summer, are visible to anyone. Will climate change affect parasites and the emergence of diseases? It is a general rule that warmer temperatures speed up growth, development, and reproduction of organisms, and this should also be true for parasites. Climate change generates new selection pressures. Parasite populations, like their hosts, will respond and adapt to the new conditions. Inevitably, some parasite groups cannot adapt as well and probably will go extinct. For example, model estimates suggest that 30 per cent of helminths will disappear, that ectoparasites are more vulnerable to extinction, and that 5–10 per cent of all parasites will disappear due to climate-induced habitat destruction. Extinctions can be direct, or because the host species disappear. Parasites with complex life cycles, vulnerable transmission stages, narrow host ranges, or narrow temperature tolerance will be especially at risk (King and Monis 2005; Cizauskas et al. 2017). Parasites infecting warm-blooded (homeothermic) hosts may perhaps be more protected to this change. At the same time, parasite species may disperse and establish in new habitats or occupy emptied niches—or new variants and species will evolve. Hence, parasite richness and diversity will not necessarily be lower but could also increase (Carlson et al. 2017).

Climate change selects for traits such as the parasite's ability to disperse, its choice of the thermal environment, or resilience to droughts and changed precipitation regimes. Effects of such factors exist in rodents and their ectoparasitic fleas (Poisot et al. 2017). The respective adaptations include altered geographical distributions or phenology, i.e. the timing of events through the seasons, which in turn can affect host range (Jeffs and Lewis 2013; Paull and Johnson 2014). Warming on Hawaii has already expanded the breeding range of mosquitoes towards higher altitudes. Therefore, avian malaria has also moved upwards in the mountains (Freed et al. 2005). Hence, climate change has impacts in potentially problematic ways. The effect of global change and warming can, furthermore, affect immune defences that enhance or counteract changes in disease prevalence (Foxman et al. 2015; Mignatti et al. 2016).

Over the last decades, the occurrence of different parasitic diseases in human populations has continuously changed. Some became less common (e.g. some 'worm' diseases, like African trypanosomiasis), others increased in frequency (e.g. certain types of leishmaniasis). However, during this time not only has climate changed, but also the socioeconomic situation for many countries and regions (Hotez 2018). To identify the effects of climate change—especially for human diseases—therefore requires accounting for such confounding variables. For example, human diseases are more diverse in lower latitudes. However, this pattern does not correlate convincingly with mean annual temperature, or the amount of precipitation that characterizes a tropical climate (which serves as the substitute for the future warm Earth) (Guernier et al. 2004). Indeed, major infectious diseases of humans have originated from the tropics just as often as from elsewhere (Wolfe et al. 2007). Similarly, the areas with the highest risk of zoonoses are not correlated with warmer climates in a straightforward manner (Han et al. 2016).

A generalized and straightforward effect of climate on disease risk is therefore not always evident; the same factor (e.g. temperature) can have opposite consequences depending on the particular parasite or host (Altizer et al. 2013; Lafferty and Mordecai 2016). For example, the capacity for transmission of malaria is expected to become lower due to changing temperature regimes for the mosquitoes (Murdock et al. 2016). At the same time, habitat suitability for mosquitoes changes. Models predict that the overall mosquito range does increase, even though the extent of the optimal habitat decreases. In combination with the predicted growth and development of human populations, the areas at highest risk of malaria will, therefore, shift from West Africa to more Central African countries (Ryan et al. 2015).

Additionally, concurrent changes such as habitat degradation and fragmentation might introduce new barriers to the dispersal of a parasite, for the vector or the reservoir host. It is therefore possible that, contrary to intuition, the geographic range of some pathogens might even become reduced as climate change unfolds (Lafferty 2009). Nevertheless, for several diseases, a link to climate changes is well founded. An example is the expected increase in disease by filarial nematodes (*Onchocerca* spp.) in the tropics, mostly through range expansion of the vectors (blackflies, Simulidae) (Gordon et al. 2016). It is also very likely that some diseases will spread to higher latitudes or altitudes as global warming continues. This amounts to a greater risk of disease emergence in areas where a particular pathogen was not present before. Climate change is particularly drastic in the Arctic, and it will also alter the abundance and distribution of parasites in many ways (Waits et al. 2018).

A particular concern with climate change is the emergence or re-emergence of diseases for the human population. For example, there is a close correlation between precipitation and the incidence of malaria because this allows vectors (mosquitoes) to breed. Changes in the weather pattern, influenced by the Southern Oscillation, have increased rainfall and triggered the re-emergence of malaria since the year 2000 in some regions of China (Anhui province, Gao et al. 2012). Weather and climate change also increase the risk of the re-emergence or geographical spread of many other vectors and their diseases. Examples are bugs that carry Chagas disease, tsetse flies in Africa, or ticks in Northern Europe (transmitting Lyme disease and encephalitis) (Short et al. 2017).

Similarly, climate change also changes the conditions in the soil, which will affect the emergence of diseases associated, for example, with soil-transmitted helminths that benefit from faster developmental times and higher survival under the new conditions. Even if it is often difficult to disentangle the effects of climate change from other effects, there are many reasons to expect that climate change will seriously change the landscape of host–parasite interactions (Cable et al. 2017). This can occur directly via the ecology and physiology of organisms or indirectly through the associated socioeconomic changes in human populations. Very likely, changes in the geographical distribution of the host communities and in the vectors will be the most significant elements of the processes that shape the future of host–parasite relationships.

Important points

- Parasitism associates with host ecology and host life history. Hosts can show fecundity compensation or gigantism in defence. Group living and sociality in hosts leads to many hosts in dense aggregations and, often, of similar genetic types.
- Host populations can be regulated in size by the effects of the parasites. Density-dependent effects on host mortality or fecundity are known, and cases from biocontrol or wildlife conservation illustrate the effects. It remains unclear how widespread parasites regulate natural populations. Epidemic outbreaks provide examples of dramatic declines in host population sizes, and also many examples of extinction.

- The structure of ecological communities of hosts is affected by parasites. Infections alter the competitive ability of species, including patterns of apparent competition. Major epidemics can change entire communities of animals and plants. With the dilution effects, more diverse host communities have lower parasite loads. In particular, diverse wildlife communities can lower the risk of human diseases, too.
- Food webs connect species by their trophic interactions in an ecosystem. Parasites are consumers themselves and also change the host phenotype (e.g. feeding rate, habitat choice). They can profoundly change the structures, and therefore are essential elements of food webs.

- Structure and composition of parasite communities relate to geographical factors such as area size or latitude. A colonization–extinction (transmission–clearance) dynamics connects a regional pool of parasites with their local presence in hosts or subpopulations. However, regularities in the communities of parasites are exceedingly difficult to identify and depend on the parasite group. Parasite communities often have variable and fleeting structure and composition.
- Hosts invading a new area can lose their parasites, or transfer them to the resident community. They can also acquire new parasites from the residents. The respective consequences are manifold. Geographic variation in immune defences might be an important determinant of parasite invasion success.
- Parasites can jump to a new host species, which leads to the emergence of new diseases. Animal reservoirs for human diseases are of particular concern, especially bacteria or viruses in bats and rodents. The process of emergence itself requires sufficient animal-to-human contacts and the subsequent evolution and maintenance of human-to-human transmission. Major risk areas vary with parasite group and reservoir species. Human habitat destruction and intensified land use, especially urbanization, increases the risk of zoonoses for human populations. Zoonoses have a long history and have caused major pandemics.
- Climate change, especially global warming, will also considerably affect host–parasite relationships. However, the exact effects are difficult to predict. The change in host communities at any one site and the shift or expansion of vector ranges are probably the most critical processes.

Bibliography

Abbate JL, Kada S, Lion S (2015). Beyond mortality: sterility as a neglected component of parasite virulence. *PLoS Pathogens*, 11: e1005299.

Abby SS, Rocha EPC (2012). The non-flagellar type III secretion system evolved from the bacterial flagellum and diversified into host-cell adapted systems. *PLoS Genetics*, 8: e1002983.

Abdallah AM, Gey van Pittius NC, DiGiuseppe Champion PA, et al. (2007). Type VII secretion: mycobacteria show the way. *Nature Reviews Microbiology*, 5: 883–91.

Abdullah MFF, Borts RH (2001). Meiotic recombination frequencies are affected by nutritional states in *Saccharomyces cerevisiae*. *Proceedings of the National Academy of Sciences USA*, 98: 14524–9.

Abel L, Fellay J, Haas DW, et al. (2018). Genetics of human susceptibility to active and latent tuberculosis: Present knowledge and future perspectives. *The Lancet Infectious Diseases*, 18: e64–75.

Abel S, Abel zur Wiesch P, Chang H-H, Davis BM, Lipsitch M, Waldor MK (2015a). Sequence tag–based analysis of microbial population dynamics. *Nature Methods*, 12: 223–9.

Abel S, Abel zur Wiesch P, Davis BM, Waldor MK (2015b). Analysis of bottlenecks in experimental models of infection. *PLoS Pathogens*, 11: e1004823.

Abeles SR, Pride DT (2014). Molecular bases and role of viruses in the human microbiome. *Journal of Molecular Biology*, 426: 3892–906.

Abraham EP, Chain E (1940). An enzyme from bacteria able to destroy penicillin. *Nature*, 146: 837.

Abram PK, Brodeur J, Urbaneja A, Tena A (2019). Nonreproductive effects of insect parasitoids on their hosts. *Annual Review of Entomology*, 64: 259–76.

Abrami L, Reig N, Gisou van der Goot F (2005). Anthrax toxin: the long and winding road that leads to the kill. *Trends in Microbiology*, 13: 72–8.

Acevedo A, Brodsky L, Andino R (2014). Mutational and fitness landscapes of an RNA virus revealed through population sequencing. *Nature*, 505: 686–90.

Adamo SA (1999). Evidence for adaptive changes in egg laying crickets exposed to bacteria and parasites. *Animal Behaviour*, 57: 117–24.

Adamo SA (2013). Parasites: Evolution's neurobiologists. *The Journal of Experimental Biology*, 216: 3–10.

Adamo SA (2014). Parasitic aphrodisiacs: Manipulation of the hosts' behavioral defenses by sexually transmitted parasites. *Integrative and Comparative Biology*, 54: 159–65.

Adamo SA, Bartlett A, Le J, Spencer N, Sullivan K (2010). Illness-induced anorexia may reduce trade-offs between digestion and immune function. *Animal Behaviour*, 79: 3–10.

Adamo SA, Roberts JL, Easy RH, Ross NW (2008). Competition between immune function and lipid transport for the protein apolipophorin III leads to stress-induced immunosuppression in crickets. *Journal of Experimental Biology*, 211: 531–8.

Adelman JS, Kirkpatrick L, Grodio JL, Hawley DM (2013). House finch populations differ in early inflammatory signaling and pathogen tolerance at the peak of *Mycoplasma gallisepticum* infection. *The American Naturalist*, 181: 674–89.

Adelman Z, Akbari O, Bauer J, et al. (2017). Rules of the road for insect gene drive research and testing. *Nature Biotechnology*, 35: 716–18.

Adema CM (2015). Fibrinogen-related proteins (FREPs) in molluscs. In Hsu E, du Pasquier L (eds) *Pathogen-host interactions: antigenic variation v. somatic adaptations*, pp. 111–29. Springer International Publishing, Cham.

Adiba S, Huet M, Kaltz O (2010). Experimental evolution of local parasite maladaptation. *Journal of Evolutionary Biology*, 23: 1195–205.

Aeschlimann PB, Häberli MA, Reusch TBH, Boehm T, Milinski M (2003). Female sticklebacks *Gasterosteus aculeatus* use self-reference to optimize MHC allele number during mate selection. *Behavioral Ecology and Sociobiology*, 54: 119–26.

Afelt A, Frutos R, Devaux C (2018). Bats, coronaviruses, and deforestation: toward the emergence of novel infectious diseases? *Frontiers in Microbiology*, 9: 702.

Ageitos JM, Sánchez-Pérez A, Calo-Mata P, Vill G (2017). Antimicrobial peptides (AMPs): ancient compounds that represent novel weapons in the fight against bacteria. *Biochemical Pharmacology*, 133: 117–38.

Agerberth B, Gudmundsson GH (2006). Host antimicrobial defence peptides in human disease. *Current Topics in Microbiology and Immunology*, 306: 67–90.

Aggarwal R, Subramanyam S, Zhao C, Chen M-S, Harris MO, Stuart JJ (2014). Avirulence effector discovery in a plant galling and plant parasitic arthropod, the Hessian fly (*Mayetiola destructor*). *PLoS ONE*, 9: e100958.

Agrawal AA (1997). Adaptation and trade-offs in fitness of mites on alternative hosts. *Ecology*, 81: 500–8.

Agrawal P, Nawadkar R, Ojha H, Kumar J, Sahu A (2017). Complement evasion strategies of viruses: an overview. *Frontiers in Microbiology*, 8: 1117.

Aidoo M, Terlouw DJ, Kolzcak MSM, et al. (2002). Protective effects of the sickle cell gene against malaria morbidity and mortality. *The Lancet*, 359: 1311–12.

Aiewsakun P, Katzourakis A (2017). Marine origin of retroviruses in the early Palaeozoic Era. *Nature Communications*, 8: 13954.

Aivelo T, Norberg A (2017). Parasite–microbiota interactions potentially affect intestinal communities in wild mammals. *Journal of Animal Ecology*, 87: 438–47.

Akçay E, Roughgarden J (2007). Extra-pair paternity in birds: review of the genetic benefits. *Evolutionary Ecology Research*, 9: 855–68.

Akopyants NS, Kimbllin N, Secundino N, et al. (2009). Demonstration of genetic exchange during cyclical development of *Leishmania* in the sand fly vector. *Science*, 324: 265–8.

Alamery S, Tirnaz S, Bayer P, et al. (2018). Genome-wide identification and comparative analysis of NBS-LRR resistance genes in *Brassica napus*. *Crop and Pasture Science*, 69: 72–93.

Albuquerque P, Casadevall A (2012). Quorum sensing in fungi: a review. *Medical Mycology*, 50: 337–45.

Alcock J, Maley CC, Aktipis CA (2014). Is eating behavior manipulated by the gastrointestinal microbiota? Evolutionary pressures and potential mechanisms. *Bioessays*, 36: 940–9.

Alizon S, Fraser C (2013). Within-host and between-host evolutionary rates across the HIV-1 genome. *Retrovirology*, 10: 49.

Alizon S, Hurford A, Mideo N, Van Baalen M (2009). Virulence evolution and the trade-off hypothesis: history, current state of affairs and the future. *Journal of Evolutionary Biology*, 22: 245–59.

Alizon S, Lion S (2011). Within-host parasite cooperation and the evolution of virulence. *Proceedings of the Royal Society London B*, 278: 3738–47.

Alizon S, Michalakis Y (2015). Adaptive virulence evolution: the good old fitness-based approach. *Trends in Ecology & Evolution*, 30: 248–54.

Alizon S, van Baalen M (2005). Emergence of a convex trade-off between transmission and virulence. *The American Naturalist*, 165: E155–67.

Alizon S, van Baalen M (2008). Acute or chronic? Within-host models with immune dynamics, infection outcome, and parasite evolution. *The American Naturalist*, 172: E244–56.

Allan BF, Langerhans RB, Ryberg WA, et al. (2009). Ecological correlates of risk and incidence of West Nile virus in the United States. *Oecologia*, 158: 699–708.

Allen GR, Seeman OD, Schmid-Hempel P, Buttermore RE (2007). Low parasite loads accompany the invading population of the bumblebee, *Bombus terrestris* in Tasmania. *Insectes Sociaux*, 54: 56–63.

Allison AC (1982). Coevolution between hosts and infectious disease agents, and its effects on virulence. In Anderson RM, May RM (eds) *Population biology of infectious diseases*, pp. 245–68. Springer-Verlag, Berlin.

Allocati N, Petrucci AG, Di Giovanni P, Masulli M, Di Ilio C, De Laurenzi V (2016). Bat–man disease transmission: zoonotic pathogens from wildlife reservoirs to human populations. *Cell Death Discovery*, 2: 16048.

Almberg ES, Cross PC, Dobson AP, Smith DW, Hudson PJ (2012). Parasite invasion following host reintroduction: a case study of Yellowstone's wolves. *Philosophical Transactions of the Royal Society London B*, 367: 2840–51.

Alonso-Alvarez C, Tella JL (2001). Effects of experimental food restriction and body-mass changes on avian T-cell mediated immune response. *Canadian Journal of Zoology*, 79: 101–5.

Altenberg L, Feldman MW (1987). Selection, generalized transmission and the evolution of modifier genes. I. The reduction principle. *Genetics*, 117: 559–72.

Altizer S, Hobson KA, Davis AK, De Roode JC, Wassenaar LI (2015). Do healthy monarchs migrate farther? Tracking natal origins of parasitized vs. uninfected monarch butterflies overwintering in Mexico. *PLoS ONE*, 10: e0141371.

Altizer S, Nunn CL, Thrall PH, et al. (2003). Social organisation and parasite risk in mammals: integrating theory and empirical studies. *Annual Review of Ecology and Systematics*, 34: 517–47.

Altizer S, Ostfeld RS, Johnson PTJ, Kutz S, Harvell CD (2013). Climate change and infectious diseases: from evidence to a predictive framework. *Science*, 341: 514–19.

Alto NM, Orth K (2012). Subversion of cell signaling by pathogens. *Cold Spring Harbor Perspectives in Biology*, 4: a006114.

Altshuler D, Donnelly P, The International HapMap C (2005). A haplotype map of the human genome. *Nature*, 437: 1299–320.

Alzaga V, Vicente J, Villanua D, Acevedo P, Casas F, Gortazar C (2007). Body condition and parasite intensity correlates with escape capacity in Iberian hares (*Lepus granatensis*). *Behavioral Ecology and Sociobiology*, 62: 769–75.

Amano H, Hayashi K, Kasuya E (2008). Avoidance of egg parasitism through submerged oviposition by tandem pairs in the water strider, *Aquarius paludum insularis* (Heteroptera: Gerridae). *Ecological Entomology*, 33: 560–3.

Amato KR, G. Sanders J, Song SJ, et al. (2018). Evolutionary trends in host physiology outweigh dietary niche in structuring primate gut microbiomes. *The ISME Journal*, 13: 576–87.

Aminov RI (2009). The role of antibiotics and antibiotic resistance in nature. *Environmental Microbiology*, 11: 2970–88.

Aminov RI (2016). History of antimicrobial drug discovery: major classes and health impact. *Biochemical Pharmacology*, 133: 4–19.

Amulic B, Cazalet C, Hayes GL, Metzler KD, Zychlinsky A (2012). Neutrophil function: from mechanisms to disease. *Annual Review of Immunology*, 30: 459–89.

Andam CP, Challagundla L, Azarian T, Hanage WP, Robinson DA (2017). Population structure of pathogenic bacteria. In Tibayrenc M (ed.) *Genetics and evolution of infectious diseases*, 2nd ed., pp. 51–70. Elsevier, London.

Andam CP, Worby CJ, Chang Q, Campana MG (2016). Microbial genomics of ancient plagues and outbreaks. *Trends in Microbiology*, 24: 978–90.

Andersen KG, Rambaut A, Lipkin WI, Holmes EC, Garry RF (2020). The proximal origin of SARS-CoV-2. *Nature Medicine*, 26: 450–2.

Anderson CJ, Kendall MM (2017). *Salmonella enterica* serovar typhimurium strategies for host adaptation. *Frontiers in Microbiology*, 8: 1983.

Anderson DP, Gormley AM, Bosson M, Livingstone PG, Nugent G (2017). Livestock as sentinels for an infectious disease in a sympatric or adjacent-living wildlife reservoir host. *Preventive Veterinary Medicine*, 148: 106–14.

Anderson JB (2005). Evolution of antifungal drug resistance: mechanisms and pathogen fitness. *Nature Reviews Microbiology*, 3: 547–56.

Anderson RM, Fraser C, Ghani AC, et al. (2004). Epidemiology, transmission dynamics and control of SARS: the 2002–2003 epidemic. *Philosophical Transactions of the Royal Society London B*, 359: 1091–105.

Anderson RM, May RM (1978). Regulation and stability of host-parasite population interactions. I. Regulatory processes. *Journal of Animal Ecology*, 47: 219–47.

Anderson RM, May RM (1980). Infectious diseases and population cycles of forest insects. *Science*, 210: 658–61.

Anderson RM, May RM (1981). The population dynamics of microparasites and their invertebrate hosts. *Philosophical Transactions of the Royal Society London B*, 291: 451–524.

Anderson RM, May RM (1982). Coevolution of hosts and parasites. *Parasitology*, 85: 411–26.

Anderson RM, May RM (1985). Helminth infections of humans: mathematical models, population dynamics and control. *Advances in Parasitology*, 24: 1–101.

Anderson RM, May RM (1991). *Infectious diseases of humans*. Oxford University Press, Oxford.

Andersson DI (2006). The biological cost of mutational antibiotic resistance: any practical conclusions? *Current Opinion in Microbiology*, 9: 461–5.

Andersson DI, Hughes D (2010). Antibiotic resistance and its cost: is it possible to reverse resistance? *Nature Reviews Microbiology*, 8: 260–71.

Andersson DI, Hughes D (2014). Microbiological effects of sublethal levels of antibiotics. *Nature Reviews Microbiology*, 12: 465–78.

Andersson DI, Hughes D, Kubicek-Sutherland JZ (2016). Mechanisms and consequences of bacterial resistance to antimicrobial peptides. *Drug Resistance Updates*, 26: 43–57.

Andersson DI, Levin BR (1999). The biological cost of antibiotic resistance. *Current Opinion in Microbiology*, 2: 489–93.

Andersson M, Simmons LW (2006). Sexual selection and mate choice. *Trends in Ecology & Evolution*, 21: 296–302.

Andrade DV, Harris E (2018). Recent advances in understanding the adaptive immune response to Zika virus and the effect of previous flavivirus exposure. *Virus research*, 254: 27–33.

André J-B, Ferdy J-B, Godelle B (2003). Within-host parasite dynamics, emerging trade-off, and evolution of virulence with immune system. *Evolution*, 57: 1489–97.

Andrews SJ (1999). The life cycle of *Fasciola hepatica*. In Dalton JP (ed.) *Fasciolosis*, pp. 1–29. CABI Publishing, Wallingford, UK.

Anthony SJ, Epstein JH, Murray KA, et al. (2013). A strategy to estimate unknown viral diversity in mammals. *mBio*, 4: e00598–13.

Antonides J, Mathur S, DeWoody JA (2019). Episodic positive diversifying selection on key immune system genes in major avian lineages. *Genetica*, 147: 337–50.

Antonovics J, Boots M, Abbate J, Baker C, McFrederick Q, Panjeti V (2011). Biology and evolution of sexual transmission. *Annals of the New York Academy of Sciences*, 1230: 12–24.

Antonovics J, Boots M, Ebert D, Koskella B, Poss M, Sadd BM (2013). The origin of specificity by means of natural selection: evolved and nonhost resistance in host–pathogen interactions. *Evolution*, 67: 1–9.

Antonovics J, Ellstrand NC (1984). Experimental studies of the evolutionary significance of sexual reproduction. I. A test of the frequency-dependent selection hypothesis. *Evolution*, 38: 103–15.

Antonovics J, Wilson AJ, Forbes MR, et al. (2017). The evolution of transmission mode. *Philosophical Transactions of the Royal Society London B*, 372: 20160083.

Anttila J, Mikonranta L, Ketola T, Kaitala V, Laakso J, Ruokolainen L (2016). A mechanistic underpinning for sigmoid dose-dependent infection. *Oikos*, 126: 910–16.

Antunes LCM, Ferreira RBR, Buckner MMC, Finlay BB (2010). Quorum sensing in bacterial virulence. *Microbiology*, 156: 2271–82.

Arama C, Maiga B, Dolo A, et al. (2015). Ethnic differences in susceptibility to malaria: what have we learned from immuno-epidemiological studies in West Africa? *Acta Tropica*, 146: 152–6.

Arandjelovic S, Ravichandran KS (2015). Phagocytosis of apoptotic cells in homeostasis. *Nature Immunology*, 16: 907–17.

Araujo SBL, Braga MP, Brooks DR, et al. (2015). Understanding host-switching by ecological fitting. *PLoS ONE*, 10: 30139225.

Arciola CR, Campoccia D, Montanaro L (2018). Implant infections: adhesion, biofilm formation and immune evasion. *Nature Reviews Microbiology*, 16: 397–409.

Aresta-Branco F, Erben E, Papavasiliou N, Stebbins E (2019). Mechanistic similarities between antigenic variation and antibody diversification during *Trypanosoma brucei* Infection. *Trends in Parasitology*, 35: 302–15.

Arkush KD, Giese AR, Mendonca HL, McBride AM, Marty GD, Hedrick PW (2002). Resistance to three pathogens in the endangered winter-run chinook salmon (*Oncorhynchus tshawytscha*): effects of inbreeding and major histocompatibilty complex genotypes. *Canadian Journal of Fisheries and Aquatic Sciences*, 59: 966–75.

Armitage SA, Freiburg RY, Kurtz J, Bravo IG (2012). The evolution of *Dscam* genes across the arthropods. *BMC Evolutionary Biology*, 12: 53.

Armitage SA, Kurtz J, Brites D, Dong Y, du Pasquier L, Wang H-C (2017). *Dscam1* in pancrustacean immunity: current status and a look to the future. *Frontiers in Immunology*, 8: 662.

Arnqvist G, Rowe L (2005). *Sexual conflict*. Princeton University Press, Princeton, NJ.

Ashby B, King KC (2017). Friendly foes: the evolution of host protection by a parasite. *Evolution Letters*, 1: 211–21.

Ashcroft P, Huisman JS, Lehtinen S, et al. (2020). COVID-19 infectivity profile correction. *Swiss Medicine Weekly*, 150: w20336.

Ashworth ST, Kennedy CR (1999). Density-dependent effects on *Anguillicola crassus* (Nematoda) within its European eel definitive host. *Parasitology*, 118: 289–96.

Atkins KE, Read AF, Savill NJ, et al. (2013). Vaccination and reduced cohort duration can drive virulence evolution: Marek's disease virus and industrialized agriculture. *Evolution*, 67: 851–60.

Ausubel FM (2005). Are innate immune signalling pathways in plants and animals conserved? *Nature Immunology*, 6: 973–9.

Ayres JS, Schneider DS (2008). A signaling protease required for melanization in *Drosophila* affects resistance and tolerance of infections. *PLoS Biology*, 6: e305.

Ayres JS, Schneider DS (2012). Tolerance of infections. *Annual Review of Immunology*, 30: 271–94.

Azevedo L, Serrano C, Amorim A, Cooper DN (2015). Trans-species polymorphism in humans and the great apes is generally maintained by balancing selection that modulates the host immune response. *Human Genomics*, 9: 21.

Babin A, Saciat C, Teixeira M, et al. (2015). Limiting immunopathology: Interaction between carotenoids and enzymatic antioxidant defences. *Developmental & Comparative Immunology*, 49: 278–81.

Bach JF (2018). The hygiene hypothesis in autoimmunity: the role of pathogens and commensals. *Nature Reviews Immunology*, 18: 105–20.

Baer B, Schmid-Hempel P (1999). Experimental variation in polyandry affects parasite loads and fitness in a bumble-bee. *Nature*, 397: 151–4.

Bah A, Vergne I (2017). Macrophage autophagy and bacterial infections. *Frontiers in Immunology*, 8: 1483.

Bahar AA, Ren D (2013). Antimicrobial peptides. *Pharmaceuticals*, 6: 1543–75.

Bakker TCM, Frommen JG, Thunken T (2017). Adaptive parasitic manipulation as exemplified by acanthocephalans. *Ethology*, 123: 779–84.

Balaban NQ, Merrin J, Chait R, Kowalik L, Leibler S (2004). Bacterial resistance as a phenotypic switch. *Science*, 305: 1622–5.

Balciunas EM, Castillo Martinez FA, Todorov SD, Gombossy de Melo Franco BD, Converti A, Pinheiro de Souza Oliveira R (2013). Novel biotechnological applications of bacteriocins: a review. *Food Control*, 32: 134–42.

Baldo L, Pretus JL, Riera JL, Musilova Z, Bitja Nyom AR, Salzburger W (2017a). Convergence of gut microbiotas in the adaptive radiations of African cichlid fishes. *The ISME Journal*, 11: 1975–87.

Baldo L, Riera JL, Mitsi K, Pretus JL (2017b). Processes shaping gut microbiota diversity in allopatric populations of the endemic lizard *Podarcis lilfordi* from Menorcan islets (Balearic Islands). *FEMS Microbiology Ecology*, 94: fix186.

Baldridge MT, Nice TJ, McCune BT, et al. (2015). Commensal microbes and interferon-λ determine persistence of enteric murine norovirus infection. *Science*, 347: 266–9.

Balenger SL, Zuk M (2014). Testing the Hamilton–Zuk hypothesis: past, present, and future. *Integrative and Comparative Biology*, 54: 601–13.

Balfour A (1914). The wild monkey as a reservoir for the virus of yellow fever. *Lancet*, 1: 1176–8.

Ballabeni P (1995). Parasite-induced gigantism in a snail: a host adaptation? *Functional Ecology*, 9: 887–93.

Balmer O, Tanner M (2011). Prevalence and implications of multiple-strain infections. *Lancet Infectious Disease*, 11: 868–78.

Bandi C, Dunn AM, Hurst GDD, Rigaud T (2001). Inherited microorganisms, sex-specific virulence and reproductive parasitism. *Trends in Parasitology*, 17: 88–94.

Banerjee S, Guedj J, Ribeiro RM, Moses M, Perelson AS (2016). Estimating biologically relevant parameters under uncertainty for experimental within-host murine West Nile virus infection. *Journal of the Royal Society Interface*, 13: 20160130.

Banyard AC, Evans JS, Luo TR, Fooks AR (2014). Lyssaviruses and bats: emergence and zoonotic threat. *Viruses*, 6: 2974–90.

Banyard AC, Hayman D, Johnson N, McElhinney L, Fooks AR (2011). Bats and lyssaviruses. In Jackson AC (ed.) *Advances in Virus Research*, pp. 239–89. Academic Press, London.

Baracos V, Whitmore W, Gale R (1987). The metabolic cost of fever. *Canadian Journal of Physiology and Pharmacology*, 65: 1248–54.

Barber I (2000). Effects of parasites on fish behaviour: a review and evolutionary perspective. *Reviews in Fish Biology and Fisheries*, 10: 131–65.

Barbi J, Pardoll DM, Pan F (2015). Ubiquitin-dependent regulation of Foxp3 and Treg function. *Immunological Reviews*, 266: 27–45.

Barclay AN (2003). Membrane proteins with immunoglobulin-like domains: a master superfamily of interaction molecules. *Seminars in Immunology*, 15: 215–23.

Barclay VC, Kennedy DA, Weaver VC, Sim D, Lloyd-Smith JO, Read AF (2014). The effect of immunodeficiency on the evolution of virulence: an experimental test with the rodent malaria, *Plasmodium chabaudi*. *The American Naturalist*, 184, Supplement: S47–57.

Barclay VC, Råberg L, Chan BHK, Brown S, Gray D, Read AF (2008). CD4+T cells do not mediate within-host competition between genetically diverse malaria parasites. *Proceedings of the Royal Society London B*, 275: 1171–9.

Barclay VC, Sim D, Chan BHK, et al. (2012). The evolutionary consequences of blood-stage vaccination on the rodent malaria *Plasmodium chabaudi. PLoS Biology*, 10: e1001368.

Barreiro LB, Quintana-Murci L (2010). From evolutionary genetics to human immunology: how selection shapes host defence genes. *Nature Reviews Genetics*, 11: 17–30.

Barrett ADT (2018). The reemergence of yellow fever. *Science*, 361: 847–8.

Barrett JA, Antonovics J (1988). Frequency-dependent selection in plant-fungal interactions. *Philosophical Transactions of the Royal Society London B*, 319: 473–83.

Barrett LG, Kniskern JM, Bodenhausen N, Zhang W, Bergelson J (2009). Continua of specificity and virulence in plant host–pathogen interactions: causes and consequences. *New Phytologists*, 183: 513–29.

Barría A, Doeschl-Wilson AB, Lhorente JP, Houston RD, Yáñez JM (2019). Novel insights into the genetic relationship between growth and disease resistance in an aquaculture strain of Coho salmon (*Oncorhynchus kisutch*). *Aquaculture*, 511: 734207.

Barribeau SM, Sadd BM, du Plessis L, et al. (2015). A depauperate immune repertoire precedes evolution of sociality in bees. *Genome Biology*, 16: 83.

Barribeau SM, Sadd BM, du Plessis L, Schmid-Hempel P (2014). Gene expression differences underlying genotype-by-genotype specificity in a host-parasite system. *Proceedings of the National Academy USA*, 111: 3496–501.

Barribeau SM, Schmid-Hempel P, Sadd BM (2016). Royal decree: gene expression in transgenerationally immune primed bumblebee workers mimics a primary immune response. *PLoS ONE*, 11: e0159635.

Barron DG, Gervasi SS, Pruitt JN, Martin LB (2015). Behavioral competence: how host behaviors can interact to influence parasite transmission risk. *Current Opinion in Behavioral Sciences*, 6: 35–40.

Bartholomay LC, Michel K (2018). Mosquito immunobiology: the intersection of vector health and vector competence. *Annual Review of Entomology*, 63: 145–67.

Bartlett MS (1957). Measles periodicity and community size. *Journal of the Royal Statistical Society A*, 120: 48–70.

Baseler L, Chertow DS, Johnson KM, Feldmann H, Morens DM (2017). The pathogenesis of Ebola virus disease. *Annual Review of Pathology: Mechanisms of Disease*, 12: 387–418.

Bashey F, Young SK, Hawlena H, Lively CM (2012). Spiteful interactions between sympatric natural isolates of *Xenorhabdus bovienii* benefit kin and reduce virulence. *Journal of Evolutionary Biology*, 25: 431–7.

Bassett MG, Popov LE, Holmer LE (2004). The oldest-known metazoan parasite? *Journal of Paleontology*, 78: 1214–16.

Battersby S (ed.) (2017). *Clay's handbook of environmental health*, 21st ed. Routledge, London and New York.

Baudoin M (1975). Host castration as a parasitic strategy. *Evolution*, 29: 335–52.

Bauernfeind F, Hornung V (2013). Of inflammasomes and pathogens: sensing of microbes by the inflammasome. *EMBO Molecular Medicine*, 5: 814–26.

Baxter NT, Wan JJ, Schubert AM, Jenior ML, Myers P, Schloss PD (2015). Intra- and interindividual variations mask interspecies variation in the microbiota of sympatric *Peromyscus populations*. *Applied and Environmental Microbiology*, 81: 396–404.

Beachboard DC, Horner SM (2016). Innate immune evasion strategies of DNA and RNA viruses. *Current Opinion in Microbiology*, 32: 113–19.

Beagley KW, Gockel CM (2003). Regulation of innate and adaptive immunity by the female sex hormones oestradiol and progesterone. *FEMS Immunology and Medical Microbiology*, 38: 13–22.

Beattie AJ, Turnbull CL, Hough T, Jobson S, Knox RB (1986). Antibiotic production: a possible function for the metapleural glands of ants (Hymenoptera: Formicidae). *Annals of the Entomological Society of America*, 79: 448–50.

Beck M, Handy J, Levander OA (2004). Host nutritional status: the neglected virulence factor. *Trends in Microbiology*, 12: 417–23.

Becker DJ, Albery GF, Kessler MK, et al. (2020). Macroimmunology: the drivers and consequences of spatial patterns in wildlife immune defence. *Journal of Animal Ecology*, 89: 972–95.

Becker DJ, Nachtmann C, Argibay HD, et al. (2019). Leukocyte profiles reflect geographic range limits in a widespread neotropical bat. *Integrative and Comparative Biology*, 59: 1176–89.

Becker DJ, Streicker DG, Altizer S (2015). Linking anthropogenic resources to wildlife–pathogen dynamics: a review and meta-analysis. *Ecology Letters*, 18: 483–95.

Beckmann JF, Ronau JA, Hochstrasser M (2017). A *Wolbachia* deubiquitylating enzyme induces cytoplasmic incompatibility. *Nature Microbiology*, 2: 17007.

Begum S, Quach J, Chadee K (2015). Immune evasion mechanisms of *Entamoeba histolytica*: progression to disease. *Frontiers in Microbiology*, 6: 1394.

Behrens HM, Six A, Walker D, Kleanthous C (2017). The therapeutic potential of bacteriocins as protein antibiotics. *Emerging Topics in Life Sciences*, 1: 65–74.

Behringer DC, Karvonen A, Bojko J (2018). Parasite avoidance behaviours in aquatic environments. *Philosophical Transactions of the Royal Society London B*, 373: 20170202.

Beier D, Gross R (2006). Regulation of bacterial virulence by two-component systems. *Current Opinion in Microbiology*, 9: 143–52.

Bell AS, De Roode JC, Sim D, Read AF (2006). Within-host competition in genetically diverse malaria infections: Parasite virulence and competitive success. *Evolution*, 60: 1358–71.

Bell G (1982). *The masterpiece of nature*. University of California Press, Berkeley.

Bell G, MacLean C (2017). The search for 'evolution-proof' antibiotics. *Trends in Microbiology*, 26: 471–83.

Bellanger X, Payot S, Leblond-Bourget N, Guedo Gr (2013). Conjugative and mobilizable genomic islands in bacteria: evolution and diversity. *FEMS Microbiological Reviews*, 38: 720–60.

Bellinvia S, Johnston PR, Reinhardt K, Otti O (2020). Bacterial communities of the reproductive organs of virgin and mated common bedbugs, *Cimex lectularius*. *Ecological Entomology*, 45: 142–54.

Belzer C, Gerber GK, Roeselers G, et al. (2014). Dynamics of the microbiota in response to host infection. *PLoS ONE*, 9: e95534.

Ben-Ami F (2017). The virulence–transmission relationship in an obligate killer holds under diverse epidemiological and ecological conditions, but where is the tradeoff? *Ecology and Evolution*, 7: 11157–66.

Ben-Ami F, Ebert D, Regoes RR (2010). Pathogen dose infectivity curves as a method to analyze the distribution of host susceptibility: a quantitative assessment of maternal effects after food stress and pathogen exposure. *The American Naturalist*, 175: 106–15.

Ben-Ami F, Mouton L, Ebert D (2008). The effects of multiple infections on the expression and evolution of virulence in a *Daphnia*-endoparasite system. *Evolution*, 62: 1700–11.

Ben-Ami F, Routtu J (2013). The expression and evolution of virulence in multiple infections: the role of specificity, relative virulence and relative dose. *BMC Evolutionary Biology*, 13: 97.

Ben-Shachar R, Koelle K (2015). Minimal within-host dengue models highlight the specific roles of the immune response in primary and secondary dengue infections. *Journal of the Royal Society Interface*, 12: 20140886.

Benesh DP, Chubb JC, Parker GA (2014). The trophic vacuum and the evolution of complex life cycles in trophically transmitted helminths. *Proceedings of the Royal Society London B*, 281: 20141462.

Bento G, Routtu J, Fields PD, Bourgeois Y, du Pasquier L, Ebert D (2017). The genetic basis of resistance and matching-allele interactions of a host-parasite system: the *Daphnia magna-Pasteuria ramosa* model. *PLoS Genetics*, 13: e1006596.

Bentur JS, Rawat N, Divya D, et al. (2016). Rice–gall midge interactions: battle for survival. *Journal of Insect Physiology*, 84: 40–9.

Benz R (2016). Channel formation by RTX-toxins of pathogenic bacteria: basis of their biological activity. *Biochimica et Biophysica Acta (BBA)—Biomembranes*, 1858: 526–37.

Berdoy M, Webster JP, MacDonald DW (2000). Fatal attraction in rats infected with *Toxoplasma gondii*. *Proceedings of the Royal Society London B*, 267: 1591–4.

Bérénos C, Schmid-Hempel P, Wegner MK (2010a). Experimental coevolution leads to a decrease in parasite-induced host mortality after initial temporal dynamics. *Journal of Evolutionary Biology*, 24: 1777–82.

Bérénos C, Wegner MK, Schmid-Hempel P (2010b). Antagonistic co-evolution with parasites maintains host genetic diversity: an experimental test. *Proceedings of the Royal Society London B*, 278: 218–24.

Bergelson J, Kreitman M, Stahl EA, Tian D (2001). Evolutionary dynamics of plant R-genes. *Science*, 292: 2281–5.

Bergstrom CT, Antia R (2006). How do adaptive immune systems control pathogens while avoiding autoimmunity? *Trends in Ecology and Evolution*, 21: 22–8.

Bernasconi G, Ashman TL, Birkhead TR, et al. (2004). Evolutionary ecology of the prezygotic stage. *Science*, 303: 971–5.

Berngruber TW, Froissart R, Choisy M, Gandon S (2013). Evolution of virulence in emerging epidemics. *PLoS Pathogens*, 9: e1003209.

Berngruber TW, Lion S, Gandon S (2015). Spatial structure, transmission modes and the evolution of viral exploitation strategies. *PLoS Pathogens*, 11: e1004810.

Bernoulli D (1766). Essai d'une nouvelle analyse de la mortalité causée par la petite vérole & des avantages de l'inoculation pour la prévenir. *Mémoires de Mathématique et de Physique, Académie Royale des Sciences, Paris*.

Bernoulli D, Blower S (2004). An attempt at a new analysis of the mortality caused by smallpox and of the advantages of inoculation to prevent it. *Reviews in Medical Virology*, 14: 275–88.

Bertels F, Marzel A, Leventhal G, et al. (2018). Dissecting HIV virulence: heritability of setpoint viral load, CD4+ t-cell decline, and per-parasite pathogenicity. *Molecular Biology and Evolution*, 35: 27–37.

Berube BJ, Wardenburg JB (2013). *Staphylococcus aureus* α-toxin: nearly a century of intrigue. *Toxins*, 5: 1140–66.

Best A, Ashby B, White A, et al. (2017). Host-parasite fluctuating selection in the absence of specificity. *Proceedings of the Royal Society London B*, 284: 20171615.

Best A, Hoyle A (2013a). The evolution of costly acquired immune memory. *Ecology and Evolution*, 3: 2223–32.

Best A, Hoyle A (2013b). A limited host immune range facilitates the creation and maintenance of diversity in parasite virulence. *Interface Focus*, 3: 20130024.

Best A, Long G, White A, Boots M (2012). The implications of immunopathology for parasite evolution. *Proceedings of the Royal Society London B*, 279: 3234–40.

Best A, Tidbury H, White A, Boots M (2013). The evolutionary dynamics of within-generation immune priming in invertebrate hosts. *Jorunal of the Royal Society Interface*, 10: 20120887.

Best A, White A, Boots M (2008). Maintenance of host variation in tolerance to pathogens and parasites. *Proceedings of the National Academy of Sciences USA*, 105: 20786–91.

Best A, White A, Boots M (2009). Resistance is futile but tolerance can explain why parasites do not always castrate their hosts. *Evolution*, 64: 348–57.

Best A, White A, Boots M (2014). The coevolutionary implications of host tolerance. *Evolution*, 68: 1426–35.

Betts A, Gray C, Zelek M, MacLean RC, King KC (2018). High parasite diversity accelerates host adaptation and diversification. *Science*, 360: 907–11.

Beutler B (2004). Innate immunity: an overview. *Molecular Immunology*, 40: 845–59.

Biek R, Drummond AJ, Poss M (2006). A virus reveals population structure and recent femographic history of its carnivore host. *Science*, 311: 538–41.

Bielby J, Fisher MC, Clare FC, Rosa GM, Garner TWJ (2015). Host species vary in infection probability, sub-lethal effects and costs of immune response when exposed to an amphibian parasite. *Scientific Reports*, 5: 10828.

Bilandžija H, Laslo M, Porter ML, Fong DW (2017). Melanization in response to wounding is ancestral in arthropods and conserved in albino cave species. *Scientific Reports*, 7: 17148.

Billal DS, Feng J, Leprohon P, Legare D, Ouellette M (2011). Whole genome analysis of linezolid resistance in *Streptococcus pneumoniae* reveals resistance and compensatory mutations. *BMC Genomics*, 12: 512.

Binning SA, Shaw AK, Roche DG (2017). Parasites and host performance: incorporating infection into our understanding of animal movement. *Integrative and Comparative Biology*, 57: 267–80.

Birget PLG, Koella JC (2015). An epidemiological model of the effects of insecticide-treated bed nets on malaria transmission. *PLoS ONE*, 10: e0144173.

Birkegård AC, Halasa T, Toft N, Folkesson A, Græsbøll K (2018). Send more data: a systematic review of mathematical models of antimicrobial resistance. *Antimicrobial Resistance and Infection Control*, 7: 117.

Biron DG, Bonhomme L, Coulon M, Øverli Ø (2015). Microbiomes, plausible players or not in alteration of host behavior. *Frontiers in Microbiology*, 5: Article 775.

Biron DG, Loxdale HD (2013). Host–parasite molecular cross-talk during the manipulative process of a host by its parasite. *The Journal of Experimental Biology*, 218: 148–60.

Bittner-Eddy PD, Crute IR, Holub EB, BEynon JL (2000). RPP13 is a simple locus in *Arabidopsis thaliana* for alleles that specify downy mildew resistance to different avirulence determinants in *Peronospora parasitica*. *Plant Journal*, 21: 177–88.

Björkman J, Nagaev I, Berg OG, Hughes DW, Andresson DI (2000). Effects of environment on compensatory mutations to ameliorate costs of antibiotic resistance. *Science*, 287: 1479–82.

Blackburn TM, Lockwood JL, Cassey P (2015). The influence of numbers on invasion success. *Molecular Ecology*, 24: 1942–53.

Blackwell AD, Tamayo MA, Beheim B, et al. (2015). Helminth infection, fecundity, and age of first pregnancy in women. *Science*, 350: 970–2.

Blair JMA, Webber MA, Baylay AJ, Ogbolu DO, Piddock LJV (2015). Molecular mechanisms of antibiotic resistance. *Nature Reviews Microbiology*, 13: 42–51.

Blander JM, Longman RS, Iliev ID, Sonnenberg GF, Artis D (2017). Regulation of inflammation by microbiota interactions with the host. *Nature Immunology*, 18: 851.

Blander JM, Sander LE (2012). Beyond pattern recognition: five immune checkpoints for scaling the microbial threat. *Nature Reviews Immunology*, 12: 215–25.

Blandin S, Levashina EA (2004). Mosquito immune responses against malaria parasites. *Current Opinion in Immunology*, 16: 16–20.

Blanford S, Thomas MB, Pugh C, Pell JK (2003). Temperature checks the Red Queen? Resistance and virulence in a fluctuating environment. *Ecology Letters*, 6: 2–5.

Blanquart F, Kaltz O, Nuismer SL, Gandon S (2013). A practical guide to measuring local adaptation. *Ecology Letters*, 16: 1195–205.

Blanquart F, Wymant C, Cornelissen M, et al. (2017). Viral genetic variation accounts for a third of variability in HIV-1 set-point viral load in Europe. *PLoS Biology*, 15: e2001855.

Blasco-Costa I, Poulin R (2013). Host traits explain the genetic structure of parasites: a meta-analysis. *Parasitology*, 140: 1316–22.

Blaser M, Schmid-Hempel P (2005). Determinants of virulence for the parasite *Nosema whitei* in its host *Tribolium castaneum*. *Journal of Invertebrate Pathology*, 89: 251–7.

Blaxter ML (2003). Nematoda: genes, genomes and the evolution of parasitism. *Advances in Parasitology*, 54: 101–95.

Blaxter ML, De Ley P, Garey JR, et al. (1998). A molecular evolutionary framework for the phylum Nematoda. *Nature*, 392: 71–5.

Blaxter ML, Koutsovoulos G (2015). The evolution of parasitism in Nematoda. *Parasitology*, 142: S26–39.

Blooi M, Laking AE, Martel A, et al. (2017). Host niche may determine disease-driven extinction risk. *PLoS ONE*, 12: e0181051.

Bloom BR (1979). Games parasites play: how parasites evade immune surveillance. *Nature*, 279: 21–6.

Blum MGB, Francois O (2005). On statistical tests of phylogenetic tree imbalance: the Sackin and other indices. *Mathematical Biosciences*, 195: 141–53.

Boe DM, Boule LA, Kovacs EJ (2017). Innate immune responses in the ageing lung. *Clinical & Experimental Immunology*, 187: 16–25.

Boehm T, Bleul CC (2007). The evolutionary history of lymphoid organs. *Nature Immunology*, 8: 131–5.

Boehm T, Hirano M, Holland SJ, Das S, Schorpp M, Cooper MD (2018). Evolution of aternative adaptive immune systems in vertebrates. *Annual Review of Immunology*, 36: 19–42.

Bogaardt C, van Tonder AJ, Brueggemann AB (2015). Genomic analyses of pneumococci reveal a wide diversity of bacteriocins—including pneumocyclicin, a novel circular bacteriocin. *BMC Genomics*, 16: 554.

Bohnhoff M, Drake BL, Miller CP (1954). Effect of streptomycin on susceptibility of intestinal tract to experimental *Salmonella* infection. *Proceedings of the Society for Experimental Biology and Medicine*, 86: 132–7.

Boillat M, Hammoudi P-M, Dogga SK, et al. (2020). Neuroinflammation-associated aspecific manipulation of mouse predator fear by *Toxoplasma gondii*. *Cell Reports*, 30: 320–34.e6.

Bolker BM, Nanda A, Shah D (2009). Transient virulence of emerging pathogens. *Journal of the Royal Society Interface*, 7: 811–22.

Bolzoni L, De Leo GA (2013). Unexpected consequences of culling on the eradication of wildlife diseases: the role of virulence evolution. *The American Naturalist*, 181: 301–13.

Bonhoeffer S, Lenksi RE, Ebert D (1996). The curse of the pharaoh: the evolution of virulence in pathogens with long living propagules. *Proceedings of the Royal Society London B*, 263: 715–21.

Bonneaud C, Mazuc J, Chastel O, Westerdahl H, Sorci G (2004). Terminal investment induced by immune challenge and fitness traits associated with major histocompatibility complex in the house sparrow. *Evolution*, 58: 2823–30.

Bons E, Regoes RR (2018). Virus dynamics and phyloanatomy: merging population dynamic and phylogenetic approaches. *Immunological Reviews*, 285: 134–46.

Boots M, Best A (2018). The evolution of constitutive and induced defences to infectious disease. *Proceedings of the Royal Society London B*, 285: 20180658.

Boots M, Donnelly R, White A (2013). Optimal immune defence in the light of variation in lifespan. *Parasite Immunology*, 35: 331–8.

Boots M, Hudson PJ, Sasaki A (2004). Large shifts in pathogen virulence relate to host population structure. *Science*, 303: 842–4.

Boots M, Mealor M (2007). Local interactions select for lower pathogen infectivity. *Science*, 315: 1284–6.

Boots M, Sasaki A (1999). Small worlds and the evolution of virulence: infection occurs locally and at a distance. *Proceedings of the Royal Society London B*, 266: 1933–8.

Boraschi D, Italiani P, Weil S, Martin MU (2017). The family of the interleukin-1 receptors. *Immunological Reviews*, 281: 197–232.

Bordes F, Blumstein DT, Morand S (2007). Rodent sociality and parasite diversity. *Biology Letters*, 3: 692–4.

Bordes F, Guégan JF, Morand S (2011). Microparasite species richness in rodents is higher at lower latitudes and is associated with reduced litter size. *Oikos*, 120: 1889–96.

Bordes F, Morand S (2011). The impact of multiple infections on wild animal hosts: a review. *Infection, Ecology and Epidemiology*, 1: 7346.

Bordes F, Morand S, Ricardo G (2008). Bat fly species richness in Neotropical bats: correlations with host ecology and host brain. *Oecologia*, 158: 109–16.

Borghans JA, De Boer RJ (2002). Memorizing innate instructions requires a sufficiently specific adaptive immune system. *International Immunology*, 14: 525–32.

Borgia G, Collins K (1990). Parasites and bright male plumage in the satin bowerbird. *American Zoologist*, 30: 251–62.

Borgia G, Egeth M, Uy JA, Patricell GL (2004). Juvenile infection and male display: testing the bright male hypothesis across individual life histories. *Behavioral Ecology*, 15: 722–8.

Bortolotti G, Mougeot F, Martinez-Padilla J, Webster L, Piertney S (2009). Physiological stress mediates the honesty of social signals. *PLoS ONE*, 4: e4983.

Bosch TCG (2014). Rethinking the role of immunity: lessons from *Hydra*. *Trends in Immunology*, 35: 495–502.

Bosio CM, Dow SW (2005). *Francisella tularensis* induces aberrant activation of pulmonary dendritic cells. *The Journal of Immunology*, 175: 6792–801.

Bostanci A (2005). A devil of a disease. *Science*, 307: 1035.

Bouchon D, Cordaux R, Grève P (2008). Feminizing *Wolbachia* and the evolution of sex determination in isopods. In Bourtzis K, Miller TA (eds) *Insect symbiosis*, pp. 273–328. CRC Press, Baton Rouge.

Boulinier T, McCoy K, Sorci G (2001). Dispersal and parasitism. In Clobert J, Danchin E, Nichols JD, Dhondt AA (eds) *Dispersal*, pp. 169–79. Oxford University Press, Oxford.

Bourguet D, Gair J, Mattice M, Whitlock MC (2003). Genetic recombination and adaptation to fluctuating environments: selection for geotaxis in *Drosophila melanogaster*. *Heredity*, 91: 78–84.

Bourke CD, Maizels RM, Mutapi F (2011). Acquired immune heterogeneity and its sources in human helminth infection. *Parasitology*, 138: 139–59.

Bourret V (2018). Avian influenza viruses in pigs: an overview. *The Veterinary Journal*, 239: 7–14.

Boutrot F, Zipfel C (2017). Function, discovery, and exploitation of plant pattern recognition receptors for broad-spectrum disease resistance. *Annual Review of Phytopathology*, 55: 257–86.

Bowden TJ, Thompson KD, Morgan AL, Gratacap RML, Nikoskelainen S (2007). Seasonal variation and the immune response: a fish perspective. *Fish and Shellfish Immunology*, 22: 695–706.

Brace AJ, Lajeunesse MJ, Ardia DR, et al. (2017). Costs of immune responses are related to host body size and lifespan. *Journal of Experimental Zoology Part A: Ecological and Integrative Physiology*, 327: 254–61.

Bradley CA, Altizer S (2006). Urbanization and the ecology of wildlife diseases. *Trends in Ecology & Evolution*, 22: 96–102.

Braga MP, Razzolini E, Boeger WA (2015). Drivers of parasite sharing among Neotropical freshwater fishes. *Journal of Animal Ecology*, 84: 487–97.

Brandvain Y, Goodnight C, Wade M (2011). Horizontal transmission rapidly erodes disequilibria between organelle and symbiont genomes. *Genetics*, 189: 397–404.

Brathwaite Dick O, San Martín JL, Montoya RH, del Diego J, Zambrano B, Dayan GH (2012). The history of dengue outbreaks in the Americas. *The American Journal of Tropical Medicine and Hygiene*, 87: 584–93.

Brauer F, Castillo-Chavez C, Feng Z (2019). *Mathematical models in epidemiology*. Springer, Cham.

Bray M (2006). Pathogenesis of viral hemorrhagic fever. *Current Opinion in Immunology*, 17: 399–403.

Bremermann HJ, Pickering J (1983). A game-theoretical model of parasite virulence. *Journal of Theoretical Biology*, 100: 411–26.

Brierley L, Vonhof MJ, Olival KJ, Daszak P, Jones KE (2016). Quantifying global drivers of zoonotic bat viruses: a process-based perspective. *The American Naturalist*, 187: E53–64.

Brisson D (2018). Negative frequency-dependent selection is frequently confounding. *Frontiers in Ecology and Evolution*, 6: 10.

Brites D, du Pasquier L (2015). Somatic and germline diversification of a putative immunoreceptor within one phylum: Dscam in arthropods. In Hsu E, du Pasquier L (eds) *Pathogen-host interactions: antigenic variation v. somatic adaptations*, pp. 131–58. Springer International Publishing, Cham.

Brites D, McTaggart S, Morris K, et al. (2008). The Dscam homologue of the crustacean *Daphnia* is diversified by alternative splicing like in insects. *Molecular Biology and Evolution*, 25: 1429–39.

Brogdon KA (2005). Antimicrobial peptides: pore formers or metabolic inhibitors in bacteria? *Nature Reviews Microbiology*, 3: 238–50.

Brook CE, Dobson AP (2015). Bats as 'special' reservoirs for emerging zoonotic pathogens. *Trends in Microbiology*, 23: 172–80.

Brooks LD (1988). The evolution of recombination rates. In Michod RE, Levin BR (eds) *The evolution of sex*, pp. 87–105. Sinauer Associates, Sunderland, MA.

Brouwer AF, Masters NB, Eisenberg JNS (2018). Quantitative microbial risk assessment and infectious disease transmission modeling of waterborne enteric pathogens. *Current Environmental Health Reports*, 5: 293–304.

Brown EA, Pilkington JG, Nussey DH, et al. (2013). Detecting genes for variation in parasite burden and immunological traits in a wild population: testing the candidate gene approach. *Molecular Ecology*, 22: 757–73.

Brown GD, Willment JA, Whitehead L (2018). C-type lectins in immunity and homeostasis. *Nature*, 18: 374–89.

Brown JKM, Tellier A (2011). Plant-parasite coevolution: bridging the gap between genetics and ecology. *Annual Review of Phytopathology*, 49: 345–67.

Brown SP (1999). Cooperation and conflict in host-manipulating parasites. *Proceedings of the Royal Society London B*, 266: 1899–904.

Brown SP, Cornforth DM, Mideo N (2012). Evolution of virulece in opportunistic pathogens: generalism, plasticity, and control. *Trends in Microbiology*, 20: 336–42.

Brown SP, Inglis RF, Taddei F (2009). Evolutionary ecology of microbial wars: within-host competition and (incidental) virulence. *Evolutionary Applications*, 2: 32–9.

Brown SP, Renaud F, Guéguan J-F, Thomas F (2001). Evolution of trophic transmission in parasites: the need to reach a mating place? *Journal of Evolutionary Biology*, 14: 815–20.

Browning GF, Citti C (eds) (2014). *Mollicutes: molecular biology and pathogenesis*. Caister Academic Press, Poole, U.K.

Broz P, Dixit VM (2016). Inflammasomes: mechanism of assembly, regulation and signalling. *Nature Reviews Immunology*, 16: 407–20.

Brune A, Dietrich C (2015). The gut microbiota of termites: digesting the diversity in the light of ecology and evolution. *Annual Review of Microbiology*, 69: 145–66.

Brunet J, Mundt CC (2000). Disease, frequency-dependent selection, and genetic polymorphisms: experiments with stripe rust and wheat. *Evolution*, 54: 406–15.

Brunham LR, Hayden MR (2013). Hunting human disease genes: lessons from the past, challenges for the future. *Human Genetics*, 132: 603–17.

Bruns E, Carson M, May G (2012). Pathogen and host genotype differently affect pathogen fitness through their effects on different life-history stages. *BMC Evolutionary Biology*, 12: 135.

Bruns E, Carson ML, May G (2014). The jack of all trades is master of none: a pathogen's ability to infect a greater number of host genotypes comes at a cost of delayed reproduction. *Evolution*, 68: 2453–66.

Brusca RC (1981). A monograph on the isopoda Cymothoidae (Crustacea) of the eastern Pacific. *Zoological Journal of the Linnean Society*, 73: 117–99.

Brütsch T, Jaffuel G, Vallat A, Turlings T, Chapuisat M (2017). Wood ants produce a potent antimicrobial agent by applying formic acid on tree-collected resin. *Ecology and Evolution*, 7: 2249–54.

Buchmann K (2014). Evolution of innate immunity: clues from invertebrates via fish to mammals. *Frontiers in Immunology*, 5: 459.

Buck JC (2019). Indirect effects explain the role of parasites in ecosystems. *Trends in Parasitology*, 35: 835–47.

Buck JC, Weinstein SB, Young HS (2018). Ecological and evolutionary consequences of parasite avoidance. *Trends in Ecology & Evolution*, 33: 619–32.

Buckley KM, Rast JP (2015). Diversity of animal immune receptors and the origins of recognition complexity in the deuterostomes. *Developmental and Comparative Immunology*, 49: 179–89.

Buckling A, Brockhurst MA (2008). Kin selection and the evolution of virulence. *Heredity*, 100: 484–8.

Bueno V, Lord JM, Jackson TA (eds) (2017). *The ageing immune system and health*. Springer, Cham.

Bull JJ (1994). Virulence. *Evolution*, 48: 1423–37.

Bull JJ, Barrick JE (2017). Arresting evolution. *Trends in Genetics*, 33: 910–20.

Bull JJ, Ebert D (2008). Invasion thresholds and the evolution of nonequilibrium virulence. *Evolutionary Applications*, 1: 172–82.

Bull JJ, Lauring AS (2014). Theory and empiricism in virulence evolution. *PLoS Pathogens*, 10: e1004387.

Bull JJ, Molineux IJ, Wilke CO (2012). Slow fitness recovery in a codon-modified viral genome. *Molecular Biology and Evolution*, 29: 2997–3004.

Bull RA, Luciani F, McElroy K, et al. (2011). Sequential bottlenecks drive viral evolution in early acute hepatitis C virus infection. *PLoS Pathogens*, 7: e1002243.

Bunker JJ, Bendelac A (2018). IgA responses to microbiota. *Immunity*, 49: 211–24.

Bunnik EM, Cook KB, Varoquaux N, et al. (2018). Changes in genome organization of parasite-specific gene families during the *Plasmodium transmission* stages. *Nature Communications*, 9: 1910.

Burgess SL, Gilchrist CA, Lynn TC, Petri WA (2017). Parasitic protozoa and interactions with the host intestinal microbiota. *Infection and Immunity*, 85: e100101–17.

Burnet M, White DO (1972). *The natural history of infectious disease*. Cambridge University Press, Cambridge.

Burns DL (2003). Type IV transporters of pathogenic bacteria. *Current Opinion in Microbiology*, 6: 29–34.

Burroughs NJ, De Boer RJ, Kesmir C (2004). Discriminating self from nonself with short peptides from large proteomes. *Immunogenetics*, 56: 311–20.

Buse HY, Schoen ME, Ashbolt NJ (2012). Legionellae in engineered systems and use of quantitative microbial risk assessment to predict exposure. *Water Research*, 46: 921–33.

Bush WS, Moore FR (2012). Genome-wide association studies. *PLoS Computational Biology*, 8: e1002822.

Busula AO, Verhulst NO, Bousema T, Takken W, de Boer JG (2017). Mechanisms of *Plasmodium*-enhanced attraction of mosquito vectors. *Trends in Parasitology*, 33: 961–73.

Byers JE, Schmidt JP, Pappalardo P, Haas SE, Stephens PR (2019). What factors explain the geographical range of mammalian parasites? *Proceedings of the Royal Society London B*, 286: 20190673.

Cabalzar AP, Fields PD, Kato Y, Watanabe H, Ebert D (2019). Parasite-mediated selection in a natural metapopulation of *Daphnia magna*. *Molecular Ecology*, 28: 4770–85.

Cable J, Barber I, Boag B, et al. (2017). Global change, parasite transmission and disease control: lessons from ecology. *Philosophical Transactions of the Royal Society London B*, 372: 20160088.

Caira JN, Jensen K, Holsinger KE (2003). On a new index of host specificity. In Combes C, Jourdane J (eds) *Taxonomy, ecology and evolution of metazoan parasites*, pp. 161–201. Presses Universitaires de Perpignan, Perpignan.

Caldwell R, Compton L, Patterson F (1958). Tolerance to cereal leaf rusts. *Science*, 128: 714–15.

Cali A, Owen RL (1988). Microsporidiosis. In Balows A, Hausler WJ, Ohashi M, Turano A, Lennete EH (eds) *Laboratory diagnosis of infectious diseases: principles and practice*, pp. 929–50. Springer, New York.

Calvete C, Blanco-Aguiar JA, Virgos E, Cabezas-Diaz S, Villafuerte R (2004). Spatial variation in helminth community structure in the red-legged partridge (*Alectoris rufa* L.): effects of definitive host density. *Parasitology*, 129: 101–13.

Cameron SA, Lim HC, Lozier JD, Duennes MA, Thorp R (2016). Test of the invasive pathogen hypothesis of bumble bee decline in North America. *Proceedings of the National Academy of Sciences USA*, 113: 4386–91.

Cameron SA, Lozier JD, Strange JP, et al. (2011). Patterns of widespread decline in North American bumble bees. *Proceedings of the National Academy of Sciences USA*, 108: 662–7.

Campbell J, Kessler B, Mayack C, Naug D (2010). Behavioural fever in infected honeybees: parasitic manipulation or coincidental benefit? *Parasitology*, 137: 1487–91.

Campoy S, Adrio JL (2017). Antifungals. *Biochemical Pharmacology*, 133: 86–96.

Canning E, Okamura B (2004). Biodiversity and evolution of the Myxozoa. *Advances in Parasitology*, 56: 43–131.

Canton J, Neculai D, Grinstein S (2013). Scavenger receptors in homeostasis and immunity. *Nature Reviews Immunology*, 13: 621–34.

Cao X (2016). Self-regulation and cross-regulation of pattern-recognition receptor signalling in health and disease. *Nature Reviews Immunology*, 16: 35–50.

Capbern A, Giroud C, Baltz T, Mattern P (1977). *Trypanosoma equiperdum*: étude des variations antigéniques au cours de la trypanosomose experimentale du lapin. *Experimental Parasitology*, 42: 6–13.

Cappuccio A, Tieri P, Castiglione F (2015). Multiscale modelling in immunology: a review. *Briefings in Bioinformatics*, 17: 408–18.

Caraco T, Wang IN (2008). Free-living pathogens: life-history constraints and competition. *Journal of Theoretical Biology*, 250: 569–79.

Carballar-Lejarazú R, Ogaugwu C, Tushar T, et al. (2020). Next-generation gene drive for population modification of the malaria vector mosquito, *Anopheles gambiae*. *Proceedings of the National Academy of Sciences USA*, 117: 22805–14.

Cardoso R, Soares H, Hemphill A, Leitao A (2016). Apicomplexans pulling the strings: manipulation of the host cell cytoskeleton dynamics. *Parasitology*, 143: 957–70.

Carius HJ, Little TJ, Ebert D (2001). Genetic variation in a host-parasite association: potential for coevolution and frequency-dependent selection. *Evolution*, 55: 1136–45.

Carlson CJ, Burgio KR, Dougherty ER, et al. (2017). Parasite biodiversity faces extinction and redistribution in a changing climate. *Science Advances*, 3: e1602422.

Carlson CJ, Hopkins S, Bell KC, et al. (2020). A global parasite conservation plan. *Biological Conservation*, 250: 108596.

Carmody RN, Gerber GK, Luevano JM, et al. (2015). Diet dominates host genotype in shaping the murine gut microbiota. *Cell Host Microbe*, 17: 72–84.

Carrai V (2003). Increase in tannin consumption by sifaka (*Propithecus verreauxi verreauxi*) females during the birth season: a case for self-medication in prosimians? *Primates*, 44: 61–6.

Carroll SB (2005). Evolution at two levels: on genes and form. *PLoS Biology*, 3: 1159–66.

Carton Y, Nappi AJ, Poirie M (2005). Genetics of anti-parasite resistance in invertebrates. *Developmental and Comparative Immunology*, 29: 9–32.

Carvalho Cabral P, Olivier M, Cermakian N (2019). The complex interplay of parasites, their hosts, and circadian clocks. *Frontiers in Cellular and Infection Microbiology*, 9: 425.

Carvalho T, Becker CG, Toledo LF (2017). Historical amphibian declines and extinctions in Brazil linked to chytridiomycosis. *Proceedings of the Royal Society London B*, 284: 20162254.

Casanova JC, Ribas A, Segovia JM (2006). Nematode zoonoses. In Morand S, Krasnov BR, Poulin R (eds) *Micromammals and macroparasites: from evolutionary ecology to management*, pp. 515–26. Springer Japan, Tokyo.

Cascales E, Buchanan SK, Duché D, et al. (2007). Colicin biology. *Microbiology and Molecular Biology Reviews*, 71: 158.

Cassat James E, Skaar Eric P (2013). Iron in Infection and Immunity. *Cell Host & Microbe*, 13: 509–19.

Castro M, Lythe G, Molina-París C, Ribeiro RM (2016). Mathematics in modern immunology. *Interface Focus*, 6: 20150093.

Catalán TP, Barceló Ma, Niemeyer HM, Kalergis AM, Bozinovic F (2011). Pathogen- and diet-dependent foraging, nutritional and immune ecology in mealworms. *Evolutionary Ecology Research*, 13: 711–23.

Catalán TP, Wozniak A, Niemeyer HM, Kalergis AM, Bozinovic F (2012). Interplay between thermal and immune ecology: effect of environmental temperature on insect immune response and energetic costs after an immune challenge. *Journal of Insect Physiology*, 58: 310–17.

Cator LJ, Pietri JE, Murdock CC, et al. (2015). Immune response and insulin signalling alter mosquito feeding behaviour to enhance malaria transmission potential. *Scientific Reports*, 5: 11947.

Caugant DA, Brynildsrud OB (2020). *Neisseria meningitidis*: using genomics to understand diversity, evolution and pathogenesis. *Nature Reviews Microbiology*, 18: 84–96.

Cayetano L, Vorburger C (2013). Genotype-by-genotype specificity remains robust to average temperature variation in an aphid/endosymbiont/parasitoid system. *Journal of Evolutionary Biology*, 26: 1603–10.

Cecconi M, Evans L, Levy M, Rhodes A (2018). Sepsis and septic shock. *The Lancet*, 392: 75–87.

Cen X, Feng Z, Zhao Y (2014). Emerging disease dynamics in a model coupling within-host and between-host systems. *Journal of Theoretical Biology*, 361: 141–51.

Cerenius L, Lee BL, Söderhäll K (2008). The proPO-system: pros and cons for its role in invertebrate immunity. *Trends in Immunology*, 29: 263–71.

Cézilly F, Perrot-Minnot M-J, Rigaud T (2014). Cooperation and conflict in host manipulation: interactions among macro-parasites and micro-organisms. *Frontiers in Microbiology*, 5.

Chakra MA, Hilbe C, Traulsen A (2014). Plastic behaviors in hosts promote the emergence of retaliatory parasites. *Scientific Reports*, 4: 4251.

Chakraborty AK (2017). A perspective on the role of computational models in immunology. *Annual Review of Immunology*, 35: 403–39.

Chalkowski K, Lepczyk CA, Zohdy S (2018). Parasite ecology of invasive species: conceptual framework and new hypotheses. *Trends in Parasitology*, 34: 655–63.

Chambers ES, Vukmanovic-Stejic M (2020). Skin barrier immunity and ageing. *Immunology*, 160: 116–25.

Chang H-D, Tokoyoda K, Radbruch A (2018). Immunological memories of the bone marrow. *Immunological Reviews*, 283: 86–98.

Chantawannakul P, de Guzman LI, Li J, Williams GR (2016). Parasites, pathogens, and pests of honeybees in Asia. *Apidologie*, 47: 301–24.

Chao A, Lee S-M (1992). Estimating the number of classes via sample coverage. *Journal of the American Statistical Association*, 87: 210–17.

Chao L, Hanley KA, Burch CL, Dahlberg C, Turner PE (2000). Kin selection and parasite evolution: higher and lower virulence with hard and soft selection. *Quarterly Review of Biology*, 75: 261–75.

Chapman JR, Dowell MA, Chan R, Unckless RL (2020). The genetic basis of natural variation in *Drosophila melanogaster* immune defense against *Enterococcus faecalis*. *Genes*, 11: 234.

Chapman JR, Hill T, Unckless RL (2019). Balancing selection drives the maintenance of genetic variation in *Drosophila* antimicrobial peptides. *Genome Biology and Evolution*, 11: 2691–701.

Chapman JW, Dumbauld BR, Itani G, Markham JC (2012). An introduced Asian parasite threatens northeastern Pacific estuarine ecosystems. *Biological Invasions*, 14: 1221–36.

Chapman SJ, Hill AVS (2012). Human genetic susceptibility to infectious disease. *Nature Reviews Genetics*, 13: 175–88.

Charbonnel N, Pemberton J (2005). A long-term genetic survey of an ungulate population reveals balancing selection acting on MHC through spatial and temporal fluctuations in selection. *Heredity*, 95: 377–88.

Charlesworth B, Barton NH (1996). Recombination load associated with selection for increased recombination. *Genetical Research*, 67: 27–41.

Chassaing B, Cascales E (2018). Antibacterial weapons: targeted destruction in the microbiota. *Trends in Microbiology*, 26: 329–38.

Che D, Hasan MS, Chen B (2014). Identifying pathogenicity islands in bacterial pathogenomics using computational approaches. *Pathogens*, 3: 36–56.

Chehoud C, Rafail S, Tyldsley AS, Seykora JT, Lambris JD, Grice EA (2013). Complement modulates the cutaneous microbiome and inflammatory milieu. *Proceedings of the National Academy of Sciences USA*, 110: 15061–6.

Chen C-Z, Schaffert S, Fragoso R, Loh C (2013). Regulation of immune responses and tolerance: the microRNA perspective. *Immunological Reviews*, 253: 112–18.

Chen L, Liu B, Yang J, Jin Q (2014). DBatVir: the database of bat-associated viruses. *Database*, 2014: bau021.

Chen Y, Liu Q, Guo D (2020). Emerging coronaviruses: genome structure, replication, and pathogenesis. *Journal of Medical Virology*, 92: 418–23.

Cheng TC (1986). *General parasitology*. Academic Press, Orlando.

Chester KS (1933). The problem of acquired physiological immunity in plants. *Quarterly Review of Biology*, 8: 275–324.

Chetouhi C, Panek J, Bonhomme L, et al. (2015). Cross-talk in host–parasite associations: what do past and recent proteomics approaches tell us? *Infection, Genetics and Evolution*, 33: 84–94.

Chiang YA, Hung HY, Lee CW, Huang YT, Wang HC (2013). Shrimp Dscam and its cytoplasmic tail splicing activator serine/arginine (SR)-rich protein B52 were both induced after white spot syndrome virus challenge. *Fish and Shellfish Immunology*, 34: 209–19.

Chichorro F, Juslén A, Cardoso P (2019). A review of the relation between species traits and extinction risk. *Biological Conservation*, 237: 220–9.

Chikindas ML, Weeks R, Drider D, Chistyakov VA, Dicks LM (2018). Functions and emerging applications of bacteriocins. *Current Opinion in Biotechnology*, 49: 23–8.

Childs LM, El Moustaid F, Gajewski Z, et al. (2019). Linked within-host and between-host models and data for infectious diseases: a systematic review. *Peer J*, 7: e7057.

Chippaux J-P, Chippaux A (2018). Yellow fever in Africa and the Americas: a historical and epidemiological perspective. *Journal of Venomous Animals and Toxins including Tropical Diseases*, 24: 20.

Chisholm PJ, Busch JW, Crowder DW (2019). Effects of life history and ecology on virus evolutionary potential. *Virus Research*, 265: 1–9.

Cho H, Kelsall BL (2014). The role of type I interferons in intestinal infection, homeostasis, and inflammation. *Immunological Reviews*, 260: 145–67.

Choisy M, Brown S, Lafferty K, Thomas F (2003). Evolution of trophic transmission in parasites: why add intermediate hosts? *The American Naturalist*, 162: 172–81.

Chong KC, Zee BCY, Wang MH (2018). Approximate Bayesian algorithm to estimate the basic reproduction number in an influenza pandemic using arrival times of imported cases. *Travel Medicine and Infectious Disease*, 23: 80–6.

Choo K, Williams PD, Day T (2003). Host mortality, predation and the evolution of parasite virulence. *Ecology Letters*, 6: 310–15.

Chorine V (1929). Immunité antitoxique chez les chenilles de *Galleria mellonella*. *Annales de l' Institut Pasteur*, 43: 954–8.

Chow J, Kagan JC (2018). The fly way of antiviral resistance and disease tolerance. *Advances in Immunology*, 140: 59–93.

Christe P, Oppliger A, Richner H (1994). Ectoparasite affects choice and use of roost sites in the great tit, *Parus major*. *Animal Behaviour*, 47: 895–8.

Christe P, Richner H, Oppliger A (1996). Begging, food provisioning, and nestling competition in great tit broods infected with ectoparasites. *Behavioral Ecology*, 7: 127–31.

Christiaansen A, Varga SM, Spencer JV (2015). Viral manipulation of the host immune response. *Current Opinion in Immunology*, 36: 54–60.

Chu W-M (2013). Tumor necrosis factor. *Cancer Letters*, 328: 222–5.

Chuong EB, Elde NC, Feschotte C (2016). Regulatory evolution of innate immunity through co-option of endogenous retroviruses. *Science*, 351: 1083.

Ciccozzi M, Lai A, Zehender G, et al. (2019). The phylogenetic approach for viral infectious disease evolution and epidemiology: an updating review. *Journal of Medical Virology*, 91: 1707–24.

Cichon M, Sendecka J, Gustafsson L (2006). Genetic and environmental variation in immune response of collared flycatcher nestlings. *Journal of Evolutionary Biology*, 19: 1701–6.

Ciupe SM, Heffernan JM (2017). In-host modeling. *Infectious Disease Modelling*, 2: 188–202.

Civitello DJ, Cohen J, Fatima H, et al. (2015). Biodiversity inhibits parasites: broad evidence for the dilution effect. *Proceedings of the National Academy of Sciences USA*, 112: 8667–71.

Cizauskas CA, Carlson CJ, Burgio KR, et al. (2017). Parasite vulnerability to climate change: an evidence-based functional trait approach. *Royal Society Open Science*, 4: 160535.

Clark EA (2014). A short history of the B-cell-associated surface molecule CD40. *Frontiers in Immunology*, 5: 472.

Clark NJ, Clegg SM (2017). Integrating phylogenetic and ecological distances reveals new insights into parasite host specificity. *Molecular Ecology*, 26: 3074–86.

Clark NJ, Olsson-Pons S, Ishtiaq F, Clegg SM (2015). Specialist enemies, generalist weapons and the potential spread of exotic pathogens: malaria parasites in a highly invasive bird. *International Journal for Parasitology*, 45: 891–9.

Clarke BC (1976). The ecological genetics of host-parasite relationships. In Taylor AER, Muller R (eds) *Genetic aspects of host-parasite interactions*, pp. 87–103. Blackwell, Oxford.

Clay CA, Lehmer EM, St. Jeor S, Dearing MD (2009). Testing mechanisms of the dilution effect: deer mice encounter rates, Sin Nombre Virus prevalence and species diversity. *EcoHealth*, 6: 250–9.

Clayton DH, Bush SE, Goates BM, Johnson KP (2003). Host defense reinforces host-parasite cospeciation. *Proceedings of the National Academy of Sciences USA*, 100: 15694–9.

Clayton DH, Bush SE, Johnson KP (2004). Ecology of congruence: past meets present. *Systematic Biology*, 53: 165–73.

Clayton DH, Johnson KP (2003). Linking coevolutionary history to ecological process: doves and lice. *Evolution*, 57: 2335–41.

Cleaveland S, Hess GR, Dobson AP, et al. (2001a). The role of pathogens in biological conservation. In *The ecology of wildlife diseases*, pp. 139–50. Oxford University Press, Oxford.

Cleaveland S, Laurenson MK, Taylor LH (2001b). Diseases of humans and their domestic mammals: pathogen characteristics, host range and the risk of emergence. *Philosophical Transactions of the Royal Society London B*, 356: 991–9.

Clutton-Brock T, Isvaran k (2007). Sex differences in ageing in natural populations of vertebrates. *Proceedings of the Royal Society London B*, 274: 3097–14.

Cobey S, Lipsitch M (2013). Pathogen diversity and hidden regimes of apparent competition. *The American Naturalist*, 18: 12–24.

Cohan FM (2006). Towards a conceptual and operational union of bacterial systematics, ecology, and evolution. *Philosophical Transactions of the Royal Society London B*, 361: 1985–96.

Cohen PE, Pollack SE, Pollard JW (2006). Genetic analysis of chromosome pairing, recombination, and cell cycle control during first meiotic prophase in mammals *Endocrine Reviews*, 27: 398–426.

Coico R, Sunshine G (2015). *Immunology: a short course*, 7th ed. Wiley Blackwell, Chichester.

Colditz IG (2008). Six costs of immunity to gastrointestinal nematode infections. *Parasite Immunology*, 30: 63–70.

Cole R, Viney M (2018). The population genetics of parasitic nematodes of wild animals. *Parasites & Vectors*, 11: 590.

Colegrave N, Kotiaho JS, Tomkins JL (2002). Mate choice or polyandry: reconciling genetic compatibility and good genes sexual selection. *Evolutionary Ecology Research*, 4: 911–17.

Coleman JR, Papamichail D, Skiena S, Futcher B, Wimmer E, Mueller S (2008). Virus attenuation by genome-scale changes in codon pair bias. *Science*, 320: 1784–7.

Colgan TJ, Carolan JC, Sumner S, Blaxter ML, Brown MJF (2019). Infection by the castrating parasitic nematode *Sphaerularia bombi* changes gene expression in *Bombus terrestris* bumblebee queens. *Insect Molecular Biology*, 29: 170–82.

Collins FWJ, O'Connor PM, O'Sullivan O, et al. (2017). Bacteriocin gene-trait matching across the complete *Lactobacillus* pan-genome. *Scientific Reports*, 7: 3481.

Colonne PM, Winchell CG, Voth DE (2016). Hijacking host cell highways: manipulation of the host actin cytoskeleton by obligate intracellular bacterial pathogens. *Frontiers in Cellular and Infection Microbiology*, 6: 107.

Coltman DW, et al. (1999). Parasite-mediated selection against inbred Soay sheep in a free-living, island population. *Evolution*, 53: 1259–67.

Combes C (2001). *Parasitism: the ecology and evolution of intimate interactions*. University of Chicago Press, London.

Conrath U, Beckers GJM, Langenbach CJG, Jaskiewicz MR (2015). Priming for enhanced defense. *Annual Review of Phytopathology*: 97–119.

Consortium (Chinese SARS Molecular Epidemiology) (2004). Molecular evolution of the SARS coronavirus during the course of the SARS epidemic in China. *Science*, 303: 1666–9.

Contreras J, Rao D (2012). MicroRNAs in inflammation and immune responses. *Leukemia*, 26: 401–13.

Contreras-Garduño J, Rodríguez MC, Rodríguez MH, Alvarado-Delgado A, Lanz-Mendoza H (2014). Cost of immune priming within generations: trade-off between infection and reproduction. *Microbes and Infection*, 16: 261–7.

Conway JM, Perelson AS (2015). Post-treatment control of HIV infection. *Proceedings of the National Academy of Sciences USA*, 112: 5467–72.

Conway Morris S (1981). Parasites and the fossil record. *Parasitology*, 82: 489–509.

Cook GC, Zumla AI (2008). *Manson's tropical diseases*, 21st ed. Saunders Elsevier Health Sciences, Edinburgh UK.

Cooling L (2015). Blood groups in infection and host susceptibility. *Clinical Microbiology Reviews*, 28: 801–70.

Coop G, Przeworski M (2007). An evolutionary view of human recombination. *Nature Reviews Genetics*, 8: 23–34.

Cooper N, Griffin R, Franz M, Omotayo M, Nunn CL (2012a). Phylogenetic host specificity and understanding parasite sharing in primates. *Ecology Letters*, 15: 1370–7.

Cooper N, Kamilar JM, Nunn CL (2012b). Host longevity and parasite species richness in mammals. *PLoS ONE*, 7: e42190–0.

Cooper P, Walker AW, Reyes J, et al. (2013). Patent human infections with the whipworm, *Trichuris trichiura*, are not associated with alterations in the faecal microbiota. *PLoS ONE*, 8: e76573.

Cooper VS, Reiskind MH, Miller JA, et al. (2002). Timing of transmission and the evolution of virulence of an insect virus. *Proceedings of the Royal Society London B*, 269: 1161–5.

Cornet S, Nicot A, Rivero A, Gandon S (2019). Avian malaria alters the dynamics of blood feeding in Culex pipiens mosquitoes. *Malaria Journal*, 18: 82.

Cornforth DM, Dees JL, Ibberson CB, et al. (2018). *Pseudomonas aeruginosa* transcriptome during human infection. *Proceedings of the National Academy of Sciences USA*, 115: E5125.

Coronaviridae Study Group (International Committee on Taxonomy of Viruses) (2020). The species 'severe acute respiratory syndrome-related coronavirus': classifying 2019-nCoV and naming it SARS-CoV-2. *Nature Microbiology*, 5: 536–44.

Cortes A, Toledo R, Cantacessi C (2018). Classic models for new perspectives: delving into helminth-microbiota-immune system interactions. *Trends in Parasitology*, 34: 640–54.

Costello MJ (2016). Parasite rates of discovery, global species richness and host specificity. *Integrative and Comparative Biology*, 56: 588–99.

Côté IM, Poulin R (1995). Parasitism and group size in social mammals: a meta-analysis. *Behavioral Ecology*, 6: 159–65.

Cottam E, Haydon D, Paton D, et al. (2006). Molecular epidemiology of the foot-and- mouth disease virus outbreak in the United Kingdom in 2001. *Journal of Virology*, 80: 11274–82.

Cotter PA, DiRita VJ (2000). Bacterial virulence gene regulations: an evolutionary perspective. *Annual Review of Microbiology*, 54: 519–65.

Cotter SC, Kilner RM (2010). Personal immunity versus social immunity. *Behavioral Ecology*, 21: 663–8.

Cotter SC, Simpson SJ, Raubenheimer D, Wilson K (2011). Macronutrient balance mediates trade-offs between immune function and life history traits. *Functional Ecology*, 25: 186–98.

Cotter SC, Wilson K (2002). Heritability of immune function in the caterpillar *Spodoptera littoralis*. *Heredity*, 88: 229–34.

Cox FEG (2001). Concomitant infections, parasites and immune responses. *Parasitology*, 122: S23–38.

Crane GM, Jeffery E, Morrison SJ (2017). Adult haematopoietic stem cell niches. *Nature Reviews Immunology*, 17: 573–90.

Cremer S, Armitage SAO, Schmid-Hempel P (2007). Social immune systems. *Current Biology*, 17: R693–702.

Cremer S, Pull CD, Fürst MA (2018). Social immunity: emergence and evolution of colony-level disease protection. *Annual Review of Entomology*, 63: 105–23.

Cremer S, Sixt M (2009). Analogies in the evolution of individual and social immunity. *Philosophical Transactions of the Royal Society London B*, 364: 129–42.

Cressler CE, McLeod DV, Rozins C, Van den Hoogen J, Day T (2016). The adaptive evolution of virulence: a review of theoretical predictions and empirical tests. *Parasitology*, 143: 915–30.

Cressler CE, Nelson WA, Day T, McCauley E (2014). Starvation reveals the cause of infection-induced castration and gigantism. *Proceedings of the Royal Society London B*, 281: 20141087.

Cribb TH, Bray RA, Olson PD, Littlewood DTJ (2003). Life cycle evolution in the Digenea: a new perspectives from phylogeny. *Advances in Parasitology*, 54: 197–254.

Criscitiello MF, Flajnik MF (2007). Four primordial immunoglobulin light chain isotypes, including λ and κ, identified in the most primitive living jawed vertebrates. *European Journal of Immunology*, 37: 2683–94.

Crook M (2014). The dauer hypothesis and the evolution of parasitism: 20 years on and still going strong. *International Journal for Parasitology*, 44: 1–8.

Crosby AW (1986). *Ecological imperialism. The biological expansion of Europe, 900–1900*. Cambridge University Press, Cambridge.

Cross GAM, Kim H-S, Wickstead B (2014). Capturing the variant surface glycoprotein repertoire (the VSGnome) of *Trypanosoma brucei* Lister 427. *Molecular & Biochemical Parasitology*, 195: 59–73.

Croze M, Wollstein A, Božicević V, Živković D, Stephan W, Hutter S (2017). A genome-wide scan for genes under balancing selection in *Drosophila melanogaster*. *BMC Evolutionary Biology*, 17: 15.

Croze M, Živković D, Stephan W, Hutter S (2016). Balancing selection on immunity genes: review of the current literature and new analysis in *Drosophila melanogaster*. *Zoology*, 119: 322–9.

Cui H, Tsuda K, Parker JE (2015). Effector-triggered immunity: from pathogen perception to robust defense. *Annual Review of Plant Biology*, 66: 487–511.

Cunningham AA, Daszak P (1998). Extinction of a species of land snail due to infection with a microsporidian parasite. *Conservation Biology*, 12: 1139–41.

Curtis VA (2007). A natural history of hygiene. *Canadian Journal of Infectious Diseases and Medical Microbiology*, 18.

Curtis VA (2014). Infection-avoidance behaviour in humans and other animals. *Trends in Immunology*, 35: 457–64.

Curtis VA, de Barra M, Aunger R (2011). Disgust as an adaptive system for disease avoidance behaviour. *Philosophical Transactions of the Royal Society London B*, 366: 389–401.

Cutts FT, Hanson M (2016). Seroepidemiology: an underused tool for designing and monitoring vaccination programmes in low- and middle-income countries. *Tropical Medicine & International Health*, 21: 1086–98.

D'Andrea MM, Arena F, Pallecchi L, Rossolini GM (2013). CTX-M-type β-lactamases: A successful story of antibiotic resistance. *International Journal of Medical Microbiology*, 303: 305–17.

D'Arc M, Ayouba A, Esteban A, et al. (2015). Origin of the HIV-1 group O epidemic in western lowland gorillas. *Proceedings of the National Academy of Sciences USA*, 112: E1343–52.

D'Costa VM, McGrann KM, Hughes DW, Wright GD (2006). Sampling the antibiotic resistome. *Science*, 311: 374–7.

Dadonaite B, Gilbertson B, Knight ML, et al. (2019). The structure of the influenza A virus genome. *Nature Microbiology*, 4: 1781–9.

Dagan Y, Liljeroos K, Jokela J, Ben-Ami F (2013). Clonal diversity driven by parasitism in a freshwater snail. *Journal of Evolutionary Biology*, 26: 2509–19.

Dallas TA, Aguirre AA, Budischak S, et al. (2018). Gauging support for macroecological patterns in helminth parasites. *Global Ecology and Biogeography*, 27: 1437–47.

Dallas TA, Huang S, Nunn C, Park AW, Drake JM (2017). Estimating parasite host range. *Proceedings of the Royal Society London B*, 284: 20171250.

Dallas TA, Presley SJ (2014). Relative importance of host environment, transmission potential and host phylogeny to the structure of parasite metacommunities. *Oikos*, 123: 866–74.

Damian RT (1997). Parasite immune evasion and exploitation: reflections and projections. *Parasitology*, 115: S169–75.

Dantas G, Sommer MOA, Oluwasegun RD, Church GM (2008). Bacteria subsisting on antibiotics. *Science*, 320: 100–3.

Dapper AL, Payseur BA (2017). Connecting theory and data to understand recombination rate evolution. *Philosophical Transactions of the Royal Society London B*, 372: 20160469.

Daversa DR, Fenton A, Dell AI, Garner TWJ, Manica A (2017). Infections on the move: how transient phases of host movement influence disease spread. *Proceedings of the Royal Society London B*, 284: 20171807.

David LA, Maurice CF, Carmody RN, et al. (2014). Diet rapidly and reproducibly alters the human gut microbiome. *Nature*, 505: 559–63.

Davies TJ, Pedersen AB (2008). Phylogeny and geography predict pathogen community similarity in wild primates and humans. *Proceedings of the Royal Society London B*, 275: 1695–701.

Davis MM, Tato CM, Furman D (2017). Systems immunology: just getting started. *Nature Immunology*, 18: 725–32.

Dawkins R, Krebs JR (1979). Arms races between and within species. *Proceedings of the Royal Society London B*, 205: 489–511.

Day T (2001). Parasite transmission modes and the evolution of virulence. *Evolution*, 55: 2389–400.

Day T (2002a). Virulence evolution via host exploitation and toxin production in spore-producing pathogens. *Ecology Letters*, 5: 471–6.

Day T (2002b). On the evolution of virulence and the relationship between various measures of mortality. *Proceedings of the Royal Society London B*, 269: 1317–23.

Day T (2003). Virulence evolution and the timing of disease life-history events. *Trends in Ecology & Evolution*, 18: 113–18.

Day T, Alizon S, Mideo N (2011). Bridging scales in the evolution of infectious disease life histories: theory. *Evolution*, 65: 3448–61.

Day T, Graham AL, Read AF (2007). Evolution of parasite virulence when host responses cause disease. *Proceedings of the Royal Society London B*, 274: 2685–92.

Day T, Proulx SR (2004). A general theory for the evolutionary dynamics of virulence. *The American Naturalist*, 163: E40–63.

de Bekker C, Merrow M, Hughes DP (2014). From behavior to mechanisms: an integrative approach to the manipulation by a parasitic fungus (*Ophiocordyceps unilateralis* s.l.) of its host ants (*Camponotu*s spp.). *Integrative and Comparative Biology*, 54: 166–76.

de Bekker C, Ohm RA, Loreto RG, et al. (2015). Gene expression during zombie ant biting behavior reflects the complexity underlying fungal parasitic behavioral manipulation. *BMC Genomics*, 16: 620.

de Boer RJ, Perelson AS (1998). Target cell limited and immune control models of HIV infection: a comparison. *Journal of Theoretical Biology*, 190: 201–14.

de Castro F, Bolker B (2005). Parasite establishment and host extinction in model communities. *Oikos*, 111: 501–13.

de Jong-Brink M, Bergamin-Sassen M, Soto M (2001). Multiple strategies of schistosomes to meet their requirements in the intermediate snail host. *Parasitology*, 35: 177–256.

de Jong-Brink M, Hoek RM, Lageweg W, Smit AB (1997). Schistosome parasites induce physiological changes in their snail host be interfering with two regulatory systems, the internal defense system and the neuroendocrine system. In Beckage NE (ed.) *Parasites and pathogens: effects on host hormones and behavior*, pp. 57–75. Chapman and Hall, New York.

de Koning-Ward TF, Dixon MWA, Tilley L, Gilson PR (2016). *Plasmodium* species: master renovators of their host cells. *Nature Reviews Microbiology*, 14: 494–507.

De Kruif P (1926). *Microbe hunters*. Harcourt, Brace, Jovanovich, Orlando, FL.

De Meeûs T, Renaud F (2002). Parasites within the new phylogeny of eukaryotes. *Trends in Parasitology*, 18: 247–51.

De Moraes CM, Stanczyk NM, Betz HS, et al. (2014). Malaria-induced changes in host odors enhance mosquito attraction. *Proceedings of the National Academy of Sciences USA*, 111: 11079–84.

De Nardo D (2015). Toll-like receptors: activation, signalling and transcriptional modulation. *Cytokine*, 74: 181–9.

De Paepe M, Taddei F (2006). Viruses' life history: towards a mechanistic basis of a trade-off between survival and reproduction. *The American Naturalist*, 163: E40–63.

de Ronde D, Butterbach P, Kormelink R (2014). Dominant resistance against plant viruses. *Frontiers in Plant Science*, 5: 307.

de Roode JC, Helinski MEH, Anwar MA, Read AF (2005a). Dynamics of multiple infection and within-host competition in genetically diverse malaria infections. *The American Naturalist*, 166: 531–42.

de Roode JC, Lefèvre T (2012). Behavioral Immunity in Insects. *Insects*, 3: 789–820.

de Roode JC, Lefèvre T, Hunter MD (2013). Self-medication in animals. *Science*, 340: 150–1.

de Roode JC, Pansini R, Cheesman SJ, et al. (2005b). Virulence and competitive ability in genetically diverse malaria infections. *Proceedings of the National Academy of Sciences USA*, 102: 7624–8.

de Roode JC, Yates AJ, Altizer S (2008). Virulence-transmission trade-offs and population divergence in virulence in a naturally occuring butterfly parasite. *Proceedings of the National Academy of Sciences USA*, 105: 7489–94.

de Vienne DM, Refrégier G, López-Villavicencio M, Tellier A, Hood ME, Giraud T (2013). Cospeciation vs host-shift speciation: methods for testing, evidence from natural associations and relation to coevolution. *New Phytologist*, 198: 347–85.

de Visser JAGM, Hermisson J, Wagner GP, et al. (2003). Perspective. Evolution and detection of genetic robustness. *Evolution*, 57: 1959–72.

de Zoete MR, Palm NW, Zhu S, Flave RA (2014). Inflammasomes. *Cold Spring Harbor Perspectives in Biology*, 6: a016287.

Dean P, Maresca M, Kenny B (2005). EPEC's weapons of mass subversion. *Current Opinion in Microbiology*, 8: 28–34.

Dearsly AL, Sinden RE, Self IA (1990). Sexual development in malarial parasites, gametocyte production, fertility and infectivity to the mosquito vector. *Parasitology*, 100: 359–68.

Debbink K, Lindesmith LC, Ferris MT, et al. (2014). Within-host evolution results in antigenically distinct GII.4 noroviruses. *Journal of Virology*, 88: 7244–55.

Decaestecker E (2002). In deep trouble: habitat selection constrained by multiple enemies in zooplankton. *Proceedings of the National Academy of Sciences USA*, 99: 5481–5.

Decaestecker E, Gaba S, Rayemaekers JAM, et al. (2007). Host-parasite 'Red Queen' dynamics archived in pond sediment. *Nature*, 450: 870–4.

Degnan PH, Moran NA (2008). Diverse phage-encoded toxins in a protective insect endosymbiont. *Applied Environmental Microbiology*, 74: 6782–91.

Deitsch KW, Dzikowski R (2017). Variant gene expression and antigenic variation by malaria parasites. *Annual Review of Microbiology*, 71: 625–41.

DeLorenze GN, Nelson CL, Scott WK, et al. (2015). Polymorphisms in HLA Class II genes are associated with susceptibility to *Staphylococcus aureus* infection in a white population. *The Journal of Infectious Diseases*, 213: 816–23.

Demandt N, Saus B, Kurvers RHJM, Krause J, Kurtz J, Scharsack1 JP (2018). Parasite-infected sticklebacks increase the risk-taking behaviour of uninfected group members. *Proceedings of the Royal Society London B*, 285: 20180956.

Deng X, Hackbart M, Mettelman RC, et al. (2017). Coronavirus nonstructural protein 15 mediates evasion of dsRNA sensors and limits apoptosis in macrophages. *Proceedings of the National Academy of Sciences USA*, 114: E4251–60.

Dennis AB, Patel V, Oliver KM, Vorburger C (2017). Parasitoid gene expression changes after adaptation to symbiont-protected hosts. *Evolution*, 71: 2599–617.

Desai PT, Porwollik S, Long F, et al. (2013). Evolutionary genomics of *Salmonella enterica* subspecies. *mBio*, 4: e00579–12.

Desai SD, Currie RW (2015). Genetic diversity within honey bee colonies affects pathogen load and relative virus levels in honey bees, *Apis mellifera* L. *Behavioral Ecology and Sociobiology*, 69: 1527–41.

Deschasaux M, Bouter KE, Prodan A, et al. (2018). Depicting the composition of gut microbiota in a population with varied ethnic origins but shared geography. *Nature Medicine*, 24: 1526–31.

Desdevises Y, Morand S, Legendre P (2002). Evolution and determinants of host specificity in the genus *Lamellodiscus* (Monogenea). *Biological Journal of the Linnean Society*, 77: 431–43.

Deshwal S, Mallon EB (2014). Antimicrobial peptides play a functional role in bumblebee anti-trypanosome defense. *Developmental & Comparative Immunology*, 42: 240–3.

Detwiler J, Janovy JJ (2008). The role of phylogeny and ecology in experimental host specificity: insights from a Eugregarine-host system. *Journal of Parasitology*, 94: 7–12.

Dewees AA (1975). Genetic modification of recombination rate in *Tribolium castaneum*. *Genetics*, 81: 537–52.

DeWitte SN (2014). Mortality risk and survival in the aftermath of the medieval Black Death. *PLoS ONE*, 9: e96513.

Dhanda SK, Mahajan S, Paul S, et al. (2019). IEDB-AR: immune epitope database: analysis resource in 2019. *Nucleic Acids Research*, 47: W502–6.

Dheilly NM, Martínez Martínez J, Rosario K, et al. (2019). Parasite microbiome project: grand challenges. *PLoS Pathogens*, 15: e1008028.

Dhinaut J, Balourdet A, Teixeira M, Chogne M, Moret Y (2017). A dietary carotenoid reduces immunopathology and enhances longevity through an immune depressive effect in an insect model. *Scientific Reports*, 7: 12429.

Dhinaut J, Chogne M, Moret Y (2018). Immune priming specificity within and across generations reveals the range of pathogens affecting evolution of immunity in an insect. *Journal of Animal Ecology*, 87: 448–63.

Dhondt AA, Altizer S, Cooch EG, et al. (2005). Dynamics of a novel pathogen in an avian host: mycoplasmal conjunctivitis in house finches. *Acta Tropica*, 94: 77–93.

Di W, Lu T-Y, Liu H-B (2009). Characterization and genetic diversity of the sturgeon *Acipenser schrenskii* Ig heavy chain. *Immunobiology*, 214: 359–66.

Dick CW, Patterson BD (2007). Against all odds: explaining high host specificity in dispersal-prone parasites. *International Journal for Parasitology*, 37: 871–6.

Didelot X, Méric G, Falush D, Darling AE (2012). Impact of homologous and non-homologous recombination in the genomic evolution of *Escherichia coli*. *BMC Genomics*, 13: 256.

Didelot X, Walker AS, Peto TE, Crook DW, Wilson DJ (2016). Within-host evolution of bacterial pathogens. *Nature Reviews Microbiology*, 14: 150–62.

Diekmann O, Heesterbeek JAP, Metz JJ (1990). On the definition and the computation of the basic reproductive rate Ro in models of infectious diseases in heterogeneous environments. *Journal of Mathematical Biology*, 28: 365–82.

Dieters MJ, Hodge GR, White TL (1996). Genetic parameter estimates for resistance to rust (*Cronartium quercuum*) infection from full-sib tests of slash pine (*Pinus elliottii*), modelled as functions of rust incidence. *Silvae Genetica*, 45: 235–42.

Dietz K (1975). Transmission and control of arbovirus diseases. *Epidemiology*, 104: 104–21.

Diez L, Lejeune P, Detrain C (2014). Keep the nest clean: survival advantages of corpse removal in ants. *Biology Letters*, 10: 20140306.

Diggle SP, Griffin AS, Campbell GS, West SA (2007). Cooperation and conflict in quorum-sensing bacterial populations. *Nature*, 450: 411–14.

Dionne MS, Pham LM, Shirasu-Hiza M, Schneider DS (2006). Akt and FOXO dysregulation contribute to infection-induced wasting in *Drosophila*. *Current Biology*, 16: 1977–85.

Dizney LJ, Ruedas LA (2009). Increased host species diversity and decreased prevalence of Sin Nombre Virus. *Emerging Infectious Diseases*, 15: 1012–18.

Dobrindt U, Hochhut B, Hentschel U, Hacker J (2004). Genomic islands in pathogenic and environmental microorganisms. *Nature Reviews Microbiology*, 2: 414–24.

Dobson AP, Hudson PJ, Lyles AM (1992). Macroparasites: worms and others. In Crawley MJ (ed.) *Natural enemies*, pp. 329–48. Blackwell, Oxford.

Doerig C, Rayner JC, Scherf A, Tobin AB (2014). Post-translational protein modifications in malaria parasites. *Nature*, 13: 160–72.

Doeschl-Wilson AB, Bishop SC, Kyriazakis I, Villanue B (2012). Novel methods for quantifying individual host response to infectious pathogens for genetic analyses. *Frontiers in Genetics*, 3: 266.

Dojiri M, Ho J-S (2018). *Systematics of the Caligidae, copepods parasitic on marine fishes*. Brill, Den Haag.

Dolan PT, Whitfield ZJ, Andino R (2018). Mechanisms and concepts in RNA virus population dynamics and evolution. *Annual Review of Virology*, 5: 69–92.

Domingo E, Perales C (2019). Viral quasispecies. *PLoS Genetics*, 15: e1008271.

Dominguez-Diaz C, Garcia-Orozco A, Riera-Leal A, Padilla-Arellano JR, Fafutis-Morris M (2019). Microbiota and its role on viral evasion: is it with us or against us? *Frontiers in Cellular and Infection Microbiology*, 9: 256.

Dong YH, Wang L-H, Xu J-L, Zhang H-B, Xi-Fen Z, Zhang L-H (2001). Quenching quorum-sensing-dependent bacterial infection by an N-acyl homoserine lactonase. *Nature*, 411: 813–17.

Dorratoltaj N, Nikin-Beers R, Ciupe SM, Eubank SG, Abbas KM (2017). Multi-scale immunoepidemiological modeling of within-host and between- host HIV dynamics: systematic review of mathematical models. *PeerJ*, 5: e3877.

Dortmans JCFM, Rottier PJM, Koch G, Peeters BPH (2011). Passaging of a Newcastle disease virus pigeon variant in chickens results in selection of viruses with mutations in the polymerase complex enhancing virus replication and virulence. *Journal of General Virology*, 92: 336–45.

Douam F, Ploss A (2018). Yellow fever virus: knowledge gaps impeding the fight against an old foe. *Trends in Microbiology*, 26: 913–28.

Doudna JA, Charpentier E (2014). The new frontier of genome engineering with CRISPR-Cas9. *Science*, 346: 1258096.

Dougherty ER, Carlson CJ, Bueno VM, et al. (2016). Paradigms for parasite conservation. *Conservation Biology*, 30: 724–33.

Douglas AE (2015). Multiorganismal insects: diversity and function of resident microorganisms. *Annual Review of Entomology*, 60: 17–34.

Dowling A, van Veller M, Hoberg E, Brooks D (2003). A priori and a posteriori methods in comparative evolutionary studies of host-parasite associations. *Cladistics*, 19: 240–53.

Doyle PS, Zhou YM, Hsieh I, Greenbaum DC, McKerrow JH, Engel JC (2011). The *Trypanosoma cruzi* protease cruzain mediates Immune evasion. *PLoS Pathogens*, 7: e1002139.

Drame PM, Poinsignon A, Besnard P, et al. (2010). Human antibody response to *Anopheles gambiae* saliva: an immuno-epidemiological biomarker to evaluate the efficacy of insecticide-treated nets in malaria vector control. *American Journal of Tropical Medicine and Hygiene*, 83: 115–21.

Drexler JF, Corman VM, Müller MA, et al. (2012). Bats host major mammalian paramyxoviruses. *Nature Communications*, 3: 796.

Drexler JF, Corman VM, Wegner T, et al. (2011). Amplification of emerging viruses in a bat colony. *Emerging Infectious Diseases*, 17: 449–56.

Drovetski SV, Aghayan SA, Mata VA, et al. (2014). Does the niche breadth or trade-off hypothesis explain the abundance–occupancy relationship in avian Haemosporidia? *Molecular Ecology*, 23: 3322–9.

Druett HA (1952). Bacterial invasion. *Nature*, 170: 16 Aug.

Drum NH, Rothenbuhler WC (1983). Non-stinging aggressive responses of worker honeybees to hive-mates, intruder bees and bees affected with chronic bee paralysis. *Journal of Apicultural Research*, 22: 256–60.

Drum NH, Rothenbuhler WC (1985). Differences in non-stinging aggressive responses of workers to diseased and healthy bees in May and July. *Journal of Apicultural Research*, 24: 184–7.

Drummond AJ, Pybus OG, Rambaut A, Forsber R, Rodrigo AG (2003). Measurably evolving populations. *Trends in Ecology & Evolution*, 18: 481–8.

Drummond RA, Brown GD (2013). Signalling C-type lectins in antimicrobial immunity. *PLoS Pathogens*, 9: e1003417.

du Pasquier L (2006). Germline and somatic diversification of immune recognition elements in Metazoa. *Immunology Letters*, 104: 2–17.

du Pasquier L (2018). Evolution of the immune system. In McQueen CA (ed.) *Comprehensive toxicology*, Vol 11, pp. 29–48.

du Plessis L, Stadler T (2015). Getting to the root of epidemic spread with phylodynamic analysis of genomic data. *Trends in Microbiology*, 23: 383–6.

Du X, King AA, Woods RJ, Pascual M (2017). Evolution-informed forecasting of seasonal influenza A (H3N2). *Science Translational Medicine*, 9: eaan5325.

Duan S, Thomas PG (2016). Balancing immune protection and immune pathology by CD8(+) T-cell responses to influenza infection. *Frontiers in Immunology*, 7: 1–16.

Dubrulle M, Mousson L, Moutailler S, Vazeille M, Failloux A (2009). Chikungunya virus and *Aedes* mosquitoes: saliva is infectious as soon as two days after oral infection. *PLoS ONE*, 4: e5895.

Dubuffet A, Zanchi C, Boutet G, Moreau J, Teixeira M, Moret Y (2015). Trans-generational immune priming protects the eggs only against gram-positive bacteria in the mealworm beetle. *PLoS Pathogens*, 11: e1005178.

Ducatez S, Shine R (2017). Drivers of extinction risk in terrestrial vertebrates. *Conservation Letters*, 10: 186–94.

Duffy S (2018). Why are RNA virus mutation rates so damn high? *PLoS Biology*, 16: e3000003.

Duffy S, Burch CL, Turner PE (2007). Evolution of host specificity drives reproductive isolation among RNA viruses *Evolution*, 61: 2614–22.

Duffy S, Shackelton LA, Holmes EC (2008). Rates of evolutionary change in viruses: patterns and determinants. *Nature Reviews Genetics*, 9: 267–76.

Dumont BL (2017). Variation and evolution of the meiotic requirement for crossing over in mammals. *Genetics*, 205: 155.

Dunbar RIM (1991). Functional significance of social grooming in primates. *Folia Primatologica*, 57: 121–31.

Duneau D, Ferdy J-B, Revah J, et al. (2017a). Stochastic variation in the initial phase of bacterial infection predicts the probability of survival in *D. melanogaster. eLife*, 6: e28298.

Duneau D, Luijckx P, Ruder LF, Ebert D (2012). Sex-specific effects of a parasite evolving in a female-biased host population. *BMC Biology*, 10: 104.

Duneau DF, Kondolf HC, Im JH, et al. (2017b). The Toll pathway underlies host sexual dimorphism in resistance to both Gram-negative and Gram-positive bacteria in mated *Drosophila*. *BMC Biology*, 15: 124.

Dunn AM, Adams L, Smith JE (1993). Transovarial transmission and sex ratio distortion by a microsporidan parasite in a shrimp. *Journal of Invertebrate Pathology*, 61: 248–52.

Dunn AM, Terry RS, Smith JE (2001). Transovarial transmission in the microsporidia. *Advances in Parasitology*, 48: 57–100.

Dunn CW, Hejnol A, Matus DQ, et al. (2008). Broad phylogenomic sampling improves resolution of the animal tree of life. *Nature*, 452: 745–9.

Dupas S, Frey F, Carton Y (1998). A single parasitoid segregating factor controls immune suppression in *Drosophila*. *Journal of Heredity*, 89: 306–11.

Duron O, Buchon D, Boutin S, et al. (2008). The diversity of reproductive parasites among arthropods: *Wolbachia* do not walk alone. *BMC Biology*, 6: 27.

Durrant WE, Dong X (2004). Systemic acquired resistance. *Annual Review of Phytopathology*, 42: 185–209.

Durrer S, Schmid-Hempel P (1994). Shared use of flowers leads to horizontal pathogen transmission. *Proceedings of the Royal Society London B*, 258: 299–302.

Duscher GG, Leschnik M, Fuehrer H-P, Joachim A (2015). Wildlife reservoirs for vector-borne canine, feline and zoonotic infections in Austria. *International Journal for Parasitology: Parasites and Wildlife*, 4: 88–96.

Dusi E, Gougat-Barbera C, Berendonk TU, Kaltz O (2015). Long-term selection experiment produces breakdown of horizontal transmissibility in parasite with mixed transmission mode. *Evolution*, 69: 1069–76.

Dwyer KG, Balent MA, Nasrallah JB, Nasrallah ME (1991). DNA sequences of self-incompatibility genes from *Brassica campestris* and *B. oleracea*: polymorphism predating speciation. *Plant Molecular Biology*, 16: 481–6.

Dybdahl MF, Jenkins CE, Nuismer SL (2014). Identifying the molecular basis of host-parasite coevolution: merging models and mechanisms. *The American Naturalist*, 184: 1–13.

Dybdahl MF, Lively CM (1995). Host-parasite interactions: infection of common clones in natural populations of a freshwater snail (*Potamopyrgus antipodarium*). *Proceedings of the Royal Society London B*, 260: 99–103.

Dybdahl MF, Lively CM (1998). Host-parasite coevolution: evidence for rare advantage and time-lagged selection in a natural population. *Evolution*, 52: 1057–66.

Dye C, Williams BG (1995). Non-linearities in the dynamics of indirectly-transmitted infections (or, does having a vector make a difference?). In Grenfell BT, Dobson AP (eds) *Ecology of infectious diseases in natural populations*, pp. 260–79. Cambridge University Press, Cambridge.

Eberhard W (2000). Spider manipulation by a wasp larva. *Nature*, 406: 255–6.

Ebert D (1994). Virulence and local adaptation of a horizontally transmitted parasite. *Science*, 265: 1084–6.

Ebert D (1998). Experimental evolution of parasites. *Science*, 282: 1432–5.

Ebert D (2008). Host–parasite coevolution: Iinsights from the *Daphnia*–parasite model system. *Current Opinion in Microbiology*, 11: 290–301.

Ebert D (2013). The epidemiology and evolution of symbionts with mixed-mode transmission. *Annual Review of Ecology Evolution and Systematics*, 44: 623–43.

Ebert D, Bull JJ (2008). The evolution and expression of virulence. In Stearns SC, Koella J (eds) *Evolution in health and disease*, 2nd ed., pp. 153–67. Oxford University Press, Oxford.

Ebert D, Carius HJ, Little TJ, Decaestecker E (2004). The evolution of virulence when parasites cause host castration and gigantism. *The American Naturalist*, 164, Suppl.: S19–32.

Ebert D, Mangin KL (1997). The influence of host demography on the evolution of virulence of a microsporidian gut parasite. *Evolution*, 51: 1828–37.

Ebert D, Weisser WW (1997). Obligate killing for obligate killers: the evolution of life histories and virulence of semelparous parasites. *Proceedings of the Royal Society London B*, 264: 965–91.

Eens M, Van Duyse E, Berghman E, Pinxten R (2000). Shield characteristics are testosterone-dependent in both male and female moorhens. *Hormones and Behaviour*, 37: 126–34.

Eftimie R, Gillard JJ, Cantrell DA (2016). Mathematical models for immunology: current state of the art and future research directions. *Bulletin of Mathematical Biology*, 78: 2091–134.

Eggert H, Kurtz J, Diddens-de Buhr MF (2014). Different effects of paternal trans-generational immune priming on survival and immunity in step and genetic offspring. *Proceedings of the Royal Society London B*, 281: 20142089.

Ehman KD, Scott ME (2001). Urinary odour preferences of MHC congenic female mice, *Mus domesticus*: implications for kin recognition and detection of parasitized males. *Animal Behaviour*, 62: 781–9.

Eigen M (1996). On the nature of virus quasispecies. *Trends in Microbiology*, 4: 216–18.

Eisenreich W, Heesemann J, Rudel T, Goebel W (2013). Metabolic host responses to infection by intracellular bacterial pathogens. *Frontiers in Cellular and Infection Microbiology*, 3: 24.

Eizaguirre C, Lenz TL, Kalbe M, Milinski M (2012a). Divergent selection on locally adapted major histocompatibility complex immune genes experimentally proven in the field. *Ecology Letters*, 15: 723–31.

Eizaguirre C, Lenz TL, Kalbe M, Milinski M (2012b). Rapid and adaptive evolution of MHC genes under parasite selection in experimental vertebrate populations. *Nature Communications*, 3: 621.

Ekblom R, French L, Slate J, Burke T (2010). Evolutionary analysis and expression profiling of zebra finch immune genes. *Genome Biology and Evolution*, 2: 781–90.

Ekenberg C, Tang M-H, Zucco AG, et al. (2019). Association between single-nucleotide polymorphisms in HLA alleles and human immunodeficiency virus type 1 viral load in demographically diverse, antiretroviral therapy–naive participants from the strategic timing of antiretroviral treatment trial. *The Journal of Infectious Diseases*, 220: 1325–34.

Eleftherianos I, Revenis C (2011). Role and importance of phenoloxidase in insect hemostasis. *Journal of Innate Immunity*, 3: 28–33.

Elena SF (2001). Evolutionary history conditions the timing of transmission in vesicular stomatitis virus. *Infection, Genetics and Evolution*, 1: 151–9.

Elinav E, Strowig T, Kau AL, et al. (2011). NLRP6 inflammasome regulates colonic microbial ecology and risk for colitis. *Cell*, 145: 745–57.

Elliot SL, Adler FR, Sabelis MW (2003). How virulent should a parasite be to its vector? *Ecology*, 84: 2568–74.

Emami SN, Lindberg BG, Hua S, et al. (2017). A key malaria metabolite modulates vector blood seeking, feeding, and susceptibility to infection. *Science*, 355: 1076–80.

Emonts M, Hazelzet JA, de Groot R, Hermans PWM (2003). Host genetic determinants of *Neisseria meningitidis* infections. *The Lancet Infectious Diseases*, 3: 565–77.

Endler JA, Basolo AL (1998). Seonsory ecology, receiver bias and sexual selection. *Trends in Ecology and Evolution*, 13: 415–20.

Engel P, Moran NA (2013). The gut microbiota of insects: diversity in structure and function. *FEMS Microbiological Reviews*, 37: 699–735.

Engelstädter J (2015). Host-parasite coevolutionary dynamics with generalized success/failure infection genetics. *The American Naturalist*, 185: E117–29.

Englehardt JD, Swartout J (2004). Predictive population dose-response assessment for *Cryptosporidium parvum*: infection endpoint. *Journal of Toxicology and Environmental Health, Part A*, 67: 651–66.

Enstrom DA, Ketterson ED, Nolan VJ (1997). Testosteron and mate choice in the dark-eyed junco. *Animal Behaviour*, 54: 1135–46.

Eraud C, Duriez O, Chastel O, Faivre B (2005). The energetic cost of humoral immunity in the Collared Dove, *Streptopelia decaocto*: is the magnitude sufficient to force energy-based trade-offs? *Functional Ecology*, 19: 110–18.

Ercolani GL (1984). Infectivity titration with bacterial plant pathogens. *Annual Review of Phytopathology*, 22: 35–52.

Esch GW, Barger MA, Fellis KJ (2002). The transmission of digenetic trematodes: style, elegance, complexity. *Integrative and Comparative Biology*, 42: 304–12.

Eshelman CM, Vouk R, Stewart JL, et al. (2010). Unrestricted migration favours virulent pathogens in experimental metapopulations: evolutionary genetics of a rapacious life history. *Philosophical Transactions of the Royal Society London B*, 365: 2503–13.

Eskew EA, Olival KJ (2018). De-urbanization and zoonotic disease risk. *EcoHealth*, 15: 707–12.

Espinaze MPA, Hellard E, Horak IG, Cumming GS (2016). Analysis of large new South African dataset using two host-specificity indices shows generalism in both adult and larval ticks of mammals. *Parasitology*, 143: 366–73.

Estrada-Reyes ZM, Tsukahara Y, Amadeu RR, et al. (2019). Signatures of selection for resistance to *Haemonchus contortus* in sheep and goats. *BMC Genomics*, 20: 735.

Evans JD, Aronstein K, Chen YP, et al. (2006). Immune pathways and defence mechanisms in honey bees *Apis mellifera. Insect Molecular Biology*, 15: 645–56.

Evans SS, Repasky EA, Fisher DT (2015). Fever and the thermal regulation of immunity: the immune system feels the heat. *Nature Reviews Immunology*, 15: 335–49.

Ewald PW (1983). Host-parasite relations, vectors, and the evolution of disease severity. *Annual Review of Ecology and Systematics*, 14: 465–85.

Ewald PW (1994). *Evolution of infectious disease*. Oxford University Press, Oxford.

Ewenson E, Zann R, Flannery G (2001). Body condition and immune response in wild zebra finches: effects of capture, confinement and captive-rearing. *Naturwissenschaften*, 88: 391–4.

Ezenwa VO (2004). Host social behavior and parasitic infection: a multifactorial approach. *Behavioral Ecology*, 15: 446–54.

Ezenwa VO, Gerardo NM, Inouye DW, Medina M, Xavier JB (2012). Animal behavior and the microbiome. *Science*, 338: 198–9.

Ezenwa VO, Ghai RR, McKay AF, Williams AE (2016). Group living and pathogen infection revisited. *Current Opinion in Behavioral Sciences*, 12: 66–72.

Ezenwa VO, Godsey MS, King RJ, Guptill SC (2006a). Avian diversity and West Nile virus: testing associations between biodiversity and infectious disease risk. *Proceedings of the Royal Society London B*, 273: 109–17.

Ezenwa VO, Price SA, Altizer S, Vitone ND, Cook KC (2006b). Host traits and parasite species richness in even and odd-toed hoofed mammals, Artiodactyla and Perissodactyla. *Oikos*, 115: 526–36.

Fabian DK, Garschall K, Klepsatel P, et al. (2018). Evolution of longevity improves immunity in *Drosophila. Evolution Letters*, 2: 567–79.

Fadista J, Manning AK, Florez JC, Groop L (2016). The (in)famous GWAS P-value threshold revisited and updated for low-frequency variants. *European Journal of Human Genetics*, 24: 1202–5.

Faillace CA, Lorusso NS, Duffy S (2017). Overlooking the smallest matter: viruses impact biological invasions. *Ecology Letters*, 20: 524–38.

Fairfax BP, Knight JC (2014). Genetics of gene expression in immunity to infection. *Current Opinion in Immunology*, 30: 63–71.

Fairlie-Clarke KJ, Shuker DM, Graham AL (2009). Why do adaptive immune responses cross-react? *Evolutionary Applications*, 2: 122–31.

Falconer DS (1989). *Introduction to quantitative genetics*, 3rd ed. John Wiley & Sons, New York.

Fallon SM, Bermingham E, Ricklefs RE (2005). Host specialization and geographic localization of avian malaria parasites: a regional analysis in the Lesser Antilles. *American Naturalist*, 165: 466–80.

Faria NR, Rambaut A, Suchard MA, et al. (2014). The early spread and epidemic ignition of HIV-1 in human populations. *Science*, 346: 56–61.

Farré D, Martínez-Vicente P, Engel P, Angulo A (2017). Immunoglobulin superfamily members encoded by viruses and their multiple roles in immune evasion. *European Journal of Immunology*, 47: 780–96.

Faruque SM, Mekalanos JJ (2003). Pathogenicity islands and phages in *Vibrio cholerae* evolution. *Trends in Microbiology*, 11: 505–10.

Fast MD (2014). Fish immune responses to parasitic copepod (namely sea lice) infection. *Developmental & Comparative Immunology*, 43: 300–12.

Faulhaber LM, Karp RD (1992). A diphasic immune response against bacteria in the american cockroach. *Immunology*, 75: 378–81.

Fazio G, Sasal P, Da Silva C, et al. (2008). Regulation of *Anguillicola crassus* (Nematoda) infrapopulations in their definitive host the European eel, *Anguilla anguilla. Parasitology*, 135: 1707–16.

Fecchio A, Bell JA, Pinheiro RBP, et al. (2019). Avian host composition, local speciation and dispersal drive the regional assembly of avian malaria parasites in South American birds. *Molecular Ecology*, 28: 2681–93.

Feng W, Zong W, Wang F, Ju S (2020). Severe acute respiratory syndrome coronavirus 2 (SARS-CoV-2): a review. *Molecular Cancer*, 19: 100.

Feng Z, Cen X, Zhao Y, Velasco-Hernandez J (2015). Coupled within-host and between-host dynamics and evolution of virulence. *Mathematical Biosciences*, 270: 204–12.

Fenner F, Ratcliffe FN (1965). *Myxomatosis*. Cambridge University Press, Cambridge.

Fenton A, Antonovics J, Brockhurst MA (2009). Inverse gene-for-gene infection genetics and coevolutionary dynamics. *The American Naturalist*, 174: E230–42.

Fenton A, Antonovics J, Brockhurst MA (2012). Two-step infection processes can lead to coevolution between functionally independent infection and resistance pathways. *Evolution*, 66: 2030–41.

Fenton A, Brockhurst MA (2008). The role of specialist parasites in structuring host communities. *Ecological Research*, 23: 975–804.

Fenton A, Rands SA (2006). The impact of parasite manipulation and predator foraging behavior on predator-prey communities. *Ecology*, 87: 2832–41.

Ferdy J, Godelle B (2005). Diversification of transmission modes and the evolution of mutualism. *The American Naturalist*, 166: 613–27.

Ferguson NM, Glavani AP, Bush RM (2003). Ecological and immunological determinants of influenza evolution. *Nature*, 422: 428–33.

Ferrandon D, Imler J-L, Hetru C, Hoffmann JA (2007). The *Drosophila* system immune response: sensing and signalling during bacterial and fungal infections. *Nature Reviews Immunology*, 7: 862–74.

Ferrer ES, García-Navas V, Bueno-Enciso J, Sanz JJ, Ortego J (2015). Multiple sexual ornaments signal heterozygosity in male blue tits. *Biological Journal of the Linnean Society*, 115: 362–75.

Ferro K, Peuß R, Yang W, Rosenstiel P, Schulenburg H, Kurtz J (2019). Experimental evolution of immunological specificity. *Proceedings of the National Academy of Sciences USA*, 116: 20598–604.

Filippitzi ME, Goumperis T, Robinson T, Saegerman C (2017). Microbiological zoonotic emerging risks, transmitted between livestock animals and humans (2007–2015). *Transboundary and Emerging Diseases*, 64: 1059–70.

Fillol-Salom A, Martínez-Rubio R, Abdulrahman RF, Chen J, Davies R, Penadés JR (2018). Phage-inducible chromosomal islands are ubiquitous within the bacterial universe. *The ISME Journal*, 12: 2114–28.

Finnegan DJ (2015). Retrotransposons. *Current Biology*, 22: R432–7.

Firmat C, Alibert P, Mutin G, Losseau M, Pariselle A, Sasal P (2016). A case of complete loss of gill parasites in the invasive cichlid *Oreochromis mossambicus*. *Parasitology Research*, 115: 3657–61.

Fisher AC (1934). A study of the schistosomiasis of the Stanleyville district of the Belgian Congo. *Transactions of The Royal Society of Tropical Medicine and Hygiene*, 28: 277–306.

Fisher RA (1915). The evolution of sexual preference. *Eugenics Review*, 7: 184–92.

Fisher RA (1930). *The genetical theory of natural selection*. Dover Publications, New York.

Fitzsimmons WJ, Woods RJ, McCrone JT, et al. (2018). A speed–fidelity trade-off determines the mutation rate and virulence of an RNA virus. *PLoS Biology*, 16: e2006459.

Flajnik MF (2002). Comparative analysis of immunoglobulin genes: surprises and portents. *Nature Reviews Immunology*, 2: 688–98.

Flajnik MF, du Pasquier L (2004). Evolution of innate and adaptive immunity: can we draw a line? *Trends in Immunology*, 25: 640–4.

Flajnik MF, du Pasquier L (2008). Evolution of the immune system. In Paul WE (ed.) *Fundamental immunology*, 6th ed., pp. 56–124. Wolters Kluver–Lippincott–Williams & Wilkins, Philadelphia.

Flajnik MF, du Pasquier L (2013). Evolution of the immune system. In Paul WE (ed.) *Fundamental immunology*, 7th ed., pp. 67–128. Lippincott Williams & Wilkins, Philadelphia.

Flemming H-C, Wingender J (2010). The biofilm matrix. *Nature Reviews Microbiology*, 8: 623–33.

Flies AS, Flies EJ, Fox S, et al. (2020). An oral bait vaccination approach for the Tasmanian devil facial tumor diseases. *Expert Review of Vaccines*, 19: 1–10.

Flor HH (1956). The complementary genic systems in flax and flax rust. *Advances in Genetics*, 8: 29–54.

Flynn J, Chan J (2003). Immune evasion by *Mycobacterium tuberculosis*: living with the enemy. *Current Opinion in Immunology*, 15: 450–5.

Foerster K, Delhey K, Johnsen A, Lifjeld JT, Kempenaers B (2003). Females increase offspring heterozygosity through extra-pair matings. *Nature*, 425: 714–17.

Foley K, Fazio G, Jensen AB, Hughes WOH (2012). Nutritional limitation and resistance to opportunistic *Aspergillus* parasites in honey bee larvae. *Journal of Invertebrate Pathology*, 111: 68–73.

Folstad I, Karter AJ (1992). Parasites, bright males, and the immunocompetence handicap. *The American Naturalist*, 139: 603–22.

Folstad I, Nilssen AC, Halvorsen O, Andersen J (1991). Parasite avoidance: the cause of post-calving migrations in *Rangifer*? *Canadian Journal of Zoology*, 69: 2423–9.

Fong IW (2017). *Emerging zoonoses. A worldwide perspective*. Springer, Cham.

Foo YZ, Nakagawa S, Rhodes G, Simmons LW (2017). The effects of sex hormones on immune function: a meta-analysis. *Biological Reviews*, 92: 551–71.

Forbes KM, Sironen T, Plyusnin A (2018). Hantavirus maintenance and transmission in reservoir host populations. *Current Opinion in Virology*, 28: 1–6.

Ford MJ (2002). Applications of selective neutrality tests to molecular ecology. *Molecular Ecology*, 11: 1245–62.

Ford SA, Kao D, Williams D, King KC (2016). Microbe-mediated host defence drives the evolution of reduced pathogen virulence. *Nature Communications*, 7: 13430.

Ford SA, Williams D, Paterson S, King KC (2017). Co-evolutionary dynamics between a defensive microbe and a pathogen driven by fluctuating selection. *Molecular Ecology*, 26: 1778–89.

Forde SE, Thompson JN, Holt RD, Bohannan BJ (2008). Coevolution drives temporal changes in fitness and diversity across environments in a bacteria-bacteriophage interaction. *Evolution*, 62: 1830–9.

Fornoni J, Nuñez-Farfan J, Valverde PL, Rausher M (2004). Evolution of mixed strategies of plant defense allocation against natural enemies. *Evolution*, 58: 1685–95.

Forsberg KJ, Patel S, Gibson MK, et al. (2014). Bacterial phylogeny structures soil resistomes across habitats. *Nature*, 509: 612–21.

Forsberg LA, Dannewitz J, Petersson E, Grahn M (2007). Influence of genetic dissimilarity in the reproductive success and mate choice of brown trout: females fishing for optimal MHC dissimilarity. *Journal of Evolutionary Biology*, 20: 1859–69.

Forsman AM, Vogel LA, Sakaluk SK, Grindstaff JL, Thompson CF (2008). Immune-challenged house wren broods differ in the relative strengths of their responses among different axes of the immune system. *Journal of Evolutionary Biology*, 21: 873–8.

Forster SC, Tate MD, Hertzog PJ (2015). MicroRNA as type I interferon- regulated transcripts and modulators of the innate immune response. *Frontiers in Immunology*, 6: 334.

Foster CSP (2019). Digest. The phylogenetic distance effect: understanding parasite host switching. *Evolution*, 73: 1494–5.

Fowler A (2007). Leaf-swallowing in Nigerian chimpanzees: evidence for assumed self-medication. *Primates*, 48: 73–6.

Foxman EF, Storer JA, Fitzgerald ME, et al. (2015). Temperature-dependent innate defense against the common cold virus limits viral replication at warm temperature in mouse airway cells. *Proceedings of the National Academy of Sciences USA*, 112: 827–32.

Fraija-Fernández N, Olson PD, Crespo EA, Raga JA, Aznar FJ, Fernández M (2015). Independent host switching events by digenean parasites of cetaceans inferred from ribosomal DNA. *International Journal for Parasitology*, 45: 167–73.

Franceschi N, Cornet S, Bollache L, et al. (2010). Variation between populations and local adaptation in acantho-cephalan-induced parasite manipulation. *Evolution*, 64: 2417–30.

Frank SA (1992). A kin selection model for the evolution of virulence. *Proceedings of the Royal Society London B*, 250: 195–7.

Frank SA (1995). George Price's contributions to evolutionary genetics. *Journal of Theoretical Biology*, 175: 373–88.

Frank SA (1996). Models of parasite virulence. *Quarterly Review of Biology*, 71: 37–78.

Frank SA (1997). The Price equation, Fisher's fundamental theorem, kin selection and causal analysis. *Evolution*, 51: 1712–29.

Frank SA (2000). Specific and non-specific defense against parasitic attack. *Journal of Theoretical Biology*, 202: 283–304.

Frank SA, Schmid-Hempel P (2008). Mechanisms of pathogenesis and the evolution of parasite virulence. *Journal of Evolutionary Biology*, 21: 396–404.

Frank SA, Schmid-Hempel P (2019). Evolution of negative immune regulators. *PLoS Pathogens*, 15: e1007913.

Frankild S, de Boer RJ, Lund O, Nielsen M, Kesmir C (2008). Amino acid similarity accounts for T-cell cross-reactivity and for "'holes'" in the T-cell repertoire. *PLoS ONE*, 3: e1831.

Franzosa EA, Huang K, Meadow JF, et al. (2015). Identifying personal microbiomes using metagenomic codes. *Proceedings of the National Academy of Sciences USA*, 112: E2930.

Fraser C, Hollingsworth TD, Chapman R, de Wolf F, Hanage WP (2007). Variation in HIV-1 set-point viral load: epidemiological analysis and evolutionary hypothesis. *Proceedings of the National Academy of Sciences USA*, 104: 17441–6.

Fredensborg BL (2014). Predictors of host specificity among behavior-manipulating parasites. *Integrative and Comparative Biology*, 54: 149–58.

Fredriksson-Ahomaa M (2018). Wild boar: a reservoir of foodborne zoonoses. *Foodborne Pathogens and Disease*, 16: 153–65.

Freed LA, Cann RL, Goff ML, Kuntz WA, Bonder GR (2005). Increase in avian malaria at upper elevation in Hawai`i. *Condor*, 105: 753–64.

Freeland WJ (1983). Parasites and the coexistence of animal host species. *The American Naturalist*, 121: 223–36.

Freely S, Kemper C, Le Friec G (2016). The 'ins and outs' of complement-driven immune responses. *Immunological Reviews*, 274: 16–32.

Freilich S, Zarecki R, Eilam O, et al. (2011). Competitive and cooperative metabolic interactions in bacterial communities. *Nature Communications*, 2: 589.

French RK, Holmes EC (2020). An ecosystems perspective on virus evolution and emergence. *Trends in Microbiology*, 28: 165–75.

Freyberg Z, Harvill ET (2017). Pathogen manipulation of host metabolism: a common strategy for immune evasion. *PLoS Pathogens*, 13: e1006669.

Fritzsche McKay A, Hoye BJ (2016). Are migratory animals superspreaders of infection? *Integrative and Comparative Biology*, 56: 260–7.

Froissart R, Doumayrou J, Vuillaume F, Alizon S, Michalakis Y (2010). The virulence-transmission trade-off in vector-borne plant viruses: a review of (non-) existing studies. *Philosophical Transactions of the Royal Society London B*, 365: 1907–18.

Fu YX, Li WH (1993). Statistical tests of neutrality of mutations. *Genetics*, 133: 693–709.

Fuentefria AM, Pippi B, Dalla Lana DF, Donato KK, de Andrade SF (2018). Antifungals discovery: an insight into new strategies to combat antifungal resistance. *Letters in Applied Microbiology*, 66: 2–13.

Fuller RC, Houle D, Travis J (2005). Sensory bias as an explanation for the evolution of mate preferences. *American Naturalist*, 166: 437–46.

Futo M, Armitage SAO, Kurtz J (2016). Microbiota plays a role in oral immune priming in *Tribolium castaneum*. *Frontiers in Microbiology*, 6: 1383.

Gaba S, Ebert D (2009). Time-shift experiments as a tool to study antagonistic coevolution. *Trends in Ecology & Evolution*, 24: 226–32.

Gahr CL, Boehm T, Milinski M (2018). Female assortative mate choice functionally validates synthesized male odours of evolving stickleback river–lake ecotypes. *Biology Letters*, 14: 20180730.

Gal-Mor O, Finlay BB (2006). Pathogenicity islands: a molecular toolbox for bacterial virulence. *Cellular Microbiology*, 8: 1707–19.

Galán JE, Lara-Tejero M, Marlovits TC, Wagner S (2014). Bacterial Type III secretion systems: specialized nanomachines for protein delivery into target cells. *Annual Review of Microbiology*, 68: 415–38.

Gallagher ME, Brooke CB, Ke R, Koelle K (2018). Causes and consequences of spatial within-host viral spread. *Viruses*, 10: 627.

Gallien L, Carboni M (2017). The community ecology of invasive species: where are we and what's next? *Ecography*, 40: 335–52.

Gammon DB (2017). *Caenorhabditis elegans* as an emerging model for virus-host interactions. *Journal of Virology*, 91: e00509–17.

Gandon S (2004). Evolution of multihost parasites. *Evolution*, 58: 1165–77.

Gandon S, Day T (2007). The evolutionary epidemiology of vaccination. *Journal of The Royal Society Interface*, 4: 803–17.

Gandon S, Mackinnon MJ, Nee S, Read AF (2001). Imperfect vaccines and the evolution of pathogen virulence. *Nature*, 414: 751–6.

Gandon S, Mackinnon MJ, Nee S, Read AF (2002). Antitoxin vaccines and pathogen virulence: reply. *Nature*, 417: 610.

Gandon S, Mackinnon MJ, Nee S, Read AF (2003). Imperfect vaccination: some epidemiological and evolutionary consequences. *Proceedings of the Royal Society London B*, 270: 1129–36.

Gandon S, Michalakis Y (2002). Local adaptation, evolutionary potential and host–parasite coevolution: interactions between migration, mutation, population size and generation time. *Journal of Evolutionary Biology*, 15: 451–62.

Gandon S, Nuismer SL (2009). Interactions between genetic drift, gene flow, and selection mosaics drive parasite local adaptation. *The American Naturalist*, 173: 212–24.

Gandon S, Otto SP (2007). The evolution of sex and recombination in response to abiotic and coevolutionary fluctuations. *Genetics*, 175: 1835–53.

Ganusov VV, Antia R, Koella J (2006). Imperfect vaccines and the evolution of pathogens causing acute infections in vertebrates. *Evolution*, 60: 957–69.

Ganyani T, Kremer C, Chen D, et al. (2020). Estimating the generation interval for coronavirus disease (COVID-19) based on symptom onset data, March 2020. *European Communicable Disease Bulletin*, 25: 2000257.

Ganz T (2003). The role of antimicrobial peptides in innate immunity. *Integrative and Comparative Biology*, 43: 300–4.

Gao H-W, Wang L-P, Liang S, et al. (2012). Change in rainfall drives malaria re-emergence in Anhui province, China. *PLoS ONE*, 7: e43686.

Garamszegi LZ (2005). Bird song and parasites. *Behavioral Ecology and Sociobiology*, 59: 167–80.

Garamszegi LZ (2006). The evolution of virulence and host specialization in malaria parasites of primates. *Ecology Letters*, 9: 933–40.

Garamszegi LZ, Møller AP (2012). The interspecific relationship between prevalence of blood parasites and sexual traits in birds when considering recent methodological advancements. *Behavioral Ecology and Sociobiology*, 66: 107–19.

Garbutt J, Bonsall MB, Wright DJ, Raymond B (2011). Antagonistic competition moderates virulence in *Bacillus thuringiensis*. *Ecology Letters*, 14: 765–72.

Garcia-Longoria L, Palinauskas V, Ilgūnas M, Valkiūnas G, Hellgren O (2020). Differential gene expression of *Plasmodium homocircumflexum* (lineage pCOLL4) across two experimentally infected passerine bird species. *Genomics*, 112: 2857–65.

García-Varela M, Pérez-Ponce de Leon G (2015). Advances in the classification of acantocephalans: evolutionary history and evolution of parasitism. In Morand S, Kransob BR, Littlewood DTJ (eds) *Parasite diversity and diversification: evolutionary ecology meets parasitology*, pp. 182–99. Cambridge University Press, Cambridge.

Gardner A, West SA, Buckling A (2004). Bacteriocins, spite and virulence. *Proceedings of the Royal Society London B*, 271: 1529–35.

Garey JR (2002). The lesser-known protostome taxa: an introduction and a tribute to Robert P. Higgins. *Integrative and Comparative Biology*, 42: 611–18.

Garira W (2018). A primer on multiscale modelling of infectious disease systems. *Infectious Disease Modelling*, 3: 176–91.

Garnier R, Graham AL (2014). Insights from parasite-specific serological tools in eco-immunology. *Integrative and Comparative Biology*, 54: 363–76.

Garrido-Olvera L, Arita HT, Perez-Ponce De León G (2012). The influence of host ecology and biogeography on the helminth species richness of freshwater fishes in Mexico. *Parasitology*, 139: 1652–65.

Gaston KJ, Blackburn TM (2000). *Pattern and process in macroecology*. Blackwell, Oxford.

Gauthier DT (2015). Bacterial zoonoses of fishes: a review and appraisal of evidence for linkages between fish and human infections. *The Veterinary Journal*, 203: 27–35.

Gay N, Olival KJ, Bumrungsri S, Siriaroonrat B, Bourgarel M, Morand S (2014a). Parasite and viral species richness of Southeast Asian bats: fragmentation of area distribution matters. *International Journal for Parasitology: Parasites and Wildlife*, 3: 161–70.

Gay NJ, Symmons MF, Gangloff M, Bryant CE (2014b). Assembly and localization of Toll-like receptor signalling complexes. *Nature Reviews Immunology*, 14: 546–58.

Gellatly SL, Hancock REW (2013). *Pseudomonas aeruginosa*: new insights into pathogenesis and host defenses. *Pathogens and Disease*, 67: 159–73.

Geoghegan JL, Senior AM, Di Giallonardo F, Holmes EC (2016). Virological factors that increase the transmissibility of emerging human viruses. *Proceedings of the National Academy of Sciences USA*, 113: 4170–5.

George TL, Harrigan RJ, LaManna JA, DeSante DF, Saracco JF, Smith TB (2015). Persistent impacts of West Nile virus on North American bird populations. *Proceedings of the National Academy of Sciences USA*, 112: 14290–4.

Gérard C, Théron A (1997). Age/size- and time-specific effects of *Schistosoma mansoni* on energy allocation patterns of its snail host, *Biomphalaria glabrata*. *Oecologia*, 112: 447–52.

Gerardo NM, Altincicek B, Anselme C, et al. (2010). Immunity and other defenses in pea aphids, *Acyrthosiphon pisum*. *Genome Biology*, 11: R21.

Getty T (2002). Signalling health versus parasites. *The American Naturalist*, 159: 363–71.

Ghiselin MT (1974). *The economy of nature and evolution of sex*. University of California Press, Berkeley.

Ghosh S, Jiang N, Farr L, Ngobeni R, Moonah S (2019). Parasite-produced MIF cytokine: role in immune evasion, invasion, and pathogenesis. *Frontiers in Immunology*, 10: 1995.

Gibb R, Redding DW, Chin KQ, et al. (2020). Zoonotic host diversity increases in human-dominated ecosystems. *Nature*, 584: 398–402.

Gibson G (2003). Population genomics: celebrating individual expression. *Heredity*, 90: 1–5.

Gibson G, Weir B (2005). The quantitative genetics of transcription. *Trends in Genetics*, 21: 616–23.

Gibson W, Lewis MD, Yeo M, Miles MA (2017). Genetic exchange in trypanosomatids and its relevance to epidemiology. In Tibayrenc M (ed.) *Genetics and evolution of infectious diseases*, 2nd ed., pp. 459–86. Elsevier, London.

Gibson W, Stevens J (1999). Genetic exchange in the Trypanosomatidae. *Advances in Parasitology*, 43: 1–46.

Gidwani K, Picado A, Rijal S, et al. (2011). Serological markers of sand fly exposure to evaluate insecticidal nets against visceral leishmaniasis in india and nepal: a cluster-randomized trial. *PLoS Neglected Tropical Diseases*, 5: e1296.

Giesecke J (2002). *Modern infectious disease epidemiology*, 2nd ed. Arnold Publishers, London.

Gilbert G, Webb C (2007). Phylogenetic signal in plant pathogen-host range. *Proceedings of the National Academy of Sciences USA*, 104: 4979–83.

Gilchrist MA, Sasaki A (2002). Modeling host-parasite coevolution: a nested approach based on mechanistic models. *Journal of Theoretical Biology*, 218: 289–308.

Gillard JJ, Laws TR, Lythe G, Molina-Paris C (2014). Modeling early events in *Francisella tularensis* pathogenesis. *Frontiers in Cellular and Infection Microbiology*, 4: 169.

Gillis CC, Hughes ER, Spiga L, et al. (2018). Dysbiosis-associated change in host metabolism generates lactate to support *Salmonella* growth. *Cell Host & Microbe*, 23: 54–64.e6.

Gingrich RD (1964). Acquired humoral immune response of the large milkweed bug, *Oncopeltus fasciatus* (Dallas), to injected materials. *Journal of Insect Physiology*, 10: 179–84.

Gire SK, Goba A, Andersen KG, et al. (2014). Genomic surveillance elucidates Ebola virus origin and transmission during the 2014 outbreak. *Science*, 345: 1369–72.

Gleeson M, Nieman DC, Pedersen BK (2004). Exercise, nutrition and immune function. *Journal of Sports Sciences*, 22: 115–25.

Glunt KD, Thomas MB, Read AF (2011). The effects of age, exposure history and malaria infection on the susceptibility of *Anopheles* mosquitoes to low concentrations of pyrethroid. *PLoS ONEOne*, 6: e24968.

Goater T, Goater CP, Esch GW (2014). *Parasitism: the diversity and ecology of animal parasites* 2nd ed. Cambridge University Press, Cambridge.

Godfray HCJ (1994). *Parasitoids: behavioral and evolutionary ecology*. Princeton University Press, Princeton, New Jersey.

Goedknegt MA, Feis ME, Wegner KM, et al. (2016). Parasites and marine invasions: ecological and evolutionary perspectives. *Journal of Sea Research*, 113: 11–27.

Goel AK (2015). Anthrax: a disease of biowarfare and public health importance. *World Journal of Clinical Cases*, 16: 20–33.

Gohar M, Gilois N, Graveline R, Garreau C, Sanchis V, Lereclus D (2005). A comparative study of *Bacillus cereus*, *Bacillus thuringiensis* and *Bacillus anthracis* extracellular proteomes. *Proetomics*, 5: 3696–711.

Gold KS, Brückner K (2015). Macrophages and cellular immunity in *Drosophila melanogaster*. *Seminars in Immunology*, 27: 357–68.

Gomes MGM, Lipsitch M, Wargo AR, et al. (2014). A missing dimension in measures of vaccination impacts. *PLoS Pathogens*, 10: e1003849.

Gomes PS, Bhardwaj J, Rivera-Correa J, Freire-De-Lima CG, Morrot A (2016). Immune escape strategies of malaria parasites. *Frontiers in Microbiology*, 7: 1617.

Gonzalez G, Sorci G, Moller AP, Ninni P, Haussy C, De LF (1999). Immunocompetence and condition-dependent sexual advertisement in male house sparrows (*Passer domesticus*). *Journal of Animal Ecology. Nov.*, 68: 1225–34.

González-González A, Wayne ML (2020). Immunopathology and immune homeostasis during viral infection in insects. In Carr JP, Roossinck MJ (eds) *Advances in virus research*, pp. 285–314. Academic Press, London.

Goodrich Julia K, Waters Jillian L, Poole Angela C, et al. (2014). Human genetics shape the gut microbiome. *Cell*, 159: 789–99.

Gordon CA, McManus DP, Jones MK, Gray DJ, Gobert GN (2016). Chapter Six: The increase of exotic zoonotic helminth infections: the impact of urbanization, climate change and globalization. *Advances in Parasitology*, 91: 311–97.

Gordon CH, Banyard AC, Hussein A, et al. (2015). Canine distemper in endangered Ethiopian wolves. *Emerging Iinfectious Ddiseases*, 21: 824–32.

Gordon DM, Whitfield PJ (1985). Interactions of the cysticercoids of *Hymenolepis diminuta* and *Raillietiena cesticillus* in their intermediate host, *Tribolium confusum*. *Parasitology*, 90: 421–31.

Gordon S (2016a). Phagocytosis: the legacy of Metchnikoff. *Cell*, 166: 1065–8.

Gordon S (2016b). Phagocytosis: an immunobiologic process. *Immunity*, 44: 463–75.

Gordy MA, Pila EA, Hanington PC (2015). The role of fibrinogen-related proteins in the gastropod immune response. *Fish and Shellfish Immunology*, 46: 39–49.

Gould SJ, Vrba ES (1982). Exaptation: a missing term in the science of form. *Palaeobiology*, 8: 4–9.

Gourbal B, Pinaud S, Beckers GJM, Van DerMeer JWM, Conrath U, Netea MG (2018). Innate immune memory: an evolutionary perspective. *Immunological Reviews*, 283: 21–40.

Grabda-Kazubska B (1976). Abbreviation of the life cycles in plagiochorid trematodes: general remarks. *Acta Parasitologica Polonica*, 24: 125–41.

Graham AL, Allen JE, Read AF (2005). Evolutionary causes and consequences of immunopathology. *Annual Review of Ecology and Systematics*, 36: 373–97.

Granato ET, Meiller-Legrand TA, Foster KR (2019). The evolution and ecology of bacterial warfare. *Current Biology*, 29: R521–37.

Grant AJ, Restif O, McKinley TJ, Sheppard M, Maskell DJ, Mastroeni P (2008). Modelling within-host spatiotemporal dynamics of invasive bacterial disease. *PLoS Biology*, 6: e74.

Graw F, Magnus C, Regoes RR (2010). Theoretical analysis of the evolution of immune memory. *BMC Evolutionary Biology*, 10: 380.

Gray RR, Parker J, Lemey P, Salemi M, Katzourakis A, Pybus OG (2011). The mode and tempo of hepatitis C virus evolution within and among hosts. *BMC Evolutionary Biology*, 11: 131.

Grech K, Chan BHK, Anders RF, Read AF, Van Baalen M (2008). The impact of immunization on competition within *Plasmodium* infections. *Evolution*, 62: 2359–71.

Green TJ, Speck P (2018). Antiviral defense and innate immune memory in the Oyster. *Viruses*, 10: 133.

Greenberg RS, Daniels SR, Flanders DW, Eley JW, Boring III JR (2004). *Medical epidemiology*, 4th ed. McGraw-Hill Professional, New York.

Greenberg S, Grinstein S (2002). Phagocytosis and innate immunity. *Current Opinion in Immunology*, 1: 136–45.

Greene NP, Kaplan E, Crow A, Koronakis V (2018). Antibiotic resistance mediated by the MacB ABC transporter family: a structural and functional perspective. *Frontiers in Microbiology*, 9: 950.

Greenwood JM, Milutinović B, Peuß R, et al. (2017). Oral immune priming with *Bacillus thuringiensis* induces a shift in the gene expression of *Tribolium castaneum* larvae. *BMC Genomics*, 18: 329.

Greischar MA, Koskella B (2007). A synthesis of experimental work on parasite local adaptation. *Ecology Letters*, 10: 418–34.

Greischar MA, Mideo N, Read AF, Bjornstad ON (2016a). Predicting optimal transmission investment in malaria parasites. *Evolution*, 70: 1542–58.

Greischar MA, Reece S, Mideo N (2016b). The role of models in translating within-host dynamics to parasite evolution. *Parasitology*, 143: 905–14.

Grenfell BT, Pybus OC, Gog JR, et al. (2004). Unifying the epidemiological and evolutionary dynamics of pathogens. *Science*, 303: 327–32.

Grenga L, Little RH, Malone JG (2017). Quick change: post-transcriptional regulation in *Pseudomonas*. *FEMS Microbiology Letters*, 364: fnx125.

Grewell BJ (2008). Parasite facilitates plant species coexistence in a coastal wetland. *Ecology*, 89: 1481–8.

Griffin AS, West SA, Buckling A (2004). Cooperation and competition in pathogenic bacteria. *Nature*, 430: 1024–7.

Griffith JW, Sokol CL, Luster AD (2014). Chemokines and chemokine receptors: positioning cells for host defense and immunity. *Annual Review of Immunology*, 32: 659–702.

Grigorian M, Hartenstein V (2013). Hematopoiesis and hematopoietic organs in arthropods. *Development Genes and Evolution*, 223: 103–15.

Grindstaff JL, Brodie EDI, Ketterson ED (2003). Immune function across generations: integrating mechanism and evolutionary process in maternal antibody transmission. *Proceedings of the Royal Society London B*, 270: 2309–19.

Grohmann E, Christie PJ, Waksman G, Backert S (2018). Type IV secretion in Gram-negative and Gram-positive bacteria. *Molecular Microbiology*, 107: 455–71.

Grosholz ED (1994). The effects of host genotype and spatial distribution on trematode parasitism in a bivalve population. *Evolution*, 48: 1514–24.

Grubaugh ND, Ladner JT, Lemey P, et al. (2019). Tracking virus outbreaks in the twenty-first century. *Nature Microbiology*, 4: 10–19.

Guégan J-F, Lambert FA, Léveque C, Combes C, Euzet L (1992). Can host body size explain the parasite species richness in tropical freshwater fish? *Oecologia*, 90: 197–204.

Guégan J-F, Morand S, Poulin R (2005). Are there general laws in parasite community ecology? The emergence of spatial parasitology and epidemiology. In Thomas F, Renaud Fc, Guégan J-F (eds) *Parasitism and ecosystems*, pp. 22–42. Oxford University Press, Oxford.

Gueguen Y, Garnier J, Robert L, et al. (2006). PenBase, the shrimp antimicrobial peptide penaeidin database: sequence-based classification and recommended nomenclature. *Developmental & Comparative Immunology*, 30: 283–8.

Guernier V, Hochberg ME, Guégan JF (2004). Ecology drives the worldwide distribution of human infectious diseases. *PLoS Biology*, 2: 740–6.

Guerra FM, Bolotin S, Lim G, et al. (2017). The basic reproduction number (R0) of measles: a systematic review. *The Lancet Infectious Diseases*, 17: e420–8.

Guizetti J, Scherf A (2013). Silence, activate, poise and switch! Mechanisms of antigenic variation in *Plasmodium falciparum*. *Cellular Microbiology*, 15: 718–26.

Gunia M, David I, Hurtaud J, Maupin M, Gilbert H, Garreau H (2018). Genetic parameters for resistance to non-specific diseases and production traits measured in challenging and selection environments; application to a rabbit case. *Frontiers in Genetics*, 9: 467.

Guo H, Callaway JB, Ting JP-Y (2015). Inflammasomes: mechanism of action, role in disease, and therapeutics. *Nature Medicine*, 21: 677–87.

Guven-Maiorov E, Tsai C-J, Nussinov R (2016). Pathogen mimicry of host protein-protein interfaces modulates immunity. *Seminars in Cell & Developmental Biology*, 58: 136–45.

Gwinn M, MacCannell DR, Khabbaz RF (2017). Integrating advanced molecular technologies into public health. *Journal of Clinical Microbiology*, 55: 703–14.

Haas CN, Rose JB, Gerba CP (2014). *Quantitative microbial risk assessment*. John Wiley & Sons, Hoboken, NJ.

Hacker J, Kaper JB (2000). Pathogenicity islands and the evolution of microbes. *Annual Review of Microbiology*, 54: 641–79.

Hadjichrysanthou C, Cauet E, Lawrence E, Vegvari C, de Wolf F, Anderson RM (2016). Understanding the within-host dynamics of influenza A virus: from theory to clinical implications. *Journal of the Royal Society Interface*, 13: 20160289.

Hafer N (2016). Conflicts over host manipulation between different parasites and pathogens: investigating the ecological and medical consequences. *Bioessays*, 38: 1027–37.

Hafer N (2017). Differences between populations in host manipulation by the tapeworm *Schistocephalus solidus*: is there local adaptation? *Parasitology*, 145: 762–9.

Hafer N, Milinski M (2015a). When parasites disagree: evidence for parasite-induced sabotage of host manipulation. *Evolution*, 69: 611–20.

Hafer N, Milinski M (2015b). An experimental conflict of interest between parasites reveals the mechanism of host manipulation. *Behavioral Ecology*, 27: 617–27.

Hafer N, Vorburger C (2019). Diversity begets diversity: do parasites promote variation in protective symbionts? *Current Opinion in Insect Science*, 32: 8–14.

Hafer-Hahmann N (2019). Experimental evolution of parasitic host manipulation. *Proceedings of the Royal Society London B*, 286: 20182413.

Hahn C, Fromm B, Bachmann L (2014). Comparative genomics of flatworms (Platyhelminthes) reveals shared genomic features of ecto- and endoparasitic Neodermata. *Genome Biology and Evolution*, 6: 1105–17.

Hahn YS (2003). Subversion of immune responses by hepatitis C virus: immunomodulatory strategies beyond evasion? *Current Opinion in Immunology*, 15: 443–9.

Haine ER (2008). Symbiont-mediated protection. *Proceedings of the Royal Society London B*, 275: 353–61.

Hairston NG, Bohonak A (1998). Copepod reproductive strategies: life-history theory, phylogenetic pattern and invasion of land waters. *Journal of Marine Systems*, 15: 23–34.

Hajek AE, Eilenberg J (2018). *Natural enemies: an introduction to biological control*, 2nd ed. Cambridge University Press, Cambridge.

Hajishengallis G, Reis ES, Mastellos DC, Ricklin D, Lambris JD (2017). Novel mechanisms and functions of complement. *Nature Immunology*, 18: 1288–98.

Haldane JBS (1949). Disease and evolution. *La Ricerca Scientifica*, 19, Suppl.: 68–76.

Haldar K, Kamoun S, Hiller NL, Bhattacharje S, van Ooij C (2006). Common infection strategies of pathogenic eukaryotes. *Nature Reviews Microbiology*, 4: 922–31.

Hall BG (2004). Predicting the evolution of antibiotic resistance genes. *Nature Reviews Microbiology*, 2: 430–5.

Hall CW, Mah T-F (2017). Molecular mechanisms of biofilm-based antibiotic resistance and tolerance in pathogenic bacteria. *FEMS Microbiology Reviews*, 41: 276–301.

Hall MD, Bento G, Ebert D (2017). The evolutionary consequences of stepwise infection processes. *Trends in Ecology & Evolution*, 32: 612–23.

Hall MD, Routtu J, Ebert D (2019). Dissecting the genetic architecture of a stepwise infection process. *Molecular Ecology*, 28: 3942–57.

Hall N, Karras M, Raine JD, et al. (2005). A comprehensive survey of the *Plasmodium* life cycle by genomic, transcriptomic, and proteomic analyses. *Science*, 307: 82–6.

Halliday FW, Heckman RW, Wilfahrt PA, Mitchell CE (2019). Past is prologue: host community assembly and the risk of infectious disease over time. *Ecology Letters*, 22: 138–48.

Halliwell B, Gutteridge JMC (1999). *Free radicals in biology and medicine*. Oxford University Press, Oxford.

Halloran ME, Longini Jr. IM, Struchiner CJ (1996). Estimability and interpretation of vaccine efficacy using frailty mixing models. *Journal of Epidemiology*, 144: 83–97.

Halsey S (2019). Defuse the dilution effect debate. *Nature Ecology & Evolution*, 3: 145–6.

Halvorson HO (1935). The effect of chance on the mortality of experimentally infected animals. *Journal of Bacteriology*, 30: 330.

Hamilton WD (1964). The genetical evolution of social behavior. *Journal of Theoretical Biology*, 7: I: 1–16; II: 17–32.

Hamilton WD (1971). Geometry for the selfish herd. *Journal of Theoretical Biology*, 31: 295–311.

Hamilton WD (1975). Gamblers since life began: barnacles, aphids, elms. *Quarterly Review of Biology*, 50: 175–80.

Hamilton WD (1980). Sex versus non-sex versus parasite. *Oikos*, 35: 282–90.

Hamilton WD (2001). *Narrow roads of gene land. Vol. 2: The evolution of sex*. Oxford University Press, Oxford.

Hamilton WD, Axelrod A, Tanese R (1990). Sexual reproduction as an adaptation to resist parasites (a review). *Proceedings of the National Academy of Sciences USA*, 87: 3566–73.

Hamilton WD, Zuk M (1982). Heritable true fitness and bright birds: a A role for parasites? *Science*, 218: 384–7.

Hamilton-Miller JMT (2004). Antibiotic resistance from two perspectives: man and microbe. *International Journal of Antimicrobial Agents*, 23: 209–12.

Hammami R, Ben Hamida J, Vergoten G, Fliss I (2008). PhytAMP: a database dedicated to antimicrobial plant peptides. *Nucleic Acids Research*, 37: D963–8.

Hamon MA, Quintin J (2016). Innate immune memory in mammals. *Seminars in Immunology*, 28: 351–8.

Han BA, Kramer AM, Drake JM (2016). Global patterns of zoonotic disease in mammals. *Trends in Parasitology*, 32: 565–77.

Han H-J, Wen H-l, Zhou C-M, et al. (2015). Bats as reservoirs of severe emerging infectious diseases. *Virus Research*, 205: 1–6.

Hancock REW, Haney EF, Gill EE (2016). The immunology of host defence peptides: beyond antimicrobial activity. *Nature Reviews Immunology*, 16: 321–34.

Handel A, Brown J, Stallknecht D, Rohani P (2013). A multi-scale analysis of influenza A virus fitness trade-offs due to temperature-dependent virus persistence. *PLoS Computational Biology*, 9: e1002989.

Handel A, Rohani P (2015). Crossing the scale from within-host infection dynamics to between-host transmission fitness: a discussion of current assumptions and knowledge. *Philosophical Transactions of the Royal Society London B*, 370.

Hanington PC, Forys MA, Loker ES (2012). A somatically diversified defense factor, FREP3, is a determinant of snail resistance to schistosome infection. *PLoS Neglected Tropical Diseases*, 6: e1591.

Hankiewicz J, Swierczek E (1974). Lysozyme in human body fluids. *Clinica Chimica Acta*, 57: 205–9.

Hansen E, Woods RJ, Read AF (2017). How to use a chemotherapeutic agent when resistance to it threatens the patient. *PLoS Biology*, 15: e2001110.

Hanski I (1982). Dynamics of regional distribution: the core and satellite hypothesis. *Oikos*, 38: 210–21.

Hanson LA (1999). Human milk and host defence: immediate and long-term effects. *Acta Paediatrica*, 88: 42–6.

Haraguchi Y, Sasaki A (2000). The evolution of parasite virulence and transmission rate in a spatially structured population. *Journal of Theoretical Biology*, 203: 85–96.

Harbison CW, Clayton DH (2011). Community interactions govern host-switching with implications for host–parasite coevolutionary history. *Proceedings of the National Academy of Sciences USA*, 108: 9525–9.

Harden LM, Kent S, Pittman QJ, Roth J (2015). Fever and sickness behavior: friend or foe? *Brain, Behavior, and Immunity*, 50: 322–33.

Hardin G (1968). The tragedy of the commons. *Science*, 162: 1243–8.

Hargreaves BJ, Yoeli M, Nussenzweig RS (1975). Immunological studies in rodent malaria. I: Protective immunity induced in mice by mild strains of *Plasmodium berghei yoelii* against a virulent and fatal line of this *Plasmodium*. *Annals of Tropical Medicine and Parasitology*, 69: 289–99.

Harnos A, Lang Z, Petrás D, Bush SE, Szabó K, Rózsa L (2017). Size matters for lice on birds: coevolutionary allometry of host and parasite body size. *Evolution*, 71: 421–31.

Harris JB, LaRocque RC, Qadri F, Ryan ET, Calderwood SB (2012). Cholera. *The Lancet*, 379: 2466–76.

Hart AR, Cloyd MW (1990). Interference patterns of human immunodeficiency viruses HIV-1 and HIV-2. *Virology*, 177: 1–10.

Hart BL (1988). Biological basis of the behavior of sick animals. *Neuroscience & Biobehavioral Review*, 12: 123–37.

Hart BL, Hart LA (2018). How mammals stay healthy in nature: the evolution of behaviours to avoid parasites and pathogens. *Philosophical Transactions of the Royal Society London B*, 373: 20170205.

Hartenstein V (2006). Blood cells and blood cell development in the animal kingdom. *Annual Review of Cell and Developmental Biology*, 22: 677–712.

Hassell JM, Begon M, Ward MJ, Fèvre EM (2017). Urbanization and disease emergence: dynamics at the wildlife-livestock-human interface. *Trends in Ecology & Evolution*, 32: 55–67.

Hassell MP (2000). *The spatial and temporal dynamics of host-parasitoid interactions*. Oxford University Press, New York.

Hassell MP, Godfray HCJ (1992). The population biology of insect parasitoids. In Crawley MJ (ed.) *Natural enemies: the population biology of predators, parasites and diseases*, pp. 265–92. Blackwell, Oxford.

Hasselquist D (2007). Comparative immunoecology in birds: hypotheses and tests. *Journal of Ornithology*, 148: S571–82.

Hata H, Sogabe A, Tada S, et al. (2017). Molecular phylogeny of obligate fish parasites of the family Cymothoidae (Isopoda, Crustacea): evolution of the attachment mode to host fish and the habitat shift from saline water to freshwater. *Marine Biology*, 164: 105.

Hatcher MJ, Dick JTA, Dunn AM (2012). Disease emergence and invasions. *Functional Ecology*, 26: 1275–87.

Hatcher MJ, Dunn AM (1995). Evolutionary consequences of cytoplasmically inherited feminizing factors. *Philosophical Transactions of the Royal Society London B*, 348: 445–56.

Hatcher MJ, Dunn AM (2011). *Parasites in ecological communities*. Cambridge University Press, Cambridge.

Hawlena H, Bashey F, Lively CM (2010). The evolution of spite: population structure and bacteriocin-mediated antagonism in two natural populations of *Xenorhabdus* bacteria. *Evolution*, 64: 23198–204.

Hawley DM, Osnas EE, Dobson AP, Hochachka WM, Ley DH, Dhondt AA (2013). Parallel patterns of increased virulence in a recently emerged wildlife pathogen. *PLoS Biology*, 11: e1001570.

Hawley DM, Sydenstricker KV, Kollias GV, Dhondt AA (2005). Genetic diversity predicts pathogen resistance and cell-mediated immunocompetence in house finches. *Biology Letters*, 1: 326–9.

Hay SI, George DB, Moyes CL, Brownstein JS (2013). Big data opportunities for global infectious disease surveillance. *PLoS Medicine*, 10: e1001413.

Haydon DT, Cleaveland S, Taylor LH, Laurenson MK (2002). Identifying reservoirs of infection: a conceptual and practical challenge. *Emerging Infectious Diseases*, 8: 1468–73.

Haydon DT, Randall DA, Matthews L, et al. (2006). Low-coverage vaccination strategies for the conservation of endangered species. *Nature*, 443: 692–5.

Hayman DTS (2016). Bats as viral reservoirs. *Annual Review of Virology*, 3: 77–99.

Hayward A, Tsuboi M, Owusu C, et al. (2017). Evolutionary associations between host traits and parasite load: insights from Lake Tanganyika cichlids. *Journal of Evolutionary Biology*, 30: 1056–67.

Hayward AD (2013). Causes and consequences of intra- and inter-host heterogeneity in defence against nematodes. *Parasite Immunology*, 35: 362–73.

Hayward AD, Garnier R, Watt KA, et al. (2014a). Heritable, heterogeneous, and costly resistance of sheep against

nematodes and potential feedbacks to epidemiological dynamics. *The American Naturalist*, 184: S58–76.

Hayward AD, Nussey DH, Wilson AJ, et al. (2014b). Natural selection on individual variation in tolerance of gastrointestinal nematode infection. *PLoS Biology*, 12: e1001917.

He X, Jing Z, Cheng G (2014). MicroRNAs: new regulators of Toll-like receptor signalling pathways. *BioMed Research International*: 945169.

Heard MJ, Smith KF, Ripp KJ, et al. (2013). The threat of disease increases as species move toward extinction. *Conservation Biology*, 27: 1378–88.

Hechinger RF (2013). A metabolic and body-size scaling framework for parasite within-host abundance, biomass, and energy flux. *The American Naturalist*, 182: 234–48.

Hechinger RF, Lafferty KD (2005). Host diversity begets parasite diversity: bird final hosts and trematodes in snail intermediate hosts. *Proceedings of the Royal Society London B*, 272: 1059–66.

Hedges AJ (1966). An examination of single-hit and multi-hit hypotheses in relation to the possible kinetics of colicin adsorption. *Journal of Theoretical Biology*, 11: 383–410.

Hedrick P (1992). Female choice and variation in the major histocompatibility complex. *Genetics*, 132: 575–81.

Heesterbeek JAP (2002). A brief history of R0 and a recipe for its calculation. *Acta Biotheoretica*, 50: 189–204.

Heffernan JM, Smith RJ, Wahl LM (2005). Perspectives on the basic reproductive ratio. *Journal of The Royal Society Interface*, 2: 281–93.

Heikkinen T, Jarvinen A (2003). The common cold. *The Lancet*, 361: 51–9.

Heimesaat MM, Alutis M, Grundmann U, et al. (2014). The role of serine protease HtrA in acute ulcerative enterocolitis and extra-intestinal immune responses during *Campylobacter jejuni* infection of gnotobiotic IL-10 deficient mice. *Frontiers in Cellular and Infection Microbiology*, 4: 77.

Heitman J (2006). Sexual reproduction and the evolution of microbial pathogens. *Current Biology*, 16: R711–25.

Heldt FS, Kupke SY, Dorl S, Reichl U, Frensing T (2015). Single-cell analysis and stochastic modelling unveil large cell-to-cell variability in influenza A virus infection. *Nature Communications*, 6: 8938.

Helgason E, Økstad OA, Caugant DA, et al. (2000). *Bacillus anthracis, Bacillus cereus*, and *Bacillus thuringiensis*: one species on the basis of genetic evidence. *Applied Environmental Microbiology*, 66: 2627–30.

Hellgren O, Pérez-Tris J, Bensch S (2009). A jack-of-all-trades and still a master of some: prevalence and host range in avian malaria and related blood parasites. *Ecology*, 90: 2840–9.

Hellriegel B (2001). Immunoepidemiology: bridging the gap between immunology and epidemiology. *Trends in Parasitology*, 17: 102–6.

Helluy S (2013). Parasite-induced alterations of sensori-motor pathways in gammarids: collateral damage of neuroinflammation? *The Journal of Experimental Biology*, 216: 67–77.

Hemachudha T, Laothamatas J, Rupprecht C (2002). Human rabies: a disease of complex neuropathogenic mechanisms and diagnostic challenges. *The Lancet Neurology*, 1: 1101–9.

Henriquez SA, Brett R, Alexander J, Pratt J, Roberts CW (2009). Neuropsychiatric disease and *Toxoplasma gondii* infection. *Neuroimmunomodulation*, 16: 122–33.

Henry LM, Roitberg BD, Gillespie DR (2008). Host range evolution in *Aphidius* parasitoids: fidelity, virulence and fitness trade-offs on an ancestral host. *Evolution*, 62: 689–99.

Hentschel U, Hacker J (2001). Pathogenicity islands: the tip of the iceberg. *Microbes and Infection*, 3: 545–8.

Herbison R, Evans S, Doherty J-F, Algie M, Kleffmann T, Poulin R (2019). A molecular war: convergent and ontogenetic evidence for adaptive host manipulation in related parasites infecting divergent hosts. *Proceedings of the Royal Society London B*, 286: 20191827.

Hernandez AD, Shukedo MVK (2008). Parasite effects on isopod feeding rates can alter the host's functional role in a natural stream ecosystem. *International Journal of Parasitology*, 38: 683–90.

Hesse E, Best A, Boots M, Hall AR, Buckling A (2015). Spatial heterogeneity lowers rather than increases host–parasite specialization. *Journal of Evolutionary Biology*, 28: 1682–90.

Hesse E, Buckling A (2016). Host population bottlenecks drive parasite extinction during antagonistic coevolution. *Evolution*, 70: 235–40.

Heun M (1987). Combining ability and heterosis for quantitative powdery mildew resistance in barley. *Plant Breeding*, 99: 24–238.

Heuverswyn FV, Li Y, Bailes E, et al. (2007). Genetic diversity and phylogeographic clustering of SIVcpzPtt in wild chimpanzees in Cameroon. *Virology*, 368: 155–71.

Hewitt EW (2003). The MHC class I antigen presentation pathway: strategies for viral immune evasion. *Immunology*, 110: 163–9.

Hibino T, Loza-Coll M, Messier C, et al. (2006). The immune gene repertoire encoded in the purple sea urchin genome. *Developmental Biology*, 300: 349–65.

Hicks O, Burthe SJ, Daunt F, et al. (2018). The energetic cost of parasitism in a wild population. *Proceedings of the Royal Society London B*, 285: 20180489.

Hilker FM, Langlais M, Malchow H (2009). The Allee effect and infectious diseases: extinction, multistability, and the (dis-)appearance of oscillations. *The American Naturalist*, 173: 72–88.

Hill NJ, Runstadler JA (2016). A bird's eye view of influenza A virus transmission: challenges with characterizing both sides of a co-evolutionary dynamic. *Integrative and Comparative Biology*, 56: 304–16.

Hille F, Richter H, Wong SP, Bratovič M, Ressel S, Charpentier E (2018). The biology of CRISPR-Cas: backward and forward. *Cell*, 172: 1239–59.

Hillgarth N (1990). Parasites and female choice in the ring-necked pheasant. *American Zoologist*, 30: 227–33.

Hillyer JnF (2016). Insect immunology and hematopoiesis. *Developmental & Comparative Immunology*, 58: 102–18.

Hirayasu K, Saito F, Suenaga T, et al. (2016). Microbially cleaved immunoglobulins are sensed by the innate immune receptor LILRA2. *Nature Microbiology*, 1: 16054.

Hird SM, Sánchez C, Carstens BC, Brumfield RT (2015). Comparative gut microbiota of 59 neotropical bird species. *Frontiers in Microbiology*, 6: 1403.

Hiscock SJ, Tabah DA (2003). The different mechanisms of sporophytic self-incompatbility. *Philosophical Transactions of the Royal Society B*, 358: 1037–45.

Hmama Z, Pena-Diaz S, Joseph S, Av-Gay Y (2015). Immunoevasion and immunosuppression of the macrophage by *Mycobacterium tuberculosis*. *Immunological Reviews*, 264: 220–32.

Ho Brian T, Dong Tao G, Mekalanos John J (2014). A view to a kill: the bacterial type VI secretion system. *Cell Host & Microbe*, 15: 9–21.

Hoang A (2001). Immune response to parasitism reduces resistance of *Drosophila melanogaster* to desiccation and starvation. *Evolution*, 55: 2353–8.

Hoberg EP, Brooks D (2008). A macroevolutionary mosaic: episodic host-switching, geographical colonization and diversification in complex host-parasite systems. *Journal of Biogeography*, 35: 1533–50.

Hoberg EP, Klassen GJ (2002). Revealing the faunal tapestry: co-evolution and historical biogeography of hosts and parasites in marine systems. *Parasitology*, 124: S3–22.

Hochberg ME, Holt RD (2002). Biogeographical perspectives on arms races. In Dieckmann U, Metz AJ, Sabelis MW, Sigmund K (eds) *Adaptive dynamics of infectious diseases: in pursuit of virulence management*, pp. 197–209. Cambridge University Press, Cambridge.

Hock K, Fefferman NH (2012). Social organization patterns can lower disease risk without associated disease avoidance or immunity. *Ecological Complexity*, 12: 34–42.

Hodgson EE, Otto SP (2012). The red queen coupled with directional selection favours the evolution of sex. *Journal of Evolutionary Biology*, 25: 797–802.

Høeg JT, Deutsch J, Chan BKK, Semmler Le H (2015). 'Crustacea': Cirripedia. In Wanninger A (ed.) *Evolutionary Developmental Biology of Invertebrates 4: Ecdysozoa II: Crustacea*, pp. 153–81. Springer Vienna, Vienna.

Høeg JT, Noever C, Rees DA, Crandall KA, Glenner H (2020). A new molecular phylogeny-based taxonomy of parasitic barnacles (Crustacea: Cirripedia: Rhizocephala). *Zoological Journal of the Linnean Society*, 190: 632–53.

Hoeksema Jason D, Forde Samantha E (2008). A meta-analysis of factors affecting local adaptation between interacting species. *The American Naturalist*, 171: 275–90.

Holmes AH, Moore LSP, Sundsfjord A, et al. (2016). Understanding the mechanisms and drivers of antimicrobial resistance. *The Lancet*, 387: 176–87.

Holmes EC (2010). The RNA virus quasispecies: fact or fiction? *Journal of Molecular Biology*, 400: 271–3.

Holmes EC, Rambaut A (2004). Viral evolution and the emergence of SARS coronavirus. *Philosophical Transactions of the Royal Society London B*, 359: 1059–965.

Holmes JC (1982). Impact of infectious disease agents on the population growth and geographical distribution of animals. In Anderson RM, May RM (eds) *Population biology of infectious diseases*, pp. 37–51. Springer Verlag, Berlin.

Holmes JC, Bethel WM (1972). Modification of intermediate host behavior by parasites. In Canning EU, Wright CA (eds) *Behavioral aspects of parasite transmission*, pp. 123–49. Academic Press, London.

Holmes JC, Hobbs RP, Leong TS (1977). Populations in perspective: community organisation and regulation of parasites populations. In Esch GW (ed.) *Regulation of parasite populations*, pp. 209–45. Academic Press, New York.

Holt III BF, Hubert DA, Dangl JL (2003). Resistance gene signalling in plants: complex similarities to animal innate immunity. *Current Opinion in Immunology*, 15: 20–5.

Holt RD, Lawton JH (1994). The ecological consequences of shared natural enemies. *Annual Review of Ecology and Systematics*, 25: 495–520.

Honda K, Littman DR (2016). The microbiota in adaptive immune homeostasis and disease. *Nature*, 535: 75–84.

Honjo T, Muramatsu M, Fagarasan S (2004). Aid: how does it aid antibody diversity? *Immunity*, 20: 659–68.

Hooper L (2009). Do symbiotic bacteria subvert host immunity? *Nature Reviews Microbiology*, 78: 367–74.

Hopkins SR, Ocampo JM, Wojdak JM, Belden LK (2016). Host community composition and defensive symbionts determine trematode parasite abundance in host communities. *Ecosphere*, 7: e01278.

Hopkins SR, Wojdak JM, Belden LK (2017). Defensive symbionts mediate host–parasite interactions at multiple scales. *Trends in Parasitology*, 33: 53–64.

Horak P, Saks L, Zilmer M, Karu U, Zilmer K (2007). Do dietary antioxidants alleviate the cost of immune activation? An experiment with greenfinches. *The American Naturalist*, 170: 625–35.

Hordiijk PL, de Jong-Brink M, ter Matt A, Penemann AW, Lodder JC, Kits KS (1992). The neuropeptide schistosomin and heamolymph from parasitized snails induce similar changes in excitability in neuroendocrine cells controlling reproduction and growth in a freshwater snail. *Neuroscience Letters*, 136: 139–97.

Horn D (2014). Antigenic variation in African trypanosomes. *Molecular & Biochemical Parasitology*, 195: 123–9.

Hornef MW, Wick MJ, Rhen M, Normark S (2002). Bacterial strategies for overcoming host innate and adaptive immune responses. *Nature Immunology*, 3: 1033–40.

Horton R, Wilming L, Rand V, et al. (2004). Gene map of the extended human MHC. *Nature Reviews Genetics*, 5: 889–99.

Horvath P, Barrangou R (2010). CRISPR/Cas, the immune system of bacteria and archaea. *Science*, 327: 167–70.

Hotchkiss RS, Moldawer LL, Opal SM, Reinhart K, Turnbull IR, Vincent J-L (2016). Sepsis and septic shock. *Nature Reviews Disease Primers*, 2: 16045.

Hotez PJ (2018). Human parasitology and parasitic diseases: heading towards 2050. In Rollinson D, Stothard JR (eds) *Advances in Parasitology*, pp. 29–38. Academic Press, London.

Hotson AG, Schneider DS (2015). *Drosophila melanogaster* natural variation affects growth dynamics of infecting *Listeria monocytogenes*. *G3: Genes Genomes Genetics*, 5: 2593–600.

Houben ENG, Korotkov KV, Bitter W (2014). Take five: type VII secretion systems of Mycobacteria. *Biochimica et Biophysica Acta (BBA): Molecular Cell Research*, 1843: 1707–16.

Houlden A, Hayes KS, Bancroft AJ, et al. (2015). Chronic *Trichuris muris* infection in C57BL/6 mice causes significant changes in host microbiota and metabolome: effects reversed by pathogen clearance. *PLoS ONE*, 10: e0125945.

Housden BE, Muhar M, Gemberling M, et al. (2017). Loss-of-function genetic tools for animal models: cross-species and cross-platform differences. *Nature Reviews Genetics*, 18: 24–40.

Howard RS, Lively CM (1994). Parasitism, mutation accumulation and the maintenance of sex. *Nature*, 367: 554–7.

Howard RS, Lively CM (2004). Good vs complementary genes for parasite resistance and the evolution of mate choice. *BMC Evolutionary Biology*, 4: 48.

Howe L, Castro IC, Schoener ER, Hunter S, Barraclough RK, Alley MR (2012). Malaria parasites (*Plasmodium* spp.) infecting introduced, native and endemic New Zealand birds. *Parasitology Research*, 110: 913–23.

Howell MJ (1985). Gene exchange between hosts and parasites. *International Journal of Parasitology*, 15: 597–600.

Howick VM, Lazzaro BP (2017). The genetic architecture of defence as resistance to and tolerance of bacterial infection in *Drosophila melanogaster*. *Molecular Ecology*, 26: 1533–46.

Hoyer BF, Radbruch A (2017). Protective and pathogenic memory plasma cells. *Immunology Letters*, 189: 10–12.

Hrdlickova R, Toloue M, Tian B (2017). RNA-Seq methods for transcriptome analysis. *WIREs RNA*, 8: e1364.

Hryckowian AJ, Pruss KM, Sonnenburg JL (2017). The emerging metabolic view of *Clostridium difficile* pathogenesis. *Current Opinion in Microbiology*, 35: 42–7.

Hsu E, Pulham N, Rumfelt LL, Flajnik MF (2006). The plasticity of immunoglobulin systems in evolution. *Immunological Reviews*, 210: 8–26.

Hsu J, Santesso N, Mustafa R, et al. (2012). Antivirals for treatment of Influenza. *Annals of Internal Medicine*, 156: 512–24.

Huang C, Liu X, Sun S, et al. (2016). Insights into the transmission of respiratory infectious diseases through empirical human contact networks. *Scientific Reports*, 6: 31484.

Huang S, Bininda-Emonds ORP, Stephens PR, Gittleman JL, Altizer S (2014). Phylogenetically related and ecologically similar carnivores harbour similar parasite assemblages. *Journal of Animal Ecology*, 83: 671–80.

Huang Y-JS, Higgs S, Vanlandingham DL (2019). Emergence and re-emergence of mosquito-borne arboviruses. *Current Opinion in Virology*, 34: 104–9.

Huang ZYX, Yu Y, Van Langevelde F, De Boer WF (2017). Does the dilution effect generally occur in animal diseases? *Parasitology*, 144: 823–6.

Hudson PJ (1986). *Red grouse: the biology and management of a wild gamebird*. The Game Conservancy Trust, Fordingbridge.

Hudson PJ, Dobson AP (1989). Population biology of *Trichostrongylus tenuis*, a parasite of economic importance for the red grouse management. *Parasitology Today*, 5: 283–91.

Hudson PJ, Dobson AP (1995). Macroparasites: observed patterns in naturally fluctuating animal populations. In Grenfell BT, Dobson AP (eds) *Ecology of infectious diseases in natural populations*, pp. 144–76. Cambridge University Press, Cambridge.

Hudson PJ, Dobson AP, Lafferty KD (2006). Is a healthy ecosystem one that is rich in parasites? *Trends in Ecology & Evolution*, 21: 381–5.

Hudson PJ, Dobson AP, Newborn D (1992a). The population dynamics of *Trichostrongylus tenuis* in red grouse. *Journal of Animal Ecology*, 61: 477–86.

Hudson PJ, Dobson AP, Newborn D (1992b). Do parasites make prey vulnerable to predation? Red Grouse and parasites. *Journal of Animal Ecology*, 61: 681–92.

Hudson PJ, Dobson AP, Newborn D (1998). Prevention of population cycles by parasite removal. *Science*, 282: 2256–8.

Hudson PJ, Rizzoli AP, Grenfell BT, Heesterbeek H, Dobson AP (2002). *Ecology of wildlife diseases*. Oxford University Press, Oxford.

Hudson RR, Kreitman RR, Aguadé M (1987). A test of neutral molecular evolution based on nucleotide data. *Genetics*, 116: 153–9.

Huffman MA (2016). Primate self-medication, passive prevention and active treatment: a brief review. *International Journal of Multidisciplinary Studies*, 3: 1–10.

Huffnagle GB, Dickson RP, Lukacs NW (2017). The respiratory tract microbiome and lung inflammation: a two-way street. *Mucosal Immunology*, 10: 299–306.

Hughes AL, Nei M (1988). Pattern of nucleotide substitution at major histocompatibility complex class I loci reveals overdominant selection. *Nature*, 335: 167–70.

Hughes DP (2013). Pathways to understanding the extended phenotype of parasites in their hosts. *Journal of Experimental Biology*, 216: 142–7.

Hughes DP, Andersen SB, Hywel-Jones NL, Himaman W, Billen J, Boomsma JJ (2011). Behavioral mechanisms and morphological symptoms of zombie ants dying from fungal infection. *BMC Ecology*, 11: 13.

Hughes DP, Kathirithamby J, Turilazzi S, Beani L (2004a). Social wasps desert colony and aggregate outside if parasitized: parasite manipulation? *Behavioral Ecology*, 15: 1037–43.

Hughes DP, Libersat F (2018). Neuroparasitology of parasite-insect Associations. *Annual Review of Entomology*, 63: 471–87.

Hughes DP, Pierce NE, Boomsma JJ (2008). Social insect symbionts: evolution in homeostatic fortresses. *Trends in Ecology & Evolution*, 23: 672–7.

Hughes J, Kennedy M, Johnson KP, Palma RL, Page RDM (2007). Multiple cophylogenetic analyses reveal frequent cospeciation between pelecaniform birds and pectinopygus lice. *Systematic Biology*, 56: 232–51.

Hughes RN, Mauriquez PH, Morley S, Craig SF, Bishop JD (2004b). Kin or self-recognition? *Evolution and Development*, 6: 431–7.

Hughes WOH, Eilenberg J, J.J. B (2002). Trade-offs in group living: transmisson and disease resistance in leaf-cutting ants. *Proceedings of the Royal Society London B*, 269: 1811–19.

Hugot JP (1999). Primates and their pinworm parasites: the Cameron hypothesis revisited. *Systematic Biology*, 48: 523–46.

Hugueny B, Guégan J-F (1997). Community nestedness and the proper way to assesss statistical significance by Monte-Carlo tests: some comments on Worthern and Rohde's (1996) paper. *Oikos*, 80: 572–4.

Huigens ME, Stouthamer R (2003). Parthenogenesis associated with *Wolbachia*. In Bourtzis K, Miler TA (eds) *Insect symbionts*, pp. 247–66. CRC Press, Baton Rouge.

Huijben S, Paaijmans KP (2018). Putting evolution in elimination: winning our ongoing battle with evolving malaria mosquitoes and parasites. *Evolutionary Applications*, 11: 415–30.

Hurd H (1998). Parasite manipulation of insect reproduction: who benefits? *Parasitology*, 116 (suppl.): S13–24.

Hurd H (2003). Manipulation of medically important insect vectors by their parasites. *Annual Review of Entomology*, 48: 141–61.

Hurd H, Warr E, Polwart A (2001). A parasite that increases host lifespan. *Proceedings of the Royal Society London B*, 268: 1749–53.

Hurford A, Day T (2013). Immune evasion and the evolution of molecular mimicry in parasites. *Evolution*, 67: 2889–904.

Hurlbert SH (1971). The nonconcept of species diversity: a A critique and alternative parameters. *Ecology*, 52: 577–86.

Hurst LD (1993). The incidence, mechanisms, and evolution of cytoplasmic sex ratio distorters. *Biological Reviews*, 68: 121–94.

Hurst LD, Smith NG (1999). Do esssential genes evolve slowly? *Current Biology*, 9: 747–50.

Hurt AC, Fouchier RAM, Vijaykrishna D (2017). Ecology and evolution of avian influenza viruses. In Tibayrenc M (ed.) *Genetics and evolution of infectious diseases*, 2nd ed., pp. 621–40. Elsevier, London.

Hussain M, Asgari S (2014). MicroRNAs as mediators of insect host–pathogen interactions and immunity. *Journal of Insect Physiology*, 70: 151–8.

Huyse T, Volckaert FAM (2005). Comparing host and parasite phylogenies: *Gyrodactylus* flatworms jumping from goby to goby. *Systematic Biology*, 54: 710–18.

Ilmonen P, Penn D, Damjanovoch K, Morrison L, Ghotbi L, Potts WK (2007). Major Histocompatibility Complex heterozygosity reduces fitness in experimentally infected mice. *Genetics*, 176: 2501–8.

Ilmonen P, Stundner G, Thoss M, Penn D (2009). Females prefer the scent of outbred males: good-genes-as-heterozygosity? *BMC Evolutionary Biology*, 9: 104.

Ilyas B, Tsai CN, Coombes BK (2017). Evolution of *Salmonella*-host cell interactions through a dynamic bacterial genome. *Frontiers in Cellular and Infection Microbiology*, 7: 428.

Imler J-L (2014). Overview of *Drosophila* immunity: a historical perspective. *Developmental and Comparative Immunology*, 42: 3–15.

Inamine H, Ellner SP, Newell PD, Luo Y, Buchon N, Douglas AE (2018). Spatiotemporally heterogeneous population dynamics of gut bacteria inferred from fecal time series data. *mBio*, 9: e01453–17.

Inglis RF, Gardner A, Cornelis P, Buckling A (2009). Spite and virulence in the bacterium *Pseudomonas aeruginosa*. *Proceedings of the National Academy of Sciences USA*, 106: 5703–7.

International Ebola Response Team, Agua-Agum J, Ariyarajah A, et al. (2016). Exposure patterns driving Ebola transmission in West Africa: a A retrospective observational study. *PLoS Medicine*, 13: e1002170.

International Helminth Genomes Consortium (2019). Comparative genomics of the major parasitic worms. *Nature Genetics*, 51: 163–74.

Ivanova E, Carpino N (2016). Negative regulation of TCR signaling by ubiquitination of Zap-70 Lys-217. *Molecular Immunology*, 73: 19–28.

Ivashkiv LB, Donlin LT (2014). Regulation of type I interferon responses. *Nature Reviews Immunology*, 14: 36–49.

Iwasaki A, Medzhitov R (2015). Control of adaptive immunity by the innate immune system. *Nature Immunology*, 16: 343–53.

Jackson AP (2015). Genome evolution in trypanosomatid parasites. *Parasitology*, 142: S40–56.

Jackson AP, Allison HC, Barry JD, Field MC, Hertz-Fowler C, Berriman M (2013). A cell-surface phylome for African trypanosomes. *PLoS Neglected Tropcial Diseases*, 7: e2121.

Jackson AP, Otto TD, Aslett M, et al. (2016). Kinetoplastid phylogenomics reveals the evolutionary innovations associated with the origins of parasitism. *Current Biology*, 26: 161–72.

Jackson MA, Verdi S, Maxan M-E, et al. (2018). Gut microbiota associations with common diseases and prescription medications in a population-based cohort. *Nature Communications*, 9: 2655.

Jackson S, Nielsen DM, Singh ND (2015). Increased exposure to acute thermal stress is associated with a non-linear increase in recombination frequency and an independent linear decrease in fitness in *Drosophila*. *BMC Evolutionary Biology*, 15: 175.

Jacob F (1977). Evolution and tinkering. *Science*, 195: 1161–6.

Jacobs AC, Fair JM, Zuk M (2015). Parasite infection, but not immune response, influences paternity in western bluebirds. *Behavioral Ecology and Sociobiology*, 69: 193–203.

Jacobs SR, Herman CE, Maciver NJ, et al. (2008). Glucose uptake is limiting in T cell activation and requires CD28-mediated Akt-dependent and independent pathways. *Journal of Immunology*, 180: 4476–86.

Jacobsen A, Hendriksen RS, Aaresturp FM, Ussery DW, Friis C (2011). The *Salmonella enterica* pan-genome. *Microbial Ecology*, 62: 487–504.

Jacquin L, Mori Q, Pause M, Steffen M, Medoc V (2014). Non-specific manipulation of gammarid behaviour by *P. minutus* parasite enhances their predation by definitive bird hosts. *PLoS ONE*, 9: e101684.

Jaenike J (1978). An hypothesis to account for the maintenance of sex in populations. *Evolutionary Theory*, 3: 191–4.

Jahnke M, Smith JE, Dubuffet A, Dunn AM (2013). Effects of feminizing microsporidia on the masculinizing function of the androgenic gland in *Gammarus duebeni*. *Journal of Invertebrate Pathology*, 112: 146–51.

James TY, Kauff F, Schoch CL, et al. (2006). Reconstructing the early evolution of Fungi using a six-gene phylogeny. *Nature*, 443: 818–22.

Jamieson AM, Pasman L, Yu S, et al. (2013). Role of tissue protection in lethal respiratory viral-bacterial coinfection. *Science Express*, 25 April 2013: 1–7.

Janeway CA, Travers P, Walport M, Shlomchik M (2001). *Immunobiology: the immune system in health and disease*. Garland Publishing, London.

Janeway CA, Travers P, Walport M, Shlomchik MJ (2005). *Immunobiology: the immune system in health and disease*. 6th ed. Garland Science Publishing, New York.

Janzen DH (1977). Why fruits rot, seeds mold, and meat spoils. *The American Naturalist*, 111: 691–713.

Jarosz AM, Burdon JJ (1991). Host-pathogen interactions in natural populations of *Linum marginale* and *Melampsora lini*. II. Local and regional variation in patterns of resistance and racial structure. *Evolution*, 45: 1618–27.

Javan RR, van Tonder AJ, King JP, Harrold CL, Brueggemann AB (2018). Genome sequencing reveals a large and diverse repertoire of antimicrobial peptides. *Frontiers in Microbiology*, 9: 2012.

Jean K, Burnside WR, Carlson L, Smith K, Guégan J-F (2016). An equilibrium theory signature in the island biogeography of human parasites and pathogens. *Global Ecology and Biogeography*, 25: 107–16.

Jeffs CT, Lewis OT (2013). Effects of climate warming on host–parasitoid interactions. *Ecological Entomology*, 38: 209–18.

Jegede AS (2007). What led to the Nigerian boycott of the polio vaccination campaign? *PLoS Medicine*, 4: e73.

Jenkins T, Delhaye J, Christe P (2015). Testing local adaptation in a natural great tit-malaria system: an experimental approach. *PLoS ONE*, 10: e0141391.

Jenner RA (2010). Higher-level crustacean phylogeny: consensus and conflicting hypotheses. *Arthropod Structure & Development*, 39: 143–53.

Jephcott TG, Sime-Ngando T, Gleason FH, Macarthur DJ (2016). Host–parasite interactions in food webs: diversity, stability, and coevolution. *Food Webs*, 6: 1–8.

Jervis MA (ed.) (2005). *Insects as natural enemies: a practical perspective*. Springer Science, New York.

Jia Q, Guo Y, Wang G, Barnes SJ (2020a). Big data analytics in the fight against major public health incidents (including Covid-19): a conceptual framework. *International Journal of Environmental Research and Public Health*, 17: 6161.

Jia Y, Shen G, Zhang Y, et al. (2020b). Analysis of the mutation dynamics of SARS-CoV-2 reveals the spread history and emergence of RBD mutant with lower ACE2 binding affinity. *bioRxiv*: 2020.04.09.034942.

Jiang F, Doudna JA (2017). CRISPR–Cas9 structures and mechanisms. *Annual Review of Biophysics*, 46: 505–29.

Jiggins FM, Kim K-W (2006). Contrasting evolutionary patterns in *Drosophila* immune receptors. *Molecular Evolution*, 63: 769–80.

Jin Z, Akao N, Ohta N (2008). Prolactin evokes lactational transmission of larvae in mice infected with *Toxocara canis*. *Parasitology International*, 57: 495–8.

Johnson B (2003). OSHA infectious dose white paper. *Applied Biosafety*, 8: 160–5.

Johnson KP, Adams RJ, Page RDM, Clayton DH (2003). When do parasites fail to speciate in response to host speciation? *Systematic Biology*, 52: 37–47.

Johnson KP, Malenke JR, Clayton DH (2009a). Competition promotes the evolution of host generalists in obligate parasites. *Proceedings of the Royal Society London B*, 276: 3921–6.

Johnson KP, Yoshizawa K, Smith VS (2004). Multiple origins of parasitism in lice. *Proceedings of the Royal Society London B*, 271: 1771–6.

Johnson PTJ, Calhoun DM, Riepe T, McDevitt-Galles T, Koprivnikar J (2019). Community disassembly and disease: realistic—but not randomized—biodiversity losses enhance parasite transmission. *Proceedings of the Royal Society London B*, 286: 20190260.

Johnson PTJ, Lund PJ, Hartson RB, Yoshino TP (2009b). Community diversity reduces *Schistosoma mansoni* transmission, host pathology and human infection risk. *Proceedings of the Royal Society London B*, 276: 1657–63.

Johnson PTJ, Ostfeld RS, Keesing F (2015). Frontiers in research on biodiversity and disease. *Ecology Letters*, 18: 1119–33.

Johnson PTJ, Preston DL, Hoverman JT, Richgels KLD (2013). Biodiversity decreases disease through predictable changes in host community competence. *Nature*, 494: 230–3.

Johnson PTJ, Wilber MQ (2017). Biological and statistical processes jointly drive population aggregation: using host–parasite interactions to understand Taylor's power law. *Proceedings of the Royal Society London B*, 284: 20171388.

Johnson PTJ, Wood CL, Joseph MB, Preston DL, Haas SE, Springer YP (2016). Habitat heterogeneity drives the host-diversity-begets-parasite-diversity relationship: evidence from experimental and field studies. *Ecology Letters*, 19: 752–61.

Johnson WE (2019). Origins and evolutionary consequences of ancient endogenous retroviruses. *Nature Reviews Microbiology*, 17: 355–70.

Johnston SE, Bérénos C, Slate J, Pemberton JM (2016). Conserved genetic architecture underlying individual recombination rate covariation in a wild population of Soay Sheep (*Ovis aries*). *Genetics*, 203: 583–98.

Jokela J, Dybdahl MF, Lively CM (2009). The maintenance of sex, clonal dynamics, and host-parasite coevolution in a mixed population of sexual and asexual snails. *The American Naturalist*, 174 Suppl: S43–53.

Jokela J, Taskinen J, Mutikainen P, Kopp K (2005). Virulence of parasites in hosts under environmental stress: experiments with anoxia and starvation. *Oikos*, 108: 156–64.

Joly DO, Messier F (2004). The distribution of *Echinococcus granulosus* in moose: evidence Evidence for parasite-induced vulnerability to predation by wolves? *Oecologia*, 140: 586–90.

Jones BD, Falkow S (1996). Salmonellosis: host immune responses and bacterial virulence determinants. *Annual Review of Immunology*, 14: 533–61.

Jones EO, White A, Boots M (2011). The evolution of host protection by vertciallyvertically transmitted parasites. *Proceedings of the Royal Society London B*, 278: 863–70.

Jones IT (1986). Inheritance of adult-plant resistance to mildew in oats. *Annals of Applied Biology*, 109: 187–92.

Jones JDG, Dangl JL (2006). The plant immune system. *Nature*, 444: 323–9.

Joo H-S, Fu C-I, Otto M (2016). Bacterial strategies of resistance to antimicrobial peptides. *Philosophical Transactions of the Royal Society London B*, 371: 20150292.

Joop G, Rolff J (2004). Plasticity of immune function and condition under the risk of predation and parasitism. *Evolutionary Ecology Research*, 6: 1051–62.

Jorgensen I, Rayamajhi M, Miao EA (2017). Programmed cell death as a defence against infection. *Nature Reviews Immunology*, 17: 151–64.

Jost S, Altfeld M (2013). Control of human viral infections by natural killer cells. *Annual Review of Immunology*, 31: 163–94.

Józefowski S (2016). The danger model: questioning an unconvincing theory. *Immunology and Cell Biology*, 94: 164–8.

Kada S, Lion S (2015). Superinfection and the coevolution of parasite virulence and host recovery. *Journal of Evolutionary Biology*, 28.

Kageyama D, Narita S, Watanebe M (2012). Insect sex determination manipulated by their endosymbionts: incidences, mechanisms and implications. *Insects*, 3: 161–99.

Kahn LP, Knox MR, Gray GD, Lea JM, Walden-Brown SW (2003). Enhancing immunity to nematode parasites in single-bearing Merino ewes through nutrition and genetic selection. *Veterinary Parasitology*, 112: 211–25.

Kaiser P, Slack E, Grant AJ, Hardt W-D, Regoes RR (2013). Lymph node colonization dynamics after oral *Salmonella typhimurium* infection in mice. *PLoS Pathogens*, 9: e1003532.

Kaltenpoth M, Engl T (2014). Defensive microbial symbionts in Hymenoptera. *Functional Ecology*, 28: 315–27.

Kaltz O, Shykoff JA (1998). Local adaptation in host-parasite systems. *Heredity*, 81: 361–70.

Kaltz O, Shykoff JA (2001). Male and female *Silene latifolia* plants differ in per-contact risk of infection by a sexually transmitted disease. *Journal of Ecology*, 89: 99–109.

Kamiya T, O'Dwyer K, Nakagawa S, Poulin R (2014a). What determines species richness of parasitic organisms?

A meta-analysis across animal, plant and fungal hosts. *Biological Reviews*, 89: 123–34.

Kamiya T, O'Dwyer K, Westerdahl H, Senior A, Nakagawa S (2014b). A quantitative review of MHC-based mating preference: the role of diversity and dissimilarity. *Molecular Ecology*, 23: 5151–63.

Kamo M, Sasaki A, Boots M (2007). The role of trade-off shapes in the evolution of parasites in spatial host populations: an approximate analytical approach. *Journal of Theoretical Biology*, 244: 588–96.

Kang X, Dong F, Shi C, et al. (2019). DRAMP 2.0, an updated data repository of antimicrobial peptides. *Scientific Data*, 6: 148.

Kao C-Y, Sheu B-S, Wu J-J (2016). *Helicobacter pylori* infection: an overview of bacterial virulence factors and pathogenesis. *Biomedical Journal*, 39: 14–23.

Kao RR (2002). The role of mathematical modelling in the control of the 2001 FMD epidemic in the UK. *Trends in Microbiology*, 6: 279–86.

Kaper JB, Nataro JP, Mobley HL (2004). Pathogenic *Escherichia coli*. *Nature Reviews Microbiology*, 21: 123–40.

Kappeler PM, Cremer S, Nunn CL (2015). Sociality and health: impacts of sociality on disease susceptibility and transmission in animal and human societies. *Philosophical Transactions of the Royal Society London B*, 370: 20140116.

Karlsson EK, Kwiatkowski DP, Sabeti PC (2014). Natural selection and infectious disease in human populations. *Nature Reviews Genetics*, 15: 379–93.

Karp RD, Rheins LA (1980). Induction of specific humoral immunity to soluble proteins in the American cockroach (*Periplaneta americana*): II. Nature of the secondary response. *Developmental & Comparative Immunology*, 4: 629–39.

Karvonen A, Seppälä O, Valtonen ET (2009). Host immunization shapes interspecific associations in trematode parasites. *Journal of Animal Ecology*, 78: 945–52.

Kathirithamby J, Hrabar M, Delgado JA, et al. (2015). We do not select, nor are we choosy: reproductive biology of Strepsiptera (Insecta). *Biological Journal of the Linnean Society*, 116: 221–38.

Kathirithamby J, Michael SE (2014). A revised key to the living and fossil families of Strepsiptera, with the description of a new family, Cretostylopidae. *Journal of the Kansas Entomological Society*, 87: 385–8.

Kaufer A, Ellis J, Stark D, Barratt J (2017). The evolution of trypanosomatid taxonomy. *Parasites & Vectors*, 10: 287.

Kaufmann J (2018). Unfinished business: evolution of the MHC and the adaptive immune system of jawed vertebrates. *Annual Review of Immunology*, 36: 383–409.

Kaunisto KM, Viitaniemi HM, Leder EH, Suhonen J (2013). Association between host's genetic diversity and parasite burden in damselflies. *Journal of Evolutionary Biology*, 26: 1784–9.

Kaushik M, Knowles SCL, Webster JP (2014). What makes a feline fatal in *Toxoplasma gondii*'s fatal feline attraction? Infected rats choose wild cats. *Integrative and Comparative Biology*, 54: 118–28.

Kavaliers M, Choleris E (2018). The role of social cognition in parasite and pathogen avoidance. *Philosophical Transactions of the Royal Society London B*, 373: 20170206.

Kavaliers M, Choleris E, Agmo A, Pfaff D (2004). Olfactory-mediated parasite recognition and avoidance: linking genes to behavior. *Hormones and Behavior*, 46: 272–83.

Kawecki TJ (2020). Sexual selection reveals a cost of pathogen resistance undetected in life-history assays. *Evolution*, 74: 338–48.

Kawecki TJ, Ebert D (2004). Conceptual issues in local adaptation. *Ecology Letters*, 7: 1225–41.

Kay GL, Millard A, Sergeant MJ, et al. (2015). Differences in the faecal microbiome in *Schistosoma haematobium* infected children vs. uninfected children. *PLoS Neglected Tropical Diseases*, 9: e0003861.

Kazmin D, Nakaya HI, Lee EK, et al. (2017). Systems analysis of protective immune responses to RTS, S malaria vaccination in humans. *Proceedings of the National Academy of Sciences USA*, 114: 2425–30.

Keck J, Gupta R, Christenson LK, Arulanandam BP (2017). MicroRNA mediated regulation of immunity against gram-negative bacteria. *International Reviews of Immunology*, 36: 287–99.

Keehnen NLP, Rolff J, Theopold U, Wheat CW (2017). Insect antimicrobial defences: a brief history, recent findings, biases, and a way forward in evolutionary studies. In Ligoxygakis P (ed.) *Advances in insect physiology*, pp. 1–33. Academic Press, London.

Keele BF, Giorgi EE, Salazar-Gonzalez JF, et al. (2008). Identification and characterization of transmitted and early founder virus envelopes in primary HIV-1 infection. *Proceedings of the National Academy of Sciences USA*, 105: 7552–7.

Keeling PJ, Fast NM (2002). Microsporidia: biology and evolution of highly reduced intracelullarintracellular parasites. *Annual Review of Microbiology*, 556: 93–116.

Keesing F, Belden LK, Daszak P, et al. (2010). Impacts of biodiversity on the emergence and transmission of infectious diseases. *Nature*, 468: 647–52.

Keiser CN, Pinter-Wollman N, Ziemba MJ, Kothamasu KS, Pruitt JN (2018). The primary case is not enough: variation among individuals, groups and social networks modify bacterial transmission dynamics. *Journal of Animal Ecology*, 87: 369–78.

Kekäläinen J, Evans JP (2018). Gamete-mediated mate choice: towards a more inclusive view of sexual selection. *Proceedings of the Royal Society London B*, 285: 20180836.

Kelley DS, Bendich A (1996). Essential nutrients and immunologic functions. *American Journal of Clinical Nutrition*, 63: S994–6.

Kelly A, Hatcher MJ, Dunn AM (2006). Intersexuality in the amphipod *Gammarus duebeni* results from incomplete feminisation by the vertically transmitted parasitic sex ratio distorter *Nosema granulosis*. *Evolutionary Ecology*, 18: 121–32.

Kelly CD, Stoehr AM, Nunn C, Smyth KN, Prokop ZM (2018). Sexual dimorphism in immunity across animals: a meta-analysis. *Ecology Letters*, 21: 1885–94.

Kelly DW, Paterson RA, Townsend CR, Poulin R, Tompkins DM (2009). Parasite spillback: a neglected concept in invasion ecology? *Ecology*, 90: 2047–56.

Kelly TR, Karesh WB, Johnson CK, et al. (2017). One Health proof of concept: bringing a transdisciplinary approach to surveillance for zoonotic viruses at the human-wild animal interface. *Preventive Veterinary Medicine*, 137: 112–18.

Kemp C, Imler J-L (2009). Antiviral immunity in *Drosophila*. *Current Opinion in Immunology*, 21: 3–9.

Kennedy CR (1996). Establishment, survival and site selection of the cestode *Eubothrium crassum* in brown trout *Salmo trutta*. *Parasitology*, 112: 347–55.

Kennedy CR (2009). The ecology of parasites of freshwater fishes: the search for patterns. *Parasitology*, 136: 1653–62.

Kennedy CR, Bush AO, Aho JM (1986). Patterns in helminth communities: why are birds and fish different? *Parasitology*, 93: 205–15.

Kennedy CR, Guégan JF (1994). Regional versus local helminth parasites richness in British freshwater fish: saturated or unsaturated parasite communities? *Parasitology*, 109: 175–85.

Kennedy CR, Moriarty C (2002). Long-term stability in the richness and structure of helminth communities in eels, *Anguilla anguilla*, in Lough Derg, River Shannon, Ireland. *Journal of Helminthology*, 76: 315–22.

Kennedy CR, Shears PC, Shears JA (2001). Long-term dynamics of *Ligula intestinalis* and roach *Rutilus rutilus*: a study of three epizootic cycles over thirty one years. *Parasitology*, 123: 257–69.

Kennedy DA, Kurath G, Brito IL, et al. (2005). Potential drivers of virulence evolution in aquaculture. *Evolutionary Applications*, 9: 344–54.

Kennedy DA, Read AF (2017). Why does drug resistance readily evolve but vaccine resistance does not? *Proceedings of the Royal Society London B*, 284: 20162562.

Kenney AD, Dowdle JA, Bozzacco L, et al. (2017). Human genetic determinants of viral diseases. *Annual Review of Genetics*, 51: 241–63.

Kepler TB, Perelson AS (1993). Cyclic re-entry of germinal center B cells and the efficiency of affinity maturation. *Immunology Today*, 14: 412–15.

Kermack WO, McKendrick AG (1927). A contribution to the mathematical theory of epidemics. *Proceedings of the Royal Society London A*, 115: 700–21.

Kerr B, Neuhauser C, Bohannan BJM, Dean AM (2006). Local migration promotes competitive restraint in a host-pathogen 'tragedy of the commons'. *Nature*, 442: 75–8.

Kerr PJ (2012). Myxomatosis in Australia and Europe: a model for emerging infectious diseases. *Antiviral Research*, 93: 387–415.

Kerr PJ, Cattadori IM, Liu J, et al. (2017a). Next step in the ongoing arms race between myxoma virus and wild rabbits in Australia is a novel disease phenotype. *Proceedings of the National Academy of Sciences USA*, 114: 9397–402.

Kerr PJ, Cattadori IM, Rogers MB, et al. (2017b). Genomic and phenotypic characterization of myxoma virus from Great Britain reveals multiple evolutionary pathways distinct from those in Australia. *PLoS Pathogens*, 2: 1006252.

Kerstes N, Bérénos C, Schmid-Hempel P, Wegner MK (2012). Antagonistic coevolution with a parasite increases host recombination frequency. *BMC Evolutionary Biology*, 12: 18.

Key FM, Teixeira JC, de Filippo C, Andrés AM (2014). Advantageous diversity maintained by balancing selection in humans. *Current Opinion in Genetics & Development*, 29: 45–51.

Khan I, Agashe D, Rolff J (2017a). Early-life inflammation, immune response and ageing. *Proceedings of the Royal Society London B*, 284: 20170125.

Khan I, Prakash A, Agashe D (2017b). Experimental evolution of insect immune memory versus pathogen resistance. *Proceedings of the Royal Society London B*, 284: 20171583.

Khan I, Prakash A, Agashe D (2019). Pathogen susceptibility and fitness costs explain variation in immune priming across natural populations of flour beetles. *Journal of Animal Ecology*, 88: 1332–42.

Kho ZY, Lai SK (2018). The human gut microbiome: a potential controller of wellness and disease. *Frontiers in Microbiology*, 9: 1835.

Khodadadi L, Cheng Q, Radbruch A, Hiepe F (2019). The maintenance of memory plasma cells. *Frontiers in Immunology*, 10: 721.

Kidwell MG (1972). Genetic change of recombination value in *Drosophila melanogaster*. II. Simulated natural selection. *Genetics*, 70: 433–43.

Kiffner C, Stanko M, Morand S, et al. (2014). Variable effects of host characteristics on species richness of flea infracommunities in rodents from three continents. *Parasitology Research*, 113: 2777–88.

Kim H, Grueneberg A, Vazquez AI, Hsu S, de los Campos G (2017). Will big data close the missing heritability gap? *Genetics*, 207: 1135–45.

Kim SY, Fargallo JA, Vergara P, Martínez-Padilla J (2013). Multivariate heredity of melanin-based coloration, body mass and immunity. *Heredity*, 111: 139–46.

Kimbrell DA, Beutler B (2001). The evolution and genetics of innate immunity. *Nature Reviews Genetics*, 2: 256–67.

Kindler E, Thiel V (2014). To sense or not to sense viral RNA: essentials of coronavirus innate immune evasion. *Current Opinion in Microbiology*, 20: 69–75.

King BJ, Monis PT (2005). Critical processes affecting *Cryptosporidium* oocyst survival in the environment. *Parasitology*, 134: 309–23.

King KC, Bonsall MB (2017). The evolutionary and coevolutionary consequences of defensive microbes for host-parasite interactions. *BMC Evolutionary Biology*, 17: 190.

King KC, Brockhurst MA, Vasieva O, et al. (2016). Rapid evolution of microbe-mediated protection against pathogens in a worm host. *The ISME Journal*, 10: 1915–24.

Kingsley DH (2018). Emerging foodborne and agriculture-related viruses. In Thakur S, Kniel KE (eds) *Preharvest food safety*, pp. 205–26. ASM Press, Washington DC.

Kirchberger S, Majdic O, Stöckl J (2007). Modulation of the immune system by Human Rhinoviruses. *International Archives of Allergy and Immunology*, 142: 1–10.

Kirchner JH (2019). The origin, evolution, and epidemiology of HIV-1 and HIV-2. In *Fundamentals of HIV medicine*, pp. 15–20. Oxford University Press, New York.

Kirchner JW, Roy BA (2002). Environmental variation mediates the deleterious effects of *Coleosporium ipomoeae* on *Ipomoea purpurea*. *Ecology*, 87: 675–85.

Kirschner D, Pienaar E, Marino S, Linderman JJ (2017). A review of computational and mathematical modeling contributions to our understanding of *Mycobacterium tuberculosis* within-host infection and treatment. *Current Opinion in Systems Biology*, 3: 170–85.

Kjemtrup S, Nimchuk Z, Dangl JL (2000). Effector proteins of phytopathogenic bacteria: bifunctional signals in virulence and host recognition. *Current Opinion in Microbiology*, 3: 73–8.

Klasing KC (1998). Nutritional modulation of resistance to infectious diseases. *Poultry Science*, 77: 1119–25.

Klein SL (2000). The effects of hormones on sex differences in infection: from genes to behavior. *Neuroscience and Biobehavioral Reviews*, 24: 627–38.

Klein SL (2003). Parasite manipulation of the proximate mechanisms that mediate social behavior in vertebrates. *Physiology & Behavior*, 79: 441–9.

Klein SL (2004). Hormonal and immunological mechanisms mediating sex differences in parasite infection. *Parasite Immunology*, 26: 247–64.

Klein SL, Flanagan KL (2016). Sex differences in immune responses. *Nature Reviews Immunology*, 16: 626–38.

Klingen TR, Reimering S, Loers J, et al. (2018). Sweep Dynamics (SD) plots: computational identification of selective sweeps to monitor the adaptation of influenza A viruses. *Scientific Reports*, 8: 373.

Klitting R, Gould EA, Paupy C, De Lamballerie X (2018). What does the future hold for yellow fever virus?(I). *Genes*, 9: 291.

Kloch A, Babik W, Bajer A, Sinski E, Radwan J (2010). Effects of an MHC-DRB genotype and allele number on the load of gut parasites in the bank vole *Myodes glareolus*. *Molecular Ecology*, 19 (Supplement 1): 255–65.

Kloch A, Wenzel MA, Laetsch DR, et al. (2018). Signatures of balancing selection in toll-like receptor (TLRs) genes: novel insights from a free-living rodent. *Scientific Reports*, 8: 8361.

Knapp S, Hacker J, Jarchau T, Goebel W (1986). Large, unstable inserts in the chromosome affect virulence properties of uropathogenic *Escherichia coli* O6 strain 536. *Journal of Bacteriology*, 168: 22–30.

Knell RJ, Webberley KM (2004). Sexually transmitted diseases of insects: distribution, evolution, ecology and host behaviour. *Biological Reviews*, 79: 557–81.

Knights D, Silverberg MS, Weersma RK, et al. (2014). Complex host genetics influence the microbiome in inflammatory bowel disease. *Genome and Medicine*, 6: 107.

Kobayashi SD, Malachowa N, Deleo FR (2018). Neutrophils and bacterial immune evasion. *Journal of Innate Immunity*, 10: 432–41.

Kobayashi T (2016). Evolving ideas about genetics underlying insect virulence to plant resistance in rice-brown planthopper interactions. *Journal of Insect Physiology*, 84: 32–9.

Koch H, Woodward J, Langat MK, Brown MJF, Stevenson PC (2019). Flagellum removal by a nectar metabolite inhibits infectivity of a bumblebee parasite. *Current Biology*, 29: 3494–500.e5.

Koch RE, Josefson CC, Hill GE (2017). Mitochondrial function, ornamentation, and immunocompetence. *Biological Reviews*, 92: 1459–74.

Koella JC (1993). Ecological correlates of chiasma frequency and recombination index of plants. *Biological Journal of the Linnean Society*, 48: 227–38.

Koella JC, Boete C (2003). A model for the coevolution of immunity and immune evasion in vector-borne diseases with implications for the epidemiology of malaria. *The American Naturalist*, 161: 698–707.

Koelle K, Cobey S, Grenfell B, Pascual M (2006). Epochal evolution shapes the phylodynamics of interpandemic influenza A (H3N2) in humans. *Science*, 314: 1898–903.

Koh LP, Dunn RR, Sodhi NS, Colwell RK, Proctor HC, Smith VS (2004). Species coextinctions and the biodiversity crisis. *Science*, 305: 1632–4.

Kolev M, Le Friec G, Kemper C (2014). Complement: tapping into new sites and effector systems. *Nature Reviews Immunology*, 14: 811–20.

Kolodny-Hirsch DM, Van Beek NAM (1997). Selection of a morphological variant of *Autographa californica* nuclear polyhedrosis virus with increased virulence following serial passage in *Plutella xylostella*. *Journal of Invertebrate Pathology*, 69: 205–11.

Kołodziej-Sobocińska M (2019). Factors affecting the spread of parasites in populations of wild European terrestrial mammals. *Mammal Research*, 64: 301–18.

Kolopack PA, Lavery JV (2017). Informed consent in field trials of gene-drive mosquitoes. *Gates Open Research*, 1: 14.

König C, Schmid-Hempel P (1995). Foraging activity and immunocompetence in workers of the bumble bee, *Bombus terrestris* L. *Proceedings of the Royal Society London B*, 260: 225–7.

Konrad M, Vyleta ML, Theis FJ, et al. (2012). Social transfer of pathogenic fungus promotes active immunisation in ant colonies. *PLoS Biology*, 10: e1001300.

Koonin EV, Dolja VV, Krupovic M (2015). Origins and evolution of viruses of eukaryotes: the ultimate modularity. *Virology*, 479–480: 2–25.

Koonin EV, Makarova KS, Zhang F (2017). Diversity, classification and evolution of CRISPR-Cas systems. *Current Opinion in Microbiology*, 37: 67–78.

Korber B, Fischer WM, Gnanakaran S, et al. (2020). Tracking changes in SARS-CoV-2 spike: evidence that D614G increases infectivity of the COVID-19 virus. *Cell*, 182: 812–27.e19.

Korol AB (1999). Selection for adaptive traits as a factor of recombination evolution: evidence from natural and experimental populations (a review). In Wasser SP (ed.) *Evolutionary theory and processes: modern modern perspectives*, pp. 31–53. Kluwer Academic Publishers, Amsterdam.

Kortet R, Rantala MJ, Hedrick A (2007). Boldness in antipredator behaviour and immune defence in field crickets. *Evolutionary Ecology Research*, 9: 185–97.

Kosiol C, Vinar T, da Fonseca RR, et al. (2008). Patterns of positive selection in six Mammalian genomes. *PLoS Genetics*, 4: e1000144–4.

Koskella B, Lively CM (2009). Evidence for negative frequency-dependent selection during experimental coevolution of a freshwater snail and a sterilizing trematode. *Evolution*, 63: 2213–21.

Kotwal GJ, Moss B (1988). Vaccinia virus encodes a secretory polypeptide structurally related to complement. *Nature*, 335: 176–8.

Kouyos RD, Althaus CL, Bonhoeffer S (2006). Stochastic or deterministic: what is the effective population size of HIV-1? *Trends in Microbiology*, 14: 507–11.

Kouyos RD, Gordon SN, Staprans SI, Silvestri G, Regoes RR (2010). Similar impact of CD8+ T Cell responses on early virus dynamics during siv infections of rhesus macaques and sooty mangabeys. *PLoS Computational Biology*, 6: e1000901.

Kraaijeveld AR, Godfray HCJ (2003). Potential life-history costs of parasitoid avoidance in *Drosophila melanogaster*. *Evolutionary Ecological Research*, 5: 1251–61.

Krasnov BR, Khokhlova IS, Shenbrot GI, Poulin R (2008). How are the host spectra of hematophagous parasites shaped over evolutionary time? Random choice vs selection of a phylogenetic lineage. *Parasitology Research*, 102: 1157–64.

Krasnov BR, Morand S, Mouillot D, Shenbrot GI, Khokhlova IS, Poulin R (2006). Resource predictability and host specificity in fleas: the effect of host body mass. *Parasitology*, 133: 81–8.

Krasnov BR, Mouillot D, Shenbrot GI, Khokhlova IS, Poulin R (2011). Beta specificity: the turnover of host species in space and another way to measure host specificity. *International Journal for Parasitology*, 41: 33–41.

Krasnov BR, Shenbrot GI, Khokhlova IS, Degen AA (2004). Flea species richness and parameters of host body, host geography and host 'milieu'. Journal *of Animal Ecology*, 73: 1121–8.

Krasnov BR, Shenbrot GI, Khokhlova IS, Poulin R (2005). Nested pattern in flea assemblages across the host's geographic range. *Ecography*, 28: 475–84.

Krause PJ, Kavathas PB, Ruddle NH (eds) (2019). *Immunoepidemiology*. Springer Nature Switzerland, Cham.

Krebs BL, Anderson TK, Goldberg TL, et al. (2014). Host group formation decreases exposure to vector-borne disease: a field experiment in a 'hotspot' of West Nile virus transmission. *Proceedings of the Royal Society London B*, 281: 20141586.

Kreisinger J, Bastien G, Hauffe HC, Marchesi J, Perkins SE (2015). Interactions between multiple helminths and the gut microbiota in wild rodents. *Philosophical Transactions of the Royal Society London B*, 370: 20140295.

Krishna Kumar S, Feldman MW, Rehkopf DH, Tuljapurkar S (2016). Limitations of GCTA as a solution to the missing heritability problem. *Proceedings of the National Academy of Sciences USA*, 113: E61.

Krist AC, Lively C (1998). Experimental exposure of juveile snails (*Potamopyrgus antipodarium*) to infection by trematode larvae (*Microphallus* sp.): infectivity, fecundity compensation and growth. *Oecologia*, 116: 575–82.

Kristensen T, Nielsen AI, Jørgensen AI, et al. (2012). The selective advantage of host feminization: a case study of the green crab *Carcinus maenas* and the parasitic barnacle *Sacculina carcini*. *Marine Biology*, 159: 2015–23.

Krueger JM, Opp MR (2016). Sleep and microbes. In Cryan JF, Clarke G (eds) *International Review of Neurobiology*, pp. 207–25. Academic Press, London.

Kruger DJ, Nesse RM (2006). An evolutionary framework for understanding sex differences in human mortality rates. *Human Nature*, 17: 74–97.

Kucharski AJ, Althaus CL (2015). The role of superspreading in Middle East respiratory syndrome coronavirus (MERS-CoV) transmission. *Eurosurveillance*, 20: 21167.

Kühnert D, Stadler T, Vaughan TG, Drummond AJ (2014). Simultaneous reconstruction of evolutionary history and epidemiological dynamics from viral sequences with the birth–death SIR model. *Journal of The Royal Society Interface*, 11: 20131106.

Kumar P, Kizhakkedathu JN, Straus SK (2018). Antimicrobial peptides: diversity, mechanism of action and strategies to improve the activity and biocompatibility in vivo. *Biomolecules*, 8: 4.

Kümmerli R, Ross-Gillespie A (2014). Exploring the sociobiology of pyoverdin-producing *Pseudomonas*: a comment on Zhang and Rainey. *Evolution*, 68: 3337–42.

Kümmerli R, Santorelli LA, Granato E, et al. (2015). Co-evolutionary dynamics between public good producers and cheats in the bacterium *Pseudomonas aeruginosa*. *Journal of Evolutionary Biology*, 28: 2264–74.

Kumpitsch C, Koskinen K, Schöpf V, Moissl-Eichinger C (2019). The microbiome of the upper respiratory tract in health and disease. *BMC Biology*, 17: 87.

Kuris AM (2012). The global burden of human parasites: who and where are they? how are they transmitted? *The Journal of Parasitology*, 98: 1056–64.

Kuris AM, Goddard JHR, Torchin ME, Muprhy N, Gruneyx R, Laffferty KD (2007). An experimental evaluation of host specificity: the role of encounter and compatibility filters for a rhizocephalan parasite of crabs. *International Journal of Parasitology*, 37: 539–45.

Kuris AM, Hechinger RF, Shaw JC, et al. (2008). Ecosystem energetic implications of parasite and free-living biomass in three estuaries. *Nature*, 454: 515–18.

Kurpiers LA, Schulte-Herbrüggen B, Ejotre I, Reeder DM (2016). Bushmeat and emerging infectious diseases: lessons from Africa. In Angelici FM (ed.) *Problematic wildlife: a cross-disciplinary approach*, pp. 507–51. Springer International Publishing, Cham.

Kurtz CC, Carey H (2007). Seasonal changes in the intestinal immune system of hibernating ground squirrels. *Developmental and Comparative Immunology*, 31: 415–28.

Kurtz J, Kalbe M, Aeschlimann PB, et al. (2004). Major histocompatibility complex diversity influences parasite resistance and innate immunity in sticklebacks. *Proceedings of the Royal Society of London Series B*, 271: 197–204.

Kurtz J, Reinhold K, Engqvist L (2000). Immunosuppression under stress: necessary for condition-dependant signalling? *Trends in Ecology & Evolution*, 15: 418–19.

Kuss SK, Best GT, Etheredge CA, et al. (2011). Intestinal microbiota promote enteric virus replication and systemic pathogenesis. *Science*, 334: 249–52.

Kussell E, Kishony R, Balaban NQ, Leibler S (2005). Bacterial persistence: a model of survival in changing environments. *Genetics*, 169: 1807–14.

Kwiatkowski DP (2005). How malaria has affected the human genome and what human genetics can teach us about malaria. *The American Journal of Human Genetics*, 77: 171–92.

Kwong WK, Engel P, Koch H, Moran NA (2014). Genomics and host specialization of honey bee and bumble bee gut symbionts. *Proceedings of the National Academy of Sciences USA*, 111: 11509–14.

Kwong WK, Medina LA, Koch H, et al. (2017). Dynamic microbiome evolution in social bees. *Science Advances*, 3: e1600513.

Kyrou K, Hammond AM, Galizi R, et al. (2018). A CRISPR–Cas9 gene drive targeting doublesex causes complete population suppression in caged *Anopheles gambiae* mosquitoes. *Nature Biotechnology*, 36: 1062–6.

La Marca A, Capuzzo M, Paglia T, Roli L, Trenti T, Nelson SM (2020). Testing for SARS-CoV-2 (COVID-19): a systematic review and clinical guide to molecular and serological in-vitro diagnostic assays. *Reproductive BioMedicine Online*, 41: 483–99.

La Marca E, Lips KR, Lötters S, et al. (2005). Catastrophic population declines and extinctions in neotropical harlequin frogs (Bufonidae: *Atelopus*). *Biotropica*, 37: 190–201.

Lacroix R, Mukabana WR, Gouagna LC, Koella JC (2005). Malaria infection increases attractiveness of humans to mosquitoes. *PLoS Biology*, 3: 1590–3.

Lafferty KD (2009). The ecology of climate change and infectious diseases. *Ecology*, 90: 888–900.

Lafferty KD, Allesina S, Arim M, et al. (2008). Parasites in food webs: the ultimate missing links. *Ecology Letters*, 11: 533–46.

Lafferty KD, Dobson AP, Kuris AM (2006). Parasites dominate food web links. *Proceedings of the National Academy of Sciences USA*, 103: 11211–16.

Lafferty KD, Kuris AM (2009). Parasites reduce food web robustness because they are sensitive to secondary extinction as illustrated by an invasive estuarine snail. *Philosophical Transactions of the Royal Society London B*, 364: 1659–63.

Lafferty KD, Mordecai EA (2016). The rise and fall of infectious disease in a warmer world. *F1000Research*, 5: F1000 Faculty Rev-2040.

Lafferty KD, Shaw JC (2013). Comparing mechanisms of host manipulation across host and parasite taxa. *The Journal of Experimental Biology*, 216: 56–66.

Lafferty KD, Thomas F, Poulin R (2000). Evolution of host phenotype-manipulation by parasites and its consequences. In Poulin R, Morand S, Skorping A (eds) *Evolutionary biology of host-parasite relationships: theory meets reality*, pp. 117–27. Elsevier, Amsterdam.

Laine A-L (2009). Role of coevolution in generating biological diversity: spatially divergent selection trajectories. *Journal of Experimental Biology*, 60: 2957–70.

Laine A-L, Burdon JJ, Nemri A, Thrall PH (2014). Host ecotype generates evolutionary and epidemiological divergence across a pathogen metapopulation. *Proceedings of the Royal Society London B*, 281: 20140522.

Lajeunesse MJ, Forbes MR (2001). Host range and local parasite adaptation. *Proceedings of the Royal Society London B*, 269: 703–10.

Lam TT-Y, Zhou B, Wang J, et al. (2015). Dissemination, divergence and establishment of H7N9 influenza viruses in China. *Nature*, 522: 102–5.

Lambeth JD (2007). Nox enzymes, ROS, and chronic disease: an example of antagonistic pleiotropy. *Free Radical Biology and Medicine*, 43: 332–47.

Lambrechts L, Scott TW (2009). Mode of transmission and the evolution of arboviurs virulence in mosquito vectors. *Proceedings of the Royal Society London B*, 276: 1369–78.

Lamiable O, Kellenberger C, Kemp C, et al. (2016). Cytokine Diedel and a viral homologue suppress the IMD pathway in *Drosophila*. *Proceedings of the National Academy of Sciences USA*, 113: 698–703.

Landry C, Garant D, Duchesne P, Bernatchez L (2001). 'Good genes as heterozygosity': the major histocompatibility complex and mate choice in Atlantic salmon (Salmo salar). *Proceedings of the Royal Society of London Series B-Biological Sciences*, 268: 1279–85.

Langridge GC, Fookes M, Connor TR, et al. (2015). Patterns of genome evolution that have accompanied host adaptation in *Salmonella*. *Proceedings of the National Academy of Sciences USA*, 112: 863.

Lanyon CV, Rushton SP, O'Donnell AG, et al. (2007). Murine scent mark microbial communities are genetically determined. *FEMS Microbiology Ecology*, 59: 576–83.

Lanz-Mendoza H, Garduño JC (2018). Insect innate immune memory. In Cooper EL (ed.) *Advances in comparative immunology*, pp. 193–211. Springer International Publishing, Cham.

Lau Clinton KY, Turner L, Jespersen Jakob S, et al. (2015). Structural conservation despite huge sequence diversity allows epcr binding by the Pfemp1 family implicated in severe childhood malaria. *Cell Host & Microbe*, 17: 118–29.

Lau MSY, Dalziel BD, Funk S, et al. (2017). Spatial and temporal dynamics of superspreading events in the 2014–2015 West Africa Ebola epidemic. *Proceedings of the National Academy of Sciences USA*, 114: 2337–42.

Lauring AS (2020). Within-host viral diversity: a window into viral evolution. *Annual Review of Virology*, 7: 63–81.

Lawniczak MKN, Barnes AI, Linklater JR, Boone JM, Wigby S, Chapman T (2006). Mating and immunity in invertebrates. *Trends in Ecology & Evolution*, 22: 48–55.

Lawniczak MKN, Begun DJ (2004). A genome-wide analysis of courting and mating responses in *Drosophila melanogaster* females. *Genome*, 47: 900–10.

Lazzaro BP, Schneider DS (2014). The genetics of immunity. *Genetics*, 197: 467–70.

Le C-F, Fang C-M, Sekaran SD (2017). Intracellular targeting mechanisms by antimicrobial peptides. *Antimicrobial Agents and Chemotherapy*, 61: e02340–16.

Leary CJ, Knapp R (2014). The stress of elaborate male traits: integrating glucocorticoids with androgen-based models of sexual selection. *Animal Behaviour*, 89: 85–92.

Lee C, Chen Y-PP, Yao T-J, et al. (2013a). GI-POP: a combinational annotation and genomic island prediction pipeline for ongoing microbial genome projects. *Gene*, 518: 114–23.

Lee GM, McGee PA, Oldroyd BP (2013b). Variable virulence among isolates of *Ascosphaera apis*: testing the parasite–pathogen hypothesis for the evolution of polyandry in social insects. *Naturwissenschaften*, 100: 229–34.

Lee J, Choi BY, Jung E (2018). Metapopulation model using commuting flow for national spread of the 2009 H1N1 influenza virus in the Republic of Korea. *Journal of Theoretical Biology*, 454: 320–9.

Lee JH, Park H, Park YH (2014a). Molecular mechanisms of host cytoskeletal rearrangements by *Shigella* invasins. *International Journal of Molecular Sciences*, 15: 18253–66.

Lee KA (2006). Linking immune defenses and life history at the levels of the individual and the species. *Integrative and Comparative Biology*, 46: 1000–15.

Lee KA, Wikelski M, Robinson WD, Robinson TR, Klasing KC (2008a). Constitutive immune defences correlate with life-history variables in tropical birds. *Journal of Animal Ecology*, 77: 356–63.

Lee KP, Simpson SJ, Wilson K (2008b). Dietary protein-quality influences melanization and immune function in an insect. *Functional Ecology*, 22: 1052–61.

Lee PH, O'Dushlaine C, Thomas B, Purcell SM (2012). INRICH: interval-based enrichment analysis for genome-wide association studies. *Bioinformatics*, 28: 1797–9.

Lee SC, Tang MS, Lim YAL, et al. (2014b). Helminth colonization is associated with increased diversity of the gut microbiota. *PLoS Neglected Tropical Diseases*, 8: e2880.

Lee T-H, Hall KN, Aguilar M-I (2016). Antimicrobial peptide structure and mechanism of action: a focus on the role of membrane structure. *Current Topics in Medicinal Chemistry*, 16: 25–39.

Lee YK, Mazmanian SK (2010). Has the microbiota played a critical role in the evolution of the adaptive immune system? *Science*, 330: 1768–73.

Leendertz SAJ, Gogarten JF, Düx A, Calvignac-Spencer S, Leendertz FH (2016). Assessing the evidence supporting fruit bats as the primary reservoirs for ebola viruses. *EcoHealth*, 13: 18–25.

Lefcort H, Blaustein AR (1995). Disease, predator avoidance, and vulnerability to predation in tadpoles. *Oikos*, 74: 469–74.

Lefèvre T, Lebarbenchon C, Gauthier-Clerc M, Missé D, Poulin R, Thomas F (2009). The ecological significance of manipulative parasites. *Trends in Ecology & Evolution*, 24: 41–8.

Lefèvre T, Thomas F (2008). Behind the scene, something else is pulling the strings: emphasizing parasitic manipulation in vector-borne diseases. *Infection, Genetics and Evolution*, 8: 504–19.

Leffler EM, Gao Z, Pfeifer S, et al. (2013). Multiple instances of ancient balancing selection shared between humans and chimpanzees. *Science*, 339: 1578–82.

Leggett HC, Benmayor R, Hodgson DJ, Buckling A (2013a). Experimental evolution of adaptive phenotypic plasticity in a parasite. *Current Biology*, 23: 139–42.

Leggett HC, Brown SP, Reece SE (2014). War and peace: social interactions in infections. *Philosophical Transactions of the Royal Society London B*, 369: 20130365.

Leggett HC, Buckling A, Long GH, Boots M (2013b). Generalism and the evolution of parasite virulence. *Trends in Ecology & Evolution*, 28: 592–6.

Leggett HC, Cornwallis CK, West SA (2012). Mechanisms of pathogenesis, infective dose and virulence in human parasites. *PLoS Pathogens*, 8: e1002512.

Leggewie M, Schnettler E (2018). RNAi-mediated antiviral immunity in insects and their possible application. *Current Opinion in Virology* 32: 108–14.

Legros M, Bonhoeffer S (2016). A combined within-host and between-hosts modelling framework for the evolution of resistance to antimalarial drugs. *Journal of the Royal Society Interface*, 13: 20160148.

Lehtinen S, Blanquart F, Croucher NJ, Turner PE, Lipsitch M, Fraser C (2017). Evolution of antibiotic resistance is linked to any genetic mechanism affecting bacterial duration of carriage. *Proceedings of the National Academy of Sciences USA*, 114: 1075–80.

Leinders-Zufall T, Brennan P, Widmayer P, et al. (2004). MHC class I peptides as chemosensory signals in the vomeronasal organ. *Science*, 306: 1033–7.

Leitão AB, Sucena É (2015). *Drosophila* sessile hemocyte clusters are true hematopoietic tissues that regulate larval blood cell differentiation. *eLife*: e01611.

Lello J, Boag B, Fenton A, Stevenson IR, Hudson PJ (2004). Competition and mutualism among the gut helminths of a mammalian host. *Nature*, 428: 840–4.

Lemaitre B, Hoffmann JA (2006). The host defense of *Drosophila melanogaster*. *Annual Review of Immunology*, 25: 697–743.

Lemke H, Hansen H, Lange H (2003). Non-genetic inheritable potential of maternal antibodies. *Vaccine* 21: 3428–31.

Lenski RE, May RM (1994). The evolution of virulence in parasites and pathogens: reconciliation between two competing hypotheses. *Journal of Theoretical Biology*, 169: 253–65.

Leulier F, MacNeil LT, Lee W-j, et al. (2017). Integrative physiology: at the crossroads of nutrition, microbiota, animal physiology, and human health. *Cell Metabolism*, 25: 522–34.

Leventhal GE, Günthard HF, Bonhoeffer S, Stadler T (2013). Using an epidemiological model for phylogenetic inference reveals density dependence in HIV transmission. *Molecular Biology and Evolution*, 31: 6–17.

Levin BR (1988). Frequency-dependent selection in bacterial populations. *Philosophical Transactions of the Royal Society London B*, 319: 459–72.

Levin BR, Bull JJ (1994). Short-sighted evolution and the virulence of pathogenic microbes. *Trends in Microbiology*, 2: 76–81.

Levin BR, Perrot V, Walker N (2000). Compensatory mutations, antibiotic resistance and the population genetics of adaptive evolution in bacteria. *Genetics*, 154: 985–97.

Levin BR, Rozen DE (2006). Non-inherited antibiotic resistance. *Nature Reviews Microbiology*, 4: 556–62.

Levin BR, Svanborg Eden C (1990). Selection and evolution of virulence in bacteria: an ecumenical excursion and a modest suggestion. *Parasitology*, 100: S103–15.

Levin DA (1973). The role of trichomes in plant defence. *Quartely Review of Biology*, 48: 3–15.

Levin S, Sela N, Chejanovsky N (2016). Two novel viruses associated with the *Apis mellifera* pathogenic mite *Varroa destructor*. *Scientific Reports*, 6: 37710.

Levy M, Kolodziejczyk AA, Thaiss CA, Elinav E (2017). Dysbiosis and the immune system. *Nature Reviews Immunology*, 17: 219–32.

Levy SB, Marshall B (2004). Antibacterial resistance worldwide: causes, challenges and responses. *Nature Medicine*, 10 (suppl.): S122–9.

Lewontin RC (1974). *The genetic basis of evolutionary change*. Columbia University Press, New York and London.

Ley DH, Hawley DM, Geary SJ, Dhondt AA (2016). House Finch (*Haemorhous mexicanus*) conjunctivitis, and *Mycoplasma* spp. isolated from North American wild birds, 1994–2015. *Journal of Wildlife Diseases*, 52: 669–73.

Li H, Li WX, Ding SW (2002). Induction and suppression of RNA silencing by an animal virus. *Science*, 296: 1319–21.

Li H, Stephan W (2006). Inferring the demographic history and rate of adaptive substitution in *Drosophila*. *PLoS Genetics*, 2: e166.

Li KS, Guan Y, Wang J, et al. (2004a). Genesis of a highly pathogenic and potentially pandemic H5N1 influenza virus in eastern Asia. *Nature*, 430: 209–13.

Li W-X, Li H, Lu R, et al. (2004b). Interferon antagonist proteins of influenza and vaccinia viruses are suppressors of RNA silencing. *Proceedings of the National Academy of Sciences USA*, 101: 1350–5.

Li X-X, Zhou XN (2013). Co-infection of tuberculosis and parasitic diseases in humans: a systematic review. *Parasites & Vectors* 6: 79.

Libersat F, Kaiser M, Emanuel S (2018). Mind control: how parasites manipulate cognitive functions in their insect hosts. *Frontiers in Psychology*, 9: 572.

Lievens EJP, Perreau J, Agnew P, Michalakis Y, Lenormand T (2018). Decomposing parasite fitness reveals the basis of specialization in a two-host, two-parasite system. *Evolution Letters*, 2–4: 390–405.

Light JE, Hafner MS (2007). Cophylogeny and disparate rates of evolution in sympatric lineages of chewing lice on pocket gophers. *Molecular Phylogenetics and Evolution*, 45: 997–1013.

Light JE, Hafner MS (2008). Codivergence in Heteromyid Rodents (Rodentia: Heteromyidae) and their sucking lice of the genus *Fahrenholzia* (Phthiraptera: Anoplura). *Systematic Biology*, 57: 449–65.

Lighten J, Papadopulos AST, Mohammed RS, et al. (2017). Evolutionary genetics of immunological supertypes reveals two faces of the Red Queen. *Nature Communications*, 8: 1294.

Ligon JD, Zwartjes PW (1995). Ornate plumage of male red junglefowl does not influence mate choice in females. *Animal Behaviour*, 49: 117–25.

Ligoxygakis P (2017). Immunity: insect immune memory goes viral. *Current Biology*, 27: R1218–20.

Lim JJ, Grinstein S, Roth Z (2017). Diversity and versatility of phagocytosis: roles in innate immunity, tissue remodeling, and homeostasis. *Frontiers in Cellular and Infection Microbiology* 7: 191.

Lin L, Hou Z (2020). Combat COVID-19 with artificial intelligence and big data. *Journal of Travel Medicine*, 27: taaa080.

Lind O, Mitkus M, Olsson P, Kelber A (2013). Ultraviolet sensitivity and colour vision in raptor foraging. *The Journal of Experimental Biology*, 216: 1819–26.

Lindsay SA, Wasserman SA (2014). Conventional and non-conventional *Drosophila* Toll signaling. *Developmental & Comparative Immunology*, 42: 16–24.

Lion S, Boots M (2010). Are parasites 'prudent' in space? *Ecology Letters*, 13: 1245–55.

Lion S, Metz JAJ (2018). Beyond Ro maximisation: on pathogen evolution and environmental dimensions. *Trends in Ecology & Evolution*, 33: 458–73.

Lion S, Van Baalen M, Wilson WG (2006). The evolution of parasite manipulation of host dispersal. *Proceedings of the Royal Society London B*, 273: 1063–71.

Lippens Cd, Guivier E, Faivre B, Sorci G (2016). Reaction norms of host immunity, host fitness and parasite performance in a mouse: intestinal nematode interaction. *International Journal for Parasitology*, 46: 133–40.

Lips KR (2016). Overview of chytrid emergence and impacts on amphibians. *Philosophical Transactions of the Royal Society London B*, 371: 20150465.

Lipsitch M, Cohen T, Cooper B, et al. (2003). Transmission dynamics and control of Severe Acute Respiratory Syndrome. *Science*, 300: 1966–70.

Lipsitch M, Dykes JK, Johnson SE, et al. (2000). Competition among *Streptococcus pneumoniae* for intranasal colonization in a mouse model. *Vaccine*, 18: 2895–901.

Lipsitch M, Herre EA, Nowak MA (1995). Host population structure and the evolution of virulence: a law of diminishing returns. *Evolution*, 49: 743–8.

Lipsitch M, Moxon ER (1997). Virulence and transmissibility of pathogens: what is the relationship? *Trends in Microbiology*, 5: 31–7.

Lipsitch M, Siller S, Nowak MA (1996). The evolution of virulence in pathogens with vertical and horizontal transmission. *Evolution*, 50: 1729–41.

Liston A, Carr EJ, Linterman MA (2016). Shaping variation in the human immune system. *Trends in Immunology*, 37: 637–46.

Little TJ (2002). The evolutionary significance of parasitism: do parasite-driven genetic dynamics occur *ex silico*? *Journal of Evolutionary Biology*, 15: 1–9.

Little TJ, Ebert D (1999). Associations between parasitism and host genotype in natural populations of *Daphnia* (Crustacea: Cladocera). *Journal of Animal Ecology*, 68: 134–49.

Little TJ, Kraaijeveld AR (2004). Ecological and evolutionary implications of immunological priming in invertebrates. *Trends in Ecology & Evolution*, 19: 58–60.

Little TJ, Shuker DM, Colegrave N, Day T, Graham AL (2010). The coevolution of virulence: tolerance in perspective. *PLoS Pathogens*, 6: e1001006.

Littlefair JE, Laughton AM, Knell RJ (2017). Maternal pathogen exposure causes diet- and pathogen-specific transgenerational costs. *Oikos*, 126: 82–90.

Littman RJ (2009). The plague of Athens: epidemiology and paleopathology. *Mount Sinai Journal of Medicine: A Journal of Translational and Personalized Medicine*, 76: 456–67.

Liu CH, Liu HY, Ge BX (2017). Innate immunity in tuberculosis: host defense vs pathogen evasion. *Cellular & Molecular Immunology*, 14: 963–75.

Liu F, Wu X, Li L, Zou Y, Liu S, Wang Z (2016a). Evolutionary characteristics of morbilliviruses during serial passages in vitro: gradual attenuation of virus virulence. *Comparative Immunology, Microbiology and Infectious Diseases*, 47: 7–18.

Liu J, Qian C, Cao X (2016b). Post-translational modification control of innate immunity. *Immunity*, 45: 15–30.

Liu Y, Gayle AA, Wilder-Smith A, Rocklöv J (2020). The reproductive number of COVID-19 is higher compared to SARS coronavirus. *Journal of Travel Medicine*, 27: 1–4.

Lively CM (1987). Evidence from a New Zealand snail for the maintenance of sex by parasitism. *Nature*, 328: 519–21.

Lively CM (1999). Migration, virulence, and the geographic mosaic of adaptation by parasites. *The American Naturalist*, 153: S34–47.

Lively CM (2006). The ecology of virulence. *Ecology Letters*, 9: 1089–95.

Lively CM, Dybdahl MF (2000). Parasite adaptation to locally common host genotypes. *Nature*, 405: 679–81.

Livermore DM, Pearson A (2007). Antibiotic resistance: location, location, location. *Clinical Microbiology and Infectious Diseases*, 13 (Suppl. 2): 7–16.

Ljubic B, Gligorijevic D, Gligorijevic J, Pavlovski M, Obradovic Z (2019). Social network analysis for better understanding of influenza. *Journal of Biomedical Informatics*, 93: 103161.

Llaurens V, Whibley A, Joron M (2017). Genetic architecture and balancing selection: the life and death of differentiated variants. *Molecular Ecology*, 26: 2430–48.

Lobet E, Letesson J-J, Arnould T (2015). Mitochondria: a target for bacteria. *Biochemical Pharmacology*, 94: 173–85.

Lochmiller RL, Deerenberg C (2000). Trade-offs in evolutionary immunology: just what is the cost of immunity? *Oikos*, 88: 87–98.

LoGiudice K, Duerr STK, Newhouse MJ, Schmidt KA, Killilea ME, Ostfeld RS (2008). Impact of host community composition on Lyme disease risk. *Ecology*, 89: 2841–9.

LoGiudice K, Ostfeldt RS, Schmidt KA, Keesing F (2003). The ecology of infectious disease: effects of host diversity and community composition on Lyme disease risk. *Proceedings of the National Academy of Sciences USA*, 100: 567–71.

Lohm J, Grahn M, Langefors A, Andersen O, Storset A, Von Schantz T (2002). Experimental evidence for major histocompatibility complex-allele-specific resistance to a bacterial infection. *Proceedings of the Royal Society London B*, 269: 2029–33.

Longdon B, Brockhurst MA, Russell CA, Welch JJ, Jiggins FM (2014). The evolution and genetics of virus host shifts. *PLoS Pathogens*, 10: e1004395.

Longdon B, Hadfield J, Webster C, Obbard D, Jiggins F (2011). Host phylogeny determines viral persistence and replication in novel hosts. *PLoS Pathogens*, 7: e1002260.

Loos-Frank B, Grencis RK (2017). Parasitic worms. In Lucius R, Loos-Frank B, Lane RP, Poulin R, Roberts C, Grencis RK (eds) *The biology of parasites*, pp. 228–338. Wiley, Berlin.

Loos-Frank B, Zimmermann G (1976). Über eine dem *Dicrocoelium*-Befall analoge Verhaltensänderung bei Ameisen der Gattung *Formica* durch einen Pilz der Gattung *Entomophthora*. *Zeitschrift für Parasitenkunde*, 49: 281–9.

Loots AK, Mitchell E, Dalton DL, Kotzé A, Venter EH (2017). Advances in canine distemper virus pathogenesis research: a wildlife perspective. *Journal of General Virology*, 98: 311–21.

Lough G, Kyriazakis I, Bergmann S, Lengeling A, Doeschl-Wilson AB (2015). Health trajectories reveal the dynamic contributions of host genetic resistance and tolerance to infection outcome. *Proceedings of the Royal Society London B*, 282: 20152151.

Louie A, Song KH, Hotson A, Thomas Tate A, Schneider DS (2016). How many parameters does it take to describe disease tolerance? *PLoS Biology*, 14: e1002435.

Lounibos LP, Kramer LD (2016). Invasiveness of *Aedes aegypti* and *Aedes albopictus* and vectorial capacity for Chikungunya virus. *The Journal of Infectious Diseases*, 214: S453–8.

Louten J (2016). Influenza viruses. In Louten J (ed.) *Essential human virology*, pp. 171–91. Academic Press, Boston.

Lozano GA, Ydenberg RC (2002). Transgenerational effects of maternal immune challenge in tree swallows (*Tachycineta bicolor*). *Canadian Journal of Zoology*, 80: 918–25.

Lu B, Leong HW (2016). Computational methods for predicting genomic islands in microbial genomes. *Computational and Structural Biotechnology Journal*, 14: 200–6.

Luca F, Kupfer SS, Knights D, Khoruts A, Blekhman R (2018). Functional genomics of host–microbiome interactions in humans. *Trends in Genetics*, 34: 30–40.

Luikart G, Pilgrim K, Visty J, Ezenwa VO, Schwartz MK (2008). Candidate gene microsatellite variation is associated with parasitism in wild bighorn sheep. *Biology Letters*, 4: 228–31.

Luk GSM, Leung CYH, Sia SF, et al. (2015). Transmission of H7N9 influenza viruses with a polymorphism at PB2 residue 627 in chickens and ferrets. *Journal of Virology*, 89: 9939–51.

Lukens S, DePasse J, Rosenfeld R, et al. (2014). A large-scale immuno-epidemiological simulation of influenza A epidemics. *BMC Public Health*, 14: 1019.

Lukeš J, Butenko A, Hashimi H, Maslov DA, Votýpka J, Yurchenko V (2018). Trypanosomatids are much more than just trypanosomes: clues from the expanded family tree. *Trends in Parasitology*, 34: 466–80.

Lukeš J, Skalický T, Týc J, Votýpka J, Yurchenko V (2014). Evolution of parasitism in kinetoplastid flagellates. *Molecular and Biochemical Parasitology*, 195: 115–22.

Łuksza M, Lässig M (2014). A predictive fitness model for influenza. *Nature*, 507: 57–61.

Lun CM, Samuel RL, Gillmor SD, Boyd A, Smith LC (2017). The recombinant sea urchin immune effector protein, rSpTransformer-E1, binds to phosphatidic acid and deforms membranes. *Frontiers in Immunology*, 8: 481.

Lundberg H, Svensson BG (1975). Studies on the behaviour of *Bombus* Latr. species (Hymenoptera: Apidae) parasitized by *Sphaerularia bombi* Dufour (Nematoda) in an alpine area. *Norwegian Journal of Entomology*, 22: 129–34.

Lundegaard C, Lamberth K, Harndahl M, Buus S, Lund O, Nielsen M (2008). NetMHC-3.0: accurate web accessible predictions of human, mouse and monkey MHC class I affinities for peptides of length 8–11. *Nucleic acids research*, 36: W509–12.

Luo S, Zhang H, Duan Y, Yao X, Clark AG, Lu J (2020). The evolutionary arms race between transposable elements and piRNAs in Drosophila melanogaster. *BMC Evolutionary Biology*, 20: 14.

Luong LT, Mathot KJ (2019). Facultative parasites as evolutionary stepping-stones towards parasitic lifestyles. *Biology Letters*, 15: 20190058.

Luong LT, Polak M (2007a). Costs of resistance in the *Drosophila-Macrocheles* system: a negative genetic correlation between ectoparasite resistance and reproduction. *Evolution*, 61: 1391–402.

Luong LT, Polak M (2007b). Environment-dependent trade-offs between ectoparasite resistance and larval competitive ability in the *Drosophila-Macrocheles* system. *Heredity*, 99: 632–40.

Lymbery AJ, Hassan M, Morgan DL, Beatty SJ, Doupé RG (2010). Parasites of native and exotic freshwater fishes in south-western Australia. *Journal of Fish Biology*, 76: 1770–85.

Lymbery AJ, Morine M, Kanani HG, Beatty SJ, Morgan DL (2014). Co-invaders: the effects of alien parasites on native hosts. *International Journal for Parasitology: Parasites and Wildlife*, 3: 171–7.

Lythgoe KA (2002). Effects of acquired immunity and mating strategy on the genetic structure of parasite populations. *The American Naturalist*, 159: 519–29.

Lythgoe KA, Pellis L, Fraser C (2013). Is HIV short-sighted? Insights from a multistrain nested model. *Evolution*, 67: 2769–82.

Ma WJ, Vavre F, Beukeboom LW (2014). Manipulation of arthropod sex determination by endosymbionts: diversity and molecular mechanisms. *Sexual Development*, 8: 59–73.

Ma Z, Damania B (2016). The cGAS-STING defense pathway and its counteraction by viruses. *Cell Host & Microbe*, 19: 150–8.

MacArthur RH, Wilson EO (1967). *The theory of island biogeography*. Princeton University Press, Princeton, NJ.

Macdonald G (1957). *The epidemiology and control of malaria*. Oxford University Press, London.

MacDonald JH, Kreitman M (1991). Adaptive protein evolution in the Adh locus in *Drosophila*. *Nature*, 351: 652–4.

Machado LR, Ottolini B (2015). An evolutionary history of defensins: a role for copy number variation in maximizing host innate and adaptive immune responses. *Frontiers in Immunology*, 6: 115.

Machovsky-Capuska GE, Senior AM, Simpson SJ, Raubenheimer D (2016). The multidimensional nutritional niche. *Trends in Ecology & Evolution*, 31: 355–65.

Mack R (1970). The great African cattle plague epidemic of the 1890's. *Tropical Animal Health and Production*, 2: 210–19.

Mackinnon MJ, Gandon S, Read AF (2008). Virulence evolution in response to vaccination: the case of malaria. *Vaccine*, 265: C42–52.

Mackinnon MJ, Mwangi TW, Snow RW, Marsh K, Williams TN (2005). Heritability of malaria in Africa. *PLoS Medicine*, 2: e340: 1253–9.

Mackinnon MJ, Read AF (2004). Immunity promotes virulence evolution in a malaria model. *PLoS Biology*, 2: 1286–92.

MacLeod CJ, Paterson AM, Tompkins DM, Duncan RP (2010). Parasites lost: do invaders miss the boat or drown on arrival? *Ecology Letters*, 13: 516–27.

Maekawa T, Krauss Jennifer L, Abe T, et al. (2014). *Porphyromonas gingivalis* manipulates complement and TLR signaling to uncouple bacterial clearance from inflammation and promote dysbiosis. *Cell Host & Microbe*, 15: 768–78.

Mahanta A, Ganguli P, Barah P, et al. (2018). Integrative approaches to understand the mastery in manipulation of host cytokine networks by protozoan parasites with emphasis on plasmodium and *Leishmania* species. *Frontiers in Immunology*, 9: 296.

Mahany S, Bray M (2004). Pathogenesis of filoviral haemorrhagic fevers. *The Lancet Infectious Diseases*, 4: 487–98.

Mahlapuu M, Håkansson J, Ringstad L, Björn C (2016). Antimicrobial peptides: an emerging category of therapeutic agents. *Frontiers in Cellular and Infection Microbiology*, 6: Article 194.

Mahmud MA, Bradley JE, MacColl ADC (2017). Abiotic environmental variation drives virulence evolution in a fish host-parasite geographic mosaic. *Functional Ecology*, 31: 2138–46.

Maizels RM, Smits HH, McSorley HJ (2018). Modulation of host immunity by helminths: the expanding repertoire of parasite effector molecules. *Immunity*, 49: 801–18.

Majerczyk C, Schneider E, Greenberg EP (2016). Quorum sensing control of Type VI secretion factors restricts the proliferation of quorum-sensing mutants. *eLife*, 5: e14712.

Manley R, Boots M, Bayer-Wilfert L (2017). Condition-dependent virulence of Slow Bee Paralysis Virus in *Bombus terrestris*: are the impacts of honeybee viruses in wild pollinators underestimated? *Oecologia*, 184: 305–15.

Marikovsky PI (1962). On some features of behavior of the ants *Formica rufa* L. infected with fungus disease. *Insectes Sociaux*, 9: 173–9.

Marino J, Sillero-Zubiri C, Deressa A, et al. (2017). Rabies and distemper outbreaks in smallest Ethiopian wolf population. *Emerging Infectious Diseases*, 23: 2102–4.

Markiewski M, Nilsson B, Nilsson Ekdahl K, Mollnes TE, Lambris JD (2007). Complement and coagulation: strangers or partners in crime? *Trends in Immunology*, 28: 184–92.

Marquardt WC, Demaree RS, Grieve RS (2000). *Parasitology and vector biology*. Harcourt-Academic Press, San Diego.

Marques JT, Imler J-L (2016). The diversity of insect antiviral immunity: insights from viruses. *Current Opinion in Microbiology*, 32: 71–6.

Marquetoux N, Stevenson MA, Wilson P, Ridler A, Heuer C (2016). Using social network analysis to inform disease control interventions. *Preventive Veterinary Medicine*, 126: 94–104.

Marsh SGE, Parham P, Barber LD (2000). *The HLA facts book*. Academic Press, San Diego.

Marston HD, Dixon DM, Knisely JMK, Palmore TN, Fauci AS (2016). Antimicrobial resistance. *Clinical Review & Education*, 316: 1193–204.

Martin CH, Johnsen S (2007). A field test of the Hamilton-Zuk hypothesis in the Trinidadian guppy (*Poecilia reticulata*). *Behavioral Ecology and Sociobiology*, 61: 1897–909.

Martin LB, Weil ZM, Nelson RJ (2007). Immune defense and reproductive pace of life in *Peromyscus* mice. *Ecology*, 88: 2516–28.

Martin LB, Weil ZM, Nelson RJ (2008). Seasonal changes in vertebrate immune activity: mediation by physiological trade-offs. *Philosophical Transactions of the Royal Society London B*, 363: 321–39.

Martin TE, Møller AP, Merin S, Clobert J (2001). Does clutch size evolve in response to parasites and immunocompetence? *Proceedings of the National Academy of Sciences USA*, 98: 2071–6.

Martínez JL (2008). Antibiotics and antibiotic resistance genes in natural environments. *Science*, 321: 365–7.

Martínez JL (2014). Short-sighted evolution of bacterial opportunistic pathogens with an environmental origin. *Frontiers in Microbiology*, 5: 239.

Martínez JL, Cogni R, Cao C, Smith S, Illingworth Christopher JR, Jiggins Francis M (2016). Addicted? Reduced host resistance in populations with defensive symbionts. *Proceedings of the Royal Society London B*, 283: 20160778.

Martinez-Bakker M, Helm B (2015). The influence of biological rhythms on host–parasite interactions. *Trends in Ecology & Evolution*, 30: 314–26.

Marvig RL, Sommer LM, Molin S, Johansen HK (2015). Convergent evolution and adaptation of *Pseudomonas aeruginosa* within patients with cystic fibrosis. *Nature Genetics*, 47: 57–65.

Marxer M, Vollenweider V, Schmid-Hempel P (2016). Insect antimicrobial peptides act synergistically to inhibit a trypanosome parasite. *Philosophical Transactions of the Royal Society London B*, 371: 20150302.

Mas-Coma S, Bargues MD (1997). Human flukes: a review. *Research and Reviews in Parasitology*, 57: 145–218.

Maslov DA, Votýpka J, Yurchenko V, Lukeš J (2013). Diversity and phylogeny of insect trypanosomatids: all that is hidden shall be revealed. *Trends in Parasitology*, 29: 43–52.

Massad E (1987). Transmission rates and the evolution of pathogenicity. *Evolution*, 41: 1127–30.

Massey RC, Buckling A, ffrench-Constant R (2004). Interference competition and parasite virulence. *Proceedings of the Royal Society London B*, 271: 785–8.

Mathee S, Krasnov BR (2009). Searching for generality in the patterns of parasite abundance and distribution: ecto-parasites of a South African rodent, *Rhabdomys pumilio*. *International Journal for Parasitology*, 39: 781–8.

Mathis D, Benoist C (2007). A decade of AIRE. *Nature Reviews Immunology*, 7: 645–50.

Matthews KR, McCulloch R, Morrison LJ (2015). The within-host dynamics of African trypanosome infections. *Philosophical Transactions of the Royal Society London B*, 370: 20140288.

Matzaraki V, Kumar V, Wijmenga C, Zhernakova A (2017). The MHC locus and genetic susceptibility to autoimmune and infectious diseases. *Genome Biology*, 18: 76.

Matzinger P (2002). The danger model: a renewed sense of self. *Science*, 296: 301–5.

Maudlin I, Welburn SC, Milligan PJ (1998). Trypanosome infections and survival in tsetse. *Parasitology*, 116: S23–8.

Maure F, Brodeur J, Ponlet N, et al. (2011). The cost of a bodyguard. *Biology Letters*, 7: 843–6.

Maure F, Thomas F, Doyon J, Brodeur J (2016). Host nutritional status mediates degree of parasitoid virulence. *Oikos*, 125: 1314–23.

Maurice CF, Knowles S, Ladau J, et al. (2015). Marked seasonal variation in the wild mouse gut microbiota. *The ISME Journal*, 9: 2423–34.

Mavridou DAI, Gonzalez D, Kim W, West SA, Foster KR (2018). Bacteria use collective behavior to generate diverse combat strategies. *Current Biology*, 28: 345–55.e4.

May RM, Anderson RM (1978). Regulation and stability of host-parasite population interactions. II. Destabilizing processes. *Journal of Animal Ecology*, 47: 249–67.

May RM, Anderson RM (1979). Population biology of infectious diseases. Part II. *Nature*, 280: 455–61.

May RM, Anderson RM (1983a). Epidemiology and genetics in the coevolution of parasites and hosts. *Proceedings of the Royal Society London B*, 219: 281–303.

May RM, Anderson RM (1983b). Parasite-host coevolution. In Futuyma DJ, Slatkin M (eds) *The study of coevolution*, pp. 186–206. Sinauer, Sunderland, MA.

Mayer A, Mora T, Rivoire O, Walczak AM (2016). Diversity of immune strategies explained by adaptation to pathogen statistics. *Proceedings of the National Academy of Sciences USA*: 201600663.

Mayer S, Raulf M-K, Lepenies B (2017). C-type lectins: their network and roles in pathogen recognition and immunity. *Histochemistry and Cell Biology*, 147: 223–37.

Maynard CL, Elson CO, Hatton RD, Weaver CT (2012). Reciprocal interactions of the intestinal microbiota and immune system. *Nature*, 489: 231–41.

Maynard Smith J (1978). *The evolution of sex*. Cambridge University Press, Cambridge.

Maynard Smith J, Haigh J (1974). The hitchhiking effect of a favorable gene. *Genetical Research*, 23: 23–35.

Maynard Smith J, Smith NH, O'Rourke M, Spratt BG (1993). How clonal are bacteria? *Proceedings of the National Academy of Sciences USA*, 90: 4384–8.

Mays HL, Hill GE (2004). Choosing mates: good genes versus genes that are a good fit. *Trends in Ecology and Evolution*, 19: 554–9.

Mazé-Guilmo E, Loot G, Páez DJ, Lefèvre T, Blanchet S (2014). Heritable variation in host tolerance and resistance inferred from a wild host–parasite system. *Proceedings of the Royal Society London B*, 281: 20132567.

Mazzoccoli G, Vinciguerra M, Carbone A, Relogio A (2020). The circadian clock, the immune system, and viral infections: the intricate relationship between biological time and host-virus interaction. *Pathogens*, 9: 83.

McCallum H (2016). Models for managing wildlife disease. *Parasitology*, 143: 805–20.

McCormack WT, Tjoelker LW, Thompson CB (1991). Avian B-cell development: generation of an immunoglobulin repertoire by gene conversion. *Annual Review of Immunology*, 9: 219–41.

McCrone JT, Lauring AS (2018). Genetic bottlenecks in intraspecies virus transmission. *Current Opinion in Virology*, 28: 20–5.

McCullers JA (2014). The co-pathogenesis of influenza viruses with bacteria in the lung. *Nature Reviews Microbiology*, 12: 252–62.

McCurdy DG, Shutler D, Mullie A, Forbes MR (1998). Sex-biased parasitism of avian hosts: relations to blood parasite taxon and mating system. *Oikos*, 82: 303–12.

McDade TW, Georgiev AV, Kuzawa CW (2016). Trade-offs between acquired and innate immune defenses in humans. *Evolution, Medicine, and Public Health*, 2016: 1–16.

McDonald MJ, Rice DP, Desai MM (2016a). Sex speeds adaptation by altering the dynamics of molecular evolution. *Nature*, 531: 233–6.

McDonald SM, Nelson MI, Turner PE, Patton JT (2016b). Reassortment in segmented RNA viruses: mechanisms and outcomes. *Nature Reviews Microbiology*, 14: 448–60.

McElroy EJ, de Buron I (2014). Host performance as a target of manipulation by parasites: a meta-analysis. *Journal of Parasitology*, 100: 399–410.

McGuinness WA, Kobayashi SD, DeLeo FR (2016). Evasion of neutrophil killing by *Staphylococcus aureus*. *Pathogens*, 5: 32.

McGuire VA, Arthur JSC (2015). Subverting Toll-Like Receptor signaling by bacterial pathogens. *Frontiers in Immunology*, 6: 607.

McKean KA, Nunney L (2001). Increased sexual activity reduces male immune function in Drosophila melanogaster. *Proc Natl Acad Sci U S A*, 98: 7904–9.

McKean KA, Yourth CP, Lazzaro BP, Clark AG (2008). The evolutionary costs of immunological maintenance and deployment. *BMC Evolutionary Biology*, 8: 76.

McKenzie FE, Smith DL, O'Meara WP, Riley EM (2008). Strain theory of malaria: the first 50 years. *Advances in Parasitology*, 66: 1–46.

McKerrow JH, Caffrey C, Kelly B, Loke P, Sajid M (2006). Proteases in parasitic diseases. *Annual Review of Pathology*, 1: 497–536.

McLean AHC, Godfray HCJ (2015). Evidence for specificity in symbiont-conferred protection against parasitoids. *Proceedings of the Royal Society London B*, 282: 20150977.

McMahon DP, Natsopoulou ME, Doublet V, et al. (2016). Elevated virulence of an emerging viral genotype as a driver of honeybee loss. *Proceedings of the Royal Society London B*, 283: 20160811.

McNab F, Mayer-Barber K, Sher A, Wack A, O'Garra A (2015). Type I interferons in infectious disease. *Nature Reviews Immunology*, 15: 87–103.

McNally L, Brown SP (2015). Building the microbiome in health and disease: niche construction and social conflict in bacteria. *Philosophical Transactions of the Royal Society London B*, 370: 20140298.

McTaggart SJ, Obbard DJ, Conlon C, Little TJ (2012). Immune genes undergo more adaptive evolution than non-immune system genes in *Daphnia pulex*. *BMC Evolutionary Biology*, 12: 63.

McVinish R, Lester RJG (2020). Measuring aggregation in parasite populations. *Journal of The Royal Society Interface*, 17: 20190886.

Mead LS, Arnold SJ (2004). Quantitative genetic models of sexual selection. *Trends in Ecology and Evolution*, 19: 264–71.

Meaden S, Koskella B (2017). Adaptation of the pathogen, *Pseudomonas syringae*, during experimental evolution on a native vs. alternative host plant. *Molecular Ecology*, 26: 1790–801.

Medeiros MCI, Ellis VA, Ricklefs RE (2014). Specialized avian Haemosporida trade reduced host breadth for increased prevalence. *Journal of Evolutionary Biology*, 27: 2520–8.

Medzhitov R, Biron CA (2003). Innate immunity. *Current Opinion in Immunology*, 15: 2–4.

Medzhitov R, Schneider DS, Soares MP (2012). Disease tolerance as a defense strategy. *Science*, 335: 936–41.

Meeus I, Pisman M, Smagghe G, Piot N (2018). Interaction effects of different drivers of wild bee decline and their influence on host–pathogen dynamics. *Current Opinion in Insect Science*, 26: 136–41.

Mehlhorn H (2015). The brain worm story. In Mehlhorn H (ed.) *Host Manipulations by Parasites and Viruses*, pp. 101–8. Springer International Publishing, Cham.

Mehta A, Baltimore D (2016). MicroRNAs as regulatory elements in immune system logic. *Nature Reviews Immunology*, 16: 279–94.

Meister H, Tammaru T, Sandre S-L, Freitak D (2017). Sources of variance in immunological traits: evidence of congruent latitudinal trends across species. *The Journal of Experimental Biology*, 220: 2606–15.

Meister S, Bogos A, Turlure F, et al. (2009). *Anopheles gambiae* PGRPLC-mediated defense against bacteria modulates infections with malaria parasites. *PLoS Pathogens*, 5: e1000542.

Melcarne C, Lemaitre B, Kurant E (2019). Phagocytosis in *Drosophila*: from molecules and cellular machinery to physiology. *Insect Biochemistry and Molecular Biology*, 109: 1–12.

Melnyk AH, Wong A, Kassen R (2014). The fitness costs of antibiotic resistance mutations. *Evolutionary Applications*, 8: 273–83.

Mendlová M, Šimková A (2014). Evolution of host specificity in monogeneans parasitizing African cichlid fish. *Parasites & Vectors*, 7: 69.

Mercereau-Puijalon O (1996). Revisiting host/parasite interactions: molecular analysis of parasites collected during longitudinal and cross-sectional surveys in humans. *Parasite Immunology*, 18: 173–80.

Merkling SH, Lambrechts L (2020). Taking insect immunity to the single-cell level. *Trends in Immunology*, 41: 190–9.

Merrell DS, Falkow S (2004). Frontal and stealth attack strategies in microbial pathogenesis. *Nature*, 430: 250–6.

Mersch D, Crespi A, Keller L (2013). Tracking individuals shows spatial fidelity is a key regulator of ant social organization. *Science*, 340: 1090–3.

Mescas J, Strauss EJ (1996). Molecular mechanisms of bacterial virulence: type III secretion and pathogenicity islands. *Emerging Infectious Diseases*, 2: 271–88.

Messenger SL, Molineux IJ, Bull JJ (1999). Virulence evolution in a virus obeys a trade-off. *Proceedings of the Royal Society London B*, 266: 397–404.

Messinger SM, Ostling A (2013). The influence of host demography, pathogen virulence, and relationships with pathogen virulence on the evolution of pathogen transmission in a spatial context. *Evolutionary Ecology*, 27: 353–80.

Metalnikov S (1921). L'immunite naturelle et acquise chez la chenille de *Galleria melonella*. *Annales de l' Institut Pasteur*, 35: 363–77.

Metcalf CJE, Birger RB, Funk S, Kouyos, R.D., Lloyd-Smith JO, Jansen VAA (2015). Five challenges in evolution and infectious diseases. *Epidemics*, 10: 40–4.

Metcalf CJE, Jones JH (2015). The evolutionary dynamics of timing of maternal immunity: evaluating the role of age-specific mortality. *Journal of Evolutionary Biology*, 28: 493–502.

Metcalf CJE, Tate AT, Graham AL (2017). Demographically framing trade-offs between sensitivity and specificity illuminates selection on immunity. *Nature Ecology & Evolution*, 1: 1766–72.

Michalska-Smith MJ, Sander EL, Pascual M, Allesina S (2018). Understanding the role of parasites in food webs using the group model. *Journal of Animal Ecology*, 87: 790–800.

Michel-Briand Y, Baysse C (2002). The pyocins of *Pseudomonas aeruginosa*. *Biochimie*, 84: 499–510.

Mideo N, Acosta-Serrano A, Aebischer T, et al. (2013a). Life in cells, hosts, and vectors: parasite evolution across scales. *Infection, Genetics and Evolution*, 13: 344–7.

Mideo N, Alizon S, Day T (2008). Linking within- and between-host dynamics in the evolutionary epidemiology of infectious diseases. *Trends in Ecology & Evolution*, 23: 511–17.

Mideo N, Nelson WA, Reece SE, Bell AS, Read AF, Day T (2011). Bridging scales in the evolution of infectious disease life histories: application. *Evolution*, 65: 3298–310.

Mideo N, Reece SE (2012). Plasticity in parasite phenotypes: evolutionary and ecological implications for disease. *Future Microbiology*, 7: 17–24.

Mideo N, Reece SE, Smith AL, Metcalf CJE (2013b). The Cinderella syndrome: why do malaria-infected cells burst at midnight? *Trends in Parasitology*, 29: 10–16.

Mignatti A, Boag B, Cattadori IM (2016). Host immunity shapes the impact of climate changes on the dynamics of parasite infections. *Proceedings of the National Academy of Sciences USA*, 113: 2970–5.

Mihaljevic JR, Hoye BJ, Johnson PTJ (2018). Parasite metacommunities: evaluating the roles of host community composition and environmental gradients in structuring symbiont communities within amphibians. *Journal of Animal Ecology*, 87: 354–68.

Mikheev VN, Pasternak AF, Valtonen ET (2015). Behavioural adaptations of argulid parasites (Crustacea: Branchiura) to major challenges in their life cycle. *Parasites & Vectors*, 8: 394.

Mikonranta L, Mappes J, Laakso J, Ketola T (2015). Within-host evolution decreases virulence in an opportunistic bacterial pathogen. *BMC Evolutionary Biology*, 15: 165.

Milinski M (1985). Risk of predation of parasitized sticklebacks (*Gasterosteus acuelatus* L.) under competition for food. *Behavior*, 93: 203–16.

Milinski M (1990). Parasites and host decision-making. In Barnard CJ, Behnke JM (eds) *Parasitism and host behaviour*, pp. 95–116. Taylor & Francis, London.

Milinski M (2006a). Fitness consequences of selfing and outcrossing in the cestode *Schistocephalus solidus*. *Integrative and Comparative Biology*, 46: 373–80.

Milinski M (2006b). The major histocompatibility complex, sexual selection, and mate choice. *Annual Review of Ecology, Evolution and Systematics*, 37: 159–86.

Milinski M (2014). Arms races, ornaments and fragrant genes: the dilemma of mate choice in fishes. *Neuroscience and Biobehavioral Reviews*, 46: 567–72.

Milinski M, Bakker TCM (1990). Female sticklebacks use male coloration in mate choice and hence avoid parasitized males. *Nature*, 344: 330–3.

Milinski M, Croy I, Hummel T, Boehm T (2013). Major histocompatibility complex peptide ligands as olfactory cues in human body odour assessment. *Proceedings of the Royal Society London B*, 280: 1–7.

Milinski M, Griffiths S, Wegner KM, Reusch TBH, Haas-Assenbaum A, Boehm T (2005a). Mate choice decisions of stickleback females predictably modified by MHC peptide ligands. *Proceedings of the National Academy of Sciences of the United States of America*, 102: 4414–18.

Miller C, Thomsen LE, Gaggero C, Mosseri R, Ingmer H, Cohen SN (2004). SOS response induction by β-lactams and bacterial defense against antibiotic lethality. *Science*, 305: 1629–31.

Miller CVL, Cotter SC (2018). Resistance and tolerance: the role of nutrients on pathogen dynamics and infection outcomes in an insect host. *Journal of Animal Ecology*, 87: 500–10.

Miller MR, White A, Boots M (2005). The evolution of host resistance: tolerance and control as distinct strategies. *Journal of Theoretical Biology*, 236: 198–207.

Miller MR, White A, Boots M (2006). The evolution of parasites in response to tolerance in their hosts: the good, the bad and apparent commensalism. *Evolution*, 60: 945–56.

Miller MR, White A, Boots M (2007). Host life span and the evolution of resistance characteristics. *Evolution*, 61: 2–14.

Milutinovic B, Kurtz J (2016). Immune memory in invertebrates. *Seminars in Immunology*, 28: 328–42.

Minchella DJ (1985). Host life-history variation in response to parasitation. *Parasitology*, 90: 205–16.

Minchella DJ, Loverde PT (1981). A cost of increased early reproductive effort in the snail *Biomphalaria glabrata*. *American Naturalist*, 118: 876–81.

Minchella DJ, Scott ME (1991). Parasitism: a cryptic determinant of animal community structure. *Trends in Ecology & Evolution*, 6: 250–4.

Minias P, Pikus E, Whittingham LA, Dunn PO (2018). A global analysis of selection at the avian MHC. *Evolution*, 72: 1278–93.

Mira A, Ochman H, Moran NA (2001). Deletional bias and the evolution of bacterial genomes *Trends in Genetics*, 17: 589–96.

Mitchell CE, Power AG (2003). Release of invasive plants from fungal and viral pathogens. *Nature*, 421: 625–7.

Mitchell CE, Reich PB, Tilman D, Groth JV (2003). Effects of elevated CO2, nitrogen deposition, and decreased species diversity on foliar fungal plant disease. *Global Change Biology*, 9: 438–51.

Mitchell J, Vitikainen EIK, Wells DA, Cant MA, Nichols HJ (2017). Heterozygosity but not inbreeding coefficient predicts parasite burdens in the banded mongoose. *Journal of Zoology*, 302: 32–9.

Mitchell SE, Rogers ES, Little TJ, Read AF (2005). Host-parasite and genotype-by-environment interactions:

temperature modifies potential for selection by a sterilizing pathogen. *Evolution*, 59: 70–80.

Mitov V, Stadler T (2018). A practical guide to estimating the heritability of pathogen traits. *Molecular Biology and Evolution*, 35: 756–72.

Miura O, Torchin ME, Kuris AMH, Ryan F., Chiba S (2006). Introduced cryptic species of parasites exhibit different invasion pathways. *Proceedings of the National Academy of Sciences USA*, 103: 19818–23.

Moayeri M, Leppla S (2004). The roles of anthrax toxin in pathogenesis. *Current Opinion in Microbiology*, 7: 19–24.

Moayeri M, Leppla SH, Vrentas C, Pomerantsev AP, Liu S (2015). Anthrax pathogenesis. *Annual Review of Microbiology*, 69: 185–208.

Mobegi VA, Duffy CW, Amambua-Ngwa A, et al. (2014). Genome-wide analysis of selection on the malaria parasite *Plasmodium falciparum* in West African populations of differing infection endemicity. *Molecular Biology and Evolution*, 31: 1490–9.

Mock M, Fouet A (2001). Anthrax. *Annual Review of Microbiology*, 55: 647–71.

Mockler BK, Kwong WK, Moran NA, Koch H (2018). Microbiome structure influences infection by the parasite *Crithidia bombi* in bumble bees. *Applied and Environmental Microbiology*, 84: e02335–17.

Mohanraju P, Makarova KS, Zetsche B, Zhang F, Koonin EV, van der Oost J (2018). Diverse evolutionary roots and mechanistic variations of the CRISPR-Cas systems. *Science*, 353: aad5147-1- aad47-12.

Moiroux N, Gomez MB, Pennetier C, et al. (2012). Changes in *Anopheles funestus* biting behavior following universal coverage of long-lasting insecticidal nets in Benin. *The Journal of Infectious Diseases*, 206: 1622–9.

Moissl-Eichinger C, Huber H (2011). Archaeal symbionts and parasites. *Current Opinion in Microbiology*, 14: 364–70.

Molina-Ochoa J, Carpenter JE, Heinrichs EA, Foster JE (2003). Parasitoids and parasites of *Spodoptera frugiperda* (Lepidoptera: Noctuidae) in the Americas and Caribbean basin: an inventory. *Florida Entomologist*, 86: 254–89.

Mollentze N, Streicker DG (2020). Viral zoonotic risk is homogenous among taxonomic orders of mammalian and avian reservoir hosts. *Proceedings of the National Academy of Sciences USA*, 117: 9423–30.

Møller AP, Cassey P (2004). On the relationship of T-cell mediated immunity in bird species and the establishment success of introduced populations. *Journal of Animal Ecology*, 73: 1035–42.

Moloo SK, Sabwa CL, Baylis M (2000). Feeding behaviour of *Glossina pallidipes* and *G. morsitans centralis* on Boran cattle infected with *Trypanosoma congolense* or *T. vivax* under laboratory conditions. *Medical Veterinary Entomology*, 14: 290–9.

Monda V, Villano I, Messina A, et al. (2017). Exercise modifies the gut microbiota with positive health effects. *Oxidative Medicine and Cellular Longevity*, 2017: 3831972.

Mondotte JA, Saleh M-C (2018). Antiviral immune response and the route of infection in *Drosophila melanogaster*. *Advances in Virus Research*, 100: 247–78.

Montaner JS, Lima VD, Barrios R, et al. (2010). Association of highly active antiretroviral therapy coverage, population viral load, and yearly new HIV diagnoses in British Columbia, Canada: a population-based study. *Lancet*, 376: 532–9.

Montgomerie R (2006). Analyzing colors. In Hill G, McGraw K (eds) *Bird coloration 1: mechanisms and measurements*, pp. 90–147. Harvard University Press, Cambridge, MA.

Monticelli LS, Outreman Y, Frago E, Desneux N (2019). Impact of host endosymbionts on parasitoid host range: from mechanisms to communities. *Current Opinion in Insect Science*, 32: 77–82.

Moon JY, Park JM (2016). Cross-talk in viral defense signaling in plants. *Frontiers in Microbiology*, 7: 2068.

Moore FR, Shuker DM, Dougherty L (2016). Stress and sexual signaling: a systematic review and meta-analysis. *Behavioral Ecology*, 27: 363–71.

Moore J (1983). Responses of an avian predator and its isopod prey to an acanthocephalan parasite. *Ecology*, 64: 1000–15.

Moore J (2002). *Parasites and the behavior of animals*. Oxford University Press, Oxford.

Moore J, Adamo SA, Thomas F (2005). Manipulation: expansion of the paradigm. *Behavioural Processes*, 68: 283–7.

Mooring MS (1992). Animal grouping for protection from parasites: selfish herd and encounter-dilution effects. *Behaviour*, 123: 173–93.

Moran N (2002). Microbial minimalism: genome reduction in bacterial pathogens. *Cell*, 108: 583–6.

Morand S (1996). Biodiversity of parasites in relation to their life cycles. In Hochberg ME, Clobert J, Barbault R (eds) *Aspects of the genesis and maintenance of biological diversity*, pp. 243–60. Oxford University Press, Oxford.

Morand S (2015). (macro-) Evolutionary ecology of parasite diversity: from determinants of parasite species richness to host diversification. *International Journal for Parasitology: Parasites and Wildlife*, 4: 80–7.

Morand S, Harvey PH (2000). Mammalian metabolism, longevity and parasite species richness. *Proceedings of the Royal Society London B*, 267: 1999–2003.

Morand S, Krasnov BR, Littlewood DTJ (eds) (2015). *Parasite diversity and diversification*. Cambridge University Press, Cambridge.

Morand S, Manning SD, Woolhouse MEJ (1996). Parasite host coevolution and geographic patterns of parasite infectivity and host susceptibility. *Proceedings of the Royal Society London B*, 263: 119–28.

Morand S, Poulin R, Rohde K, Hayward C (1999). Aggregation and species coexistence of ectoparasites of marine fishes. *International Journal for Parasitology*, 29: 663–72.

Morand S, Robert F, Connors VA (1995). Complexity in parasite life cycles: population biology of cestodes in fish. *Journal of Animal Ecology*, 64: 256–64.

Morand S, Rohde K, Hayward C (2002). Order in ectoparasite communities of marine fish is explained by epidemiological processes. *Parasitology*, 124: S57–63.

Moravec F (1994). *Parasitic nematodes of freshwater fish in Europe*. Academic Press, Kluwer, Prag. Dordrecht.

Moreau J, Girgis D, Hume E, Dajes J, Austin M, O'Callaghan R (2001). Phospholipase A(2) in rabbit tears: a host defense against *Staphylococcus aureus*. *Investigations in Ophthalmology and Visual Sciences*, 42: 2347–54.

Moreira-Soto A, Torres MC, Lima de Mendonça MC, et al. (2018). Evidence for multiple sylvatic transmission cycles during the 2016–2017 yellow fever virus outbreak, Brazil. *Clinical Microbiology and Infection*, 24: 1019.e1–19.e4.

Moret Y, Schmid-Hempel P (2000). Survival for immunity: the price of immune system activation for bumblebee workers. *Science*, 290: 1166–8.

Moret Y, Schmid-Hempel P (2001). Immune defence in bumble-bee offspring. *Nature*, 414: 506.

Morley D, Broniewski JM, Westra ER, Buckling A, van Houte S (2017). Host diversity limits the evolution of parasite local adaptation. *Molecular Ecology*, 26: 1756–63.

Morozinska-Gogol J (2006). A checklist of parasites recorded on sticklebacks (Actinopterygii: Gasterosteidae) from Poland. *Parasitology International*, 55: 69–73.

Morozov AY, Adamson MW (2011). Evolution of virulence driven by predator–prey interaction: possible consequences for population dynamics. *Journal of Theoretical Biology*, 276: 181–91.

Morrill A, Dargent F, Forbes MR (2017). Explaining parasite aggregation: more than one parasite species at a time. *International Journal for Parasitology*, 47: 185–8.

Mougari S, Sahmi-Bounsiar D, Levasseur A, Colson P, La Scola B (2019). Virophages of giant viruses: an update at eleven. *Viruses*, 11: 733.

Mouillot D, Krasnov BR, Shenbrot GI, Gaston KJ, Poulin R (2006). Conservatism of host specificity in parasites. *Ecography*, 29: 596–602.

Mouritsen KN, Poulin R (2010). Parasitism as a determinant of community structure on intertidal flats. *Marine Biology*, 157: 201–13.

Muegge BD, Kuczynski J, Knights D, et al. (2011). Diet drives convergence in gut microbiome functions across mammalian phylogeny and within humans. *Science*, 332: 970–4.

Muehlenbein MP (2017). Primates on display: potential disease consequences beyond bushmeat. *American Journal of Physical Anthropology*, 162: 32–43.

Mueller S, Gebhardt T, Carbone F, Heath W (2013). Memory T cell subsets, migration patterns, and tissue residence. *Annual Review of Immunology*, 31: 137–61.

Mugnier MR, Cross GAM, Papavasiliou FN (2015). The in vivo dynamics of antigenic variation in *Trypanosoma brucei*. *Science*, 347: 1470–3.

Mühlenkamp M, Oberhettinger P, Leo JC, Linke D, Schütz MS (2015). Yersinia adhesin A (YadA): beauty & beast. *International Journal of Medical Microbiology*, 305: 252–8.

Müller CB, Schmid-Hempel P (1992). Variation in life-history pattern in relation to worker mortality in the bumblebee, *Bombus lucorum*. *Functional Ecology*, 6: 48–56.

Müller CB, Schmid-Hempel P (1993). Exploitation of cold temperature as defense against parasitoids in bumblebees. *Nature*, 363: 65–7.

Müller V, Bonhoeffer S (2003). Quantitative constraints on the scope of negative selection. *Trends in Immunology*, 3: 132–5.

Munro AD, Smallman-Raynor M, Algar AC (2020). Long-term changes in endemic threshold populations for pertussis in England and Wales: a A spatiotemporal analysis of Lancashire and South Wales, 1940–69. *Social Science & Medicine*: 113295.

Munson E, Carroll KC (2017). What's in a name? New bacterial species and changes to taxonomic status from 2012 through 2015. *Journal of Clinical Microbiology*, 55: 24.

Murdock CC, Sternberg ED, Thomas MB (2016). Malaria transmission potential could be reduced with current and future climate change. *Scientific reportsReports*, 6: 27771–1.

Murillo LN, Murillo MS, Perelson AS (2013). Towards multiscale modeling of influenza infection. *Journal of Theoretical Biology*, 332: 267–90.

Muskett JC, Reed NE, Thornton DH (1985). Increased virulence of an infectious bursal disease live virus vaccine after passage in chicks. *Vaccine*, 3: 309–12.

Mussabekova A, Daeffler, Imler JL (2017). Innate and intrinsic antiviral immunity in *Drosophila*. *Cellular and Molecular Life Sciences*, 74: 2039–54.

Mutanen M, Ovaskainen O, Várkonyi G, et al. (2020). Dynamics of a host–parasitoid interaction clarified by modelling and DNA sequencing. *Ecology Letters*, 23: 851–9.

Myers JH (2018). Population cycles: generalities, exceptions and remaining mysteries. *Proceedings of the Royal Society London B*, 285: 20172841.

Myers JH, Cory JS (2016). Ecology and evolution of pathogens in natural populations of Lepidoptera. *Evolutionary Applications*, 9: 231–47.

N'Goran EK, Diabate S, Utzinger J, Sellin B (1997). Changes in human schistosomiasis levels after the construction of two large hydroelectric dams in central Côte d'Ivoire. *Bulletin of the World Health Organization*, 75: 541–5.

Nadell CD, Drescher K, Foster KR (2016). Spatial structure, cooperation and competition in biofilms. *Nature Reviews Microbiology*, 14: 589–600.

Nakjang S, Williams TA, Heinz E, et al. (2013). Reduction and expansion in microsporidian genome evolution: new insights from comparative genomics. *Genome Biology and Evolution*, 5: 2285–303.

Näpflin K, Schmid-Hempel P (2016). Immune response and gut microbial community structure in bumblebees after microbiota transplants. *Proceedings of the Royal Society London B*, 283: 20160312.

Nappi AJ, Christensen BM (2005). Melanogenesis and associated cytotoxic reactions: applications to insect innate immunity. *Insect Biochemistry and Molecular Biology*, 35: 443–59.

Nappi AJ, Ottaviani E (2000). Cytotoxicity and cytotoxic molecules in invertebrates. *Bioessays*, 22: 469–80.

Nascimento MTC, Garcia MCF, da Silva KP, et al. (2010). Interaction of the monoxenic trypanosomatid *Blastocrithidia culicis* with the *Aedes aegypti* salivary gland. *Acta Tropica*, 113: 269–78.

Natsopoulou ME, McMahon DP, Paxton RJ (2016). Parasites modulate within-colony activity and accelerate the temporal polyethism schedule of a social insect, the honey bee. *Behavioral Ecology and Sociobiology*, 70: 1019–31.

Nazario-Toole AE, Wu LP (2017). Phagocytosis in insect immunity. In Ligoxygakis P (ed.) *Advances in insect physiology*, pp. 35–82. Academic Press, London.

Neethling LAM, Avenant-Oldewage A (2016). Branchiura: a compendium of the geographical distribution and a summary of their biology. *Crustaceana*, 89: 1243–446.

Neff BD, Pitcher TE (2005). Genetic quality and sexual selection: an integrated framework for good genes and compatible genes. *Molecular Ecology*, 14: 19–38.

Neiman M, Lively CM, Meirmans S (2017). Why sex? a pluralist approach revisited. *Trends in Ecology & Evolution*, 32: 589–600.

Nelson RJ, Demas GE, Klein SL (1998). Photoperiodic mediation of seasonal breeding and immune function in rodents: a multifactorial approach. *American Zoologist*, 38: 226–37.

Ness JH, Foster SA (1999). Parasite-associated phenotype modifications in threespine sticklebacks. *Oikos*, 85: 127–34.

Netea MG, Joosten LAB, Latz E, et al. (2016). Trained immunity: a program of innate immune memory in health and disease. *Science*, 352: aaf1098.

Netea MG, Latz E, Mills KHG, O'Neill LAJ (2015). Innate immune memory: a paradigm shift in understanding host defense. *Nature Immunology*, 16: 675–9.

Netea MG, Quintin J, van der Meer JW (2011). Trained immunity: a memory for innate host defense. *Cell Host Microbe*, 9: 355–61.

Netea MG, Schlitzer A, Placek K, Joosten LAB, Schultze JL (2019). Innate and adaptive immune memory: an evolutionary continuum in the host's response to pathogens. *Cell Host & Microbe*, 25: 13–26.

Netea MG, Wijmenga C, O'Neill LAJ (2012). Genetic variation in Toll-like receptors and disease susceptibility. *Nature Immunology*, 13: 535–42.

Neuenschwander P, Hammond WNO, Gutierrez AP, et al. (1989). Impact assessment of the biological control of the cassava mealybug, *Phenacoccus manihoti* (Matile-Ferrero) (Hemiptera: Pseudococcidae), by the introduced parasitoid *Epidinocarsis lopezi* (De Santis) (Hymenoptera: Encyrtidae). *Bulletin of Entomological Research*, 79: 579–94.

Newbold LK, Burthe SJ, Oliver AE, et al. (2016). Helminth burden and ecological factors associated with alterations in wild host gastrointestinal microbiota. *The ISME Journal*, 11: 663–75.

Newell EW, et al. (2013). Combinatorial tetramer staining and mass cytometry analysis facilitate T-cell epitope mapping and characterization. *Nature Biotechnology*, 31: 623–9.

Newton AH, Cardani A, Braciale TJ (2016). The host immune response in respiratory virus infection: balancing virus clearance and immunopathology. *Seminars in Immunopathology*, 38: 471–82.

Ng ACY, Eisenberg JM, Heath RJW, et al. (2011). Human leucine-rich repeat proteins: a genome-wide bioinformatic categorization and functional analysis in innate immunity. *Proceedings of the National Academy of Sciences USA*, 108 Suppl 1: 4631–8.

Ng ACY, Xavier RJ (2011). Leucine-rich repeat (LRR) proteins: integrators of pattern recognition and signaling in immunity. *Autophagy*, 7: 1082–4.

Ngo TTN, Senior AM, Culina A, Santos ESA, Vlak JM, Zwart MP (2018). Quantitative analysis of the dose-response of white spot syndrome virus in shrimp. *Journal of Fish Diseases*, 41: 1733–44.

Nguyen VK, Klawonn F, Mikolajczyk R, Hernandez-Vargas EA (2016). Analysis of practical identifiability of a viral infection model. *PLoS ONE*, 11: e0167568.

Ní Cheallaigh C, Sheedy FJ, Harris J, et al. (2016). A common variant in the adaptor mal regulates interferon gamma signaling. *Immunity*, 44: 368–79.

Ni J, Wang D, Wang S (2018). The CCR5-Delta32 genetic polymorphism and HIV-1 infection susceptibility: a meta-analysis. *Open Medicine*, 13: 467–74.

Nicholson AJ, Bailey VA (1935). The balance of animal populations. Part I. *Proceedings of the Zoological Society London*, 3: 551–98.

Niehl A, Heinlein M (2019). Perception of double-stranded RNA in plant antiviral immunity. *Molecular Plant Pathology*, 20: 1203–10.

Nielsen R, Williamson S, Kim Y, Hubisz MJ, Clark AG, Bustamante C (2005). Genomic scans for selective sweeps using SNP data. *Genome*, 15: 1566–75.

Nieto PA, Pardo-Roa C, Salazar-Echegarai FJ, et al. (2016). New insights about excisable pathogenicity islands in *Salmonella* and their contribution to virulence. *Microbes and Infection*, 18: 302–9.

Niezgoda M, A. HC, Rupprecht CE (2002). Animal rabies. In Jackson AC, Wunner WH (eds) *Rabies*, pp. 163–218. Academic Press, New York.

Nikbakht R, Baneshi MR, Bahrampour A, Hosseinnataj A (2019). Comparison of methods to estimate basic reproduction number (R (0)) of influenza, using Canada 2009 and 2017–18 A (H1N1) data. *Journal of Research in Medical Sciences*, 24: 67–7.

Nishida AH, Ochman H (2019). A great-ape view of the gut microbiome. *Nature Reviews Genetics*, 20: 195–206.

Nishiura H, Linton NM, Akhmetzhanov AR (2020). Serial interval of novel coronavirus (COVID-19) infections. *International Journal of Infectious Diseases*, 93: 284–6.

Noble ER, Noble GA, Schad GA, MacInnes AJ (1989). *Parasitology: Tthe biology of animal parasites*. Lee and Febiger, Philadelphia.

Noguiera T, Rankin DJ, Touchon M, Taddei F, Brown SP, Rocha EPC (2009). Horizontal gene transfer of the secretome drives the evolution of bacterial cooperation and virulence. *Current Biology*, 19: 1683–91.

Nokes D-J (1992). Microparasites: viruses and bacteria. In Crawley E (ed.) *Natural enemies: the population biology of predators, parasites and diseases*, pp. 349–74. Blackwell's, Oxford.

Noland D, Drisko JA, Wagner L (eds) (2020). *Integrative and functional medical nutrition therapy: principles and practices*. Springer, Cham.

Noland J, Noland D (2020). Nutritional influences on immunity and infection. In Noland D, Drisko JA, Wagner L (eds) *Integrative and functional medical nutrition therapy: principles and practices*, pp. 303–21. Springer, Cham.

Nonaka M (2001). Evolution of the complement system. *Current Opinion in Immunology*, 13: 69–73.

Nonaka M (2014). Evolution of the complement system. In Anderluh G, Gilbert R (eds) *MACPF/CDC proteins: agents of defence, attack and invasion*, pp. 31–43. Springer, Dordrecht.

Norouzitallab P, Biswas P, Baruah K, Bossier P (2015). Multigenerational immune priming in an invertebrate parthenogenetic *Artemia* to a pathogenic *Vibrio campbellii*. *Fish and Shellfish Immunology*, 42: 426–9.

Norton J, Lewis JW, Rollinson D (2004). Temporal and spatial patterns of nestedness in eel macroparasite communities. *Parasitology*, 129: 203–11.

Nothelfer K, Sansonetti PJ, Phalipon A (2015). Pathogen manipulation of B cells: the best defence is a good offence. *Nature Reviews Microbiology*, 13: 173–84.

Novick RP, Ram G (2016). The floating (pathogenicity) island: a genomic dessert. *Trends in Genetics*, 32: 114–26.

Novotny V, Basset Y (1998). Seasonality of sap-sucking insects (Auchenorrhyncha, Hemiptera) feeding on *Ficus* (Moraceae) in a lowland rain forest in New Guinea. *Oecologia*, 115: 514–22.

Nowak J, Pawłowski B, Borkowska B, Augustyniak D, Drulis-Kawa Z (2018). No evidence for the immunocompetence handicap hypothesis in male humans. *Scientific Reports*, 8: 7392.

Nowak MA, May RM (2000). *Virus dynamics: mathematical principles of immunology and virology*. Oxford University Press, New York.

Nuismer SL (2006). Parasite local adaptation in a geographic mosaic. *Evolution*, 60: 24–30.

Nuismer SL (2017). Rethinking conventional wisdom: are locally adapted parasites ahead in the coevolutionary race? *The American Naturalist*, 190: 584–93.

Nunn CL, Altizer S (2006). *Infectious diseases in primates*. Oxford University Press, Oxford.

Nunn CL, Altizer SM, Jones KE, Sechrest W (2003). Comparative tests of parasite species richness in primates. *The American Naturalist*, 162: 597–614.

Nunn CL, Lindenfors P, Pursall ER, Rolff J (2009). On sexual dimorphism in immune function. *Philosophical Transactions of the Royal Society London B*, 364: 61–9.

Nussenzweig PM, McGinn J, Marraffini LA (2019). Cas9 cleavage of viral genomes primes the acquisition of new immunological memories. *Cell Host & Microbe*, 26: 515–26.e6.

O'Brien SJ, Evermann JF (1988). Interactive influence of infectious disease and genetic diversity in natural populations. *Trends in Ecology & Evolution*, 3: 254–9.

O'Connell RM, Rao DS, Baltimore D (2012). microRNA Regulation of Inflammatory Responses. *Annual Review of Immunology*, 30: 295–312.

O'Connor EA, Cornwallis CK, Hasselquist D, Nilsson J-Å, Westerdahl H (2018). The evolution of immunity in relation to colonization and migration. *Nature Ecology & Evolution*, 2: 841–9.

O'Connor MI, Bernhardt JR (2018). The metabolic theory of ecology and the cost of parasitism. *PLoS Biology*, 16: e2005628.

O'Grady E, Mulcahy H, Admas C, Morrissey JP, O'Gara F (2007). Manipulation of host Kruppel-like factor (KLF) function by exotoxins from diverse bacterial pathogens. *Nature Reviews Microbiology*, 5: 337–41.

O'Hanlon SJ, Rieux A, Farrer RA, et al. (2018). Recent Asian origin of chytrid fungi causing global amphibian declines. *Science*, 360: 621–7.

O'Loughlin CT, Miller LC, Siryaporn A, Drescher K, Semmelhack MF, Bassler BL (2013). A quorum-sensing inhibitor blocks *Pseudomonas aeruginosa* virulence and biofilm formation. *Proceedings of the National Academy of Sciences USA*, 110: 17981–6.

O'Shea B, Rebollar-RTellez E, Ward RD, Hamilton JGC, El Naiem D, Polwart A (2002). Enhanced sandfly attraction to *Leishmania*-infected hosts. *Transactions of the Royal Society of Tropical Medicine and Hygiene*, 96: 117–18.

Obadia B, Güvener ZT, Zhang V, et al. (2017). Probabilistic invasion underlies natural gut microbiome stability. *Current Biology*, 27: 1999–2006.

Obbard DJ, Callister DM, Jiggins FM, Soares DC, Yan G, Little TJ (2008). The evolution of TEP1, an exceptionally polymorphic immunity gene in *Anopheles gambiae*. *BMC Evolutionary Biology*, 8: 274.

Obbard DJ, Jiggins FM, Halligan DL, Little TJ (2006). Natural selection drives extremely rapid evolution in antiviral RNAi genes. *Current Biology*, 16: 580–5.

Ochman H, Wilson AC (1987). Evolution in bacteria: evidence for a universal substitution rate in cellular genomes. *Journal of Molecular evolutionEvolution*, 26: 74–86.

Ödeen A, Håstad O (2013). The phylogenetic distribution of ultraviolet sensitivity in birds. *BMC Evolutionary Biology*, 13: 36.

Odegard VH, Schatz DG (2006). Targeting of somatic hypermutation. *Nature Reviews Immunology*, 6: 573–83.

Ojha CR, Rodriguez M, Dever SM, Mukhopadhyay R, El-Hage N (2016). Mammalian microRNA: an important modulator of host-pathogen interactions in human viral infections. *Journal of Biomedical Science*, 23: 74.

Okin D, Medzhitov R (2012). Evolution of inflammatory diseases. *Current Biology*, 22.

Oldstone MBA (1998). *Viruses, plagues and history*. Oxford University Press, Oxford.

Oleksiak MF, Churchill GA, Crawford DL (2002). Variation in gene expression within and among natural populations. *Nature Genetics*, 32: 261–6.

Oli AN, Obialor WO, Ifeanyichukwu MO, et al. (2020). Immunoinformatics and vaccine development: an overview. *ImmunoTargets and Therapy*, 9: 13–30.

Olival KJ, Hosseini PR, Zambrana-Torrelio C, Ross N, Bogich TL, Daszak P (2017). Host and viral traits predict zoonotic spillover from mammals. *Nature*, 546: 646–50.

Onyilagha C, Uzonna JE (2019). Host immune responses and immune evasion strategies in African trypanosomiasis. *Frontiers in Immunology*, 10: 2738.

Onzima RB, Upadhyay MR, Doekes HP, et al. (2018). Genome-wide characterization of selection signatures and runs of homozygosity in ugandan goat breeds. *Frontiers in Genetics*, 9: 318.

Oren M, Rosental B, Hawley TS, et al. (2019). Individual sea urchin coelomocytes undergo somatic immune gene diversification. *Frontiers in Immunology*, 10: 1298.

Org E, al. e (2015). Genetic and environmental control of host– gut microbiota interactions. *Genome Research*, 25: 1558–69.

Orlandi-Pradines V, Penhoat K, Durand C, et al. (2006). Antibody responses to several malaria pre-erythrocytic antigens as a marker of malaria exposure among travelers. *The American Journal of Tropical Medicine and Hygiene*, 74: 979–85.

Orozco-Solis R, Aguilar-Arnal L (2020). Circadian regulation of immunity through epigenetic mechanisms. *Frontiers in Cellular and Infection Microbiology*, 10: 96.

Orzalli MH, Knipe DM (2014). Cellular sensing of viral DNA and viral evasion mechanisms. *Annual Review of Microbiology*, 68: 477–92.

Ostfeld RS, Keesing F (2012). Effects of host diversity on infectious disease. *Annual Review of Ecology, Evolution, and Systematics*, 43: 157–82.

Otterstatter MC, Gegear RJ, Colla S, Thomson JD (2005). Effects of parasitic mites and protozoa on the flower constancy and foraging rate of bumble bees. *Behavioural Ecology and ScoiobiologySociobiology*, 58: 383–9.

Otto SP, Lenormand T (2002). Resolving the paradox of sex and recombination. *Nature Reviews Genetics*, 3: 252–61.

Otto SP, Payseur BA (2019). Crossover interference: shedding light on the evolution of recombination. *Annual Review of Genetics*, 53: 19–44.

Overton K, Barrett LT, Oppedal F, Kristiansen TS, Dempster T (2020). Sea lice removal by cleaner fish in salmon aquaculture: a review of the evidence base. *Aquaculture Environment Interactions*, 12: 31–44.

Overton K, Dempster T, Oppedal F, Kristiansen TS, Gismervik K, Stien LH (2019). Salmon lice treatments and salmon mortality in Norwegian aquaculture: a review. *Reviews in Aquaculture*, 11: 1398–417.

Oyebola KM, Idowu ET, Olukosi YA, Awolola TS, Amambua-Ngwa A (2017). Pooled-DNA sequencing identifies genomic regions of selection in Nigerian isolates of *Plasmodium falciparum*. *Parasites & Vectors*, 10: 320.

Page RDM (2003). *Tangled trees: phylogeny, cospeciation and coevolution*. Chicago University Press, Chicago.

Paillot A (1920). L'immunite acquise chez les insectes. *Comptes Rendues de la Societé Biologique de Paris*, 83: 278–80.

Palmer WH, Hadfield JD, Obbard DJ (2018). RNA-Interference pathways display high rates of adaptive protein evolution in multiple invertebrates. *Genetics*, 208: 1585–99.

Pancer Z (2000). Dynamic expression of multiple scavenger receptor cysteine-rich genes in coelomocytes of the purple sea urchin. *Proceedings of the National Academy of Sciences USA*, 97: 13161.

Panda AK, Das BK, Panda A, et al. (2016). Heterozygous mutants of TIRAP (S180L) polymorphism protect adult patients with *Plasmodium falciparum* infection against severe disease and mortality. *Infection, Genetics and Evolution*, 43: 146–50.

Pandit A, de Boer RJ (2014). Reliable reconstruction of HIV-1 whole genome haplotypes reveals clonal interference and genetic hitchhiking among immune escape variants. *Retrovirology*, 11: 56.

Pandit PS, Doyle MM, Smart KM, Young CCW, Drape GW, Johnson CK (2018). Predicting wildlife reservoirs and global vulnerability to zoonotic Flaviviruses. *Nature Communications*, 9: 5425.

Pang X, Xiao X, Liu Y, et al. (2016). Mosquito C-type lectins maintain gut microbiome homeostasis. *Nature Microbiology*, 1: 16023.

Papavasilou FN, Schatz DG (2002). Somatic hypermutation of immunoglobulin genes: merging mechanisms for genetic diversity. *Cell*, 109: S35–44.

Papkou A, Guzella T, Yang W, et al. (2019). The genomic basis of Red Queen dynamics during rapid reciprocal host–pathogen coevolution. *Proceedings of the National Academy of Sciences USA*, 116: 923–8.

Park T (1948). Experimental studies of interspecific competition. I. Competition between populations of the flour beetles *Tribolium confusum* Duval and *Tribolium castaneum* Herbst. *Ecological Monographs*, 18: 267–307.

Parker BJ, Barribeau SM, Laughton AM, de Roode JC, Gerardo NM (2011). Non-immunological defense in an evolutionary framework. *Trends in Ecology & Evolution*, 26: 242–8.

Parker BJ, Hrcek J, McLean AHC, Godfray HCJ (2017). Genotype specificity among hosts, pathogens, and beneficial microbes influences the strength of symbiont-mediated protection. *Evolution*, 71: 1222–31.

Parker GA (1979). Sexual selection and sexual conflict In Blum MS, Blum NA (eds) *Sexual selection and reproductive competition in insects*, pp. 123–66. Academic Press, New York.

Parker GA, Ball MA, Chubb JC (2009). To grow or not to grow? Intermediate and paratenic hosts as helminth life cycle strategies. *Journal of Theoretical Biology*, 258: 135–47.

Parker GA, Ball MA, Chubb JC (2015a). Evolution of complex life cycles in trophically transmitted helminths. I. Host incorporation and trophic ascent. *Journal of Evolutionary Biology*, 28: 267–91.

Parker GA, Chubb JC, Ball MA, Roberts GN (2003). Evolution of complex life cycles in helminth parasites. *Nature*, 425: 480–4.

Parker IM, Saunders M, Bontrager M, et al. (2015b). Phylogenetic structure and host abundance drive disease pressure in communities. *Nature*, 520: 542–4.

Parks T, Elliott K, Lamagni T, et al. (2020). Elevated risk of invasive group A streptococcal disease and host genetic variation in the human leucocyte antigen locus. *Genes & Immunity*, 21: 63–70.

Paro S, Imler J-L, Meignin C (2015). Sensing viral RNAs by Dicer/RIG-I like ATPases across species. *Current Opinion in Immunology*, 32: 106–13.

Parret AHA, De Mot R (2002). Bacteria killing their own kind: novel bacteriocins of *Pseudomonas* and other γ-proteobacteria. *Trends in Microbiology*, 10: 107–12.

Parsons B, Foley E (2016). Cellular immune defenses of *Drosophila melanogaster*. *Developmental & Comparative Immunology*, 58: 95–101.

Patel MK, Gacic-Dobo M, Strebel Pv, et al. (2016). Progress toward regional measles elimination: worldwide, 2000–2015. *Morbidity and Mortality Weekly Report (CDC)*, 65: 1228–33.

Paterson S, Wilson K, Pemberton JM (1998). Major histocompatibility complex variation associated with juvenile survival and parasite resistance in a large unmanaged population (*Ovis aries* L.). *Proceedings of the National Academy of Sciences USA*, 95: 3714–19.

Patterson JEH, Ruckstuhl KE (2013). Parasite infection and host group size: a meta-analytical review. *Parasitology*, 140: 803–13.

Paules CI, Marston HD, Fauci AS (2019). Measles in 2019: going backward. *New England Journal of Medicine*, 380: 2185–7.

Paull SH, Johnson PTJ (2014). Experimental warming drives a seasonal shift in the timing of host-parasite dynamics with consequences for disease risk. *Ecology Letters*, 17: 445–53.

Pawluk A, Davidson AR, Maxwell KL (2018). Anti-CRISPR: discovery, mechanism and function. *Nature*, 16: 12–17.

Payne S (2017). Family Coronaviridae. *Viruses*: 149–58.

Pearson JP, Feldman M, Iglewski BH, Prince A (2000). *Pseudomonas aeruginosa* cell-to-cell signaling is required for virulence in a model of acute pulmonary infection. *Infection and Immunity*, 68: 4331–4.

Pechous RD (2017). With friends like these: the complex role of neutrophils in the progression of severe pneumonia. *Frontiers in Cellular and Infection Microbiology*, 7: 160.

Pedersen AB, Altizer S, Poss M, Cunningham AA, Nunn CL (2005). Patterns of host specificity and transmission among parasites of wild primates. *International Journal for Parasitology*, 35: 647–57.

Pedersen AB, Fenton A (2006). Emphasizing the ecology in parasite community ecology. *Trends in Ecology & Evolution*, 22: 133–9.

Pedersen AB, Jones KE, Nunn CL, Altizer S (2007). Infectious diseases and extinction risk in wild mammals. *Conservation Biology*, 21: 1269–79.

Peeri NC, Shrestha N, Rahman MS, et al. (2020). The SARS, MERS and novel coronavirus (COVID-19) epidemics, the newest and biggest global health threats: what lessons have we learned? *International Journal of Epidemiology*, 49: 717–26.

Pees B, Yang WT, Zarate-Potes A, Schulenburg H, Dierking K (2016). High innate immune specificity through diversified C-type lectin-like domain proteins in invertebrates. *Journal of Innate Immunity*, 8: 129–42.

Penadés JR, Christie GE (2015). The phage-inducible chromosomal islands: a family of highly evolved molecular parasites. *Annual Review of Virology*, 2: 181–201.

Peng Y, Grassl J, Millar AH, Baer B (2016). Seminal fluid of honeybees contains multiple mechanisms to combat infections of the sexually transmitted pathogen *Nosema apis*. *Proceedings of the Royal Society London B*, 283: 20151785.

Penley MJ, Greenberg AB, Khalid A, Namburar SR, Morran LT (2018). No measurable fitness cost to experimentally evolved host defence in the *Caenorhabditis elegans–Serratia marcescens* host–parasite system. *Journal of Evolutionary Biology*, 31: 1976–81.

Penn D (2002). The scent of genetic compatibility: sexual selection and the major histocompatibility complex. *Ethology*, 108: 1–21.

Pennings PS, Kryazhimskiy S, Wakeley J (2014). Loss and recovery of genetic diversity in adapting populations of HIV. *PLoS Genetics*, 10: e1004000.

Peraro MD, van der Goot FG (2016). Pore-forming toxins: ancient, but never really out of fashion. *Nature Reviews Microbiology*, 14: 77–92.

Perazzo P, Eguibar N, González RH, Nusblat AD, Cuestas MaL (2015). Hepatitis B virus (HBV) and S-escape mutants: from the beginning until now. *Journal of Human Virology & Retrovirology*, 2: 00046.

Perelson AS (2018). *Theoretical immunology, Part One*. Taylor & Francis, Boca Raton.

Perelson AS, Ribeiro RM (2018). Introduction to modeling viral infections and immunity. *Immunological Review*, 285: 5–8.

Pérez-Ponce de León G, Hernández-Mena DI (2019). Testing the higher-level phylogenetic classification of Digenea (Platyhelminthes, Trematoda) based on nuclear rDNA sequences before entering the age of the 'next-generation' Tree of Life. *Journal of Helminthology*, 93: 260–76.

Perlman SJ, Hunter MS, Zchori-Fein E (2006). The emerging diversity of *Rickettsia*. *Proceedings of the Royal Society London B*, 273: 2097–106.

Perlman SJ, Jaenike J (2003). Infection success in novel hosts: an experimental and phylogenetic study of *Drosophila*-parasitic nematodes. *Evolution*, 57: 544–57.

Perron GG, Gonzalez A, Buckling A (2007). Source–sink dynamics shape the evolution of antibiotic resistance and its pleiotropic fitness cost. *Proceedings of the Royal Society London B*, 274: 2351–6.

Peterson E, Kaur P (2018). Antibiotic resistance mechanisms in bacteria: relationships between resistance determinants of antibiotic producers, environmental bacteria, and clinical pathogens. *Frontiers in Microbiology*, 9: 2928.

Petrosillo N, Viceconte G, Ergonul O, Ippolito G, Petersen E (2020). COVID-19, SARS and MERS: are they closely related? *Clinical Microbiology and Infection*, 26: 729–34.

Pham LN, Dionne MS, Shirasu-Hiza M, Schneider DS (2007). A specific primed immune response in *Drosophila* is dependent on phagocytes. *PLoS Pathogens*, 3: e26.

Phillips AN (1996). Reduction of HIV concentration during acute infection: independence from a specific immune response. *Science*, 271: 497–9.

Phillips CD, Hanson J, Wilkinson JE, et al. (2017). Microbiome structural and functional interactions across host dietary niche space. *Integrative and Comparative Biology*, 57: 743–55.

Phillips ZN, Tram G, Seib KL, Atack JM (2019). Phase-variable bacterial loci: how bacteria gamble to maximise fitness in changing environments. *Biochemical Society Transactions*, 47: 1131–41.

Phongsisay V (2016). The immunobiology of *Campylobacter jejuni*: innate immunity and autoimmune diseases. *Immunobiology*, 221: 535–43.

Piddock LJV (2006). Clinically relevant chromosomally encoded multidrug resistance efflux pumps in bacteria. *Clinical Microbiology Reviews*, 19: 382–402.

Pierini F, Lenz TL (2018). Divergent allele advantage at human MHC genes: signatures of past and ongoing selection. *Molecular Biology and Evolution*, 35: 2145–58.

Pietrocola G, Nobile G, Rindi S, Speziale P (2017). *Staphylococcus aureus* manipulates innate immunity through own and host-expressed proteases. *Frontiers in Cellular and Infection Microbiology*, 7: 166.

Pigeault R, Garnier R, Rivero A, Gandon S (2018). Evolution of transgenerational immunity in invertebrates. *Proceedings of the Royal Society London B*, 283: 20161136.

Pike VL, Ford SA, King KC, Rafaluk-Mohr C (2019). Fecundity compensation is dependent on the generalized stress response in a nematode host. *Ecology and Evolution*, 9: 11957–61.

Pila EA, Li H, Hambrook JR, Wu X, Hanington PC (2017). Schistosomiasis from a snail's perspective: advances in snail immunity. *Trends in Parasitology*, 33: 845–57.

Pinaud S, Portela J, Duval D, et al. (2016). A shift from cellular to humoral responses contributes to innate immune memory in the vector snail *Biomphalaria glabrata*. *PLoS Pathogens*, 12: e1005361.

Pirtskhalava M, Gabrielian A, Cruz P, et al. (2016). DBAASP v.2: an enhanced database of structure and antimicrobial/cytotoxic activity of natural and synthetic peptides. *Nucleic Acids Research*, 44 (D1): D1104–12.

Pitombo FB (2020). Class Cirripedia. In Rogers DC, Damborenea C, Thorp J (eds) *Thorp and Covich's freshwater invertebrates* (4th ed.), pp. 579–84. Academic Press, London.

Plasterk RHA (2002). RNA silencing: the genome's immune system. *Science*, 296: 1263–5.

Plotkin SA (ed.) (2011). *History of vaccine development.* Springer, New York.

Plowright RK, Becker DJ, McCallum H, Manlove KR (2019). Sampling to elucidate the dynamics of infections in reservoir hosts. *Philosophical Transactions of the Royal Society London B*, 374: 20180336.

Plowright RK, Parrish CR, McCallum H, et al. (2017). Pathways to zoonotic spillover. *Nature Reviews Microbiology*, 15: 502–10.

Plowright RK, Sokolow SH, Gorman ME, Daszak P, Foley JE (2008). Causal inference in disease ecology: investigating ecological drivers of disease emergence. *Frontiers in Ecology and the Environment*, 6: 420–9.

Poiani A, Goldsmith AR, Evans MR (2000). Ectoparasites of house sparrows (*Passer domesticus*): an experimental test of the immunocompetence handicap hypothesis and a new model. *Behavioral Ecology and Sociobiology*, 47: 230–42.

Poirie M, Frey F, Hita M, et al. (2000). *Drosophila* resistance genes to parasitoids: chromosomal location and linkage analysis. *Proceedings of the Royal Society London B*, 267: 1417–21.

Poirotte C, Kappeler PM (2019). Hygienic personalities in wild grey mouse lemurs vary adaptively with sex. *Proceedings of the Royal Society London B*, 286: 20190863.

Poirotte C, Massol F, Herbert A, et al. (2017). Mandrills use olfaction to socially avoid parasitized conspecifics. *Science Advances*, 3: e1601721.

Poisot T, Guéveneux-Julien C, Fortin M-J, Gravel D, Legendre P (2017). Hosts, parasites and their interactions respond to different climatic variables. *Global Ecology and Biogeography*, 26: 942–51.

Pojmańska T, Niewiadomska K (2012). New trends in research on parasite host specificity: a survey of current parasitological literature. *Annals of Parasitology*, 58: 57–61.

Polak M (2003). Heritability of resistance against ectoparasitism in the *Drosophila- Macrocheles* system. *Journal of Evolutionary Biology*, 16: 74–82.

Poland J, Rutkoski J (2016). Advances and challenges in genomic selection for disease resistance. *Annual Review of Phytopathology*, 54: 79–98.

Pollitt EJG, West SA, Crusz SA, Burton-Chellew MN, Diggle SP (2014). Cooperation, quorum sensing, and evolution of virulence in *Staphylococcus aureus*. *Infection and Immunity*, 82: 1045–51.

Pomiankowski A (1987). Sexual selection: the handicap principle does work—sometimes. *Proceedings of the Royal Society London B*, 231: 123–45.

Pomiankowski A, Iwasa Y (1998). Runaway ornament diversity caused by Fisherian sexual selection. *Proceedings of the National Academy of Sciences USA*, 95: 5106–11.

Ponton F, Lalubin F, Fromont C, Wilson K, Behm C, Simpson SJ (2011a). Hosts use altered macronutrient intake to circumvent parasite-induced reduction in fecundity. *International Journal for Parasitology*, 41: 43–60.

Ponton F, Wilson K, Cotter S, Raubenheimer D, Simpson S (2011b). Nutritional immunology: a multi-dimensional approach. *PLoS Pathogens*, 7: e1002223.

Ponton F, Wilson K, Holmes AJ, Cotter SC, Raubenheimer D, Simpson SJ (2013). Integrating nutrition and immunology: a new frontier. *Journal of Insect Physiology*, 59: 130–7.

Poore GCB, Bruce NL (2012). Global diversity of marine Isopods (except Asellota and Crustacean symbionts). *PLoS ONE*, 7: e43529.

Popat R, Harrison F, da Silva Ana C, et al. (2017). Environmental modification via a quorum sensing molecule influences the social landscape of siderophore production. *Proceedings of the Royal Society London B*, 284: 20170200.

Portnoy DA (2005). Manipulation of innate immunity by bacterial pathogens. *Current Opinion in Immunology*, 17: 1–4.

Portnoy DA, Auerbuch V, Glomski I (2002). The cell biology of *Listeria monocytogenes* infection: the intersection of bacterial pathogenesis and cell-mediated immunity. *Journal of Cell Biology*, 126: 869–80.

Pott J, Hornef M (2012). Innate immune signalling at the intestinal epithelium in homeostasis and disease. *EMBO Reports*, 13 684–98.

Poulin R (1992). Determinants of host-specificity in parasites of freshwater fishes. *International Journal for Parasitology*, 22: 753–8.

Poulin R (1994). The evolution of parasite manipulation of host behaviour: a theoretical analysis. *Parasitology*, 109: S109–18.

Poulin R (1995). 'Adaptive' change in the behaviour of parasitized animals: a critical review. *International Journal for Parasitology*, 25: 1371–83.

Poulin R (1996). Richness, nestedness, and randomness in parasite infracommunity structure. *Oecologia*, 105: 545–51.

Poulin R (1997). Species richness of parasite assemblages: evolution and patterns. *Annual Review of Ecology and Systematics*, 28: 341–58.

Poulin R (1998). Large-scale patterns of host use by parasites of freshwater fishes. *Ecology Letters*, 1: 118–28.

Poulin R (1999). The intra- and interspecific relationships between abundance and distribution of helminth parasites of birds. *Journal of Animal Ecology*, 68: 719–25.

Poulin R (2000). Manipulation of host behaviour by parasites: a weakening paradigm? *Proceedings of the Royal Society London B*, 267: 787–92.

Poulin R (2004). *Parasites in marine systems*. Cambridge University Press, Cambridge.

Poulin R (2007a). Are there general laws in parasite ecology? *Parasitology*, 134: 763–76.

Poulin R (2007b). *Evolutionary ecology of parasites*. 2nd ed. Princeton University Press, Princeton, NJ.

Poulin R (2010). Parasite manipulation of host behavior: an update and frequently asked questions. In Brockmann HJ, Roper TJ, Naguib M, Wynne-Edwards KE, Mitani JC, Simmons LW (eds) *Advances in the study of behavior*, pp. 151–86. Academic Press, London.

Poulin R (2014). Parasite biodiversity revisited: frontiers and constraints. *International Journal for Parasitology*, 44: 581–9.

Poulin R, Cribb T (2002). Trematode life cycles: short is sweet? *Trends in Parasitology*, 18: 176–83.

Poulin R, FitzGerald GJ (1989). Risk of parasitism and microhabitat selection in juvenile sticklebacks. *Canadian Journal of Zoology*, 67: 14–18.

Poulin R, Fredensborg BL, Hansen E, Leung TLF (2005). The true cost of manipulation by parasites *Behavioural Processes*, 68: 241–4.

Poulin R, Keeney DB (2008). Host specificity under molecular and experimental scrutiny. *Trends in Parasitology*, 24: 24–8.

Poulin R, Krasnov BR, Mouillot D (2011). Host specificity in phylogenetic and geographic space. *Trends in Parasitology*, 27: 355–61.

Poulin R, Krasnov BR, Shenbrot GI, Mouillot D, Khokhlova IS (2006). Evolution of host specificity in fleas: is it directional and irreversible? *International Journal for Parasitology*, 36: 185–91.

Poulin R, Maure F (2015). Host manipulation by parasites: a look back before moving forward. *Trends in Parasitology*, 31: 563–70.

Poulin R, Morand S (1999). Geographical distance and the similarity among parasite communities of conspecific host populations. *Parasitology*, 119: 369–74.

Poulin R, Morand S (2004). *Parasite biodiversity*. Smithsonian Institution Press, Washington DC.C.

Poulin R, Mouillot D (2003). Parasite specialization from a phylogenetic perspective: a new index of host specificity. *Parasitology*, 126: 473–80.

Poulin R, Mouillot D (2005). Combining phylogenetic and ecological information into a new index of host specificity. *Journal of Parasitology*, 91: 511–14.

Poulin R, Randhawa HS (2015). Evolution of parasitism along convergent lines: from ecology to genomics. *Parasitology*, 142: S6–15.

Poulin R, Valtonen ET (2001). Nested assemblages resulting from host-size variation: the case of endoparasite communities in fish hosts. *International Journal for Parasitology*, 31: 1194–204.

Pounds JA, Bustamante MR, Coloma LA, et al. (2006). Widespread amphibian extinctions from epidemic disease driven by global warming. *Nature*, 439: 161–7.

Povey S, Cotter SC, Simpson SJ, Lee KP, Wilson K (2009). Can the protein costs of bacterial resistance be offset by altered feeding behaviour? *Journal of Animal Ecology*, 78: 437–46.

Povey S, Cotter SC, Simpson SJ, Wilson K (2014). Dynamics of macronutrient self-medication and illness-induced anorexia in virally infected insects. *Journal of Animal Ecology*, 83: 245–55.

Powell JR, Gloria-Soria A, Kotsakiozi P (2018). Recent history of *Aedes aegypti*: vector genomics and epidemiology records. *BioScience*, 68: 854–60.

Pradeu T, Cooper EL (2012). The danger theory: 20 years later. *Frontiers im in Immunology*, 3: 287.

Pradeu T, du Pasquier L (2018). Immunological memory: what's in a name? *Immunological Reviews*, 283: 7–20.

Prenter J, MacNeil C, Dick JTA, Dunn AM (2004). Roles of parasites in animal invasions. *Trends in Ecology & Evolution*, 19: 385–90.

Presley SJ, Dallas T, Klingbeil BT, Willig MR (2015). Phylogenetic signals in host–parasite associations for Neotropical bats and Nearctic desert rodents. *Biological Journal of the Linnean Society*, 116: 312–27.

Presloid JB, Novella IS (2015). RNA viruses and RNAi: quasispecies implications for viral escape. *Viruses*, 7: 3226–40.

Pressman A, Blanco C, Chen Irene A (2015). The RNA world as a model system to study the origin of life. *Current Biology*, 25: R953–63.

Previtali MA, Ostfeld RS, Keesing F, Jolles AE, Hanselmann R, Martin LB (2012). Relationship between pace of life and immune responses in wild rodents. *Oikos*, 121: 1483–92.

Price PW (1980). *Evolutionary biology of parasites*. Princeton University Press, Princeton, NJ.

Price PW, Westoby M, Rice B (1988). Parasite-mediated competition: some predictions and tests. *The American Naturalist*, 13: 544–55.

Prior KF, Rijo-Ferreira F, Assis PA, et al. (2020). Periodic parasites and daily host rhythms. *Cell Host & Microbe*, 27: 176–87.

Pruett-Jones SG, Pruett-Jones MA, Jones HI (1991). Parasites and sexual selection in a New Guinea avifauna. *Current Ornithology*, 7: 22–43.

Prugnolle F, Manica A, Charpentier M, Guégan JF, Guernier V, Balloux F (2005). Pathogen-driven selection and worldwide HLA Class I diversity. *Current Biology*, 15: 1022–7.

Pujol JM, Eisenberg JE, Haas CN, Koopman JS (2009). The effect of ongoing exposure dynamics in dose response relationships. *PLoS Computational Biology*, 5: e100399.

Pull CD, Ugelvig LV, Wiesenhofer F, et al. (2018). Destructive disinfection of infected brood prevents systemic disease spread in ant colonies. *eLife*: 32073.

Pulliam JRC (2008). Viral host jumps: moving toward a predictive framework. *Ecohealth*, 5: 80–91.

Punt J, Strandford S, Jones P, Owen JA (2018). *Kuby Immunology, 8th ed.* Macmillan Education Elt, London.

Pybus OG, Rambaut A, Harvey PH (2000). An integrated framework for the inference of viral population history from reconstructed genealogies. *Genetics*, 155: 1429.

Qian C, Cao X (2013). Regulation of Toll-like receptor signaling pathways in innate immune responses. *Annals of the New York Academy of Science*, 1283: 67–74.

Qin J, Li R, Raes J, et al. (2010). A human gut microbial gene catalogue established by metagenomic sequencing. *Nature*, 464: 59–67.

Qiu P, al e (2011). Extracting a cellular hierarchy from high-dimensional cytometry data with SPADE. *Nature Biotechnology*, 29: 886–91.

Qiu X, Duvvuri VR, Gubbay JB, Webby RJ, Kayali G, Bahl J (2017). Lineage-specific epitope profiles for HPAI H5 pre-pandemic vaccine selection and evaluation. *Influenza and Other Respiratory Viruses*, 11: 445–56.

Quinnell RJ, Bethony J, Pritchard DI (2004). The immunoepidemiology of human hookworm infection. *Parasite Immunology*, 26: 443–54.

Qureshi ST, Larivière L, Leveque G, et al. (1999). Endotoxin-tolerant mice have mutations in Toll-like receptor 4 (Tlr4). *The Journal of Experimental Medicine*, 189: 615–25.

Råberg L (2014). How to live with the enemy: understanding tolerance to parasites. *PLoS Biology*, 12: e1001989.

Råberg L, de Roode JC, Bell AS, Stamou P, Gray D, Read AF (2006). The role of immune-mediated apparent competition in genetically diverse malaria infections. *The American Naturalist*, 168: 41–53.

Råberg L, Grahn M, Hasselquist D, Svensson E (1998). On the adaptive significance of stress-induced immunosuppression. *Proceedings of the Royal Society of London B*, 265: 1637–41.

Råberg L, Sim D, Read AF (2007). Disentangling genetic variation for resistance and tolerance to infectious diseases in animals. *Science*, 318: 812–14.

Råberg L, Stjernman M (2003). Natural selection on immune responsiveness in blue tits *Parus caeruleus*. *Evolution*, 57: 1670–8.

Råberg L, Vestberg M, Hasselquist D, Homdahl R, Svensson E, Nilsson J-A (2002). Basal metabolic rate and the evolution of the adaptive immune system. *Proceedings of the Royal Society London B*, 269: 817–21.

Rafaluk C, Jansen G, Schulenburg H, Joop G (2015). When experimental selection for virulence leads to loss of virulence. *Trends in Parasitology*, 31: 426–34.

Rahnamaeian M, Cytryńska M, Zdybicka-Barabas A, et al. (2015). Insect antimicrobial peptides show potentiating functional interactions against Gram-negative bacteria. *Proceedings of the Royal Society London B*, 282: 20150293.

Rakus K, Ronsmans M, Vanderplasschen A (2017). Behavioral fever in ectothermic vertebrates. *Developmental & Comparative Immunology*, 66: 84–91.

Rall GF (2003). Measles virus 1998–2002: progress and controversy. *Annual Review of Microbiology*, 57: 343–67.

Ramirez JL, de Almeida Oliveira G, Calvo E, et al. (2015). A mosquito lipoxin/lipocalin complex mediates innate immune priming in *Anopheles gambiae*. *Nature Communications*, 6: 7403.

Ramirez-Prado JS, Abulfaraj AA, Rayapuram N, Benhamed M, Hirt H (2018). Plant immunity: from signaling to epigenetic control of defense. *Trends in Plant Science*, 23: 833–44.

Randhawa HS, Saunders GW, Burt MDB (2007). Establishment of the onset of host specificity in four phyllobothriid tapeworm species (Cestoda: Tetraphyllidea) using a molecular approach. *Parasitology*, 134: 1291–300.

Randolph SE, Dobson ADM (2012). Pangloss revisited: a critique of the dilution effect and the biodiversity-buffers-disease paradigm. *Parasitology*, 139: 847–63.

Ranjeva S, Subramanian R, Fang VJ, et al. (2019). Age-specific differences in the dynamics of protective immunity to influenza. *Nature Communications*, 10: 1660.

Rao CR (1982). Diversity and dissimilarity coefficients: a unified approach. *Theoretical Population Biology*, 21: 24–43.

Rasmussen E (1959). Behaviour of sacculinized shore crabs. *Nature*, 183: 479–80.

Raubenheimer D, Simpson SJ (2016). Nutritional ecology and human health. *Annual Review of Nutrition*, 36: 603–26.

Razakandrainibe FG, Duran P, Koella JC, et al. (2005). 'Clonal' population structure of malaria agent, *Plasmodium falciparum*, in high-infection regions. *Proceedings of the National Academy of Sciences USA*, 102: 17388–99.

Read AF (1987). Comparative evidence supports the Hamilton and Zuk hypothesis on parasites and sexual selection. *Nature*, 382: 68–70.

Read AF (1988). Sexual selection and the role of parasites. *Trends in Ecology & Evolution*, 3: 97–102.

Read AF (1994). The evolution of virulence. *Trends in Microbiology*, 2: 73–6.

Read AF, Baigent SJ, Powers C, et al. (2015). Imperfect vaccination can enhance the transmission of highly virulent pathogens. *PLoS Biology*, 13: e1002198.

Read AF, Lynch PA, Thomas MB (2009). How to make evolution-proof insecticides for malaria control. *PLoS Biology*, 7: e1000058.

Read AF, Taylor LH (2001). The ecology of genetically diverse infections. *Science*, 292: 1099–102.

Read AF, Weary DM (1990). Sexual selection and the evolution of bird song: a test of the Hamilton-Zuk hypothesis. *Behavioral Ecology and Sociobiology*, 26: 47–56.

Redfield RJ (2001). Do bacteria have sex? *Nature Reviews Genetics*, 2: 634–9.

Redman EM, Wilson K, Cory JS (2016). Trade-offs and mixed infections in an obligate-killing insect pathogen. *Journal of Animal Ecology*, 85: 1200–9.

Reed P, Francis-Floyd R, Klinger RE (2002). *Monogenean parasites of fish (FA28)*. Fisheries and Aquatic Sciences Department Series. University of Florida, Institute of Food and Agricultural Sciences, Gainesville, FL.

Regoes RR, Bonhoeffer S, Nowak MA (2000). Evolution of virulence in a heterogeneous host population. *Evolution*, 54: 64–71.

Regoes RR, Hamblin S, Tanaka MM (2013). Viral mutation rates: modelling the roles of within-host viral dynamics and the trade-off between replication fidelity and speed. *Proceedings of the Royal Society London B*, 280: 20122047.

Regoes RR, Hottinger JW, Sygnarski L, Ebert D (2003). The infection rate of *Daphnia magna* by *Pasteuria ramosa* conforms with the mass-action principle. *Epidemiology and Infection*, 131: 957–66.

Regoes RR, McLaren P, Battegay M, et al. (2014). Disentangling human tolerance and resistance against HIV. *PLoS Biology*, 12: e1001951.

Reid JM, Arcese P, Cassidy ALEV, Marr AB, Smith JNM, Keller LF (2005). Hamilton and Zuk meet heterozygosity? Song repertoire size indicates inbreeding and immunity in song sparrows (*Melospiza melodia*). *Proceedings of the Royal Society London B*, 272: 481–7.

Reimer-Michalski E, Conrath U (2016). Innate immune memory in plants. *Seminars in Immunology*, 28: 319–27.

Reinhardt K, Siva-Jothy MT (2007). Biology of the bed bugs (Cimicidae). *Annual Review of Entomology*, 52: 351–74.

Reis ES, Mastellos DC, Hajishengallis G, Lambris JD (2019). New insights into the immune functions of complement. *Nature Reviews Immunology*, 19: 503–16.

Reiss ER, Drinkwater LE (2018). Cultivar mixtures: a meta-analysis of the effect of intraspecific diversity on crop yield. *Ecological Applications*, 28: 62–77.

Rejmanek D, Vanwormer E, Mazet JAK, Packham AE, Aguilar B, Conrad PA (2010). Congenital transmission of *Toxoplasma gondii* infection in deer mice (*Peromyscus maniculatus*) after oral oocyst infection. *Journal of Parasitology*, 96: 516–20.

Remoue F, Cisse B, Ba F, et al. (2006). Evaluation of the antibody response to *Anopheles* salivary antigens as a potential marker of risk of malaria. *Transactions of The Royal Society of Tropical Medicine and Hygiene*, 100: 363–70.

Restif O, Grenfell BT (2006). Integrating life history and cross-immunity into the evolutionary dynamics of pathogens. *Proceedings of the Royal Society London B*, 273: 409–16.

Restif O, Koella J (2004). Concurrent evolution of resistance and tolerance to pathogens. *The American Naturalist*, 164: E90–102.

Reuter JA, Spacek DV, Snyder Michael P (2015). High-throughput sequencing technologies. *Molecular Cell*, 58: 586–97.

Reverter M, Tapissier-Bontemps N, Lecchini D, Banaigs B, Sasal P (2018). Biological and ecological roles of external fish mucus: a review. *Fishes*, 3: 41.

Reynolds LA, Finlay BB, Maizels RM (2015). Cohabitation in the intestine: interactions among helminth parasites, bacterial microbiota, and host immunity. *Journal of Immunology*, 195: 4059–66.

Rice WR (2002). Experimental tests of the adaptive significance of sexual reproduction. *Nature Reviews Genetics*, 3: 241–51.

Richard FJ, Aubert A, Grozinger CM (2008). Modulation of social interactions by immune stimulation in honey bee, *Apis mellifera*, workers. *BMC Biology*, 6: 50.

Richard FJ, Holt HL, Grozinger CM (2012). Effects of immunostimulation on social behavior, chemical communication and genome-wide gene expression in honey bee workers (*Apis mellifera*). *BMC Genomics*, 13: 558.

Richards EL, van Oosterhout C, Cable J (2010). Sex-specific differences in shoaling affect parasite transmission in guppies. *PLoS ONE*, 5: e13285.

Richner H, Christie P, Oppliger A (1995). Paternal investment affects prevalence of malaria. *Proceedings of the National Academy of Sciences USA*, 92: 1192–4.

Richter S (1993). Phoretic association between the dauerjuveniles of *Rhabditis stammeri* (Rhabditidae) and life history stages of the burying beetle *Nicrophorus vespilloides* (Coleoptera: Silphidae). *Nematologica*, 39: 346–55.

Ricklefs RE, Outlaw DC, Svensson-Coelho M, Medeiros MCI, Ellis VA, Latta S (2014). Species formation by host shifting in avian malaria parasites. *Proceedings of the National Academy of Sciences USA*, 111: 14816–21.

Ricklin D, Reis ES, Mastellos DC, Gros P, Lambris JD (2016). Complement component C3: the 'Swiss Army Knife' of innate immunity and host defense. *Immunological Reviews*, 274: 33–58.

Riddell CR, Admas S, Schmid-Hempel P, Mallon EB (2009). Differential expression of immune defences underlies specific host-parasite interactions in insects. *PLoS ONE*, 4: e7621.

Riedel S (2005). Edward Jenner and the history of smallpox and vaccination. *Baylor University Medical Center Proceedings* 18: 21–5.

Rifkin JL, Nunn CL, Garamszegi LZ (2012). Do animals living in larger groups experience greater parasitism? a meta-analysis. *The American Naturalist*, 180: 70–82.

Riley MA, Gordon DM (1999). The ecological role of bacteriocins in bacterial competition. *Trends in Microbiology*, 7: 129–33.

Riley MA, Wertz JA (2002). Bacteriocins: evolution, ecology, and application. *Annual Review of Microbiology*, 56: 117–37.

Robert F, Renaud F, Mathieu E, Gabrion C (1988). Importance of the paratenic host in the biology of *Bothriocephalus gregarius* (Cestoda: Pseudophylllidae), a parasite of the turbot. *International Journal for Parasitology*, 24: 1099–115.

Roberts MAJ (2019). Recombinant DNA technology and DNA sequencing. *Essays in Biochemistry*, 63: 457–68.

Roberts MG, Heesterbeek JAP (2020). Characterizing reservoirs of infection and the maintenance of pathogens in ecosystems. *Journal of The Royal Society Interface*, 17: 20190540.

Roberts ML, Buchanan KL, Evans MR (2004). Testing the immunocompetence handicap hypothesis: a review of the evidence. *Animal Behaviour*, 68: 227–39.

Roberts SC, Gosling LM (2003). Genetic similarity and quality interact in mate choice decisions by female mice. *Nature Genetics*, 35: 103-.

Robinson M, Donnelly S, Dalton J (2013). Helminth defence molecules: immunomodulators designed by parasites! *Frontiers in Microbiology*, 4: 296.

Roche DG, Leung B, Franco EF, Torchin ME (2010). Higher parasite richness, abundance and impact in native versus introduced cichlid fishes. *International Journal for Parasitology*, 40: 1525–30.

Rochlin I, Ninivaggi DV, Hutchinson ML, Farajollahi A (2013). Climate change and range expansion of the Asian tiger mosquito (*Aedes albopictus*) in Northeastern USA: implications for public health practitioners. *PLoS ONE*, 8: e60874.

Rockman MV, Kruglyak L (2006). Genetics of global gene expression. *Nature*, 7: 862–72.

Rodell CF, Schipper MR, Keenan DK (2004). Modes of selection and recombination response in *Drosophila melanogaster*. *Journal of Heredity*, 95: 70–5.

Rodgers-Gray TP, Smith JE, Ashcroft AE, Isaac RE, Dunn AM (2004). Mechanisms of parasite-induced sex reversal in *Gammarus duebeni*. *International Journal for Parasitology*, 34: 747–53.

Rogawa A, Ogata S, Mougi A (2018). Parasite transmission between trophic levels stabilizes predator–prey interaction. *Scientific Reports*, 8: 12246.

Rohde K (1980). Host-specificity indices of parasites and their application. *Experientia*, 36: 1369–71.

Rohde K (2002). Ecology and biogeography of marine parasites. In *Advances in marine biology*, pp. 1–83. Academic Press, London.

Rohde K (2005a). *Marine parasitology*. CSIRO Publishing, 592 pp., Collingwood, Victoria.

Rohde K (2005b). *Nonequilibrium ecology*. Cambridge University Press, Cambridge.

Rohde K, Heap M (1998). Latitudinal differences in species and community richness and in community structure of metazoan endo- and ectoparasites of marine teleost fish. *International Journal for Parasitology*, 28: 461–74.

Rohr JR, Barrett CB, Civitello DJ, et al. (2019). Emerging human infectious diseases and the links to global food production. *Nature Sustainability*, 2: 445–56.

Rolff J (2007). Why did the acquired immune system of vertebrates evolve? *Developmental and Comparative Immunology*, 1: 467–82.

Rolff J, Kraaijeveld AR (2003). Selection for parasitoid resistance alters mating success in *Drosophila*. *Proceedings of the Royal Society London B*, 270: S154–5.

Rolff J, Schmid-Hempel P (2016). Perspectives on the evolutionary ecology of arthropod antimicrobial peptides *Philosophical Transactions of the Royal Society London B*, 371: 20150297.

Rolff J, Siva-Jothy MT (2002). Copulation corrupts immunity: A mechanism for a cost of mating in insects. *Proceedings of the National Academy of Sciences, USA*, 99: 9916–18.

Rolff J, Siva-Jothy MT (2004). Selection on insect immunity in the wild. *Proceedings of the Royal Society London B*, 271: 2157–60.

Romero-Meza G, Mugnier MR (2020). *Trypanosoma brucei*. *Trends in Parasitology*, 36: 571–2.

Rosenfeld Y, Barra D, Simmaco M, Shai Y, Mangoni ML (2006). A synergism between temporins toward Gram-negative bacteria overcomes resistance imposed by the lipopolysaccharide protective layer. *Journal of Biological Chemistry*, 281: 28565–74.

Rosental B, Kowarsky M, Seita J, et al. (2018). Complex mammalian-like haematopoietic system found in a colonial chordate. *Nature*, 564: 425–9.

Rosselló-Móra R, Amann R (2015). Past and future species definitions for Bacteria and Archaea. *Systematic and Applied Microbiology*, 38: 209–16.

Rossignol PA, Ribeiro JMC, Jungery M, Turell MJ, Spielman A, Bailey CL (1985). Enhanced mosquito blood-finding success on parasitemic hosts: evidence of vector parasite mutualism. *Proceedings of the National Academy of Sciences USA*, 82: 7725–7.

Rossinelli S, Bacher S (2014). Higher establishment success in specialized parasitoids: support for the existence of trade-offs in the evolution of specialization. *Functional Ecology*, 229: 277–84.

Rossolini GM, Arena F, Pecile P, Pollini S (2014). Update on the antibiotic resistance crisis. *Current Opinion in Pharmacology*, 18: 56–60.

Roth O, Beemelmanns A, Barribeau SM, Sadd BM (2018). Recent advances in vertebrate and invertebrate transgenerational immunity in the light of ecology and evolution. *Heredity*, 121: 225–38.

Roth O, Joop G, Eggert H, et al. (2010). Paternally derived immune priming for offspring in the red flour beetle, *Tribolium castaneum*. *Journal of Animal Ecology*, 79: 403–13.

Roth O, Sadd BM, Schmid-Hempel P, Kurtz J (2009). Strain-specific priming of resistance in the red flour beetle, *Tribolium castaneum*. *Proceedings of the Royal Society London B*, 276: 145–51.

Rothschild DE, McDaniel DK, Ringel-Scaia VM, Allen IC (2018). Modulating inflammation through the negative regulation of NF-κB signaling. *Journal of Leukocyte Biology*, 103: 1131–50.

Routtu J, Ebert D (2015). Genetic architecture of resistance in *Daphnia* hosts against two species of host-specific parasites. *Heredity*, 114: 241–8.

Roved J, Westerdahl H, Hasselquist D (2017). Sex differences in immune responses: hormonal effects, antagonistic selection, and evolutionary consequences. *Hormones and Behavior*, 88: 95–105.

Roy BA, Kirchner JW (2000). Evolutionary dynamics of pathogen resistance and tolerance. *Evolution*, 54: 51–63.

Roy HE, Steinkraus DC, Eilenberg J, Hajek AE, Pell JK (2006). Bizarre interactions and endgames: entomopathogenic fungi and their arthropod hosts. *Annual Review of Entomology*, 51: 331–57.

Rückert C, Ebel GD (2018). How do virus–mosquito interactions lead to viral emergence? *Trends in Parasitology*, 34: 310–21.

Ruiz-Gonzalez MX, Brown MJF (2006). Honey bee and bumblebee trypanosomatids: specificity and potential for transmission. *Ecological Entomology*, 31: 616–22.

Ruiz-López MJ, Monello RJ, Gompper ME, Eggert LS (2012). The effect and relative importance of neutral genetic diversity for predicting parasitism varies across parasite taxa. *PLoS ONE*, 7: e45404–4.

Rukambile E, Sintchenko V, Muscatello G, Kock R, Alders R (2019). Infection, colonization and shedding of *Campylobacter* and *Salmonella* in animals and their contribution to human disease: a review. *Zoonoses and Public Health*, 66: 562–78.

Rumbaugh KP, Trivedi U, Watters C, Burton-Chellew MN, Diggle SP, West SA (2012). Kin selection, quorum sensing and virulence in pathogenic bacteria. *Proceedings of the Royal Society London B*, 279: 3584–8.

Rupprecht C, Hamlon C, Hemachudha T (2002). Rabies re-examined. *The Lancet Infectious Disease*, 2: 327–43.

Rushmore J, Bisanzio D, Gillespie TR (2017). Making new connections: insights from primate-parasite networks. *Trends in Parasitology*, 33: 547–60.

Russell CA, Fonville JM, Brown AEX, et al. (2012). The Potential for Respiratory Droplet–Transmissible A/H5N1 Influenza Virus to Evolve in a Mammalian Host. *Science*, 336: 1541–7.

Ryan MJ (1998). Sexual selection, receiver biases, and the evolution of sex differences. *Science*, 281: 1999–2003.

Ryan SJ, McNally A, Johnson LR, et al. (2015). Mapping physiological suitability limits for malaria in Africa under climate change. *Vector-Borne and Zoonotic Diseases*, 15: 718–25.

Rybicki J, Kisdi E, Anttila JV (2018). Model of bacterial toxin-dependent pathogenesis explains infective dose. *Proceedings of the National Academy of Sciences USA*, 115: 10690–5.

Saavedra I, Amo L (2018). Insectivorous birds eavesdrop on the pheromones of their prey. *PLoS ONE*, 13: e0190415.

Sachs J, Skophammer R, Regus J (2011). Evolutionary transitions in bacterial symbiosis. *Proceedings of the National Academy of Sciences USA*, 108: 10800–7.

Sacks D, Sher A (2002). Evasion of innate immunity by parasitic protozoa. *Nature Immunology*, 3: 1041–7.

Sackton TB, Lazzaro BP, Clark AG (2010). Genotype and gene expression associations with immune function in *Drosophila*. *PLoS Genetics*, 6: e1000797.

Sadd BM, Kleinlogel Y, Schmid-Hempel R, Schmid-Hempel P (2005). Trans-generational immune priming in a social insect. *Biology Letters*, 1: 386–8.

Sadd BM, Schmid-Hempel P (2006). Insect immunity shows specificity in protection upon secondary pathogen exposure. *Current Biology*, 16: 1206–10.

Sadd BM, Schmid-Hempel P (2007). Facultative but persistent trans-generational immunity via the mother's eggs in bumblebees. *Current Biology*, 17: R1046–7.

Sadd BM, Schmid-Hempel P (2009). A distinct infection cost associated with trans-generational immune priming of antibacterial immunity in bumble-bees. *Biology Letters*, 5: 798–801.

Sadd BM, Siva-Jothy MT (2006). Self-harm caused by an insect's innate immunity. *Proceedings of the Royal Society London B*, 273: 2571–4.

Sagna A, Poinsignon A, Remoue F (2017). Epidemiological applications of assessing mosquito exposure in a malaria-endemic area. In Wikel SK, Aksoy S, Dimopoulos G (eds) *Arthropod vector: controller of disease transmission, Volume 2*, pp. 209–29. Academic Press, London.

Sah P, Mann J, Bansal S (2018). Disease implications of animal social network structure: a synthesis across social systems. *Journal of Animal Ecology*, 87: 546–58.

Saini J, McPhee JS, Al-Dabbagh S, Stewart CE, Al-Shanti N (2016). Regenerative function of immune system: modulation of muscle stem cells. *Ageing Research Reviews*, 27: 67–76.

Saintenac C, Lee W-S, Cambon F, et al. (2018). Wheat receptor-kinase-like protein Stb6 controls gene-for-gene resistance to fungal pathogen *Zymoseptoria tritici*. *Nature Genetics*, 50: 368–74.

Saitou M, Gokcumen O (2020). An evolutionary perspective on the impact of genomic copy number variation on human health. *Journal of Molecular Evolution*, 88: 104–19.

Salathé M, Kouyos RD, Bonhoeffer S (2008a). The state of affairs in the kingdom of the Red Queen. *Trends in Ecology & Evolution*, 23: 439–45.

Salathé M, Kouyos RD, Regoes RR, Bonhoeffer S (2008b). Rapid parasite adaptation drives selection for high recombination rates. *Evolution*, 62: 295–300.

Salathé M, Soyer O (2008). Parasites lead to evolution of robustness against gene loss in host signaling networks. *Molecular Systems Biology*, 4: 1–9.

Salkeld DJ, Padgett KA, Jones JH (2013). A meta-analysis suggesting that the relationship between biodiversity and risk of zoonotic pathogen transmission is idiosyncratic. *Ecology Letters*, 16: 679–86.

Salles TS, da Encarnacao Sa-Guimaraes T, de Alvarenga ESL, et al. (2018). History, epidemiology and diagnostics of dengue in the American and Brazilian contexts: a review. *Parasites & Vectors*, 11: 264.

Salmela H, Amdam GV, Freitak D (2015). Transfer of immunity from mother to offspring is mediated via egg-yolk protein vitellogenin. *PLoS Pathogens*, 11: e1005015.

Salyers AA, Whitt DD (2002). *Bacterial pathogenesis*, 2nd ed. ASM Press, Washington D.C.

San Filippo J, Sung P, Klein H (2008). Mechanism of eukaryotic homologous recombination. *Annual Review of Biochemistry*, 77: 229–57.

Sánchez Barranco V, Van der Meer MTJ, Kagami M, et al. (2020). Trophic position, elemental ratios and nitrogen transfer in a planktonic host–parasite–consumer food chain including a fungal parasite. *Oecologia*: https://doi.org/10.1007/s00442-020-4721-w.

Sánchez-Romero MA, Casadesús J (2020). The bacterial epigenome. *Nature Reviews Microbiology*, 18: 7–20.

Sanchez-Villagra MR, Pope TR, Salas V (1998). Relation of intergroup variation in allogrooming to group social structure and ectoparasite loads in red howlers (*Alouatta seniculus*). *International Journal of Primatology*, 19: 473–91.

Sandmeier FC, Tracy RC (2014). The metabolic pace-of-life model: incorporating ectothermic organisms into the theory of vertebrate ecoimmunology. *Integrative and Comparative Biology*, 54: 387–95.

Sanjuán R (2010). Mutational fitness effects in RNA and single-stranded DNA viruses: common patterns revealed by site-directed mutagenesis studies. *Philosophical Transactions of the Royal Society London B*, 365: 1975–82.

Sanjuán R, Domingo-Calap P (2016). Mechanisms of viral mutation. *Cellular and Molecular Life Sciences*, 73: 4433–48.

Sanjuán R, Nebot MR, Chirico N, Mansky LM, Belshaw R (2010). Viral mutation rates. *Journal of Virology*, 84: 9733–48.

Saranathan R, Sathyamurthi P, Thiruvengadam K, et al. (2020). MAL adaptor (TIRAP) S180L polymorphism and severity of disease among tuberculosis patients. *Infection, Genetics and Evolution*, 77: 104093.

Sardelis MR, Turell MJ, O'Guinn ML, Andre RG, Roberts DR (2002). Vector competence of three North American strains of *Aedes albopictus* for West Nile virus. *Journal of the American Mosquito Control Association*, 18: 284–9.

Sarvetnick N, Ohashi P (2003). Autoimmunity. *Current Opinion in Immunology*, 15: 647–50.

Sasal P, Desdevises Y, Morand S (1998). Host-specialization and species diversity in fish parasites: phylogenetic conservationism? *Ecography*, 21: 639–43.

Sasal P, Trouvé S, Müller-Graf C, Morand S (1999). Specificity and host predictability: a comparative analysis among monogenean parasites of fish. *Journal of Animal Ecology*, 68: 437–44.

Sato T, Egusa T, Fukushima K, et al. (2012). Nematomorph parasites indirectly alter the food web and ecosystem function of streams through behavioural manipulation of their cricket hosts. *Ecology Letters*, 15: 786–93.

Sato T, Iritani R, Sakura M (2019). Host manipulation by parasites as a cryptic driver of energy flow through food webs. *Current Opinion in Insect Science*, 33: 69–76.

Saulnier E, Gascuel O, Alizon S (2017). Inferring epidemiological parameters from phylogenies using regression-ABC: a comparative study. *PLoS Computational Biology*, 13: e1005416.

Schafer F (1971). Tolerance to plant diseases. *Annual Review of Phytopathology*, 9: 235–52.

Schaller M (2011). The behavioural immune system and the psychology of human sociality. *Philosophical Transactions of the Royal Society London B*, 366: 3418–26.

Schatz DG (2007). DNA deaminases converge on adaptive immunity. *Nature Immunology*, 8: 551–3.

Scheele BC, Pasmans F, Skerratt LF, et al. (2019). Amphibian fungal panzootic causes catastrophic and ongoing loss of biodiversity. *Science*, 363: 1459–63.

Scherer C, Radchuk V, Franz M, et al. (2020). Moving infections: individual movement decisions drive disease persistence in spatially structured landscapes. *Oikos*, 129: 651–67.

Schieber AMP, Ayres JS (2016). Thermoregulation as a disease tolerance defense strategy. *Pathogens and Disease*, 74: ftw106.

Schieber AMP, Lee YM, Chang MW, et al. (2015). Disease tolerance mediated by microbiome *E. coli* involves inflammasome and IGF-1 signaling. *Science*, 350: 558–63.

Schijven J, Rijs GBI, de Roda Husman AM (2005). Quantitative risk assessment of FMD virus transmission via water. *Risk Analysis*, 25: 13–21.

Schlenke TA, Begun, D, J, (2003). Natural selection drives *Drosophila* immune system evolution. *Genetics*, 164: 1471–80.

Schmid-Hempel P (1998). *Parasites in social insects*. Princeton University Press, Princeton, New Jersey.

Schmid-Hempel P (2003). Variation in immune defence as a question of evolutionary ecology. *Proceedings of the Royal Society London B*, 270: 357–66.

Schmid-Hempel P (2008). Immune evasion by parasites: a momentous molecular war. *Trends in Ecology & Evolution*, 23: 318–26.

Schmid-Hempel P (2017). Parasites and their social hosts. *Trends in Parasitology*, 33: 453–62.

Schmid-Hempel P, Crozier RH (1999). Polyandry versus polygyny versus parasites. *Philosophical Transactions of the Royal Society London B*, 354: 507–15.

Schmid-Hempel P, Ebert D (2003). On the evolutionary ecology of specific immune defence. *Trends in Ecology & Evolution*, 18: 27–32.

Schmid-Hempel P, Frank SA (2007). Pathogenesis, virulence, and infective dose. *PLoS Pathogens*, 3: 1372–3.

Schmid-Hempel P, Schmid-Hempel R (1990). Endoparasitic larvae of conopid flies alter pollination behaviour of bumblebees. *Naturwissenschaften*, 27: 50–2.

Schmid-Hempel P, Schmid-Hempel R (1993). Transmission of a pathogen in *Bombus terrestris*, with a note on division of labour in social insects. *Behavioural Ecology and Sociobiology*, 33: 319–27.

Schmid-Hempel P, Stauffer HP (1998). Parasites and flower choice of bumblebees. *Animal Behaviour*, 55: 819–25.

Schmid-Hempel R, Tognazzo M, Salathé R, Schmid-Hempel P (2011). Genetic exchange and emergence of novel strains in directly transmitted trypanosomatids. *Infection, Genetics, and Evolution*, 11: 564–71.

Schmidt O, Theopold U, Strand MR (2001). Innate immunity and its evasion and suppression by hymenopteran endoparasitoids. *BioEssays*, 23: 344–51.

Schneider DS (2011). Tracing personalized health curves during infections. *PLoS Biology*, 9: e1001158.

Schneider P, Rund SSC, Smith NL, Prior KF, O'Donnell AJ, Reece SE (2018). Adaptive periodicity in the infectivity of malaria gametocytes to mosquitoes. *Proceedings of the Royal Society London B*, 285: 20181876.

Schrag SJ, Perrot V, Levin BR (1997). Adaptation to fitness costs of antibiotic resistance in *Escherichia coli*. *Proceedings of the Royal Society of London B*, 264: 1287–91.

Schrauwen EJA, Fouchier RAM (2014). Host adaptation and transmission of influenza A viruses in mammals. *Emerging Microbes & Infections*, 3: e9.

Schröder G, Dehio C (2005). Virulence-associated type UIV secretion systems of *Bartonella*. *Trends in Microbiology*, 13: 336–42.

Schroeder BO, Bäckhed F (2016). Signals from the gut microbiota to distant organs in physiology and disease. *Nature Medicine*, 22: 1079–89.

Schrom EC, Prada JM, Graham AL (2017). Immune signaling networks: sources of robustness and constrained evolvability during coevolution. *Molecular Biology and Evolution*, 35: 676–87.

Schubert K, Karousis ED, Jomaa A, et al. (2020). SARS-CoV-2 Nsp1 binds the ribosomal mRNA channel to inhibit translation. *Nature Structural & Molecular Biology*, 27: 959–66.

Schulte RD, Makus C, Hasert B, Michiels NK, Schulenburg H (2010). Multiple reciprocal adaptations and rapid genetic change upon experimental coevolution of an animal host and its microbial parasite. *Proceedings of the National Academy of Sciences USA*, 91: 7184–8.

Schuster A, Erasimus H, Fritah S, et al. (2019a). RNAi/CRISPR screens: from a pool to a valid hit. *Trends in Biotechnology*, 37: 38–55.

Schuster S, Miesen P, van Rij RP (2019b). Antiviral RNAi in insects and mammals: parallels and differences. *Viruses*, 11: 448.

Schwenke RA, Lazzaro BP (2017). Juvenile hormone suppresses resistance to infection in mated female *Drosophila melanogaster*. *Current Biology*, 27: 596–601.

Schwenke RA, Lazzaro BP, Wolfner MF (2015). Reproduction–immunity trade-offs in insects. *Annual Review of Entomology*, 61: 239–56.

Screaton G, Mongkolsapaya J, Yacoub S, Roberts C (2015). New insights into the immunopathology and control of dengue virus infection. *Nature Reviews Immunology*, 15: 745–59.

Scudellari M (2019). Hacking evolution. *Nature*, 571: 160–2.

Seebens H, Schwartz N, Schuppa PJ, Blasius B (2016). Predicting the spread of marine species introduced by global shipping. *Proceedings of the National Academy of Sciences USA*, 113: 5646–51.

Seeger C, Mason WS (2015). Molecular biology of hepatitis B virus infection. *Virology*, 479–480: 672–86.

Seeley TD, Tarpy DR (2007). Queen promiscuity lowers disease within honeybee colonies. *Proceedings of the Royal Society London B*, 274: 67–72.

Seet BT, Jonhnston JB, Brunetti CR, et al. (2003). Poxviruses and immune evasion. *Annual Review of Immunology*, 21: 377–423.

Seghal RNM, Jones HI, Smith TB (2005). Molecular evidence for host specificity of parasitic nematode microfilariae in some African rainforest birds. *Molecular Ecology*, 14: 3977–88.

Ségurel L, Thompson EE, Flutre T, et al. (2012). The ABO blood group is a trans-species polymorphism in primates. *Proceedings of the National Academy of Sciences USA*, 109: 18493.

Seppälä O, Karvonen A, Valtonen ET (2008a). Shoaling behaviour of fish under parasitism and predation risk. *Animal Behaviour*, 75: 145–50.

Seppälä O, Valtonen ET, Benesh DP (2008b). Host manipulation by parasites in the world of dead-end predators: adaptation to enhance transmission? *Proceedings of the Royal Society London B*, 275: 1611–15.

Serhan CN, Savill J (2005). Resolution of inflammation: the beginning programs the end. *Nature Immunology*, 6: 1191–7.

Sewell DL (1995). Laboratory-associated infections and biosafety. *Clinical Microbiology Reviews*, 8: 389–405.

Shakeel M, Xu X, De Mandal S, Ji F (2019). Role of serine protease inhibitors in insect-host-pathogen interactions. *Insect Biochemistry and Physiology*, 102: e21556.

Shao S, Sun XM, Chen Y, Zhan B, Zhu XP (2019). Complement evasion: an effective strategy that parasites utilize to survive in the host. *Frontiers in Microbiology*, 10: 532.

Shapiro JW, Turner PE (2014). The impact of transmission mode on the evolution of benefits provided by microbial symbionts. *Ecology and Evolution*, 4: 3350–61.

Sharp NP, Otto SP (2016). Evolution of sex: using experimental genomics to select among competing theories. *BioEssays*, 38: 751–7.

Sharp PM, Hahn BH (2011). Origins of HIV and the AIDS pandemic. *Cold Spring Harbor Perspectives in Medicine*, 1: a006841.

Sheehan G, Garvey A, Croke M, Kavanagh K (2018). Innate humoral immune defences in mammals and insects: the same, with differences? *Virulence*, 9: 1625–39.

Sheldon BC, Verhulst S (1996). Ecological immunology: costly parasite defences and trade offs in evolutionary ecology. *Trends in Ecology & Evolution*, 11: 317–21.

Sheldon JR, Heinrichs DE (2015). Recent developments in understanding the iron acquisition strategies of gram positive pathogens. *FEMS Microbiology Reviews*, 39: 592–630.

Shen H-M, Chen S-B, Wang Y, Xu B, Abe EM, Chen J-H (2017). Genome-wide scans for the identification of *Plasmodium vivax* genes under positive selection. *Malaria Journal*, 16: 238.

Shi W, Guo Y, Xu C, et al. (2014). Unveiling the mechanism by which microsporidian parasites prevent locust swarm behavior. *Proceedings of the National Academy of Sciences USA*, 111: 1343–8.

Shim E, Galvani AP (2009). Evolutionary repercussions of avian culling on host resistance and influenza virulence. *PLoS ONE*, 4: e5503.

Shinya K, Ebina M, Yamada S, Ono M, Kasai N, Kawaoka Y (2006). Avian flu: influenza virus receptors in the human airway. *Nature*, 440: 435–6.

Shinzawa N, Nelson B, Aonuma H, et al. (2009). p38 MAPK-dependent phagocytic encapsulation confers infection tolerance in *Drosophila*. *Cell Host & Microbe*, 6: 244–52.

Shmakov S, Smargon A, Scott D, et al. (2017). Diversity and evolution of class 2 CRISPR–Cas systems. *Nature Reviews Microbiology*, 15: 169.

Shoemaker KL, Parsons NM, Adamo SA (2006). Mating enhances parasite resistance in the cricket *Gryllus texensis*. *Animal Behaviour*, 71: 371–80.

Shore AC, Deasy EC, Slickers P, et al. (2011). Detection of staphylococcal cassette chromosome mec type XI carrying highly divergent mecA, mecI, mecR1, blaZ, and ccr genes in human clinical isolates of clonal complex 130 methicillin-resistant *Staphylococcus aureus*. *Antimicrobial Agents and Chemotherapy*, 55: 3765–73.

Short EE, Caminade C, Thomas BN (2017). Climate change contribution to the emergence or re-emergence of parasitic diseases. *Infectious Diseases: Research and Treatment*, 10: 1–7.

Shudo E, Iwasa Y (2001). Inducible defence against pathogens and parasites: optimal choice among multiple options. *Journal of Theoretical Biology*, 209: 233–47.

Shudo E, Iwasa Y (2004). Dynamic optimization of host defense, immune memory, and post-infection pathogen levels in mammals. *Journal of Theoretical Biology*, 228: 1–9.

Shutt DP, Manore CA, Pankavich S, Porter AT, Del Valle SY (2017). Estimating the reproductive number, total outbreak size, and reporting rates for Zika epidemics in South and Central America. *Epidemics*, 21: 63–79.

Sia AK, Allred BE, Raymond KN (2013). Siderocalins: siderophore binding proteins evolved for primary pathogen host defense. *Current Opinion in Chemical Biology*, 17: 150–7.

Sikora S, Strongin A, Godzik A (2005). Convergent evolution as a mechanism for pathogenic adaptation. *Trends in Microbiology*, 13: 522–7.

Sillero-Zubiri C, Marino J, Gordon CH, et al. (2016). Feasibility and efficacy of oral rabies vaccine SAG2 in endangered Ethiopian wolves. *Vaccine*, 34: 4792–8.

Silverman JM, Brunet YR, Cascales E, Mougous JD (2012). Structure and regulation of the Type VI secretion system. *Annual Review of Microbiology*, 66: 453–72.

Šimková A, Verneau O, Gelnar M, Morand S (2006). Specificity and specialization of congeneric monogeneans parasitizing cyprinid fish. *Evolution*, 60: 1023–37.

Simms EL, Triplett J (1994). Costs and benefits of plant responses to disease: resistance and tolerance. *Evolution*, 48: 1973–85.

Simone-Finstrom M, Walz M, Tarpy DR (2016). Genetic diversity confers colony-level benefits due to individual immunity. *Biology Letters*, 12: 20151007.

Simpson RJ, Kunz H, Agha N, Graff R (2015a). Exercise and the regulation of immune functions. In Bouchard C (ed.) *Progress in molecular biology and translational science*, pp. 355–80. Academic Press, London.

Simpson SJ, Clissold FJ, Lihoreau M, Ponton F, Wilder SM, Raubenheimer D (2015b). Recent advances in the integrative nutrition of arthropods. *Annual Review of Entomology*, 60: 293–311.

Simpson SJ, Le Couteur DG, Raubenheimer D, et al. (2017). Dietary protein, aging and nutritional geometry. *Ageing Research Reviews*, 39: 78–86.

Sironi M, Cagliani R, Forni D, Clerici M (2015). Evolutionary insights into host–pathogen interactions from mammalian sequence data. *Nature Reviews Genetics*, 16: 224–36.

Siva-Jothy MT (2000). A mechanistic link between parasite resistance and expression of a sexually selected trait in a damselfly. *Proceedings of the Royal Society London B*, 267: 2523–7.

Siva-Jothy MT (2009). Reproductive immunity. In Rolff J, Reynolds SE (eds) *Insect infection and immunity*, pp. 241–50. Oxford University Press, Oxford.

Skevaki C, Pararas M, Kostelidou K, Tsakris A, Routsias JG (2015). Single nucleotide polymorphisms of Toll-like receptors and susceptibility to infectious diseases. *Clinical and Experimental Immunology*, 180: 165–77.

Slatko BE, Gardner AF, Ausubel FM (2018). Overview of next-generation sequencing technologies. *Current Protocols in Molecular Biology*, 122: e59.

Smallbone W, Van Oosterhout C, Cable J (2016). The effects of inbreeding on disease susceptibility: *Gyrodactylus turnbulli* infection of guppies, *Poecilia reticulata*. *Experimental Parasitology*, 167: 32–7.

Smallegange RC, van Gemert G-J, van de Vegte-Bolmer M, et al. (2013). Malaria infected mosquitoes express enhanced attraction to human odor. *PLoS ONE*, 8: e63602.

Smit NJ, Bruce NL, Hadfield KA (2014). Global diversity of fish parasitic isopod crustaceans of the family Cymothoidae. *International Journal for Parasitology: Parasites and Wildlife*, 3: 188–97.

Smith AM, Perelson AS (2011). Influenza A virus infection kinetics: Qquantitative data and models. *WIREs Systems Biology and Medicine*, 3: 429–45.

Smith DR, Steele KE, Shamblin J, et al. (2010). The pathogenesis of Rift Valley fever virus in the mouse model. *Virology*, 407: 256–67.

Smith JE, Dunn AM (1991). Transovarial transmission. *Parasitology Today*, 7: 146–8.

Smith KF, Sax DF, Lafferty KD (2006). Evidence for the role of infectious disease in species extinction and endangerment. *Conservation Biology*, 20: 1349–57.

Smith LC, Arizza V, Barela Hudgell MA, et al. (2018a). Echinodermata: the complex immune system in echinoderms. In Cooper EL (ed.) *Advances in comparative immunology*, pp. 409–501. Springer International Publishing, Cham.

Smith NR, Trauer JM, Gambhir M, et al. (2018b). Agent-based models of malaria transmission: a systematic review. *Malaria Journal*, 17: 299.

Smith RA, M'Ikanatha NM, Read AF (2015). Antibiotic resistance: a primer and call to action. *Health Communication*, 30: 309–14.

Smith T, Felger I, Tanner M, Beck HP (1999). The epidemiology of multiple *Plasmodium falciparum* infections. 11. Premunition in *Plasmodium falciparum* infection: insights from the epidemiology of multiple infections. *Transactions of the Royal Society of Tropical Medicine and Hygiene*, 93 (suppl.), S1: 59- 64.

Smythe AB, Font WF (2001). Phylogenetic analysis of *Alloglossidium* (Digenea: Macroderoidiidae) and related genera: life cycle evolution and taxonomic revision. *Journal of ParsitologyParasitology*, 87: 386–91.

Snaith TV, Chapman CA, Rothman JM, Wasserman MD (2008). Bigger groups have fewer parasites and similar cortisol levels: a multi-group analysis in red colobus monkeys. *American Journal of Primatology*, 70: 1072–80.

Soares MP, Teixeira L, Moita LF (2017). Disease tolerance and immunity in host protection against infection. *Nature Reviews Immunology*, 17: 83–96.

Söderhäll I (2016). Crustacean hematopoiesis. *Developmental and Comparative Immunology*, 58: 129–41.

Sofonea MT, Alizon S, Michalakis Y (2017). Exposing the diversity of multiple infection patterns. *Journal of Theoretical Biology*, 419: 278–89.

Solter LF (2006). Transmission as a predictor of ecological host specificity with a focus on vertical transmission of microsporidia. *Journal of Invertebrate Pathology*, 92: 132–40.

Sommer F, Anderson JM, Bharti R, Raes J, Rosenstiel P (2017). The resilience of the intestinal microbiota influences health and disease. *Nature Reviews Microbiology*, 15: 630–8.

Sommer F, Nookaew I, Sommer N, Fogelstrand P, Backhed F (2015). Site-specific programming of the host epithelial transcriptome by the gut microbiota. *Genome Biology*, 16: 62.

Sondo P, Derra K, Lefevre T, et al. (2019). Genetically diverse *Plasmodium falciparum* infections, within-host competition and symptomatic malaria in humans. *Scientific Reports*, 9: 127.

Sorbara MT, Pamer EG (2019). Interbacterial mechanisms of colonization resistance and the strategies pathogens use to overcome them. *Mucosal Immunology*, 12: 1–9.

Sorci G, Faivre B (2009). Inflammation and oxidative stress in vertebrate host-parasite systems. *Philosophical Transactions of the Royal Society London B*, 364: 71–83.

Sorci G, Lippens C, Lechenault C, Faivre B (2017). Benefits of immune protection versus immunopathology costs: a synthesis from cytokine KO models. *Infection Genetics and Evolution*, 54: 491–5.

Sørensen JH, Mackay DKJ, Jensen CØ, Donaldson AI (2000). An integrated model to predict the atmospheric spread of foot-and-mouth disease virus. *Epidemiology and Infection*, 124: 577–90.

Sorensen RE, Minchella DJ (2001). Snail-trematode life history interactions: past trends and future directions. *Parasitology*, 123: S3-18.

Sousa WP (1983). Host life history and the effect of parasitic castration on growth: a field study on *Cerithidea californica* Haldenau (Gastropoda: Prosobranchia) and its trematode parasites. *Journal of Experimental Marine Biology*, 73: 273–96.

Southgate VR (1997). Schistosomiasis in the Senegal river basin: before and after the construction of the dams at Diama, Senegal and Manantali, Mali and future prospects. *Journal of Helminthology*, 71: 125–32.

Soyer OS, Bonhoeffer S (2006). Evolution of complexity in signaling pathways. *Proceedings of the National Academy of Sciences USA*, 103: 16337–42.

Spor A, Koren O, Ley R (2011). Unravelling the effects of the environment and host genotype on the gut microbiome. *Nature Reviews Microbiology*, 9: 279–90.

Spottiswoode C (2008). Cooperative breeding and immunity: a comparative study of PHA response in African birds. *Behavioral Ecology and Sociobiology*, 62: 963–74.

Spyrou MA, Bos KI, Herbig A, Krause J (2019). Ancient pathogen genomics as an emerging tool for infectious disease research. *Nature Reviews Genetics*, 20: 323–40.

Stadler T (2010). Sampling-through-time in birth–death trees. *Journal of Theoretical Biology*, 267: 396–404.

Stadler T, Kouyos R, von Wyl V, et al. (2011). Estimating the basic reproductive number from viral sequence data. *Molecular Biology and Evolution*, 29: 347–57.

Stahl EA, Dwyer G, Mauricio R, Kreitmann M, Bergelson J (1999). Dynamics of disease resistance polymorphism at the Rpm1 locus of *Arabidopsis*. *Nature*, 400: 667–71.

Stahlschmidt ZR, Adamo SA (2013). Context dependency and generality of fever in insects. *Naturwissenschaften*, 100: 691–6.

Staley JT (2006). The bacterial species dilemma and the genomic-phylogenetic species concept. *Philosophical Transactions of the Royal Society London B*, 361: 1899–909.

Stanczyk NM, Mescher MC, De Moraes CM (2017). Effects of malaria infection on mosquito olfaction and behavior: extrapolating Extrapolating data to the field. *Current Opinion in Insect Science*, 20: 7–12.

Stapley J, Feulner PGD, Johnston SE, Santure AW, Smadja CM (2017). Variation in recombination frequency and distribution across eukaryotes: Ppatterns and processes. *Philosophical Transactions of the Royal Society London B*, 372: 20160455.

Starks PT, Blackie CA, Seeley TD (2000). Fever in honeybee colonies. *Naturwissenschaften*, 87: 229–31.

Staszewski V, Siitari H (2010). Antibody injection in the egg yolk: maternal antibodies affect humoral immune response of the offspring. *Functional Ecology*, 24: 1333–41.

Stearns SC (1977). The evolution of life history traits: a critique of the theory and a review of the data. *Annual Review of Ecology and Systematics*, 8: 145–71.

Stearns SC (1992). *Life history evolution*. Oxford University Press, Oxford.

Stephens PR, Altizer S, Smith KF, et al. (2016). The macroecology of infectious diseases: a new perspective on global-scale drivers of pathogen distributions and impacts. *Ecology Letters*, 19: 1159–71.

Stephenson EB, Peel AJ, Reid SA, Jansen CC, McCallum H (2018). The non-human reservoirs of Ross River virus: a systematic review of the evidence. *Parasites & Vectors*, 11: 188.

Sternberg Samuel H, Richter H, Charpentier E, Qimron U (2016). Adaptation in CRISPR-Cas systems. *Molecular Cell*, 61: 797–808.

Stetzenbach LD (2009). Airborne Infectious microorganisms. *Encyclopedia of Microbiology*, 2009: 175–82.

Stevens JR, Noyes HA, Schofield CI, Gibson W (2001). The molecular evolution of Trypanosomatidae. *Advances in Parasitology*, 48: 1–56.

Stevison L, S., Sefick S, Rushton C, Graze Rita M (2017). Recombination rate plasticity: revealing mechanisms by design. *Philosophical Transactions of the Royal Society London B*, 372: rp.

Stewart AD, Logsdon Jr. JM, Kelley SE (2005). An empirical study of the evolution of virulence under both horizontal and vertical transmission. *Evolution*, 59: 730–9.

Stewart PS (2002). Mechanisms of antibiotic resistance in bacterial biofilms. *International Journal for Medical Microbiology*, 292: 107–13.

Stijlemans B, Caljon G, Van den Abbeele J, Van Ginderachter JA, Magez S, De Trez C (2016). Immune evasion strategies of *Trypanosoma brucei* within the mammalian host: progression to pathogenicity. *Frontiers in Immunology*, 7: 233.

Stites EC, Wilen CB (2020). The Interpretation of SARS-CoV-2 diagnostic tests. *Med*: https://doi.org/10.1016/j.medj.2020.08.001.

Stotz HU, Bishop JG, Bergmann CW, et al. (2000). Identification of target amino acids that affect interactions of fungal polygalacturonases and their plant inhibitors. *Physiological and Molecular Plant Pathology*, 56: 117–30.

Stow A, Briscoe D, Gillings M, et al. (2007). Antimicrobial defences increase with sociality in bees. *Biology Letters*, 3: 422–4.

Stowe K, Maqruis R, Hochwender C, Simms EL (2000). The evolutionary ecology of tolerance to consumer damage. *Annual Review of Ecology and Systematics*, 31: 565–95.

Strand MR (2008). The insect cellular immune response. *Insect Science*, 15: 1–14.

Strassmann JE, Gilbert OM, Queller DC (2011). Kin discrimination and cooperation in microbes. *Annual Review of Microbiology*, 65: 349–67.

Straub L, Williams GR, Pettis J, Fries I, Neumann P (2015). Superorganism resilience: eusociality and susceptibility of ecosystem service providing insects to stressors. *Current Opinion in Insect Science*, 12: 109–12.

Streicker DG, Turmelle AS, Vonhof MJ, Kuzmin IV, McCracken GF, Rupprecht CE (2010). Host phylogeny constrains cross-species emergence and establishment of rabies virus in bats. *Science*, 329: 676–9.

Strich JR, Chertow DS (2019). CRISPR-Cas biology and its application to infectious diseases. *Journal of Clinical Microbiology*, 57: e01307–18.

Stroeymeyt N, Casillas-Pérez B, Cremer S (2014). Organisational immunity in social insects. *Current Opinion in Insect Science*, 5: 1–15.

Stroeymeyt N, Grasse AV, Crespi A, Mersch DP, Cremer S, Keller L (2018). Social network plasticity decreases disease transmission in a eusocial insect. *Science*, 362: 941–5.

Struzik J, Szulc-Dąbrowska L, Papiernik D, Winnicka A, Niemiałtowski M (2015). Modulation of proinflammatory NF-κB signaling by ectromelia virus in RAW 264.7 murine macrophages. *Archives of Virology*, 160: 2301–14.

Stuart LM, Ezekowitz RA (2008). Phagocytosis and comparative immunity: learning on the fly. *Nature Reviews Immunology*, 8: 131–41.

Stuart LM, Paquette N, Boyer L (2013). Effector-triggered versus pattern-triggered immunity: how animals sense pathogens. *Nature Reviews Immunology*, 13: 199–206.

Stuart P, Paredis L, Henttonen H, Lawton C, Ochoa Torres CA, Holland CV (2020). The hidden faces of a biological invasion: parasite dynamics of invaders and natives. *International Journal for Parasitology*, 50: 111–23.

Stuart SN, Chanson JS, Cox NA, et al. (2004). Status and trends of amphibian declines and extinctions worldwide. *Science*, 306: 1783–6.

Stylianou A, Hadjichrysanthou C, Truscott JE, Anderson RM (2017). Developing a mathematical model for the evaluation of the potential impact of a partially efficacious vaccine on the transmission dynamics of *Schistosoma mansoni* in human communities. *Parasites & Vectors*, 10: 294.

Su S, Wong G, Shi W, et al. (2016). Epidemiology, genetic recombination, and pathogenesis of Coronaviruses. *Trends in Microbiology*, 24: 490–502.

Suárez-Morales E (2020). Class Branchiura. In Rogers DC, Damborenea C, Thorp J (eds) *Thorp and Covich's freshwater invertebrates*, (4th eEd.), pp. 797–807. Academic Press, London.

Sugumaran M, Nellaiappan K, Valivattan K (2000). A new mechanism for the control of phenoloxidase activity: inhibition and complex formation with quinone isomease. *Archives of Biochemistry and Biophysics*, 379: 252–60.

Sullivan JT, Ronson CW (1998). Pathogenicity islands and other mobile virulence determinants. *Proceedings of the National Academy of Sciences USA*, 95: 5145–9.

Sun JC, Ugolini S, Vivier E (2014). Immunological memory within the innate immune system. *The EMBO Journal*, 33: 1295–303.

Sutter GR, Rothenbuhler WC, Raun ES (1968). Resistance to American foulbrood in honey bees. VII. Growth of resistant and susceptible larvae. *Journal of Invertebrate Pathology*, 12: 25–8.

Suurväli J, Jouneau L, Thépot D, et al. (2014). The Proto-MHC of placozoans, a region specialized in cellular stress and ubiquitination/proteasome pathways. *The Journal of Immunology*, 193: 2891–901.

Suzán G, Marcé E, Giermakowski JT, et al. (2009). Experimental evidence for reduced rodent diversity causing increased hantavirus prevalence. *PLoS ONE*, 4: e5461.

Suzuki MM, Dird A (2008). DNA methylation landscapes: provocative insights from epigenomics. *Nature Reviews Genetics*, 9: 465–76.

Svensson-Coelho M, Loiselle BA, Blake JG, Ricklefs RE (2016). Resource predictability and specialization in avian malaria parasites. *Molecular Ecology*, 25: 4377–91.

Swaddle JP, Calos SE (2008). Increased avian diversity is associated with lower incidence of human west nile infection: observation of the dilution effect. *PLoS ONE*, 3: e2488.

Sweet AD, Boyd BM, Allen JM, et al. (2017). Integrating phylogenomic and population genomic patterns in avian lice provides a more complete picture of parasite evolution. *Evolution*, 72: 95–112.

Switzer WM, Salemi M, Shanmugam V, et al. (2005). Ancient co-speciation of simian foamy viruses and primates. *Nature*, 434: 376–80.

Syed Musthaq SK, Kwang J (2014). Evolution of specific immunity in shrimp: a vaccination perspective against white spot syndrome virus. *Developmental & Comparative Immunology*, 46: 279–90.

Tanner JR, Kingsley RA (2018). Evolution of *Salmonella* within hosts. *Trends in Microbiology*, 26: 986–98.

Tanser F, Vandormael A, Cuadros D, et al. (2017). Effect of population viral load on prospective HIV incidence in a hyperendemic rural African community. *Science Translational Medicine*, 9: eaam8012.

Tarpy DR (2003). Genetic diversity within honeybee colonies prevents severe infections and promotes colony growth. *Proceedings of the Royal Society London B*, 270: 99–103.

Tas JMJ, Mesin L, Pasqual G, et al. (2016). Visualizing antibody affinity maturation in germinal centers. *Science*, 351: 1048–54.

Tashiro M, Ciborowski P, Klenk HD, Pulverer G, Rott R (1987). Role of *Staphylococcus* protease in the development of influenza pneumonia. *Nature*, 325: 536–7.

Taubenberger JK, Kash JC (2010). Influenza virus evolution, host adaptation and pandemic formation. *Cell Host Microbe*, 7: 440–51.

Taylor J, Purvis A (2003). Have mammals and their chewing lice diversified in parallel? In Page RDM (ed.) *Tangled trees: phylogeny, cospeciation and coevolution*, pp. 240–61. University of Chicago Press, Chicago.

Taylor KE, Mossman KL (2013). Recent advances in understanding viral evasion of type I interferon. *Immunology*, 138: 190–7.

Taylor LH, Walliker D, Read AF (1997). Mixed-genotype infections of malaria parasites: within-host dynamics and transmission success of competing clones. *Proceedings of the Royal Society London B*, 264: 927–35.

Taylor P, Hurd H (2001). The influence of host hematocrit on the blood feeding success of *Anopheles stephensi*: implications for enhanced malaria transmission. *Parasitology*, 122: 491–6.

Tebit DM, Arts EJ (2011). Tracking a century of global expansion and evolution of HIV to drive understanding and to combat disease. *The Lancet Infectious Diseases*, 11: 45–56.

Teixeira L, Ferreira A, Ashburner M (2008). The bacterial symbiont *Wolbachia* induces resistance to RNA viral infections in *Drosophila melanogaster*. *PLoS Biology*, 6: e2.

Tella J, Scheuerlein A, Ricklefs R (2002). Is cell-mediated immunity related to the evolution of life-history strategies in birds? *Proceedings of the Royal Society London B*, 269: 1059–66.

Tellier A, Brown JKM (2011). Spatial heterogeneity, frequency-dependent selection and polymorphism in host-parasite interactions. *BMC Evolutionary Biology*, 11: 319.

Templeton AR (2000). Epistasis and complex traits. In Wolf JB, Brodie EB, Wade MJ (eds) *Epistasis and the evolutionary process*, pp. 41–57. Oxford University Press, Oxford.

Tenover FC, Goering RV (2009). Methicillin-resistant *Staphylococcus aureus* strain USA300: origin and epidemiology. *Journal of Antimicrobial Chemotherapy*, 64: 441–6.

Těšický M, Vinkler M (2015). Trans-species polymorphism in immune genes: general pattern or MHC-restricted phenomenon? *Journal of Immunology Research*, 2015: 838035.

Tetreau G, Dhinaut J, Gourbal B, Moret Y (2019). Transgenerational immune priming in invertebrates: current knowledge and future prospects. *Frontiers in Immunology*, 10: 1938.

Teunis PFM, Havelaar AH (2000). The Beta Poisson dose-response model is not a single-hit model. *Risk Analysis*, 20: 513–20.

Thaiss CA, Zmora N, Levy M, Elinav E (2016). The microbiome and innate immunity. *Nature*, 535: 65–74.

Thangamani S, Huang J, Hart CE, Guzman H, Tesh RB (2016). Vertical transmission of Zika Virus in *Aedes aegypti* mosquitoes. *The American Journal of Tropical Medicine and Hygiene*, 95: 1169–73.

Theopold U, Schmidt O, Söderhall K, Dushay MS (2004). Coagulation in arthropods: defence, wound closure and healing. *Trends in Immunology*, 25: 289–94.

Théron A (1984). Early and late shedding patterns of *Schistosoma mansoni* cercariae: ecological significance in transmission to human and murine hosts. *Journal of Parasitology*, 70: 652–5.

Théron A, Combes C (1988). Genetic analysis of cercarial emergence rhythms of *Schistosoma mansoni*. *Behavior Genetics*, 18: 201–9.

Thieltges DW, Amundsen P-A, Hechinger RF, et al. (2013). Parasites as prey in aquatic food webs: implications for predator infection and parasite transmission. *Oikos*, 122: 1473–82.

Thogmartin WE, Sanders-Reed CA, Szymanski JA, et al. (2013). White-nose syndrome is likely to extirpate the endangered Indiana bat over large parts of its range. *Biological Conservation*, 160: 162–72.

Thomas F, Adamo SA, Moore J (2005). Parasitic manipulation: where are we and where should we go? *Behavioural Processes*, 68: 185–99.

Thomas F, Mete K, Helluy S, et al. (1997). Hitch-hike parasites or how to benefit from the strategy of another parasite. *Evolution*, 51: 1316–18.

Thomas F, Renaud F, Poulin R (1998). Exploitation of manipulators: 'hitch-hiking' as a parasite transmission strategy. *Animal Behaviour*, 56: 199–206.

Thomas MB, Blanford S (2003). Thermal biology in insect-parasite interactions. *Trends in Ecology & Evolution*, 18: 344–50.

Thornhill JA, Jones JT, Kusel JR (1986). Increased oviposition and growth in immature *Biomphalaria glabrata* after exposure to *Schistosoma mansoni*. *Parasitology*, 93: 443–50.

Thrall PH, Barrett LG, Dodds PN, Burdon JJ (2016). Epidemiological and evolutionary outcomes in gene-for-gene and matching allele models. *Frontiers in Plant Science*, 6: 1084.

Thrall PH, Burdon JJ (2003). Evolution of virulence in a plant host-pathogen metapopulation. *Science*, 299: 1735–7.

Thrall PH, Laine A-L, Ravensdale M, et al. (2012). Rapid genetic change underpins antagonistic coevolution in a natural host-pathogen metapopulation. *Ecology Letters*, 15: 425–35.

Tian C, Hromatka BS, Kiefer AK, et al. (2017). Genome-wide association and HLA region fine-mapping studies identify susceptibility loci for multiple common infections. *Nature Communications*, 8: 599.

Tibayrenc M (1999). Toward an integrated genetic epidemiology of parasitic protozoa and other pathogens. *Annual Review of Genetics*, 33: 449–77.

Tibayrenc M, Ayala FJ (2002). The clonal theory of parasitic protozoa: 12 years on. *Trends in Parasitology*, 18: 405–10.

Tibayrenc M, Kjellberg F, Ayala FJ (1990). A clonal theory of parasitic protozoa: the population structures of *Entamoeba*, *Giardia*, *Leishmania*, *Naegleria*, *Plasmodium*, *Trichomonas*, and *Trypanosoma*, and their medical and taxonomical consequences. *Proceedings of the National Academy of Sciences USA*, 87: 2414–18.

Tibbetts SA, Loh J, van Berkel V, et al. (2003). Establishment and maintenance of Gammaherpesvirus latency are independent of infective dose and route of infection. *Journal of Virology*, 77: 7696–701.

Tidbury HJ, Best A, Boots M (2012). The epidemiological consequences of immune priming. *Proceedings of the Royal Society London B*, 279: 4505–12.

Tieleman BI, Williams JB, Ricklefs RE, Klasing KC (2005). Constitutive innate immunity is a component of the pace-of-life syndrome in tropical birds. *Proceedings of the Royal Society London B*, 272: 1715–20.

Tiensin T, Chaitaweesub P, Songserm T, et al. (2005). Highly pathogenic avian influenza H5N1, Thailand, 2004. *Emerging Infectious Diseases*, 11: 1664–72.

Timme RE, Pettengill JB, Allard MW, et al. (2013). Phylogenetic diversity of the enteric pathogen *Salmonella enterica* subsp. *enterica* inferred from genome-wide reference-free snp characters. *Genome Biology and Evolution*, 5: 2109–23.

Timpson NJ, Greenwood CMT, Soranzo N, Lawson DJ, Richards JB (2018). Genetic architecture: the shape of the genetic contribution to human traits and disease. *Nature Reviews Genetics*, 19: 110–24.

Tinbergen N (1951). *The study of instinct.* Oxford University Press, Oxford.

Tinsley M, Majerus M (2007). Small steps or giant leaps for male-killers? Phylogenetic constraints to male-killer host shifts. *BMC Evolutionary Biology*, 7: 238.

Tinsley RC (1990). Host behaviour and opportunism in parasite life cycles. In Barnard CJ, Behnke JM (eds) *Parasitism and host behaviour*, pp. 158–92. Taylor and Francis, London.

Tisoncik JR, Korth MJ, Simmons CP, Farrar J, Martin TR, Katze MG (2012). Into the eye of the cytokine storm. *Microbiology and Molecular Biology Reviews*, 76: 16–32.

Toft CA, Karter AJ (1990). Parasite-host coevolution. *Trends in Ecology & Evolution*, 5: 326–9.

Tompkins DM, Dobson AP, Arneberg P, et al. (2001a). Parasites and host population dynamics. In Hudson PJ (ed.) *The ecology of wildlife diseases*, pp. 45–62. Oxford University Press, Oxford.

Tompkins DM, Sainsbury AW, Nettleton P, Buxton D, Gurnell J (2001b). Parapoxvirus causes a deleterious disease in red squirrels associated with UK population declines. *Proceedings of the Royal Society London B*, 269: 529–33.

Torchin ME, Lafferty KD (2009). Escape from parasites. In Rilov G, Crooks JA (eds) *Biological invasions in marine ecosystems*, pp. 203–14. Springer, Berlin, Heidelberg.

Torchin ME, Lafferty KD, Dobson AP, McKenzie VJ, Kuris M (2003). Introduced species and their missing parasites. *Nature*, 421: 628–30.

Torres BY, Oliveira JHM, Tate, Ann Thomas, Rath P, Cumnock K, Schneider DS (2016). Tracking resilience to infections by mapping disease space. *PLoS Biology*, 14: e1002436.

Tortorella D, Gewurz BE, Furman MH, Schust DJ, Ploegh HL (2000). Viral subversion of the immune system. *Annual Review of Immunology*, 18: 861–926.

Trabalon M, Plateaux L, Péru L, Bagnères A-G, Hartmann N (2000). Modification of morphological characters and cuticular compounds in worker ants *Leptothorax nylanderi* induced by endoparasites *Anomotaenia brevis*. *Journal of Insect Physiology*, 46: 169–78.

Trachtenberg S (2005). Mollicutes. *Current Biology*, 15: R483–4.

Trancoso I, Morimoto R, Boehm T (2020). Co-evolution of mutagenic genome editors and vertebrate adaptive immunity. *Current Opinion in Immunology*, 65: 32–41.

Trang TT, Hung NH, Ninh NH, Knibb W, Nguyen NH (2019). Genetic variation in disease resistance against White Spot Syndrome Virus (WSSV) in *Liptopenaeus vannamei*. *Frontiers in Genetics*, 10: 264.

Tranter C, LeFevre L, Evison SEF, Hughes WOH (2015). Threat detection: contextual recognition and response to parasites by ants. *Behavioural Ecology*, 26: 396–405.

Trauer-Kizilelma U, Hilker M (2015). Impact of transgenerational immune priming on the defence of insect eggs against parasitism. *Developmental & Comparative Immunology*, 51: 126–33.

Treganza T, Wedell N (2002). Polyandrous females avoid costs of inbreeding. *Nature*, 415: 71–3.

Trimpert J, Osterrieder N (2019). Herpesvirus DNA polymerase mutants: how important is faithful genome replication? *Current Clinical Microbiology Reports*, 6: 240–8.

Tripet F, Christe P, Møller AP (2002). The importance of host spatial distribution for parasite specialization and speciation: a comparative study of bird fleas (Siphonaptera: Ceratophyllidae). *Journal of Animal Ecology*, 71: 735–48.

Trivers RL (1972). Parental investment and sexual selection. In Campbell E (ed.) *Sexual conflict and the descent of man*, pp. 136–79. Aladine, Chicago.

Tsai IJ, Zarowiecki M, Holroyd N, et al. (2013). The genomes of four tapeworm species reveal adaptations to parasitism. *Nature*, 496: 57–63.

Tsairidou S, Allen AR, Pong-Wong R, et al. (2018). An analysis of effects of heterozygosity in dairy cattle for bovine tuberculosis resistance. *Animal Genetics*, 49: 103–9.

Tscharntke T, Karp DS, Chaplin-Kramer R, et al. (2016). When natural habitat fails to enhance biological pest control: five hypotheses. *Biological Conservation*, 204: 449–58.

Tseng T-T, Tyler BM, Setubal JC (2009). Protein secretion systems in bacterial-host associations, and their description in the Gene Ontology. *BMC Microbiology*, 9: S2.

Tuncer N, Gulbudak H, Cannataro VL, Martcheva M (2016). Structural and practical identifiability issues of immuno-epidemiological vector–host models with application to Rift Valley Fever. *Bulletin of Mathematical Biology*, 78: 1796–827.

Tuomanen E, Cozens R, Tosch W, Zak O, Toamsz A (1986). The rate of killing of *Escherichia coli* by β-lactam antibiotics is strictly proportional to the rate of bacterial growth. *Journal of Genetical Microbiology*, 132: 1297–304.

Turner CMR (2002). A perspective on clonal phenotypic (antigenic) variation in protozoan parasites. *Parasitology*, 125: S17-S32.

Turner DL, Farber DL (2014). Mucosal resident memory CD4 T cells in protection and immunopathology. *Frontiers in Immunology*, 5: 331.

Turner PE, Chao L (1999). Prisoner's dilemma in an RNA virus. *Nature*, 398: 441–3.

Turner VM, Mabbott NA (2017). Influence of ageing on the microarchitecture of the spleen and lymph nodes. *Biogerontology*, 18: 723–38.

Turney S, Gonzalez A, Millien V (2014). The negative relationship between mammal host diversity and Lyme disease incidence strengthens through time. *Ecology*, 95: 3244–50.

Ulrich Y, Sadd BM, Schmid-Hempel P (2011). Strain filtering and transmission of a mixed infection in a social insect. *Journal of Evolutionary Biology*, 24: 354–62.

Ulrich Y, Schmid-Hempel P (2012). Host modulation of parasite competition in multiple infections. *Proceedings of the Royal Society London B*, 279: 2982–9.

Underhill DM, Ozinsky A (2002). Phagocytosis of microbes: complexity in action. *Annual Review of Immunology*, 20: 825–52.

United Nations Department of Economic and Social Affairs (2003). *World population prospects*. United Nations, New York.

Unterholzner L, Almine JF (2019). Camouflage and interception: how pathogens evade detection by intracellular nucleic acid sensors. *Immunology*, 156: 217–27.

Upton C, Macen JL, Schreiber M, McFadden G (1991). Myxoma virus expresses a secreted protein with homology to the tumor necrosis factor receptor gene family that contributes to viral virulence. *Virology*, 184: 370–82.

Urban CF, Lourido S, Zychlinsky A (2006). How do microbes evade neutrophile killing? *Cellular Microbiology*, 8: 1687–96.

Vale PF, Fenton A, Brown SP (2014). Limiting damage during infection: lessons from infection tolerance for novel therapeutics. *PLoS Biology*, 12: e1001769.

Valtonen ET, Holmes JC, Koskivaara M (1997). Eutrophication, pollution and fragmentation: effects on parasite communities in roach (*Rutilus rutilus*) and perch (*Perca fluviatilis*) in four lakes in Central Finland. *Canadian Journal of Fisheries and Aquatic Sciences*, 54: 572–85.

Van Acker H, Van Dijck P, Coenye T (2014). Molecular mechanisms of antimicrobial tolerance and resistance in bacterial and fungal biofilms. *Trends in Microbiology*, 22: 326–33.

van Baarlen P, van Belkum A, Summerbell RC, Crous PW, Thomma BPHJ (2007). Molecular mechanisms of pathogenicity: how do pathogenic microorganisms develop cross-kingdom jumps? *FEMS Microbiology Reviews*, 31: 239–77.

van Ballegooijen W, Boerlijst MC (2004). Emergent trade-offs and selection for outbreak frequency in spatial epidemics. *Proceedings of the National Academy of Sciences USA*, 101: 18246–50.

Van den Berg HA, Rand DA (2007). Quantitative theories of T-cell responsiveness. *Immunological Reviews*, 216: 81–92.

Van den Berg HA, Rand DA, Borroughs NJ (2001). A reliable and safe T-cell repertoire based on low-affinity T-cell receptors. *Journal of Theoretical Biology*, 209: 465–96.

van der Meer J (2006). Metabolic theories in ecology. *Trends in Ecology & Evolution*, 21: 136–40.

van der Most P, de Jong B, Parmentier H, Verhulst S (2010). Trade-off between growth and immunocompetence: a meta-analysis of selection experiments. *Functional Ecology*, 25: 74–80.

van der Poll T, van de Veerdonk FL, Scicluna BP, Netea MG (2017). The immunopathology of sepsis and potential therapeutic targets. *Nature Reviews Immunology*, 17: 407–20.

van der Woude MW (2017). Epigenetic phase variation in bacterial pathogens. In Doerfler W, Casadesús J (eds) *Epigenetics of infectious diseases*, pp. 159–73. Springer International Publishing, Cham.

van Diepen LTA, Olson Å, Ihrmark K, Stenlid J, James TY (2013). Extensive trans-specific polymorphism at the mating type locus of the root decay fungus *Heterobasidion*. *Molecular Biology and Evolution*, 30: 2286–301.

van Dijk EL, Jaszczyszyn Y, Naquin D, Thermes C (2018). The third revolution in sequencing technology. *Trends in Genetics*, 34: 666–81.

van Dijk JGB, Matson KD (2016). Ecological immunology through the lens of exercise immunology: new perspective on the links between physical activity and immune function and disease susceptibility in wild animals. *Integrative and Comparative Biology*, 56: 290–303.

van Mierlo JT, Overheul GJ, Obadia B, et al. (2014). Novel *Drosophila* viruses encode host-specific suppressors of RNAi. *PLoS Pathogens*, 10: e1004256.

Van Noordwijk AJ, De Jong G (1986). Acquisition and allocation of resources: their influence on variation in life history tactics. *The American Naturalist*, 128: 137–42.

Van Valen L (1973). A new evolutionary law. *Evolutionary Theory*, 1: 1–30.

Vance SA (1996). Morphological and behavioural sex reversal in mermithid-infected mayflies. *Proceedings of the Royal Society London B*, 263: 907–12.

VanderWaal KL, Ezenwa VO (2016). Heterogeneity in pathogen transmission: mechanisms and methodology. *Functional Ecology*, 30: 1606–22.

Vanhove MPM, Hablützel PI, Pariselle A, Šimková A, Huyse T, Raeymaekers JAM (2016). Cichlids: a host of opportunities for evolutionary parasitology. *Trends in Parasitology*, 32: 820–32.

Vantaux A, Dabiré KR, Cohuet A, Lefèvre T (2014). A heavy legacy: offspring of malaria-infected mosquitoes show reduced disease resistance. *Malaria Journal*, 13: 442.

Varela M-L, Koffi D, White M, et al. (2020). Practical example of multiple antibody screening for evaluation of malaria control strategies. *Malaria Journal*, 19: 117.

Vaughan TG, Leventhal GE, Rasmussen DA, Drummond AJ, Welch D, Stadler T (2019). Estimating epidemic incidence and prevalence from genomic data. *Molecular Biology and Evolution*, 36: 1804–16.

Vazquez DP, Poulin R, Krasnov BR, Shenbrot GI (2005). Species abundance and the distribution of specialization in host-parasite interaction networks. *Journal of Animal Ecology*, 74: 946–55.

Vega-Rúa A, Zouache K, Girod R, Failloux A-B, Lourenço-de-Oliveira R (2014). High level of vector competence of *Aedes aegypti* and *Aedes albopictus* from ten american countries as a crucial factor in the spread of Chikungunya Virus. *Journal of Virology*, 88: 6294–306.

Vergara D, Jokela J, Lively CM (2014). Infection dynamics in coexisting sexual and asexual host populations: support for the red queen hypothesis. *The American Naturalist*, 184: S22–30.

Verhey TB, Castellanos M, Chaconas G (2019). Antigenic variation in the Lyme spirochete: detailed functional assessment of recombinational switching at vlsE in the JD1 strain of *Borrelia burgdorferi*. *Molecular Microbiology*, 111: 750–63.

Verhulst S, Riedsrea B, Wiersma P (2005). Brood size and immunity costs in zebra finches (*Taeniopygia guttata*). *Journal of Avian Biology*, 36: 22–30.

Vernon JG, Okamura B, Jones CS, Noble LR (1996). Temporal patterns of clonality and parasitism in a population of freshwater bryozoans. *Proceedings of the Royal Society London B*, 263: 1313–18.

Versteeg GA, Benke S, García-Sastre A, Rajsbaum R (2014). InTRIMsic immunity: positive and negative regulation of immune signaling by tripartite motif proteins. *Cytokine & Growth Factor Reviews*, 25: 563–76.

Versteegh M, Schwabl I, Jaquier S, Tieleman B (2012). Do immunological, endocrine and metabolic traits fall on a single Pace-of-Life axis? Covariation and constraints among physiological systems. *Journal of Evolutionary Biology*, 25: 1864–76.

Viana M, Mancy R, Biek R, et al. (2014). Assembling evidence for identifying reservoirs of infection. *Trends in Ecology & Evolution*, 29: 270–9.

Victora G, Nussenzweig M (2012). Germinal centers. *Annual Review of Immunology*, 30: 429–57.

Vigant F, Santos NC, Lee B (2015). Broad-spectrum antivirals against viral fusion. *Nature Reviews Microbiology*, 13: 426–37.

Vignuzzi M, Stone JK, Arnold JJ, Cameron CE, Andino R (2006). Quasispecies diversity determines pathogenesis through cooperative interactions in a viral population. *Nature*, 439: 344–8.

Villeneuve C, Kou HH, Eckermann H, et al. (2018). Evolution of the hygiene hypothesis into biota alteration theory: what are the paradigms and where are the clinical applications? *Microbes and Infection*, 20: 147–55.

Viney M (2017). How can we understand the genomic basis of nematode parasitism? *Trends in Parasitology*, 33: 444–52.

Viney M (2018). The genomic basis of nematode parasitism. *Briefings in Functional Genomics*, 17: 8–14.

Viney ME, Riley EM, Buchanan KL (2005). Optimal immune responses: immuncompetence revisited. *Trends in Ecology & Evolution*, 20: 665–9.

Vita R, Mahajan S, Overton JA, et al. (2019). The Immune Epitope Database (IEDB): 2018 update. *Nucleic Acids Research*, 47: D339–43.

Vittor AY, Gilman RH, Tielsch J, et al. (2006). The effect of deforestation on the human-biting rate of *Anopheles darlingi*, the primary vector of Falciparum malaria in the Peruvian Amazon. *American Journal of Tropical Medicine and Hygiene*, 74: 3–11.

Vlisidou I, Wood W (2015). *Drosophila* blood cells and their role in immune responses. *FEBS Journal*, 82: 1368–82.

Voegeli B, Saladin V, Wegmann M, Richner H (2012). Parasites as mediators of heterozygosity–fitness correlations in the Great Tit (*Parus major*). *Journal of Evolutionary Biology*, 25: 584–90.

Vogwill T, MacLean RC (2014). The genetic basis of the fitness costs of antimicrobial resistance: a meta-analysis approach. *Evolutionary Applications*, 8: 284–95.

Volz EM (2012). Complex population dynamics and the coalescent under neutrality. *Genetics*, 190: 187.

Von Bonsdorff B, Bylund G (1982). The ecology of *Diphyllobothrium latum*. *Ecology and Disease*, 1: 21–6.

von Schantz T, Bensch S, Grahn M, Hasselquist D, Wittzell H (1999). Good genes, oxidative stress and condition-dependent sexual signals. *Proceedings of the Royal Society London B*, 266: 1–12.

Vonaesch P, Anderson M, Sansonetti PJ (2018). Pathogens, microbiome and the host: emergence of the ecological Koch's postulates. *FEMS Microbiology Reviews*, 42: 273–92.

Voordouw MJ, Lambrechts L, Koella J (2008). No maternal effects after stimulation of the melanization response in the yellow fever mosquito *Aedes aegypti*. *Oikos*, 117: 1269–79.

Vorburger C, Perlman SJ (2018). The role of defensive symbionts in host–parasite coevolution. *Biological Reviews*, 93: 1747–64.

Waddington KD, Rothenbuhler WC (1976). Behaviour associated with hairless-black syndrome of adult honeybees. *Journal of Apicultural Research*, 15: 35–41.

Waddington SN, Privolizzi R, Karda R, O'Neill HC (2016). A broad overview and review of CRISPR--Cas technology and stem cells. *Current Stem Cell Reports*, 2: 9–20.

Wade MJ (2007). The co-evolutionary genetics of ecological communities. *Nature Reviews Genetics*, 8: 185–95.

Waghu FH, Barai RS, Gurung P, Idicula-Thomas S (2015). CAMPR3: a A database on sequences, structures and signatures of antimicrobial peptides. *Nucleic Acids Research*, 44: D1094–7.

Waits A, Emelyanova A, Oksanen A, Abass K, Rautio A (2018). Human infectious diseases and the changing climate in the Arctic. *Environment International*, 121: 703–13.

Wajnberg E, Bernstein C, van Alphen J (eds) (2008). *Behavioral ecology of insect parasitoids: Ffrom theoretical approaches to field applications*. John Wiley & Sons, Chichester, U.K.

Wajnberg E, Colazza S (eds) (2013). *Chemical ecology of insect parasitoids*. John Wiley & Sons, Chichester, U.K.

Wale N, Sim DG, Jones MJ, Salathé R, Day T, Read AF (2017). Resource limitation prevents the emergence of drug resistance by intensifying within-host competition. *Proceedings of the National Academy of Sciences USA*, 114: 13774–9.

Walker JG, Hurford A, Cable J, Ellison AR, Price SJ, Cressler CE (2017). Host allometry influences the evolution of parasite host-generalism: theory and meta-analysis. *Philosophical Transactions of the Royal Society London B*, 372: 20160089.

Wallace RL (2002). Rotifers: exquisite metazoans. *Integrative and Comparative Biology*, 42: 660–7.

Walsh NP, Oliver SJ (2016). Exercise, immune function and respiratory infection: an update on the influence of training and environmental stress. *Immunology & Cell Biology*, 94: 132–9.

Walter DE, Proctor HC (1999). *Mites: ecology, evolution and behaviour*. CABI Publishing, Sydney & Wallingford, UK.

Walther BA, Ewald PW (2004). Pathogen survival in the external environment and the evolution of virulence. *Biological Reviews of the Cambridge Philosophical Society*, 79: 849–69.

Wang CJ, Ng CY, Brook RH (2020). Response to COVID-19 in Taiwan: Bbig data analytics, new technology, and proactive testing. *JAMA*, 323: 1341–2.

Wang G, Li X, Wang Z (2016). APD3: Tthe antimicrobial peptide database as a tool for research and education. *Nucleic Acids Research*, 44: D1087–93.

Wang G, Zhang S, Wang Z (2009). Responses of alternative complement expression to challenge with different combinations of *Vibrio anguillarum*, *Escherichia coli* and *Staphylococcus aureus*: Eevidence for specific immune priming in amphioxus *Branchiostoma belcheri*. *Fish and Shellfish Immunology*, 26: 33–9.

Wang J, Dou X, Song J, et al. (2019a). Antimicrobial peptides: Ppromising alternatives in the post feeding antibiotic era. *Medicinal Research Reviews*, 39: 831–59.

Wang L, Liu Z, Dai S, Yan J, Wise MJ (2017). The sit-and-wait hypothesis in bacterial pathogens: Aa theoretical study of durability and virulence. *Frontiers in Microbiology*, 8: 2167.

Wang L, Yue F, Song X, Song L (2015a). Maternal immune transfer in mollusc. *Developmental & Comparative Immunology*, 48: 354–9.

Wang LD, Wagers AJ (2011). Dynamic niches in the origination and differentiation of haematopoietic stem cells. *Nature Reviews Molecular Cell Biology*, 12: 643–56.

Wang P, Zhu S, Yang L, et al. (2015b). Nlrp6 regulates intestinal antiviral innate immunity. *Science*, 350: 826–30.

Wang SYS, Tattersall GJ, Koprivnikar J (2018). Trematode parasite infection affects temperature selection in aquatic host snails. *Physiological and Biochemical Zoology*, 92: 71–9.

Wang X, Zhang Y, Zhang R, Zhang J (2019b). The diversity of pattern recognition receptors (PRRs) involved with insect defense against pathogens. *Current Opinion in Insect Science*, 33: 105–10.

Wang Y, Oberley LW, Murhammer DW (2001). Antioxidant defence systems of two Lepidopteran insect cell lines. *Free Radical Biology & Medicine*, 30: 1254–62.

Ward AJW, Duff AJ, Krause J, Barber I (2005). Shoaling behaviour of sticklebacks infected with the microsporidian parasite, *Glugea anomala*. *Environmental Biology of Fishes*, 72: 155–60.

Ward PI (1988). Sexual dichromatism and parasitism in British and Irish freshwater fishes. *Animal Behaviour*, 36: 1210–15.

Warner RE (1968). The role of introduced diseases in the extinction of the endemic Hawaiian avifauna. *Condor*, 70: 101–20.

Warren RL, al. e (2011). Exhaustive T-cell repertoire sequencing of human peripheral blood samples reveals signatures of antigen selection and a directly measured repertoire size of at least 1 million clonotypes. *Genome Research*, 21: 7890–797.

Washburn JO, Gross ME, Mercer DR, Anderson JR (1988). Predator-induced trophic shift of a free-living ciliate: parasitism of mosquito larvae by their prey. *Science*, 240: 1193–5.

Waterhouse RM, Kriventseva EV, Meister S, et al. (2007). Evolutionary dynamics of immune-related genes and pathways in disease-vector mosquitoes. *Science*, 316: 1738–43.

Waters CM, Bassler BL (2005). Quroum sensing: cell-to-cell communication in bacteria. *Annual Review of Cell and Developmental Biology*, 21: 319–46.

Waters E, Hohn M, Ahel I, et al. (2003). The genome of *Nanoarchaeum equitans*: insights into early archaeal evolution and derived parasites. *Proceedings of the National Academy of Sciences USA*, 100: 12984–8.

Waxman D, Weinert LA, Welch JJ (2014). Inferring host range dynamics from comparative data: the protozoan parasites of New World monkeys. *The American Naturalist*, 184: 65–74.

Weatherhead PJ, Bennett GF, Schluter D (1990). Sexual selection and parasites in wood warblers. *Auk*, 108: 147–52.

Wegner KM, Kalbe M, Kurtz J, Reusch TBH, Milinski M (2003a). Parasite selection for immunogenetic optimality. *Science*, 301: 1343.

Wegner KM, Kalbe M, Reusch TBH (2007). Innate versus adaptive immunity in sticklebacks: evidence for trade-offs from a selection experiment. *Evolutionary Ecology*, 21: 473–83.

Wegner KM, Reusch TBH, Kalbe M (2003b). Multiple parasites are driving major histocompatibility complex polymorphism in the wild. *Journal of Evolutionary Biology*, 16: 224–32.

Weiberg A, Jin H (2015). Small RNAs: the secret agents in the plant–pathogen interactions. *Current Opinion in Plant Biology*, 26: 87–94.

Weinersmith K, Faulkes Z (2014). Parasitic manipulation of hosts' phenotype, or how to make a zombie:an introduction to the symposium. *Integrative and Comparative Biology*, 54: 93–100.

Weinstein S, Titcomb G, Agwanda B, Riginos C, Young H (2017). Parasite responses to large mammal loss in an African savanna. *Ecology*, 98: 1839–48.

Weinstock GM (2012). Genomic approaches to studying the human microbiota. *Nature*, 489: 250–6.

Weiss RA (2002). Virulence and pathogenesis. *Trends in Microbiology*, 10: 314–17.

Welling MM, Lupetti A, Balter HS, et al. (2001). 99mTc-labeled antimicrobial peptides for detection of bacterial and *Candida albicans* infections. *Journal of Nuclear Medicine*, 42: 788–94.

Wenzel MA, Webster LMI, Paterson S, Mougeot F, Martínez-Padilla J, Piertney SB (2013). A transcriptomic investigation of handicap models in sexual selection. *Behavioral Ecology and Sociobiology*, 67: 221–34.

Werden L, Barker IK, Bowman J, et al. (2014). Geography, deer, and host biodiversity shape the pattern of lyme disease emergence in the thousand islands archipelago of Ontario, Canada. *PLoS ONE*, 9: e85640.

Werren JH (1997). Biology of *Wolbachia*. *Annual Review of Entomology*, 42: 587–609.

West SA, Buckling A (2003). Cooperation, virulence and siderophore production in bacterial parasites. *Proceedings of the Royal Society London B*, 270: 37–44.

West SA, Gardner A (2010). Altruism, spite, and greenbeards. *Science*, 327: 1341–4.

West SA, Griffin AS, Gardner A, Diggle SP (2006). Social evolution theory for microorganisms. *Nature Reviews Microbiology*, 4: 597–607.

West SA, Lively CM, Read AF (1999). A pluralist approach to sex and recombination. *Journal of Evolutionary Biology*, 12: 1003–12.

West SA, Pen I, Griffin AS (2002). Cooperation and competition between relatives. *Science*, 296: 72–5.

West SA, Smith TG, Read AF (2000). Sex allocation and population structure in apicomplexan (protozoa) parasites. *Proceedings of the Royal Society London B*, 267: 257–63.

Westneat DF, Birkhead TR (1998). Alternative hypotheses linking the immune system and mate choice for good genes. *Proceedings of the Royal Society London B*, 265: 1065–73.

Westwood ML, O'Donnell AJ, de Bekker C, Lively CM, Zuk M, Reece SE (2019). The evolutionary ecology of circadian rhythms in infection. *Nature Ecology & Evolution*, 3: 552–60.

Wetterer JK (1995). Forager size and ecology of *Acromyrmex coronatus* and other leaf-cutting ants in Costa Rica. *Oecologia*, 104: 409–15.

White EC, Houlden A, Bancroft AJ, et al. (2018a). Manipulation of host and parasite microbiotas: survival strategies during chronic nematode infection. *Science Advances*, 4: eaap7399.

White LA, Forester JD, Craft ME (2018b). Disease outbreak thresholds emerge from interactions between movement behavior, landscape structure, and epidemiology. *Proceedings of the National Academy of Sciences USA*, 115: 7374–9.

White LA, Forester JD, Craft ME (2018c). Dynamic, spatial models of parasite transmission in wildlife: their structure, applications and remaining challenges. *Journal of Animal Ecology*, 87: 559–80.

Whiteley M, Diggle SP, Greenberg P (2017). Progress in and promise of bacterial quorum sensing research. *Nature*, 551: 313–20.

Whiteman NK, Matson KD, Bollmer JL, Parker PG (2005). Disease ecology in the Galapagos Hawk (*Buteo galapagoensis*): host genetic diversity, parasite load and natural antibodies. *Proceedings of the Royal Society London B*, 273: 797–804.

Whiteman NK, Santiago-Alarcon D, Johnson KP, Parker PG (2004). Differences in straggling rates between two genera of dove lice (Insecta: Phthiraptera) reinforce population genetic and cophylogenetic patterns. *International Journal for Parasitology*, 34: 1113–19.

Whitfield JB (2003). Phylogenetic insights into the evolution of parasitism in Hymenoptera. *Advances in Parasitology*, 54: 69–100.

WHO (2005). Evolution of H5N1 avian influenza viruses in Asia. *Emerging Infectious Diseases*, 11: 1515–21.

WHO (2015). Avian influenza weekly update. *http://www.who.int/influenza/*, 480.

Wikramaratna PS, Gupta S (2009). Influenza outbreaks. *Cellular Microbiology*, 11: 1016–24.

Wilber MQ, Johnson PTJ, Briggs CJ (2017). When can we infer mechanism from parasite aggregation? A constraint-based approach to disease ecology. *Ecology*, 98: 688–702.

Wilfert L, Jiggins FM (2012). The dynamics of reciprocal selective sweeps of host resistance and a parasite counter-adaptation in *Drosophila*. *Evolution*, 67: 761–73.

Wilfert L, Long G, Leggett HC, et al. (2016). Deformed Wing Virus is a recent global epidemic in honeybees driven by Varroa mites. *Science*, 351: 594–7.

Wilfert L, Schmid-Hempel P (2008). The genetic architecture of susceptibility to parasites. *BMC Evolutionary Biology*, 8: 187.

Willem L, Verelst F, Bilcke J, Hens N, Beutels P (2017). Lessons from a decade of individual-based models for infectious disease transmission: a A systematic review (2006–2015). *BMC Infectious Diseases*, 17: 612.

Williams A, Antonovics J, Rolff J (2011). Dioecy, hermaphrodites and pathogen load in plants. *Oikos*, 120: 657–60.

Williams CG (1975). *Sex and evolution*. Princeton University Press, Princeton, N.J.

Williams JD, Boyko CB (2012). The global diversity of parasitic Isopods associated with Crustacean hosts (Isopoda: Bopyroidea and Cryptoniscoidea). *PLoS ONE*, 7: e35350.

Williams PD, Day T (2001). Interactions between sources of mortality and the evolution of parasite virulence. *Proceedings of the Royal Society London B*, 268: 2331–7.

Williams PD, Day T (2008). Epidemiological and evolutionary consequences of targeted vaccination. *Molecular Ecology*, 17: 485–99.

Williams TD, Christians JK, Aiken JJ, Evanson M (1999). Enhanced immune function does not depress reproductive output. *Proceedings of the Royal Society London B*, 266: 753–7.

Wills C (1996). *Plagues: their origin, history and future*. Harper Collins, London.

Wilson JW, Schurr MJ, LeBlanc CL, Ramamurthy R, Buchanan KL, Nickerson CA (2002). Mechanisms of bacterial pathogenicity. *Postgraduate Medical Journal*, 78: 216–24.

Wilson K (2003). Group living and investment in immune defence: an interspecific analysis. *Journal of Animal Ecology*, 72: 133–43.

Wilson K, Graham RI (2015). Transgenerational effects modulate density-dependent prophylactic resistance to viral infection in a lepidopteran pest. *Biology Letters*, 11: 20150012.

Wilson K, Thomas MB, Blanford S, Doggett M, Simpson SJ, Moore SL (2002). Coping with crowds: density-dependent disease resistance in desert locusts. *Proceedings of the National Academy of Sciences USA*, 99: 5471–5.

Wilson SN, Sindi SS, Brooks HZ, et al. (2020). How emergent social patterns in allogrooming combat parasitic infections. *Frontiers in Ecology and Evolution*, 8: 54.

Wilson-Rich N, Spivak M, Fefferman NH, Starks PT (2008). Genetic, individual, and group facilitation of disease resistance in insect societies. *Annual Review of Ecology and Systematics*, 54: 405–23.

Windsor DA (1988). Most of the species on earth are parasites. *International Journal for Parasitology*, 28: 1939–41.

Winstead CR, Zhai S-K, Sethupathi P, Knight KL (1999). Antigen-induced somatic diversification of rabbit IgH genes: gene conversion and point mutation. *The Journal of Immunology*, 162: 6602–12.

Winterhalter WE, Fedorka KM (2009). Sex-specific variation in the emphasis, inducibility and timing of the post-mating immune response in *Drosophila melanogaster*. *Proceedings of the Royal Society London B*, 276: 1109–17.

Withenshaw SM, Devevey G, Pedersen AB, Fenton A (2016). Multihost *Bartonella* parasites display covert host specificity even when transmitted by generalist vectors. *Journal of Animal Ecology*, 85: 1442–52.

Witter RL (1997). Increased virulence of Marek's disease virus field isolates. *Avian Diseases*, 41: 149–63.

Wittkopp PJ, Haerum BK, Clark AG (2004). Evolutionary changes in *cis* and *trans* gene regulation. *Nature*, 430: 85–8.

Woese CR, Fox GE (1977). Phylogenetic structure of the prokaryotic domain: the primary kingdoms. *Proceedings of the National Academy of Sciences USA*, 74: 5088–90.

Wolfe ND, Dunavan CP, Diamond J (2007). Origins of major human infectious diseases. *Nature*, 447: 279–83.

Wolowczuk I, Verwaerde C, Viltart O, et al. (2008). Feeding our immune system: impact on metabolism. *Clinical and Developmental Immunology*, 2008: 639803.

Wong J, Hamel MJ, Drakeley CJ, et al. (2014). Serological markers for monitoring historical changes in malaria transmission intensity in a highly endemic region of Western Kenya, 1994–2009. *Malaria Journal*, 13: 451.

Wong V, Cooney D, Bar-Yam Y (2016). Beyond contact tracing: community-based early detection for ebola response. *PLoS Currents*, 8: 1–42.

Wood CL, Lafferty KD (2013). Biodiversity and disease: a synthesis of ecological perspectives on Lyme disease transmission. *Trends in Ecology & Evolution*, 28: 239–47.

Wood CL, Lafferty KD, DeLeo G, Young HS, Hudson PJ, Kuris AM (2014). Does biodiversity protect humans against infectious disease? *Ecology*, 95: 817–32.

Wood CL, McInturff A, Young HS, Kim D, Lafferty KD (2017). Human infectious disease burdens decrease with urbanization but not with biodiversity. *Philosophical Transactions of the Royal Society London B*, 372: 20160122.

Woolhouse MEJ (1998). Patterns in parasite epidemiology: the peak shift. *Parasitology Today*, 14: 428–34.

Woolhouse MEJ, Hagan PE (1999). Seeking the ghost of worms past. *Nature Medicine*, 5: 1225–7.

Woolhouse MEJ, Taylor LH, Haydon DT (2001). Population biology of multihost parasites. *Science*, 292: 1109–12.

World Health Organization (2016). *Quantitative microbial risk assessment: application Application for water safety management.* WHO Press, Geneva.

Worley J, Meng J, Allard MW, Brown EW, Timme RE (2018). *Salmonella enterica* phylogeny based on whole-genome sequencing reveals two new clades and novel patterns of horizontally acquired genetic elements. *mBio*, 9: e02303–18.

Wu G, Li M, Liu Y, Ding Y, Yi Y (2015). The specificity of immune priming in silkworm, *Bombyx mori*, is mediated by the phagocytic ability of granular cells. *Journal of Insect Physiology*, 81: 60–8.

Wuilmart C, Urbain J, Givol D (1977). On the location of palindromes in immunoglobulin genes. *Proceedings of the National Academy of Sciences USA*, 74: 2526–30.

Xiao Y, Dolan PT, Goldstein EF, Li M, Farkov M, etal. (2017). Poliovirus intrahost evolution is required to overcome tissue-specific innate immune responses. *Nature Communications*, 8: 375.

Xie M, Chen Q (2020). Insight into 2019 novel coronavirus: an updated interim review and lessons from SARS-CoV and MERS-CoV. *International Journal of Infectious Diseases*, 94: 119–24.

Xu J, Cherry S (2014). Viruses and antiviral immunity in *Drosophila*. *Developmental & Comparative Immunology*, 42: 67–84.

Xu X, Zhao H, Gong Z, Han G-Z (2018). Endogenous retroviruses of non-avian/mammalian vertebrates illuminate diversity and deep history of retroviruses. *PLoS Pathogens*, 14: e1007072.

Xue KS, Moncla LH, Bedford T, Bloom JD (2018). Within-host evolution of human influenza virus. *Trends in Microbiology*, 26: 781–93.

Yan H, Hancock REW (2001). Synergistic interactions between mammalian antimicrobial defense peptides. *Antimicrobial Agents and Chemotherapy*, 45: 1558–60.

Yang S, Ruuhola T, Rantala MJ (2007). Impact of starvation on immune defence and other life-history traits of an outbreaking geometrid, *Epirrita autumnat*a: a A possible causal trigger for the crash phase of population cycles. *Annales Zoologici Fennici*, 44: 89–96.

Yang S, Sleight SC, Sauro HM (2012). Rationally designed bidirectional promoter improves the evolutionary stability of synthetic genetic circuits. *Nucleic Acids Research*, 41: e33–3.

Yang Z, Xian H, Hu J, et al. (2015). USP18 negatively regulates NF-κB signaling by targeting TAK1 and NEMO for deubiquitination through distinct mechanisms. *Scientific Reports*, 5: 12738.

Ye Z, Vasco DA, Carter TC, Brilliant MH, Schrodi SJ, Shukla SK (2014). Genome wide association study of SNP-, gene-, and pathway-based approaches to identify genes influencing susceptibility to *Staphylococcus aureus* infections. *Frontiers in Genetics*, 5: 125.

Yin S, Kleijn D, Müskens GJDM, et al. (2017). No evidence that migratory geese disperse avian influenza viruses from breeding to wintering ground. *PLoS ONE*, 12: e0177790.

Yoshida GM, Bangera R, Carvalheiro R, et al. (2018). Genomic prediction accuracy for resistance against *Piscirickettsia salmonis* in farmed rainbow trout. *G3: Genes, Genomes, Genetics*, 8: 719–26.

You C, Deng Y, Hu W, et al. (2020). Estimation of the time-varying reproduction number of COVID-19 outbreak in China. *International Journal of Hygiene and Environmental Health*, 228: 113555.

Young BC, Golubchik T, Batty EM, et al. (2012). Evolutionary dynamics of *Staphylococcus aureus* during progression from carriage to disease. *Proceedings of the National Academy of Sciences USA*, 109: 4550–5.

Young BC, Wu C-H, Gordon NC, et al. (2017a). Severe infections emerge from commensal bacteria by adaptive evolution. *eLife*, 6: e30637.

Young D, Hussell T, Dougan G (2002). Chronic bacterial infections: living with unwanted guests. *Nature Immunology*, 3: 1026–32.

Young HS, Griffin RH, Wood CL, Nunn CL (2013). Does habitat disturbance increase infectious disease risk for primates? *Ecology Letters*, 16: 656–63.

Young HS, Parker IM, Gilbert GS, Sofia Guerra A, Nunn CL (2017b). Introduced species, disease ecology, and biodiversity-disease relationships. *Trends in Ecology & Evolution*, 32: 41–54.

Yu J, An J, Li Y, Boyko CB (2018). The first complete mitochondrial genome of a parasitic isopod supports Epicaridea Latreille, 1825 as a suborder and reveals the less conservative genome of isopods. *Systematic Parasitology*, 95: 465–78.

Yuan H-Y, Baguelin M, Kwok KO, Arinaminpathy N, van Leeuwen E, Riley S (2017). The impact of stratified immunity on the transmission dynamics of influenza. *Epidemics*, 20: 84–93.

Yuan S, Tao X, Huang S, Chen S, Xu A (2014). Comparative immune systems in animals. *Annual Review of Animal Biosciences*, 2: 235–58.

Yuen K-S, Ye ZW, Fung S-Y, Chan C-P, Jin D-Y (2020). SARS-CoV-2 and COVID-19: the most important research questions. *Cell & Bioscience*, 10: 40.

Zacharof MP, Lovitt RW (2012). Bacteriocins produced by lactic acid bacteria: a review article. *APCBEE Procedia*, 2: 50–6.

Zahavi A (1975). Mate selection: selection for a handicap. *Journal of Theoretical Biology*, 53: 205–14.

Zambon M (2014). Influenza and other emerging respiratory viruses. *Medicine*, 42: 45–51.

Zani IA, Stephen SL, Mughal NA, et al. (2015). Scavenger receptor structure and function in health and disease. *Cells*, 4: 178–201.

Zarkadis IK, Mastellos D, Lambris JD (2001). Phylogenetic aspects of the complement system. *Developmental and Comparative Immunology*, 25: 745–62.

Zasloff M (2002). Antimicrobial peptides of multicellular organisms. *Nature*, 415: 389–95.

Zeh JA, Zeh DW (1996). The evolution of polyandry I: intragenomic conflict and genetic incompatibility. *Proceedings of the. Royal. Society. London. B*, 263: 1711–17.

Zelano B, Edwards SV (2002). An MHC component to kin recognition and mate choice in birds: predictions, progress, and prospects. *The American Naturalist*, 160: S225–37.

Zhang D, Zou H, Wu SG, et al. (2017). Sequencing of the complete mitochondrial genome of a fish-parasitic flatworm *Paratetraonchoides inermis* (Platyhelminthes: Monogenea): tRNA gene arrangement reshuffling and implications for phylogeny. *Parasites & Vectors*, 10: 462.

Zhang L-J, Gallo RL (2016). Antimicrobial peptides. *Current Biology*, 26: R14–19.

Zhang Q, Yin J, Zhang Y, et al. (2013). HLA-DP polymorphisms affect the outcomes of chronic Hepatitis B virus infections, possibly through interacting with viral mutations. *Journal of Virology*, 87: 12176–86.

Zhang Q, Zmasek C, Godzik A (2010). Domain architecture evolution of pattern-recognition receptors. *Immunogenetics*, 62: 263–72.

Zhang S-M, Adema CM, Kepler TB, Loker ES (2004). Diversification of Ig superfamily genes in an invertebrate. *Science*, 305: 251–4.

Zhang S-M, Loker ES (2003). The FREP gene family in the snail *Biomphalaria glabrata*: additional members, and evidence consistent with alternative splicing and FREP retrosequences. *Developmental & Comparative Immunology*, 27: 175–87.

Zhang T, Wu Q, Zhang Z (2020). Probable pangolin origin of SARS-CoV-2 associated with the COVID-19 outbreak. *Current Biology*, 30: 1346–51.e2.

Zhang XR, Rainey PB (2013). Exploring the sociobiology of pyoverdin-producing *Pseudomonas*. *Evolution*, 67: 3161–74.

Zhang Y-Z, Holmes EC (2020). A genomic perspective on the origin and emergence of SARS-CoV-2. *Cell*, 181: 223–7.

Zhang Y-Z, Shi M, Holmes EC (2018a). Using metagenomics to characterize an expanding virosphere. *Cell*, 172: 1168–72.

Zhang Y-Z, Wu W-C, Shi M, Holmes EC (2018b). The diversity, evolution and origins of vertebrate RNA viruses. *Current Opinion in Virology*, 31: 9–16.

Zhang Y, Cheng TC, Huang G, et al. (2019). Transposon molecular domestication and the evolution of the RAG recombinase. *Nature*, 569: 79–84.

Zhao S, Chen H (2020). Modeling the epidemic dynamics and control of COVID-19 outbreak in China. *Quantitative Biology*, 8: 11–19.

Zhao W, Caro F, Robins W, Mekalanos JJ (2018). Antagonism toward the intestinal microbiota and its effect on *Vibrio cholerae* virulence. *Science*, 359: 210–13.

Zhao Z, Li H, Wu X, et al. (2004). Moderate mutation rate in the SARS coronavirus genome and its implications. *BMC Evolutionary Biology*, 4: 21.

Zhao Z, Wu G, Wang J, Liu C, Qiu L (2013). Next-generation -based transcriptome analysis of *Helicoverpa armigera* larvae immune-primed with *Photorhabdus luminescens* TT01. *PLoS ONE*, 8: e80146.

Zhou X, Li X, Wu M (2018). miRNAs reshape immunity and inflammatory responses in bacterial infection. *Signal Transduction and Targeted Therapy*, 3: 14.

Zinkernagel RM, Hengartner H (2001). Regulation of the immune response by antigen. *Science*, 293: 252–3.

Zipfel C, Oldroyd GED (2017). Plant signalling in symbiosis and immunity. *Nature*, 543: 328–36.

Zipfel PF, Skerka C (2009). Complement regulators and inhibitory proteins. *Nature Reviews Immunology*, 9: 729–40.

Zipperer A, Konnerth MC, Laux C, et al. (2016). Human commensals producing a novel antibiotic impair pathogen colonization. *Nature*, 535: 511–16.

Zlotnik A, Yoshie O (2012). The chemokine superfamily revisited. *Immunity*, 36: 705–16.

Zueva KJ, Lumme J, Veselov AE, Kent MP, Primmer CR (2018). Genomic signatures of parasite-driven natural selection in north European Atlantic salmon (*Salmo salar*). *Marine Genomics*, 39: 26–38.

Zug R, Hammerstein P (2015). Bad guys turned nice? A critical assessment of *Wolbachia* mutualisms in arthropod hosts. *Biological Reviews*, 90: 89–111.

Zuk M, McKean KA (1996). Sex differences in parasite infections: patterns and processes. *International Journal of Parasitology*, 26: 1009–24.

Zuk M, Stoehr AM (2010). Sex differences in susceptibility to infection: an evolutionary perspective. In Klein SL, Roberts CW (eds) *Sex hormones and immunity to infection*, pp. 1–17. Springer, Berlin, Heidelberg.

Zuk M, Thornhill R, Ligon JD (1990). Parasites and mate choice in red jungle fowl. *American Zoologist*, 30: 235–44.

Zuk O, Schaffner SF, Samocha K, et al. (2014). Searching for missing heritability: designing rare variant association studies. *Proceedings of the National Academy of Sciences USA*, 111: E455–64.

Zurita-Gutierrez YH, Lion S (2015). Spatial structure, host heterogeneity and parasite virulence: implications for vaccine-driven evolution. *Ecology Letters*, 18: 779–89.

Subject index

dominance (genetics), 121, 200, 264–265, 269, 271

dominance (immunology, response), 82, 93, 96, 265, 331

dose,
  effective dose, 223, 225–227
  infective dose, 130, 218–219, 221–223, 226, 228–229, 240, 355
  lethal dose 50 (LD50), 223, 235, 375

droplet, 38–39, 213–214, 216, 263, 309–310, 351

drug resistance, 277, 335, 338

Down syndrome cell adhesion molecule (Dscam), 74–75, 78, 88, 91, 97, 106

durable stage, 34, 39, 167, 172, 304, 330

dysbiosis, 235–236, 347

dysregulation, 188, 207

ectoparasite, 20, 31–33, 38–39, 41, 45–46, 56, 121, 149, 169–170, 177, 206, 264, 392, 418, 421–422, 435, 437, 448, 450

effective population size, 274, 277, 332, 402

effector, 31, 59–60, 63, 69, 74, 76, 78, 85–88, 98, 102, 104, 131, 134, 141, 176, 180, 185, 187–195, 211, 232, 234, 236, 242, 249, 255, 272, 276, 302, 387

emerging disease, 163, 372, 433

encapsulation, 58, 63, 67, 114, 118, 120, 122–125, 127, 129, 139, 145, 151, 273

endangered species, 260, 425, 428

endemic, 4, 215, 283, 285–290, 292–294, 299, 312, 316, 322, 343, 371–372, 425, 440, 445, 447

endocytosis, 5, 67, 215, 233

endoparasite, 20, 32, 38–39, 41, 56, 126, 258, 273, 421

energy, 72, 113, 116, 120, 122, 124–128, 132–133, 135, 142, 149, 173, 195, 208, 249, 319

epidemic, 3, 6, 17, 22, 25, 97, 135, 182, 213, 217, 224, 236, 250, 256, 277–278, 280–284, 286–287, 290–299, 301, 303–308, 310–312, 314, 316, 339, 358, 362, 366, 369, 371–372, 376, 384, 390, 424–427, 431, 443–444, 447–448, 451

epigenetic, 98–99, 178, 180, 262

epistasis (genetics), 11, 247–249, 264–267, 409–411

epithelium, 25, 40, 47, 88, 103, 190, 232, 235, 238, 347

epitope, 62–63, 68–69, 74–75, 77–79, 81, 89, 91, 93–94, 96, 104, 129, 137, 173–174, 181–182, 184, 231, 241–242, 250, 262, 264, 272, 315, 319

erythrocyte, 64, 195, 203, 239, 241

escape mutant, 186, 191, 239, 252, 351, 355, 360

establishment (of infection), 13, 15, 37–39, 49, 52–53, 98, 134, 159, 164, 166, 196, 209, 213, 218, 225, 227, 230, 236, 240, 250, 263, 303, 314, 317, 330, 342, 376, 385, 435, 441

evolutionarily stable strategy (ESS), 16–18, 135–136, 211, 299, 364, 380–382

Europe, 6, 151, 276, 282–283, 301, 305, 336, 343, 356, 426–427, 443, 448–449, 451

evasion (by parasite), 17, 183–188, 191–192, 194, 196, 211, 233–234, 238–240, 277, 385–387

evolution-proof (strategy), 340, 400, 415

experimental evolution, 348, 372–374, 377

exposure (to parasites), 38, 51–52, 54, 59, 93, 96, 98–99, 109, 111, 113–114, 143, 174–175, 179, 224, 227, 229, 282–283, 314, 316, 335, 433, 447

expression (of genes), 5, 16, 21, 37, 66, 68, 70, 73, 75, 78–79, 83, 85–86, 90–92, 95, 98–99, 103–104, 115, 129, 132–133, 137, 143, 145–148, 152–154, 157, 172, 174–175, 177, 179, 184–186, 188, 191–192, 194–195, 200, 205, 209, 231–232, 236, 241–242, 244–247, 255, 257, 261–266, 274, 276, 279, 315, 321, 327, 331, 334–335, 338, 341, 346–347, 349, 376, 378, 381, 383–384, 402

extinction (of species), 137, 260, 273, 276, 307, 313, 331, 365, 391, 393, 398, 401, 423, 425–429, 432, 434, 436–437, 440, 442, 450–452

extraction (of host resources), 5, 32–33, 39, 41, 342, 344, 353, 360, 362, 376–377, 381, 383

faecal-oral route, 39, 213, 216, 218, 355, 445

fat body, 61, 186

fatality, 4, 6, 14, 43, 72, 183, 218, 222–223, 231, 235–236, 238–239, 243, 276, 281, 284, 290, 294, 309–310, 323, 354

fecundity, 42–43, 54, 56, 114, 116–119, 121, 124, 129, 134–135, 147, 171, 173, 179, 196–198, 204–205, 207–208, 302–305, 312, 353, 358–359, 365, 367, 369, 383, 407, 414, 417–419, 422, 424

fecundity compensation, 56, 196, 198, 204, 418–419

feminization, 33, 200, 203, 205, 365

fever, 1, 3–6, 52, 56–57, 102, 126, 144, 216–218, 222, 230, 235, 237, 258, 282, 288, 300–301, 314, 355, 372, 374, 427, 442, 444, 446, 448

fitness, 3, 5–6, 12, 14–16, 36–37, 39, 44, 51, 54, 69, 115–118, 120, 122, 127–134, 139, 142, 147–149, 154–155, 159, 166, 171–172, 180, 183, 187, 196–197, 209, 211, 220, 229–230, 260, 266–267, 286, 299, 319, 326, 329, 331, 340–342, 353–356, 358, 361–362, 365, 373, 381–386, 397, 401, 409–412, 415, 417–418, 439

food web, 49, 430–432, 451

foodborne, 52, 216, 222, 256, 445, 449

force of infection, 135, 285, 294, 314, 418, 439, 445

free-living, 5, 15, 20, 24–26, 28–29, 31, 34–36, 38, 45–46, 88, 144, 156, 170–171, 217, 304, 372, 405, 442

fibrinogen-related protein (FREP), 74–75, 78–81, 88, 97, 106

gametocyst, 27, 278

gastrointestinal, 25, 101, 184, 215, 308, 344, 421, 437

gene,
  gene conversion, 78, 80–82, 89, 105–106, 335, 398
  gene duplication, 11, 78, 104
  gene expression, 5, 16, 21, 37, 66, 68, 70, 73, 75, 78–79, 83, 85–86, 90–92, 95, 98–99, 103–104, 115, 129, 132–133, 137, 143, 145–148, 152–154, 157, 172, 174–175, 177, 179, 184–186, 188, 191–192, 194–195, 200, 205, 209, 231–232, 236, 241–242, 244–247, 255, 257, 261–266, 274, 276, 279, 315, 321, 327, 331, 334–335, 338, 341, 346–347, 349, 376, 378, 381, 383–384, 402
  gene family, 35, 79, 92, 262
  gene flow, 11, 167, 277, 399, 412–413
  gene transfer, 39, 252, 256–257, 305, 330–331, 337, 351
  jumping gene, 35, 252–253

Taxonomic index